Recent Advances in Poultry Nutrition and Production

Recent Advances in Poultry Nutrition and Production

Editors

Natalie Morgan
Youssef A. Attia

MDPI

Basel • Beijing • Wuhan • Barcelona • Belgrade • Novi Sad • Cluj • Manchester

Editors

Natalie Morgan
Animal Science
Curtin University
Perth
Australia

Youssef A. Attia
Agriculture Department
Faculty of Environmental Sciences
King Abdulaziz University
Jeddah
Saudi Arabia

Editorial Office
MDPI
St. Alban-Anlage 66
4052 Basel, Switzerland

This is a reprint of articles from the Special Issue published online in the open access journal *Animals* (ISSN 2076-2615) (available at: www.mdpi.com/journal/animals/special_issues/Poultry_Nutrition_Production).

For citation purposes, cite each article independently as indicated on the article page online and as indicated below:

Lastname, A.A.; Lastname, B.B. Article Title. *Journal Name* **Year**, *Volume Number*, Page Range.

ISBN 978-3-0365-9421-7 (Hbk)
ISBN 978-3-0365-9420-0 (PDF)
doi.org/10.3390/books978-3-0365-9420-0

Cover image courtesy of Home for Hooves Farm Sanctuary

Contents

About the Editors ix

Preface xi

Carlos Alexandre Granghelli, Carrie Louise Walk, Gilson Alexandre Gomes, Tiago Tedeschi dos Santos, Paulo Henrique Pelissari and Brunna Garcia de Souza Leite et al.
Influence of Dietary Phytase Inclusion Rates on Yolk Inositol Concentration, Hatchability, Chick Quality, and Early Growth Performance
Reprinted from: *Animals* **2023**, *13*, 1000, doi:10.3390/ani13061000 1

Hani H. Al-Baadani, Rashed A. Alhotan, Abdulaziz A. Al-Abdullatif, Ibrahim A. Alhidary, Abdulrahman S. Alharthi and Saud I. Al-Mufarrej et al.
The Effect of Gum Arabic Supplementation on Growth Performance, Blood Indicators, Immune Response, Cecal Microbiota, and the Duodenal Morphology of Broiler Chickens
Reprinted from: *Animals* **2022**, *12*, 2809, doi:10.3390/ani12202809 12

Jose R. Hernandez, Joseph P. Gulizia, John B. Adkins, Martha S. Rueda, Samuel I. Haruna and Wilmer J. Pacheco et al.
Effect of Phytase Level and Form on Broiler Performance, Tibia Characteristics, and Residual Fecal Phytate Phosphorus in Broilers from 1 to 21 Days of Age
Reprinted from: *Animals* **2022**, *12*, 1952, doi:10.3390/ani12151952 28

Maged A. Al-Garadi, Hani H. Al-Baadani and Abdulmohsen H. Alqhtani
Growth Performance, Histological Changes and Functional Tests of Broiler Chickens Fed Diets Supplemented with *Tribulus Terrestris* Powder
Reprinted from: *Animals* **2022**, *12*, 1930, doi:10.3390/ani12151930 38

Magdalena Krauze, Katarzyna Ognik, Dariusz Mikulski and Jan Jankowski
Assessment of Neurodegenerative Changes in Turkeys Fed Diets with Different Proportions of Arginine and Methionine Relative to Lysine
Reprinted from: *Animals* **2022**, *12*, 1535, doi:10.3390/ani12121535 53

Mona H. Hafez, Sara E. El-Kazaz, Badr Alharthi, Heba I. Ghamry, Mohammed A. Alshehri and Samy Sayed et al.
The Impact of Curcumin on Growth Performance, Growth-Related Gene Expression, Oxidative Stress, and Immunological Biomarkers in Broiler Chickens at Different Stocking Densities
Reprinted from: *Animals* **2022**, *12*, 958, doi:10.3390/ani12080958 72

Hanan Al-Khalaifah, Afaf Al-Nasser, Tahani Al-Surrayai, Hanan Sultan, Dalal Al-Attal and Rawan Al-Kandari et al.
Effect of Ginger Powder on Production Performance, Antioxidant Status, Hematological Parameters, Digestibility, and Plasma Cholesterol Content in Broiler Chickens
Reprinted from: *Animals* **2022**, *12*, 901, doi:10.3390/ani12070901 88

Hong T. Nguyen, Michael R. Bedford, Shu-Biao Wu and Natalie K. Morgan
Dietary Soluble Non-Starch Polysaccharide Level Influences Performance, Nutrient Utilisation and Disappearance of Non-Starch Polysaccharides in Broiler Chickens
Reprinted from: *Animals* **2022**, *12*, 547, doi:10.3390/ani12050547 102

Panpan Lu, Thanidtha Morawong, Amonrat Molee and Wittawat Molee
Influences of L-Arginine In Ovo Feeding on the Hatchability, Growth Performance, Antioxidant Capacity, and Meat Quality of Slow-Growing Chickens
Reprinted from: *Animals* **2022**, *12*, 392, doi:10.3390/ani12030392 119

Mohammed M. Qaid, Saud I. Al-Mufarrej, Mahmoud M. Azzam, Maged A. Al-Garadi, Abdulmohsen H. Alqhtani and Abdulaziz A. Al-abdullatif et al.
Dietary Cinnamon Bark Affects Growth Performance, Carcass Characteristics, and Breast Meat Quality in Broiler Infected with *Eimeria tenella* Oocysts
Reprinted from: *Animals* **2022**, *12*, 166, doi:10.3390/ani12020166 130

Filip Larsberg, Maximilian Sprechert, Deike Hesse, Gudrun A. Brockmann and Susanne Kreuzer-Redmer
Chicken Immune Cell Assay to Model Adaptive Immune Responses In Vitro
Reprinted from: *Animals* **2021**, *11*, 3600, doi:10.3390/ani11123600 147

Mukti Barua, Mohammad Reza Abdollahi, Faegheh Zaefarian, Timothy J. Wester, Channarayapatna Krishnegowda Girish and Peter V. Chrystal et al.
Influence of Age on the Standardized Ileal Amino Acid Digestibility of Corn and Barley in Broilers
Reprinted from: *Animals* **2021**, *11*, 3575, doi:10.3390/ani11123575 164

Magdalena Kopacz, Aleksandra Alicja Drażbo, Katarzyna Śmiecińska and Katarzyna Ognik
Performance and Egg Quality of Laying Hens Fed Diets Containing Raw, Hydrobarothermally-Treated and Fermented Rapeseed Cake
Reprinted from: *Animals* **2021**, *11*, 3083, doi:10.3390/ani11113083 184

Yinglu Liu, Jia Feng, Yamin Wang, Jing Lv, Jinghe Li and Lijuan Guo et al.
Fermented Corn–Soybean Meal Mixed Feed Modulates Intestinal Morphology, Barrier Functions and Cecal Microbiota in Laying Hens
Reprinted from: *Animals* **2021**, *11*, 3059, doi:10.3390/ani11113059 199

Ahsan Mustafa, Shiping Bai, Qiufeng Zeng, Xuemei Ding, Jianping Wang and Yue Xuan et al.
Limitation and Potential Effects of Different Levels of Aging Corn on Performance, Antioxidative Capacity, Intestinal Health, and Microbiota in Broiler Chickens
Reprinted from: *Animals* **2021**, *11*, 2832, doi:10.3390/ani11102832 213

Ahmed A. Saleh, Abdulrahman S. Alharthi, Rashed A. Alhotan, Mustafa Shukry Atta and Abdel-Moneim Eid Abdel-Moneim
Soybean Oil Replacement by Poultry Fat in Broiler Diets: Performance, Nutrient Digestibility, Plasma Lipid Profile and Muscle Fatty Acids Content
Reprinted from: *Animals* **2021**, *11*, 2609, doi:10.3390/ani11092609 231

Dimitrios Galamatis, Georgios A. Papadopoulos, Diamanto Lazari, Dimitrios Fletouris, Evanthia Petridou and Georgios I. Arsenos et al.
Effects of Dietary Supplementation of *Salvia officinalis* L. in Organic Laying Hens on Egg Quality, Yolk Oxidative Stability and Eggshell Microbiological Counts
Reprinted from: *Animals* **2021**, *11*, 2502, doi:10.3390/ani11092502 244

Abdul Jabbar, Muhammad Tahir, Ibrahim A. Alhidary, Mutassim A. Abdelrahman, Hani Albadani and Rifat Ullah Khan et al.
Impact of Microbial Protease Enzyme and Dietary Crude Protein Levels on Growth and Nutrients Digestibility in Broilers over 15–28 Days
Reprinted from: *Animals* **2021**, *11*, 2499, doi:10.3390/ani11092499 261

Mervat A. Abdel-Latif, Ahmed R. Elbestawy, Ali H. El-Far, Ahmed E. Noreldin, Mohamed Emam and Roua S. Baty et al.
Quercetin Dietary Supplementation Advances Growth Performance, Gut Microbiota, and Intestinal mRNA Expression Genes in Broiler Chickens
Reprinted from: *Animals* **2021**, *11*, 2302, doi:10.3390/ani11082302 268

Akshat Goel, Beom-June Kim, Chris-Major Ncho, Chae-Mi Jeong, Vaishali Gupta and Ji-Young Jung et al.
Dietary Supplementation of Shredded, Steam-Exploded Pine Particles Decreases Pathogenic Microbes in the Cecum of Acute Heat-Stressed Broilers
Reprinted from: *Animals* **2021**, *11*, 2252, doi:10.3390/ani11082252 . . . . **283**

Mohamed E. Abd El-Hack, Bothaina A. Alaidaroos, Reem M. Farsi, Diaa E. Abou-Kassem, Mohamed T. El-Saadony and Ahmed M. Saad et al.
Impacts of Supplementing Broiler Diets with Biological Curcumin, Zinc Nanoparticles and *Bacillus licheniformis* on Growth, Carcass Traits, Blood Indices, Meat Quality and Cecal Microbial Load
Reprinted from: *Animals* **2021**, *11*, 1878, doi:10.3390/ani11071878 . . . . **299**

Ahmed A. Al-Sagan, Abdullah H. Al-Yemni, Alaeldein M. Abudabos, Abdulaziz A. Al-Abdullatif and Elsayed O. Hussein
Effect of Different Dietary Betaine Fortifications on Performance, Carcass Traits, Meat Quality, Blood Biochemistry, and Hematology of Broilers Exposed to Various Temperature Patterns
Reprinted from: *Animals* **2021**, *11*, 1555, doi:10.3390/ani11061555 . . . . **323**

Xiumei Li, Minhong Zhang, Jinghai Feng and Ying Zhou
Myostatin and Related Factors Are Involved in Skeletal Muscle Protein Breakdown in Growing Broilers Exposed to Constant Heat Stress
Reprinted from: *Animals* **2021**, *11*, 1467, doi:10.3390/ani11051467 . . . . **336**

Guangju Wang, Xiumei Li, Ying Zhou, Jinghai Feng and Minhong Zhang
Effects of Heat Stress on Gut-Microbial Metabolites, Gastrointestinal Peptides, Glycolipid Metabolism, and Performance of Broilers
Reprinted from: *Animals* **2021**, *11*, 1286, doi:10.3390/ani11051286 . . . . **346**

Mohammed M. Qaid and Maged A. Al-Garadi
Protein and Amino Acid Metabolism in Poultry during and after Heat Stress: A Review
Reprinted from: *Animals* **2021**, *11*, 1167, doi:10.3390/ani11041167 . . . . **359**

Adel I. Attia, Fayiz M. Reda, Amlan K. Patra, Shaaban S. Elnesr, Youssef A. Attia and Mahmoud Alagawany
Date (*Phoenix dactylifera* L.) by-Products: Chemical Composition, Nutritive Value and Applications in Poultry Nutrition, an Updating Review
Reprinted from: *Animals* **2021**, *11*, 1133, doi:10.3390/ani11041133 . . . . **372**

Dariusz Lisiak, Piotr Janiszewski, Eugenia Grześkowiak, Karol Borzuta, Beata Lisiak and Łukasz Samardakiewicz et al.
Research on the Effects of Gender and Feeding Geese Oats and Hybrid Rye on Their Slaughter Traits and Meat Quality
Reprinted from: *Animals* **2021**, *11*, 672, doi:10.3390/ani11030672 . . . . **385**

About the Editors

Natalie Morgan

Dr. Natalie Morgan is a Senior Lecturer in Animal Science at Curtin University, Perth, Western Australia. Her primary interest areas are poultry nutrition, mainly feed enzymes and carbohydrate chemistry. Her current research focuses on dietary fiber and non-starch polysaccharides (NSPs), improving the efficacy of NSP-degrading enzymes, and the effects of xylo-oligomers as prebiotics in broiler chicken and laying hen diets. Dr Morgan completed a PhD and post-doctoral research fellow position in poultry nutrition at Nottingham Trent University, UK, and focused on the efficacy of phytase in meat chicken diets. She moved to the University of New England, Australia, 2015, where she was a Lecturer in Animal Science and a Post-Doctoral Research Fellow in Poultry Nutrition before moving to Curtin University, Australia, in 2022. She has published over 70 peer-reviewed journal articles and is an Associate Editor for three high-impact animal nutrition and poultry science journals. She is regularly invited to present at domestic conferences in Australia and international conferences in Europe, Asia, and America on the topic of NSP and prebiotics in poultry diets. She has research collaborations with multiple universities and industry bodies around the world and regularly supervises Ph.D., Master, and Honours student projects in poultry nutrition.

Youssef A. Attia

Professor Dr. Youssef A. Attia is a professor of Poultry Nutrition at King Abdulaziz University, Jeddah, Saudi Arabia. He gained a Doctor of Philosophy in Poultry Production at the University of Zagazig, Egypt, as well as the University of Georgia, USA. His titles have included Department Head, Vice Dean, and Dean in Charge. Professor Attia has published 330 articles (including 16 books and book chapters) and has translated books and student books. Dr. Attia is a reviewer and editorial board member and has gained a lot of awards.

Preface

As global consumer demand for poultry meat and eggs continues to rise, it is necessary to improve diet precision and utilisation among birds to keep the costs of poultry products low and reduce the industry's environmental impact. One possible approach is to feed cheaper, more readily available, and less environmentally harmful feed ingredients. However, these alternative ingredients are often rich in anti-nutritional factors, negatively impacting bird health and productivity. This Special Issue explores the potential of various alternative ingredients for poultry diets and identifies approaches to overcoming barriers to their adoption. The digestibility of common feed ingredients, such as corn and wheat, can also be enhanced, reducing the wastage of valuable nutrients. This can be achieved by applying feed additives, such as enzymes, and improving the accuracy of feed formulations. This Special Issue features 27 articles to provide insight into the inter-relations between nutrition, metabolism, gut microflora, poultry performance, production, and product quality. The impact of diet on several different parameters has been examined, including bird performance in terms of bird growth, chick health, and meat and egg quality, as well as immunity, gut morphology and gene expression, bone characteristics, blood and neuron health, and microbiota composition, thus presenting a holistic and dynamic overview of current research in this field.

Natalie Morgan and Youssef A. Attia
Editors

animals

Article

Influence of Dietary Phytase Inclusion Rates on Yolk Inositol Concentration, Hatchability, Chick Quality, and Early Growth Performance

Carlos Alexandre Granghelli [1,*], Carrie Louise Walk [2], Gilson Alexandre Gomes [3], Tiago Tedeschi dos Santos [3], Paulo Henrique Pelissari [4], Brunna Garcia de Souza Leite [1], Fabricia Arruda Roque [1], Mário Henrique Scapin Lopes [1], Lúcio Francelino Araujo [1,4] and Cristiane Soares da Silva Araujo [1,4]

1 Department of Animal Science, University of Sao Paulo, Pirassununga 13635-900, SP, Brazil
2 DSM Nutritional Products, Heanor DE75 7SG, UK
3 AB Vista, Marlborough SN8 4AN, UK
4 Department of Animal Nutrition and Production, University of Sao Paulo, Pirassununga 13635-900, SP, Brazil
* Correspondence: carlosgranghelli@usp.br; Tel.: +55-19-99709859

Simple Summary: Appropriate broiler nutrition is essential to guarantee good body development and a uniform growth rate. However, broiler breeder nutrition is also key for progeny success because the development of embryos and chicks depends directly on the nutrients in the yolk. The physiological status of embryos is influenced by the nutritional status of the breeder hens, which is reported to have a significant impact on progeny quality, development, and hatchability. Although phytase levels in broiler production are already consensual, there are no studies reporting the effects of phytase supplementation for broiler breeders on their progeny. The aim of this study was to evaluate how different levels of phytase supplementation, including superdosing, influence the performance and growth of progeny. In addition, the extra phosphoric effects of using high phytase supplementation to overcome the anti-nutritional effects of phytate were studied.

Abstract: The aim of this study was to determine the influence of dietary phytase in breeder hens on yolk nutrients, hatchability, chick quality, and growth rate of their progeny, and their subsequent performance to 42 d post-hatch when fed diets with the same phytase concentrations. Breeder hens (n = 216) were divided into 3 groups receiving nutrient-adequate diets with reduced calcium (Ca) and phosphorus (P) (by 0.16% and 0.15%, respectively), supplemented with either 500, 1500, or 4500 FTU/kg phytase from 27 to 50 weeks of age. Eggs were collected at 38 weeks of age and incubated. On the day of hatch, the chick quality and hatchability were determined, and 18 chicks/group were euthanized for yolk sac collection and the determination of inositol and glycerol concentrations. The remaining chicks were divided into three groups, receiving different diets with reduced Ca and P (by 0.16% and 0.15%, respectively), supplemented with 0, 500, or 1500 FTU/kg phytase to 42 d post-hatch. Increasing the phytase concentration in the breeder hen diet linearly ($p < 0.05$) increased the number of early embryo deaths and decreased the number of late deaths and pips. The inositol concentration in the yolk sac at day of hatch increased (quadratic; $p < 0.05$) as the phytase dose increased in the breeder hen diet. The breeder hen diet ($p < 0.05$) influenced the body weight (BW), feed intake (FI), and feed conversion ratio (FCR) up to 21 days of age. The supplementation of breeder hen diets with 1500 FTU/kg phytase increased the concentration of sodium (Na), magnesium (Mg), potassium (K), manganese (Mn), and zinc (Zn) in the yolk sac. The inclusion of phytase doses up to 4500 FTU/kg appeared to influence embryo mortality, chick feed intake, and BW gain to 21 days and the FCR throughout the entire production phase.

Keywords: broiler breeder; glycerol; mineral; superdosing; yolk sac

check for updates

Citation: Granghelli, C.A.; Walk, C.L.; Gomes, G.A.; Santos, T.T.d.; Pelissari, P.H.; Leite, B.G.d.S.; Roque, F.A.; Lopes, M.H.S.; Araujo, L.F.; Araujo, C.S.d.S. Influence of Dietary Phytase Inclusion Rates on Yolk Inositol Concentration, Hatchability, Chick Quality, and Early Growth Performance. *Animals* **2023**, *13*, 1000. https://doi.org/10.3390/ani13061000

Academic Editors: Natalie Morgan, Youssef A. Attia and Alireza Seidavi

Received: 23 January 2023
Revised: 19 February 2023
Accepted: 8 March 2023
Published: 9 March 2023

1. Introduction

Breeder hen nutrition has been reported to have an influence on progeny quality, hatchability, and early growth rates [1–3]. Thus, the development of embryos and chicks depends directly on the nutrients in the yolk, and their physiological status is influenced by the nutritional status of the breeder hens [4]. In this context, an improvement in the digestibility of feed ingredients could improve the capacity for the transfer of nutrients from the hen's diet to the egg. The use of exogenous enzymes in poultry feed has beneficial effects by increasing the bioavailability of nutrients and digestibility as well as helping to eliminate several anti-nutritional factors [5,6].

The poultry industry has used phytase supplementation for decades as a nutritional strategy to make available additional nutrients, including phytate phosphorus, which has secondary advantages of reducing environmental pollution [7–9] and decreasing the cost of feed formulation. However, the extra phosphoric effects are progressively evident and desired [10,11]. These extra phosphoric effects have been associated with increased nutrient availability through the degradation of phytate to inositol [12].

Inositol plays an important role in many physiological processes, including lipid transport and the function of coenzyme Q10. Coenzyme Q10 is an important antioxidant that aids mitochondrial functions, and thus improves not only energy metabolism, but also the nutrient transfer to the embryo through the egg yolk [13]. Studies have already shown that the use of high doses of dietary phytase increases free inositol concentrations in the gizzard, and improves weight gain and feed conversion in broilers [14]. In addition, previous research has shown that myo-inositol administered orally improves the performance of broiler chicks [15,16].

Furthermore, previous studies evaluating increasing doses of phytase in diets for laying hens [17] and catfish [18] have reported a significant increase in the mineral concentration in the liver, plasma, and, ultimately, egg yolk. The authors of these studies hypothesized that the mineral availability was increased due to an almost complete destruction of phytate. The improved performance associated with phytase supplementation for both broilers and laying hens is clear [6,19–22]. However, further studies are necessary to demonstrate how the supplementation of phytase in the diet of broiler breeders can influence chick performance and quality. Therefore, the aim of this study was to determine the influence of different dietary phytase inclusion rates for breeders on yolk inositol concentration, hatchability, chick quality, and early growth performance in progeny.

2. Materials and Methods

The study was conducted at the Poultry Science Laboratory of the School of Veterinary Medicine and Animal Science at the University of Sao Paulo, Pirassununga, SP, Brazil. The animal experiment was approved, and the experimental procedure followed the Institutional Animal Care and Use Committee Guidelines of the University of Sao Paulo (CEUA n. 9196040614).

2.1. Broiler Breeder Trial

The birds were housed in a shed with a negative ventilation system to allow the control of temperature and humidity. The facility consisted of pens with a capacity for 4 birds each (density of 2250 cm^2 per bird), containing a trough feeder, nipple drinker, nesting box, and wood shavings as bedding.

A total of 216 AP95 Aviagen broiler breeder hens were randomly distributed between 3 treatment groups consisting of 18 replicates of 4 birds each. The birds were reared in a commercial environment up to 20 weeks of age, then transferred to the experimental facility. The management, photostimulation, and feeding practices during the rearing, breeding, and production phases followed the recommendations in the genetics manual [23]. From 20 to 26 weeks of age, all breeder hens received a common diet during the adaptation period in the new environment. After this period, the hens were reallocated to ensure similar average weights of birds in each pen. According to the lineage manual [23], the hens were expected to reach 5% production at 25 weeks and reach peak laying by 31 weeks. The experimental diets were offered at the beginning of week 27 until the end of the experiment, at 50 weeks of age. In addition, the hens were weighed at the beginning and at the end of the study. Mash feed was served once a day and each replicate received the same feed allocation, adjusted monthly in line with the egg production of each experimental unit, following the genetic performance manual (Table 1) [23]. During the production phase, the birds experienced 16 h of light/day.

Table 1. Feed intake of breeder hens from 27 to 50 weeks of laying.

Feed Intake/Pen (g)						
Week 27–30	**Week 31–34**	**Week 35–38**	**Week 39–42**	**Week 43–46**	**Week 47–50**	**Week 27–50**
676.0	676.0	670.0	662.0	654.0	646.0	664.0

The experimental treatments consisted of a basal diet based on corn and soybean meal supplemented with graded levels of phytase (500, 1500, and 4500 FTU/kg), with corresponding reductions in calcium and available phosphorus (by 0.16% and 0.15%, respectively). The diets were formulated to provide similar nutrient profiles that met the recommendations proposed in the manual of nutritional specifications [24], regardless of the enzyme treatment. The basal diet, containing the same macronutrient and micronutrient formulations (Table 2), varied only in the concentration of phytase.

Table 2. Ingredients and calculated composition in an experimental diet for broiler breeders fed superdoses of phytase.

Ingredients (%)	Basal Diet
Corn	63.70
Soybean meal	22.92
Wheat middlings	3.12
Soybean oil	1.44
Dicalcium phosphate	1.06
Vitamin premix [1]	0.10
Mineral premix [2]	0.10
Salt	0.41
DL-Methionine (99%)	0.11
L-Lysine HCL (78.4%)	0.06
Limestone	6.98
Total	100.00
Calculated Composition	
Metabolizable energy (kcal/kg)	2.82
Crude protein (%)	16.50
Calcium [3] (%)	3.04
Available phosphorus [3] (%)	0.30
Digestible lysine (%)	0.77
Digestible methionine + cysteine (%)	0.59
Digestible methionine (%)	0.58
Digestible threonine (%)	0.54
Digestible valine (%)	0.67
Linoleic acid (%)	2.20

[1] Vitamin premix provided per kg of diet: vitamin A, 9000 UI; vitamin D3, 2600 UI; vitamin E, 14 UI; vitamin K3, 1.6 UI; vitamin B1, 2.2 mg; vitamin B2, 6 mg; vitamin B6, 3 mg; vitamin B12, 10 mcg; nicotinic acid, 0.03 g; pantothenic acid, 0.005 g; folic acid, 0.6 mg; biotin, 0.1 mg. [2] Mineral premix provided per kg of diet: Zn (ZnO), 0.126 g; Cu ($CuSO_4$), 0.0126 g; I ($Ca(IO_3)_2$), 2.52 mg; Fe ($FeSO_4$), 0.105 g; Mn (M_nSO_4), 0.126 g. [3] Matrix estimated by phytase: 0.16% Ca and 0.15% available P. The mean value analyzed was 3.18% and 0.50% for Ca and total P, respectively.

The phytase used was an enhanced *E. coli* phytase (Quantum Blue), provided by AB Vista Feed Ingredients (Marlborough, UK), with an expected activity of 5000 FTU/g. The experimental diets were analyzed for phytase enzyme activity, calcium, and total phosphorus. Phytase enzyme activity was analyzed via a colorimetric enzymatic method, as previously described [25]. The calcium and total phosphorus determination followed the official guidelines [26].

A total of 40 roosters, also AP95 Aviagen and the same age as the hens, were used for artificial insemination and were kept in the same environment as the broiler breeders. However, they received a common diet without phytase because the male effect was not being considered in this study. At 37 weeks of age, semen was manually collected from the roosters. A small amount of feather trimming was performed around the cloacal area to facilitate visualization and avoid contact with excreta or debris. A week later, at 38 weeks of age and when the hens were in the mid-production cycle, semen was collected from the roosters by abdominal massage. The semen was pooled from 3 roosters, and the broiler breeder hens were immediately inseminated with fresh semen using a dose of 0.5 mL per hen. The eggs were collected from day 3 until day 10 after insemination and temporarily placed in a holding room at 18 °C. Subsequently, the eggs were grouped by treatment and incubated using standard procedures. Any soiled, cracked, and/or deformed eggs were not incubated. Eggs that failed to hatch were broken to determine the fertility (the number of fertile eggs divided by the number of incubated eggs multiplied by 100), and embryonic mortality was then classified as early (1 to 7 d), intermediate (8 to 14 d), and late (15 to 21 d). Additionally, the egg hatchability (the number of chicks hatched divided by the number of fertile eggs multiplied by 100) was evaluated.

2.2. *Progeny Trial*

The chicks were housed in an experimental pen facility in a masonry shed with open sides bound by screens and equipped with fans and foggers for temperature control. The temperature and humidity of the experimental house were assessed throughout the trial. On the day of hatching, 18 chicks/treatment were euthanized for yolk sac collection. The yolk sacs were freeze-dried and analyzed for the inositol and glycerol content using the methods described by [27] for the blood samples, adjusting the parameters and the molecular weight of the two analyzed components. The concentration of the minerals calcium (Ca), phosphorus (P), sodium (Na), magnesium (Mg), potassium (K), copper (Cu), iron (Fe), manganese (Mn), and zinc (Zn) was determined by optical emission spectrometry with inductively coupled plasma [26].

The remaining chicks were divided by sex and allotted into a 50:50 ratio. For each treatment, the chicks were allocated with 100 × 120 cm surface floor pens equipped with nipple drinkers and a tubular feeder, with rice husk as bedding. Feed and water were provided ad libitum for the duration of the study. Heating was provided using an infrared heating lamp when necessary. The light program was set as specified by the recommendations of the genetics manual [28] as follows: 23 h light and 1 h darkness until the chicks were 3 days old and then 18 h light and 6 h darkness until slaughtering age.

A total of 648 mixed chicks were equally divided into the experimental treatments to create a 3 × 3 factorial (breeder hen diet × progeny diet) (Table 3). Each treatment was replicated 6 times (12 birds each, including 6 of each sex). The animals were fed from hatching to day 42. The progeny diets consisted of a corn and soybean meal-based diet supplemented with graded levels of phytase containing 0, 500, or 1500 FTU/kg (Quantum Blue, AB Vista, Marlborough, UK).

The diets were formulated to provide a similar nutrient profile that met the recommendations proposed in the nutrition specifications manual [29] regardless of the enzyme treatment. Ca and available P were reduced by 0.16% and 0.15%, respectively. The diets were fed in a two-phase feeding program: starter (1 to 21 d) and grower/finisher (22 to 42 d) (Table 4). In the groups where the effect of the maternal diet on the progeny was evaluated, the birds received a common basal progeny diet (without the inclusion of phytase) to verify these effects.

Table 3. Progeny experiment design: completely randomized 3 × 3 factorial design.

Treatment	Breeder Diets	Progeny Diets
1	Basal diet + 500 FTU/kg	Basal progeny diet (BP)
2	Basal diet + 500 FTU/kg	BP + 500 FTU/kg
3	Basal diet + 500 FTU/kg	BP + 1500 FTU/kg
4	Basal diet + 1500 FTU/kg	BP
5	Basal diet + 1500 FTU/kg	BP + 500 FTU/kg
6	Basal diet + 1500 FTU/kg	BP + 1500 FTU/kg
7	Basal diet + 4500 FTU/kg	BP
8	Basal diet + 4500 FTU/kg	BP + 500 FTU/kg
9	Basal diet + 4500 FTU/kg	BP + 1500 FTU/kg

Table 4. Progeny trial: composition of the basal progeny diets.

Ingredients (%)	Initial	Grower/Final
Corn	61.20	64.73
Soybean meal	34.53	29.53
Dicalcium phosphate	1.00	1.64
Soybean oil	1.20	2.19
Limestone	0.94	0.79
Salt	0.42	0.42
L-Lysine HCl (78.4%)	0.20	0.25
DL-Methionine (99%)	0.08	0.24
Vitamin premix [1]	0.05	0.05
Mineral premix [2]	0.10	0.10
Threonine	0.09	0.06
Total	100.00	100.00
Composition Calculated		
Metabolizable energy (kcal/kg)	2983	3100
Crude protein (%)	21.27	19.41
Calcium [3] (%)	0.74	0.82
Available phosphorus [3] (%)	0.30	0.41
Digestible methionine + cysteine (%)	0.87	0.77
Digestible lysine (%)	1.21	1.07
Digestible methionine (%)	0.47	0.77
Digestible threonine (%)	0.79	0.70

[1] Vitamin premix provided per kg of diet: vitamin A, 6000 UI; vitamin D3, 2000 UI; vitamin E, 10 UI; vitamin K3, 1.6 UI; vitamin B1, 1.4 mg; vitamin B2, 4 mg; vitamin B6, 2 mg; vitamin B12, 10 mcg; niacin, 0.03 g; pantothenic acid, 0.011 g; folic acid, 0.6 mg. [2] Mineral premix provided per kg of diet: Zn (ZnO), 0.126 g; Cu ($CuSO_4$), 0.0126 g; I ($Ca(IO_3)_2$), 2.52 mg; Fe ($FeSO_4$), 0.105 g; Mn (M_nSO_4), 0.126 g. [3] Matrix estimated by phytase: 0.16% Ca and 0.15% available P. The mean value analyzed was 0.77% Ca and 0.53% total P (initial diet), and 0.78% Ca and 0.63% total P (grower/final diet).

Birds and feed were weighed at 1, 7, 21 and 42 d of age to determine BW, FI and FCR (calculated from FI divided by BW gain) on a pen basis.. The incidence of mortality was recorded daily.

2.3. Statistical Analysis

The data were analyzed using the fit model platform in JMP Pro v. 14.0 [30]. Each replicate was considered to be an experimental unit in both the parent and progeny experiments. The experimental model included the breeder hen diet for the chick quality, with a statistical analysis performed on the data for fertility (%); hatchability (%); early, mid, and late deaths (%); pips (%); and inositol and glycerol concentrations in the yolk content (µmol/g). The hatchability data were transformed using Box–Cox transformations and the untransformed means were presented. The experimental model also included the breeder hen diet, progeny diet, and interactions for the progeny performance parameters, including the body weight (g/bird), feed intake (g/bird), and feed conversion rate (g/g). All model factors were considered to be nominal variables. When the model effects were significant at $p < 0.05$, the means were separated using linear and quadratic orthogonal contrast statements.

3. Results

There was no effect of diet ($p > 0.10$) on hen weight at week 27 or 50. The weight on day 0 (189 days old; 27 weeks) was 2.6 kg and 4.1 kg at week 50. Phytase activity in the breeder diets was 336, 1160, and 4610 FTU/kg for 500, 1500, or 4500 FTU/kg, respectively. The average phytase activities in the progeny diets were < 50, 360, and 1510 FTU/kg (initial) and < 50, 670, and 1310 FTU/kg (grower/finisher) for 0, 500, and 1500 FTU/kg, respectively. Hen mortality was 0%, and egg production and total eggs/hen/week were as expected according to the breed guidelines from 27 to 50 weeks [23].

There was no significant effect of the breeder diet on the percentage of fertile eggs or hatchability (Table 5). The percentage of early embryo death linearly ($p < 0.05$) increased as the phytase dose increased in the breeder diet whereas the percentage of late deaths and pips linearly ($p < 0.05$) decreased (Table 5).

Table 5. Influence of phytase supplementation in 38-week-old breeder diets on progeny quality and yolk sac nutrient concentration on day of hatching.

Variable	Breeder Diet (FTU/kg)			SEM [3]	*p*	Contrasts [1]	
	500	1500	4500			L	Q
Fertility (%)	97.0	95.9	96.5	1.3	0.92	-	-
Hatchability (%)	81.5	88.0	84.6	3.1	0.40	-	-
Early death [2] (%)	3.44	6.01	10.8	1.7	0.02 [4]	0.006 [4]	0.61
Intermediate death [2] (%)	0.00	1.12	0.64	0.5	0.23	-	-
Late death [2] (%)	6.37	1.62	1.46	1.4	0.03 [4]	0.02 [4]	0.19
Pip (%)	7.46	2.32	1.91	1.9	0.08	0.04 [4]	0.31
Inositol (µmol/g)	1.36	1.22	1.52	0.07	0.02 [4]	0.13	0.02 [4]
Glycerol (µmol/g)	1.58	1.48	1.50	8.17	0.66	-	-

[1] L: linear contrast statement; Q: quadratic contrast statement. [2] Early death: incubation days 1–7; intermediate death: incubation days 8–14; late death: incubation days 15–21. [3] Standard error of the mean. [4] Results with significant ($p < 0.05$) values.

The inositol content in the yolk sac was greater in the progeny from hens fed 4500 FTU/kg and the progeny from hens fed 500 FTU/kg, resulting in a quadratic ($p < 0.05$) influence of breeder diet on yolk sac inositol concentration (Table 5). There was no effect of breeder diet on the glycerol content of the yolk sac on the day of hatching.

The breeder hen diet influenced the progeny performance to a greater extent than the chick diet post-hatching. For example, the initial BW and the BW at day 7 post-hatching linearly increased ($p < 0.05$) as the phytase dose increased from 500 to 4500 FTU/kg in the breeder hen diet (Table 6). No significant effect of the breeder diet on the BW of broiler chicks at day 21 or 42 post-hatching was observed (Table 6). Conversely, the progeny diet supplementation with phytase had no significant effect on the BW throughout the production phase. The breeder diet affected the feed intake from 7 to 21 days post-hatch ($p < 0.05$), with a linear increase as the phytase dose increased from 500 to 4500 FTU/kg (Table 6).

There was no significant effect on the FI during the phytase supplementation of the progeny diet from 1 to 42 days of age. The breeder diet continued to influence the FCR up to 21 days post-hatching (Table 6). There was a quadratic effect at 7 and 21 days of age ($p < 0.02$ and $p < 0.01$, respectively), and the FCR increased with the use of 4500 FTU/kg (Table 6). In addition, there was a significant quadratic effect of phytase supplementation on the progeny diet at 21 d, with an increase in the FCR when at a dose of 500 FTU/kg.

Table 6. Influence of phytase dose in diets of 38-week-old breeder hens and the subsequent effect of phytase dose in the diets of the progeny from hatching to day 42.

Phytase (FTU/kg)	BW (g/bird)				FI (g/bird)			FCR (g/g)		
	1 d	7 d	21 d	42 d	7 d	21 d	42 d	7 d	21 d	42 d
				Breeder Diet						
500	47.7	168	901	2858	126	1138	4713	1.05	1.33	1.70
1500	47.9	173	926	2872	131	1154	4753	1.04	1.31	1.68
4500	48.6	181	929	2862	146	1200	4830	1.08	1.37	1.71
SEM [1]	0.29	2.07	12.3	39.3	1.96	16.9	48.3	0.01	0.01	0.01
				Progeny Diet						
0	47.9	175	908	2848	133	1140	4693	1.04	1.32	1.68
500	48.3	172	910	2893	134	1169	4781	1.07	1.37	1.70
1500	48.0	175	938	2852	135	1182	4821	1.06	1.33	1.72
SEM [1]	0.29	2.08	12.2	39.3	1.96	16.9	48.4	0.01	0.01	0.01
				Probability						
Breeder diet	0.08	<0.01 [2]	0.22	0.97	<0.01 [2]	0.03 [2]	0.24	0.02 [2]	<0.01 [2]	0.09
Linear	0.03 [2]	<0.01 [2]	-	-	<0.01 [2]	0.01 [2]	-	0.02 [2]	0.06	0.21
Quadratic	0.53	0.70	-	-	0.03 [2]	0.47	-	0.06	<0.01 [2]	0.07
Progeny diet	0.70	0.39	0.17	0.66	0.70	0.22	0.18	0.31	0.03 [2]	0.07
Linear	-	-	-	-	-	-	-	-	0.66	0.02 [2]
Quadratic	-	-	-	-	-	-	-	-	0.01 [2]	0.79
Interaction	0.39	0.60	0.68	0.62	0.84	0.91	0.51	0.39	0.80	0.85

[1] Standard error of the mean. [2] Results with significant ($p < 0.05$) values.

There was no interaction between the breeder diet and the progeny diet and the BW, FI, or FCR.

The supplementation of phytase in the breeder diet influenced the concentration of macrominerals in the yolk sac. There was a quadratic effect on the concentrations of Mg and K ($p < 0.01$) with an increase in the concentration of these minerals up to the dose of 1500 FTU/kg phytase (Table 7). There was a linear increase in the Na concentration ($p < 0.05$) with increasing doses of phytase in the maternal diet (Table 7). There was no significant effect on the Ca and P mineral concentrations (Table 7). Phytase supplementation in the maternal diet also influenced the concentrations of microminerals in the yolk sac. There was a significant and quadratic effect on the Mn concentration ($p < 0.05$), increasing up to the dose of 1500 FTU/kg.

Table 7. Mineral concentration of yolks sac collected from one day-old chicks of broiler breeders fed superdoses of phytase from 27 to 50 weeks of age.

Phytase (FTU/kg)	Ca (%)	Na (%)	Mg (%)	K (%)	P (%)	Cu (ppm)	Fe (ppm)	Mn (ppm)	Zn (ppm)
500	0.975	0.116	0.036	0.101	0.534	7.39	51.7	3.28	38.4
1500	1.040	0.129	0.046	0.139	0.539	11.70	46.9	4.45	44.5
4500	1.105	0.136	0.036	0.118	0.507	18.83	46.4	2.98	48.9
SEM [1]	0.078	0.006	0.002	0.006	0.015	5.24	3.1	0.37	3.2
				Probability					
Breeder diet	0.49	0.05	<0.01	<0.01	0.28	0.29	0.40	0.02	0.06
Linear	-	0.01	0.83	0.05	-	-	-	0.58	0.02
Quadratic	-	0.66	<0.01	<0.01	-	-	-	<0.01	0.82

[1] Standard error of the mean.

4. Discussion

There are numerous causes of embryo mortality, such as insufficient time of egg storage, incorrect incubation temperatures, inadequate humidity or ventilation, mechanical impact during transport, contamination, or nutritional deficiencies. Early embryonic mortality on average was 6.75%; this was slightly lower than that observed by [31], but greater than the guidelines at 3.5% [23]. In our study, there was a significant reduction in late embryonic mortality as well as pips in the egg shell with an increasing phytase supplementation in breeder diets. This may have been due to the increase in inositol, which can be used by the chick as a nutrient to complete the hatching process. The increased early embryonic mortality in the diets with increased doses of phytase in the breeder diet was not expected. Nonetheless, the increase in early deaths combined with the decrease in late deaths could indicate an increase in carbohydrate metabolism during the embryonic growth cycle of the chick, which could result in toxic byproducts such as CO_2 or lactic acid very early in the embryonic stage before the establishment of adequate respiratory surfaces, as stated by [31].

The progeny performance was influenced by the breeder hen diet to a greater extent than the diet the chicks were fed post-hatching for the BW, FI, and FCR, with the post-hatching diet only significant for the FCR during the final 21 days of rearing (Table 6). Differing from the results found in the present study, a study evaluating two levels of phytic phosphorus (0.22% and 0.44%) and three concentrations of phytase (0, 500, and 1000 FTU/kg of feed) in broiler diets showed that phytase improved the weight gain of broilers [32], although there was no difference between the two levels of phytic phosphorus studied. In agreement with the results in the current experiment, a study using two diets supplemented with phytase (750 FTU/kg and 1500 FTU/kg of feed) showed no effect on the body weight gain of broilers [33].

The increase in the FI as the phytase dosages in the breeder diet increased may have been due to the intrinsic relationship between the initial BW and FI. Birds with a greater body development will have a better development of internal organs, especially the intestine. The growth rate is known to be partially mediated by the development of different organs [34]—namely, the duodenum—the weight of which increases as the BW gain increases [35]. The effects of the hen diet did not persist beyond 21 days of age. This difference may have been more noticeable in the initial feeding phase because broilers have a greater food voracity in the finishing stage. Corroborating the results of our study, no differences in the FI were observed when broilers diets were supplemented with phytase at 750 FTU/kg or 1500 FTU/kg [33].

The performance results obtained in this experiment contradicted the hypothesis that the performance of birds may continue to improve with increasing the phytase supplementation of diets above the standards recommended by the industry. The absence of effects on the BW and FI and the increase in the FCR at 42 d after the supplementation with higher doses of phytase in the broiler diet may have been due to an insufficient reduction in the levels of calcium and, more particularly, available phosphorus in the diet. The performance of chickens and the supplementation of phytase in the diet depends on the levels of non-phytic phosphorus [36]. According to these authors, a supplementation of 600 FTU/kg in broiler diets resulted in an increase in weight gain, FI, and tibia ash percentage and a reduction in mortality. The authors showed that phytase was able to reduce the requirements for non-phytic phosphorus by 8.5% for weight gain, 3.5% for FI, and 6.5% for the tibia ash percentage. The phytase effect was more evident in diets with low, rather than with high, concentrations of non-phytic phosphorus. However, another study showed that broilers consuming diets containing 12,500 FTU/kg of feed had a 6% greater weight gain and a 9.4% lower FCR when compared with broilers that consumed a positive control diet [37]. Therefore, improvements in broiler performance have been observed with higher phytase doses than those added to the progeny diet in the current study.

In contrast, improvements in the BW (hatch day and 7 d), FI (7 and 21 d), and FCR (7 and 21 d) were observed in the progeny with a maternal supplementation of phytase, suggesting that it might be better to focus phytase supplementation on the breeder hen rather than broiler chickens, or that there may be benefits to re-evaluating the ratio between the supplementation of both diets. Therefore, further studies are needed to better understand the relationship between phytase supplementation for breeders and the effects on the progeny.

The improvements in the availability of Na, Mg, and K could be explained by the greater absorption of minerals due to the reduction in the concentration of phytates in the diets, as observed by others in other species [17,18,38]. In the present study, this was more evident in the diets containing 1500 and 4500 FTU/kg phytase. Ca and P concentrations varied in a manner different from the other macrominerals. This could be explained by the increasing demand for Ca and P in embryonic development during the early growth of the chick.

The result observed for Zn may have been the effect of the reduction in phytate in the diet due to the destruction of the molecule by greater phytase concentrations; hence, the availability of microminerals increased, as was also observed to a lesser extent in the Mn concentrations. The supplementation of diets with higher levels of phytase in the breeder diet did not influence the concentrations of the minerals Cu and Fe. Differences in the concentrations of the minerals Na, K, Mn, and Zn in the yolk sac indicated that phytate not only interfered with the absorption of Ca and P, but also had a much broader effect on the nutrition of broiler breeders. The authors, therefore, suggest that a higher mineral concentration in the yolk sac may have a direct effect on hatchability and initial chick development.

There have been few studies investigating the effects of breeder hen dietary phytase on progeny, and there remains a degree of uncertainty as to the dose that will best support broiler development. Arguelles-Ramos et al. [39] aimed to analyze the effects of phytase inclusion on broiler breeders, although this study did not focus on superdosing or progeny performance and development. Berry et al. [40] studied the supplementation of 300 FTU/kg phytase for breeder hens, analyzing the effect of phytase on the bone mineral content and density as well as the breeder performance, but without considering the progeny. Therefore, this is the first known report analyzing superdoses of phytase in broiler breeder hens, and following the effect on the hatchability, performance, and growth of progeny.

5. Conclusions

In summary, breeder hen nutrition influences progeny hatchability and early growth rates. Increasing phytase doses up to 4500 FTU/kg significantly increases the inositol concentration of the yolk sac, which may be used by the embryo as an energy source during hatching. This may decrease the number of late deaths during incubation and pips as well as increasing the chick body weight on the day of hatching. However, an increase in early embryo deaths was seen as the phytase doses increased. The supplementation of superdoses of phytase for breeder diets also resulted in an improvement in the BW, FI, and FCR up to 21 days post-hatching in chicks in addition to increasing the availability of Na, K, Mn, and Zn in the yolk sac on the day of hatching.

Author Contributions: Conceptualization, C.L.W., C.S.d.S.A. and L.F.A.; methodology, C.L.W., C.S.d.S.A. and L.F.A.; validation, C.A.G., C.L.W., C.S.d.S.A. and L.F.A., formal analysis, C.A.G. and C.L.W.; investigation, C.A.G., C.L.W., C.S.d.S.A., L.F.A., P.H.P., B.G.d.S.L. and F.A.R.; resources, C.L.W., C.S.d.S.A., L.F.A., G.A.G. and T.T.d.S.; data curation, C.A.G., C.L.W., C.S.d.S.A. and L.F.A.; writing—original draft preparation, C.A.G. and C.S.d.S.A.; writing—review and editing, C.A.G., C.S.d.S.A., C.L.W., M.H.S.L., G.A.G. and T.T.d.S.; visualization, C.A.G., C.L.W., C.S.d.S.A. and L.F.A.; supervision, C.A.G., C.L.W., C.S.d.S.A. and L.F.A.; project administration, C.S.d.S.A. and L.F.A.; funding acquisition, C.S.d.S.A. and L.F.A. All authors have read and agreed to the published version of the manuscript.

Funding: This research was funded by AB Vista, UK. Funding Number QB A23.

Institutional Review Board Statement: The animal experiment was approved and the experimental procedure followed the Institutional Animal Care and Use Committee Guidelines of the University of Sao Paulo (CEUAn. 9196040614).

Informed Consent Statement: Not applicable.

Data Availability Statement: The data presented in this study are available on request from the corresponding author.

Acknowledgments: The authors gratefully acknowledge the funding support for this study from AB Vista and CAPES for the scholarship to researcher Carlos A. Granghelli (Finance Code 001).

Conflicts of Interest: The authors declare no conflict of interest.

References

1. Chang, A.; Halley, J.; Silva, M. Can feeding the broiler breeder improve chick quality and offspring performance? *Anim. Prod. Sci.* **2016**, *56*, 1254–1262. [CrossRef]
2. Nusairat, B.; Qudsieh, R.; Brake, J. Use of phytase to enhance nutritional value of broiler breeder diets. *Int. J. Poult. Sci.* **2018**, *17*, 211–220. [CrossRef]
3. Araújo, C.S.S.; Hermes, R.G.; Bittencourt, L.C.; Silva, C.C.; Araújo, L.F.; Granghelli, C.A.; Pelissari, P.H.; Roque, F.A.; Leite, B.G.S. Different dietary trace mineral sources for broiler breeders and their progenies. *Poult. Sci.* **2019**, *98*, 4716–4721. [CrossRef]
4. Kenny, M.; Kemp, C. Breeder nutrition and chick quality. *Int. Hatch. Pract.* **2005**, *19*, 7–11.
5. Selle, P.H.; Ravindran, V. Microbial phytase in poultry nutrition. *Anim. Feed Sci. Technol.* **2007**, *135*, 1–41. [CrossRef]
6. Walk, C.L.; Rama Rao, S.V. Increasing dietary phytate has a significant anti-nutrient effect on apparent ileal amino acid digestibility and digestible amino acid intake requiring increasing doses of phytase as evidenced by prediction equations in broilers. *Poult. Sci.* **2020**, *99*, 290–300. [CrossRef] [PubMed]
7. Gautier, A.E.; Walk, C.L.; Dilger, R.N. Effects of a high level of phytase on broiler performance, bone ash, phosphorus utilization, and phytate dephosphorylation to inositol. *Poult. Sci.* **2018**, *97*, 211–218. [CrossRef] [PubMed]
8. Babatunde, O.O.; Cowieson, A.J.; Wilson, J.W.; Adeola, O. Influence of age and duration of feeding low-phosphorus diet on phytase efficacy in broiler chickens during the starter phase. *Poult. Sci.* **2019**, *98*, 2588–2597. [CrossRef]
9. Viveros, A.; Brenes, A.; Arija, I.; Centeno, C. Effects of microbial phytase supplementation on mineral utilization and serum enzyme activities in broiler chicks fed different levels of phosphorus. *Poult. Sci.* **2002**, *81*, 1172–1183. [CrossRef]
10. Selle, P.H.; Cowieson, A.J.; Ravindran, V. Consequences of calcium interactions with phytate and phytase for poultry and pigs. *Livest. Sci.* **2009**, *124*, 126–141. [CrossRef]
11. Kriseldi, R.; Walk, C.L.; Bedford, M.R.; Dozier, W.A. Inositol and gradient phytase supplementation in broiler diets during a 6-week production period: 1. effects on growth performance and meat yield. *Poult. Sci.* **2021**, *100*, 964–972. [CrossRef] [PubMed]
12. Beeson, L.A.; Walk, C.L.; Bedford, M.R.; Olukosi, O.A. Hydrolysis of phytate to its lower esters can influence the growth performance and nutrient utilization of broilers with regular or super doses of phytase. *Poult. Sci.* **2017**, *96*, 2243–2253. [CrossRef] [PubMed]
13. Huber, K. Cellular *myo*-inositol metabolism. In *Phytate Destruction—Consequences for Precision Animal Nutrition*; Walk, C.L., Kühn, I., Stein, H.H., Kidd, M.T., Rodehutscord, M., Eds.; Wageningen Academic Publishers: Wageningen, The Netherlands, 2016; pp. 53–60.
14. Walk, C.L.; Santos, T.T.; Bedford, M.R. Influence of superdoses of a novel microbial phytase on growth performance, tibia ash, and gizzard phytate and inositol in young broilers. *Poult. Sci.* **2014**, *93*, 1172–1177. [CrossRef]
15. Cowieson, A.J.; Ptak, A.; Maćkowiak, P.; Sassek, M.; Pruszyńska-Oszmałek, E.; Żyła, K.; Świątkiewicz, S.; Kaczmarek, S.; Józefiak, D. The effect of microbial phytase and *myo*-inositol on performance and blood biochemistry of broiler chickens fed wheat/corn-based diets. *Poult. Sci.* **2013**, *92*, 2124–2134. [CrossRef]
16. Żyła, K.; Duliński, R.; Pierzchalska, M.; Grabacka, M.; Józefiak, D.; Świątkiewicz, S. Phytases and *myo*-inositol modulate performance, bone mineralization and alter lipid fractions in the serum of broilers. *J. Anim. Feed Sci.* **2013**, *22*, 56–62. [CrossRef]
17. Soto-Salanova, M.; Santos, T.T. High phytase levels increase mineral deposition in egg yolks. In Proceedings of the XV European Symposium on the Quality of Eggs and Egg Products, Bergamo, Italy, 15–19 September 2013.
18. Peatman, E.; Beck, B.H. From floor sweepings to fish flesh phytase superdosing in the US catfish industry. In *Phytate Destruction—Consequences for Precision Animal Nutrition*; Walk, C.L., Kühn, I., Stein, H.H., Kidd, M.T., Rodehutscord, M., Eds.; Wageningen Academic Publishers: Wageningen, The Netherlands, 2016; pp. 237–250.
19. Denbow, D.M.; Ravindran, V.; Kornegay, E.T.; Yi, Z.; Hulet, R.M. Improving phosphorus availability in soybean meal for broilers by supplemental phytase. *Poult. Sci.* **1995**, *74*, 1831–1842. [CrossRef] [PubMed]
20. Costa, F.G.P.; Jácome, I.M.T.D.; da Silva, J.H.V.; de Araújo, M.J.; de Campos, K.M.F.; Barbosa, J.G.; Peixoto, J.P.N.; da Silva, J.C.A.; do Nascimento, G.A.J.; Clementino, R.H. Níveis de fósforo disponível e de fitase na dieta de poedeiras de ovos de casca marrom. *Ciênc. Anim. Bras.* **2004**, *5*, 73–81.
21. Manangi, M.K.; Coon, C.N. Phytate phosphorus hydrolysis in broilers in response to dietary phytase, calcium, and phosphorus concentrations. *Poult. Sci.* **2008**, *87*, 1577–1586. [CrossRef]
22. Walk, C.L.; Olukosi, O.A. Influence of graded concentrations of phytase in high-phytate diets on growth performance, apparent ileal amino acid digestibility, and phytate concentration in broilers from hatch to 28 D post-hatch. *Poult. Sci.* **2019**, *98*, 3884–3893. [CrossRef]
23. Aviagen. Manual de Objetivos de Desempenho Ross (308) AP (AP95). Available online: https://pt.aviagen.com/assets/Tech_Center/BB_Foreign_Language_Docs/portuguese/Ross308AP-PS-PO-PT-2017.pdf (accessed on 8 November 2022).

24. Aviagen. Manual de Especificações Nutricionais Ross (308) AP (AP95). Available online: https://pt.aviagen.com/assets/Tech_Center/BB_Foreign_Language_Docs/Portuguese/Ross308AP-PS-NS-2016-PT.pdf (accessed on 21 January 2023).
25. Engelen, A.J.; van der Heeft, F.C.; Randsdorp, P.H.; Somers, W.A.; Schaefer, J.; van der Vat, B.J. Determination of phytase activity in feed by a colorimetric enzymatic method: Collaborative interlaboratory study. *J. AOAC Int.* **2001**, *84*, 629–633. [CrossRef] [PubMed]
26. *Official Methods of Analysis of AOAC International*, 18th ed.; Horwitz, W.; Latimer, G.W., Jr. (Eds.) AOAC International: Gaithersburg, MD, USA, 2005.
27. Lee, S.A.; Dunne, J.; Febery, E.; Brearley, C.A.; Mottram, T.; Bedford, M.R. Exogenous phytase and xylanase exhibit opposing effects on real-time gizzard pH in broiler chickens. *Br. Poult. Sci.* **2018**, *59*, 568–578. [CrossRef]
28. Aviagen. Manual de Manejo de Frangos de Corte Ross (308) AP (AP95). Available online: https://pt.aviagen.com/assets/Tech_Center/BB_Foreign_Language_Docs/Portuguese/Ross-BroilerHandbook2018-PT.pdf (accessed on 19 January 2023).
29. Aviagen. Manual de Especificações Nutricionais de frangos de corte Ross (308) AP (AP95). Available online: https://pt.aviagen.com/assets/Tech_Center/BB_Foreign_Language_Docs/Portuguese/Ross-BroilerNutritionSpecifications2022-PT.pdf (accessed on 21 January 2023).
30. SAS Institute. *User's Guide, Statistics Version JMP Pro v. 14.0*; SAS Inst, Inc.: Cary, NC, USA, 2014.
31. Peñuela, S.A.; Hernandez, V.A. Caracterización de mortalidad embrionaria en pollos de engorde. *Rev. MVZ Córdoba* **2018**, *23*, 6500–6513. [CrossRef]
32. Liu, N.; Ru, Y.J.; Cowieson, A.J.; Li, F.D.; Cheng, X.C. Effects of phytate and phytase on the performance and immune function of broilers fed nutritionally marginal diets. *Poult. Sci.* **2008**, *87*, 1105–1111. [CrossRef]
33. Brunelli, S.R.; Pinheiro, J.W.; Bridi, A.M.; Fonseca, N.A.N.; Silva, C.A.; Oba, A. Efeitos da Fitase no desempenho e na qualidade da carne de frangos de corte. *Semina Ciênc. Agrár.* **2012**, *33*, 3279–3286. [CrossRef]
34. Lilja, C. A comparative study of postnatal growth and organ development in some species of birds. *Growth* **1983**, *47*, 317–339.
35. Wijtten, P.J.A.; Hangoor, E.; Sparla, J.K.W.M.; Verstegen, M.W.A. Dietary amino acid levels and feed restriction affect small intestinal development, mortality, and weight gain of male broilers. *Poult. Sci.* **2010**, *89*, 1424–1439. [CrossRef]
36. Persia, M.E.; Saylor, W.W. Effects of broiler strain, dietary nonphytate phosphorus, and phytase supplementation on chick performance and tibia ash. *J. Appl. Poult. Res.* **2006**, *15*, 72–81. [CrossRef]
37. Karadas, F.; Pirgozliev, V.; Pappas, A.C.; Acamovic, T.; Bedford, M.R. Effects of different dietary phytase activities on the concentration of antioxidants in the liver of growing broilers. *J. Anim. Physiol. Anim. Nutr.* **2010**, *94*, 519–526. [CrossRef]
38. Martinez Rojas, I.Y.; Ávila González, E.; Arce Menocal, J.; Dos Santos, T.T.; Rubio Arguello, J.; López Coello, C. Assessment of a phytase included with lactic acid on productive parameters and on deposition of phosphorus, calcium, and zinc in laying hens fed with sorghum–soybean-meal-based diets. *J. Appl. Anim. Res.* **2018**, *46*, 314–321. [CrossRef]
39. Arguelles-Ramos, M.; Nusairat, B.; Qudsieh, R.; Brake, J. Effects of phytase inclusion in broiler breeder diets during early lay on their fecal and egg characteristics. *Braz. J. Poult. Sci.* **2020**, *22*, 1–6. [CrossRef]
40. Berry, W.D.; Hess, J.B.; Lien, R.J.; Roland, D.A. Egg production, fertility, and hatchability of breeder hens receiving dietary phytase. *J. Appl. Poult. Res.* **2003**, *12*, 264–270. [CrossRef]

animals

MDPI

Article

The Effect of Gum Arabic Supplementation on Growth Performance, Blood Indicators, Immune Response, Cecal Microbiota, and the Duodenal Morphology of Broiler Chickens

Hani H. Al-Baadani [1,*], Rashed A. Alhotan [1], Abdulaziz A. Al-Abdullatif [1], Ibrahim A. Alhidary [1], Abdulrahman S. Alharthi [1], Saud I. Al-Mufarrej [1], Maged A. Al-Garadi [1], Mohammed M. Qaid [1], Ahmed A. Al-Sagan [2], Khalid E. Ibrahim [3] and Mahmoud M. Azzam [1,4,*]

1 Department of Animal Production, College of Food and Agriculture Science, King Saud University, P.O. Box 2460, Riyadh 11451, Saudi Arabia
2 King Abdul-Aziz City for Science and Technology, Riyadh 11451, Saudi Arabia
3 Department of Zoology, College of Science, King Saud University, P.O. Box 2460, Riyadh 11451, Saudi Arabia
4 Poultry Production Department, Faculty of Agriculture, Mansoura University, Mansoura 35516, Egypt
* Correspondence: hsaeed@ksu.edu.sa (H.H.A.-B.); mazzam@ksu.edu.sa (M.M.A.); Tel.: +966-541047445 (M.M.A.)

Simple Summary: Gum Arabic (GA) is considered a natural prebiotic because it contains soluble and indigestible fibers that can stimulate the growth and activity of commensal bacteria through fermentation. Administration of GA in broiler diets may promote early growth as well as the development and health of the intestine and immune functions. However, studies on the mechanism of GA on broiler chickens to improve growth performance and intestinal health are scarce. This study aimed to evaluate the growth performance, internal organs, immune-related gene expression, microbiota, and histological changes of broiler chickens fed with increasing levels of GA. We suggest that GA (0.25 to 0.75%) positively affects the performance, microbiota, immune response, morphology, and gut health of post-hatched chickens.

Abstract: Gum Arabic (GA) belongs to the Fabaceae family and contains indigestible soluble fibers (80–85%) that could be fermented by commensal bacteria to enhance performance, immune response, and intestinal integrity. This study aimed to investigate the effects of GA on performance, serum biochemical indicators, microbiota, immune-related gene expression, and histological changes in chickens. Six GA levels (0.0, 0.12, 0.25, 0.5, 0.75, 1.0%) were allocated using a total of 432 1-day-old male chickens (12 replicates with 6 chickens each). Growth performance was evaluated on days 10 and 24 of age. Blood parameters, organ pH levels, and intestinal health were determined on day 10 of age. Results showed that GA at 0.12% increased weight gain and 0.12 to 1.0% decreased feed intake but was best in feed conversion ratio and production efficiency except for 1.0% on day 1–10 of age. There was an increase in the thymus weight at GA level 0.25 to 0.75%. GA decreased the pH value of the proventriculus (at 0.50 and 1.0%) as well as the duodenum and cecum (at 0.12 and 1.0%). Chickens fed GA between 0.25 to 1.0% had higher protein and HDL, but lower cholesterol, LDL, and creatinine. Globulin was increased at 0.50% GA, while glucose and triglycerides were decreased (at 0.25 and 0.75% GA, respectively). The immune-related gene expression was reduced, except for 0.25% GA, which increased IL-10. Furthermore, chickens fed GA (0.25 to 0.75%) had higher *Lactobacillus* spp. and lower *Salmonella* spp. and *Escherichia coli*. When chickens received GA, the villus length and length to crypt ratio were higher, which also improved the integrity of intestinal epithelial cells and early duodenal development. We conclude that using GA (0.25 to 0.75%) as a natural prebiotic positively affects the performance, microbiota, immune response, morphology, and gut health of post-hatched chickens. More studies are needed to determine the potential mechanism of GA on broiler chickens.

Keywords: broiler chickens; *Acacia Senegal*; performance, gene expression; microbiota; morphology

Citation: Al-Baadani, H.H.; Alhotan, R.A.; Al-Abdullatif, A.A.; Alhidary, I.A.; Alharthi, A.S.; Al-Mufarrej, S.I.; Al-Garadi, M.A.; Qaid, M.M.; Al-Sagan, A.A.; Ibrahim, K.E.; et al. The Effect of Gum Arabic Supplementation on Growth Performance, Blood Indicators, Immune Response, Cecal Microbiota, and the Duodenal Morphology of Broiler Chickens. *Animals* **2022**, *12*, 2809. https://doi.org/10.3390/ani12202809

Academic Editor: Ilias Giannenas

Received: 2 August 2022
Accepted: 10 October 2022
Published: 17 October 2022

1. Introduction

The inclusion of antibiotic growth promoters (AGP) in poultry feed has long been practiced worldwide due to their capacity to enhance intestinal health, prevent pathogenic bacteria and promote growth [1]. However, the use of AGP has declined substantially due to bacterial resistance, an imbalance in microflora, and increased consumer demand for poultry products free of antibiotics [2]. Schokker et al. [3] reported that AGP application in the first days post-hatching affects microbial colonization negatively in broilers. In response, the European Union banned the use of AGP in poultry feed [4]. All of these concerns prompted researchers to look for the safest and most natural diet supplements, such as prebiotics, medicinal plants, and herbal products [5,6]. On the other hand, dietary supplements could be a useful strategy for promoting early growth and gastrointestinal development in broiler chickens [7].

Gum Arabic or GA (*Acacia Senegal*) is a natural supplement obtained from the exudate of tubers or tears of Acacia species [8]. The U.S. Food and Drug Administration has recognized GA as one of the safest dietary fibers for humans [9]. GA is considered to be a natural prebiotic as a result of containing indigestible fibers such as polysaccharides, neutral sugars (rhamnose, arabinose, and galactose), and glucuronic acid that selectively stimulate the growth and activity of beneficial bacteria by fermentation of the caecum besides containing organic matter, amino acids, and minerals [10]. Previous research has shown that GA at levels up to 6% as a broiler feed ingredient effectively acts as a prebiotic in improving growth performance and gut health [11]. Dietary supplementation with GA (0.1 to 2%) improved biochemical serum indicators in rabbits [12] and rats [13]. Sharma et al. [14] reported that GA at 0.25 to 2% has antimicrobial activity. However, increasing beneficial bacteria and eliminating pathogens by GA leads to a healthy gut, which may appear in the growth performance of broiler chickens [15]. A study by Teng and Kim [16] indicated that GA as a natural prebiotic lead to improved gut health by promoting lactobacilli in young chickens. Moreover, using GA in the chicken diet after hatching (1 to 10 days old) successfully enhances the gastrointestinal tract's early growth and development. [17].

To our knowledge, published reports on the efficacy of GA supplements in broiler chickens are scarce, and their effects on the microbiota, development, and health of the chicken intestine have not been elucidated. The present study hypothesized that dietary supplementation with GA as a natural prebiotic could have beneficial effects on gut health by activating beneficial bacteria and promoting early gastrointestinal tract development, which could be reflected in improved growth performance of broiler chickens. Therefore, this study aimed to investigate the effects of GA on growth performance, weight of internal organs, pH levels of gut segments, immune-related gene expression by real-time quantitative PCR, counting microbiota colonies of cecal, and duodenal morphology of broiler chickens.

2. Materials and Methods

The King Saud University's Scientific Research Ethics Committee (SREC) gave its approval to the current study and the use of all chickens (Ethics reference number: KSU-SE-20-39).

2.1. Chemical Composition Analysis

Gum Arabic or GA (*Acacia Senegal*) was purchased as natural product from local company (Abnaa Sayed Elobied Agro Export, Khartoum State, Sudan). It was ground into a fine powder at the College of Food and Agriculture Sciences, King Saud University, Saudi Arabia. Nutrient composition analysis of GA powder and feed samples (starter and grower) was carried out according to the methods of AOAC International [18]. Amino acid content was analyzed using high-performance liquid chromatography (HPLC) according to the method described previously [19]. All minerals were determined by an atomic absorption spectrometer system (PerkinElmer, instruments, AAnalyst, Shelton, USA).

Sugar derivatives of GA were analyzed by the procedures described by Gashua et al. [20] and Elsayed et al. [21] using gas chromatography-mass spectrometry (GC-MS; Agilent Technologies, Palo Alto, CA, USA). Bioactive compounds were expressed as a percentage of extracted GA.

2.2. Housing Chickens and Experimental Design

A total of 432 one-day-old male broiler chickens (Ross 308) purchased from a hatchery near Riyadh City were used in this study. The chickens were randomly divided into 6 GA groups (basal diet was supplemented with 0.0, 0.12, 0.25, 0.5, 0.75, 1.0% GA) with 12 replicates (6 chickens per replicate). The basal diet was formulated in the mash form at two stages: starter (1–10 days) and grower (11–24 days), according to the nutrient requirements of the Ross 308 Management Guide recommendations (Aviagen, 2019, New York, NY, USA). The feed ingredients and nutrient composition of the basal diets are listed in Table 1. Feed and water were provided ad libitum to the chickens until the end of the study period. All chickens were reared in environmentally controlled battery cages where the temperature and humidity were 35 °C and 65% on the first day and then gradually decreased (2 °C per 3 days) until they reached 24 °C and 50% after 24 days, according to standard management practices. The lighting program was continuous for 24 h during the first week after which 23 h of light and 1 h of darkness were maintained throughout the study. All chickens were vaccinated against NDV, IBV, and IBDV (Fort Dodge Animal Health-USA).

Table 1. Feed ingredients and nutrient content of the basal diets (%).

Ingredients	Basal Diet	
	Starter (1–10 Day)	**Grower (11–24 Day)**
Corn	52.66	57.38
Soybean meal 48%	39.10	33.98
Corn oil	3.72	4.41
Dicalcium phosphate	1.82	1.63
Limestone	1.00	0.92
Salt	0.42	0.32
DL-Methionine	0.35	0.32
L-Lysin HCL	0.20	0.19
L-Threonine	0.13	0.11
Premix Blank [a]	0.50	0.50
Choline CL 60%	0.09	0.09
Sodium bicarbonate	0.01	0.15
Total	100	100
Calculated nutrient, %		
Metabolizable energy, kcal/kg	3000	3100
Crude protein	23.29	21.15
Crude fiber	2.83	2.72
Calcium	0.96	0.87
Non-phytate P	0.48	0.44
Digestible lysine	1.28	1.15
Digestible methionine and cysteine	0.95	0.87
Digestible threonine	0.86	0.77
Digestible arginine	1.43	1.28

[a] Containing by kg of diets: Vit. A = 2,400,000 IU; Vit. D = 1,000,000 IU; Vit. E = 16,000 IU; Vit. K = 800 mg; Vit. B1 = 600 mg; Vit. B2 = 1600 mg; Vit. B6 = 1000 mg; Vit. B12 = 6 mg; Biotin = 40 mg; Folic Acid = 400 mg; Niacin; 8000 mg; Pantothenic Acid = 3000 mg; Cobalt = 80 mg; Copper = 2000 mg; Iodine = 400 mg; Iron = 1200 mg; Manganese = 18,000 mg; Selenium = 60 mg; Zinc = 14,000 mg.

2.3. Performance Measurements

At one day of age, chickens were weighed individually to obtain an equal initial body weight per replicate in each level (group). Chickens and feed were weighed at the end of the nutritional stages after 10 days (starters) and after 24 days (growers) to calculate body weight gain, total feed intake, and feed conversion ratio [22]. In addition, the European index for production efficiency was evaluated as (the sum of live weight (%) multiplied by the live weight (kg), then divided by age in days multiplied by feed conversion ratio (g/g)) $\times$ 100 according to Goiri et al. [23].

2.4. Relative Weight of Some Internal Organs

At 10 days of age, the internal organs, such as proventriculus, gizzard, thymus, bursa, spleen, liver, heart, pancreas, kidney, and small intestine, were removed from 12 chickens of each treatment group selected for slaughter and weighted separately. The relative weights of the different organs were calculated as a percentage based on live weight [24].

2.5. Determine the pH Values of Various Gastrointestinal Tract Segments

The pH values of the segments of the gastrointestinal tract (proventriculus, gizzard, duodenum, jejunum, ileum, and cecum) were determined directly in the lumen without touching the walls during the slaughter period of the chicken in duplicate using a method previously described by Zanu et al. [25]. The pH was measured using a microprocessor-controlled digital pH meter (model pH 211; Hanna Instruments, Woonsocket, RI, USA).

2.6. Serum Biochemical Analysis

Blood samples from 12 chickens per treatment group were collected in tubes without EDTA at 10 days of age. Serum was separated by centrifugation at 3000$\times$ g for 15 min for biochemical analysis. Total protein, albumin, glucose, total cholesterol, high-density lipoprotein (HDL), triglycerides, and creatinine were determined spectrophotometrically (Randox, London, U.K.) using reagent kits (Randox, London, UK) according to the manufacturer's instructions. Serum globulin concentration was determined by subtracting albumin concentration from total protein [26]. To determine low-density lipoprotein (LDL), the following formula was used: LDL = triglycerides – HDL − (triglycerides/5), according to Panda et al. [27].

2.7. Immune Response

Total mRNA was extracted from 10 samples (proximal upper jejunum) per treatment group according to the protocol of the ZymoQuick mRNA kit (Quick-RNA Miniprep, CA, USA). The quantity and quality of mRNA extracted were measured by a Nanodrop spectrophotometer (Thermo Scientific, 2000 Nanodrop, Waltham, MA, USA). Synthesis of complementary DNA (cDNA) was performed using the cDNA Reverse Transcription Kit (Applied Biosystems, Thermo Fisher Scientific, Foster, CA, USA). Gene expression of cDNA was determined using a Real-time quantitative PCR system (7300 Real-Time PCR system, Applied Biosystems) with primer sequence of target genes: IL-4 (F: GCCACCATGAGAAG-GACACT and R: ACTCTGGTTGGCTTCCTTCA), IL-6 (F: CTGCTCCTCGTGATGGCTAC and R: CCGAGGATGTACTTAATGTGCTG) and IL-10 (F: GACGTAATGCCGAAGGCAGA and R: TGCTCTTGTTTTCACAGGGC) according to the Basic Local Alignment Search Tool. Each reaction was performed in duplicate for each target gene with added SYBR Green PCR Master Mix (Applied Biosystems, Thermo Fisher Scientific, Foster, CA, USA) and calculated for fold change in gene expression using the $2^{-\Delta\Delta Ct}$ method [28].

2.8. Cecal Microbiota

Cecal digesta samples were collected from 10 chickens per treatment group in a sterile 1.5-mL Eppendorf tube and stored (−20 °C) until analysis to count microbiota colonies according to the method of Azzam et al. [6]. Approximately one gram of the cecal content was serially diluted in 9 mL of buffered peptone water (1:10) until the desired dilution was achieved. The colonies were clear and easy to count (50 to 300 colonies). From each dilution, 0.1 mL was cultured on selective media for the bacterial species studied. Selective agar media were used for the enumeration of bacterial target groups such as *Lactobacillus* spp. on de Man, Rogosa, and Sharpe agar (MRS, Himedia, Mumbai, India), *Clostridium perfringens* on Brain Heart Infusion (BHI, Oxoid, Milan, Italy) agar, and anaerobic bacteria by plate counting agar (incubated at 37 °C with 5% CO_2 for 24 h), while total aerobic bacteria by plate counting agar (Himedia, Mumbai, India), *Salmonella* spp. and *Escherichia coli* by eosin methylene blue agar (EMB, Hardy Diagnostics, Santa Maria, CA, USA) to distinguish the two microbes (incubated at 37 °C for 24 h) according to the method of Qaid et al. [29]. Colonies were counted using a colony counter, and results were expressed as $\log^{10}$ colony forming units per gram.

2.9. Duodenum Histomorphometry and Histopathological

Histomorphometric of the duodenum were prepared according to the method described by Daneshmand et al. [30]. A sample approximately 2 cm long was taken from the medial part of the duodenum (sample/chicken) to obtain 12 samples for each treatment group. After cutting, the samples were rinsed with sterile normal saline (0.9% NaCl) without stressing the tissue wall and placed in neutral buffered formalin (10%) for 72 h to fix the samples. All samples were washed with distilled water to remove excess fixative and then dehydrated in ascending gradations of ethyl alcohol (70–95%) for 60 min each and finally in two changes of absolute ethyl alcohol (100%) for 60 min each and then in two changes of xylene I and II for 60 min each. The fabrics were then impregnated with melted paraffin wax at 60 °C for 60 min each and embedded in paraffin wax. All previous processing steps were performed automatically using a sample processor (Tissue-Tek VIP 5 Jr, Sakura, Japan). Samples of 5 µm were cut with the Rotary Microtome (Leica Biosystems, RM 2255, Wetzlar, Germany), mounted on slides, and stained with hematoxylin and eosin (Leica, CV5030, Wetzlar, Germany).

Morphometric analysis of villi length and depth of crypts was measured [31] using at least five villi from the sample at 100× using a light microscope (Nikon, Corp., Tokyo, Japan) in conjunction with camera software for image analysis (AmScope digital camera-attached Ceti England microscope, Irvine, CA, USA). The percentage of villus length/crypt depth is calculated [6]. In addition, the histopathological examination in the duodenal of broiler chickens was evaluated using a light microscope according to the method described by Diler et al. [22].

2.10. Statistical Analysis

For statistical analysis, the cage mean (6 chickens/cage) was used as the experimental unit for the parameters of growth performance, whereas one chicken per cage was used as the experimental unit for internal organs, pH, serum biochemical profile, cecal microbiota, and histometrics, based on a completely randomized block design. All data were statistically analyzed using one-way ANOVA in SAS software 2008 (Cary, NC, USA) [32]. The Dunnett test ($p < 0.05$) was used to compare GA levels with the basal diet (0.0% GA; control group). It was also examined whether the responses to increasing amounts of GA were linear or quadratic by applying regression analysis. All values were expressed as mean ± standard error of the means (SEM). Pearson A correlation between cecal pH and the microbiota was calculated [33].

3. Results

The nutrient content analysis of the basal diet and GA powder (*Acacia Senegal*) are shown in Table 2. GA was rich in gross energy, organic matter, starch, and ash while less in fat. Vital amino acids such as threonine and valine were the most abundant residues. Minerals such as calcium, phosphorus, potassium, magnesium, and iron were the most abundant in GA. In addition, GA extract consists of complex polysaccharides, oligosaccharides, glycoproteins, arabinogalactans, and monosaccharides (Table 3).

Table 2. Chemical composition analysis of basal diet and gum Arabic (GA) on a dry matter basis [1].

Nutrient Analysis, %	Basal Diet		GA
	Starter	Grower	
Gross energy Kcal/kg	3740	3850	4157
Dry matter	92.49	93.23	90.68
Organic matter	86.36	87.27	85.50
Crude Protein	22.21	20.78	2.30
Starch	37.21	37.79	9.32
Crude Fat	7.17	8.70	0.10
Ash	6.13	5.96	5.18
Insoluble fiber	-	-	2.93
Soluble fiber	-	-	80.22
Calcium	1.06	1.15	1.10
Total phosphate	0.79	0.77	0.60
Magnesium	0.21	0.2	0.46
Potassium	1.27	1.21	0.98
Sodium	0.2	0.23	0.02
Iron, ppm	182	205	750
Copper, ppm	21	18	10
Zinc, ppm	97	109	2
Lysine	1.34	1.20	0.06
Methionine	0.68	0.63	-
Threonine	0.92	0.83	0.14
Valine	1.13	1.03	0.13

[1] The chemical composition analysis was performed in duplicate.

Table 3. Analyses of sugar derivatives compounds in gum Arabic (GA) by GC-MS as a percentage of extracted.

RT (min)	Compound Name	% by Area	Mol Weight
20.08	Methyl. alpha. -Arabinofuranoside, 3TMS derivative	10.0	380.1
20.63	D-(-)-Ribofuranose, tetrakis(trimethylsilyl) ether (isomer 2)	12.8	394.2
20.99	D-Ribose, 4TMS derivative	6.4	438.2
21.78	Arabinofuranose, 1,2,3,5-tetrakis-O-(trimethylsilyl)-	9.8	438.2
23.08	Propanoic acid, 2-[4-(1-buten-3-yl) phenyl]-, methyl ester	8.6	218.1
24.49	D-(+)-Talofuranose, pentakis(trimethylsilyl) ether (isomer 2)	6.8	540.2
24.64	D-Arabinopyranose, 4TMS derivative (isomer 2)	38.2	438.2
25.91	Lactose, 8TMS derivative	7.2	918.4

The effects of GA on the general growth performance of male broiler chickens from 1 to 24 days of age are shown in Table 4. According to Dunnett's test, the results show that adding GA at 0.12% of the basal diet had the highest weight gain during the starter phase (p = 0.004) compared with the basal diet (0.0% of GA). The feed intake and feed conversion ratio were reduced by 0.12, 0.25, 0.75, and 1.0% of GA (p = 0.001) except for 1.0% GA which did not affect feed conversion ratio compared with the basal diet. Additionally, there was an observed quadratic response (p = 0.007) of GA on weight gain and linear response on feed intake, feed conversion ratio, and production efficiency index upon increasing GA levels (p = 0.004, p = 0.001, p = 0.007, respectively). For the grower phase (11 to 24 days of age), the 0.12 and 0.25% GA groups consumed less feed, gained more weight and had a reduced feed conversion ratio compared with the control group. Feed intake and weight gain of the chickens responded in a quadratic fashion to increasing GA levels. In contrast, the feed conversion ratio responded linearly with increasing GA levels (p = 0.001). The production efficiency index was increased for the groups that received 0.12 up to 0.50% of GA compared to the control group during the starter and grower phases. During the overall period (1 to 24 days of age), both 0.12 and 0.25% of GA resulted in lower feed intake and feed conversion ratio and higher production efficiency index (for 0.12% inclusion only) compared with chickens fed a basal diet ($p < 0.05$). On the contrary, weight gain was not influenced nor showed any linear or quadratic response to GA levels at 1 to 24 days ($p > 0.05$).

Table 4. Effect of gum Arabic (GA) on general growth performance of male broiler chickens from 1 to 24 days of age.

Parameters	GA Levels, %						SEM [1]	p-Value [2]		
	0.00	0.12	0.25	0.50	0.75	1.00		GA	L	Q
Body weight gain, g										
1–10 d	215 [b]	231 [a]	222 [b]	227 [b]	220 [b]	212 [b]	3.80	0.004	0.122	0.007
11–24 d	889 [b]	951 [a]	946 [a]	945 [b]	925 [b]	911 [b]	15.3	0.040	0.007	0.002
1–24 d	1166	1192	1160	1182	1168	1163.3	18.1	0.866	0.767	0.664
Total feed intake, g										
1–10 d	260 [a]	245 [b]	245 [b]	256 [a]	249 [b]	247 [b]	2.95	0.001	0.004	0.333
11–24 d	1263 [a]	1176 [b]	1196 [b]	1222 [a]	1233 [a]	1222 [a]	14.8	0.003	0.004	0.034
1–24 d	1524 [a]	1421 [b]	1442 [b]	1478 [a]	1484 [a]	1469 [a]	16.2	0.001	0.001	0.033
Feed conversion ratio, g/g										
1–10 d	1.22 [a]	1.06 [b]	1.11 [b]	1.13 [b]	1.14 [b]	1.17 [a]	0.02	0.001	0.001	0.850
11–24 d	1.42 [a]	1.23 [b]	1.26 [b]	1.30 [b]	1.34 [b]	1.35 [b]	0.02	0.001	0.001	0.773
1–24 d	1.31 [a]	1.19 [b]	1.24 [b]	1.25 [a]	1.27 [a]	1.26 [a]	0.01	0.006	0.004	0.678
European Production Efficiency Index										
1–10 d	214 [b]	260 [a]	240 [a]	241 [a]	233 [b]	220 [b]	5.63	0.001	0.007	0.818
11–24 d	341 [b]	416 [a]	403 [a]	396 [a]	376 [b]	367 [b]	10.3	0.001	0.001	0.106
1–24 d	387 [b]	433 [a]	404 [b]	409 [b]	398 [b]	398 [b]	10.0	0.047	0.067	0.911

[a,b] Means that do not share a common superscripted letter within a row for each parameter differ significantly from those of the basal diet (0.0%), as determined by the Dunnett test ($p < 0.05$). [1] SEM = Standard error of means for diet effect. [2] GA = gum Arabic response; L = linear response; Q = quadratic response.

The effects of GA on the internal organ relative weight and pH values of the gastrointestinal tract segments are shown in Table 5. The relative weights of all internal organs were not influenced nor showed any linear or quadratic responses by GA levels ($p > 0.05$) except for the relative weight of the thymus, which was increased for the chickens that received GA at 0.25 to 0.75% compared to the control. Thymus weight also increased ($p = 0.002$) gradually in a quadratic way to reach a maximum response at 0.75% GA, then dropped thereafter. The pH values of gizzard, jejunum, and ileum segments were not influenced ($p > 0.05$) nor did they show any linear or quadratic responses when GA levels were increased in the diets. On the contrary, pH was lowered when chickens received GA at 0.50 and 1.0% in the proventriculus, 1.0% in the duodenum, and 0.12 and 1.0% in the cecum compared to the control group. Additionally, there was an observed linear response to GA levels of in the duodenum and cecum ($p = 0.006$, $p = 0.005$; respectively).

The effects of GA on the measurement of blood parameters in male broiler chickens are shown in Table 6, including 0.25% or more of GA increased total serum protein and HDL concentrations ($p = 0.001$, $p = 0.002$; respectively) but lowered total cholesterol and LDL ($p = 0.001$) compared to the basal diet. The 0.50% GA group was associated with a higher serum globulin concentration compared to the control. The groups 0.25% and 0.75% GA were found to have lower glucose and triglyceride concentrations, respectively ($p < 0.05$). Serum creatinine concentrations decreased ($p = 0.001$) when chickens received 0.12 or more GA compared to the control. Additionally, there was a quadratic response in total protein, globulin, glucose, total cholesterol, LDL, triglycerides, and creatinine concentrations with increasing GA levels. Positive linear relationships in HDL and albumin concentrations were observed due to GA supplementation.

Table 5. Effect of gum Arabic (GA) powder on the relative weight of internal organs and pH values of gastrointestinal tract segments in male broiler chickens on d 10 of age.

Parameters	GA Levels, %						SEM [2]	p-Value [3]		
	0.00	0.12	0.25	0.50	0.75	1.00		GA	L	Q
Internal organs [1]										
Proventriculus	0.87	0.84	0.82	1.04	0.89	0.90	0.05	0.105	0.634	0.696
Gizzard	2.99	2.74	3.06	3.13	2.97	2.82	0.10	0.176	0.678	0.374
Thymus	0.29 [b]	0.39 [b]	0.46 [a]	0.52 [a]	0.56 [a]	0.43 [b]	0.03	0.002	0.001	0.002
Bursa	0.19	0.19	0.18	0.21	0.18	0.20	0.01	0.857	0.738	0.565
Spleen	0.08	0.09	0.09	0.08	0.09	0.07	0.004	0.284	0.392	0.051
Liver	3.23	3.09	3.08	3.42	3.48	3.31	0.09	0.057	0.664	0.734
Heart	0.66	0.69	0.64	0.67	0.71	0.70	0.02	0.587	0.403	0.621
Pancreas	0.52	0.52	0.46	0.48	0.54	0.46	0.02	0.225	0.322	0.710
Kidney	0.72	0.69	0.75	0.80	0.84	0.76	0.05	0.558	0.458	0.662
SmallIntestine	9.6	9.2	8.2	9.2	8.7	8.9	0.33	0.130	0.040	0.040
PH values of gastrointestinal tract segments										
Proventriculus	4.2 [a]	4.5 [a]	4.2 [a]	3.3 [b]	4.0 [a]	3.7 [b]	0.17	0.013	0.230	0.880
Gizzard	2.8	3.1	2.6	2.7	2.9	3.1	0.12	0.061	0.440	0.210
Duodenum	6.5 [a]	6.3 [b]	6.6 [a]	6.3 [a]	6.3 [a]	6.2 [b]	0.05	0.002	0.006	0.590
Jejunum	6.2	6.2	6.3	6.0	6.1	6.2	0.13	0.801	0.610	0.960
Ileum	6.2	6.0	6.4	5.9	6.0	6.4	0.13	0.145	0.680	0.330
Cecum	6.6 [a]	6.1 [b]	6.3 [a]	6.2 [a]	6.3 [a]	6.0 [b]	0.10	0.007	0.005	0.403

[a,b] Means that do not share a common superscripted letter within a row for each parameter differ significantly from those of the basal diet (0.0%), as determined by the Dunnett test ($p < 0.05$). [1] (g/100 g of live BW). [2] SEM = Standard error of means for diet effect. [3] GA = gum Arabic response; L = linear response; Q = quadratic response.

Table 6. Effect of gum Arabic (GA) on serum biochemical profile of male broiler chickens on d 10 of age.

Parameters	GA Levels, %						SEM [1]	p-Value [2]		
	0.00	0.12	0.25	0.50	0.75	1.00		GA	L	Q
Total protein, g/dl	3.14 [b]	3.36 [b]	3.51 [a]	3.96 [a]	3.60 [a]	3.65 [a]	0.07	0.001	0.001	0.001
Albumin, g/dl	1.74	1.78	1.94	1.96	2.00	2.12	0.11	0.100	0.038	0.902
Globulin, g/dl	1.39 [b]	1.57 [b]	1.56 [b]	1.99 [a]	1.60 [b]	1.53 [b]	0.10	0.011	0.020	0.015
Albumin/Globulin	1.33	1.15	1.35	1.07	1.40	1.41	0.14	0.626	0.727	0.342
Glucose, mg/dl	253 [a]	239 [a]	222 [b]	239 [a]	232 [a]	241 [a]	6.20	0.050	0.011	0.011
Cholesterol, mg/dl	151 [a]	158 [a]	119 [b]	122 [b]	99 [b]	130 [b]	3.11	0.001	0.001	0.001
HDL, mg/dl	45.3 [b]	50.0 [b]	63.7 [a]	67.6 [a]	66.9 [a]	72.7 [a]	4.14	0.002	0.002	0.234
LDL, mg/dl	73.6 [a]	77.1 [a]	24.4 [b]	25.5 [b]	7.9 [b]	24.0 [b]	3.48	0.001	0.001	0.007
Triglycerides, mg/dl	160 [a]	155 [a]	152 [a]	144 [a]	118 [b]	164 [a]	5.12	0.001	0.026	0.024
Creatinine, mg/dl	0.47 [a]	0.37 [b]	0.31 [b]	0.36 [b]	0.30 [b]	0.35 [b]	0.02	0.001	0.001	0.001

[a,b] Means that do not share a common superscripted letter within a row for each parameter differ significantly from those of the basal diet (0.0%), as determined by the Dunnett test ($p < 0.05$). [1] SEM = Standard error of means for diet effect. [2] GA = gum Arabic response; L = linear response; Q = quadratic response.

The effects of GA on immune-related gene expression of male broiler chickens are shown in Figure 1. The 0.12 and 1.0% groups had a lower fold change in IL-4 expression than the control group ($p = 0.020$) and an observed negative linear response with levels of GA ($p = 0.002$). The fold change in the expression of IL-6 was reduced when chickens received GA compared to their counterparts in the control group ($p = 0.001$), with a quadratic response observed ($p = 0.008$). Furthermore, the expression of IL-10 was reduced in chickens that received 0.12, 0.50, and 1.0% GA and increased in chickens that received 0.25% GA compared to the control ($p = 0.001$) but did not show any linear or quadratic responses to GA levels ($p > 0.05$).

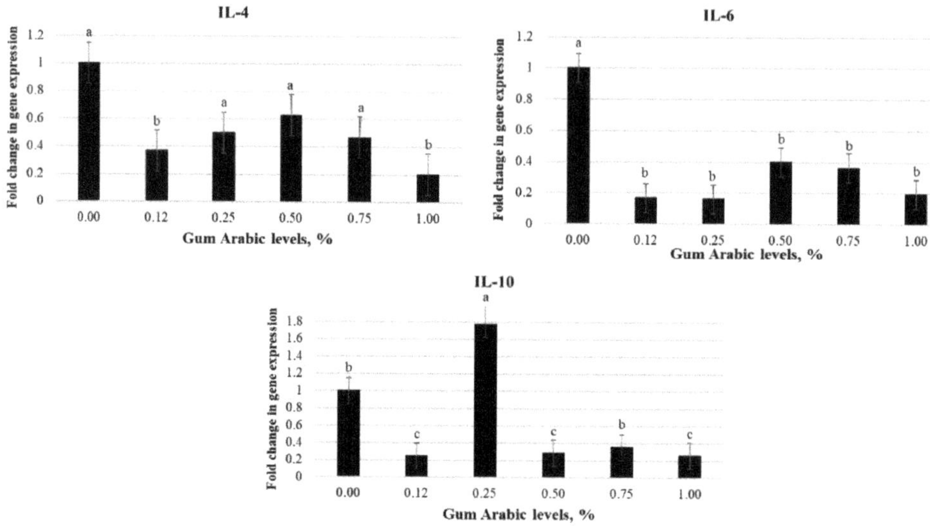

Figure 1. Effect of gum Arabic (GA) on immune-related gene expression of male broiler chickens on day 10 of age. IL-4 = Interleukin 4 (p-value: GA = 0.020; L = 0.002; Q = 0.662). IL-6 = Interleukin 6 (p-value: GA = 0.001; L = 0.001; Q = 0.008). IL-10 = Interleukin 10 (p-value: GA = 0.001; L = 0.051; Q = 0.321).

Aerobic and anaerobic bacteria and *Clostridium perfringens* were not influenced by GA supplementation ($p > 0.05$), as shown in Table 7. The inclusion of 0.25 or more of GA resulted in a higher *Lactobacillus* spp. count and lower *Salmonella* spp. compared to the control ($p = 0.001$, $p = 0.009$, respectively). Only the 1.0% GA group had a higher *Escherichia coli* ($p = 0.010$), and the 0.25% GA had a higher *Lactobacillus* to *Escherichia coli* ratio ($p < 0.030$) than the control. In addition, there was an observed linear response of GA levels on *Salmonella* spp. count and a quadratic response on *Lactobacillus* and *Lactobacillus* to *Escherichia coli* ratio upon increasing GA levels ($p < 0.05$). The results of correlations between the cecal microbiota and pH values are shown in Table 8. A strong negative correlation was observed between *Lactobacillus* count and cecal pH values ($p = 0.010$). The other bacterial populations (*Clostridium perfringens, Escherichia coli and Salmonella Typhimurium*) showed no correlations ($p > 0.05$) with the pH values of the cecum.

Table 7. Effect of gum Arabic (GA) on cecal microbiota of male broiler chickens on d 10 of age.

Parameters [1]	GA Levels, %						SEM [2]	p-Value [3]		
	0.00	0.12	0.25	0.50	0.75	1.00		GA	L	Q
Aerobic	11.4	11.4	11.4	11.5	11.5	11.6	0.25	0.970	0.770	0.790
Escherichia coli	7.9 [b]	8.2 [b]	7.7 [b]	8.4 [b]	8.5 [b]	8.6 [a]	0.20	0.010	0.060	0.440
Salmonella Typh.	8.9 [a]	9.0 [a]	7.9 [b]	8.0 [b]	7.9 [b]	8.0 [b]	0.27	0.009	0.020	0.110
Anaerobic	10.9	11.1	11.7	12.0	11.9	11.9	0.37	0.180	0.040	0.220
Clostridium perfringens	11.1	11.3	10.9	11.5	11.5	11.6	0.32	0.620	0.410	0.760
Lactobacillus spp.	9.9 [c]	10.5 [b]	11.0 [a]	11.1 [a]	11.3 [a]	11.1 [a]	0.19	0.001	0.001	0.001
Lactobacillus/Escherichia coli	1.3 [b]	1.3 [b]	1.5 [a]	1.3 [b]	1.3 [b]	1.3 [b]	0.03	0.030	0.050	0.010

[a,b] Means that do not share a common superscripted letter within a row for each parameter differ significantly from those of the basal diet (0.0%), as determined by the Dunnett test ($p < 0.05$). [1] colonies were counted using a colony counter and the results were expressed as log10 colony forming units per gram. [2] SEM = Standard error of means for diet effect. [3] GA = gum Arabic response; L = linear response; Q = quadratic response.

Table 8. Pearson correlation between cecal pH and bacterial populations.

Cecal Microbiota	Cecal pH Values	
	Correlation Coefficient	
	r_{xy}	*p*-Value
Lactobacillus spp.	−0.420	0.010
Clostridium perfringens	−0.318	0.058
Escherichia coli	−0.302	0.073
Salmonella Typhimurium	0.125	0.467

The effects of GA on the duodenal histometric measurements of male broiler chickens are shown in Table 9. Chickens that received 0.12 to 1.0% GA had higher villus length and villus length to crypt depth ratio compared to the basal diet (p = 0.001, p = 0.001, respectively). Only the 0.25% group had a higher crypt depth than the control group (p = 0.001). Additionally, there was an observed quadratic response of GA levels to villus length, crypt depth, and villus length to crypt depth ratio upon increasing GA levels ($p < 0.05$).

Table 9. Effect of gum Arabic (GA) on duodenal histomorphometric of male broiler chickens on d 10 of age.

Parameters	GA Levels, %						SEM [1]	*p*-Value [2]		
	0.00	0.12	0.25	0.50	0.75	1.00		GA	L	Q
Villus length, μm	833 [b]	1077 [a]	904 [a]	975 [a]	1054 [a]	1025 [a]	16.9	0.001	0.001	0.047
Crypt depth, μm	148 [b]	159 [b]	162 [a]	154 [b]	157 [b]	137 [b]	3.80	0.001	0.161	0.001
Length to crypt depth	8.8 [b]	13.2 [a]	14.0 [a]	10.2 [a]	12.7 [a]	12.2 [a]	0.30	0.001	0.001	0.001

[a,b] Means that do not share a common superscripted letter within a row for each parameter differ significantly from those of the basal diet (0.0%), as determined by the Dunnett test ($p < 0.05$). [1] SEM = Standard error of means for diet effect. [2] GA = gum Arabic response; L = linear response; Q = quadratic response.

The effects of GA on the duodenal histopathological examination of broiler chickens are shown in Figure 2A–F. In chickens fed the basic diet (A), proliferative enterocytes are seen raised besides the goblet cells, which also include hyperplastic association and activated crypts with mild thickening of some villi. In addition, a mucinous exudate mixed with desquamated epithelial sheets was noticed in the intestinal lumen within samples. For the chickens that received 0.12% GA (B), the villi structure was improved compared with the control chickens, which characterized by many criteria including thickening of villi was seen and desquamated epithelial sheets inside the intestinal lumen also seen in some samples. The intestinal crypts showed hyperplastic due to increase regenerative. At 0.25% GA (C), most of the villi retained their morphological structure, except still a little mucus seen in the lumen without desquamated sheets. The intestinal villi were shorter and lined by only one layer of enterocytes than the control chickens. Regeneration villi due to more active deeper intestinal crypts were noticed with a fusion of some villi. The duodenum villi structures showed length and thickness in chicken groups receiving 0.50, 0.75, and 1.0% GA (D–F) were consistent with morphological appearance. All duodenal crypts exhibited numerous mitotic activations (marked regenerative index) compared with the basal diet. Local and slight changes in some villi maintained their height and width, but they still atrophied with little separation of villous sheets into the lumen in chickens fed GA at level of 0.50% (D).

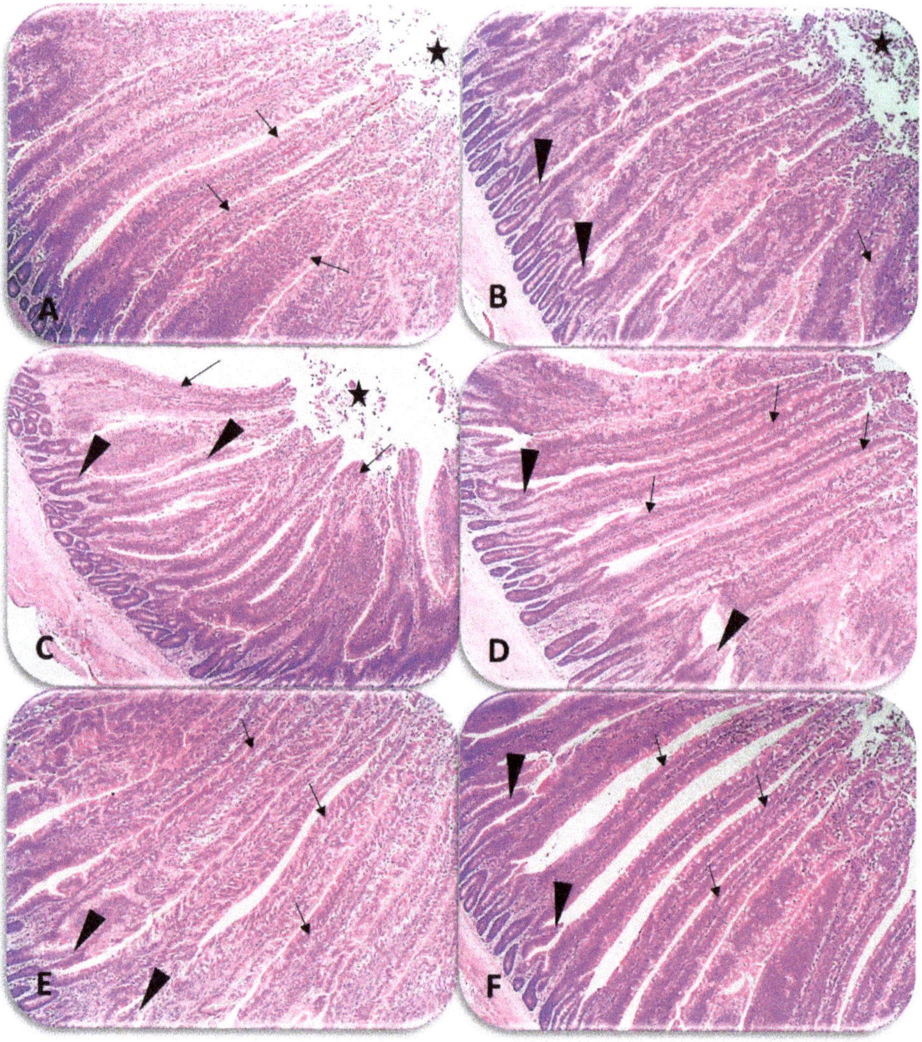

Figure 2. Representative photomicrographs of histopathological examination of broiler chicken's duodenum samples on d 10 of age stained with hematoxylin and eosin (100×). (**A**): duodenum of chickens (0.00% GA) showing proliferative and some denuded villus enterocytes with abundant goblet cells (arrow), desquamated epithelial sheets inside the lumen (star). (**B**): 0.12% GA, villus structure was nearly high with abundant goblet cells, still found desquamated epithelial sheets in the lumen (star), and proliferative enterocytes (arrow) with crypts appeared to have high compensatory regenerative activity (arrowhead). (**C**): 0.25% GA showing the enterocytes and all villus coats were nearly better than 0.0 and 0.12% GA (arrow) with still faint mucus and desquamate in the lumen (star). Clear regeneration events from deeper intestinal crypts (arrowhead) was observed. (**D–F**): levels 0.50, 0.75, and 1.0% GA, respectively, morphological appearance, including villus coats and enterocytes layers were improved than chickens of other GA levels (arrow) and all crypts have regenerative activity (arrowhead).

4. Discussion

Previous studies have confirmed that GA (*Acacia Senegal*) is rich in soluble fiber (galactose, rhamnose, and arabinose), essential amino acids, and minerals [12,34]. Usually, GA especially *Acacia Senegal* is used as a traditional medicine to treat many diseases such as intestinal infections, diabetes, inflammation, and antibacterial agents in humans [35]. The mechanism of action of GA was studied by Kishimoto et al. [36]; Calame et al. [37]; Adil et al. [38]; and Gultermirian et al. [39] in pigs, humans, laying hens, and broilers. They indicated that GA is not degraded in the gastrointestinal tract and thus fermented by the microbiota, which may be reflected in the improved performance and healthy gut of broilers. Our study's results show adding GA to a basal diet improved overall growth performance, with maximum optimization achieved in chickens fed 0.12% GA during the starter phase. The results agree with those of Tabidi and Ekram [40], who found that chickens receiving GA powder at 0.6% exhibited a higher growth rate and a lower feed conversion ratio. However, cumulative feed intake was lower in the groups fed GA at 0.12 and 0.25%. In contrast, Al-Fadil et al. [41] reported that adding GA at levels up to 6.0% did not affect total feed intake, suggesting that GA acts as a prebiotic when incorporated into the diet and improves the growth performance of broiler chickens. The relative weight of the internal organs of chickens that received GA after 10 days were not affected, except for the thymus gland. This is in contrast to a study by Sato et al. [42], who found that the relative weights of the spleen and bursa were higher in chickens fed GA for 10 days. The pH values were lower in chickens fed GA levels (0.50 and 1.0%) in the proventriculus (0.12 and 1.0%) in the duodenum and cecum compared to the control, which could be due to the physical and chemical properties of GA (relative viscosity = 14.98, pH = 4.69, solubility = 25.66%), resulting in higher acidity in these segments, in addition to its dependence on the amount and buffering capacity of the available feed, which affects the pH range [34].

The biochemical changes in serum are metabolic indicators of health and nutritional status [43]. Total protein (0.25 to 1.0% of GA) and globulin (0.50% of GA) were increased in concentrations compared to the control, which indicates that GA can improve body protein anabolism in chickens. These results agree with Amber et al. [44], who reported that GA at 0.2 to 2.0% of a rabbit's diet increased serum protein and globulin. The glucose concentration in serum was decreased by GA at a level of 0.25%. This decrease may be attributed to the inhibition of glucose uptake by sodium-glucose transporter 1 in the intestine [45]. The results obtained in this study indicated an improvement in serum lipid profile by GA. Cholesterol and LDL were decreased while HDL was increased at 0.25 to 1.0% GA levels and triglycerides at 0.75% GA compared to the control. In agreement with these findings, Musa et al. [46] reported that GA reduced cholesterol and LDL while increasing HDL levels. This could be due to bile acid absorption disruption in the intestinal (impairing bile acid circulation), resulting in lower serum cholesterol. Another possibility may be that the viscosity increases in duodenum contents with GA levels, which reduces intestinal lipid absorption [47]. The creatinine was decreased with GA levels (0.12 to 1.0%), possibly due to reduced intestinal fluid absorption and enhanced renal functions. Ali et al. [48] indicated that GA was associated with decreased creatinine in healthy mice.

IL-4 and IL-6 function as pre-inflammatory cytokines, while IL-10 is an anti-inflammatory cytokine. This study indicated that GA leads to down-regulation of pro and anti-inflammatory cytokine expressions in chickens through the lowest fold changes in IL-4 expression (at 0.12 and 1.0% GA) and IL-6 expression (at 0.12 to 1.0% GA) and the highest fold change in IL-10 expression (0.25% GA). These results indicate that GA can act as antigens through recognition by immune cell receptors, which beneficially modulate host immunity. Kamal et al. [49] reported that GA increased IL10 and decreased IL-4 and IL-6 in humans. Stabilizing the gut microbiota by increasing beneficial bacteria and eliminating pathogens can promote gut health and modulate host immunity, which may be reflected in broiler chicken growth performance [16]. The microbiota of the cecum is more abundant than the gastrointestinal tract segments and plays a role in the fermentation of indigestible fibers [50]. Teng and Kim [16] indicated that GA improves gut health by promoting *lactobacilli* spp. in young chickens. In our study, chickens that received GA levels (0.25 to 1.0%) had higher *Lactobacillus* spp. and lower *Salmonella* spp. content. Therefore, the pH values of the cecum of chickens were lower than the control. These results agree with the study's findings that prebiotics increase lactic acid through fermentation and growth of bacterial populations, especially lactobacilli, in the cecum, thereby lowering pH values. Lactic acid is a major byproduct of Lactobacillus bacteria [51]. Our results confirmed this, which showed a strong negative correlation between Lactobacillus count and cecal pH. Pelicano et al. [52] found that the low pH of the lumen inhibited acid-sensitive pathogenic bacteria, such as *Salmonella Typhimurium*, *Clostridium perfringens*, and *Escherichia coli*.

The most commonly used standards to assess nutrient absorption and gut health are villus length, crypt depth, and villus length to crypt ratio [53]. Moreover, broiler chickens have a strong relationship with increased villus height, gut health, and absorption efficiency [54]. High crypt height may indicate increased proliferative activity to compensate for villus height loss [55]. The ratio of villus height to crypt height is a useful measurement for estimating the absorptive capacity of the small intestine, which correlates with increased epithelial cell turnover, and longer villi are associated with activated cell mitosis [56]. In the current study, villus length and villus length to crypt depth ratio were higher in the duodenum, while the crypt depth of villi decreased when chickens were fed GA powder (0.12 to 1.0%) compared to the control group. In agreement with Macari and Maiorka [57], it was shown that the use of fermented prebiotics by bacteria in the cecum increases the height of the duodenal villi of chickens aged 1 to 7 days after hatching. Moreover, the increase in absorbent surface area of villi in the duodenum occurs more rapidly with the age of chickens up to 7 days. The use of GA in chicken diets (1 to 10 days old) is an effective strategy to improve the early growth and development of the gastrointestinal tract. Moreover, the early development of duodenal morphological and functional characteristics of broiler chickens in the initial stage leads to an improvement in early growth performance [17]. GA could be altering the gut microbiota and improving the integrity of intestinal epithelial cells, leading to better absorption of nutrients and hence better growth performance.

5. Conclusions

We concluded that using GA (0.25 to 0.75%) as a natural prebiotic in the diet of post-hatching chickens could be an effective strategy to improve the early growth and development of the gastrointestinal tract by altering the gut microbiota, modulating the immune response, and improving the intestinal epithelial.

Author Contributions: Conceptualization, methodology, formal analysis and writing—original draft: H.H.A.-B. Supervision, conceptualization, visualization, investigation, writing—review, and editing: R.A.A., M.M.A. and S.I.A.-M. Methodology: I.A.A., M.A.A.-G., K.E.I. and A.S.A. Visualization: A.A.A.-A., A.A.A.-S. and M.M.Q. All authors have read and agreed to the published version of the manuscript.

Funding: This work was funded by the Deanship of Scientific Research at King Saud University, through research group No (RG-1436-021).

Institutional Review Board Statement: The current study and the use of all chickens were approved by the Scientific Research Ethics Committee (SREC), King Saud University, Saudi Arabia (Ethics reference number: KSU-SE-20-39).

Informed Consent Statement: Not applicable.

Data Availability Statement: The data and analyses presented in this paper are freely available in the thesis of the first author (HHA).

Acknowledgments: The authors would like to extend their sincere appreciation to the Deanship of Scientific Research at King Saud University for funding this work through research project number RG-1436-021. The authors thank the Deanship of Scientific Research and RSSU at King Saud University for their technical support.

Conflicts of Interest: The authors declare no conflict of interest.

References

1. Smith, J.A. Broiler production without antibiotics: United States field perspectives. *Anim. Feed Sci. Technol.* **2019**, *250*, 93–98. [CrossRef]
2. Xiong, W.; Wang, Y.; Sun, Y.; Ma, L.; Zeng, Q.; Jiang, X.; Zhang, T. Antibiotic-mediated changes in the fecal microbiome of broiler chickens define the incidence of antibiotic resistance genes. *Microbiome* **2018**, *6*, 34. [CrossRef] [PubMed]
3. Schokker, D.; Jansman, A. J.; Veninga, G.; de Bruin, N.; Vastenhouw, S. A.; de Bree, F.M.; Smits, M.A. Perturbation of microbiota in one-day old broiler chickens with antibiotic for 24 hours negatively affects intestinal immune development. *BMC Genom.* **2017**, *18*, 241–255. [CrossRef] [PubMed]
4. Belal, A.S.; Ismail, A.; Elnaggar, M.M.; Belal, T.S. Click chemistry inspired copper sulphidenanoparticle-based fluorescence assay of kanamycin using DNA aptamer. *Spectrochim. Acta A Mol. Biomol. Spectrosc.* **2018**, *205*, 48–54. [CrossRef] [PubMed]
5. Amein, S.M.; Mosaad, G.M.; Hussein, M.K. Effect of some medicinal plants as feed additives on growth performance, blood constituents and carcass characteristics of broilers. *J. Adv. Vet. Res.* **2019**, *9*, 170–177.
6. Azzam, M.M.; Qaid, M.M.; Al-Mufarrej, S.I.; Al-Garadi, M.A.; Albaadani, H.H.; Alhidary, I.A. Rumex nervosus leaves meal improves body weight gain, duodenal morphology, serum thyroid hormones, and cecal microflora of broiler chickens during the starter period. *Poult. Sci.* **2020**, *99*, 5572–5581. [CrossRef]
7. Berrocoso, J.D.; Kida, R.; Singh, A.K.; Kim, Y.S.; Jha, R. Effect of in ovo injection of raffinose on growth performance and gut health parameters of broiler chicken. *Poult. Sci.* **2017**, *96*, 1573–1580. [CrossRef]
8. Abdelkareem, A.A.; Jaafar, S.F.; Hassan, H.M.; Taha, H.M.; Amal, Z.S. Gum Arabic supplementation improved antioxidant status and alters expression of oxidative stress gene in ovary of mice fed high fat diet. Middle. *East. Fertil. Soc. J.* **2016**, *21*, 101–108.
9. Babiker, R.; Merghani, T.H.; Elmusharaf, K.; Badi, R.M.; Lang, F.; Saeed, A.M. Effects of gum Arabic ingestion on body mass index and body fat percentage in healthy adult females: Two-arm randomized, placebo controlled, double blind trial. *Nutr. J.* **2012**, *11*, 111. [CrossRef] [PubMed]
10. Khalid, S.A.; Musa, A.M.; Saeed, A.M.; Abugroun, E.A.; Ahmed, E.O.S.; Ghalib, M.B.; Phillips, G.O. Manipulating dietary fibre: Gum Arabic making friends of the colon and the kidney. *Bioact. Carbohyd. Diet Fiber.* **2014**, *3*, 71–76. [CrossRef]
11. Abdulla, S.A.; Abdel-Atti, K.A.; Malik, H.E.; Dousa, B.M.; ElAmin, K.M. Effect of dietary inclusion of gum Arabic (*Acasia senegal*) on performance and blood chemistry of broiler chicks. *Global J. Anim. Sci. Res.* **2015**, *3*, 305–310.
12. El-Ratel, I.T.; Ismail, R.F.; Fouda, S.F. Productive performance, carcass traits, lipid profile, antioxidants and immunity of growing rabbits treated with gum Arabic under Egyptian summer condition. *Egyptian J. Nutr. Feeds.* **2019**, *22*, 143–154. [CrossRef]
13. Longdet, I.Y.; Eyibo, A.S.; Istifanus, G.; Blessing, O.E.; Bogolnaan, A.D.; Denkok, Y. Determination of the effect of gum arabic on body weight and some bio-chemical parameters on albino wistar rat. *Eur. J. Nutr. Food Safety* **2018**, *8*, 14–19.
14. Sharma, C.; Anejam, K.R.; Surain, P.; Dhiman, R. In vitro, evaluation of anti-microbial spectrum of *Acacia nilotica* leaves and bark extracts against pathogens causing otitis infection. *J. Innov. Biol.* **2014**, *1*, 51–56.
15. Lockyer, S.; Stanner, S. Prebiotics—An added benefit of some fiber types. *Nutr. Bull.* **2019**, *44*, 74–91. [CrossRef]
16. Teng, P.Y.; Kim, W.K. Roles of prebiotics in intestinal ecosystem of broilers. *Front. Vet. Sci.* **2018**, *5*, 245–263. [CrossRef]
17. Jha, R.; Kim, S.W. Nutritional Intervention for the Intestinal Health of Young Monogastric Animals. *Front. Vet. Sci.* **2021**, *8*, 1–9. [CrossRef]
18. AOAC. *Official Methods of Analysis*, 19th ed.; Association of Official Analytical Chemists: Washington, DC, USA, 2012.
19. Saleh, A.A.; Amber, K.A.; Soliman, M.M.; Soliman, M.Y.; Morsy, W.A.; Shukry, M.; Alzawqari, M.H. Effect of low protein diets with amino acids supple-mentation on growth performance, carcass traits, blood parameters and muscle amino acids profile in broiler chickens under high ambient temperature. *Agriculture* **2021**, *11*, 185. [CrossRef]
20. Gashua, I.B.; Williams, P.A.; Yadav, M.P.; Baldwin, T.C. Characterization and molecular association of Nigerian and Sudanese Acacia gum exudates. *Food Hydrocoll.* **2015**, *51*, 405–413. [CrossRef]
21. Elsayed, T.R.; Galil, D.F.; Sedik, M.Z.; Hassan, H.M.; Sadik, M.W. Anti-microbial and anticancer activities of actinomycetes isolated from Egyptian soils. *Int. J. Curr. Microbiol. App. Sci.* **2020**, *9*, 1–12. [CrossRef]
22. Diler, Ö.; Özil, Ö.; Bayrak, H.; Yiğit, N. Ö.; Özmen, Ö.; Saygın, M.; Aslankoç, R. Effect of dietary supplementation of sumac fruit powder (*Rhuscoriaria* L.) on growth performance, serum biochemistry, intestinal morphology and antioxidant capacity of rainbow trout (*Oncorhynchus mykiss*, Walbaum). *Anim. Feed Sci. Technol.* **2021**, *278*, 114993–115002. [CrossRef]
23. Goiri, I.; Ruiz, R.; Atxaerandio, R.; Lavin, J.L.; de Otálora, X.D.; García-Rodríguez, A. Assessing the potential use of a feed additive based on biochar on broilers feeding upon productive performance, pH of digestive organs, cecum fermentation and bacterial community. *Anim. Feed Sci. Technol.* **2021**, *279*, 115039. [CrossRef]
24. Shahryari, M.; Tabeidian, S.A.; Shahraki, A.D.F.; Tabatabaei, S.N.; Toghyani, M.; Forouzmand, M.; Habibian, M. Using soybean acid oil or its calcium salt as the energy source for broiler chickens: Effects on growth performance, carcass traits, intestinal morphology, nutrient digestibility, and immune responses. *Anim. Feed. Sci. Technol.* **2021**, *276*, 114919. [CrossRef]
25. Zanu, H.K.; Kheravii, S.K.; Morgan, N.K.; Bedford, M.R.; Swick, R.A. Over-processed meat and bone meal and phytase effects on broilers challenged with subclinical necrotic enteritis: Part 1. Performance, intestinal lesions and pH, bacterial counts and apparent ileal digestibility. *Anim. Nutr.* **2020**, *6*, 313–324. [CrossRef] [PubMed]
26. Al-Baadani, H.H.; Abudabos, A.M.; Al-Mufarrej, S.I.; Al-Baadani, A.A.; Alhidary, I.A. Dietary supplementation of *Bacillus subtilis*, *Saccharomyces cerevisiae* and their symbiotic effect on serum biochemical parameters in broilers challenged with *Clostridium perfringens*. *J. Appl. Anim. Res.* **2018**, *46*, 1064–1072. [CrossRef]

27. Panda, A.K.; Rao, S.V.R.; Raju, M.V.; Sharma, S.R. Dietary supplementation of *Lactobacillus sporogenes* on performance and serum biochemico-lipid profile of broiler chickens. *J. Poult. Sci.* **2006**, *43*, 235–240. [CrossRef]
28. Elnagar, R.; Elkenany, R.; Younis, G. Interleukin gene expression in broiler chickens infected by different *Escherichia coli* serotypes. *Vet. World.* **2021**, *14*, 2727. [CrossRef] [PubMed]
29. Qaid, M.M.; Al-Mufarrej, S.I.; Azzam, M.M.; Al-Garadi, M.A.; Albaadani, H.H.; Alhidary, I.A.; Aljumaah, R.S. Growth performance, serum biochemical indices, duodenal histomorphology, and cecal microbiota of broiler chickens fed on diets supplemented with cinnamon bark powder at pre-starter and starter phases. *Animals* **2021**, *11*, 94. [CrossRef]
30. Daneshmand, A.; Kermanshahi, H.; King, A.J.; Ibrahim, S.A. Combination of purine and pyrimidine nucleosides influences growth performance, gut morphology, digestive enzymes, serum biochemical indices and immune functions in broiler chickens. *Anim. Feed Sci. Technol.* **2017**, *228*, 186–193. [CrossRef]
31. Pour, A.R.A.; Kermanshahi, H.; Golian, A. Effects of conditioning time and activated sodium bentonite on pellet quality, performance, intestinal morphology, and nutrients retention in growing broilers fed wheat-soybean meal diets. *Anim. Feed Sci. Technol.* **2021**, *277*, 114955. [CrossRef]
32. SAS Institute. *SAS Users Guide: Statistics*; SAS Institute Inc.: Cary, NC, USA, 2008.
33. Belote, B.L.; Soares, I.; Tujimoto-Silva, A.; Sanches, A.W.; Kraieski, A.L.; Santin, E. Applying I see inside histological methodology to evaluate gut health in broilers challenged with *Eimeria*. *Vet. Parasitol.* **2019**, *276*, 100004. [CrossRef] [PubMed]
34. Khojah, E.Y. Biological effects of low protein diet with gum *Arabic* on rat's chronic kidney disease. *Adv. Environ. Biol.* **2017**, *11*, 60–69.
35. Musa, H.H.; Ahmed, A.A.; Fedail, J.S.; Musa, T.H.; Sifaldin, A.Z. Gum Arabic attenuates the development of nephropathy in type 1 diabetes rat. In *Gums and Stabilisers for the Food Industry*; RSC: London, OH, USA, 2016; Volume 18, pp. 245–255.
36. Kishimoto, A.; Ushida, K.; Phillips, G.O.; Ogasawara, T.; Sasaki, Y. Identification of intestinal bacteria responsible for fermentation of gum *Arabic* in pig model. *Curr. Microbiol.* **2006**, *53*, 173–177. [CrossRef] [PubMed]
37. Calame, W.; Weseler, A.R.; Viebke, C.; Flynn, C.; Siemensma, A.D. Gum Arabic establishes prebiotic functionality in healthy human volunteers in a dose-dependent manner. *Br. J. Nutr.* **2008**, *100*, 1269–1275. [CrossRef]
38. Adil, S.; Banday, T.; Bhat, G.A.; Mir, M.S.; Rehman, M. Effect of dietary supplementation of organic acids on performance, intestinal histomorphology, and serum biochemistry of broiler chicken. *Vet. Med. Int.* **2010**, *2010*, 479485. [CrossRef] [PubMed]
39. Gultermirian, M.L.; Corti, H.R.; Perez Chaia, A.; Apella, M.C. Fermentation of in vitro of a mixture of dietary fibers and cane molasses by the cecal microbiota: Application on mineral absorption through the laying hen's colonic epithelium. *Anim. Feed Sci. Technol.* **2014**, *191*, 76–82. [CrossRef]
40. Tabidi, M.H.; Ekram, K.A. Effect of feeding gum Arabic with or without commercial xylem enzyme 500 on the performance of broiler chicks. *World J. Pharm. Pharm. Sci.* **2015**, *4*, 1863–1872.
41. Al-Fadil, S.; Mukhtar, M.A.M.; Tabidi, M.H. Response of broiler chicks to diets containing gum *Arabic* as a natural prebiotic. *J. Curr. Res. Sci.* **2013**, *1*, 247–253.
42. Sato, K.; Takahashi, K.; Tohno, M.; Miura, Y.; Kamada, T.; Ikegami, S.; Kitazawa, H. Immunomodulation in gut-associated lymphoid tissue of neonatal chicks by immunobiotic diets. *Poult. Sci.* **2009**, *88*, 2532–2538. [CrossRef]
43. Raju, S.; Gurram, S.; Bora, S.; Kumar, B.S. Effect of herbal immunomodulators on immune status, haematological and serum biochemical parameters of Japanese quails (*Coutrinixjapanica*) during summer stress. *Indian J. Anim. Sci.* **2019**, *89*, 575–577.
44. Amber, K.; Abd El-Nabi, F.M.; Morsy, W.A.; Shama, S.H.A. Gum Arabic as prebiotic in growing rabbit's diet. *Glob. Vet.* **2017**, *19*, 465–471.
45. Nasir, O.; Babiker, S.; Salim, A. M. M. Protective effect of gum Arabic supplementation for type 2 diabetes mellitus and its complications. *Int. J. Multidiscip. Curr. Res.* **2016**, *4*, 288–294.
46. Musa, H.H.; Ahmed, A.A.; Musa, T.H. Chemistry, biological, and pharmacological properties of gum *Arabic*. In *Bioactive Molecules in Food.*; Springer International Publishing AG: Cham, Switzerland, 2018; pp. 1–18.
47. Lattimer, J.M.; Haub, M.D. Effects of dietary fiber and its components on metabolic health. *Forum. Nutr.* **2010**, *2*, 1266–1289. [CrossRef]
48. Ali, B.H.; Ziada, A.; Blunden, G. Biological effects of gum Arabic: A review of some recent research. *Food Chem. Toxicol.* **2009**, *47*, 1–8. [CrossRef] [PubMed]
49. Kamal, E.; Kaddam, L.A.; Dahawi, M.; Osman, M.; Salih, M.A.; Alagib, A.; Saeed, A. Gum arabic fibers decreased inflammatory markers and disease severity score among rheumatoidarthritis patients, Phase II Trial. *Int. J. Rheumatol.* **2018**, *2018*, 4197537. [CrossRef]
50. Pourabedin, M.; Zhao, X. Prebiotics and gut microbiota in chickens. *FEMS. Microb. Letters.* **2015**, *362*, 122–129. [CrossRef] [PubMed]
51. Alvarez-Sieiro, P.; Montalbán-López, M.; Mu, D.; Kuipers, O.P. Bacteriocins of lactic acid bacteria: Extending the family. *Appl. Microbiol. Biotechnol.* **2016**, *100*, 2939–2951. [CrossRef]
52. Pelicano, E.R.L.; Souza, P.A.; Souza, H.B.A.; Figueiredo, D.F.; Boiago, M.M.; Carvalho, S.R.; Bordon, V.F. Intestinal mucosa development in broiler chickens fed natural growth promoters. *Braz. J. Poult. Sci.* **2005**, *7*, 221–229. [CrossRef]
53. Ducatelle, R.; Goossens, E.; Meyer, F.; Eeckhaut, V.; Antonissen, G.; Haesebrouck, F.; Immerseel, F. Biomarkers for monitoring intestinal health in poultry: Present status and future perspectives. *Vet. Res.* **2018**, *49*, 1–9. [CrossRef] [PubMed]

54. Alfaro, D.M.; Silva, A.V.F.; Borges, S.A.; Maiorka, F.A.; Vargas, S.; Santin, E. Use of *Yucca schidigera* extract in broiler diets and its effects on performance results obtained with different coccidiosis control methods. *J. Appl. Poult. Res.* **2007**, *16*, 248–254. [CrossRef]
55. Chiang, G.; Lu, W.Q.; Piao, X.S.; Hu, J.K.; Gong, L.M.; Thacker, P.A. Effects of feeding solid-state fermented rapeseed meal on performance, nutrient digestibility, intestinal ecology, and intestinal morphology of broiler chickens. *Asian-Austr. J. Anim. Sci.* **2010**, *23*, 263–271. [CrossRef]
56. Awad, W.A.; Ghareeb, K.; Abdel-Raheem, S. Effects of dietary inclusion of probiotic and symbiotic on growth performance, organ weights and intestinal histomorphology of broiler chickens. *Poult. Sci.* **2009**, *88*, 49–56. [CrossRef] [PubMed]
57. Macari, M.; Maiorka, A. Gastrointestinal function and its impact on poultry performance. In Proceedings of the Apinco Conference on Poultry Science and Technology. *Campinas: Poult. Foun. Poult. Sci. Techno.* **2000**, *2*, 161–174.

animals

MDPI

Article

Effect of Phytase Level and Form on Broiler Performance, Tibia Characteristics, and Residual Fecal Phytate Phosphorus in Broilers from 1 to 21 Days of Age

Jose R. Hernandez [1], Joseph P. Gulizia [1], John B. Adkins [2], Martha S. Rueda [1], Samuel I. Haruna [2], Wilmer J. Pacheco [1] and Kevin M. Downs [2,*]

[1] Department of Poultry Science, Auburn University, Auburn, AL 36849, USA; jzh0237@auburn.edu (J.R.H.); jzg0120@auburn.edu (J.P.G.); msr0036@auburn.edu (M.S.R.); wjp0010@auburn.edu (W.J.P.)

[2] School of Agriculture, Middle Tennessee State University, Murfreesboro, TN 37132, USA; jba0030@auburn.edu (J.B.A.); samuel.haruna@mtsu.edu (S.I.H.)

* Correspondence: kevin.downs@mtsu.edu

Simple Summary: The demand for poultry products is increasing at the same time as the prices of raw materials and other ingredients used in feed manufacturing, leading to the implementation of strategies to improve nutrient utilization in broiler diets and reduce feeding costs. One strategy used in broiler production is the dietary supplementation of phytase, an enzyme that assists in phytate degradation. Phytate is the primary storage form of phosphorus (P) in seeds and it accounts for two-thirds of the P in vegetable feedstuffs. Dietary phytase supplementation has become a common practice because the activity of endogenous phytase in broilers does not allow for the adequate utilization of phytate-bound P. Consequently, the effects of phytase have been widely studied and the beneficial effects on broiler growth performance and reduced P excretion into the environment have been previously described. In the present study, broilers were fed a diet with a combination of two phytase forms (coated and uncoated) to assess the effects on broiler performance, tibia characteristics, and residual fecal phytate P. The results indicated no distinct advantage of combining phytase forms, but this information may be useful for the direction of future research.

Abstract: The present study evaluated the individual and combined effects of coated and uncoated phytase on broiler performance, tibia characteristics, and residual phytate phosphorus (P) in manure. Two repeated studies were conducted using 240-day-old Cobb 500 by-product male broilers per trial. For each trial, birds were assigned to four treatments with four replicate battery cages per treatment (60 birds/trt) and grown for 21 days. Treatments included: (1) negative control (NC), (2) NC + 1000 phytase units (FTU) coated phytase (C), (3) NC + 1000 FTU uncoated phytase (U), and (4) NC + 500 FTU coated + 500 FTU uncoated phytase (CU). Data were analyzed with a one-way ANOVA and means were separated using Tukey's HSD. In the pooled data for both trials, all treatments with dietary phytase had a higher body weight (BW) and feed consumption (FC) than the NC on day 21 ($p < 0.05$). Similarly, a six-point reduction was observed for day 1 to 21 feed conversion (FCR) for U and CU ($p < 0.05$). All treatments with phytase inclusion differed from the NC in every evaluated parameter for bone mineralization ($p < 0.05$) and had significantly lower fecal phytate P concentrations compared to the NC ($p < 0.05$). Overall, bird performance was essentially unaffected by phytase form, indicating that combining phytase forms does not appear to offer any advantage to the evaluated parameters from day 1 to 21.

Keywords: phytase; broilers; growth performance; bone mineralization; phytate phosphorus

check for updates

Citation: Hernandez, J.R.; Gulizia, J.P.; Adkins, J.B.; Rueda, M.S.; Haruna, S.I.; Pacheco, W.J.; Downs, K.M. Effect of Phytase Level and Form on Broiler Performance, Tibia Characteristics, and Residual Fecal Phytate Phosphorus in Broilers from 1 to 21 Days of Age. *Animals* **2022**, *12*, 1952. https://doi.org/10.3390/ani12151952

Academic Editor: Sylwester Świątkiewicz

Received: 3 July 2022
Accepted: 29 July 2022
Published: 1 August 2022

1. Introduction

The increasing demand for poultry meat and higher feed prices have led to the implementation of nutritional strategies that optimize nutrient utilization and reduce feeding costs.

Among these strategies is the dietary inclusion of phytase, which reduces the nutrient variability of feedstuffs and counteracts the antinutritional effects of phytate [1]. Phytase inclusion has become a common practice in broiler feeding as a means of increasing the hydrolysis of phytate, improving phytate phosphorus (P) utilization, and reducing P excretion [2].

Phytate (myo-inositol (1,2,3,4,5,6) hexakisphosphate; IP_6) is the salt form of phytic acid (PA) and the primary storage form, typically contributing 50 to 80% of total P, in plant seeds [3]. Chemically, it consists of a myo-inositol ring associated with up to six phosphate anions [4]. Since broiler diets are mainly manufactured with feedstuffs of plant origin, the antinutritive effects of phytate in broiler production have been extensively reviewed [5,6]. Approximately two-thirds of P in vegetable feedstuffs is poorly digested by poultry because it is bound to PA; birds do not produce sufficient amounts of endogenous phytase to effectively hydrolyze PA [7]. Additionally, since phytate is negatively charged under many pH conditions, it can form complexes with positively charged molecules, which may reduce the bioavailability of minerals, amino acids, and protein [8–10]. There are reports that phytate also reduces endogenous enzyme activity in broilers [11]. Ultimately, phytate can negatively affect bird performance and increase the amount of P excreted into the environment.

Phytases (myo-inositol (1,2,3,4,5,6) hexakisphosphate phosphohydrolases) represent a subgroup of phosphatases that are capable of initiating phytate dephosphorylation, thus releasing phosphate groups from PA [7,9]. These enzymes are often sourced from bacteria and fungi and can be classified as 3-phytases or 6-phytases depending on the carbon site of hydrolysis in the phytate molecule [12]. Phosphate group removal starts with a fully phosphorylated PA, followed by penta-, tetra-, tri-, di- and mono-esters of inositol in descending order of preference [5,13,14]. In addition, the release of inorganic P depends on several factors, including, but not limited to, dietary phytate concentration, source of phytate, species and age of animals, mineral concentrations in the diet, phytase sources, and phytase dosing [15].

Inorganic P sources are expensive and represent the third most expensive component of poultry diets after energy and amino acids [7]. Furthermore, P excretion into the environment is a hazard to water quality because P is a predisposing factor for the eutrophication of freshwater bodies [16]. Therefore, the use of phytase is beneficial economically and for sustainable production. Previous studies have determined that the inclusion of phytase improves broiler performance by increasing feed consumption (FC) and body weight (BW) and decreasing feed conversion ratio (FCR) [17–22]. Similarly, phytase supplementation has proven beneficial to bone mineralization from improved P digestibility [23,24].

Commercially available phytases may differ in several characteristics, including form and origin. Some phytases are encapsulated by lipid or carbohydrate coating to minimize oxidation at high temperatures, particularly during pelleting conditions [25]. Thermostability is an important characteristic of phytases, as animal diets are frequently pelleted at temperatures up to 90 °C [26], but evaluating this attribute is not within the scope of this study. Previous studies have assessed the activity of coated and uncoated phytases on broiler performance and tibia mineralization. Ayres et al. [12] evaluated the activity of a coated phytase and an uncoated phytase developed to be intrinsically heat-stable on 21-day broiler performance and tibia mineralization. They concluded that the uncoated and intrinsically heat-stable phytase was more efficacious than coated phytase. However, there is limited information regarding the effect of combining different phytase forms (coated and uncoated) in the same diet. Therefore, the objective of this study was to determine the effects associated with combining coated and uncoated phytase sources on broiler performance, tibia characteristics, and residual phytate P in broiler manure from 1 to 21 days of age (starter phase).

2. Materials and Methods

2.1. Diet Preparation

Two repeated trials were conducted to fully elucidate the effect of phytase form and inclusion on broiler performance, tibia characteristics, and residual fecal phytate P. Coated (OptiPhos®; Huvepharma Inc., Peachtree City, GA, USA) and uncoated (OptiPhos® Plus; Huvepharma Inc., Peachtree City, GA, USA) phytase was incorporated into a basal broiler starter diet (mash form) as either a single inclusion or a combination to create four experimental diets (Table 1). Both products are 6-phytase derived from *E. coli.* The basal diet was manufactured as a negative control (NC) with reduced calcium (Ca) and available P (aP) levels; however, all remaining nutrients represented a typical broiler starter diet. The NC diet was formulated with 0.20% less Ca and aP compared to industry standards [27] and National Research Council (NRC) requirements [28]. Three additional treatments were manufactured using the NC with different combinations and forms of phytase products. Two treatments contained either a single inclusion of a coated (1000 FTU/kg complete feed; FTU = phytase units) or uncoated (1000 FTU/kg complete feed) phytase, whereas one treatment contained a combination of coated (500 FTU/kg complete feed) and uncoated (500 FTU/kg complete feed) phytase. The addition of 1000 FTU of phytase activity per kg of complete feed was expected to increase Ca and P availability by 0.20%.

Table 1. Ingredient and analyzed nutrient compositions of the negative control (NC) diet (as-fed basis).

Ingredient, % of Diet (Unless Otherwise Noted)	NC
Corn	51.60
Soybean meal, 46% crude protein	37.94
Corn oil	3.31
Distillers dried grains with solubles	4.00
Dicalcium phosphate	0.55
Ground limestone	1.45
Salt (NaCl)	0.38
DL-Methionine	0.33
L-Lysine	0.18
Trace mineral premix [A]	0.10
Vitamin premix [B]	0.10
Choline chloride	0.07
Phytase, g supplement/kg diet	0.00 [C]
Calculated Nutrients, % (Unless Otherwise Noted)	
AMEn, kcal/kg	3000
Crude protein	23.17
Calcium	0.80
Available phosphorus	0.20
Digestible methionine	0.64
Digestible methionine + cysteine	0.93
Digestible lysine	1.23
Digestible threonine	0.73
Digestible valine	0.96
Digestible tryptophan	0.25

[A] Trace Mineral premix source and amount provided per kg of diet: Mn (manganese sulfate), 120 mg; Zn (zinc sulfate), 100 mg; Fe (iron sulfate monohydrate), 30 mg; Cu (tri-basic copper chloride), 8 mg; I (ethylenediamine dihydriodide), 1.4 mg; and Se (sodium selenite), 0.3 mg. [B] Vitamin premix source and amount provided per kg of diet: Vitamin A (Vitamin A acetate), 18,739 IU; Vitamin D (cholecalciferol), 6614 IU; Vitamin E (DL-alpha tocopherol acetate), 66 IU; menadione (menadione sodium bisulfate complex), 4 mg; Vitamin B12 (cyanocobalamin), 0.03 mg; folacin (folic acid), 2.6 mg; D-pantothenic acid (calcium pantothenate), 31 mg; riboflavin (riboflavin), 22 mg; niacin (niacinamide), 88 mg; thiamine (thiamine mononitrate), 5.5 mg; D-biotin (biotin), 0.18 mg; and pyridoxine (pyridoxine hydrochloride), 7.7 mg. [C] Using the NC, 3 treatments were supplemented with either OptiPhos® (0.126 g/kg; Huvepharma Inc., Peachtree City, GA, USA), OptiPhos® Plus (0.090 g/kg; Huvepharma Inc., Peachtree City, GA, USA), or a combination of both phytase products (0.108 g/kg) to achieve 1000 FTU of phytase activity per kg of diet.

Dietary treatments were prepared at the Auburn University Animal Nutrition Center as described in Downs et al. [29]. The phytase supplement was premixed with ground corn prior to its addition to the whole feed batch. After mixing, coated, uncoated, or the combination of both phytase products were added to the basal diet and mixed for 5 min to prepare each treatment with dietary phytase inclusion. The starter feed used in each treatment met or exceeded Cobb 500 nutrient recommendations [27] and NRC requirements [28], except for Ca and aP.

2.2. Phytase Analysis

Feed samples were collected during feed manufacture and analyzed for phytase activity levels by an external laboratory. The analyzed phytase activity for each treatment is shown in Table 2.

Table 2. Analysis of phytase activity in the negative control (NC) and phytase-supplemented diets.

Item	NC	NC + Coated [A]	NC + Uncoated [B]	NC + Combined [C]
Expected value, FTU/kg	0	1000	1000	1000
Analyzed value, FTU/kg [D]	<13	1180	1090	1150

[A] OptiPhos® (Huvepharma Inc., Peachtree City, GA, USA) added to provide 1000 FTU/kg of phytase activity per kg of diet. [B] OptiPhos® Plus (Huvepharma Inc., Peachtree City, GA, USA) added to provide 1000 FTU/kg of phytase activity per kg of diet. [C] OptiPhos® and OptiPhos® Plus combined to provide 1000 FTU/kg of phytase activity per kg of diet. [D] Phytase was analyzed by an external laboratory (Huvepharma; BIOVET laboratory for feed analysis, Peshtera, Bulgaria).

2.3. Bird Management and Data Collection

Animal handling procedures were approved by the Middle Tennessee State University Institutional Animal Care and Use Committee (IACUC) under proposal 21–2001 and conformed to accepted animal welfare standards [30,31]. Both trials were conducted using 240-day-old Cobb 500 by-product males obtained from a local hatchery. Chicks were randomly allocated to 16 battery cages (15 birds per cage; 527 cm^2/bird) and each cage was randomly assigned to a treatment (4 treatments; 4 replicate cages/treatment; 60 birds/treatment). Birds were brooded at 35 °C with the temperature reduced by 3 °C every 7 days. The temperature of the room housing battery cages was maintained at 27 °C and continuous light was provided (24 L:0 D; 25 lux), with feed and water offered *ad libitum*. Bird spacing for feeders (9.6 cm/bird) and drinkers (4.8 cm/bird) met or exceeded the recommendations for broilers grown to 21 days of age [30,31].

All birds and feed were weighed at days 1, 14, and 21 to determine average BW and FC. Feed conversion was calculated using FC and BW gain data and adjusted for mortality. Mortality weights were included in the FCR calculation to determine mortality-adjusted values.

On day 21, left tibias were collected to assess shear strength and ash content. For this procedure, 5 birds per cage were randomly selected and euthanized according to AVMA guidelines [32]. Tibias were evaluated for shear strength using a TA.XT plus texture analyzer (Stable Microsystems, Surrey, UK) according to the official method (ANSI/ASAE method s459) [33]. A test speed of 5 mm/sec and a trigger force of 5 g were used. Applied fracture force was recorded in newtons (N) and determined as the peak force at the initial fracture. Peak 1 was considered the force to crack the bone and peak 2 was considered the force to break the bone. After breaking tibias and analyzing for shear strength, each bone was exposed to sequential extractions using 200-proof ethanol and anhydrous ether and ashed in a muffle furnace, as described by Hall et al. [34].

Excreta samples were collected on day 21 to assess phytate P level. An aliquot of fresh excreta was collected from the refuge tray of each battery cage, screened to remove non-fecal material, and frozen (−4 °C) for further analysis. Excreta samples were analyzed at the University of Georgia Feed and Environmental Water Laboratory (Agricultural and Environmental Services Laboratories, Athens, GA, USA) for phytate P level determination using a standard extraction-colorimetric procedure.

2.4. Statistical Analysis

Data were analyzed as a completely randomized block design, with a battery cage representing an experimental unit and trials treated as replicates. Data presented were pooled from both trials. Mortality data were arcsine-transformed. Data were analyzed using the general linear model (GLM) procedure of the SAS statistical package [35]. Least square means were compared using post hoc Tukey's HSD procedure, and all data were analyzed for normality using the Shapiro–Wilk test. Statistical significance was established at $p \leq 0.05$. Statistical tendency was considered as $0.05 < p < 0.10$.

3. Results

Phytase activity was determined in the four dietary treatments (Table 2). Diets without phytase inclusion had less than 13 FTU/kg. The treatments with phytase supplementation had an expected value of 1000 FTU/kg and the analyzed value had a slight variation between treatments. Phytase recoveries were 118, 109, and 115% for NC + Coated (C), NC + Uncoated (U), and NC + Combined (CU), respectively.

Performance results (Table 3) and other evaluated parameters were pooled from trials 1 and 2. For day 14 BW, CU was the only treatment not different from NC ($p > 0.05$). All phytase-treated birds had higher day 21 BW than NC ($p < 0.05$) regardless of the phytase form used. Feed consumption was the only parameter that did not differ across treatments from day 1 to 14 ($p > 0.05$), but from day 1 to 21, all three treatments containing phytase presented a higher FC than the NC ($p < 0.05$). In accordance with the results for BW, the data for FCR suggest that the use of phytase reduced FCR between 4 and 6 points. From day 1 to 14, C and CU both reduced FCR by 5 points compared to the NC ($p < 0.05$). Similarly, a 6-point reduction was observed for day 1 to 21 FCR for U and CU ($p < 0.05$). However, days 1 to 14 FCR of U did not differ from the NC and the same was observed for the C treatment from 1 to 21 days of age ($p > 0.05$). In addition, the pooled data indicate that treatments did not have an effect on bird mortality ($p > 0.05$).

Table 3. Influence of negative control (NC), coated phytase supplementation (NC + 1000 FTU [A] coated phytase), uncoated phytase supplementation (NC + 1000 FTU uncoated phytase), and combined phytase supplementation (NC + 500 FTU coated + 500 FTU uncoated) on broiler live performance from 1 to 21 days of age (pooled data from trial 1 and 2).

Item	NC	NC + Coated [B]	NC + Uncoated [C]	NC + Combined [D]	$P_r > F$ [E]	Pooled SEM [E]
Body weight, g/bird						
Day 1	43.1	43.4	43.3	43.0	0.614	0.25
Day 14	451 ^{b}	487 ^{a}	488 ^{a}	480 ^{ab}	0.019	17.4
Day 21	909 ^{b}	993 ^{a}	995 ^{a}	1002 ^{a}	<0.001	27.2
Feed consumption, g/bird						
Day 1 to 14	524	548	548	543	0.225	18.0
Day 1 to 21	1151 ^{b}	1231 ^{a}	1206 ^{a}	1216 ^{a}	0.002	28.1
Mortality adjusted FCR, g:g						
Day 1 to 14	1.29 ^{a}	1.24 ^{b}	1.25 ^{ab}	1.24 ^{b}	0.016	0.029
Day 1 to 21	1.34 ^{a}	1.30 ^{ab}	1.28 ^{b}	1.28 ^{b}	0.009	0.033
Mortality, %						
Day 1 to 14	0.00	2.50	4.17	1.67	0.212	3.88
Day 1 to 21	0.83	2.50	5.00	2.50	0.400	4.82

[A] Phytase unit. OptiPhos® (Huvepharma Inc., Peachtree City, GA, USA) provided 1000 FTU/kg of phytase activity per kg of diet. [B] OptiPhos® Plus (Huvepharma Inc., Peachtree City, GA, USA) provided 1000 FTU/kg of phytase activity per kg of diet. [C] OptiPhos® and [D] OptiPhos® Plus combined to provide 1000 FTU/kg of phytase activity per kg of diet. [E] P-value of F statistic; SEM = standard error of the mean. [a,b] Means in the same row with different superscript letters are significantly different ($p < 0.05$).

Tibias were assessed for bone weight, shear strength, and tibia ash percentage (Table 4). The obtained results indicate that all treatments with phytase inclusion differed from the NC in every bone parameter evaluated ($p < 0.05$). Tibia ash increased around 5% in diets with phytase inclusion and bone weight increase was 8, 10, and 12% for C, U, and CU, respectively ($p < 0.05$).

Table 4. Influence of negative control (NC), coated phytase supplementation (NC + 1000 FTU [A] coated phytase), uncoated phytase supplementation (NC + 1000 FTU uncoated phytase), and combined phytase supplementation (NC + 500 FTU coated + 500 FTU uncoated) on day 21 broiler tibia shear strength and tibia ash (pooled data from trials 1 and 2).

Item	NC	NC + Coated [B]	NC + Uncoated [C]	NC + Combined [D]	$P_r > F$	Pooled SEM
Bone Wt., g	5.0 ^{b}	5.4 ^{a}	5.5 ^{a}	5.6 ^{a}	<0.001	0.10
Shear strength, peak 1 [E], N	246 ^{b}	365 ^{a}	365 ^{a}	343 ^{a}	<0.001	11.1
Shear strength, peak 2 [F], N	261 ^{b}	410 ^{a}	398 ^{a}	384 ^{a}	<0.001	11.5
Tibia ash, %	47.16 ^{b}	52.38 ^{a}	52.12 ^{a}	52.59 ^{a}	0.003	0.310

[A] Phytase unit. [B] OptiPhos® (Huvepharma Inc., Peachtree City, GA, USA) provided 1000 FTU/kg of phytase activity per kg of diet. [C] OptiPhos® Plus (Huvepharma Inc., Peachtree City, GA, USA) provided 1000 FTU/kg of phytase activity per kg of diet. [D] OptiPhos® and OptiPhos® Plus combined to provide 1000 FTU/kg of phytase activity per kg of diet. [E] Force required to crack the bone. [F] Force required to break the bone. [a,b] Means in the same row with different superscript letters are significantly different ($p < 0.05$).

For fecal phytate P, significant differences were observed across treatments in the pooled data for both trials (Table 5). All treatments with dietary phytase inclusion had significantly lower fecal phytate P concentration compared to the NC ($p < 0.05$). The reduction in phytate P concentrations in feces were 74, 79, and 77% for C, U, and CU, respectively. In addition, no differences in fecal phytate P concentrations were observed between the phytase-containing treatments ($p > 0.05$).

Table 5. Influence of negative control (NC), coated phytase supplementation (NC + 1000 FTU [A] coated phytase), uncoated phytase supplementation (NC + 1000 FTU uncoated phytase), and combined phytase supplementation (NC + 500 FTU coated + 500 FTU uncoated) on phytate phosphorus concentrations in feces from 21-day old broilers (pooled data from trials 1 and 2).

Item	NC	NC + Coated [B]	NC + Uncoated [C]	NC + Combined [D]	$P_r > F$	Pooled SEM
Fecal phytate P, mg/kg	2.423 ^{a}	630 ^{b}	510 ^{b}	552 ^{b}	< 0.001	94.9

[A] Phytase unit. [B] OptiPhos® (Huvepharma Inc., Peachtree City, GA, USA) provided 1000 FTU/kg of phytase activity per kg of diet. [C] OptiPhos® Plus (Huvepharma Inc., Peachtree City, GA, USA) provided 1000 FTU/kg of phytase activity per kg of diet. [D] OptiPhos® and OptiPhos® Plus combined to provide 1000 FTU/kg of phytase activity per kg of diet. [a,b] Means in the same row with different superscript letters are significantly different ($p < 0.05$).

4. Discussion

Birds poorly utilize phytate-bound P due to their limited endogenous phytase production. Previous studies have reported that the inclusion of exogenous phytase can reduce the negative effects of phytate on nutrient utilization, resulting in phytase products being commonly utilized in poultry diets. As mentioned previously, coated and uncoated phytases have been evaluated to determine their effects on broiler performance, bone mineralization, and phytate hydrolysis. However, there is limited information on the effects of combining phytase forms. In this regard, it is important to mention that bird age can influence the utilization of nutrients and the efficacy of phytase in broiler chickens [36,37]. Broilers are able to utilize P more efficiently and phytase is more efficacious during the starter phase, with a higher sensitivity during the end of the second week of the growing period [37]. This period of rapid growth and development serves as a foundation for the deposition of meat in the subsequent phases [38], hence the importance of reporting results from the starter phase.

The BW results of this study were similar to the data obtained by Leyva-Jimenez et al. [39]. The authors reported that broilers fed diets supplemented with regular (250 FTU/kg) and superdose (1500 FTU/kg) levels of Optiphos 2000 (Huvepharma Inc. Peachtree, GA, USA) had a significantly higher day-22 BW in comparison to birds fed no phytase. Ayres et al. [12] used the same two products evaluated in the present study with increasing inclusion levels. In two of their experiments, they observed higher day-21 BW for birds fed diets with 1000 FTU of an uncoated phytase, compared to birds fed diets with the same inclusion of a coated phytase and a negative control with no phytase inclusion. Several studies have obtained similar results of increased body weight and feed intake with phytase supplementation at low and high inclusion levels [40,41]. This may be caused by the activity of phytase in the gastrointestinal tract, which contributes to the breakdown of IP_6 and lesser phytate esters, therefore increasing the availability of phytate-bound nutrients [41,42]. Additionally, these data indicate that apart from improving nutrient utilization, phytase also stimulates a feed intake response and, consequently, improves BW. It has been suggested that the effects of phytase in stimulating digestible nutrient intake may be associated with its impact on phytate degradation since phytate reduces feed intake in broilers [43,44].

The results for FCR that were obtained in this study suggest that phytase inclusion in diets with reduced Ca and aP is able to prevent the P deficiency effects on growth performance parameters, thus allowing birds to match the performance of birds fed nutrient-adequate diets [41]. Dersjant-Li et al. [45] obtained similar results when using a dose of 1000 FTU/kg of a coated phytase in diets with different reduced Ca levels. The authors reported a lower day 11 to 21 FCR in broilers fed phytase-supplemented diets, compared to an NC with no phytase inclusion and no Ca reduction. Similarly, Broch et al. [46] evaluated increasing levels of phytase inclusion (1000, 2000, and 3000 FTU/kg) in broiler diets deficient in aP and Ca and observed a linear response for all growth performance parameters from 1 to 21 days of age. In the present study, combining two different phytase forms did not appear to offer any distinct performance advantages. The benefits of combining different phytase forms could have been observed if they were released at different rates throughout the gastrointestinal tract of broilers. Exogenous phytase is, for the most part, active in the proximal segments of the gastrointestinal tract (crop, proventriculus, and gizzard) of poultry [47]. If combining phytase forms could extend phytase activity to the upper part of the small intestine, it would reduce the binding of phytate to dietary protein and the formation of calcium phytate complexes, thus reducing its antinutritional effects and endogenous amino acid losses [3,10]. Nevertheless, the results obtained in this study do not indicate any clear benefits of combining a coated and uncoated phytase product in the growth performance of broilers.

According to Cardoso Junior et al. [48], bone mineralization reflects adequate bone quality, which is associated with beneficial effects on broiler performance and is important to support muscular development. Furthermore, bone mineralization data is commonly used to estimate and validate inorganic P release by phytase and is an efficient parameter to quantify phytate P released in corn and soybean meal-based diets [3,49]. It is worth noting that Ca and P are closely related in bone mineralization results because they are both stored together in bone and Ca is stored almost entirely as hydroxyapatite crystals of Ca phosphate in bone [50]. In this study, the three treatments with phytase supplementation had improved results for all bone parameters in comparison to the NC. Similarly, Chung et al. [51] reported improvements in bone mineral content and bone mineral density in broilers fed diets with two inclusion levels (500 and 1000 FTU/kg) of a coated phytase compared to broilers fed low-P diets. In addition, Walk et al. [20] and several other authors have observed that tibia ash and other bone parameters improve with phytase addition [36,39,52]. This indicates a positive effect of phytase in bone mineralization, which is associated with an increase in the amount of aP [41]. Moreover, these results suggest that phytase supplementation is beneficial for bone mineralization and is in accordance with the data obtained for growth performance measurements. However, no clear effects were observed for the combination of different phytase forms.

The results for phytate P concentrations in feces from 21-day-old broilers demonstrate the beneficial effect of phytase on phytate degradation. The feces from broilers fed phytase-supplemented diets had reduced fecal phytate P concentrations compared to the NC and no differences were observed between treatments combining phytase forms. Martins et al. [53] observed a reduction in residual fecal phytate P of broilers fed phytase-supplemented diets (750 FTU/kg) compared to broilers fed diets with no phytase addition. Kriseldi et al. [54]

reported similar results when evaluating two doses of phytase (400 and 1200 FTU/kg) on phytate concentration in the ileal digesta of 28-day-old birds. The authors reported that both concentrations of phytase effectively increased phytate degradation and this led to increased inositol liberation. It is essential to achieve the maximum degradation of phytate to obtain the extra-phosphoric effects of phytase and observe improvements in the growth performance of broilers [55]. In addition, the degradation of IP_3, IP_4, and lower esters is critical, as the anti-nutritive effects of these lower IP esters may still be present to chelate nutrients [56]. Therefore, more research is required to assess the effect of combining different phytase forms in the degradation of IP esters and the liberation of inositol. Furthermore, reductions in the P excretion of broilers can reduce the environmental pollution burden, since P from poultry manure can pollute soil and is a significant eutrophication agent [57].

5. Conclusions

The results obtained in this study are consistent with previous research on phytase and its effects on broiler performance, tibia characteristics, and residual fecal phytate P. Improvements in BW, FC, and FCR were observed when using 1000 FTU/kg of dietary phytase, regardless of its form. Phytase addition also improved tibia characteristics and reduced residual phytate P in broiler feces. However, the combination of phytase forms does not appear to offer any distinct advantage to bird performance or bone mineralization from day 1 to 21. Although the phytase research area has been widely studied, a paucity of research exists that evaluates the effects of combining two different phytase forms (coated and uncoated) in the same broiler diet. Therefore, further research is needed to better understand if combining different phytase forms represents potential benefits for broiler growth.

Author Contributions: Conceptualization, K.M.D., J.P.G., W.J.P. and S.I.H.; methodology, J.B.A., K.M.D., J.P.G., S.I.H. and M.S.R.; formal analysis, J.P.G.; investigation, J.B.A. and K.M.D.; resources, W.J.P. and K.M.D.; data curation, K.M.D.; writing—original draft preparation, J.R.H. and J.P.G.; writing—review and editing, J.R.H., J.P.G., K.M.D. and W.J.P.; supervision, K.M.D.; project administration, K.M.D. and J.B.A. All authors have read and agreed to the published version of the manuscript.

Funding: This research received no external funding.

Institutional Review Board Statement: This experiment was approved by the Middle Tennessee State University Institutional Animal Care and Use Committee (IACUC: PRN 21–2001) and conformed to the accepted practices of the American Society of Animal Science/American Dairy Science Association/Poultry Science Association. Birds were euthanized by cervical dislocation in accordance with American Veterinary Medical Association Guidelines.

Informed Consent Statement: Not applicable.

Data Availability Statement: All data are contained within the article.

Acknowledgments: Authors would like to express appreciation to Mark Jackson, Huvepharma, Inc., for his assistance in the phytase analysis of experimental diets.

Conflicts of Interest: The authors declare no conflict of interest.

References

1. De Sousa, J.P.; Albino, L.F.; Vaz, R.; Rodrigues, K.F.; Da Silva, G.F.; Renno, L.N.; Barros, V.; Kaneko, I.N. The effect of dietary phytase on broiler performance and digestive, bone, and blood biochemistry characteristics. *Rev. Bras. Cienc. Avic.* **2015**, *17*, 69–76. [CrossRef]
2. Selle, P.H.; Ravindran, V. Microbial phytase in poultry nutrition. *Anim. Feed Sci. Technol.* **2007**, *135*, 1–41. [CrossRef]
3. Dersjant-Li, Y.; Awati, A.; Schulze, H.; Partridge, G. Phytase in non-ruminant animal nutrition: A critical review on phytase activities in the gastrointestinal tract and influencing factors. *J. Sci. Food Agric.* **2015**, *95*, 878–896. [CrossRef] [PubMed]
4. Cowieson, A.J.; Acamovic, T.; Bedford, M.R. The effects of phytase and phytic acid on the loss of endogenous amino acids and minerals from broiler chickens. *Br. Poult. Sci.* **2004**, *45*, 101–108. [CrossRef]
5. Woyengo, T.A.; Nyachoti, C.M. Review: Supplementation of phytase and carbohydrases to diets for poultry. *Can. J. Anim. Sci.* **2011**, *91*, 177–192. [CrossRef]
6. Woyengo, T.A.; Nyachoti, C.M. Review: Anti-nutritional effects of phytic acid in diets for pigs and poultry–current knowledge and directions for future research. *Can. J. Anim. Sci.* **2013**, *93*, 9–21. [CrossRef]

7. Walk, C.L.; Rama Rao, S.V. Increasing dietary phytate has a significant anti-nutrient effect on apparent ileal amino acid digestibility and digestible amino acid intake requiring increasing doses of phytase as evidenced by prediction equations in broilers. *Poult. Sci.* **2020**, *99*, 290–300. [CrossRef] [PubMed]
8. Dersjant-Li, Y.; Davin, R.; Christensen, T.; Kwakernaak, C. Effect of two phytases at two doses on performance and phytate degradation in broilers during 1-21 days of age. *PLoS ONE* **2021**, *16*, e0247420. [CrossRef] [PubMed]
9. Broch, J.; dos Santos, E.C.; Damasceno, J.L.; Nesello, P.D.O.; de Souza, C.; Eyng, C.; Pesti, G.M.; Nunes, R.V. Phytase and phytate interactions on broilers' diet at 21 days of age. *J. Appl. Poult. Res.* **2020**, *29*, 240–250. [CrossRef]
10. Selle, P.H.; Cowieson, A.J.; Ravindran, V. Consequences of calcium interactions with phytate and phytase for poultry and pigs. *Livest. Sci.* **2009**, *124*, 126–141. [CrossRef]
11. Cowieson, A.J.; Acamovic, T.; Bedford, M.R. Phytic acid and phytase: Implications for protein utilization by poultry. *Poult. Sci.* **2006**, *85*, 878–885. [CrossRef]
12. Ayres, V.E.; Jackson, M.E.; Cantley, S.A.; Rochell, S.J.; Crumpacker, C.D.; Lee, D.T.; Bodle, B.C.; Pacheco, W.J.; Rueda, M.S.; Bailey, C.A.; et al. Multiexperiment evaluation of increasing phytase activity from Optiphos® and Optiphos Plus® on 21-day broiler performance and tibia mineralization. *J. Appl. Poult. Res.* **2021**, *30*, 100210. [CrossRef]
13. Yu, S.; Cowieson, A.; Gilbert, C.; Plumstead, P.; Dalsgaard, S. Interactions of phytate and myo-inositol phosphate esters (IP1-5) including IP5 isomers with dietary protein and iron and inhibition of pepsin. *J. Anim. Sci.* **2012**, *90*, 1824–1832. [CrossRef] [PubMed]
14. Vats, P.; Banerjee, U.C. Production studies and catalytic properties of phytases (myo-inositolhexakisphosphate phosphohydrolases): An overview. *Enzym. Microb. Technol.* **2004**, *35*, 3–14. [CrossRef]
15. Adeola, O.; Cowieson, A.J. BOARD-INVITED REVIEW: Opportunities and challenges in using exogenous enzymes to improve nonruminant animal production. *J. Anim. Sci.* **2011**, *89*, 3189–3218. [CrossRef] [PubMed]
16. Singh, P.K. Significance of phytic acid and supplemental phytase in chicken nutrition: A review. *World's Poult. Sci. J.* **2008**, *64*, 553–580. [CrossRef]
17. Cowieson, A.J.; Acamovic, T.; Bedford, M.R. Supplementation of corn-soy-based diets with an Eschericia coli-derived phytase: Effects on broiler chick performance and the digestibility of amino acids and metabolizability of minerals and energy. *Poult. Sci.* **2006**, *85*, 1389–1397. [CrossRef] [PubMed]
18. Walk, C.L.; Santos, T.T.; Bedford, M.R. Influence of superdoses of a novel microbial phytase on growth performance, tibia ash, and gizzard phytate and inositol in young broilers. *Poult. Sci.* **2014**, *93*, 1172–1177. [CrossRef] [PubMed]
19. Olukosi, O.A.; Kong, C.; Fru-Nji, F.; Ajuwon, K.M.; Adeola, O. Assessment of a bacterial 6-phytase in the diets of broiler chickens. *Poult. Sci.* **2013**, *92*, 2101–2108. [CrossRef]
20. Walk, C.L.; Bedford, M.R.; McElroy, A.P. Influence of limestone and phytase on broiler performance, gastrointestinal pH, and apparent ileal nutrient digestibility. *Poult. Sci.* **2012**, *91*, 1371–1378. [CrossRef]
21. Ptak, A.; Bedford, M.R.; Świątkiewicz, S.; Żyła, K.; Józefiak, D. Phytase modulates ileal microbiota and enhances growth performance of the broiler chickens. *PLoS ONE* **2015**, *10*, e0119770. [CrossRef]
22. Campasino, A.; York, T.; Wyatt, C.; Bedford, M.R.; Dozier, W.A., III. Effect of increasing supplemental phytase concentration in diets fed to Hubbard × Cobb 500 male broilers from 1 to 42 days of age. *J. Appl. Poult. Res.* **2014**, *23*, 705–714. [CrossRef]
23. Woyengo, T.A.; Slominski, B.A.; Jones, R.O. Growth performance and nutrient utilization of broiler chickens fed diets supplemented with phytase alone or in combination with citric acid and multicarbohydrase. *Poult. Sci.* **2010**, *89*, 2221–2229. [CrossRef]
24. Wu, D.; Wu, S.B.; Choct, M.; Swick, R.A. Comparison of 3 phytases on energy utilization of a nutritionally marginal wheat-soybean meal broiler diet. *Poult. Sci.* **2015**, *94*, 2670–2676. [CrossRef]
25. Sulabo, R.C.; Jones, C.K.; Tokach, M.D.; Goodband, R.D.; Dritz, S.S.; Campbell, D.R.; Ratliff, B.W.; DeRouchey, J.M.; Nelssen, J.L. Factors affecting storage stability of various commercial phytase sources. *J. Anim. Sci.* **2011**, *89*, 4262–4271. [CrossRef]
26. Nováková, J.; Vértesi, A.; Béres, E.; Petkov, S.; Niederberger, K.E.; van Gaver, D.; Hirka, G.; Balázs, Z. Safety assessment of a novel thermostable phytase. *Toxicol. Rep.* **2021**, *8*, 139–147. [CrossRef] [PubMed]
27. Cobb-Vantress, Inc. Cobb 500: Broiler Performance and Nutrition Supplement. 2022. Available online: https://www.cobb-vantress.com/assets/Cobb-Files/product-guides/5502e86566/2022-Cobb500-Broiler-Performance-Nutrition-Supplement.pdf (accessed on 15 February 2020).
28. National Research Council. *Nutrient Requirements of Poultry*, 9th ed.; The National Academies Press: Washington, DC, USA, 1994.
29. Downs, K.M.; Gulizia, J.P.; Stafford, E.K.; Pacheco, W.J. Influence of varying dietary kudzu leaf meal particle size on performance, breast weight, and organ weight of broiler chickens from 1 to 21 days of age. *Poultry* **2022**, *1*, 30–39. [CrossRef]
30. ASAS/ADSA/PSA. Husbandry, housing, and biosecurity. In *Guide for the Care and Use of Agricultural Animal in Agricultural Research and Teaching*; American Society of Animal Science/American Dairy Science Association/Poultry Science Association: Champaign, IL, USA, 2020; pp. 17–19.
31. ASAS/ADSA/PSA. Environmental enrichment. In *Guide for the Care and Use of Agricultural Animal in Agricultural Research and Teaching*; American Society of Animal Science/American Dairy Science Association/Poultry Science Association: Champaign, IL, USA, 2020; pp. 30–53.
32. American Veterinary Medical Association. *AVMA Guidelines for the Euthanasia of Animals*; American Veterinary Medical Association: Schaumberg, IL, USA, 2020.

33. ASABE. *Shear and Three-Point Bending Test of Animal Bone*; American Society of Agricultural and Biological Engineers: St. Joseph, MI, USA, 2017.
34. Hall, L.E.; Shirley, R.B.; Bakalli, R.I.; Aggrey, S.E.; Pesti, G.M.; Edwards, H.M., Jr. Power of two methods for the estimation of bone ash of broilers. *Poult. Sci.* **2003**, *82*, 414–418. [CrossRef]
35. SAS Institute. *SAS/STAT®User's Guide Version 14.3*; SAS Institute, Inc.: Cary, NC, USA, 2017.
36. Babatunde, O.O.; Cowieson, A.J.; Wilson, J.W.; Adeola, O. Influence of age and duration of feeding low-phosphorus diet on phytase efficacy in broiler chickens during the starter phase. *Poult. Sci.* **2019**, *98*, 2588–2597. [CrossRef]
37. Babatunde, O.O.; Cowieson, A.J.; Wilson, J.W.; Adeola, O. The impact of age and feeding length on phytase efficacy during the starter phase of broiler chickens. *Poult. Sci.* **2019**, *98*, 6742–6750. [CrossRef]
38. Batal, A.B.; Parsons, C.M. Effects of Age on Nutrient Digestibility in Chicks fed Different Diets. *Poult. Sci.* **2002**, *81*, 400–407. [CrossRef] [PubMed]
39. Leyva-Jimenez, H.; Alsadwi, A.M.; Gardner, K.; Voltura, E.; Bailey, C.A. Evaluation of high dietary phytase supplementation on performance, bone mineralization, and apparent ileal digestible energy of growing broilers. *Poult. Sci.* **2019**, *98*, 811–819. [CrossRef]
40. Kozlowski, K.; Nollet, L.; Lanckriet, A.; Vanderbeke, E.; Mielnik, P.; Outchkourov, N.; Petkov, S. Effect of different phytases derived from E. coli AppA gene on the performance, bone mineralisation and nutrient digestibility of broiler chicken. *J. Appl. Anim. Nutr.* **2019**, *7*, 1–9. [CrossRef]
41. Sens, R.F.; Bassi, L.S.; Almeida, L.M.; Rosso, D.F.; Teixeira, L.V.; Maiorka, A. Effect of different doses of phytase and protein content of soybean meal on growth performance, nutrient digestibility, and bone characteristics of broilers. *Poult. Sci.* **2021**, *100*, 100917. [CrossRef] [PubMed]
42. Beeson, L.A.; Walk, C.L.; Bedford, M.R.; Olukosi, O.A. Hydrolysis of phytate to its lower esters can influence the growth performance and nutrient utilization of broilers with regular or super doses of phytase. *Poult. Sci.* **2017**, *96*, 2243–2253. [CrossRef]
43. Kriseldi, R.; Walk, C.L.; Bedford, M.R.; Dozier, W.A., III. Inositol and gradient phytase supplementation in broiler diets during a 6-week production period: 1. effects on growth performance and meat yield. *Poult. Sci.* **2021**, *100*, 964–972. [CrossRef] [PubMed]
44. Dos Santos, T.T.; Srinongkote, S.; Bedford, M.R.; Walk, C.L. Effect of high phytase inclusion rates on performance of broilers fed diets not severely limited in available phosphorus. *Asian-Aust. J. Anim. Sci.* **2013**, *26*, 227–232. [CrossRef] [PubMed]
45. Dersjant-Li, Y.; Evans, C.; Kumar, A. Effect of phytase dose and reduction in dietary calcium on performance, nutrient digestibility, bone ash and mineralization in broilers fed corn-soybean meal-based diets with reduced nutrient density. *Anim. Feed Sci. Technol.* **2018**, *242*, 95–110. [CrossRef]
46. Broch, J.; Nunes, R.V.; Eyng, C.; Pesti, G.M.; de Souza, C.; Sangalli, G.G.; Fascina, V.; Teixeira, L. High levels of dietary phytase improves broiler performance. *Anim. Feed Sci. Technol.* **2018**, *244*, 56–65. [CrossRef]
47. Campbell, G.L.; Bedford, M.R. Enzyme applications for monogastric feeds: A review. *Can. J. Anim. Sci.* **1992**, *72*, 449–466. [CrossRef]
48. Cardoso Júnior, A.; Rodrigues, P.B.; Bertechini, A.G.; de Freitas, R.T.F.; de Lima, R.R.; Lima, G.F.R. Levels of available phosphorus and calcium for broilers from 8 to 35 days of age fed rations containing phytase. *R. Bras. Zootec.* **2010**, *39*, 1237–1245. [CrossRef]
49. Pereira, R.; Menten, J.; Romano, G.G.; Silva, C.; Zavarize, K.C.; Barbosa, N. Efficiency of a bacterial phytase to release phytate phosphorus in broiler chicken diets. *Arq. Bras. Med. Vet. Zootec.* **2012**, *64*, 137–144. [CrossRef]
50. Bougouin, A.; Appuhamy, J.A.D.R.N.; Kebreab, E.; Dijkstra, J.; Kwakkel, R.P.; France, J. Effects of phytase supplementation on phosphorus retention in broilers and layers: A meta-analysis. *Poult. Sci.* **2014**, *93*, 1981–1992. [CrossRef] [PubMed]
51. Chung, T.K.; Rutherfurd, S.M.; Thomas, D.V.; Moughan, P.J. Effect of two microbial phytases on mineral availability and retention and bone mineral density in low-phosphorus diets for broilers. *Br. Poult. Sci.* **2013**, *54*, 362–373. [CrossRef] [PubMed]
52. Babatunde, O.O.; Bello, A.; Dersjant-Li, Y.; Adeola, O. Evaluation of the responses of broiler chickens to varying concentrations of phytate phosphorus and phytase. II. Grower phase (day 12-23 post hatching). *Poult. Sci.* **2022**, *101*, 101616. [CrossRef] [PubMed]
53. Martins, B.A.B.; Borgatti, L.M.D.O.; Souza, L.W.D.O.; Robassini, S.L.D.A.; de Albuquerque, R. Bioavailability and poultry fecal excretion of phosphorus from soybean-based diets supplemented with phytase. *R. Bras. Zootec.* **2013**, *42*, 174–182. [CrossRef]
54. Kriseldi, R.; Johnson, J.A.; Walk, C.L.; Bedford, M.R.; Dozier, W.A., III. Influence of exogenous phytase supplementation on phytate degradation, plasma inositol, alkaline phosphatase, and glucose concentrations of broilers at 28 days of age. *Poult. Sci.* **2021**, *100*, 224–234. [CrossRef]
55. Gautier, A.E.; Walk, C.L.; Dilger, R.N. Effects of a high level of phytase on broiler performance, bone ash, phosphorus utilization, and phytate dephosphorylation to inositol. *Poult. Sci.* **2018**, *97*, 211–218. [CrossRef]
56. Kriseldi, R.; Walk, C.L.; Bedford, M.R.; Dozier, W.A., III. Inositol and gradient phytase supplementation in broiler diets during a 6-week production period: 2. Effects on phytate degradation and inositol liberation in gizzard and ileal digesta contents. *Poult. Sci.* **2021**, *100*, 100899. [CrossRef]
57. Abbasi, F.; Fakhur-un-Nisa, T.; Liu, J.; Luo, X.; Abbasi, I.H.R. Low digestibility of phytate phosphorus, their impacts on the environment, and phytase opportunity in the poultry industry. *Environ. Sci. Pollut. Res.* **2019**, *26*, 9469–9479. [CrossRef]

animals

MDPI

Article

Growth Performance, Histological Changes and Functional Tests of Broiler Chickens Fed Diets Supplemented with *Tribulus Terrestris* Powder

Maged A. Al-Garadi , Hani H. Al-Baadani and Abdulmohsen H. Alqhtani *

Department of Animal Production, College of Food and Agriculture Science, King Saud University, P.O. Box 2460, Riyadh 11451, Saudi Arabia; malgaradi@ksu.edu.sa (M.A.A.-G.); hanee7811@gmail.com (H.H.A.-B.)
* Correspondence: ahalqahtani@ksu.edu.sa; Tel.: +96-65-4000-0232

Simple Summary: *Tribulus Terrestris* (TT) is extremely rich in substances with potential biological activity and is therefore considered to be a promising ingredient in industrial medicinal preparations based on its saponin fraction. On the other hand, studies regarding broiler chickens are limited. However, the aim of this study was to verify the safe amount of TT powder that can be used as a medicinal plant in the diet of broiler chickens by determining the growth performance and histological changes and functional tests of some target organs. In conclusion, 0.75 g/kg TT is the best dosage for the health of broiler chickens. There is a need for low dose studies to identify the mechanism of the pharmacological effect of TT on broilers.

Abstract: The current experiment aimed to investigate the effects of TT powder on performance parameters and functional tests, as well as on morphological and histological changes in the liver, kidney and ileum in broiler chickens. Commercial broilers (total = 168 females) were used, equally divided into three dietary treatments (C = 0.0, T1 = 0.75, and T2 = 1.5 g/kg diet). The growth performance (1–35 days of age), absolute and relative weight, liver and kidney functional tests, intestinal morphology (14 and 35 days of age), and histomorphology of the ileum (35 days of age) were evaluated. At 35 days of age, histopathological changes in the ileum, liver, and kidney were also examined. The results showed that the growth performance and absolute and relative weights of the liver and kidney had no negative effects when dietary supplementation with TT powder was given at 0.75 g/kg diet (T1), whereas a decrease was observed at T2 ($p < 0.05$). Liver and kidney functional tests showed no significant effects in all feed treatments (14 days), while T1 showed lower ($p < 0.05$) ALT and AST levels (35 days). T1 exhibited higher weights, lengths, and weight-to-length ratios of the small intestine, and relative lengths of the duodenum ($p < 0.05$). Histomorphometric measurements of the ileum were higher ($p < 0.05$) in chickens fed the 0.75 g TT/kg diet, and except for in the goblet cell count and epithelial thickness, there were no differences between treatments ($p > 0.05$). In T1, hepatocytes were more normal but hepatic sinusoids were dilated, whereas in T2, lymphocytes had infiltrated around the central vein and lining endothelial cells had been lost. The kidney was improved in T1 and T2 compared with the control group. Ileal villi were shorter in T2, and some villi fused with enterocyte necrosis and inflammatory cells accumulated in the lumen. We concluded that TT powder (0.75 g/kg feed) has a safe effect and is healthy for broilers.

Keywords: *Gallus domesticus*; *Tribulus Terrestris*; histology; liver; kidney; intestine

check for updates

Citation: Al-Garadi, M.A.; Al-Baadani, H.H.; Alqhtani, A.H. Growth Performance, Histological Changes and Functional Tests of Broiler Chickens Fed Diets Supplemented with *Tribulus Terrestris* Powder. *Animals* **2022**, *12*, 1930. https://doi.org/10.3390/ani12151930

Academic Editors: Ilias Giannenas and Vincenzo Tufarelli

Received: 24 May 2022
Accepted: 18 July 2022
Published: 28 July 2022

1. Introduction

Medicinal herbal and aromatic plants have been a gaining interest in recent years as feed additives in poultry. Medicinal herbs can alter the active compounds in final products depending on the part used (e.g., seed, leaf, root, or bark) and the processing technique (e.g., extraction with non-aqueous solvents) [1,2]. Several studies in recent years have shown that the use of medicinal herbs as feed additives can improve the health, performance, and nutrient digestibility of broiler chickens [3–5].

Tribulus Terrestris (TT) belongs to the Zygophyllaceae family [6]. It is widely distributed in many areas (tropical and temperate regions of the world), especially in northeastern Yemen and southern Saudi Arabia, and it competes with most crops in the early stage of development [7]. TT is a medicinal plant accepted and recognized by humans as a traditional remedy for many diseases in the form of the whole plant or especially the fruits [4]. Rathod et al. [8]; Daniyal et al. [9] reported that the phytotherapeutic potential of TT is present in the leaves, roots, seeds, and fruits. TT is extremely rich in substances with potential biological activity such as steroid glycosides (saponins), protodioscin, glycosides, alkaloids and flavonoids [10]. Additionally, Khalid et al. [1] showed that the pharmacologically active components of TT belong to flavonoids and phenols. However, it is considered as an ingredient in the industrial production of medicinal preparations based on its saponin fraction [11]. Therefore, TT and its extracts are used as medicinal plants to enhance performance and digestibility, to remove kidney stones, as a diuretic, to treat urinary tract infections, hypercholesterolemia, and diabetes, and to improve liver health [12–15]. In addition, the fruit of TT powder is widely and effectively used as a feed additive to improve the reproductive and health potential of roosters [16], mice [17], and humans [18]. Vincent et al. [19] reported that saponins extracted from TT had a beneficial effect on the control of poultry diseases. Feeding 1 TT powder/kg feed [20] or 0.06 and 0.12 g TT extracts/kg [21] had no significant effect on the growth performance of broiler chickens. In contrast, a 0.8 g TT powder/kg diet can improve the growth performance of broilers [22]. In another study, TT fruit powder was safe for the growth performance of mice up to a dose of 1 g/kg [2]. Sharma et al [23] found that the use of TT extract reduced functional renal disorders and cellular damage in rats. Liver function tests were increased as a result of cellular damage and degenerative changes with the increased permeability of the plasma membrane in the liver of broilers that received an overdose of TT [24]. Other studies reported that TT fruit powder can be used to treat liver regeneration and digestive problems and diarrhea in humans [25,26].

Although studies have shown some effects of TT, there is still much debate about possible mechanisms of action and therapeutic applications, as well as a significant lack of research data on this medicinal plant. To our knowledge, there is a limited number of studies in which the effects of adding TT powder as a whole plant to the diet on the performance and health status of broiler chickens has been examined. All previous findings are existing hypotheses from which this study was created. Therefore, the aim of the current study was to evaluate the use of TT powder on the performance, histological changes, and functional markers of the liver, kidney, and small intestine in broilers.

2. Materials and Methods

The use of birds, sampling and analyses were performed in accordance with the recommendations of the Scientific Research Ethics Committee of King Saud University, Riyadh, Saudi Arabia (approval number: KSU-SE-21-47).

2.1. Preparation of Tribulus Terrestris (TT) Powder

The TT was collected as a whole plant (flowers, seeds, leaves, and stems) in the valleys of the Republic of Yemen. The plants were classified in the herbarium of the College of Science-King Saud University (NO. 24519). Then, the plants were dried at a normal temperature (27 °C) until they reached a constant weight and were then ground into powder at the Research Department of Animal Production-Food and Agriculture Sciences at King Saud University.

2.2. Chemical Composition Profiles of Tribulus Terrestris (TT) Powder

The analysis of TT powder was performed in duplicate to determine the content of nutrients (dry matter, crude protein, ash, and total fiber) according to the methods of

AOAC [27]. The fatty acids composition of TT powder was performed using the method of extraction with a mixture of chloroform and methanol, and then the total fat (g/100 g of TT) was calculated [28]. Fatty acid profiles were analyzed using a gas chromatograph mass spectrometer (GC-MS) [29]. Fatty acids were calculated based on the peak areas of the chromatogram and expressed as the percentage of fatty acid methyl esters. For the analysis of phenolic compounds, ethanol and methanol TT powder extracts were prepared according to the method of Al-Fatimi et al. [30]. Phenolic compounds were analyzed using high performance liquid chromatography (HPLC). Resorcinol, dinitrobenzene, chlorogenic acid, caffeic acid, vanillin, acetylsalicylic acid, salicylic acid, and quercetin were used as external standards. The amounts were expressed as ng/µL of extracted TT.

2.3. Birds and Experimental Design

One-day-old commercial Ross chickens (168 females in total) were used for this experiment. Birds were weighed individually and randomly divided equally into 3 treatments with 8 replicates per treatment (7 chicks per replicate) in an environmentally controlled battery with automatic electric heating. The initial temperature of 35 °C at one day of age was gradually reduced to 22 °C at 21 days of age and then maintained at this temperature and 55% humidity for the remainder of the 35-day period. Dietary treatments were supplemented with TT at three doses (C = 0.0, T1 = 0.75, and T2 = 1.5 g/kg feed). The basal diet for the chickens during the starter and finisher periods was formulated according to the recommendations from the commercial practice (Saudi Arabia) (see Table 1). Feed and water were offered ad libitum to the chickens during the experimental period. All chickens were vaccinated against NDV, IBV, and IBDV according to the manufacturer's instructions (Fort Dodge Animal Health, Overland Park, KS, USA).

Table 1. Feed ingredients and calculated content of the basal diet.

Ingredients	**Basal Diet (%)**	
	Starter (Days 1–14)	**Finisher (Days 15–35)**
Corn	54.89	60.13
Soybean meal	38.77	31.92
Palm oil	1.96	4.21
Dicalcium phosphate	1.74	1.38
Limestone	1.19	1.05
Salt	0.40	0.40
Min. Vit. Premix, 0.5% [a]	0.50	0.50
DL-Methionine	0.31	0.25
L-Lysine-HCL	0.13	0.08
L-Threonine	0.07	0.02
Choline CL-70%	0.05	0.05
Total	100	100
Calculated composition		
Metabolizable energy, Kcal/kg	3000	3200
Crude protein	23.0	20.0
Crude fiber	2.21	2.10
Available P	0.48	0.41
Lysine	1.28	1.06
Methionine	0.64	0.55
Methionine + cysteine	0.95	0.83
Threonine	0.86	0.71
Tryptophan	0.27	0.23
Arginine	1.50	1.29
Valine	1.13	0.99

[a] Containing by kg of diets: 2,400,000 IU of vitamin A; 1,000,000 IU of vitamin D; 16,000 IU of vitamin E; 800 mg of vitamin K; 600 mg of vitamin B1; 1600 mg of vitamin B2; 1000 mg of vitamin B6; 6 mg of vitamin B12; 8000 mg of niacin; 400 mg of folic acid; 3000 mg of pantothenic acid; 40 mg of biotin; 80 mg of cobalt; 2000 mg of copper; 400 mg of iodine; 1200 mg of iron; 18,000 mg of manganese; 60 mg of selenium and 14,000 mg of zinc.

2.4. Live Performance Measurements

All chicks were weighed individually to determine their body weight at days 1 and 35 and to calculate body weight gain per treatment. Feed intake was recorded during the experimental periods to calculate the feed conversion ratio.

2.5. Measurement of Enzyme Activity

At 14 and 35 days of age, blood samples were collected from 16 birds in each dietary treatment in tubes without EDTA to measure liver function tests (alanine aminotransferase (ALT) and aspartate aminotransferase (AST)) and kidney function tests (uric acid and creatinine). Serum was separated via centrifugation at 3000× *g* for 20 min and then frozen at −80 °C until analysis. These functional tests were analyzed spectrophotometrically (RANDOX, London, UK) using reagent kits (Randox, London, UK) according to the manufacturer's instructions.

2.6. Morphology, Histological Changes

At 14 and 35 days of age, the entire small intestine, liver, and kidney were randomly collected from 8 birds per treatment, weighed and calculated as a percentage based on live weight. The measurement of small intestine length was recorded and the relative length of intestinal segments was calculated based on the total small intestine length. The length (L) and weight (W) of the small intestine were used to calculate the W/L ratio [31].

The mid-portions of the ileum (1.5 cm) between the Meckel's diverticulum and the cecal junction and the liver and kidneys (0.75 cm) of each bird were taken without stressing the tissue wall. Tissue samples were washed in phosphate-buffered saline (pH = 7.4) and fixed in 10% phosphate-buffered formalin for 72 h, then dehydrated in graded ethyl alcohol (50% to 100%) and paraffin-embedded into the action sections (Tissue-Tek, Sakura, Tokyo, Japan). Approximately 4 μm was cut using the Microtome System (Leica, RM 2245, Wetzlar, Germany). Sections were deparaffinized, hydrated and stained in hematoxylin and eosin (H&E) with Alcian blue stain to count the goblet cells (Leica, CV5030, Germany) using a protocol based on the modified method of Winsor and Sluys [32].

Tissue sections were examined under a light microscope (Nikon, Eclipse i80, Tokyo, Japan), and the required images were taken at various magnifications using a Nikon digital camera (OXM 1200C, Nikon, Tokyo, Japan). Measurements of villus length, width, depth, goblet cells, epithelial thickness, mucosa, submucosa, and lamina propria were based on at least ten intact villi from one section (bird) with a total of 80 measurements per treatment. The following equation based on villus length and width was used to calculate villus surface area (surface area = $2\pi \times (\text{width}/2) \times \text{length}$) according to Al-Baadani et al. [33]. In addition, the ratio of villus length to crypt height was calculated [31]. Goblet cell density per 100 μm of villus area was determined according to the method of Qaid et al. [34]. Histopathological changes in the ileum, liver, and kidney in the different treatments were examined microscopically (100 to 400×) according to the method described by Belote et al. [35].

2.7. Statistical Analysis

The data obtained from the measurements of performance, enzymes, and morphology in the different treatments were statistically analyzed using the GLM procedures of SAS software [36] via one-way analysis ANOVA for a completely randomized design. Means were analyzed using Duncan's multiple range test and treatments, with statistical significance being based on $p < 0.05$. All mean values for each parameter within the different treatments were reported as ± standard error of the means (SEM).

3. Results

The nutritional values and fatty acids composition of TT are presented in Table 2. The phytochemical compositions of TT included dry matter (92.84%), crude protein (15.52%), ash (17.19%), crude fiber (35.55%), and total fat (2.80%). As a result of the fatty acid profile, the saturated fatty acids and unsaturated fatty acid amounts of TT oil extract were found to be 19.38% and 80.62%, respectively. According to the measurement results, a total of nine various fatty acids for each saturated and unsaturated were determined in TT oil extract, including palmitic acid (C16:0; 11.44%), oleic acid (C18:1; 13.76%), and linoleic acid (C18:2; 64.32%).

Table 2. Nutritional values and fatty acids composition of *Tribulus Terrestris* (TT).

Item	%
Chemical composition [1]	
Dry matter	92.84
Crude protein	15.52
Ash	17.19
Crude fiber	35.55
Total fat	2.80
Fatty acids of total fat [2]	
C6:0	0.15
C8:0	0.34
C12:0	0.08
C14:0	0.21
C16:0	11.44
C17:0	0.17
C18:0	6.86
C22:0	0.03
C20:0	0.10
Saturated fatty acids	19.38
C16:1	0.11
C18:1	13.76
C18:2	64.32
C18:3 (gamma)	0.54
C18:3 (α-linoliec)	1.42
C20:2	0.09
C20:3	0.03
C20:4	0.16
C22:5	0.19
Unsaturated fatty acids	80.62

[1] The chemical composition analysis was performed in duplicate on a dry matter basis; [2] The fatty acids composition was analyzed via GC-MS in duplicate.

The results of HPLC analysis (Table 3) showed that the phenolic content of TT is resorcinol, 1,2-Dinitrobenzene, chlorogenic acid, caffeic acid, vanillin, acetyl-salicylic acid, and salicylic acid, and the flavonoid content is quercetin. The major compound was acetylsalicylic acid at the rate of 160.99 or 524.86 ng/µL either in the ethanol or methanol extract, respectively.

Table 3. Phenolic and flavonoid compounds of *Tribulus Terrestris* (TT). The phenolic compounds were analyzed via HPLC in duplicate.

Compounds	Ethanol Extract			Methanol Extract		
	Area (mAU*s)	Ret Time (min)	Amount (ng/µL)	Area (mAU*s)	Ret Time (min)	Amount (ng/µL)
Resorcinol	117.7	7.33	10.24	341.40	7.36	29.58
1,2-Dinitrobenzene	492.3	9.34	31.16	98.90	8.99	6.30
Chlorogenic acid	388.4	12.02	23.23	126.90	11.77	7.45
Caffeic acid	539.6	14.13	12.10	1008.50	14.19	22.37
Vanillin acid	1065.4	16.30	19.65	1773.10	16.37	32.53
Acetyl salicylic acid	278.3	22.17	160.99	915.80	22.27	524.86
Salicylic acid	470.5	24.95	32.70	487.30	24.77	36.17
Quercetin	92.6	34.80	33.37	235.74	34.89	91.98

The results of the effect of TT on growth performance during the experimental period (1 to 35 days) are presented in Table 4. The present study showed that the intake of the highest level of TT at 1.5 g/kg diet (T2) resulted in lower ($p \leq 0.05$) body weight gain and total feed intake compared to the basal diet (control), while the intake of TT at 0.75 g/kg diet (T1) had no negative effect. The feed conversion ratio was not affected by any of the treatments ($p > 0.05$).

Table 4. Effect of *Tribulus Terrestris* (TT) supplementation on growth performance in broiler chickens from 1 to 35 days of age.

Variables	Treatments [1]			SEM [2]	*p*-Value
	Control	T1	T2		
Bodyweight gain, g	1714.02 [a]	1716.76 [a]	1630.41 [b]	14.12	0.0003
Total feed intake, g	2435.97 [a]	2350.63 [a,b]	2323.31 [b]	32.58	0.050
Feed conversion ratio, g/g	1.42	1.36	1.42	0.018	0.072

[a,b] Means values within a row for each variable with clarification of the significant difference in the form of superscripts ($p < 0.05$). [1] Treatments: control, birds fed basal diet; T1, basal diet with 0.75 g *Tribulus Terrestris*/kg diet; T2, basal diet with 1.5 g *Tribulus Terrestris*/kg diet. [2] SEM = standard error of means for treatment effect.

The body weight and absolute and relative weights of the liver and kidney in broiler chickens fed TT powder are shown in Table 5. The body weight during slaughter data obtained showed that the chickens fed 0.75 g TT/kg diet (T1) displayed no effects ($p > 0.05$) compared to the control, while the chickens fed 1.5 g/kg had weights lower than those in the other treatments at 14 days of age ($p < 0.05$). At 35 days of age, the live body displayed no effects in all treatments ($p > 0.05$). The absolute and relative weights of the liver and kidney displayed no effects ($p > 0.05$) between all treatments during the starter and finisher phase (14 and 35 days).

Table 5. Effect of *Tribulus Terrestris* (TT) supplementation on absolute and relative weight of liver and kidney in broiler chickens.

Variables	Treatments [1]			SEM [2]	*p*-Value
	Control	T1	T2		
At14 days of age					
Body weight (g)	438.75 [a]	429.88 [a]	369.38 [b]	15.47	0.009
Liver (g)	12.74	10.81	9.86	0.89	0.090
Liver (%)	2.94	2.50	2.67	0.21	0.363
Kidney (g)	2.26	2.90	2.60	0.25	0.238
Kidney (%)	0.528	0.672	0.710	0.06	0.149
At 35 days of age					
Body weight (g)	1672.63	1728.00	1693.75	21.66	0.238
Liver (g)	35.68	38.33	35.72	1.25	0.250
Liver (%)	2.13	2.21	2.11	0.09	0.672
Kidney (g)	2.32	2.40	1.99	0.19	0.319
Kidney (%)	0.14	0.13	0.11	0.01	0.391

[a,b] Means values within a row for each variable with clarification of the significant difference in the form of superscripts ($p < 0.05$). [1] Treatments: control, birds fed basal diet; T1, basal diet with 0.75 g *Tribulus Terrestris*/kg diet; T2, basal diet with 1.5 g *Tribulus Terrestris*/kg diet. [2] SEM = standard error of means for treatment effect.

The effects of TT supplementation on liver and kidney function tests in broiler chickens are shown in Table 6. The measured liver function tests (ALT; alanine aminotransferase and AST; aspartate aminotransferase) and kidney function tests (uric acid and creatinine) showed no significant effect ($p > 0.05$) between the dietary treatments and the control group at 14 days of age. At 35 days of age, the chickens receiving the 0.75 g TT/kg diet (T1) had lower values ($p < 0.05$) on ALT and AST compared to those in the other dietary treatments. In contrast, there were no effects on uric acid and creatinine for all treatments ($p > 0.05$).

Table 6. Effect of *Tribulus Terrestris* (TT) supplementation on function tests of liver and kidney in broiler chickens.

Variables	Treatments [1]			SEM [2]	*p*-Value
	Control	T1	T2		
At 14 days of age					
AST U/L	171.5	172.5	173.8	3.13	0.874
ALT U/L	14.6	13.2	12.3	2.09	0.732
Uric acid mg/dl	2.12	2.43	2.46	0.18	0.362
Creatinine mg/dl	0.444	0.452	0.363	0.04	0.387
At 35 days of age					
AST U/L	245.2 ^{a}	208.7 ^{c}	227.9 ^{b}	3.39	<0.0001
ALT U/L	17.5 ^{a}	9.3 ^{c}	11.1 ^{b}	2.29	0.050
Uric acid mg/dl	1.82	1.63	1.48	0.22	0.584
Creatinine mg/dl	0.365	0.414	0.359	0.04	0.647

$^{a–c}$ Means values within a row for each variable with clarification of the significant difference in the form of superscripts ($p < 0.05$). [1] Treatments: control, birds fed basal diet; T1, basal diet with 0.75 g *Tribulus Terrestris*/kg diet; T2, basal diet with 1.5 g *Tribulus Terrestris*/kg diet. [2] SEM = standard error of means for treatment effect.

The results of histopathological changes in the livers of broiler chickens are shown in Figure 1, stained with (H&E) 100× to 400×. The livers of chickens fed a basal diet (control) showed the normal distribution of hepatocytes and dilation of the central vein. However, the examination of many chicken tissues revealed focal inflammatory infiltration by lymphocytes between the hepatocytes and around the central vein. The livers of chickens fed the 0.75 g TT powder/kg diet (T1) showed more normal hepatocytes compared with the control group, but dilation of the hepatic sinusoids located between the hepatic cords and narrowing of the central vein. In addition, the livers of T2 chickens (1.5 g TT powder/kg diet) showed normal hepatocytes but swollen hepatocytes in some areas, with lymphocyte infiltration around the central vein and the loss of lining endothelial cells.

Histopathological changes in the kidneys of broiler chickens are shown in Figure 2, stained with (H&E) 100× to 400×. Many renal tissues showed interstitial hemorrhage and congestion in renal blood vessels with the degeneration of the epithelial lining of renal tubules with the accumulation of inflammatory cells in chickens fed the basal diet (control). However, all histopathological changes in the control group improved in the tissues of the chickens when they received the 0.75 and 1.5 g TT powder/kg diet (T1 and T2, respectively).

The intestinal morphology, such as the weight and length of the small intestine, the ratio between the weight and length of the intestine, and the relative length of the small intestinal fragments in broilers fed diets enriched with TT powder, are shown in Table 7. At 14 days of age, the data revealed that the chickens fed a 0.75 g TT/kg diet (T1) had higher weights and lengths of small intestine ($p < 0.05$) than the control group. In addition, T1 and the control group had higher intestinal weight to intestinal length ratios and relative lengths of the duodenum ($p < 0.05$) compared to the chickens fed the 1.5 g/kg diet (T2). In contrast, the relative length of the ileum was lower ($p < 0.05$) in T1 and the control group. The small intestine weight and intestine weight to length ratio (W/L) were higher ($p < 0.05$) in chickens fed the 0.75 g TT/kg diet (T1) than in the control group, but there was no significant difference from T2 at 35 days of age.

Table 8 shows that the histomorphometric measurements of the ileum in broiler chickens were affected by the treatments ($p < 0.001$), while the number of goblet cells and epithelial thickness showed no differences between the treatments compared to the control ($p > 0.05$). The length and width of ileal villi and villus surface area were higher ($p < 0.05$) in chickens fed the 0.75 g TT/kg diet (T1) than T2 but were not significantly different between the T1 and control groups. In addition, the results showed that the ratio of villus length to crypt depth and the thickness of mucosa and lamina propria were higher ($p < 0.05$) in T1 chickens than in the other diets (T2 and control). In contrast, the crypt depth was lower in T1 compared to the other feed treatments, and the number of goblet cells per 100 µm villus area was also lower ($p < 0.05$) but not significantly different from the chickens fed the basal diet (control).

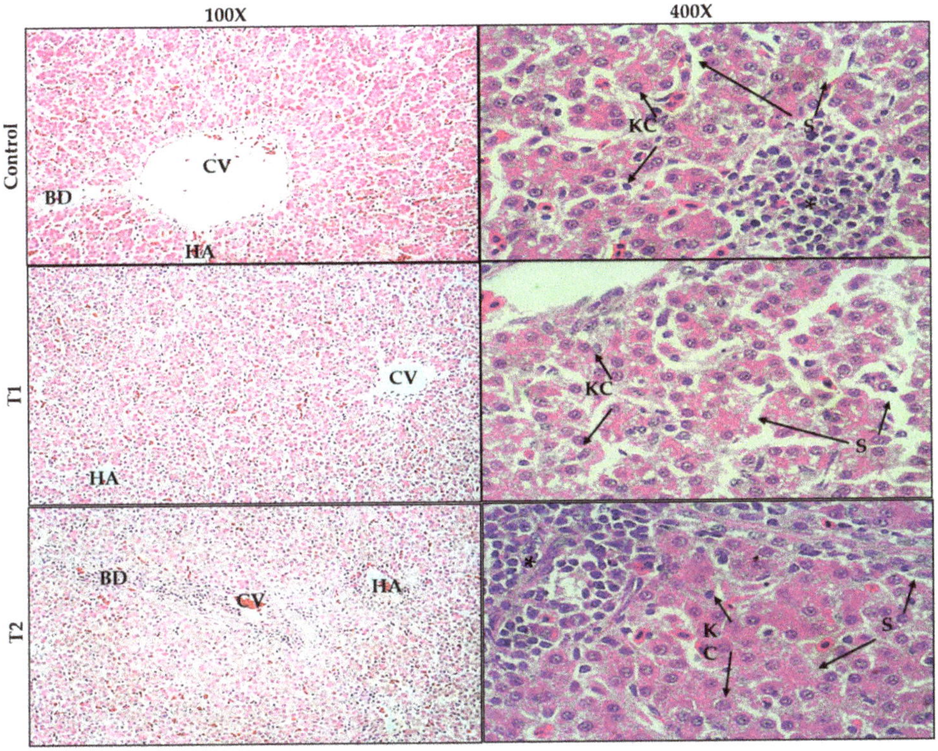

Figure 1. Histological changes in the liver of broiler chickens stained with (H&E) 100× to 400×. Control, birds fed basal diet; T1, basal diet containing 0.75 g *Tribulus Terrestris*/kg feed; T2, basal diet containing 1.5 g *Tribulus Terrestris*/kg feed. CV = central vein; S = sinusoids; KC = Kupffer cells; HA = hepatic artery; BD = bile duct. Chicken liver (control): shows normal morphology of hepatocytes with focal inflammatory infiltration by leukocytes (lymphocytes) (star) and dilatation of the CV. Chicken liver (T1): hepatocytes are more normal but sinusoidal spaces are dilated (S) and CV is narrower than control group. The chicken liver (T2): hepatocytes normal, but swollen hepatocytes in some areas with lymphocyte infiltration around the central vein (star) and loss of lining endothelial cells.

The histopathological changes in the ileum in broiler chickens are shown in Figure 3. In the chickens fed the basic diet (control), the height and width of the villi were within the normal range, except for some cases where proliferative enterocytes with metaplasia of goblet cells were seen with disturbances in the villous spaces and the lumen contained sheets of epithelial cells, mucus, and denuded tips (Figure 3; control—100× and 300×). In birds fed a basal diet containing 0.75 g TT/kg feed (T1), normal heights and widths of the intestinal villi and a free lumen without any exudate, as well as a normal appearance of the submucosal and muscular sheaths were observed (Figure 3; T1—100× and 300×). In chicken intestines at T2 (1.5 g TT/kg feed), villi were shorter, and some villi fused due to the proliferation of enterocytes. Some villi showed the necrosis of enterocytes with numerous inflammatory cells. The lumen contained layers of epithelial cells that are mucinous (Figure 3; T2—100× and 300×).

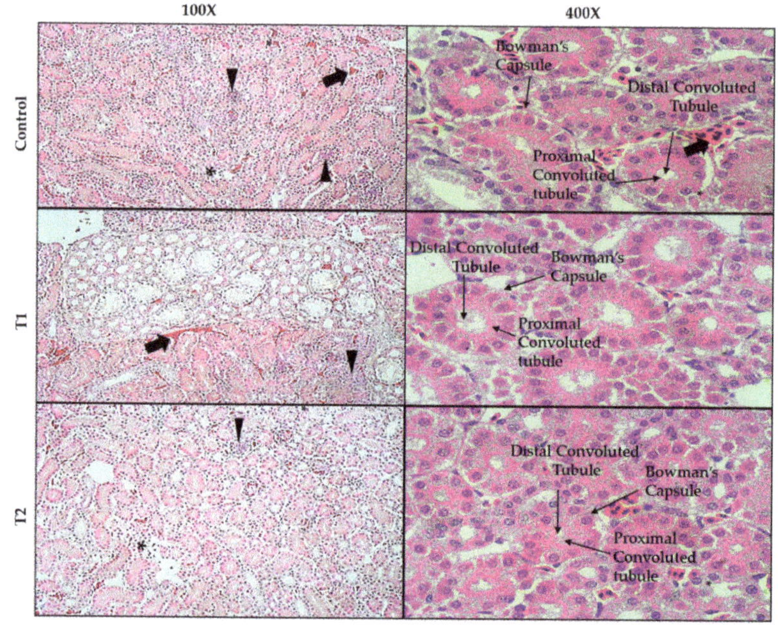

Figure 2. Histological changes in the kidney of broiler chickens stained with (H&E) 100× to 400×. Control, birds receiving basal diet; T1, basal diet containing 0.75 g *Tribulus Terrestris*/kg feed; T2, basal diet containing 1.5 g *Tribulus Terrestris*/kg feed. The kidney of the chickens (control) showed interstitial hemorrhage and congestion in the renal blood vessels (striped arrow) with degeneration of the epithelial lining of the renal tubules (star) with accumulation of inflammatory cells (arrowhead). In contrast, all histopathological changes improved in chickens fed diets T1 to T2 (0.75 or 1.5 g TT powder per kg diet).

Table 7. Effect of *Tribulus Terrestris* (TT) supplementation on the intestinal morphology in broiler chickens.

Variables	Treatments [1] Control	T1	T2	SEM [2]	*p*-Value
At14 days of age					
Small intestine weight (g)	19.33 ^{b}	23.61 ^{a}	18.09 ^{b}	1.08	0.004
Small intestine length (cm)	118.75 ^{b}	131.56 ^{a}	130.04 ^{a}	2.26	0.001
Small intestine length (%)					
Duodenum	17.93 ^{a}	16.66 ^{a,b}	16.13 ^{b}	0.50	0.050
Jejunum	44.00	44.19	41.90	0.93	0.183
Ileum	38.06 ^{b}	39.14 ^{b}	41.96 ^{a}	0.92	0.019
W/L ratio [3] (g/cm)	0.163 ^{a}	0.177 ^{a}	0.138 ^{b}	0.007	0.007
At 35 days of age					
Small intestine weight (g)	55.91 ^{b}	63.31 ^{a}	60.26 ^{a,b}	1.79	0.027
Small intestine length (cm)	177.38	176.21	180.53	3.53	0.676
Small intestine length (%)					
Duodenum	16.25	16.91	15.97	0.42	0.303
Jejunum	42.38	42.28	43.32	0.43	0.197
Ileum	41.37	40.80	40.69	0.53	0.638
W/L ratio [3] (g/cm)	0.313 ^{b}	0.360 ^{a}	0.333 ^{a,b}	0.009	0.009

^{a,b} Means values within a row for each variable with clarification of the significant difference in the form of superscripts ($p < 0.05$). [1] Treatments: control, birds fed basal diet; T1, basal diet with 0.75 g *Tribulus Terrestris*/kg diet; T2, basal diet with 1.5 g *Tribulus Terrestris*/kg diet. [2] SEM = standard error of means for treatment effect. [3] W/L ratio: intestine weight to length ratio.

Table 8. Effect of *Tribulus Terrestris* (TT) supplementation on histomorphometric measurements of ileum in broiler chickens.

Variables	Treatments [1]			SEM [2]	*p*-Value
	Control	T1	T2		
Length (μm)	419.17 [a]	403.55 [a]	307.47 [b]	15.92	<0.0001
Width (μm)	49.76 [a,b]	52.43 [a]	44.87 [b]	1.89	0.007
Villus surface area (mm^2)	0.065 [a]	0.066 [a]	0.042 [b]	0.002	<0.0001
Crypt-depth (μm)	26.96 [a]	19.99 [b]	26.12 [a]	0.91	<0.0001
Villus length/crypt depth	15.67 [b]	21.01 [a]	11.85 [c]	0.93	<0.0001
Goblet cells (NO.)	87.58	86.25	80.35	3.35	0.182
Goblet cells/100 μm Villi area	10.55 [b]	10.73 [b]	13.87 [a]	0.87	0.004
Epithelial thickness (μm)	5.76	5.34	6.55	0.44	0.084
Mucosa (μm)	28.84 [b]	33.61 [a]	23.90 [c]	0.76	<0.0001
Sub-mucosa (μm)	13.48 [a,b]	15.25 [a]	12.25 [b]	0.67	0.003
Lamina propria thickness (μm)	19.33 [b]	23.52 [a]	14.78 [c]	1.26	<.0001

[a–c] Means values within a row for each variable with clarification of the significant difference in the form of superscripts ($p < 0.05$). [1] Treatments: control, birds fed basal diet; T1, basal diet with 0.75 g *Tribulus Terrestris*/kg diet; T2, basal diet with 1.5 g *Tribulus Terrestris*/kg diet. [2] SEM = standard error of means for treatment effect.

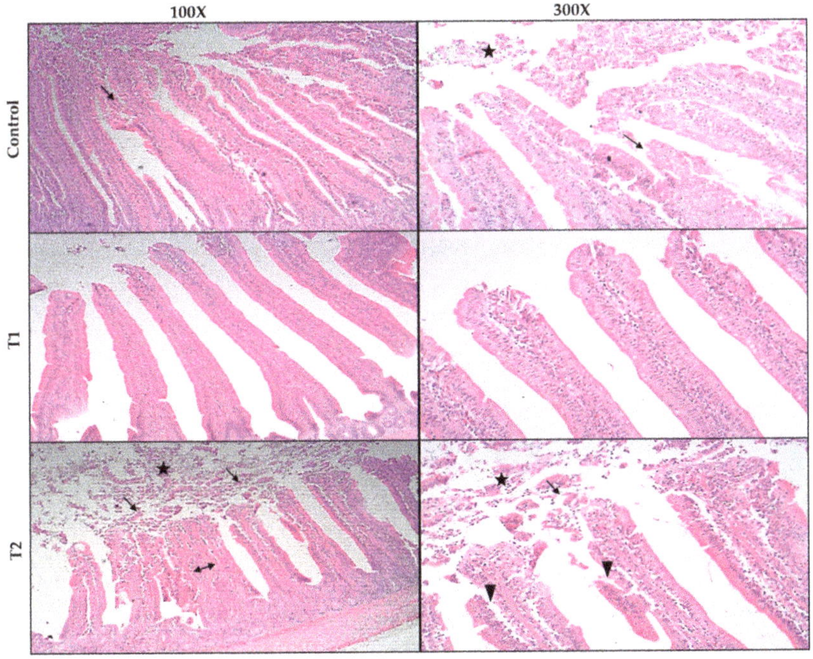

Figure 3. Histological changes in the ileum of broiler chickens stained with (H&E) 100× to 300×. Control, birds fed basal diet; T1, basal diet containing 0.75 g *Tribulus Terrestris*/kg feed; T2, basal diet containing 1.5 g *Tribulus Terrestris*/kg feed. Chicken ileum (control): showing normal height and width of intestinal villi with proliferative enterocytes (arrows). The lumen contains layers of epithelial cells (star). The chicken ileum (T1): normal height and width of the intestinal villi and the lumen free of any exudate and greater improvement than observed in the other treatment groups. Chicken ileum (T2): short villi with fusion of some villi (double arrow). Some villi show necrosis of enterocytes (arrowhead) and contain sheets of epithelial cells and mucus in the lumen (star) as well as proliferative enterocytes (arrows).

4. Discussion

TT is a pharmaceutical herb that has long been used by people as a traditional medicine to treat many diseases, including urinary tract infections, to remove kidney urolithiasis, hypercholesterolemia, and liver health, and as a diuretic medication. These traditional uses would be due to the substances it contains with potential biological activity, including steroidal glycosides (saponins), protodeacon, glycosides, alkaloids, flavonoids, and tannins, as shown by previous studies [8,10,14,37]. The results of the current study show that the chemical composition of TT powder as a whole plant was rich in dry matter, crude protein, ash, and especially crude fiber (35.55%). Moreover, HPLC analysis showed that the major compound of TT was acetylsalicylic acid, in both ethanol and methanol extracts. The results of HPLC analysis of the chemical compounds in the extracts of TT in the current study agree with [1,2], who found that TT contains many pharmacologically active components belonging to flavonoids and phenols. However, the chemical composition of the extract varies depending on the extraction method and plant parts [38,39]. Several studies suggest that TT has preventive and therapeutic effects on many diseases, possibly due to the components it contains, such as phenolics and flavonoids [40,41]. Despite the diversity of saturated and unsaturated fatty acids in TT extract, but surprisingly, the majority of oils from TT seem to be composed of unsaturated fatty acids. Therefore, analysis revealed that palmitic acid (11.44%), oleic acid (13.76%), and linoleic acid (64.32%) were the major fatty acid components in TT extracts. These results are in agreement with those of Çömlekçioğlu and Çırak [42], who reported that the major fatty acids in TT plant are palmitic acid, oleic acid, and linoleic acid. These main fatty acids extracted from TT have many health benefits, anti-inflammatory properties, and promote metabolism [43,44]. Linoleic acid is an omega-6 polyunsaturated fatty acid and one of the essential fatty acids that should be consumed in the diet [45].

The results obtained show that the addition of TT powder (0.75 g/kg) to a basal diet did not affect overall growth (weight gain, feed intake and the feed conversion ratio) compared to chickens fed a basal diet from 1 to 35 days of age, while performance decreased when they received a high amount of TT powder (1.5 g/kg). Our results are in agreement with those of Ammar et al. [2], who showed that TT was safe for the growth performance of mice up to a dose of 1 g/kg. Another study by Şahin [22] reported that TT powder (0.80 g/kg feed) can be used to improve the growth performance of broiler chickens. In addition, the aqueous extract or powder of TT (5 and 10 mg/kg body weight) had positive effects on the growth performance of laying hens [11]. In contrast, dietary supplementation with TT powder (1 g/kg) or TT extracts (0.06 and 0.12 g/kg) did not affect the growth performance of broilers [3,21]. These results indicate that the over dosage of TT powder has a negative effect on growth performance in broiler chickens. This may be due to side effects of the high level of TT supplementation powder on the intestine, especially the ileum surface and measurements. Hayirli et al. [46] reported that TT powder at a dosage of 0.2 or 0.4 mg/kg had no positive effect on feed intake in broilers. The lower feed intake in T2 (1.5 g TT powder/kg diet) could be due to the high content of dietary fiber and phenolic compounds in TT or other factors.

There were no effects on the absolute and relative weights of liver and kidney in broilers fed diets supplemented with TT powder (T1 and T2). Our results agree with the findings of Martino-Andrade et al. [47], who reported that the absolute and relative weights of liver and kidney were not different in all treated male rats compared to the control. On the other hand, Şahin and Duru [48] reported that dietary supplementation with TT extract (0.36 g/kg) decreased the weight of liver in broiler chickens. In liver function tests (ALT and AST) and kidney function tests (uric acid and creatinine), TT showed no effect at 14 days of age compared with the control group. The current results show that there was no difference in the function tests of the liver and kidney. At 35 days of age, chickens receiving T1 had lower levels of ALT and AST, while uric acid and creatinine showed no effect in all treatments. These results may be attributed to the concept that the diet supplemented with TT powder is not sufficient to induce changes in hepatocytes and cause renal dysfunction. Meanwhile, we found that examining histological changes in liver and kidney tissue of chickens that received TT powder (0.75 g/kg) showed normal morphological results, so it could be that TT powder at 0.75 g/kg has beneficial health effects in broiler chickens. The above results are consonant with those of Yazdi et al. [49], who reported that TT reduced histological damage and improved kidney function in rats. Furthermore, acetylsalicylic acid has no effect on human kidney function [50]. The results of this study are not in agreement with those of Anand et al. [19], who concluded that chickens fed a TT overdose had higher levels of ALT and AST, which were attributed to cellular damage, degenerative changes in the liver, and the increased permeability of the plasma membrane. In addition, acetylsalicylic acid contained in TT may play a beneficial role by reducing liver inflammation and oxidative stress [46]. However, lower levels of ALT and AST in chickens that received TT, especially in T1 (0.75 TT/kg feed may be attributed to a more normal liver histological structure compared with the control group. These were detected in the histological changes in the liver, although it seems that the changes in liver cells and structure did not affect the health of the birds so much, and the kidneys also showed improvement when the chickens received TT powder. Miranda et al. [51] concluded that the inclusion of TT (0.25, 0.50, and 0.75 g/kg diet) had safe effects on the liver and kidneys since no histopathological changes were observed in rats.

According to our results on intestinal morphology, such as the weight and length of the small intestine and the ratio of weight to length of the intestine, the absorption area was higher when the diet was supplemented with TT, especially at T1 (0.75 g/kg). Moreover, histomorphometric measurements of the ileum in broiler chickens were improved when they received TT powder, whereas ileal villus length and width, surface area, villus length to crypt height ratio, mucosa, and lamina propria thickness were lower in chickens receiving 1.5 g TT/kg feed (T2). Crypt height is an indicator of the multiplication and absorption capacity of the intestine [52]. Previous reports have shown that as villus height increases, both the digestive and absorptive functions of the intestine increase, resulting in an increased absorptive surface area, the increased expression of brush border enzymes, nutrient transport systems, and body weight [53]. However, many substances can affect the development of the inner villi, making enzymatic activity and enterocyte structure two essential features of the physiology of the intestinal mucosa [54]. The intake of the 1.5 g TT/kg diet resulted in a decrease in the length, width, and surface area of villi, which may be due to the suppression of beneficial bacteria in the intestine, thus reflected in the decreased growth performance. Mucus production is related to the density of goblet cells and is an important component of the intestinal barrier [55]. However, the results obtained showed no differences in goblet cell density between all treatments. Changes in the histopathology of the ileum of broiler chickens fed the basal diet and 1.5 g TT/kg diet (T2) were observed; shorter villi were observed with the fusion of some villi, which could be due to proliferation of enterocytes with enterocyte necrosis, and the lumen contains sheets of epithelial cells that are slimy. The use of TT decreased the crypt depth and increased the ratio between villus height and crypt depth in the ileum of the broiler. The higher this ratio, the greater the ability to digest and absorb nutrients [56]. These results related to changes in gut histology, which may have led to decreased growth performance in chickens fed the basal diet containing 1.5 g TT/kg feed (T2). Supplementing the diet with a high dose of medicinal plants may have negative effects on some beneficial microbial populations [57]. In addition, birds fed a 1.5 g TT/kg diet may suffer from the negative effects on the intestinal epithelium, resulting in mucosal damage. Therefore, the most common standards for assessing abdominal mucosal injury and health in poultry are villus length, crypt depth, and villus/crypt ratio [58,59]. The villus crypt is considered a villus factory, and deeper crypts indicate rapid tissue turnover, allowing villi to be renewed as needed in response to normal shedding [60]. A reduction in villus height and deeper crypts may result in poor nutrient uptake, increased secretion into the internal tract, and decreased performance [61]. In contrast, Awad et al. [62] reported that an increase in villus length and the villus-to-crypt ratio correlated with increased epithelial cell turnover, and longer villi were associated with activated cell mitosis.

5. Conclusions

The current results conclude that the use of TT powder as a whole plant contains nutritional compositions and saturated fatty acids as well as pharmacologically active components. Therefore, the addition of TT in a 0.75 g/kg diet had no negative effects on growth performance and liver and kidney morphology and improved intestinal morphology and ileum histomorphometry. Furthermore, adding TT powder (0.75 g/kg) to the diet improved the histological changes in the liver, kidneys, and ileum. The dosage of 0.75 g/kg diet used in the current study is likely to be safe and beneficial for the health of broiler chickens. However, further studies are needed to discover the proper dosages and mechanism of pharmacological effect of TT powder as a whole plant on the physiological and health characteristics of broiler chickens.

Author Contributions: Conceptualization, methodology, formal analysis, writing—original draft: H.H.A.-B. Data curation, investigation, writing—review, and editing: M.A.A.-G., H.H.A.-B., A.H.A. All authors have read and agreed to the published version of the manuscript.

Funding: Researchers Supporting Project number (RSP2022R439), King Saud University, Riyadh, Saudi Arabia.

Institutional Review Board Statement: The study complied with the King Saudi Arabia standards on animal use (KSU-SE-21-47) and was approved by the local animal care and welfare committee of King Saud University.

Informed Consent Statement: Not applicable.

Data Availability Statement: All data sets collected and analyzed during the current study are available from the corresponding author on fair request.

Acknowledgments: The authors extend their appreciation to the King Saud University, Riyadh, Saudi Arabia for funding this work through research project group No. (RSP2022R439).

Conflicts of Interest: The authors declare no conflict of interest.

References

1. Khalid, A.; Algarni, A.S.; Homeida, H.E.; Sultana, S.; Javed, S.A.; Abdalla, H.; Abdalla, A.N. Phytochemical, cytotoxic, and antimicrobial evaluation of *Tribulus terrestris L.*, Typha domingensis pers., and ricinus communis L.: Scientific evidences for folkloric uses. *J. Evid. Based Complementary Altern. Med.* **2022**, *2022*, 6519712. [CrossRef] [PubMed]
2. Ammar, N.M.; El-Hawary, S.S.E.D.; Mohamed, D.A.; Afifi, M.S.; Ghanem, D.M.; Awad, G. Phytochemical and biological studies of *Tribulus terrestris L.* growing in Egypt. *Int. J. Pharmacol.* **2018**, *14*, 248–259.
3. Yazdi, F.F.; Ghalamkari, G.; Toghyani, M.; Modaresi, M.; Landy, N. Efficiency of Tribulus terrestris L. as an antibiotic growth promoter substitute on performance and immune responses in broiler chicks. *Asian Pac. J. Trop. Dis.* **2014**, *4*, 1014–1018. [CrossRef]
4. Bravo, D.; Pirgozliev, V.; Rose, S.P. A mixture of carvacrol, cinnamaldehyde, and capsicum oleoresin improves energy utilization and growth performance of broiler chickens fed maize-based diet. *J. Anim. Sci.* **2014**, *92*, 1531–1536. [CrossRef] [PubMed]
5. Pirgozliev, V.; Mansbridge, S.C.; Rose, S.P.; Lillehoj, H.S.; Bravo, D. Immune modulation, growth performance, and nutrient retention in broiler chickens fed a blend of phytogenic feed additives. *Poult. Sci.* **2019**, *98*, 3443–3449. [CrossRef]
6. Guo, W.; Wang, Y.; Fan, M.; Xie, S.; Zhao, H.; Wang, J.; Xu, Y. Integrating metabolomics and network pharmacology to explore the protective effect of gross saponins of *Tribulus terrestris L.* fruit against ischemic stroke in rat. *J. Ethnopharmacol.* **2020**, *263*, 113202. [CrossRef]
7. Zhang, J.D.; Xu, Z.; Cao, Y.B.; Chen, H.S.; Yan, L.; An, M.M.; Gao, P.H.; Wang, Y.; Jia, X.M.; Jiang, Y.Y. Antifungal activities and action mechanisms of compounds from *Tribulus terrestris* L. *J. Ethnopharmacol.* **2006**, *103*, 76–84. [CrossRef]
8. Rathod, P.; Hajare, S.W.; Jamgade, S.; Ingole, R.S.; Ingawale, M.V.; Pooja, P.; Waghmare, S.P. Ameliorative effect of *Tribulus terrestris* on haematological parameters against acephate induced subacute toxicity in Wistar rats. *Pharma. Innov. J.* **2020**, *9*, 23–26.
9. Daniyal, M.; Akram, M.; Zainab, R.; Munir, N.; Sharif, A.; Shah, S.M.A.; Wang, W. Prevalence and current therapy in chronic liver disorders. *Inflammopharmacologya* **2019**, *27*, 213–231. [CrossRef]
10. Chhatre, S.; Nesari, T.; Somani, G.; Kanchan, D.; Sathaye, S. Phytopharmacological overview of *Tribulus terrestris*. *Pharmacogn. Rev.* **2014**, *8*, 45–51. [CrossRef]

11. Amirshekari, T.; Ziaei, N.; Ghoreishi, S.M.; Esfandiarpour, E. The effects of adding aqueous extract and dried aerial part powder of *Tribulus terrestris* on productive performance and blood parameters of laying hens. *J. Appl. Poult. Res.* **2016**, *25*, 145–155. [CrossRef]
12. Nikbakht, S.A.; Mohammadabadi, T.; Mirzadeh, K. The effect of feeding *Tribulus terrestris* plant powder on growth performance, digestibility, rumen and blood parameters of Iranian arabic lambs. *Iran. J. Appl. Anim. Sci.* **2021**, *11*, 781–788.
13. Grigorova, S.; Kashamov, B.; Sredkova, V.; Surdjiiska, S.; Zlatev, H. Effect of *Tribulus terrestris* extract on semen quality and serum total cholesterol content in White Plymouth Rock-mini cocks. *Biotechnol. Anim. Husb.* **2008**, *24*, 139–146. [CrossRef]
14. Shahid, M.; Riaz, M.; Talpur, M.M.; Pirzada, T. Phytopharmacology of *Tribulus terrestris*. *J. Biol. Regul. Homeost. Agents* **2016**, *30*, 785–788. [PubMed]
15. Azam, F.; Munier, S.; Batoo, M.; Ahmad, B.; Abbas, G. A review on advancements in ethnomedicine and phytochemistry of *Tribulus terrestris*–a plant with multiple health benefits. *Int. J. Biosci.* **2019**, *14*, 21–37.
16. Masoudi, R.; Javaheri, B.H.; Hosseini, S.A.; Zarei, F.; Abdollahi, Z. Effect of *Tribulus terrestris*, Ceratonia silique and Zingiber officinale on reproductive potential of Ross broiler breeder roosters. *Vet. Res. Biol. Prod.* **2021**, *34*, 177–186.
17. Kumari, M.; Kumar, P.; Singh, P. Safety evaluation of *Tribulus terrestris* on the male reproductive health of laboratory mouse. *Int. J. Pharm. Phytopharm. Res.* **2015**, *4*, 281–287.
18. Kamenov, Z.; Fileva, S.; Kalinov, K.; Jannini, E.A. Evaluation of the efficacy and safety of *Tribulus terrestris* in male sexual dysfunction-A prospective, randomized, double-blind, placebo-controlled clinical trial. *Maturitas* **2017**, *99*, 20–26. [CrossRef]
19. Vincent, J.L.; Patel, G.H.; Fox, M.D.; Snyder, A.Z.; Baker, J.T.; Van Essen, D.C.; Raichle, M.E. Intrinsic functional architecture in the anaesthetized monkey brain. *Nature* **2007**, *447*, 83–86. [CrossRef]
20. Duru, M.; Şahin, A. Effects of dietary puncture vine (*Tribulus Terrestris*) powder in different carriers on growth performance, carcass characteristics and blood parameters of broiler chicks. *Kafkas Univ. Vet. Fak. Derg.* **2012**, *18*, 359–365. [CrossRef]
21. Duru, M.; Şahin, A. Effects of dietary yohimbe (Pausinystalia Yohimbe) and puncture vine (*Tribulus Terrestris*) extracts for growth performance, body composition and digestive parts of broiler chicks. *Europ. Int. J. Sci. Technol.* **2016**, *5*, 2304–9693.
22. Şahin, A. Effects of dietary *Tribulus Terrestris L.* Powder on growth performance, body components and digestive system of broiler chicks. *J. Appl. Anim. Res.* **2009**, *35*, 193–195. [CrossRef]
23. Sharma, U.S.; Yadav, H.N.; Singh, S.; Gupta, Y.K. Hydroalcoholic extract of *Tribulus terrestris* protects against accumulation of platinum in rat kidney. *Innov. Pharm. Pharmacother.* **2019**, *7*, 7–14.
24. Anand, V.P.; Lakkawar, A.W.; Narayanaswamy, H.D.; Sathyanarayana, M.L. Serum biochemical changes in ochratoxin induced toxicity and its amelioration using diatomaceous earth in broiler chicken. *Int. J. Chem. Stud.* **2018**, *6*, 673–679.
25. Chu, S.; Qu, W.; Pang, X.; Sun, B.; Huang, X. Effect of saponin from *Tribulus terrestris* on hyperlipidemia. *Zhong. Vao Cai.* **2003**, *26*, 341–344.
26. Kostova, I.; Dinchev, D. Saponins in *Tribulus Terrestris*—Chemistry and bioactivity. *Phytochem. Rev.* **2005**, *4*, 111–137. [CrossRef]
27. AOAC. *Official Methods of Analysis*, 19th ed.; Association of Official Analytical Chemists: Rockville, MD, USA, 2012.
28. Karadağoğlu, Ö.; Şahin, T.; Ölmez, M.; Yakan, A.; Özsoy, B. Changes in serum biochemical and lipid profile, and fatty acid composition of breast meat of broiler chickens fed supplemental grape seed extract. *Turk. J. Vet. Anim. Sci* **2020**, *44*, 182–190. [CrossRef]
29. Biesek, J.; Kuźniacka, J.; Banaszak, M.; Maiorano, G.; Grabowicz, M.; Adamski, M. The effect of various protein sources in goose diets on meat quality, fatty acid composition, and cholesterol and collagen content in breast muscles. *Poult. Sci.* **2020**, *99*, 6278–6286. [CrossRef]
30. Al-Fatimi, M.; Wurster, M.; Schroder, G.; Lindequist, U. Antioxidant, antimicrobial and cytotoxic activities of selected medicinal plants from Yemen. *J. Ethnopharmacol.* **2007**, *111*, 657–666. [CrossRef]
31. Al-Mufarrej, S.I.; Al-Baadani, H.H.; Fazea, E.H. Effect of level of inclusion of clove (*Syzygium aromaticum*) powder in the diet on growth and histological changes in the intestines and livers of broiler chickens. *S. Afr. J. Anim. Sci.* **2019**, *49*, 166–175. [CrossRef]
32. Winsor, L.; Sluys, R. Basic histological techniques for planarians. In *Planarian Regeneration*; Humana Press: New York, NY, USA, 2018; pp. 285–351.
33. Al-Baadani, H.H.; Abudabos, A.M.; Al-Mufarrej, S.I.; Alzawqari, M. Effects of dietary inclusion of probiotics, prebiotics and synbiotics on intestinal histological changes in challenged broiler chickens. *S. Afr. J. Anim. Sci.* **2016**, *46*, 157–165. [CrossRef]
34. Qaid, M.M.; Al-Mufarrej, S.I.; Azzam, M.M.; Al-Garadi, M.A.; Albaadani, H.H.; Alhidary, I.A.; Aljumaah, R.S. Growth performance, serum biochemical indices, duodenal histomorphology, and cecal microbiota of broiler chickens fed on diets supplemented with cinnamon bark powder at prestarter and starter phases. *Animals* **2021**, *11*, 94. [CrossRef] [PubMed]
35. Belote, B.L.; Soares, I.; Tujimoto-Silva, A.; Sanches, A.W.; Kraieski, A.L.; Santin, E. Applying I see inside histological methodology to evaluate gut health in broilers challenged with Eimeria. *Vet. Parasitol.* **2019**, *276*, 100004–100010. [CrossRef] [PubMed]
36. SAS Institute. *SAS Users Guide: Statistics*; SAS Institute Inc.: Cary, NC, USA, 2008.
37. Hamed, A.I.; Janda, B.; Mahalel, U.A.; Stochmal, A.; Oleszek, W. Profiles of steroidal saponins from the aerial parts of *Tribulus pentandrus*, *T. megistopterus subsp.* pterocarpus and *T. parvispinus* by LC-ESI-MS/MS. *Phytochem. Anal.* **2012**, *23*, 613–621. [CrossRef]
38. Zhu, W.; Du, Y.; Meng, H.; Dong, Y.; Li, L. A review of traditional pharmacological uses, phytochemistry, and pharmacological activities of *Tribulus Terrestris*. *Chem. Cent. J.* **2017**, *11*, 60. [CrossRef]

39. Ştefănescu, R.; Tero-Vescan, A.; Negroiu, A.; Aurică, E.; Vari, C. A comprehensive review of the phytochemical, pharmacological, and toxicological properties of *Tribulus Terrestris* L. *Biomolecules* **2020**, *10*, 752. [CrossRef]
40. Majidinia, M.; Bishayee, A.; Yousefi, B. Polyphenols: Major regulators of key components of DNA damage response in cancer. *DNA Repair* **2019**, *82*, 102679. [CrossRef]
41. Sanagoo, S.; Oskouei, B.S.; Abdollahi, N.G.; Salehi-Pourmehr, H.; Hazhir, N.; Farshbaf-Khalili, A. Effect of *Tribulus terrestris* L. on sperm parameters in men with idiopathic infertility: A systematic review. *Complementary Ther. Med.* **2019**, *42*, 95–103. [CrossRef]
42. Çömlekçioğlu, N.; Çırak, R. Comparison of fatty acid composition and antioxidant contents of *Tribulus Terrestris L.* collected from different localities. *Turk. J. Agric. Food Sci. Technol.* **2021**, *9*, 1448–1454. [CrossRef]
43. Jabeur, I.; Pereira, E.; Barros, L.; Calhelha, R.C.; Soković, M.; Oliveira, M.B.P.; Ferreira, I.C. Hibiscus sabdariffa L. as a source of nutrients, bioactive compounds and colouring agents. *Food Res. Int.* **2017**, *100*, 717–723. [CrossRef]
44. Buckley, M.T.; Racimo, F.; Allentoft, M.E.; Jensen, M.K.; Jonsson, A.; Huang, H.; Hormozdiari, F.; Sikora, M.; Marnetto, D.; Eskin, E.; et al. Selection in Europeans on fatty acid desaturases associated with dietary changes. *Mol. Biol. Evol.* **2017**, *34*, 1307–1318. [CrossRef] [PubMed]
45. Whitney, E.; Rolfes, S.R. *Understanding Nutrition*, 11th ed.; Thomson Wadsworth: Belmont, CA, USA, 2008; p. 154.
46. Hayirli, A.; Esenbuga, N.; Macit, M.; Acit, E.; Lacin, M.; Karaoglu, H.; Karaca, L.; Yidiz, L. Nutrition practice to alleviate the adverse effects of stress on laying performance, metabolic profile, and egg quality in peak producing hens: I. The humate supplementation. *Asian-Austr. J. Anim. Sci.* **2005**, *18*, 1310–1319. [CrossRef]
47. Martino-Andrade, A.J.; Morais, R.N.; Spercoski, K.M.; Rossi, S.C.; Vechi, M.F.; Golin, M.; Dalsenter, P.R. Effects of *Tribulus terrestris* on endocrine sensitive organs in male and female Wistar rats. *J. Ethnopharmacol.* **2010**, *127*, 165–170. [CrossRef] [PubMed]
48. Şahin, A.; Duru, M. Effects of *Tribulus Terrestris* (puncture vine) supplementation on performance and digestive system of broiler chicks. *J. Agric. Sci.* **2010**, *16*, 271–277.
49. Najafi, H.; Firouzifar, M.R.; Shafaat, O.; Ashtiyani, S.C.; Hosseini, N. Protective effects of Tribulus Terrestris L extract against acute kidney injury induced by reperfusion injury in rats. *Iran. J. Kidney Dis.* **2014**, *8*, 292. [PubMed]
50. Hansen, H.P.; Gaede, P.H.; Jensen, B.R.; Parving, H.H. Lack of impact of low-dose acetylsalicylic acid on kidney function in type 1 diabetic patients with microalbuminuria. *Diabetes Care* **2000**, *23*, 1742–1745. [CrossRef]
51. Miranda, D.R.; Chaves, B.R.; Vilela Soares, L.; Konig, I.F.M.; Pinto, J.T.; Miranda, J.R.; Zangeronimo, M.G. Use of *Tribulus terrestris* and Lepidium meyenii extract in rats: Reproductive, biochemical and body parameters. *Andrologia.* **2021**, *54*, e14358. [CrossRef]
52. Franco, J.R.G.; Murakami, A.E.; Natali, M.R.M.; Garcia, E.R.M.; Furlan, A.C. Influence of delayed placement and dietary lysine levels on small intestine morphometrics and performance of broilers. *Braz. J. Poult. Sci.* **2006**, *8*, 233–241. [CrossRef]
53. Mohamed, M.A.; El-Daly, E.F.; El-Azeem, N.A.; Youssef, A.W.; Hassan, H.M.A. Growth performance and histological changes in ileum and immune related organs of broilers fed organic acids or antibiotic growth promoter. *Int. J. Poult. Sci.* **2014**, *13*, 602–610. [CrossRef]
54. Awad, W.; Ghareeb, K.; Böhm, J. Intestinal structure and function of broiler chickens on diets supplemented with a synbiotic containing Enterococcus faecium and oligosaccharides. *Int. J. Mol. Sci.* **2008**, *9*, 2205–2216. [CrossRef]
55. Jeurissen, S.H.; Lewis, F.; van der Klis, J.D.; Mroz, Z.; Rebel, J.M.; Huurne, A.A. Parameters and techniques to determine intestinal health of poultry as constituted by immunity, integrity, and functionality. *Curr. Issues Intest. Microbiol.* **2002**, *3*, 1–14. [PubMed]
56. Hu, Y.; Wang, Y.; Li, A.; Wang, Z.; Zhang, X.; Yun, T.; Qiu, L.; Yin, Y. Effects of fermented rapeseed meal on antioxidant functions, serum biochemical parameters and intestinal morphology in broilers. *Aliment. Agric. Immunol.* **2016**, *27*, 182–193. [CrossRef]
57. Toghyani, M.; Tohidi, M.; Gheisari, A.A.; Tabeidian, S.A. Performance, immunity, serum biochemical and hematological parameters in broiler chicks fed dietary thyme as alternative for an antibiotic growth promoter. *Afr. J. Biotechnol.* **2010**, *9*, 6819–6825.
58. Budd, G.R.; Aitchison, A.; Day, A.S.; Keenan, J.I. The effect of polymeric formula on enterocyte differentiation. *Innate Immun.* **2017**, *23*, 240–248. [CrossRef] [PubMed]
59. Ducatelle, R.; Goossens, E.; Meyer, F.; Eeckhaut, V.; Antonissen, G.; Haesebrouck, F.; Immerseel, F. Biomarkers for monitoring intestinal health in poultry: Present status and future perspectives. *Vet. Res.* **2018**, *49*, 43. [CrossRef] [PubMed]
60. Yason, C.V.; Summers, B.A.; Schat, K.A. Pathogenesis of rotavirus infection in various age groups of chickens and turkeys: Pathology. *Am. J. Vet. Res.* **1987**, *6*, 927–938.
61. Xu, Z.R.; Hu, C.H.; Xia, M.S.; Zhan, X.A.; Wang, M.Q. Effects of dietary fructooligosaccharide on digestive enzyme activities, intestinal microflora and morphology of male broilers. *Poult. Sci.* **2003**, *82*, 1030–1036. [CrossRef]
62. Awad, W.A.; Ghareeb, K.; Abdel-Raheem, S. Effects of dietary inclusion of probiotic and symbiotic on growth performance, organ weights and intestinal histomorphology of broiler chickens. *Poult. Sci.* **2009**, *88*, 49–56. [CrossRef]

animals

Article

Assessment of Neurodegenerative Changes in Turkeys Fed Diets with Different Proportions of Arginine and Methionine Relative to Lysine

Magdalena Krauze [1,*], Katarzyna Ognik [1], Dariusz Mikulski [2] and Jan Jankowski [2]

1 Department of Biochemistry and Toxicology, University of Life Sciences, 20-950 Lublin, Poland; katarzyna.ognik@up.lublin.pl
2 Department of Poultry Science, University of Warmia and Mazury in Olsztyn, 10-719 Olsztyn, Poland; dariusz.mikulski@uwm.edu.pl (D.M.); janj@uwm.edu.pl (J.J.)
* Correspondence: magdalena.krauze@up.lublin.pl

Simple Summary: It is important to take care of a properly balanced amino acid composition in the diet in order to inhibit or delay the occurrence of processes and changes related to the destruction of nervous tissue. Therefore, an attempt was made in this manuscript to evaluate the effect of different ratios of the key amino acids arginine and methionine, relative to lysine, in relation to two turkey feeding standards. The amino acid guidelines formulated by British United Turkeys (BUT) suggest higher levels of lysine (Lys) in turkey diets than those recommended by the National Research Council (NRC). In order to assess the impact of such supplementation, we analyzed the level of indicators informing the presence or degree of advancement of neurodegenerative processes in the nervous tissue (the level of acetylcholinesterase and amyloid-β; the concentration of AChE complexes with amyloid-β and Tau protein, called glycosylated acetylcholinesterase (GAChE), indicative of the destruction of neurons). The level of low-density lipoprotein receptor-related protein 1, or LRP-1, which facilitates the breakdown of toxic amyloid-β, was also assessed. In addition, the effect of different doses of these amino acids on neurodegenerative changes in DNA, especially the degree of methylation of histone proteins resulting from covalent modifications was compared between lysine and arginine residues.

Abstract: We postulated that the use of optimal levels and proportions of Arg and Met relative to a low or high concentration of Lys in diets for meat turkeys would reduce the occurrence of metabolic disturbances in the nervous tissue that can lead to neurodegenerative changes. The aim of the study was to determine the effect of various proportions of Lys, Arg, and Met in diets for turkeys, with a low content of Lys in accordance with NRC (Experiment 1) recommendations, and in diets with high Lys levels that are close to the recommendations of breeding companies (Experiment 2) on selected indicators of potential neurodegenerative effects in the brain and liver of turkeys. The Experiment 1 and Experiment 2 was conducted using 864 day-old turkey chicks randomly assigned to six groups, in eight replicates (6 groups × 18 birds × 8 replicates). A full description of the methodology can be found in previously published papers using the same experimental design. Indicators informing about the presence or advancement of neurodegenerative processes in the nervous tissue were determined in the brain and liver (level of: AChE, amyloid-β, GAChE, Tau protein, LRP1, and the degree of DNA methylation). It was established that in the case of both a low (National Research Council, NRC) and a high (British United Turkeys, BUT) level of Lys in the diet of turkeys, the Arg level can be reduced to 90% of the Lys level and Met to 30% of the Lys level, because this does not cause neurodegenerative changes in turkeys. Unfavorable neurodegenerative changes may appear if the Arg level is increased from 100 to 110% of the Lys level recommended by the NRC. However, due to the lack of such a relationship when Arg is increased from 100 to 110% of the Lys level recommended by BUT, at this stage of research no definitive conclusions can be drawn regarding the risk of neurodegenerative changes caused by increasing Arg in the diet of turkeys.

Citation: Krauze, M.; Ognik, K.; Mikulski, D.; Jankowski, J. Assessment of Neurodegenerative Changes in Turkeys Fed Diets with Different Proportions of Arginine and Methionine Relative to Lysine. *Animals* **2022**, *12*, 1535. https://doi.org/10.3390/ani12121535

Academic Editor: Ilias Giannenas

Received: 27 April 2022
Accepted: 9 June 2022
Published: 14 June 2022

Keywords: turkey; amino acid; neurodegeneration; methionine; lysine; arginine

1. Introduction

A diet with an appropriate amino acid profile plays a key role in exploiting the genetic potential of fast-growing turkeys [1]. The main amino acids limiting the biological value of dietary protein in poultry are methionine (Met) and arginine (Arg) [2,3]. Lysine (Lys), on the other hand, may determine the digestibility and absorption of Arg in the intestines [4]. Requirements for the intake of amino acids in the diet of turkeys formulated by NRC, as seen in [5,6], differ substantially in the proportions of individual amino acids. According to the NRC [5], the Arg level in the diet of turkeys should be 90–100% of the Lys content, whereas BUT [6] recommends a higher Arg level (102–105% Lys). The Met level established by the NRC [5] is 30–38% of the content of Lys, while BUT [6] recommends a level of 36–41%. Supplementation of the diet with selected amino acids can have highly varied effects on the functioning of the brain, liver, and nervous tissue, and the biosynthesis of amino acids in the brain and liver is strictly regulated by the concentration of amino acids in the plasma [7]. According to Morales et al. [8], optimal levels of amino acids can play a significant role in maintaining the integrity and functioning of brain cells and hepatocytes, take part in the synthesis of neurotransmitters, and—owing to their ability to regenerate and reorganize neurons—play an active role in the treatment of neurological diseases. Excessive amounts of these compounds, on the other hand, can have toxic effects on nervous tissue, disturb its metabolism, and lead to neurodegenerative changes [7,9,10].

According to Toue et al. [11] and Koladiya et al. [12], one of the amino acids essential to the normal functioning of nervous tissue is methionine. Methionine and cysteine, which is derived from it, are key amino acids with an important role in the cellular metabolism of neurons. Some other authors report that excessive methionine intake in the diet may induce neurodegenerative changes in the brain and accelerate the ageing process, which may be caused by methionine sulfoxide, an intermediate product of the methionine cycle [7,13–16]. On the other hand, according to Toue et al. [11], a deficiency of methionine can cause dementia. In addition, excessive amounts of cysteine, derived from methionine, can also have neurotoxic effects. The most neurotoxic form of cysteine, S-nitroso cysteine, is responsible for inhibition of mitochondrial respiration, caused by metabolic disturbances and inhibition of ATP synthesis in the neurons. This compound is produced in a reaction between methionine and reactive oxygen species, mainly nitric oxide (NO) [17]. Another amino acid with an important role in determining the biological value of fodder protein is arginine (Arg). There is no urea cycle in birds, and as a result, Arg is not synthesized endogenously, therefore, it must be added to their diet [18].

According to Rajagopal et al. [19], Arg stimulates nerve signal transmission between neurons, activates the growth and differentiation of new nerve cells, and delays neuron apoptosis. According to Balnave and Brake [4], both the deficiency and surplus of Arg can adversely affect the concentrations of other amino acids in the blood plasma and muscles, which impedes not only nerve cell metabolism but also the growth and development of birds. This is because Arg is a precursor of NO, ornithine, glutamate, creatine, proline, glutamine, and dimethylarginine, and therefore plays an important role in the metabolism of birds [20].

Lysine is another amino acid that is essential for the growing turkeys, and a Lys deficiency or defects associated with catabolism of this amino acid can result in neurodegenerative and other physiological changes [21]. Nutritional studies conducted by Jankowski et al. [2,3] and Ognik et al. [22] have shown that a diet with low Lys content, recommended by the NRC [5], with an Arg level of 100% relative to Lys and a Met level of 45% of the Lys level, can stimulate immune and antioxidant defenses, as well as eliminate oxidative changes and neutralize biologically important compounds. Another study by the same authors showed that in growing turkeys, which were fed a diet with a high Lys content (according to the recommendations of BUT [6], Arg and Met can reach levels of 90 and 45% of the Lys content with no negative effect on metabolic parameters or the growth of the birds [23]). In addition, the inclusion of Met at a level of 45% of Lys levels increases the antioxidant potential of turkeys and reduces the risk of oxidative changes in important biomolecules, especially lipids and proteins, as well as DNA.

Diets for turkeys with a high Arg content (110% Lys) were found to be unfavorable, due to the risk of lipid oxidation, protein nitration, and adverse changes in the concentrations of hormones regulating metabolism, especially of sugars [22]. Due to the potential neurotoxic effects of a diet with excessive Met, which must be correctly correlated with Arg, relative to Lys [5,6], in order to choose the optimal levels of these amino acids, it seems important to determine indicators of the appearance of neurodegenerative changes in the brain resulting from the use of these amino acids. For this purpose, it may be useful to analyze the level of acetylcholinesterase (AChE), a key enzyme for conduction of nerve impulses and protection of neurons, as well as the levels of neurotoxic amyloid-β and the concentration of AChE complexes with amyloid-β and Tau protein, called glycosylated acetylcholinesterase (GAChE), which are responsible for neuronal apoptosis. Both a deficiency of acetylcholinesterase and the formation of complexes of this enzyme with amyloid-β (glycosylated acetylcholinesterase, GAChE) are conducive to mitochondrial dysfunction and the generation of large amounts of reactive oxygen species, which initiate neuronal apoptosis [24–27]. Moreover, it may be useful to determine the level of low-density lipoprotein receptor-related protein 1 (LRP1), which facilitates the degradation of toxic amyloid-β [28]. Evaluation of the degree of methylation of histone proteins, resulting from the covalent modifications of lysine and arginine residues, may also be helpful in recognizing the intensification of neurodegenerative changes [29].

A further suggestion to explore the question of the effect of different levels and proportions of lysine, arginine, and methionine in diets for meat turkeys was the findings of other researchers, who evaluated the rearing efficiency, as well as biochemical and immune parameters of the blood, intestinal morphology, dressing percentage, and meat quality [30–32]. According to Ghamari Monavvar et al. [33], additional supplementation with arginine improves growth performance and the quality traits of poultry (especially meat tenderness and meat fat content). What is more, Arg supplementation improves the intestinal morphology (especially the ratio of villus height to crypt depth), enzyme activity, and the composition of the intestinal microbiota. It also stimulates the immune system, a significant portion of which is located in the intestines, and increases levels of insulin and thyroid hormones, thus activating metabolism. A new aspect of the study is focused on the problem of choosing optimum levels and proportions of Arg and Met relative to a low or high concentration of Lys in diets for meat turkeys, in the context of minimizing the occurrence of metabolic disorders in nervous tissue. Efforts to improve rearing efficiency in turkeys by modifying their diet may adversely affect health parameters, disturb the general metabolism, increase the risk of oxidative stress, and importantly, affect nervous tissue as well.

We found no reports in the world literature of research conducted in poultry in the context of neurodegenerative changes. Because Arg, Met, and Lys are essential amino acids for poultry and determine the biological value of fodder protein, it seems useful to observe such changes, especially when we attempt to improve the growth performance of turkeys by changing the proportions of these amino acids. We took an interest in potential neurodegenerative changes after an in-depth analysis of the results obtained by the co-authors of this study in their research on growth performance and the indicators of sugar metabolism, peptide nitration, lipid oxidation, and systemic antioxidant defenses [2,3,22,23].

Due to the fact that the nervous tissue plays a superior role in the functioning of the system, any interference leading to the improvement of its condition is justified, provided that the portion of amino acids used does not constitute a toxic dose, and we tried to avoid this problem by assessing the impact of individual proportions within the acceptable standards for poultry. We considered that, due to the fact that two different nutritional standards for poultry (NRC, BUT) that are commonly used in poultry farming give slightly different values for the nutritional recommendations of Arg, Met, and Lys, and that these amino acids limit the biological value of the dietary protein in poultry and may affect the metabolism of nervous tissue in a different way, it is important to balance them carefully and in our opinion there is a need for a thorough verification of this problem.

We postulated that the use of optimal levels and proportions of Arg and Met relative to a low or high concentration of Lys in diets for meat turkeys would reduce the occurrence of metabolic disturbances in the nervous tissue that can lead to neurodegenerative changes. The aim of the study was to determine the effect of various proportions of Lys, Arg, and Met in diets for turkeys with a low content of Lys, in accordance with NRC [5] recommendations, and in diets with high Lys levels, close to the recommendations of breeding companies [6], on selected indicators of potential neurodegenerative effects in the brain and liver of turkeys.

2. Materials and Methods

2.1. Experiment 1

The experiment was conducted using 864 day-old turkey chicks randomly assigned to 6 groups, in 8 replicates (6 groups × 18 birds × 8 replicates). A full description of the methodology can be found in previously published papers using the same experimental design [2,3,23,30]. During each of 4 feeding periods (4 weeks each), the birds received ad libitum isocaloric diets containing 1.60, 1.50, 1.30, and 1.00% Lys, in accordance with recommendations for turkeys specified in Nutrient Requirements of Poultry [5]. The factors differentiating the experimental groups were the level of Arg, which was 90, 100, and 110% of the level of Lys in the diet, and for Met it was 30 or 45% of the level of Lys in the diet. For each of the 4 feeding periods, basal diets were prepared without the addition of Lys, Met, or Arg (Table 1). The content of amino acids in the basal diets was determined (Table 2), and then they were mixed with the appropriate amounts of Lys, Met, and Arg. The total content of amino acids in all experimental diets was determined analytically (Table 2). Starter diets (days 1–28) and grower and finisher diets (days 29–112), with no feed additives, were provided as crumbles and pellets (3 mm pellets at 65 °C for 45 s), respectively.

Table 1. Ingredient composition and nutrient content of basal diets (g/kg, as-fed basis) fed to turkeys at 1–4, 5–8, 9–12, and 13–16 weeks of age, Experiment 1 [1].

Item	Feeding Period, Weeks			
	1–4	**5–8**	**9–12**	**13–16**
Ingredients				
Wheat	463.9	486.7	537.4	656.3
Maize	100.0	100.0	100.0	100.0
Soybean meal	250.5	232.7	187.3	79.1
Rapeseed meal	30.0	50.0	71.8	70.0
Potato protein	55.2	30.1	-	-
Soybean oil	2.0	23.2	35.3	32.2
Maize gluten meal	55.0	35.0	35.0	35.0
Sodium bicarbonate	2.0	2.0	2.0	2.0
Sodium chloride	1.5	1.6	1.6	1.4
Limestone	22.0	18.6	16.4	13.8
Monocalcium phosphate	14.6	12.9	9.0	5.0
L-Threonine	-	0.7	0.7	1.7

Table 1. *Cont.*

Item	Feeding Period, Weeks			
	1–4	5–8	9–12	13–16
Choline chloride	1.0	1.0	1.0	1.0
Vitamin-mineral premix [2]	2.5	2.5	2.5	2.5
Titanium oxide	-	3.0	-	-
Calculated nutrient content				
Metabolizable energy, kcal/kg	2820	2900	3000	3100
Crude protein	26.5	23.5	20.5	17.0
Arginine total [3]	14.4	13.5	11.7	8.9
Lysine total [3]	12.8	11.2	8.9	6.4
Methionine total [3]	4.5	3.9	3.4	2.9
Methionine + Cysteine total	9.2	8.2	8.0	7.5
Threonine total	10.2	9.5	7.4	6.5
Calcium	12.5	11.0	9.5	7.5
Available phosphorus	6.5	5.5	4.7	3.8

[1] Source: This table was published in Poultry Science [2]. [2] Provided per kg diet (feeding periods: weeks 0–4, 5–8, 9–12 and 13–16): mg: retinol 3.78, 3.38, 2.88 and 2.52; cholecalciferol 0.13, 0.12, 0.10 and 0.09; α-tocopheryl acetate 100, 90, 80 and 70; vit. K_3 5.8, 5.6, 4.8 and 4.2; thiamine 5.4, 4.7, 4.0 and 3.5; riboflavin 8.4, 7.5, 6.4 and 5.6; pyridoxine 6.4, 5.6, 4.8 and 4.2; cobalamin 0.032, 0.028, 0.024 and 0.021′ biotin 0.32, 0.28, 0.24 and 0.21; pantothenic acid 28, 24, 20 and 18; nicotinic acid 84, 75, 64 and 56; folic acid 3.2, 2.8, 2.4 and 2.1; Fe 64, 60, 56, 48 and 42; Mn 120, 112, 96 and 84; Zn 110, 103, 88 and 77; Cu 23, 19, 16 and 14; I 3.2, 2.8, 2.4 and 2.1; Se 0.30, 0.28, 0.24 and 0.21, respectively. [3] Actual levels of supplementary Lys, Arg, and Met in experimental diets were obtained by adding supplementary L-Lys HCl, L-Arg HCl and DL-Met to the basal feed. L-Lys HCl was added to the basal diet to obtain 1.60, 1.50, 1.30, and 1.00 g of Lys per 100 g of feed in four successive feeding periods, according to the nutrient requirements of turkeys [5]. L-Arg HCl was added to the basal diet to obtain 90, 100, and 110% Arg relative to the content of dietary Lys. DL-Met was added to obtain 30 and 45% Met relative to the content of dietary Lys. Lys—lysine, Arg—arginine, Met—methionine.

Table 2. Amino acid content (g/kg) of basal diets of turkeys, Experiment 1 [1].

Item	Feeding Period, Weeks			
	1–4	5–8	9–12	13–16
Crude protein	270.8	246.3	209.3	177.0
Alanine	13.27	11.04	9.87	9.16
Arginine	14.81	13.70	11.93	9.16
Aspartic acid	25.40	21.22	17.88	11.66
Cysteine	4.62	4.09	3.77	3.34
Glutamic acid	53.73	46.06	41.84	37.12
Glycine	11.42	9.72	8.69	6.69
Histidine	6.49	5.67	5.18	4.44
Isoleucine	11.93	9.83	8.68	6.23
Leucine	24.24	19.76	17.33	13.25
Lysine	12.96	11.89	9.62	6.22
Mathionine	4.56	3.94	3.36	2.47
Methionine + Cysteine	9.18	8.03	7.13	5.81
Phenyloalanine	14.56	11.91	10.35	7.81
Proline	18.15	15.89	14.92	14.60
Serine	13.66	11.47	10.13	7.72
Threonine	10.61	9.09	7.68	6.76
Tyrosine	8.70	8.19	7.51	5.41
Valine	13.50	11.34	9.84	7.34

[1] Source: This table was published in Poultry Science [2].

2.2. Experiment 2

The experiment was conducted using 864 day-old turkey chicks, which were randomly assigned to 6 groups, in 8 replicates (6 groups × 18 birds × 8 replicates). A full description of the methodology can be found in previously published papers using the same experimental design [3,24]. During each of 4 feeding periods (4 weeks each), the birds received ad libitum isocaloric diets containing 1.83, 1.67, 1.48, and 1.20% Lys, in accordance with recommendations for turkeys by Hybrid Turkeys [6]. The factors differentiating the experimental groups were the level of Arg, which was 90, 100, and 110% of the level of Lys in the diet, and for Met it was 30 or 45% of the level of Lys in the diet. For each of the 4 feeding periods, basal diets were prepared without the addition of Lys, Met, or Arg (Table 3). The content of amino acids in the basal diets was determined (Table 4), and then they were mixed with the appropriate amounts of Lys, Met, and Arg. The total content of amino acids in all experimental diets was determined analytically (Table 4). Starter diets (days 1–28) and grower and finisher diets (days 29–112), with no feed additives, were provided as crumbles and pellets (3 mm pellets at 65 °C for 45 s), respectively.

Table 3. Ingredient composition and nutrient content of basal diets (g/kg, as-fed basis) fed to turkeys at 1–4, 5–8, 9–12, and 13–16 weeks of age, Experiment 2 [1].

Item	Feeding Period, Weeks			
	1–4	5–8	9–12	13–16
Ingredients				
Wheat	439.8	471.2	519.9	617.1
Maize	100.0	100.0	100.0	100.0
Soybean meal, 48% CP	287.7	265.4	238.5	152.4
Rapeseed meal	30.0	30.0	30.0	30.0
Potato protein	50.0	29.6	-	-
Soybean oil	9.5	28.5	47.8	42.2
Maize gluten meal	35.0	30.0	30.0	30.0
Sodium bicarbonate	2.0	2.0	2.0	2.0
Sodium chloride	1.5	1.6	1.6	1.2
Limestone	20.7	18.7	16.4	14.5
Monocalcium phosphate	19.4	15.5	9.6	6.5
L-Threonine	0.9	1.0	0.7	0.6
Choline chloride	1.0	1.0	1.0	1.0
Vitamin-mineral premix [2]	2.5	2.5	2.5	2.5
Titanium oxide	-	3.0	-	-
Calculated nutrient content				
Metabolizable energy, kcal/kg	2820	2950	3000	3150
Crude protein	27.0	24.5	21.5	18.5
Arginine total [3]	1.58	1.44	1.27	1.04
Lysine total [3]	1.36	1.19	0.97	0.76
Methionine total [3]	0.44	0.39	0.34	0.30
Methionine + Cysteine total	0.91	0.83	0.74	0.67
Threonine total	1.02	1.01	0.83	0.70
Calcium	1.30	1.15	0.95	0.80
Available phosphorus	0.70	0.60	0.47	0.40

[1] Source: This table was published in Poultry Science [3]. [2] Provided per kg diet (feeding periods: weeks 0–4, 5–8, 9–12 and 13–16): mg: retinol 3.78, 3.38, 2.88, and 2.52; cholecalciferol 0.13, 0.12, 0.10, and 0.09; α-tocopherol acetate 100, 90, 80, and 70; vit. K_3 5.8, 5.6, 4.8, and 4.2; thiamine 5.4, 4.7, 4.0, and 3.5; riboflavin 8.4, 7.5, 6.4, and 5.6; pyridoxine 6.4, 5.6, 4.8, and 4.2; cobalamin 0.032, 0.028, 0.024, and 0.021; biotin 0.32, 0.28, 0.24, and 0.21; pantothenic acid 28, 24, 20, and 18; nicotinic acid 84, 75, 64, and 56; folic acid 3.2, 2.8, 2.4, and 2.1; Fe 64, 60, 56, 48, and 42; Mn 120, 112, 96, and 84; Zn 110, 103, 88, and 77; Cu 23, 19, 16, and 14; I 3.2, 2.8, 2.4, and 2.1; Se 0.30, 0.28, 0.24, and 0.21, respectively. [3] Actual levels of supplementary Lys, Arg, and Met in experimental diets were obtained by adding supplementary L-Lys HCl, L-Arg HCl, and DL-Met to the basal feed. L-Lys HCl was added to the basal diet to obtain 1.60, 1.50, 1.30, and 1.00 g of Lys per 100 g of feed in four successive feeding periods, according to the nutrient requirements of turkeys [6]. L-Arg HCl was added to the basal diet to obtain 90, 100, and 110% Arg relative to the content of dietary Lys. DL-Met was added to obtain 30 and 45% Met relative to the content of dietary Lys. Lys—lysine, Arg—arginine, Met—methionine.

Table 4. Amino acid content (g/kg) of basal diets of turkeys, Experiment 2 [1].

Item	Feeding Period, Weeks			
	1–4	5–8	9–12	13–16
CP	265.7	248.0	221.0	195.50
Alanine	11.9	12.3	10.0	8.0
Arginine	15.6	14.2	13.7	11.1
Aspartic acid	24.1	23.4	17.7	14.4
Cysteine	4.2	3.9	2.9	2.6
Glutamic acid	45.0	51.0	43.8	43.0
Glycine	10.2	11.3	9.0	8.3
Histidine	6.5	6.7	5.4	5.2
Isoleucine	10.5	10.6	8.6	7.7
Leucine	20.4	21.4	17.9	15.7
Lysine	13.7	12.1	10.2	8.1
Methionine	4.2	3.9	2.9	2.6
Methionine + Cysteine	8.4	7.9	6.5	5.9
Phenylalanine	11.6	12.9	10.9	9.0
Proline	13.8	19.0	14.4	15.7
Serine	11.7	13.0	9.8	8.3
Threonine	10.6	11.1	8.4	7.3
Tyrosine	9.6	10.1	7.5	6.6
Valine	11.7	12.1	10.5	9.0

[1] Source: This table was published in Poultry Science [3].

2.3. Laboratory Analyses

At the end of the experiment the birds were weighed after an 8 h feed deprivation, and one bird from each replicate representing the group average BW was selected and euthanized after electrical stunning. Birds were then hung on a processing line and bled out for 3 min by a unilateral neck cut, severing the right carotid artery and jugular vein. The non-edible viscera, including the intestines, proventriculus, gall bladder, spleen, esophagus, and full crop, were manually excised after scalding at 61 °C for 60 s and defeathering in a rotary drum picker for 25 s. After removing the head, the brain and liver were collected for further analysis.

2.4. Determination of Indicators of Potential Neurodegenerative Effects and Epigenetic DNA Damage

The activity of acetylcholinesterase (AChE) was determined in homogenates of the brain tissue (Experiments 1 and 2) and liver tissue (Experiment 2) of turkeys using the Chicken Acetylcholinesterase ELISA Kit (Bioassay Technology Laboratory, Inc., Shanghai, China), and the amyloid-β level was determined using the Chicken Total β amyloid Protein (βAP) ELISA Kit (Blue Gene Biotech, Shanghai, China). The levels of glycosylated acetylcholinesterase (GAChE) in the brain (Experiments 1 and 2) and liver (Experiment 2) of the turkeys were determined using the Chicken Glycosylated Acetylcholinesterase ELISA kit (Blue Gene Biotech, Shanghai, China). Low-density lipoprotein receptor-related protein 1 (LRP1) in the homogenates of brain tissue (Experiments 1 and 2) and liver tissue (Experiment 2) were determined using the Chicken Low Density Lipoprotein Receptor Related Protein 1 ELISA kit (Blue Gene Biotech, Shanghai, China). The level of phosphorylated Tau protein was determined using the Chicken Phosphorylated tau 231 (p Tau231) ELISA kit (Blue Gene Biotech, Shanghai, China). The levels of epigenetic changes in the brain (Experiments 1 and 2) and liver (Experiment 2) were determined by analyzing global DNA methylation (methylome), using diagnostic kits from Sigma Aldrich.

2.5. Statistical Analysis

This experiment was performed in a completely randomized 3 × 2 factorial design, and the data (presented as the mean ± standard error of the mean) were subjected to a

2-way ANOVA to examine the effect of 3 levels of Arg (90, 100, and 110%) and 2 levels of Met (30 and 45%). The Shapiro–Wilk and Levene tests were applied to test the model assumptions of normality and homogeneity of variance. When a significant interaction effect was noted (F test), treatment means were separated using the post-hoc Tukey's test. The significance level was set at $p < 0.05$, and statistical calculations were performed using the GLM procedures of the STATISTICA software system ver. 12.0 (Stat Soft Inc., Tulsa, OK, USA, 2014).

3. Results

3.1. Effect of Different Levels of Arg and Met Relative to Lys, According to the NRC

Differences in Arg content relative to Lys [5] caused no significant changes in the GAChE level in the brain of turkeys (Table 5). The use of a higher proportion of Arg (110% of the Lys level) in the diet caused an increase in the levels of AChE ($p = 0.019$), amyloid-β ($p = 0.032$), and Tau protein ($p = 0.002$) and a decrease in the level of LRP 1 ($p = 0.035$) in the brain of the turkeys. Decreasing Arg content to 90% of the Lys level caused a decrease in AChE ($p = 0.019$) in the brain. Increasing the Met content from 30 to 45% of the Lys level (NRC, 1994) [5] had no effect on the level of AChE, GAChE, amyloid-β, LRP 1, or Tau protein in the brain (Table 5). In the case of Tau protein, an Arg × Met interaction ($p = 0.05$) was noted: when the intermediate Arg level (100% of the Lys level) was applied, increasing the Met content from 30 to 45% of the Lys level caused an increase in the content of Tau protein, which was not noted in the cases of the lowest and highest Arg levels used (90 and 110% of the Lys level) (Table 5).

Table 5. Indicators of neurodegenerative changes in the brain of turkeys, Experiment 1.

Treatment [1]	AChEng/mL	GAChEng/mL	Amyloid-β pg/mL	LRP 1 pg/mL	Taung/mL	% Methylation
$Arg_{90}Met_{30}$	23.21	0.323	128.89	210.85	34.56 ^{ab}	61.37
$Arg_{90}Met_{45}$	20.91	0.337	145.82	254.25	34.93 ^{ab}	69.41
$Arg_{100}Met_{30}$	26.52	0.253	137.29	212.86	31.67 ^{b}	65.45
$Arg_{100}Met_{45}$	26.72	0.29	140.38	265.25	35.92 ^{a}	67.22
$Arg_{110}Met_{30}$	31.54	0.261	174.86	214.66	31.65 ^{b}	66.44
$Arg_{110}Met_{45}$	26.51	0.266	190.02	219.29	33.39 ^{ab}	69.24
SEM	0.015	0.004	0.145	0.526	0.047	0.022
Arg level, %						
90	22.06 ^{c}	0.330	137.36 ^{b}	232.55 ^{a}	34.75 ^{b}	65.39
100	26.62 ^{b}	0.272	138.84 ^{b}	239.01 ^{a}	33.79 ^{b}	66.34
110	30.21 ^{a}	0.296	158.56 ^{a}	205.14 ^{b}	48.08 ^{a}	67.84
Met level, %						
30	27.85	0.27	161.03	218.33	32.75	64.42
45	26.18	0.286	154.04	227.93	33.55	68.62

Table 5. *Cont.*

Treatment [1]	AChEng/mL	GAChEng/mL	Amyloid-β pg/mL	LRP 1 pg/mL	Taung/mL	% Methylation
			p-value			
Arg	0.019	0.129	0.032	0.035	0.002	0.147
Met	0.062	0.235	0.075	0.127	0.225	0.221
Arg × Met	0.057	0.092	0.235	0.357	0.051	0.187

[abc] values in same column with no common superscript denote a significant difference ($p \leq 0.05$). Abbreviations: Arg, arginine; Lys, lysine; Met, methionine; SEM, standard error of mean; AChE, acetylcholinesterase; GAChE, glycosylated acetylcholinesterase; LRP 1, low-density lipoprotein receptor-related protein 1; Tau, Tau protein. [1] Treatment: $Arg_{90}Met_{30}$ received 90% Arg and 30% Met relative to the content of dietary Lys; $Arg_{90}Met_{45}$ received 90% Arg and 45% Met relative to the content of dietary Lys; $Arg_{100}Met_{30}$ received 100% Arg and 30% Met relative to the content of dietary Lys; $Arg_{100}Met_{45}$ received 100% Arg and 45% Met relative to the content of dietary Lys; $Arg_{110}Met_{30}$ received 110% Arg and 30% Met relative to the content of dietary Lys; $Arg_{110}Met_{45}$ received 110% Arg level and 45% Met level relative to the content of dietary Lys.

3.2. *Effect of Different Levels of Arg and Met Relative to Lys, According to the BUT*

Turkeys that were fed diets with the highest Arg content (110%) relative to Lys [6] had the lowest AChE and GAChE levels ($p < 0.001$, both) and the lowest percentage of DNA methylation ($p < 0.001$) in the brain (Table 6). In comparison to the turkeys receiving diets with the intermediate (100% of the Lys level) and highest (110% of the Lys level) content of Arg, the diet with its lowest content (90% of the Lys level) resulted in an increase in LRP 1 ($p = 0.047$) in the brain. Increasing the Met content from 30 to 45% of the Lys level [6] had no effect on the level of AChE, GAChE, amyloid-β, LRP 1, or Tau protein or on the percentage of DNA methylation in the brain of turkeys (Table 6). In the case of amyloid-β there was also an Arg × Met interaction ($p = 0.042$), as the amyloid-β level in the brain decreased when the Met content was increased from 30 to 45% of the Lys level while maintaining the 1:1 Arg to Lys ratio, which was not observed in the case of the reduced Arg content (90% Lys) or increased Arg content (110% Lys). The two-way ANOVA showed an Arg × Met interaction for Tau protein in the brain ($p = 0.035$); in the case of the lowest Arg level (90% of the Lys level), increasing the Met content from 30 to 45% of the Lys level reduced the level of Tau protein, which was not observed in the case of the intermediate and highest Arg content (100 and 110% of the Lys level, respectively) (Table 6). Differences in Arg content relative to Lys [6] caused no statistically significant changes in the LRP1 or Tau protein levels in the liver of turkeys (Table 7). In comparison with the intermediate Arg content (100% of the Lys level) in the diet, reducing Arg to 90% of the Lys level caused an increase in AChE content ($p = 0.022$) in the liver. The two-way ANOVA showed Arg × Met interactions for GAChE and amyloid-β in the liver ($p = 0.008$, $p = 0.008$, $p = 0.025$; respectively). The interaction for GAChE resulted from the fact that, in the case of the highest Arg content (110% of the Lys level), increasing the Met content from 30 to 45% of the Lys level caused an increase in GAChE in the liver, which was not observed in the case of the intermediate and the lowest levels of Arg (100 and 90% of the Lys level, respectively). In the case of amyloid-β, the interaction resulted from the fact that when the lowest Arg level was applied (110% of the Lys level), increasing the Met content from 30 to 45% of the Lys level caused a decrease in the level of amyloid-β in the liver, which was not observed in the case of the intermediate and lowest levels of Arg (100 and 90% of the Lys level, respectively) (Table 7).

Table 6. Indicators of neurodegenerative changes in the brain of turkeys, Experiment 2.

Treatment [1]	AChEng/mL	GAChEng/mL	Amyloid-β pg/mL	LRP 1 pg/mL	Taung/mL	% Methylation
$Arg_{90}Met_{30}$	32.83	0.279	164.92 ^{a}	250.62	104.11 ^{a}	75.54 ^{ab}
$Arg_{90}Met_{45}$	27.70	0.316	176.95 ^{a}	229.86	37.19 ^{b}	69.36 ^{b}
$Arg_{100}Met_{30}$	27.82	0.262	168.38 ^{a}	196.14	33.42 ^{b}	65.16 ^{b}
$Arg_{100}Met_{45}$	28.80	0.329	140.72 ^{b}	186.03	32.38 ^{b}	86.88 ^{ab}
$Arg_{110}Met_{30}$	30.96	0.335	133.53 ^{b}	177.65	35.59 ^{b}	105.61 ^{a}
$Arg_{110}Met_{45}$	34.98	0.253	158.46 ^{ab}	170.46	38.42 ^{b}	99.33 ^{a}
SEM	0.022	0.012	0.245	0.078	0.069	0.124
Arg level, %						
90	30.27 ^{a}	0.298 ^{a}	170.94 ^{a}	240.24 ^{a}	70.65 ^{a}	72.45 ^{a}
100	28.31 ^{a}	0.296 ^{a}	154.55 ^{b}	190.09 ^{b}	32.90 ^{b}	76.02 ^{a}
110	17.08 ^{c}	0.156 ^{c}	151.12 ^{b}	201.96 ^{b}	36.25 ^{b}	17.08 ^{b}
Met level, %						
30	30.87	0.293	153.13	195.39	46.87	86.94
45	30.57	0.295	156.58	200.88	46.85	84.12
			p-value			
Arg	≤0.001	≤0.001	0.032	0.047	0.033	≤0.001
Met	0.235	0.182	0.075	0.069	0.082	0.066
Arg × Met	0.125	0.236	0.042	0.075	0.035	0.079

^{abc} values in same column with no common superscript denote a significant difference ($p \leq 0.05$). Abbreviations: Arg, arginine; Lys, lysine; Met, methionine; SEM, standard error of mean; AChE, acetylcholinesterase; GAChE, glycosylated acetylcholinesterase; LRP 1, low-density lipoprotein receptor-related protein 1; Tau, Tau protein. [1] Treatment: $Arg_{90}Met_{30}$ received 90% Arg and 30% Met relative to the content of dietary Lys; $Arg_{90}Met_{45}$ received 90% Arg and 45% Met relative to the content of dietary Lys; $Arg_{100}Met_{30}$ received 100% Arg and 30% Met relative to the content of dietary Lys; $Arg_{100}Met_{45}$ received 100% Arg and 45% Met relative to the content of dietary Lys; $Arg_{110}Met_{30}$ received 110% Arg and 30% Met relative to the content of dietary Lys; $Arg_{110}Met_{45}$ received 110% Arg level and 45% Met level relative to the content of dietary Lys.

Table 7. Indicators of neurodegenerative changes in the liver of turkeys, Experiment 2.

Treatment [1]	AChEng/mL	GAChEng/mL	Amyloid-β pg/mL	LRP 1 pg/mL	Taung/mL	% Methylation
$Arg_{90}Met_{30}$	20.61	0.183 ^{b}	219.54 ^{a}	185.51	35.29	75.85
$Arg_{90}Met_{45}$	17.82	0.134 ^{b}	195.24 ^{a}	209.19	37.41	69.77
$Arg_{100}Met_{30}$	15.76	0.103 ^{c}	75.22 ^{c}	181.07	36.98	65.71
$Arg_{100}Met_{45}$	14.99	0.090 ^{c}	79.79 ^{c}	219.39	37.36	81.01
$Arg_{110}Met_{30}$	14.75	0.075 ^{c}	185.59 ^{c}	203.4	33.71	75.78
$Arg_{110}Met_{45}$	18.12	0.406 ^{a}	168.71 ^{b}	219.54	35.69	79.47
SEM	0.03	0.008	0.145	0.167	0.047	0.095

Table 7. *Cont.*

Treatment [1]	AChEng/mL	GAChEng/mL	Amyloid-β pg/mL	LRP 1 pg/mL	Taung/mL	% Methylation
Arg level, %						
90	19.22 [a]	0.159 [a]	207.39 [a]	197.35	36.35	72.81
100	15.38 [b]	0.097 [b]	77.51 [c]	200.23	37.17	73.36
110	17.08 [a]	0.156 [a]	151.12 [b]	201.96	36.25	77.63
Met level, %						
30	16.64	0.156	151.89	199.30	35.54	72.45
45	16.96	0.164	153.71	202.49	35.99	76.75
			p-value			
Arg	0.022	0.018	≤0.001	0.682	0.799	0.574
Met	0.326	0.095	0.228	0.072	0.083	0.087
Arg × Met	0.093	0.008	0.025	0.374	0.189	0.121

[abc] values in same column with no common superscript denote a significant difference ($p \leq 0.05$). Abbreviations: Arg, arginine; Lys, lysine; Met, methionine; SEM, standard error of mean; AChE, acetylcholinesterase; GAChE, glycosylated acetylcholinesterase; LRP 1, low-density lipoprotein receptor-related protein 1; Tau, Tau protein. [1] Treatment: $Arg_{90}Met_{30}$ received 90% Arg and 30% Met relative to the content of dietary Lys; $Arg_{90}Met_{45}$ received 90% Arg and 45% Met relative to the content of dietary Lys; $Arg_{100}Met_{30}$ received 100% Arg and 30% Met relative to the content of dietary Lys; $Arg_{100}Met_{45}$ received 100% Arg and 45% Met relative to the content of dietary Lys; $Arg_{110}Met_{30}$ received 110% Arg and 30% Met relative to the content of dietary Lys; $Arg_{110}Met_{45}$ received 110% Arg level and 45% Met level relative to the content of dietary Lys.

4. Discussion

Due to the lack of similar studies in the world literature in animal models or data on human patients, it is difficult to compare our results on neurodegenerative changes in turkeys fed diets with different proportions of arginine and methionine relative to lysine. It is also difficult to explain the more specific mechanism of physiological reactions and changes in the biochemical parameters, due to the diametrically different effects in the two experiments, depending on the level of Arg.

In Experiment 1, where the Lys content of turkey diets was based on NRC [5] guidelines, an increase in Arg and Met inclusion rates to Arg 100% and Met 45% of Lys content, respectively, improved BWG [2]. In the 16th week of rearing, the body weight of the turkeys was very similar between the groups, but the lowest value was recorded for the turkeys from the Arg 90%, Met 30% (10.2 kg) treatment, and the highest for the Arg 90% and Met 45% and also Arg 100% and Met 45% (10.7 kg) treatments. In Experiment 2, differences in Arg and Met inclusion rates in diets, relative to Lys content (which was close to BUT [6] recommendations, i.e., high) did not affect the final BW of turkeys. In the 16th week of rearing, the body weight of the turkeys was very similar between the groups, however, the final body weight was slightly higher than in Experiment 1. The lowest value was recorded for the turkeys from the Arg 100%, Met 45% (11.38 kg) treatment, and the highest was recorded for the Arg 110%, Met 45% (11.49 kg) treatment [3].

Among the pathological changes taking place in neurodegenerative diseases, in both humans and laboratory animals, the first symptoms are the excessive deposition of amyloid-β in the brain and neurofibrillary degeneration in the form of hyperphosphorylated Tau protein. According to Całyniuk et al. [24] Arg promotes growth performance in turkeys because it acts as a substrate for creatine biosynthesis. On the other hand, methionine is also directly implicated in creatine synthesis, because it donates a methyl group to glycocyamine (it is a biological precursor for creatine synthesis in birds) which is synthesized from Arg and Gly [34]. What is more, the neuroprotective effect of creatine is related to the intensification of the processes of neurogenesis, supporting the biosynthesis of serotonin and dopamine, and inhibiting the aggregation of amyloid and Tau protein [35].

According to Li et al. [36], amyloid-β fibers can also accumulate in the liver. Amyloid-β additionally circulates in the plasma, the cerebrospinal fluid, and the brain interstitial fluid [37,38], and amyloid deposits can also accumulate near the blood vessels, impairing the functioning of the blood–brain barrier. Complexes of amyloid-β and highly phosphorylated Tau protein with acetylcholinesterase are toxic as well. These complexes are an inactive, glycosylated form of AChE called glycosylated acetylcholinesterase (GAChE). Another pathological symptom of neurodegenerative diseases is a decreasing level of the protein LRP1 or of AChE itself [39–43]. In our study, the highest proportion of Arg (110%) relative to the Lys level recommended by the NRC [5] caused a beneficial increase in the AChE level in combination with a very unfavorable increase in the levels of amyloid-β and Tau protein, as well as an unfavorable decrease in LRP1. The use of the intermediate level of Arg (100% of the Lys level) in combination with increased Met from 30 to 45% also caused an unfavorable increase in the level of Tau protein [44,45].

The decrease in the AChE level following the use of 90% Arg relative to Lys [5] may suggest the initiation of neurodegenerative changes. According to many researchers, a decrease in the level of this enzyme is an early symptom of neurodegenerative disease, which leads to the accumulation of large amounts of acetylcholine in the synaptic spaces and stimulates hyperphosphorylation of Tau protein, which is toxic [46]. A 110% share of Arg relative to Lys [5] cannot be considered to be more beneficial than a 90% share in terms of neurodegenerative changes, despite the increase in the AChE level in the brain tissue. The increase in the AChE level is accompanied by an increase in the level of toxic amyloid-β and Tau protein clusters and a decrease in the level of LRP1 (lipoprotein receptor 1), which is responsible for the removal of amyloid-β from the brain to the blood vessels by active transport [47].

Pathological amyloid-β, produced by fragmentation of its precursor APP (amyloid precursor protein), which is a component of the cell membrane of neurons, is present in tissues in the form of insoluble clusters, and is toxic for nerve cells, called amyloid-β [37,38]. The marked increase in the deposition of amyloid-β in the brain tissue of turkeys receiving a level of Arg that was increased to 110% of the Lys level recommended by NRC [5] is an unfavorable phenomenon. It is likely that excessive Arg stimulates production of amyloid-β, which disturbs calcium balance, damages the mitochondria, and contributes to the release of free radicals that degenerate DNA and cellular proteins. Toxic amyloid-β can cause an increase in the level of Ca^{2+} in the cell, stimulating calmodulin-dependent protein kinase II (CaMKII). Activation of CaMKII in turn stimulates the highly unfavorable phenomenon of hyperphosphorylation of Tau protein. These processes together can lead to oxidative stress, to which nervous tissue is particularly sensitive. This is because neurons use up large amounts of oxygen and have the highest mitochondrial activity of all cells of the body. An increasing amount of research indicates a relationship between oxidative stress of the nervous system and neurodegenerative diseases [48]. In the present study, an excessive amyloid-β deposition in the brain was observed following the use of 110% Arg (relative to the Lys level, recommended by NRC [5], which was not influenced by the Met level. In the liver, the amyloid-β level was lower following the use of 100% Arg and 45% Met (relative to the Lys level, recommended by BUT [6]). The liver plays an important role in removing excess amyloid-β from the blood, which can exacerbate neurodegenerative changes in this organ. In the hepatocytes, amyloid-β is degraded in a reaction with LRP-1, but bile from the liver may prevent this and allow the compound to be excreted unchanged [49,50]. In the nervous tissue of the liver, similarly as in the brain, pathological deposits of proteins (amyloid-β and Tau protein) may accumulate and cause damage to neurons. The presence of significant interactions between Arg and Met supplementation levels in turkey diets with the least Lys (90% of the Lys level, Met 45%) and high Lys (110% of the Lys level, Met 45%) could be attributed to the fact that Tau protein and amyloid-β levels were affected by both Arg and Met content, but in some cases, Arg and Met exerted a different influence on the same parameter, as demonstrated by the results of the one-way ANOVA. According to Ghamari Monavvar et al. [33], arginine, as a precursor of polyamines and nitric oxide, stimulates protein synthesis and the proliferation and migration of intestinal cells, thereby improving intestinal morphometry. Nitric oxide also stimulates glucose uptake from the digesta, systemic immunity, and hormone secretion. Liu et al. [47] showed that in arginine deficiency, the enzyme nitric oxide synthase (EC 1.14.13.39) (physiologically responsible for nitric oxide production) can generate the production of free radicals with neurodegenerative effects. Notably, nitric oxide is derived from arginine; nitric oxide synthase itself is dependent on Ca^{2+}/calmodulin, and its coenzyme is $NADPH^+$ H^+. The efficiency of this enzyme depends on an appropriate concentration of Ca^{2+} ions in the neuron, and excessive calcium inhibits the activity of nitric oxide synthase. Interestingly, research by Keith et al. [48] shows that amyloid-β can increase nitric oxide production by this enzyme and at the same time stimulate synthesis of the protein component of the enzyme. Zeng et al. [49] report that the neurotoxicity of amyloid-β worsens the dysfunction of the blood–brain barrier, increasing oxidative stress and inflammation of the nervous system, which can lead to the impairment of blood vessel function [49]. According to the vascular theory of amyloid plaque formation, impairment of the blood–brain barrier may allow blood proteins to penetrate the brain, including components of hemoglobin. Blood proteins generated from metabolic transformations and hemoglobin proteins in the brain tissue can both stimulate production of amyloid β [50].

According to Butterfield et al. [51], neurodegenerative changes induced by the presence of amyloid-β and free radicals involve oxidative damage to key enzymes taking part in glycolysis, the tricarboxylic acid cycle, and the biosynthesis of ATP. This type of damage to processes associated with biological oxidation in neurons adversely affects the metabolism of glucose, a key energy source for the brain, and then results in a characteristic decrease in cerebral glucose metabolism [50]. On the other hand, glucogenic amino acids such as arginine and methionine can be converted into one of the intermediate compounds of the Krebs cycle, which during successive reactions can be utilized for glucose synthesis. In addition, metabolism of arginine and methionine, similar to lysine metabolism, generates large amounts of ATP in the cell. Moreover, each molecule of the glucogenic amino acid donates one molecule of $NADPH^+$ H^+, which is used for the synthesis of fatty acids or cholesterol. A high concentration of cholesterol stimulates the production of amyloid-β from APP by activating β-secretase, which is much more efficient in a cholesterol-rich environment. It may be that a 110% proportion of Arg relative to Lys [5] is conducive to these conditions, resulting in an increased production of amyloid-β and a decrease in the level of LRP1. No such relationship was observed in the case of 110% Arg relative to the Lys level recommended by BUT [6]. As reported by Chen et al. [35] the neuroprotective action of Arg consists in inhibiting inflammatory processes, preventing oxidative stress, and improving cerebral circulation thanks to nitric oxide, for which Arg is a substrate. Chen et al. [35] argue that the neuroprotective effects of Arg can be enhanced by stimulating the metabolism of this amino acid by adding Met to a poultry diet. According to this author, the neuroprotection is possible thanks to the regulation of the transcription factor HIF-1α/LDHA, which is formed from arginine. The molecular mechanism of arginine-mediated neuroprotection is by suppression of the HIF-1α/LDHA signaling pathway during hypoxia, resulting in an inhibition of the inflammatory response. Inhibition of the inflammatory neuron response by attenuating HIF-1α/LDHA signaling exerts neuroprotective effects and reduces the occurrence of degenerative changes. Such reactions also regulate glycolysis; by improving this process in neurons, the inflammatory reaction is inhibited. In turn, arginine deficiencies inhibit glycolysis and redirects the cell's metabolism, which stimulates the formation of toxic protein deposits, i.e., amyloid and the Tau protein deposits [52]. According to Kremer et al. [53] arginine may influence the synergistic effect between the urea cycle and glycolysis. Arginine is hydrolyzed by the enzyme arginase, leading to the production of urea and ornithine. According to Całyniuk et al. [24] urea production increases with a rise in the Arg content of bird diets, but excess Arg has no significant effect on Lys metabolism, whereas excess Lys strongly antagonizes the metabolism of Arg.

Neurodegenerative changes in liver cells are also associated with disturbances of mitochondrial function. On the one hand, toxic substances for neurons are produced, and on the other hand, processes that are associated with energy acquisition for neurons are disturbed [54]. In our study, a beneficial reduction in amyloid-β in the brain tissue of turkeys was obtained in Experiment 2 by increasing the Met level from 30 to 45% of the Lys level while maintaining the 1:1 ratio for Lys and Arg [6]. It is likely that this is the optimum dose for protecting brain and liver tissue, because it does not initiate neurodegenerative changes. Among Arg levels proposed in BUT [6] guidelines for turkey diets, 90 or 110% of the Lys level in combination with 30 or 45% Met, can be considered optimal in terms of neuroprotective effects. A study by Tapia-Rojas et al. [7] in mice demonstrated that a diet with too much Met can increase levels of amyloid-β and phosphorylated Tau protein, which are indicative of neurodegenerative changes. According to Zhang et al. [55], an increase in amyloid-β levels may be due to intensification of the methionine cycle reaction, which causes an increase in the activity of γ-secretase and APP cleavage, resulting in an increase in amyloid-β protein in the brain tissue [54].

Unfortunately, as the highest GAChE level was noted in the liver of turkeys from the treatment using 100% Arg and 45% Met (relative to the Lys level recommended by BUT [6], it seems likely that large amounts of amyloid-β were involved in the formation of these toxic complexes, which consist of amyloid-β, Tau protein, and inactive AChE. According to Reitz and Mayeux [56] and Kamal at al. [57], the deposition of amyloid-β in the central nervous system takes place in the neuronal axonal membranes, disturbing homeostasis in nerve cells and leading to abnormal APP metabolism and the overproduction of amyloid-β peptides [58]. Metabolic disturbances in liver hepatocytes and neurons also cause the glycosylation of acetylcholinesterase, i.e., complexes of acetylcholinesterase with amyloid-β and Tau protein. These toxic complexes, called glycosylated acetylcholinesterase (GAChE), stimulate apoptosis of both hepatocytes and neurons. In the case of a change in the level of Arg relative to the Lys level recommended by BUT [6], the variant with a 110% share of Arg relative to Lys seems to be much more favorable, because it reduces the concentration of GAChE and methylation changes in the brain tissue. The 90% share of Arg relative to Lys [6] caused a very promising increase in the LRP1 level in the brain tissue. Another cause of metabolic disturbances in the neurons is the formation of complexes of AChE and Tau protein, known as GAChE. According to Lesné et al. [59], GAChE causes an activation of glutamate receptor NMDA, inducing the mass influx of Ca^{2+} and a non-physiological elevation of the intracellular concentration of this ion in the nerve cell. This leads to a reduction in the rate of oxidative phosphorylation and the ATP level in the neuron [25–27] disturbing the metabolism of this cell. The most neurotoxic effects of GAChE include the activation of the glutaminergic NMDA receptor (N-methyl-D-aspartate receptor), the opening of calcium channels, activation of calcium/calmodulin-dependent kinase II, and the activation of caspases [25–27]. Activation of glutamate NMDA receptor induces a mass influx of Ca^{2+} and a non-physiological elevation of the intracellular concentration of this ion in the cell, which leads to a decrease in the rate of oxidative phosphorylation and the ATP level. Disruption of cellular respiration is conducive to the opening of mitochondrial mega channels, which can lead to the release of cytochrome C from the intermembrane space and activation of caspase 3. The disruption of cellular homeostasis leads to apoptotic cell death [25–27]. Another result of the neurodegenerative effects of both amyloid-β itself and GAChE is the generation of reactive oxygen species, especially in the presence of Cu^{2+} and Zn^{2+} ions. This results in oxidative stress and a decrease in ATP production in the neurons, which in combination with disturbances of Ca^{2+} metabolism induces a number of pathological changes leading to the death of neurons [60].

In both Experiments 1 and 2, the varied levels of Met relative to Lys had no negative effect on indicators of neurodegeneration in the brain. Therefore, a Met level amounting to 30 or 45% of the lysine level can be considered to be beneficial, causing no neurodegenerative changes. In the present study, the level of Tau protein in the brain of turkeys decreased when the Met level in the diet was increased to 45% of the Lys level and Arg content was reduced to 90% of the Lys level (recommended by BUT) [6]. According to Lesné et al. [59], Met reduces the activity of Tau kinase, thereby promoting the dissociation of Tau protein from microtubules, which results in the formation of neurofibrillary tangles, manifesting the onset of neurodegenerative changes. Kinase and phosphatase are responsible for the phosphorylation of Tau protein, and an imbalance between these enzymes leads to the hyperphosphorylation of Tau protein. Tau protein may also undergo other post-translation modifications, including acetylation, which magnifies its role in the development of neurodegenerative changes [61]. The results of our study indicate that increasing the proportion of Arg to 110% of the Lys level proposed by BUT [6] causes an unfavorable decrease in the AChE level, but this is accompanied by a highly favorable decrease in the concentration of GAChE molecules and the percentage of DNA methylation in the brain tissue. Reducing Arg to 90% of BUT [6] guidelines caused a highly favorable increase in the LRP1 level in the brain tissue.

According to Sharma [62], the declining level of AChE in neurodegenerative diseases leads to the accumulation of large amounts of acetylcholine and to overstimulation of the cholinergic system. In our study, the higher AChE level in the liver of turkeys receiving an increased addition of Arg to 110% relative to Lys (in comparison to BUT guidelines) [6] should be considered to be highly favorable. In contrast, the consequences of using Arg in the amount of 90% of the Lys level (NRC) [5] were clearly unfavorable, as the level of Tau protein in the brain increased. According to Méndez et al., Elufioye et al., and Ikonomovic et al. [63–65], a decrease in AChE activity results in mitochondrial dysfunction and inflammation in brain cells. In addition, reduced AChE activity inhibits certain complexes of the respiratory chain, leading to an increased production of toxic free radicals, especially hydroxyl radical and superoxide ions, inducing neuroinflammatory changes [66]. According to Day and Greenfield [25–27], a peptide can detach from the C-terminal of the AChE molecule and attach to a glycosylated fragment of the amyloid-β precursor or Tau protein [34], forming a molecule of glycosylated acetylcholinesterase (GAChE). A decrease in the level of AChE, the main enzyme of cholinergic synapses in the brain and neuromuscular connections, may therefore be an early symptom of neuropathological changes [67–69]. According to Méndez et al. [63], the most severe changes of this type are observed in the cerebral cortex and hippocampus and are accompanied by a significant increase in the cholesterol level in these tissues [68]. The high level of Lys and Met used in Experiment 1 (110 and 45% in relation to Lys), in relation to the nutritional recommendations of the NRC, according to Ognik et al. [22], can stimulate the synthesis of cholesterol, while a high level of Arg in the diet may partially counteract such reactions. Due to the fact that high cholesterol according to the NRC [5] clearly stimulates the production of amyloid-β, only its supplementation in accordance with the BUT [6] requirements can be considered neuroprotective. An additional premise supporting this fact may be the ability of Arg to stimulate the urea cycle. The product of this process, i.e., urea, has the ability to change the conformational structure of amyloid-β, which causes the unfolding of the structure of this molecule [70]. Xiang et al. [70] suggest that the level of this neurotransmitter may be decreased by acetylcholine-binding amyloid-β. In Experiment 2, increasing the Met level to 45% and the Arg level to 110% of the Lys level recommended by BUT [6] had highly diverse effects. It caused an unfavorable increase in the GAChE level accompanied by a very promising decrease in the level of amyloid-β itself in the liver tissue of the turkeys. Unfortunately, because GAChE is a complex of amyloid-β and Tau protein with molecules of the enzyme AChE, this result should probably be considered unfavorable, as this type of complex causes the blockage of active AChE, a valuable enzyme for normal neuron function.

To verify the hypotheses put forward in this paper, further research should include other useful indicators of neurodegenerative changes, especially levels of APP, crucial enzymes (calcium/calmodulin-dependent protein kinase type II (CAMK2), calcium/calmodulin-dependent protein kinase type IV (CAMK4) and glucose 6 phosphate dehydrogenase), calcium, vitamins, especially B6 and B12, thyroid hormones, and biomarkers of inflammation and oxidative stress (cytokines and isoprostanes). Microscopic examination may prove useful as well (observation of plaques formed of Tau protein and clusters of amyloid β).

5. Conclusions

It was established that in the case of both a low (NRC) [5] and a high (BUT) [6] level of Lys in the diet of turkeys, the Arg level can be reduced to 90% of the Lys level and Met to 30% of the Lys level, because this does not cause neurodegenerative changes in turkeys. Unfavorable neurodegenerative changes may appear if the Arg level is increased from 100 to 110% of the Lys level recommended by the NRC [5].

Among the tested proportions of Lys and Arg in the diet of meat chickens, the safest variant is the higher level of lysine recommended by BUT [6] and an increase in the Arg level to 110% of that Lys level. In our opinion, these proportions of amino acids in the diet, among those tested, are the most effective at limiting neurodegenerative changes.

Author Contributions: Conceptualization, J.J. and K.O.; methodology, J.J., K.O. and M.K.; sample collection, D.M. and K.O., formal analysis, J.J., K.O. and M.K.; investigation, M.K.; resources, J.J. and D.M.; writing—original draft preparation, M.K.; writing—review and editing, K.O. and D.M.; supervision, J.J.; project administration, M.K. All authors have read and agreed to the published version of the manuscript.

Funding: This work was supported by the National Science Centre, Grant No. 2017/27/B/NZ9/01007.

Institutional Review Board Statement: The study was conducted according to the guidelines of the Declaration of Helsinki, and approved by the Ethics Committee of Local Ethics Committee (decision No. 82/2017), and the animals were cared for under guidelines comparable to those laid down by EU Directive 2010/63/EU.

Informed Consent Statement: Not applicable.

Data Availability Statement: The data presented in this study are available in the article.

Conflicts of Interest: The authors declare no conflict of interest.

References

1. Fouad, A.M.; El-Senousey, H.K.; Yang, X.J.; Yao, J.H. Role of dietary L-arginine in poultry production. *Int. J. Poult. Sci.* **2012**, *11*, 718–729. [CrossRef]
2. Jankowski, J.; Mikulski, D.; Mikulska, M.; Ognik, K.; Całyniuk, Z.; Mróz, E.; Zduńczyk, Z. The effect of different dietary ratios of arginine, methionine, and lysine on the performance, carcass traits, and immune status of turkeys. *Poult. Sci.* **2019**, *99*, 1028–1037. [CrossRef] [PubMed]
3. Jankowski, J.; Ognik, K.; Konieczka, P.; Mikulski, D. Effects of different levels of arginine and methionine in a high-lysine diet on the immune status, performance, and carcass traits of turkeys. *Poult. Sci.* **2020**, *99*, 4730–4740. [CrossRef]
4. Balnave, D.; Brake, J. Re-evaluation of the classical dietary arginine: Lysine interaction for modern poultry diets: A review. *Worlds Poult. Sci. J.* **2002**, *58*, 275–289. [CrossRef]
5. NRC: Research Council National. *Nutrient Requirements of Poultry*, 9th ed.; The National Academies Press: Washington, DC, USA, 1994.
6. British United Turkeys (BUT); Aviagen Turkeys. Management Guidelines for Raising Commercial Turkeys. Available online: https://www.aviagenturkeys.com/media/183481/aviagencommercialguide.pdf (accessed on 30 August 2016).
7. Tapia-Rojas, C.; Lindsay, C.B.; Montecinos-Oliva, C.; Arrazola, M.S.; Retamales, R.M.; Bunout, D.; Hirsch, S.; Inestrosa, N.C. Is L-methionine a trigger factor for Alzheimer's-like neurodegeneration? Changes in Aβ oligomers, tau phosphorylation, synaptic proteins, Wnt signaling and behavioral impairment in wild-type mice. *Mol. Neurodegener.* **2015**, *10*. [CrossRef]
8. Morales, I.; Guzmán-Martínez, L.; Cerda-Troncoso, C.; Farías, G.A.; Maccioni, R.B. Neuroinflammation in the pathogenesis of Alzheimer's disease. A rational framework for the search of novel therapeutic approaches. *Front. Cell Neurosci.* **2014**, *8*, 112–130. [CrossRef]
9. Tu, R.; Grover, H.M.; Kotra, L.P. Peptidyl Arginine Deiminases and Neurodegenerative Diseases. *Curr. Med. Chem.* **2016**, *23*, 104–114. [CrossRef]
10. Tóthová, B.; Kovalska, M.; Kalenská, D.; Tomascova, A.; Lehotský, J. Effect of methionine-induced hyperhomocysteinemia on neurodegeneration in experimental conditions. *Act. Nerv. Super. Rediviva* **2017**, *59*, 57–63.
11. Toue, S.; Kodama, R.; Amao, M.; Kawamata, Y.; Kimura, T.; Sakai, R. Screening of toxicity biomarkers for methionine excess in rats. *J. Nutr.* **2006**, *136*, 1716S–1721S. [CrossRef]

12. Koladiya, R.U.; Jaggi, A.S.; Singh, N.; Sharma, B.K. Ameliorative role of Atorvastatin and Pitavastatin in L-Methionine induced vascular dementia in rats. *BMC Pharm.* **2008**, *8*, 258–263. [CrossRef]
13. Goren, J.L.; Stoll, A.L.; Damico, K.E.; Sarmiento, I.A.; Cohen, B.M. Bioavailability and lack of toxicity of S-adenosyl-L-methionine (SAMe) in humans. *Pharmacotherapy* **2004**, *24*, 1501–1507. [CrossRef] [PubMed]
14. Ringman, J.M.; Coppola, G. New genes and new insights from old genes: Update on Alzheimer disease. *Contin. Lifelong Learn. Neurol.* **2013**, *19*, 35–8371. [CrossRef] [PubMed]
15. Stadtman, E.R.; Van Remmen, H.; Richardson, A.; Wehr, N.B.; Levine, R.L. Methionine oxidation and aging. *Biochim. Biophys. Acta* **2005**, *1703*, 135–140. [CrossRef] [PubMed]
16. Vuaden, F.C.; Savio, L.E.; Piato, A.L.; Pereira, T.C.; Vianna, M.R.; Bogo, M.R.; Bonan, C.D.; Wyse, A.T. Long-term methionine exposure induces memory impairment on inhibitory avoidance task and alters acetylcholinesterase activity and expression in zebrafish (Danio rerio). *Neurochem. Res.* **2012**, *37*, 1545–1553. [CrossRef] [PubMed]
17. Cudic, P.; Joshi, N.; Sagher, D.; Williams, B.T.; Stawikowski, M.J.; Weissbach, H. Identification of activators of methionine sulfoxide reductases A and B. *Biochem. Biophys. Res. Commun.* **2015**, *469*, 863–867. [CrossRef] [PubMed]
18. Olney, J.W.; Zorumski, C.; Price, M.T.; Labruyere, J. L-cysteine, a bicarbonate-sensitive endogenous excitotoxin. *Science* **1990**, *248*, 596–599. [CrossRef]
19. Khajali, F.; Wideman, R. Dietary arginine: Metabolic, environmental, immunological and physiological interrelationships. *Worlds Poult. Sci. J.* **2010**, *66*, 751–766. [CrossRef]
20. Rajagopal, S.; Joginapally, V.R.; Srinivasan, T.S.; Ramanathan, V. *Beneficial Effects of Dietary Amino Acids on Brain Health*; Essa, M.M., Memon, M.A., Eds.; Food and brain function; Nova Publications: New York, NY, USA, 2014; pp. 221–224.
21. Oso, A.O.; Williams, G.A.; Oluwatosin, O.O.; Bamgbose, A.M.; Adebayo, A.O.; Olowofeso, O.V.; Pirgozliev, A.A.; Adegbenjo, S.O.; Osho, J.O.; Alabi, F.; et al. Effect of dietary supplementation with arginine on hematological indices, serum chemistry, carcass yield, gut microflora, and lymphoid organs of growing turkeys. *Livest. Sci.* **2017**, *198*, 58–64. [CrossRef]
22. Bouchereau, J.; Schiff, M. Inherited Disorders of Lysine Metabolism: A Review. *J. Nutr.* **2020**, *1*, 2556S–2560S. [CrossRef] [PubMed]
23. Ognik, K.; Całyniuk, Z.; Mikulski, D.; Stępniowska, A.; Konieczka, P.; Jankowski, J. The effect of different dietary ratios of lysine, arginine and methionine on biochemical parameters and hormone secretion in turkeys. *J. Anim. Physiol. Anim. Nutr.* **2021**, *105*, 108–118. [CrossRef]
24. Całyniuk, Z.; Cholewińska, E.; Konieczka, P.; Ognik, K.; Mikulski, D.; Jankowski, J. The effect of the application of diets with varied proportions of arginine and lysine on biochemical and antioxidant status in turkeys. *Ann. Anim. Sci.* **2021**. [CrossRef]
25. Day, T.; Greenfield, S.A. A non-cholinergic, trophic action of acetylcholinesterase on hippocampal neurons in vitro: Molecular mechanisms. *Neuroscience* **2002**, *111*, 649–656. [CrossRef]
26. Day, T.; Greenfield, S.A. Bioactivity of a peptide derived from acetylcholinesterase in hippocampal organotypic cultures. *Exp. Brain Res.* **2003**, *155*, 500–508. [CrossRef]
27. Day, T.; Greenfield, S.A. A peptide derived from acetylcholinesterase induces neuronal cell death: Characterization of possible mechanisms. *Exp. Brain Res.* **2003**, *153*, 334–342. [CrossRef] [PubMed]
28. Hasani, S.A.; Mayeli, M.; Salehi, M.A.; Barzegar Parizi, R. A Systematic Review of the Association between Amyloid-β and τ Pathology with Functional Connectivity Alterations in the Alzheimer Dementia Spectrum Utilizing PET Scan and rsfMRI. *Dement. Geriatr. Cogn. Disord. Extra.* **2021**, *11*, 78–90. [CrossRef] [PubMed]
29. Rowe, E.M.; Xing, V.; Biggar, K.K. Lysine methylation: Implications in neurodegenerative disease. *Brain Res* **2019**, *1707*, 164–171. [CrossRef] [PubMed]
30. Jankowski, J.; Ognik, K.; Całyniuk, Z.; Stępniowska, A.; Konieczka, P.; Mikulski, D. The effect of different dietary ratios of lysine, arginine and methionine on protein nitration and oxidation reactions in turkey tissues and DNA. *Animal* **2021**, *15*, 100–183. [CrossRef]
31. Tufarelli, V.; Mehrzad-Gilmakek, H.; Booyeh, M.; Qotbi, A.; Seidavi, A.; Paz, A.E.; Laudadio, V. Effect of different levels of L-carnitine and excess lysine-methionine on broiler performance, carcass characteristics, blood constituents, immunity and triiodothyronine hormone. *Agriculture* **2020**, *10*, 138. [CrossRef]
32. Ghasemabad, B. Effects of copper sulfate and arginine supplements on performance and carcass traits in broiler chickens fed with canola meal based diet. *Iran. J. Appl. Anim. Sci.* **2017**, *7*, 647–654.
33. Ghamari Monavvar, H.; Moghaddam, G.; Ebrahimi, M. A review on the effect of arginine on growth performance, meat quality, intestine morphology, and immune system of broiler chickens. *Iran. J. Appl. Anim. Sci.* **2020**, *10*, 587–594.
34. Ghoreyshi, S.M.; Omri, B.; Chalghoumi, R.; Bouyeh, M.; Seidavi, A.R.; Dadashbeiki, M.; Lucarini, M.; Durrazo, A.; van den Hoven, R.; Santini, A. Effects of Dietary Supplementation of L-Carnitine and Excess Lysine-Methionine on Growth Performance, Carcass Characteristics, and Immunity Markers of Broiler Chicken. *Animals* **2019**, *9*, 362. [CrossRef] [PubMed]
35. Chen, S.F.; Pan, M.X.; Tang, J.C. Arginine is neuroprotective through suppressing HIF-1α/LDHA-mediated inflammatory response after cerebral ischemia/reperfusion injury. *Mol. Brain* **2020**, *13*, 63. [CrossRef] [PubMed]
36. Li, Y.; Li, K.; Wang, X.; Cui, M.; Ge, P.; Zhang, J.; Qiu, F.; Zhong, C. Conformable self-assembling amyloid protein coatings with genetically programmable functionality. *Sci. Adv.* **2020**, *6*, 1425–1435. [CrossRef]
37. Ghiso, J.; Tagliavini, F.; Frederik Timmers, W.; Frangione, B. Alzheimer's disease amyloid precursor protein is present in senile plaques and cerebrospinal fluid: Immunohistochemical and biochemical characterization. *Biochem. Biophys. Res. Commun.* **2002**, *16*, 430–437. [CrossRef]

38. Zlokovic, B.V.; Frangione, B. Transport-clearance hypothesis for Alzheimer's disease and potential therapeutic implications. *Landes Biosci.* **2003**, 114–122. Available online: https://www.ncbi.nlm.nih.gov/books/NBK5975/ (accessed on 7 June 2022).
39. Madra, M.; Sturley, S.L. Niemann-Pick type C pathogenesis and treatment: From statins to sugars. *Clin. Lipidol.* **2010**, *5*, 387–395. [CrossRef]
40. Block, R.C.; Dorsey, E.R.; Beck, C.A.; Brenna, J.T.; Shoulson, I. Altered cholesterol and fatty acid metabolism in Huntington disease. *J. Clin. Lipidol.* **2010**, *4*, 17–23. [CrossRef]
41. Di Paolo, G.; Kim, T.W. Linking lipids to Alzheimer's disease: Cholesterol and beyond. *Nat. Rev. Neurosci.* **2011**, *12*, 284–296. [CrossRef]
42. Wang, Q.; Yan, J.; Chen, X.; Li, J.; Yang, Y.; Weng, J.; Deng, C.; Yenari, M.A. Statins: Multiple neuroprotective mechanisms in neurodegenerative diseases. *Exp. Neurol.* **2011**, *230*, 27–34. [CrossRef]
43. Estrada, L.D.; Ahumada, P.; Cabrera, D.; Arab, J.P. Liver Dysfunction as a Novel Player in Alzheimer's Progression: Looking Outside the Brain. *Front. Aging Neurosci.* **2019**, *11*, 174. [CrossRef]
44. Underhill, S.M.; Amara, S.G. Acetylcholine Receptor Stimulation Activates Protein Kinase C Mediated Internalization of the Dopamine Transporter. *Front. Cell Neurosci.* **2021**, *15*, 2216. [CrossRef] [PubMed]
45. Nikolakopoulou, A.M.; Wang, Y.; Ma, Q.; Sagare, A.P.; Montagne, A.; Huuskonen, M.T.; Sanket, V.; Kisler, R.K.; Dai, Z.; Körbelin, J.; et al. Endothelial LRP1 protects against neurodegeneration by blocking cyclophilin A. *J. Exp. Med.* **2021**, *218*, e20202207. [CrossRef] [PubMed]
46. Oka, M.; Fujisaki, N.; Maruko-Otake, A. Ca^{2+}/calmodulin-dependent protein kinase II promotes neurodegeneration caused by tau phosphorylated at Ser262/356 in a transgenic Drosophila model of tauopathy. *J. Biochem.* **2017**, *162*, 335–342. [CrossRef] [PubMed]
47. Liu, C.; Liang, M.C.; Soong, T.W. Nitric Oxide, Iron and Neurodegeneration. *Front. Neurosci.* **2019**, *13*. [CrossRef]
48. Keith, T.; Akama, L.J.; Eldik, V. β-Amyloid Stimulation of Inducible Nitric-oxide Synthase in Astrocytes Is Interleukin-1β- and Tumor Necrosis Factor-α (TNFα)-dependent, and Involves a TNFα Receptor-associated Factor- and NFκB-inducing Kinase-dependent Signaling Mechanism. *J. Biol. Chem.* **2020**, *11*, 7918–7924. [CrossRef]
49. Zeng, P.; Shi, Y.; Wang, X.M.; Lin, L.; Du, Y.J.; Tang, N.; Wang, O.; Fang, Y.Y.; Wang, J.Z.; Zhou, X.W.; et al. Emodin Rescued Hyperhomocysteinemia-Induced Dementia and Alzheimer's Disease-Like Features in Rats. *Int. J. Neuropsychopharmacol.* **2019**, *2*, 57–70. [CrossRef]
50. Wook, K.J.; Soo, B.M.; Dahyun, Y.; Ho, L.Y.; Yeon, J.S.; Kang, K.; Haejung, J.; Gijung, J.; Jun-Young, L.; Chul-Ho, S.; et al. Blood Hemoglobin, in-vivo Alzheimer Pathologies, and Cognitive Impairment: A Cross-Sectional Study. *Front. Aging Neurosci.* **2021**, *13*. [CrossRef]
51. Butterfield, D.A.; Halliwell, B. Oxidative stress, dysfunctional glucose metabolism and Alzheimer disease. *Nat. Rev. Neurosci.* **2019**, *20*, 148–160. [CrossRef]
52. Baik, S.H.; Kang, S.; Lee, W.; Choi, H.; Chung, S.; Kim, J.I.; Mook-Jung, I. A Breakdown in Metabolic Reprogramming Causes Microglia Dysfunction in Alzheimer's Disease. *Cell Metab.* **2019**, *3*, 493–507. [CrossRef]
53. Kremer, J.C.; Prudner, B.C.; Lange, S.E.S.; Bean, G.R.; Schultze, M.B.; Brashears, C.B.; Van Tine, B.A. Arginine deprivation inhibits the Warburg effect and upregulates glutamine anaplerosis and serine biosynthesis in ASS1-deficient cancers. *Cell Rep.* **2017**, *18*, 991–1004. [CrossRef]
54. Liu, Y.H.; Wang, Y.R.; Xiang, Y.; Zhou, H.D.; Giunta, B.; Mañucat-Tan, N.B. Clearance of amyloid-β in Alzheimer's disease: Shifting the action site from center to periphery. *Mol. Neurobiol.* **2015**, *51*, 1–7. [CrossRef] [PubMed]
55. Zhang, C.; Wei, W.; Liu, Y.; Peng, J.; Tian, Q.; Liu, G. Hyperhomocysteinemia increases beta-amyloid by enhancing expression of gamma-secretase and phosphorylation of amyloid precursor protein in rat brain. *Am. J. Pathol.* **2009**, *174*, 1481–1491. [CrossRef] [PubMed]
56. Reitz, C.; Mayeux, R. Alzheimer Disease: Epidemiology, Diagnostic Criteria, Risk Factors and Biomarkers. *Biochem. Pharm.* **2014**, *88*, 640–651. [CrossRef] [PubMed]
57. Kamal, A.; Almenar-Queralt, A.; LeBlanc, J.F.; Roberts, E.A.; Goldstein, L.S. Kinesin-mediated axonal transport of a membrane compartment containing beta-secretase and presenilin-1 requires APP. *Nature* **2001**, *414*, 643–648. [CrossRef]
58. Gómez, X.; Sanon, S.; Zambrano, K. Key points for the development of antioxidant cocktails to prevent cellular stress and damage caused by reactive oxygen species (ROS) during manned space missions. *Microgravity Sci. Tech.* **2021**, *7*. [CrossRef] [PubMed]
59. Lesné, S.; Sherman, M.; Grant, M.; Kuskowski, M.; Schneider, J.; Bennett, D. Brain amyloid-β oligomers in ageing and Alzheimer's disease. *Brain* **2013**, *136*, 1383–1398. [CrossRef]
60. Cohen, T.J.; Guo, J.L.; Hurtado, D.E.; Kwong, L.K.; Mills, I.P.; Trojanowski, J.Q.; Lee, V.M. The acetylation of tau inhibits its function and promotes pathological tau aggregation. *Nat. Commun.* **2011**, *2*, 252. [CrossRef]
61. Irwin, D.J.; Cohen, T.J.; Grossman, M.; Arnold, S.E.; Xie, S.X.; Lee, V.M.Y.; Trojanowski, J.Q. Acetylated tau, a novel pathological signature in Alzheimer's disease and other tauopathies. *Brain* **2012**, *135*, 807–818. [CrossRef]
62. Sharma, K. Cholinesterase inhibitors as Alzheimer's therapeutics (Review). *Mol. Med. Rep.* **2019**, *20*, 1479–1487. [CrossRef]
63. Méndez, M.; Méndez-López, M.; López, L.; Aller, M.A.; Arias, J.; Arias, J.L. Acetylcholinesterase activity in an experimental rat model of Type C hepatic encephalopathy. *Acta Histochem.* **2011**, *113*, 358–362. [CrossRef]
64. Elufioye, T.O.; Chinaka, C.G.; Oyedeji, A.O. Antioxidant and anticholinesterase activities of macrosphyra longistyla (dc) hiern relevant in the management of Alzheimer's Disease. *Antioxidants* **2019**, *8*, 400. [CrossRef] [PubMed]

65. Ikonomovic, M.D.; Mufson, E.J.; Wuu, J.; Bennett, D.A.; DeKosky, S.T. Reduction of choline acetyltransferase activity in primary visual cortex in mild to moderate Alzheimer's disease. *Arch. Neurol.* **2005**, *62*, 425–430. [CrossRef] [PubMed]
66. Jamshidnejad-Tosaramandani, T.; Kashanian, S.; Babaei, M.; Al-Sabri, M.H.; Schiöth, H.B. The potential effect of insulin on AChE and its interactions with rivastigmine in vitro. *Pharmaceuticals* **2021**, *14*, 1136. [CrossRef] [PubMed]
67. Perera, T.D.; Dwork, A.J.; Keegan, K.A.; Thirumangalakudi, L.; Lipira, C.M.; Joyce, N.; Lange, C.; Higley, J.D.; Rosoklija, G.; Hen, R.; et al. Necessity of hippocampal neurogenesis for the therapeutic action of antidepressants in adult nonhuman primates. *PLoS ONE* **2011**, *6*, e17600. [CrossRef] [PubMed]
68. Anschev, A.S. Small intestine acetylcholinesterase activity in experimental animals exposed to low doses of ionizing radiation. Journal of Education. *Health Sport* **2020**, *10*, 373–380.
69. Dułak, D.; Gadzała, M.; Banach, M.; Roterman, I. Analysis of alternative conformations of the AB(1–40) amyloid protein. In *From Globular Proteins to Amyloids*; Irena, R.-K., Ed.; Elsevier: Amsterdam, The Netherlands; Oxford, UK; Cambridge, MA, USA, 2020.
70. Xiang, Y.; Bu, X.L.; Liu, Y.H. Physiological amyloid-beta clearance in the periphery and its therapeutic potential for Alzheimer's disease. *Acta Neuropathol.* **2015**, *130*, 487–499. [CrossRef]

animals

MDPI

Article

The Impact of Curcumin on Growth Performance, Growth-Related Gene Expression, Oxidative Stress, and Immunological Biomarkers in Broiler Chickens at Different Stocking Densities

Mona H. Hafez [1,*], Sara E. El-Kazaz [2], Badr Alharthi [3], Heba I. Ghamry [4], Mohammed A. Alshehri [5], Samy Sayed [6], Mustafa Shukry [7,*] and Yasser S. El-Sayed [8]

1 Physiology Department, Faculty of Veterinary Medicine, Alexandria University, Alexandria 22758, Egypt
2 Department of Animal Husbandry and Animal Wealth Development, Faculty of Veterinary Medicine, Alexandria University, Alexandria 22758, Egypt; saraelkazaz@alexu.edu.eg
3 Department of Biology, University College of Al Khurmah, Taif University, P.O. Box 11099, Taif 21944, Saudi Arabia; b.harthi@tu.edu.sa
4 Department of Home Economics, College of Home Economics, King Khalid University, P.O. Box 960, Abha 61421, Saudi Arabia; hgmry@kku.edu.sa
5 Biology Department, College of Science, University of Tabuk, Tabuk 71491, Saudi Arabia; ma.alshehri@ut.edu.sa
6 Department of Science and Technology, University College-Ranyah, Taif University, P.O. Box 11099, Taif 21944, Saudi Arabia; samy_mahmoud@hotmail.com
7 Physiology Department, Faculty of Veterinary Medicine, Kafrelsheikh University, Kafrelsheikh 33516, Egypt
8 Department of Forensic Medicine and Toxicology, Faculty of Veterinary Medicine, Damanhour University, Damanhour 22511, Egypt; elsayed-ys@vetmed.dmu.eduu.eg
* Correspondence: mona.hafez@alexu.edu.eg (M.H.H.); mostafa.ataa@vet.kfs.edu.eg (M.S.)

Simple Summary: The primary goal of global poultry production is to optimize the amount of chicken produced per square meter of floor area. Consequently, stocking density (SD) and curcumin supplementation on broiler performance were investigated. Our results revealed that supplemental curcumin improved birds' growth, behaviours, and immunity by lowering oxidative stress, enhancing humoral immune response, and modulating the suppression of growth-related gene expressions in broilers raised in high stocking density circumstances.

Abstract: Curcumin's antioxidant properties reduce free radicals and may improve broiler growth. Therefore, the influence of stocking density (SD) and administration of curcumin in the diet on broiler performance was explored to clarify the impact of HSD and curcumin on the performance of growth, behavioural patterns, haematological, oxidant/antioxidant parameters, immunity markers, and the growth-related genes expression in broiler chickens. A total of 200 broiler chickens (Cobb 500, 2-weeks old) were allotted into 4 groups; SD (moderate and high) and curcumin (100 and 200 mg/kg diet)-supplemented HSD, respectively. Behavioural observations were performed. After a 28-day experimental period, tissue and blood samples were collected for analysis. Expressions of mRNA for insulin-like growth factor-1 (IGF-1), growth hormone receptor (GHR), myostatin (MSTN), and leptin in liver tissues were examined. HSD birds exhibited lower growth performance measurements, haematological parameters, circulating 3,5,3-triiodothyronine and thyroxine levels, antioxidant activities (GSH-Px, catalase, superoxide dismutase), immunoglobulins (A, G, M), and hepatic GHR and IGF-1 expression values. However, HSD birds even had an increment of serum corticosterone, malondialdehyde, pro-inflammatory cytokine (TNF-a, IL-2, IL-6) levels, hepatic leptin and MSTN expression. Moreover, HSD decreased drinking, feeding, crouching, body care, and increased standing and walking behaviour. The addition of curcumin, particularly at a 200 mg/kg diet, alleviated the effect of HSD through amending growth-related gene expression in the chickens. In conclusion, curcumin can enhance birds' growth performance, behavioural patterns, and immunity by reducing oxidative stress and up-regulating the growth-related gene expressions of broilers under stressful conditions due to a high stocking density.

check for updates

Citation: Hafez, M.H.; El-Kazaz, S.E.; Alharthi, B.; Ghamry, H.I.; Alshehri, M.A.; Sayed, S.; Shukry, M.; El-Sayed, Y.S. The Impact of Curcumin on Growth Performance, Growth-Related Gene Expression, Oxidative Stress, and Immunological Biomarkers in Broiler Chickens at Different Stocking Densities. *Animals* **2022**, *12*, 958. https://doi.org/10.3390/ani12080958

Academic Editors: Peter F. Surai and Ilias Giannenas

Received: 30 January 2022
Accepted: 31 March 2022
Published: 8 April 2022

Keywords: broilers; curcumin; physiological efficiency; poultry welfare; stress

1. Introduction

The main target of worldwide poultry production is to maximize the bulk of chicken made in each floor space square meter. To gain an appropriate advantage, overloading-related manufacturing damages have been reduced [1]. In the majority of the modern livestock industry, animals' physiological status is being compromised under several stressful circumstances, such as high rearing density, high ambient temperature, low sanitation, disease challenges, and improper management, thereby threatening health status, productive performance, and the well-being of the animals [2,3]. In addition, stressful circumstances could produce an imbalance between oxygen demand and supply [4]; thus, hypoxia subsequently happens, which enhances the production of free radicals and, as a result, interrupts the normal body functions because of the increased activities of circulating enzymes [5]. Consequently, scientific studies have made numerous efforts to improve the physiological procedures regarding the stress reactions in animals exposed to several stressors.

Stocking density (SD) is considered one of the central alarms in the poultry production industry due to the health consequences and its financial viability. SD is the body's mass (kg) or the reared quantity of birds in each area (m^2). High stocking density (HSD) is assumed to decline poultry production due to the high temperature of the environment and decreased flow of air between the bird [6]. An HSD is financially profitable for producers. Recently, HSD has been frequently practised, improving the economic outcomes by gaining other broilers' meat for each fixed rearing area [2].

The influence of stocking density (20 to 40 kg/m^2) on broiler production and performance has been studied. However, the majority of these investigations were not always definitive and yielded mixed results. Reduced stocking density has improved broiler performance in some trials [7]. Broiler welfare minimum requirements were announced by the European Commission in 2007, with a maximum stocking density of 30 kg/m^2 (0.073 m^2/bird) of broiler chickens across the EU. Furthermore, the National Chicken Council has established a voluntary welfare audit programme for broiler producers. This programme recommends a density range of 31.8 kg/m^2 for light broilers to 41.6 kg/m^2 for roosters based on final body weight [8]. Broiler output is heavily influenced by stocking density. Increasing space allowances in the production systems can have a significant negative economic impact on the sector. High-density flocks of birds exhibit physiological adaptations indicative of stress at the transcriptional and telomere levels of the genes they carry [9]. Extension of clutch size experimentally inflicted on the lesser blackback gull [10] has shown that reproductive females deposit more female eggs under dietary stress. This is assumed to be a reproductive adaption. Male gull chicks, more prominent but not heavier at hatching, have fewer reserves and a lower survival rate [11]. In addition, the performance of broiler chickens has been shown to be impacted by the increase in stocking density from 28 to 40 kg BW/m^2, depending on the system, deep litter or free-range, which could put their welfare at risk [8]. In addition to affecting broiler chickens' ability to absorb nutrients, the high density also impacts their productivity, the production of stress hormones [4], and their ability to fight off disease [12].

On the other hand, HSD could enhance stress to birds, which is the primary cause of suffering in poultry. This stress induces pathological lesions and deteriorates the broilers' immune status. On the other hand, broilers need to be delivered ideal environmental circumstances that prompt their genetic perspective.

Li et al. [13] recorded that bodyweight reduces with high stocking density. Nonetheless, HSD has been documented to decline the broiler productivity by decreasing meat quality and growth development and deteriorating the physiological health condition of broilers through the escalating status of oxidative stress [5,14]. There are numerous expensive approaches available to decrease the harmful effects of HSD as outdoor access, but dietary management, through antioxidants supplementation, is a more accessible methodology that is supported by some experimental works [5,14,15].

Numerous experiments on broilers supplemented with phytogenic mixtures of thyme, turmeric, coriander, and others have shown substantial progress in immune indicators, growth performance, and the carcass features in mallard ducks (*Anas platyrhynchos*) [16,17]. Phytogenic mixtures are given individually or in groups as dietary supplements in poultry rations [18]. Among numerous flavours, turmeric (*Curcuma longa* L.), recognized as the golden spice, has exceptional awareness due to its various useful pharmacological properties supporting the well-being and health of poultry and animals [19]. As a member of the Zingiberaceae family, turmeric (*Curcuma longa*) is one of the spices with a member element known as curcumin [20]. Curcumin [1,7-bis (4-hydroxy 3- methoxyphenyl)- 1,6-heptadiene-3,5-dione; diferulylmethane] retains its economic value because of its weird bulb [21]. Curcumin is frequently utilized as a food condiment and a tint due to its defence in contradiction with oxidative stress. It can remove free radicals, guarding tissues and organs, contrary to peroxidation lipids [22]. Furthermore, various experimental and clinical studies have verified that, pharmacologically, curcumin is a harmless substance with antioxidant, anti-inflammatory, and antimicrobial activities [23–27]. It also possesses immune-modulatory and hepatoprotective properties [28].

Nonetheless, the information associated with curcumin usage under HSD circumstances on the physiological responses of broiler birds is scarce. Therefore, the existing work was executed to clarify the impact of HSD and food administration of curcumin on the performance of growth, behavioural patterns, haematological, oxidant/antioxidant parameters, immunity markers, and the growth-related genes expression in broiler chickens.

2. Materials and Methods

2.1. Experimental Animals

2.1.1. Animal Care

The current experiment was approved and, complying to the rules for maintaining animals and poultry supplied by the National Research Council, was permitted (2021/013/28) by the Resident Commission of Ethics for the Use and Care of Lab Animals at Alexandria University, Egypt. Every phase of the current study plan was performed with insignificant distress or pain for the broilers. The research work of our study was performed at the Department of Animal Husbandry and Animal Wealth Development, Faculty of Veterinary Medicine, Alexandria University. The birds were allocated to ground litter cages with completely controlled temperature and humidity.

2.1.2. Birds and Housing Conditions

The current study was conducted on Cobb 500 chicks (n = 200 chicks) from a broilers farm and assigned by weight (to minimize the difference in average body weight) into 4 experimental groups. The ground was cemented and shielded in a fresh straw litter with a depth of 5 cm. The drinking space and food holders were placed so the broilers in every pen acquired approximately identical feeding and drinking area regardless of the stocking density. Broilers were nourished ad libitum consuming a commercial ration. Chicks were fed on starter crumbles, as shown in Table 1 [29], until reaching two weeks of age, and then, for one week, they were provided grower pellets. They were then provided with finisher pellets until the end of the experiment. The temperature of the pens was sustained at 32 °C through the first seven days, and then slowly reduced by 3 °C for each week, until it reached 24 °C by the time the study finished. Controlled artificial light was sustained for 23 h per day during the whole study duration.

Table 1. Composition of experimental starter, grower, and finisher diets (g/kg diet) and the calculated chemical analysis of the basal diet.

Ingredients	Diet		
	Starter	Grower	Finisher
Yellow corn	542	558.8	606
Soybean meal (44%)	319	281	253.3
Corn gluten meal (60%)	71	81	48.1
Vegetable oil	29.8	41	54.4
Limestone [1]	15	15	15
Monocalcium phosphate [2]	14	14	14
Common salt	3	3	3
Mineral Premix [3]	1.5	1.5	1.5
Vitamin Premix [3]	1.5	1.5	1.5
Methionine [4]	1	1	1
Lysine [5]	1	1	1
Anti-coccidial [6]	0.2	0.2	0.2
Antimould [7]	1	1	1
Calculated analysis			
Crude protein (CP, %)	23.1	22.18	19.39
Metabolizable energy (ME, Kcal/kg diet) [8]	3053	3160.7	3252.6
Calorie/protein ratio [9]	132.16	142.5	167.7

[1] Limestone contains 36% calcium. [2] Monocalcium phosphate contains 22% phosphorus and 16% calcium. [3] Mineral and Vitamin premix was produced by Heropharm and composed (per 3 kg) of vitamin A 12,000,000 IU, vitamin D3 2,500,000 IU, vitamin E 15,000 IU, vitamin K3 1000 mg, vitamin B1 1000 mg, vitamin B2 3000 mg, vitamin B6 1500 mg, vitamin B12 13.3 mg, niacin 30,000mg, biotin 50 mg, folic acid 600 mg, pantothenic acid 10,000 mg, Mn: 60,000 mg, Zn: 50,000 mg, Fe: 30,000 mg, Cu: 4000 mg, I: 300 mg, Se: 100 mg and Co:100 mg. [4] DL-Methionine (Produced by Evonic Co. Birmingham, AL, Germany and contain 99% methionine), [5] Lysine (lysine hydrochloride contains 98 % Lysine). [6] Kill Cox® (produced by Arabian Company for Pharmaceutical Industries, [7] Antimould produced by EL-TOBA CO. For Premixes & Feed El-Sadat city Egypt). [8] ME calculated according to NRC (1994), [9] Calorie/protein ratio = ME Kcal/CP%.

MSD (10 birds/m^2) and HSD (20 birds/m^2) was accomplished by keeping a different number of broilers for each floor pen [5] with the exact size of the floor (2.0×2.4 m^2). Before beginning the study, all birds were reared with similar environmental circumstances and were fed similar starter rations for 14 days (Table 1). Space unavailable by permanent substances (i.e., one bell drinker and two feeders) was not involved when the floor space was measured.

At 2 weeks of age, all birds were weighed and divided into 4 groups, including 5 replicates per each group (10 birds/ replicate) for every group of an exact number of birds, as follows: Group 1 (MSD, negative control) (n = 50); without curcumin supplementation; Group 2 (HSD, positive control) (n = 50); without curcumin supplementation; Group 3 (HSD + curcumin 100 mg kg^{-1} diet) (n = 50) [25]; bought from Sigma–Aldrich, St. Louis, MO, USA), curcumin mixed into pellets; Group 4 (HSD + curcumin 200 mg kg^{-1} diet) (n = 50) [28].

2.2. Growth Performance

Throughout the study progression, broilers were individually weighed weekly (g) to assess the weight gain in body weight, feed conversion ratio (FCR), and to record the intake of food. FCR was measured by dividing the total intake of food (g) by the total gaining of weight (g) of the birds [30] from the beginning of the experiment (14 days old) until the finish of the study (day 42 of age).

2.3. Behavioral Observations

The behavioural observation was conducted via continuous visual screening of the flock. The behaviour of all birds was recorded during predetermined time intervals.

Experienced behavioural observations were started at two weeks old until the finish of the experiment (6 weeks of age). Observations were undertaken during the daytime (6:00 a.m.–6.00 p.m.) with 12 h of observation divided into 4 h per day through 3 consecutive days. Every day was allocated into two stages, the morning (6:00 a.m.–12:00 p.m.) and the afternoon (12:00 p.m.–6:00 p.m.). The observation was completed for 2 h for every stage of the day, i.e., 2 h during the morning and 2 h during the afternoon with alternation. During day one, the observations were performed from 6:00–8:00 a.m. and 12:00–2:00 p.m.; on day two, it was from 8:00–10:00 a.m. and 2:00–4:00 p.m., and the same for day three [31]. Each observation hour was divided into 5-min intermission scanning of broilers, which was followed with a new 5-min scan of behaviour until the observation time was finished. Behavioural configurations perceived were drinking, feeding, walking, crouching, standing, and behaviour of body care (ruffling, shaking, and preening). The documented behaviours are presented in Table 2. Results were presented as % of broilers doing the categorized behaviour/whole number of broilers scanned [31].

Table 2. Ethogram of the recorded behaviours.

Behaviour	Description
Feeding	Pecking at feed-on-feed troughs.
Drinking	Obtaining water from the cup.
Standing idle	The abdomen is not touching the litter, and the bird is motionless with no other behaviour.
Crouching	The bird remains sitting and lying on the litter, looking around or with closed eyes, no other behaviour.
Walking	Take at least two successive steps.
Preening	The bird cleans and aligns its feathers using the beak.
Ruffling	The action of ruffle or shackling all body feathers.
Shaking	The action of body trembling.

2.4. Blood Sampling

On day 42 of age, 6 birds of each replicate were arbitrarily chosen and slaughtered. Two blood samples were taken from the brachial vein, and one was into a heparinized tube for haematological measurements. At the same time, the other sample was collected into plain tubes and centrifuged at 1968× g for 15 min. Collected serum samples were retained at −20 °C till examination.

2.5. Hematological and Immunological Parameters Measurements

At 1 to 2 h after the collection, the samples of blood were examined for Hb (haemoglobin) level, PCV (packed cell volume), RBCs (red blood cells), and WBCs (white blood cells) count. In addition, a differential leukocytic count was completed. Erythrocyte sedimentation rate (ESR) and RBC settlement rate at uncoagulated blood during an hour inside the Wintrobe tube were then performed [32]. Heterophil/lymphocyte (H/L) ratio was determined by dividing the number of heterophils by lymphocytes. Immunoglobulins in the serum (IgG, IgA, and IgM), pro-inflammatory cytokines (IL-2, IL-6), and tumour necrosis factor-A (TNF-α) were estimated via kits obtained from Elabscience Biotechnology Co., Ltd. (Houston, TX, USA) using a reader for enzyme-linked immunosorbent assay (ELISA).

2.6. Biochemical and Hormonal Analysis

Using the specific commercial assay kits (Bio-diagnostic Co., Cairo, Egypt), total serum protein, albumin, total cholesterol, alanine aminotransferase (ALT), and aspartate aminotransferase (AST) were determined using Robert R. GmbH Photometer (5010 VST, Berlin, Germany) [33]. Analysis of thyroid hormones [thyroxine (T4) and triiodothyronine (T3)] and corticosterone concentration according to the photometric recognition principle was accomplished using thyroid hormones and corticosterone ELISA kit (through IBL international GmbH, 22335 Hamburg, Germany).

2.7. Serum Malondialdehyde Level and Anti-Oxidative Enzymes Activity

Serum malondialdehyde (MDA) value and superoxide dismutase (SOD) activity were determined using kits supplied from Ransod Diagnostics (London, UK). Meanwhile, glutathione peroxidase (GPx) and catalase (CAT) enzymatic activities were determined via provided test kits (Bio-Diagnostic Co., Cairo, Egypt) following the manufacturer's instruction.

2.8. Gene Expression Analysis

The liver samples (approximately one cm^3) ($n = 10$) were gathered, then positioned in liquid nitrogen, and kept at −80 °C until investigation. According to the manufacturer's procedures, total RNA was isolated with TRI reagent (easyRED™, iNtRON Biotechnology, Seongnam-Si, Korea). Its quality was confirmed by gel electrophoresis with 2% agarose. RNA was measured via Nanodrop (UV–VIS Spectrophotometer Q5000, Qua-Well, San Jose, CA, USA). RNA was conversely transcribed to the first-strand cDNA by the SensiFAST™ cDNA synthesis kit (Bioline, London, UK) and kept at −20 °C until the examination. Sequences of gene-specific primers (Table 3) were utilized to strengthen the designated genes. Primer sequences were definite via UCSC In Silico PCR (https://genome.ucsc.edu/cgibin/hgPcr, accessed on 10 March 2021) and NCBI primer blast. Real-time was accomplished by consuming the Stratagene MX300P Realtime PCR machine (Agilent Technologies, Santa Clara, CA, USA), with SensiFast™ SYBR Lo-Rox kit (Bioline, London, UK) according to manufacturer's recommendations. The intensification technique includes denaturing at 95 °C for 15 min, then by 40 sequences at 95 °C for 15 s and an annealing temperature (specific for every primer, Table 2) for 1 min [15]. Dissociation curves were inspected to approve that only one peak was recognized for each exact melting temperature, therefore presenting that PCR products were propagated specifically. Samples were examined in replicas. Target gene Ct levels were primarily standardized in contradiction of the Ct levels of the house-keeping gene (β actin), then utilized to estimate the gene expression levels (fold change) [34].

2.9. Statistical Analysis

Data collected were subjected to ANOVA by applying the General Linear Model Procedure of SAS software (SAS Institute Inc., Cary, NC, USA) [35]. Shapiro-Wilk and Levene tests confirmed normal distribution and homogeneity of variance [36]. Tukey's multiple comparison test is based on $p < 0.05$. The following formula was exhausted for the examination of variance:

Table 3. Gene-specific primers sequence used in the experiment.

Gene	Accession No.	Primer	References	Amplicon Size (bp)	Annealing Temp (°C)
Growth hormone receptor (GHR)	NM_001001293	R: AGAAGTCAGTGTTTGTCAGGG F: AACACAGATACCCAACAGCC	[37]	93	60
Insulin-like growth factor-1 (IGF-1)	NM_001004384	R: CTTGTGGATGGCATGATCT F: CACCTAAATCTGCACGCT	[34]	91	60
Leptin	KT970642.1	R: CCAGGACGC-CATCCAGGCTCTCTGGC F: TCCGCCAAGCAGAGGGGT	[38]	261	58
Myostatin (MSTN)	NM_001001461	F: TACCCGCTGACAGTGGATTTC R: GCCTCTGGGATTTGCTTGG	[37]	173	58
β-actin	NM_205518	F: ACCTGAGCGCAAGTACTCTGTCT R: CATCGTACTCCTGCTTGCTGAT	[34]	160	60

Statistical model

$$Xij = \mu + Ti + eij,$$

where, Xij is the value of ith observation (variables as weight of the body or behaviour or biochemical measurements) of ith treatment, μ is the overall mean, Ti is the effect of the treatment (treatment: four different treatments with two stocking density)), and eij is the random error.

3. Results

3.1. Productive Performance

In Table 4, the HSD group revealed a marked decrease ($p < 0.0001$) in body weight, weight gain, and the intake of food, and an increase in feed conversion ratio (FCR) than the MSD control group. Meanwhile, curcumin with HSD shows a significant increase in body weight, weight gain, and food intake, with a substantial decrease in FCR. However, the higher dose of curcumin (200 mg) displays a significant increase in all productive parameters compared to a lower dose of curcumin (100 mg) as the best bodyweight, weight gain, intake of food, and FCR were observed in the MSD control birds and in the group supplemented with 200 mg of curcumin.

Table 4. Effect of curcumin on productive performance of broilers reared under high stocking density.

	Initial Body Weight (g)	Final Body Weight (g)	Body Weight Gain (g)	Feed Intake (g)	Feed Conversion Ratio (FCR)
MSD control	412.15 ± 15.23	2101.28 ± 165.23 ^{a}	1689.13 ± 161.32 ^{a}	2785.26 ± 62.52 ^{a}	1.65 ± 0.03 ^{b}
HSD control	419.96 ± 18.52	1558.23 ± 150.23 ^{c}	1118.27 ± 149.25 ^{c}	2099.35 ± 38.11 ^{d}	1.88 ± 0.05 ^{a}
HSD + curcumin 100	429.88 ± 20.14	1723.77 ± 193.42 ^{b}	1293.89 ± 185.96 ^{b}	2302.67 ± 26.33 ^{c}	1.77 ± 0.06 ^{ab}
HSD + curcumin 200	428.23 ± 23.18	1989.87 ± 201.78 ^{ab}	1561.64 ± 198.23 ^{ab}	2623.25 ± 27.27 ^{b}	1.68 ± 0.04 ^{b}
P-Tukey	0.41	<0.0001	<0.0001	<0.0001	<0.0001

The data shown are the mean and standard deviation using ANOVA. $^{a–d}$ Means bearing different superscript letters within the same row are significantly different ($p < 0.05$). MSD = low stocking density, HSD = high stocking density.

3.2. Behavioural Observation

In Figure 1, HSD markedly ($p < 0.0001$) reduced the ingestive behaviour of birds. It is considered an allelomimetic behaviour, as two or more birds showed similar behaviour simultaneously with some degree of mutual stimulation and coordination. Therefore, the results revealed a pronounced influence of stocking density on feeding and drinking of engaged broilers. Moreover, crouching and body care behaviour (preening, ruffling, shaking) significantly decreased in birds reared in HSD. In contrast, birds stocked in high density show a significant increase in walking and standing behaviour.

Supplementation of curcumin powder at (200 mg) shows a significant enhancement in the proportion of birds that exhibit standard behaviour patterns than the low dose (100 mg). It was observed that there was a marked enhancement in ingestive behaviour, crouching, and body care behaviour, while there was a substantial decrease in walking and standing behaviour. In comparison, curcumin supplementation shows a considerable enhancement in exhibiting standard patterns of birds' behaviour.

3.3. Haematological Parameters

In Table 5, Hb, PCV, and RBCs count were significantly ($p < 0.05$) decreased in the HSD reared birds relative to their levels in the MSD birds. However, the H/L ratio and ESR were enhanced markedly in the HSD group comparable to the MSD one. However, a marked ($p < 0.0001$) improvement was documented in curcumin (100 or 200 mg/kg diet)-treated HSD chicks Hb, PCV, and RBCs count, and a marked decrease in ESR and H/L ratio relative to HSD-reared broilers. The better values were recorded in a dose-dependent effect. In contrast, the WBCs count revealed a non-significant difference between the groups.

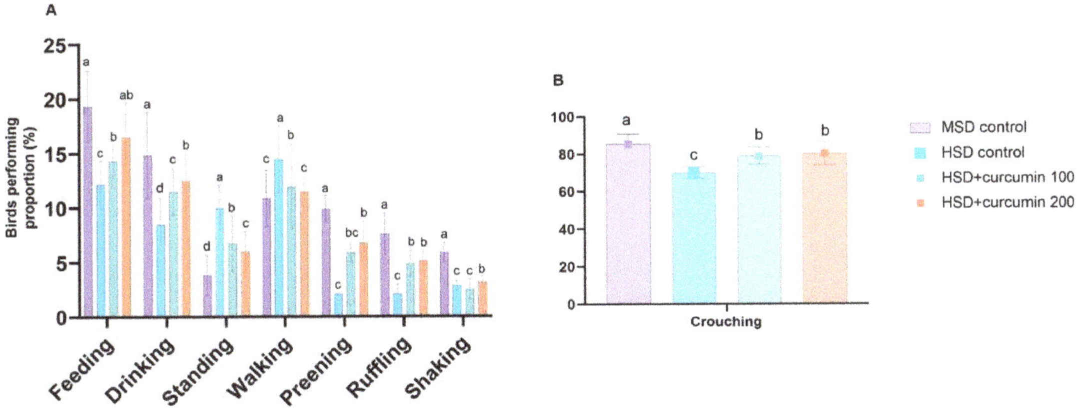

Figure 1. Effect of curcumin on the proportion (%) of birds performing different behavioural patterns (**A**). Feeding, Drinking, Standing, Walking, Preening, Ruffling and Shaking. (**B**). Crouching reared under high stocking density. The data shown are the mean and standard deviation. [a–d] Means bearing different superscript letters within the same row are significantly different ($p < 0.05$). MSD = low stocking density, HSD = high stocking density.

Table 5. Effect of curcumin on haematological parameters of broilers reared under high stocking density.

	PCV (%)	Hb (g/dL)	RBCs (10^{12}/L)	WBCs (10^9/L)	ESR (mm/h)	H/L Ratio (%)
MSD control	24.42 ± 0.27 [a]	12.23 ± 0.04 [a]	2.49 ± 0.01 [a]	24.1 ± 0.26	3.15 ± 0.02 [c]	0.263 ± 0.008 [c]
HSD control	17.35 ± 0.31 [d]	8.35 ± 0.08 [c]	1.92 ± 0.01 [b]	24.5 ± 0.38	5.52 ± 0.05 [a]	0.452 ± 0.007 [a]
HSD + curcumin 100	20.61 ± 0.24 [c]	10.87 ± 0.06 [b]	2.11 ± 0.02 [ab]	23.6 ± 0.32	4.17 ± 0.01 [b]	0.329 ± 0.006 [b]
HSD + curcumin 200	22.61 ± 0.22 [ab]	11.87 ± 0.06 [a]	2.31 ± 0.03 [a]	24.2 ± 0.26	3.97 ± 0.01 [c]	0.289 ± 0.006 [c]
P-Tukey	<0.0001	<0.0001	<0.0001	0.51	<0.0001	<0.0001

The data shown are the mean and standard deviation. [a–d] Means bearing different superscript letters within the same row are significantly different ($p < 0.05$). MSD = low stocking density, HSD = high stocking density, PCV = packed cell volume, Hb = haemoglobin, RBCs = red blood cells, WBCs = white blood cells, ESR = erythrocyte sedimentation rate, H/L = heterophil/lymphocyte.

3.4. Immunological Parameters

Treatment of the HSD broilers with curcumin (100 or 200 mg/kg diet) led to marked ($p < 0.05$) improvement in IgG and IgA concentrations compared to the HSD group. On the other hand, the HSD broilers showed a significant ($p < 0.05$) reduction in IgA, IgM, and IgG compared to the MSD reared broilers. Moreover, IL-2, IL-6, and TNF-a revealed a marked enhancement ($p > 0.05$) with HSD birds relative to MSD ones. Simultaneously, treatment with curcumin reduced the pro-inflammatory cytokines values to nearly average physiological importance of the MSD group (Table 6).

3.5. Biochemical and Hormonal Analysis

The data presented in Table 7 revealed that total serum protein and albumin levels were not varied between the groups. Relative to MSD birds, total cholesterol, ALT, and AST values were markedly enhanced with HSD broilers; however, these parameters were dramatically ($p < 0.05$) decreased with curcumin-treated HSD birds. Relative to 100 mg curcumin kg^{-1} diet HSD treated broilers, the 200 mg curcumin kg^{-1} diet treated birds markedly ($p < 0.05$) enhanced the alternations of these measurements.

Table 6. Effect of curcumin on immune measurements in the serum of broilers reared under high stocking density.

	IgA (g/L)	IgG (g/L)	IgM (g/L)	IL-2 (pg/mL)	IL-6 (pg/mL)	TNF-a (pg/mL)
MSD control	0.062 ± 0.002 ^{a}	0.089 ± 0.003 ^{a}	0.092 ± 0.005 ^{a}	110.14 ± 5.41 ^{d}	15.62 ± 0.87 ^{d}	69.92 ± 3.54 ^{c}
HSD control	0.015 ± 0.003 ^{b}	0.034 ± 0.008 ^{c}	0.048 ± 0.009 ^{d}	166.67 ± 8.11 ^{a}	28.58 ± 2.74 ^{a}	98.02 ± 7.22 ^{a}
HSD + curcumin 100	0.048 ± 0.002 ^{ab}	0.055 ± 0.006 ^{bc}	0.066 ± 0.007 ^{c}	141.67 ± 6.33 ^{b}	22.54 ± 1.66 ^{b}	77.37 ± 5.81 ^{b}
HSD + curcumin 200	0.061 ± 0.003 ^{a}	0.065 ± 0.006 ^{b}	0.079 ± 0.007 ^{b}	122.14 ± 7.27 ^{c}	18.54 ± 0.96 ^{c}	71.16 ± 4.54 ^{c}
P-Tukey	<0.0001	<0.0001	<0.0001	<0.0001	<0.0001	<0.0001

The data shown are the mean and standard deviation. $^{a–d}$ Means bearing different superscript letters within the same row are significantly different ($p < 0.05$). MSD = low stocking density, HSD = high stocking density, Ig = immunoglobulin, IL = interleukin, TNF = tumour necrosis factor.

Table 7. Effect of curcumin on biochemical parameters in the serum of broilers reared under high stocking density.

	Total Protein (mg/dL)	Albumin (mg/dL)	Total Cholesterol (mg/dL)	AST (U/L)	ALT (U/L)
MSD control	3.85 ± 0.25	2.205 ± 0.038	120.0 ± 4.8 ^{d}	212.3 ± 16.3 ^{d}	24.75 ± 5.04 ^{b}
HSD control	3.42 ± 0.31	2.045 ± 0.027	159.8 ± 9.04 ^{a}	302.8 ± 19.3 ^{a}	32.08 ± 4.22 ^{a}
HSD + curcumin 100	3.65 ± 0.21	2.155 ± 0.029	139.8 ± 12.14 ^{b}	242.3 ± 12.9 ^{b}	29.57 ± 3.73 ^{ab}
HSD + curcumin 200	3.81 ± 0.18	2.179 ± 0.038	126.8 ± 8.23 ^{c}	226.7 ± 14.7 ^{c}	25.75 ± 3.04 ^{b}
P-Tukey	0.27	0.12	<0.0001	<0.0001	<0.0001

The data shown are the mean and standard deviation. $^{a–d}$ Means bearing different superscript letters within the same row are significantly different ($p < 0.05$). MSD = low stocking density, HSD = high stocking density, AST = aspartate aminotransferase, ALT = alanine aminotransferase.

As revealed in Table 8, relative to the MSD-raised broilers, the HSD chickens expressed significantly ($p < 0.05$) enhanced corticosterone contents and decreased T3 and T4 levels. Curcumin treatment significantly reduced corticosterone and enhanced T3 and T4 serum levels close to the HSD birds. The 200 mg curcumin/kg diet-treated HSD broilers nearly reached their respective MSD values.

Table 8. Effect of curcumin on hormonal concentrations and oxidant/antioxidant parameters in the serum of broilers reared under high stocking density.

	Hormonal Concentrations			Oxidant/Antioxidant Parameters			
	T3 (ng/mL)	T4 (ng/mL)	Corticosterone (ng/mL)	MDA (μmol/L)	SOD (U/mL)	GPx (U/L)	CAT (U/L)
MSD control	3.57 ± 0.13 ^{a}	55.99 ± 3.84 ^{a}	2.95 ± 0.081 ^{d}	1.45 ± 0.027 ^{c}	2.93 ± 0.031 ^{a}	109.57 ± 8.64 ^{a}	69.57 ± 4.73 ^{a}
HSD control	2.03 ± 0.08 ^{c}	29.74 ± 1.77 ^{d}	5.74 ± 0.027 ^{a}	3.02 ± 0.031 ^{a}	1.77 ± 0.017 ^{c}	62.22 ± 4.18 ^{d}	32.08 ± 2.22 ^{c}
HSD + curcumin 100	2.76 ± 0.11 ^{b}	41.21 ± 2.03 ^{c}	4.06 ± 0.065 ^{b}	2.14 ± 0.021 ^{b}	2.23 ± 0.029 ^{b}	83.41 ± 5.04 ^{c}	55.57 ± 3.34 ^{b}
HSD + curcumin 200	3.17 ± 0.09 ^{a}	49.65 ± 2.22 ^{ab}	3.26 ± 0.082 ^{c}	1.74 ± 0.021 ^{c}	2.72 ± 0.031 ^{a}	95.49 ± 4.19 ^{b}	63.75 ± 3.84 ^{ab}
P-Tukey	<0.0001	<0.0001	<0.0001	<0.0001	<0.0001	<0.0001	<0.0001

The data shown are the mean and standard deviation. $^{a–d}$ Means bearing different superscript letters within the same row are significantly different ($p < 0.05$). MSD = low stocking density, HSD = high stocking density, T3 = triiodothyronine, T4 = thyroxine, MDA = malondialdehyde, SOD = superoxide dismutase, GPx = glutathione peroxidase, CAT= catalase.

3.6. Serum Oxidant/Antioxidant Measurements

In addition, Table 8 showed that curcumin supplementation with the HSD groups markedly ($p < 0.05$) improved SOD, GPx, and CAT activities in birds relative to the HSD broilers. However, MDA was decreased significantly ($p < 0.05$) with curcumin-treated groups close to the HSD broilers. Also, better values were recorded with the 200 mg curcumin/kg diet-treated HSD birds. However, the HSD birds revealed a marked ($p < 0.05$)

reduction in SOD, GPx, and CAT activities and a marked increase in MDA relative to MSD-reared birds.

3.7. Gene Expression of Growth Regulatory Factors

Growth regulatory genes, IGF-1, and GHR mRNA expressions were downregulated with the HSD broilers ($p < 0.05$) relative to the MSD birds (Figure 2). In contrast, curcumin treatment (100 or 200 mg kg^{-1} diet) with the HSD up-regulated the mRNA expression of GHR and IGF-1 genes ($p < 0.001$); in addition, gene expression of GHR and IGF-1 up-regulation revealed a dose-dependent impact. HSD significantly increased the expression of MSTN and leptin genes relative to the MSD control group. However, curcumin HSD treated groups markedly decreased leptin and MSTN ($p < 0.001$) close to HSD birds; on the contrary, the better values of growth regulatory genes were obtained by 200 mg curcumin kg^{-1} diet treated HSD group.

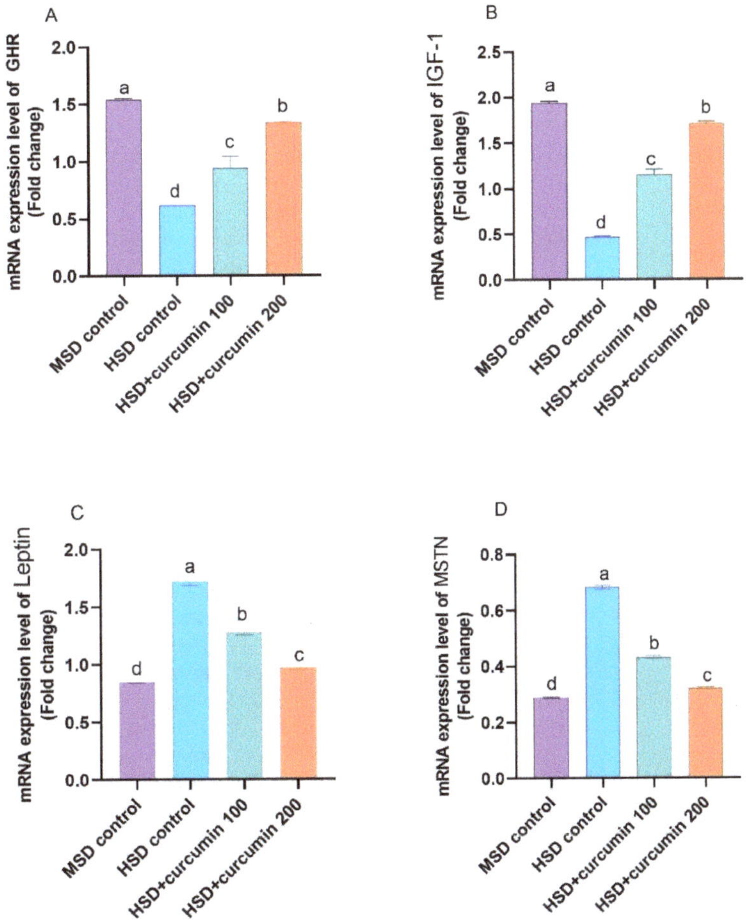

Figure 2. The relative mRNA expression of growth-related regulatory factors of broilers-reared under high stocking density. MSD = low stocking density, HSD = high stocking density, GHR = growth hormone receptor, IGF-1 = insulin-like growth factor 1, MSTN = myostatin. $^{a–d}$ Columns Means bearing different superscript letters are significantly different ($p < 0.05$).

4. Discussion

The existing experiment studied the haematological, immunological, behavioural, and growth performance combined with HSD for broilers given a control ration and two curcumin supplemented diets: 100 and 200 mg/kg. The growth performance with HSD broilers was markedly reduced, and FCR was enhanced relative to MSD birds. In contrast, the curcumin-treated HSD birds were more similar to the MSD group parameters. Similarly, HSD significantly decreased growth performance and compromised intestinal morphology, and metabolized energy results were recorded [5]. Yadav et al. [39] revealed no variation in the growth performance throughout the starter stage of age (0–21 d) with curcumin supplementation at given concentrations.

Nevertheless, broilers served curcumin-supplemented diets for 42 days had a marked enhancement in the body's weight and food conversion efficiency through a finisher phase (22–42 d). The increased gastric digestion liquor, intestinal villi length, caecal width, and bile production may have improved the digestion of fats. All these benefits are recorded by feeding curcumin for a longer time, enhancing the absorption of nutrients in the advanced stage [22]. Besides, the upregulation of pancreatic lipase and increased trypsin, amylase, and chymotrypsin improved intestinal maltase and sucrose activities [21], reflecting mainly in a higher feed intake and digestion FCR in the curcumin-supplemented birds. Moreover, the negative impacts of these findings with the HSD group may be attributed to the higher levels of serum corticosterone revealed in the current study, which indicate an enhanced stressful condition in the broilers. Those findings were comparable to previous data, in which they reported that the density of stockings had increased reduced growth, FCR, and carcass characteristics; induced oxidative stress and immunodeficiency [13]. For example, increasing corticosterone (catabolic hormone) levels markedly decreases the intake of food and body weight gain in chickens [30]. Meanwhile, curcumin supplementation significantly reduces the corticosterone hormone level, which may enhance the productive performance parameters in birds. Stress can impair immunity and egg production in layers by increasing the level of the stress hormone corticosterone [40].

Behavioural observation revealed that HSD leads to a reduced space allowance, reduced drinking and feeding space, enhanced fighting between the broilers, and significantly reduced ingestive behaviour (feeding and drinking) and crouching behaviour. Moreover, increasing stocking density adversely affects body care behaviour (preening, ruffling, shaking) as birds are under stress when exhibiting body care behaviour [30]. In contrast, increasing stocking density led to a significant increase in standing and walking behaviour. These findings show that as stocking density enhanced, the broilers stand more due to the absence of floor area for lying or crouching. Therefore, HSD limits the floor area accessible for broilers to exhibit normal behaviour, putting the birds under stress conditions. Generally, all features of welfare were harmfully impacted at HSD. These findings agree with [31,41–43]; they reported that higher stocking densities resulted in a rise in mortality during the rearing phase, an increase in the leg and contact dermatitis and carcass bruising, as well as a decrease in resting, locomotion, and ground pecking patterns. Stocking density harmed several aspects of welfare. Moreover, HSD induces tissue lipid peroxidation in birds, which increases MDA, reduces serum SOD, GPx, and CAT activity [9,11], and increases the ratio of H/L, the level of corticosterone hormone, and stress indicators in the blood [14]. All these physiological changes are reflected in the birds, as they cannot exhibit their standard behavioural patterns, and it also adversely affects the welfare of the birds.

In comparison, curcumin supplementation of the HSD group shows a significant increase in the welfare of birds as it caused a marked increase in ingestive behaviour and body care behaviour. In contrast, it caused a reduction in walking and standing behaviour, as the curcumin has a significant impact on the inhibition of oxidative stress in exposed chickens via increasing antioxidants enzyme activity and reducing the H/L ratio [44] and corticosterone hormonal level. Therefore, HSD increases stress and disruption to the HPA axis, which induces oxidative stress [45].

The variations in the haematological measurements were recorded in research studies on poultry to evaluate the physiological condition of the birds under stressful conditions [23]. Our results approve with earlier results of Goo et al. [46], who recorded a reduction in Hb, PCV, and RBCs count in the blood of chickens under the effect of HSD; which might be due to the deterioration of the overall health condition accompanied by the reduced feed intake and growth rate. In addition, with environmental or physiological stress (due to HSD), the temperature might rise, which may negatively impact birds' oxidative condition and the process of erythropoiesis [5]. In the current study, the HSD did not alter the count of WBCs, which agrees with the previous findings [47]. Curcumin treatment improved all the haematological parameters, with concurrent improvement in the antioxidant defence system of broiler under HSD. As a result, the erythropoiesis process is usually resumed [21].

In the current study, a significant impact of SD on other inflammatory markers such as H/L ratio and ESR was recorded. Comparable findings were reported in the ratio of H/L and ESR amongst various SD [14], which might indicate a stressful condition inside the broilers and might be the outcome of a mixture of environmental factors and genotypes. The stressors have rapidly decreased or may be due to the anti-inflammatory impact of curcumin [33]. Conversely, the ratio of H/L and ESR was reduced under curcumin treatment.

The immune system is highly impactful to the animal's health status. The immune system primarily comprises cell-mediated immunity and humoral immunity. The current study's marked reduction in serum globulin titers (IgG, IgA, and IgM), suggesting that chickens raised at HSD might be immunocompromised. Likewise, Li et al. [13] recorded that HSD inhibited humoral immunity and enhanced the rate of mortality. Furthermore, the IgG, IgA, and IgM level significantly improved with curcumin supplementation with HSD. Those findings propose that SD might disturb the humoral immune response, particularly curcumin, preventing HSD-mediated Ig synthesis deficiency. In the existing work, HSD enhanced the value of pro-inflammatory cytokine IL-6, IL-2, and TNF-a, which stimulated an inflammatory reaction and therefore was not accepted by the birds.

In an earlier experiment, which enhanced the administration of dietary curcumin, a decrease was reported in IL-6 value in broiler's blood under the stressful condition of high environmental temperature [22]. The pro-inflammatory cytokines increase with any inflammatory response, encouraging acute-phase protein production, i.e., C-reactive protein, which helps restore homeostasis [48]. In addition, variations occurred in quantities of distinct serum proteins, i.e., an enhancement in fibrinogen and globulins with a reduction in albumins, leading to faster ESR and increased H/L ratio. Subsequently, the administration of curcumin to the diet of broilers under HSD reduced IL-2, IL-6, and TNF-a, so the H/L ratio and ESR value decreased additionally, indicating the potential anti-inflammatory properties of curcumin against HSD [23].

It is believed that during normal physiological circumstances, equilibrium between pro-oxidant production and antioxidant defences is essential in living animals. Disproportioning the compensation raises reactive oxygen species (ROS) and persuades lipid peroxidation. An enhanced value of ROS will further increase the oxidation state and harm the component inside the cell [49]. Consequently, the body is prepared to succeed in the oxidation stress by manufacturing antioxidant enzymes to restore the physiological systems. As GPx, SOD, and CAT, these enzymes play a prominent role in antioxidant defence systems [21]. Our findings relate with findings of Li, Zhang et al. [15] and Selvam, et al. [14], who revealed that HSD caused the initiation of the oxidative stress status in birds under the impact of insufficient floor space, reduced air-flow, and the over-crowding of the birds, as was evident by the higher MDA and lower activates of GPx, SOD, and CAT in the bird's serum. A lipid peroxidation key end product, MDA, is continuously measured as an oxidative stress biomarker [49]. In the present experiment, the MDA level in the HSD bird's serum was markedly increased compared to that in MSD birds. This outcome might be owed to over-gathering enhancing the competition between the birds and leading to stress, thus causing elevated lipid peroxidation. Likewise, Li et al. [13] revealed that over-crowding increased oxidative damage and higher MDA production. So, it might be proposed that higher oxidative stress might result from reduced antioxidant enzyme activities with an HSD.

Curcumin can inhibit oxidative stress by improving the activities of antioxidants enzymes containing GPx, CAT, and SOD [43], which was also recorded in the current study in a dose-dependent manner. Consequently, a decrease in lipid peroxidation becomes more vital for keeping the regular production of the animal. Parallel to our results, Salah et al. [21] stated that the administration of curcumin in the ration decreased the TBA (thiobarbituric acid) indicator and enhanced the activities of antioxidant enzymes in chickens relative to those fed a standard ration. The presence of phenolic groups in curcumin construction has an essential role in inhibiting oxidative stress. Those groups can eliminate the superoxide ion, hydroxyl free radicals, and nitric oxides [22]. Akbarian et al. [50] recorded the HSD impact on enhanced activities of AST/ALT in broilers. Higher activities of AST/ALT might indicate a progressive injury to the liver cells, followed by enhanced free radicals, leading to liver lipid peroxidation and injury to various organs [5]. On the contrary, dietary supplementation of the HSD group with curcumin decreased activities of ALT and AST in birds, which indicates that the ability of curcumin to scavenge and neutralize free radicals can shield the liver cells in front of any free radical attack [23]. Consequently, reduced activities of liver enzymes in the current experiment might indicate enhanced functions of the liver succeeding curcumin antioxidant impacts, in order to inhibit any damage to the liver cells by free radicals.

A non-significant variation in blood concentrations of total protein and albumin was revealed in our study; on the contrary, there was a marked enhancement in the whole cholesterol level with HSD. The effect of HSD on higher serum cholesterol in the existing experiment agreed with Wang et al. [51]. They recorded that plasma cholesterol levels enhance with higher stress conditions in broilers. We detected that administration of curcumin to diet markedly reduced the total cholesterol in the blood of broilers relative to those fed a regular diet. Other findings on the impact of curcumin on total cholesterol were recorded by others and a reduction in total cholesterol and LDL in Sprague-Dawley rats [52]. HSD negatively impacted blood lipids, which could be consistent with higher corticosterone release in blood circulation. It is a fact that elevated corticosterone levels in the blood stimulate lipolysis and gluconeogenesis [53], clarifying the increased total cholesterol concentrations.

High corticosterone (catabolic hormone) values can impact the growth of bone and markedly decrease the intake of food and body weight gain in poultry [30]. Serum corticosterone concentrations were increased with HSD birds in the present work. Those findings were comparable to Li et al. [13], who revealed that corticosterone values at 42 D were elevated with HSD (15, 20, 25 per m^2), which might indicate the enhanced stressful condition of broilers. Several types of research have recorded that stressful situations decreased the serum levels of T4 and T3. The thyroid hormones impact nearly every physiological procedure in the organism and are considered essential to improving broilers' growth [53]. Nevertheless, data on the impact of SD on thyroid hormones in HSD reared birds is limited. On the contrary, curcumin supplementation to the HSD-raised broilers seems to retain the hormonal profile of the birds to nearly their average physiological values, indicating an ameliorative effect of curcumin against the harmful impacts of HSD [16].

Muscular growth is controlled by various elements, including GHR, IGF-1, MSTN, and leptin. IGF-1 contributes to the growth mechanism, and numerous researchers confirmed that IGF-1 controls the growth of skeletal muscle, which is produced under the GH and GHR control [45]. IGF-1 stimulates skeletal muscle hypertrophy and inhibits muscles atrophy [54]. Leptin plays a critical role in regulating the growth of muscles, and it is produced by adipocytes to signal the brain to decrease the intake of food and improve the expenditure of energy. Hence, it harms growth [55]. MSTN is considered a crucial element for the growth and improvement of muscle, and it also performs as an adverse regulator of muscular development. Before birth, MSTN negatively impacts muscular enlargement, and MSTN inhibits muscle hypertrophy in broilers [55]. In existing work, expression values of mRNA of IGF-1 and GHR in the liver samples were markedly reduced compared to the other groups. However, the expression of mRNA values of MSTN and leptin at HSD was enhanced markedly. Those findings propose that HSD could impact muscular differentiation and hypertrophy by controlling the GHR, IGF-I, leptin, and MSTN gene expression. On the contrary, curcumin supplementation in the diet of HSD reared broilers up-regulated the growth-enhancing genes expression (GHR and IGF-1) while it down-regulated the expression of the growth-inhibitory genes (leptin and MSTN), which might indicate a growth-promoting effect of curcumin [55].

5. Conclusions

The supplementation of dietary curcumin had a positive impact on the growth performance. The welfare of the HSD stressed birds, through modulation of the growth-related genes GHR and IGF-1 expression, in addition to the mitigative effect on the inflammatory cytokines (TNF-a, IL-2, IL-6) with increased antioxidant activity and the immunity status was accompanied by improvement of the behavioural performance of the HSD stressed birds Therefore, our study exposed the importance of curcumin supplementation in the poultry diet to relieve the stressful conditions and increase the birds' growth performance and immune status.

Author Contributions: Conceptualization, M.H.H., S.E.E.-K., S.S., M.S., Y.S.E.-S., H.I.G. and M.A.A.; data curation, M.H.H., S.E.E.-K., S.S., M.S. and Y.S.E.-S.; formal analysis, M.H.H., S.E.E.-K., B.A., M.S. and Y.S.E.-S.; funding acquisition, M.H.H.,. S.E.E.-K., H.I.G. and M.A.A.; investigation, M.H.H., S.E.E.-K., S.S., M.S. and Y.S.E.-S.; methodology, M.H.H., S.E.E.-K., B.A., S.S., M.S. and Y.S.E.-S.; supervision, M.H.H., S.E.E.-K., B.A., S.S., M.S. and Y.S.E.-S.; writing—original draft, M.H.H., S.E.E.-K., B.A., S.S., M.S. and Y.S.E.-S. All authors have read and agreed to the published version of the manuscript.

Funding: This study was supported by Taif University Researchers Supporting Project number (TURSP-2020/92), Taif University, Taif, Saudi Arabia.

Institutional Review Board Statement: The current research protocol was approved by the "Resident Commission of Ethics for the Use and Care of Lab Animals at Alexandria University" (2021/013/28). All efforts were made to minimize animal suffering.

Informed Consent Statement: Not applicable.

Data Availability Statement: Data are available upon request.

Acknowledgments: We thank Taif University Researchers Supporting Project number (TURSP-2020/92), Taif University, Saudi Arabia.

Conflicts of Interest: The authors declare no conflict of interest.

References

1. Petek, M.; Çİbİk, R.; Yİldİz, H.; Sonat, F.A.; Gezen, S.S.; Orman, A.; Aydİn, C. The influence of different lighting programs, stocking densities and litter amounts on the welfare and productivity traits of a commercial broiler line. *Vet. Zootech.* **2010**, *51*, 36–43.
2. Pulina, G.; Francesconi, A.H.D.; Stefanon, B.; Sevi, A.; Calamari, L.; Lacetera, N.; Dell'Orto, V.; Pilla, F.; Ajmone Marsan, P.; Mele, M. Sustainable ruminant production to help feed the planet. *Ital. J. Anim. Sci.* **2017**, *16*, 140–171. [CrossRef]
3. Sejian, V.; Silpa, M.V.; Reshma Nair, M.R.; Devaraj, C.; Krishnan, G.; Bagath, M.; Chauhan, S.S.; Suganthi, R.U.; Fonseca, V.F.; König, S. Heat stress and goat welfare: Adaptation and production considerations. *Animals* **2021**, *11*, 1021. [CrossRef] [PubMed]
4. Lara, L.J.; Rostagno, M.H. Impact of heat stress on poultry production. *Animals* **2013**, *3*, 356–369. [CrossRef]
5. Goo, D.; Kim, J.H.; Park, G.H.; Delos Reyes, J.B.; Kil, D.Y. Effect of heat stress and stocking density on growth performance, breast meat quality, and intestinal barrier function in broiler chickens. *Animals* **2019**, *9*, 107. [CrossRef]
6. Feddes, J.J.; Emmanuel, E.J.; Zuidhoft, M.J. Broiler performance, body weight variance, feed and water intake, and carcass quality at different stocking densities. *Poult. Sci.* **2002**, *81*, 774–779. [CrossRef]
7. Škrbić, Z.; Pavlovski, Z.; Lukić, M.; Perić, L.; Milošević, N. The effect of stocking density on certain broiler welfare parameters. *Biotechnol. Anim. Husb.* **2009**, *25*, 11–21. [CrossRef]
8. Abudabos, A.M.; Samara, E.M.; Hussein, E.O.; Al-Ghadi, M.a.Q.; Al-Atiyat, R.M. Impacts of stocking density on the performance and welfare of broiler chickens. *Ital. J. Anim. Sci.* **2013**, *12*, e11. [CrossRef]

9. Beloor, J.; Kang, H.; Kim, Y.; Subramani, V.; Jang, I.; Sohn, S.; Moon, Y.S. The effect of stocking density on stress related genes and telomeric length in broiler chickens. *Asian-Australas. J. Anim. Sci.* **2010**, *23*, 437–443. [CrossRef]
10. Nager, R.; Monaghan, P.; Griffiths, R.; Houston, D.; Dawson, R. Experimental demonstration that offspring sex ratio varies with maternal condition. *Proc. Natl. Acad. Sci. USA* **1999**, *96*, 570–573. [CrossRef]
11. Griffiths, R. Sex-biased mortality in the lesser black-backed gull Larus fuscus during the nestling stage. *Ibis* **1992**, *134*, 237–244. [CrossRef]
12. Kamel, N.; Hady, M.; Ragaa, N.; Mohamed, F. Effect of nucleotides on growth performance, gut health, and some immunological parameters of broiler chicken exposed to high stocking density. *Livest. Sci.* **2021**, *253*, 104703. [CrossRef]
13. Li, W.; Wei, F.; Xu, B.; Sun, Q.; Deng, W.; Ma, H.; Bai, J.; Li, S. Effect of stocking density and alpha-lipoic acid on the growth performance, physiological and oxidative stress and immune response of broilers. *Asian-Australas. J. Anim. Sci.* **2019**, *32*, 1914–1922. [CrossRef] [PubMed]
14. Selvam, R.; Saravanakumar, M.; Suresh, S.; Sureshbabu, G.; Sasikumar, M.; Prashanth, D. Effect of vitamin E supplementation and high stocking density on the performance and stress parameters of broilers. *Rev. Bras. Ciência Avícola* **2017**, *19*, 587–594. [CrossRef]
15. Li, X.M.; Zhang, M.H.; Liu, S.M.; Feng, J.H.; Ma, D.D.; Liu, Q.X.; Zhou, Y.; Wang, X.J.; Xing, S. Effects of stocking density on growth performance, growth regulatory factors, and endocrine hormones in broilers under appropriate environments. *Poult. Sci.* **2019**, *98*, 6611–6617. [CrossRef]
16. Qasem, M.A.A.; Alhajj, M.S.; Ger El Nabi, A.R.; Al-Mufarrej, S.I. Effect of turmeric powder as a dietary supplement on performance indicators and immune responses in broiler chickens. *J. Anim. Vet. Adv.* **2015**, *14*, 30–35.
17. Hady, M.M.; Zaki, M.M.; EL-Ghany, W.A.; Korany, M.S.R. Assessment of the broilers performance, gut healthiness and carcass characteristics in response to dietary inclusion of dried coriander, turmeric and thyme. *Int. J. Environ. Agric. Res.* **2016**, *2*, 153–159.
18. Valenzuela-Grijalva, N.V.; Pinelli-Saavedra, A.; Muhlia-Almazan, A.; Dominguez-Diaz, D.; Gonzalez-Rios, H. Dietary inclusion effects of phytochemicals as growth promoters in animal production. *J. Anim. Sci. Technol.* **2017**, *59*, 8. [CrossRef]
19. Prasad, S.; Tyagi, A.K.; Aggarwal, B.B. Recent developments in delivery, bioavailability, absorption and metabolism of curcumin: The golden pigment from golden spice. *Cancer Res. Treat.* **2014**, *46*, 2–18. [CrossRef]
20. Al-Mashhadani, H.E. Effect of different levels of turmeric (*Curcuma longa*) supplementation on broiler performance, carcass characteristic and bacterial count. *Egypt. Poult. Sci. J.* **2015**, *35*, 25–39.
21. Salah, A.S.; Mahmoud, M.A.; Ahmed-Farid, O.A.; El-Tarabany, M.S. Effects of dietary curcumin and acetylsalicylic acid supplements on performance, muscle amino acid and fatty acid profiles, antioxidant biomarkers and blood chemistry of heat-stressed broiler chickens. *J. Therm. Biol.* **2019**, *84*, 259–265. [CrossRef] [PubMed]
22. Nawab, A.; Li, G.; Liu, W.; Lan, R.; Wu, J.; Zhao, Y.; Kang, K.; Kieser, B.; Sun, C.; Tang, S.; et al. Effect of dietary curcumin on the antioxidant status of laying hens under high-temperature condition. *J. Therm. Biol.* **2019**, *86*, 102449. [CrossRef] [PubMed]
23. Laganá, C.; Saldanha, E.S.P.B.; Sartori, J.R.; Turco, P.H.N.; Gonzales, E.; Luciano, R.L.; Zanatta, G.; Fascina, V.B. Turmeric on poultry production: A Review. *Agric. Sci.* **2019**, *10*, 1592–1601. [CrossRef]
24. Moniruzzaman, M.; Min, T. Curcumin, curcumin nanoparticles and curcumin nanospheres: A review on their pharmacodynamics based on monogastric farm animal, poultry and fish nutrition. *Pharmaceutics* **2020**, *12*, 447. [CrossRef]
25. Nawab, A.; Tang, S.; Li, G.; An, L.; Wu, J.; Liu, W.; Xiao, M. Dietary curcumin supplementation effects on blood immunological profile and liver enzymatic activity of laying hens after exposure to high temperature conditions. *J. Therm. Biol.* **2020**, *90*, 102573. [CrossRef]
26. Paw, M.; Gogoi, R.; Sarma, N.; Pandey, S.K.; Borah, A.; Begum, T.; Lal, M. Study of anti-oxidant, anti-inflammatory, genotoxicity, and antimicrobial activities and analysis of different constituents found in rhizome essential oil of *Curcuma caesia* Roxb., collected from north east India. *Curr. Pharm. Biotechnol.* **2020**, *21*, 403–413. [CrossRef]
27. Ruan, D.; Wang, W.; Lin, C.; Fouad, A.; Chen, W.; Xia, W.; Wang, S.; Luo, X.; Zhang, W.; Yan, S. Effects of curcumin on performance, antioxidation, intestinal barrier and mitochondrial function in ducks fed corn contaminated with ochratoxin A. *Animal* **2019**, *13*, 42–52. [CrossRef]
28. Rajput, N.; Muhammad, N.; Yan, R.; Zhong, X.; Wang, T. Effect of dietary supplementation of curcumin on growth performance, intestinal morphology and nutrients utilization of broiler chicks. *J. Poult. Sci.* **2013**, *50*, 44–52. [CrossRef]
29. Saleh, A.A.; Alhotan, R.A.; Alharthi, A.S.; Nassef, E.; Kassab, M.A.; Farrag, F.A.; Hendam, B.M.; Abumnadour, M.; Shukry, M. Insight view on the role of in ovo feeding of clenbuterol on hatched chicks: Hatchability, growth efficiency, serum metabolic profile, muscle, and lipid-related markers. *Animals* **2021**, *11*, 2429. [CrossRef]
30. El-Kazaz, S.E.; Hafez, M.H. Evaluation of copper nanoparticles and copper sulfate effect on immune status, behavior, and productive performance of broilers. *J. Adv. Vet. Anim. Res.* **2020**, *7*, 16–25. [CrossRef]
31. Reiter, K.; Bessei, W. The behaviour of broilers in response to group size and stocking density. *Arch. Geflügelk* **1999**, *64*, 93–98.
32. Schalm, O.W.; Jain, N.C.; Carroll, E.J. *Veterinary Hematology*; Lea & Febiger: Columbus, OH, USA, 1975.
33. Reitman, S.; Frankel, S. A colorimetric method for the determination of serum glutamic oxalacetic and glutamic pyruvic transaminases. *Am. J. Clin. Pathol.* **1957**, *28*, 56–63. [CrossRef] [PubMed]
34. Kirrella, A.A.; Abdo, S.E.; El-Naggar, K.; Soliman, M.M.; Aboelenin, S.M.; Dawood, M.A.O.; Saleh, A.A. Use of corn silk meal in broiler diet: Effect on growth performance, blood biochemistry, immunological responses, and growth-related gene expression. *Animals* **2021**, *11*, 1170. [CrossRef] [PubMed]

35. SAS Institute. *Using JMP Student Edition for Windows and Macintosh: The User's Guide to Statistics with JMP Student Edition*; SAS Institute: Cary, NC, USA, 2009.
36. Levene, H. Robust tests for equality of variances. In *Contributions to Probability and Statistics. Essays in Honor of Harold Hotelling*; Stanford University Press: Palo Alto, CA, USA, 1961; pp. 279–292.
37. Ran, L.; Wang, X.; Mi, A.; Liu, Y.; Wu, J.; Wang, H.; Guo, M.; Sun, J.; Liu, B.; Li, Y. Loss of adipose growth hormone receptor in mice enhances local fatty acid trapping and impairs brown adipose tissue thermogenesis. *Iscience* **2019**, *16*, 106–121. [CrossRef]
38. Farkašová, H.; Hron, T.; Pačes, J.; Pajer, P.; Elleder, D. Identification of a GC-rich leptin gene in chicken. *Agri. Gene* **2016**, *1*, 88–92. [CrossRef]
39. Yadav, S.; Teng, P.-Y.; Dos Santos, T.S.; Gould, R.L.; Craig, S.W.; Fuller, A.L.; Pazdro, R.; Kim, W.K. The effects of different doses of curcumin compound on growth performance, antioxidant status, and gut health of broiler chickens challenged with *Eimeria* species. *Poult. Sci.* **2020**, *99*, 5936–5945. [CrossRef]
40. Sokół, R.; Koziatek-Sadłowska, S.; Michalczyk, M. The influence of *Dermanyssus gallinae* and different lighting regimens on selected blood proteins, corticosterone levels and egg production in layer hens. *Vet. Res. Commun.* **2019**, *43*, 31–36. [CrossRef]
41. Fouad, M.A.; Abdel Razek, A.H.; Badawy, E.M. Broilers welfare and economics under two management alternatives on commercial scale. *Int. J. Poult. Sci.* **2008**, *7*, 1167–1173. [CrossRef]
42. Hall, A.L. The effect of stocking density on the welfare and behaviour of broiler chickens reared commercially. *Anim. Welf.* **2001**, *10*, 23–40.
43. Thomas, D.G.; Son, J.-H.; Ravindran, V.; Thomas, D.V. The Effect of stocking density on the behaviour of broiler chickens. *Korean J. Poult. Sci.* **2011**, *38*, 1–4. [CrossRef]
44. Pimson, C.; Bakban, P.; Suwanrat, S.; Chanutsa, N. The effect of curcumin on growth performance, blood biochemistry and antioxidant activities in broiler chickens. *Vet. Integr. Sci.* **2018**, *16*, 95–107.
45. Lin, H.; Decuypere, E.; Buyse, J. Oxidative stress induced by corticosterone administration in broiler chickens (*Gallus gallus domesticus*): 1. Chronic exposure. *Comp. Biochem. Physiol. Part B Biochem. Mol. Biol.* **2004**, *139*, 737–744. [CrossRef] [PubMed]
46. Goo, D.; Kim, J.H.; Park, G.H.; Delos Reyes, J.B.; Kil, D.Y. Effect of stocking density and dietary tryptophan on growth performance and intestinal barrier function in broiler chickens. *Poult. Sci.* **2019**, *98*, 4504–4508. [CrossRef] [PubMed]
47. El-Gogary, M.; Abo El-Maaty, H. Impact of zinc supplementation and stocking density on performance, physiological and immune responses in broiler chickens. *J. Anim. Poult. Prod.* **2020**, *11*, 95–102. [CrossRef]
48. Polinska, B.; Matowicka-Karna, J.; Kemona, H. The cytokines in inflammatory bowel disease. *Postepy Hig. Med. Dosw. (Online)* **2009**, *63*, 389–394.
49. Rahmani, M.; Golian, A.; Kermanshahi, H.; Reza Bassami, M. Effects of curcumin or nanocurcumin on blood biochemical parameters, intestinal morphology and microbial population of broiler chickens reared under normal and cold stress conditions. *J. Appl. Anim. Res.* **2017**, *46*, 200–209. [CrossRef]
50. Akbarian, A.; Golian, A.; Kermanshahi, H.; Gilani, A.; Moradi, S. Influence of turmeric rhizome and black pepper on blood constituents and performance of broiler chickens. *Afr. J. Biotechnol.* **2012**, *11*, 8606–8611.
51. Wang, B.; Min, Z.; Yuan, J.; Zhang, B.; Guo, Y. Effects of dietary tryptophan and stocking density on the performance, meat quality, and metabolic status of broilers. *J. Anim. Sci. Biotechnol.* **2014**, *5*, 44. [CrossRef]
52. Kim, M.; Kim, Y. Hypocholesterolemic effects of curcumin via up-regulation of cholesterol 7a-hydroxylase in rats fed a high fat diet. *Nutr. Res. Pract.* **2010**, *4*, 191–195. [CrossRef]
53. Xiao, Y.; Wu, C.; Li, K.; Gui, G.; Zhang, G.; Yang, H. Association of growth rate with hormone levels and myogenic gene expression profile in broilers. *J. Anim. Sci. Biotechnol.* **2017**, *8*, 43. [CrossRef]
54. Wen, C.; Chen, Y.; Wu, P.; Wang, T.; Zhou, Y. MSTN, mTOR and FoxO4 are involved in the enhancement of breast muscle growth by methionine in broilers with lower hatching weight. *PLoS ONE* **2014**, *9*, e114236. [CrossRef] [PubMed]
55. Seroussi, E.; Cinnamon, Y.; Yosefi, S.; Genin, O.; Smith, J.G.; Rafati, N.; Bornelov, S.; Andersson, L.; Friedman-Einat, M. Identification of the long-sought leptin in chicken and duck: Expression pattern of the highly GC-rich avian leptin fits an autocrine/paracrine rather than endocrine function. *Endocrinology* **2016**, *157*, 737–751. [CrossRef] [PubMed]

animals

MDPI

Article

Effect of Ginger Powder on Production Performance, Antioxidant Status, Hematological Parameters, Digestibility, and Plasma Cholesterol Content in Broiler Chickens

Hanan Al-Khalaifah *, Afaf Al-Nasser, Tahani Al-Surrayai, Hanan Sultan, Dalal Al-Attal, Rawan Al-Kandari, Haya Al-Saleem, Aisha Al-Holi and Fatma Dashti

Environment and Life Sciences Research Center, Kuwait Institute for Scientific Research, P.O. Box 24885, Safat 13109, Kuwait; anasser@kisr.edu.kw (A.A.-N.); tsuraia@kisr.edu.kw (T.A.-S.); hsultan@kisr.edu.kw (H.S.); dattal@kisr.edu.kw (D.A.-A.); rkandari@kisr.edu.kw (R.A.-K.); hsaleem@kisr.edu.kw (H.A.-S.); aholi@kisr.edu.kw (A.A.-H.); fadashti@kisr.edu.kw (F.D.)

* Correspondence: hkhalifa@kisr.edu.kw

Simple Summary: Chicken meat is a popular food item all over the wrld. Improving the nutrition of broilers is important for producing high-quality broiler meat. The inclusion of natural effective ingredients, such as ginger, in the diet of broilers did not adversely affect the palatability of the diet, nor did it cause anemia in the broilers. Rather, ginger enhanced the oxidative status and growth rate of broiler chickens.

Abstract: The effect of dietary ginger powder on the production performance, digestibility, hematological parameters, antioxidant status, dietary oxidation stability, and plasma cholesterol content of broiler chickens was investigated. Ginger powder was included in the diet at 0, 5, 10, or 15 g/kg. Total antioxidant capacity and malondialdehyde in sera samples, superoxide dismutase activity, glutathione peroxidase, catalase, and malondialdehyde in liver samples, and the peroxide value and acid value of the stored diets were evaluated. The results showed that ginger inclusion significantly improved antioxidation indices in broiler sera and liver. Total body weight gain in ginger-supplemented birds was higher than that of control birds ($p < 0.048$). Supplementing the broiler chickens with ginger powder reduced total feed consumption ($p < 0.031$). White blood cell counts and the percentage of heterophils in the blood were increased in birds that received ginger supplementation ($p < 0.001$). The inclusion of ginger in the diet improved dry matter digestibility, crude protein utilization, crude fiber utilization, and ether extract utilization. In addition, blood cholesterol, triglyceride, and very low-density lipoprotein levels were decreased ($p < 0.001$), and high-density lipoprotein and levels were increased, following the inclusion of ginger in the diet ($p < 0.001$).

Keywords: antioxidant; broilers; ginger; hematological parameters

check for updates

Citation: Al-Khalaifah, H.; Al-Nasser, A.; Al-Surrayai, T.; Sultan, H.; Al-Attal, D.; Al-Kandari, R.; Al-Saleem, H.; Al-Holi, A.; Dashti, F. Effect of Ginger Powder on Production Performance, Antioxidant Status, Hematological Parameters, Digestibility, and Plasma Cholesterol Content in Broiler Chickens. *Animals* **2022**, *12*, 901. https://doi.org/10.3390/ani12070901

Academic Editor: Ilias Giannenas

Received: 24 January 2022
Accepted: 14 March 2022
Published: 31 March 2022

1. Introduction

There is interest in elevating the production performance of broiler chickens using effective nutritional additives in the feed rations, especially after the COVID-19 crisis. Medicinal plants are used as natural feed additives in poultry diets to enhance the performance, anti-oxidative status, and immune response of chickens [1–8]. One of these additives is ginger powder. Ginger is the rhizome of the plant *Zingiber officinale*. It belongs to the family *Zingibeaceae*, which includes aromatic herbs with fleshy, tuberous or non-tuberous rhizomes, that often have tuber-bearing roots [9]. It has long served as a popular culinary and traditional medicinal herb. Ginger contains several effective compounds, such as gingerol and gingerdione that exert strong antioxidant activity. In addition, it has antibacterial properties and is immunomodulatory in laboratory animals [10–13]. Plant-derived additives used in animal feed to improve production performance are known as phytogenic feed additives, and ginger is one such additive [14]. Ginger powder has lipid-reducing effects and can also be used as a growth promoter. When included in chicken feed, it has properties similar to those of antibiotics. These natural feed additives lower enteric pathogen microbial loads and improve nutrient digestion and absorption, which improve poultry production and broiler performance [15].

Antioxidants can impact the health status of poultry [16]. The inclusion of ginger in the diet at 5–6 g/kg is thought to increase total protein and lower cholesterol concentration. Studies have also found that ginger in chicken feed enhances immunity against Newcastle disease and bacterial bursal infections [17]. Because of its antioxidant properties, and ability to enhance immune function and inflammatory responses, ginger can improve both chicken production performance and the immune system [18]. Furthermore, Sahoo, Mishra [19] reported that feed rations supplemented with ginger, either alone or with 1% turmeric, significantly enhanced the antioxidative status and gut health of broiler chickens. An, Liu [18] investigated the effect of ginger supplementation on antioxidative indices in broiler chickens. The results of their study revealed that neither the activity of glutathione peroxidase nor total antioxidant capacity were affected by ginger extract supplementation, but plasma dismutase activity and malondialdehyde (MDA) content were significantly decreased. In addition, the antioxidant stability of the feed ration was increased.

The effects of ginger have also been investigated in laying hens. For example, Wen, Gu [20] reported increased superoxide dimutase (TSOD) activity, and decreased yolk MDA and cholesterol content in hens supplemented with ginger, compared to control hens. These findings were consistent with those of Zhao, Yang [21], who found that layers fed a diet rich in ginger powder had a reduced concentration of MDA and increased TSOD activity in the yolk. In addition, Yang, Ding [22] observed increased blood antioxidant enzymes and improved egg quality in ginger root-fed laying hens.

In addition to its effect as an antioxidant, several research studies have reported that poultry diets supplemented with ginger powder have positive effects on broiler performance. As a natural feed additive, ginger may have great benefit and value in poultry nutrition—especially for broilers—due to its antibacterial, anti-inflammatory, antioxidant, antiseptic, antiparasitic, and immunomodulatory properties [23]. Ginger is a natural plant that can be used as a phytobiotic to improve the performance of broilers. This improved performance may be attributed to two types of digestive enzymes found in ginger—protease and lipase—which are part of the plant's natural protective mechanisms [24]. Diets enriched with ginger may have the potential to improve production performance and modulate the biological properties of the blood in broiler chickens.

Sa'aci, Alabi [25] studied the effect of aqueous ginger extract (AGE) on growth, nutrient digestibility, and the economy of Marshal broiler chicks. The AGE supplementation of the diet had no effect on dry matter, crude protein, ether extract, or nitrogen-free extract, but crude fiber digestibility was significantly affected.

Shewita and Taha [17] reported that ethanolic extract of ginger significantly lowered serum total cholesterol and triglyceride levels and increased high-density lipoprotein (HDL) cholesterol levels, preventing tissue damage due to lipid peroxidation. It also showed lipid-lowering activity. In their study on broiler chicks, the authors fed the chicks diets supplemented with 2, 4, and 6 g/kg ginger powder and observed that serum very low-density lipoprotein (VLDL) and triglyceride levels were reduced significantly in all ginger-supplemented groups. Another study used ginger (5, 10, and 15 g/kg ginger powder) and thymol (200 and 400 mg/kg) as feed supplements in the diet of broiler chicks, which led to significantly decreased levels of serum total cholesterol and triacylglycerol [26].

The aim of this study was to investigate the effects of ginger on the antioxidant status, production performance, and hematological parameters of broiler chickens fed a ginger-enriched diet, as well as the oxidation stability of ginger-supplemented feed. The hypothesis was that ginger powder would improve the aforementioned parameters but would have dose-dependent effects. Although previous studies have investigated the effect of ginger on productive performance parameters in broiler chickens, relatively limited data exist in the literature on the direct effects of ginger on the antioxidative status of blood and liver tissues in these birds and the effect on the oxidative stability of the stored diets at different ginger levels.

2. Materials and Methods

2.1. Chickens, Experimental Design, and Diet

This research study was approved by the department committee of the Environment and Life Sciences Research Center in Kuwait Institute for Scientific Research under project No. FA157K (2017). These procedures and protocols followed the official animal welfare guidelines and regulations encoded with reference No. PMO/PV/RP/032/2017. This protocol recommends humane treatment of experimental animals with no pain, stress, or harm. The vitality rate was 100% and no abnormal signs were observed during the experiment.

Fresh ginger (*Zingiber officinale*) roots were purchased from a reliable local supermarket, originated from India. The roots were washed, sliced, freeze-dried, and milled into a powder that was used in the broiler feed rations. This study used 1-day-old, male, Cobb 500 broiler chicks that were vaccinated against infectious bronchitis and Newcastle disease. Water and feed were provided ad libitum. Four experimental diets/treatments were used, with 0, 5, 10, or 15 g/kg ginger in the diet. For each treatment, 340 birds were randomly housed in a battery cage with five levels. Each level contained 17 birds, for a total of 85 birds in the battery. This density provided a space of 0.05 m^2/bird. Each level was considered as a replicate (total of 5 replicates). The broiler chicks were fed a corn/soy-based diet that met Cobb 500 guidelines [27]. The chicks received a starter diet from hatching until 7 days of age, a grower diet from 8 to 21 days of age, and a finisher diet from 22 to 35 days of age. All diets were prepared as needed. Diet formulation, as well as chemical analyses of the control diet and ginger supplement are shown in Table 1. The control birds received no ginger. The ambient temperature for the broilers was kept at 30 °C for 14 days, and then gradually reduced to 21 °C by 21 days. A proximate analysis of the ginger was performed for crude protein, ash, dry matter, and crude fiber.

Table 1. Formulation and chemical analyses of the basal broiler diets.

Ingredient%	Starter	Grower	Finisher
	0–7 Day	8–21 Day	22–35 Day
Corn	55.6	57.6	61.2
Soyabean Meal	39.4	35.6	32.25
Soya oil	1.35	3.2	3.35
Limestone	1.45	1.45	1.3
DiCalcium Phosphate	1.4	1.4	1.2
Salt	0.21	0.21	0.2
L-Lysine	0.12	0.12	0.1
DL-Methionine	0.27	0.27	0.26
Vitamin-Mineral Premix *	0.2	0.2	0.2
Total%	100	100	100
Nutrient Composition Chemical Analysis			
Crude Protein (CP) (%)	22	20	18
Metabolizable Energy (kcal/kg)	2975	3025	3150
Fat (g/kg)	3.86	5.75	6
Calculated analysis			
Calcium (g\kg)	0.9	0.84	0.76
Phosphorus (g\kg)	0.45	0.42	0.38

Table 1. *Cont.*

Ingredient%	Starter	Grower	Finisher
	0–7 Day	8–21 Day	22–35 Day
Sodium (g\kg)	0.19	0.19	0.19
Lysine (g\kg CP)	1.22	1.12	0.97
Methionine (g\kg CP)	0.46	0.45	0.4
Choline (mg\kg)	1420	1329	1260
Proximate analysis of the ginger used			
Crude protein 6.0			
Ash 3.0			
Crude fiber 3.0			
Dry matter 91.2			

* Supplied per kg of premix: trans-retinol (A), 12,500,000 IU; cholecalciferol (D3), 500,000 IU; α-tocopherol acetate (E), 75,000 mg; thiamine (B1), 4500 mg; riboflavin (B2), 8000 mg; pyridoxine (B6), 5000 mg; vitamin B12, 22,000 mg; pantothenic acid, 20,000 mg; folic acid, 2000 mg; biotin, 200,000 µg; Fe, 100,000 mg; Co, 250 mg; Mn, 100 mg; Cu, 10,000 mg; Zn, 80,000 mg; I, 1000 mg; Se, 300 mg; Mo, 0.5 mg; Ca, 7.7%; P, 0.01%; Na, 0.18%.

2.2. Sample Collection

At 5 weeks of age, blood samples were collected from the branchial vein of birds in vacutainer tubes (K2EDTA). Blood samples were collected from five chickens from each treatment, and 8–10 mL of blood was collected in each tube. Meat tissue and liver samples were also collected from five chickens per treatment. Analysis was completed in triplicates.

2.3. Production Performance Parameters

Body weight and feed consumption were recorded; broiler chicks were weighed at hatch, at one week of age, and at the end of every two weeks afterward until 35 days [28]. The temperature and relative humidity were recorded daily and adjusted to minimize stress surroundings in the poultry house. Mortality was recorded daily.

2.4. Apparent Digestibility Coefficient

The apparent digestibility coefficient was evaluated, using a biological assay, as affected by the different levels of ginger powder in the experimental diets. Dry matter digestibility (DMD), crude protein utilization (CPU), crude fiber utilization (CFU), and ether extract utilization (EEU) were determined according to Horwitz [29]. The apparent digestibility coefficient was estimated and expressed as a percentage [30,31].

The experimental birds were individually raised in cages and fed ad libitum on diets with ginger solution for four days as the adjusting period. During this period, the excreta was discarded. Water was available ad libitum throughout the experimental period. After the adjusting period, birds were fasted for 24 h to empty all remaining contents in the alimentary canal. In one group, birds were continuously starved for 24 h to obtain data on endogenous nitrogen excretion. Birds in the other group were force-fed with the ginger solution at a dose rate of 60 g/day/bird for three consecutive days. The excreta was collected using clean, rigid trays placed under the clean wire cages in which the birds were housed. Droppings retained on the wire screen floor of the cage were also collected. The contamination of excreta with feathers, fed, scale, and vomit was avoided. The excreta of the first group was collected after the starving period for determination of endogenous nitrogen content. Excreta of the other group was individually collected every 6 h during the three subsequent days and for approximately 24 h after the birds were fed for three subsequent days. The excreta samples were individually weighed and sprayed with 2% boric acid solution to fix the nitrogen in the samples and stored in the freezer at 20 °C until analyses. Excreta was then dried in the freeze drier for 2 days, weighed, homogenized, and ground for the estimation of nutrients. Crude protein was detected using Macrokjeldahl method that involves the transformation of nitrogen into ammonium sulphate by acid digestion using boiling sulphuric acid. The ammonia was trapped by boric acid and was titrated using standard hydrochloric acid solution. The percentage of nitrogen was calculated from the percentage of crude protein. The crude fiber was determined using the Weende method and the Fiber Tec system. A moisture-free and fat-free sample was first digested with a weak acid solution, and then with the weak base solution. The digested residue was then collected in a filter crucible, dried and ignited. The loss of weight on ignition was the crude fiber. For crude fat determination, an organic solvent was used to extract the crude fat from a known weight of the sample. The dissolved fat was then recovered by the evaporation and condensation of the solvent. The fat extracted was a representative of the crude fat of the sample. The digestibility coefficients of the nutrients were calculated as follows:

$$\text{Digestibility (\%)} = (\text{NF} - \text{NE} + \text{NENC}) \times 100$$

where NF = nutrient in feed, NE = Nutrient in Excreta, and NENC = Nutrient in Excreta of negative control.

2.5. Antioxidant Status

Antioxidant status was investigated by measuring the antioxidant indices in the sera and livers of broiler chickens supplemented with different concentrations of ginger powder. The TSOD activity in the liver was measured using a Ransod kit from Randox Laboratories, UK, as described by Habibi, Sadeghi [32]. Liver glutathione peroxidase (Gpx) was indicated based on the protocol used by Paglia, Valentine [33]. Catalase (CAT) enzyme activity in the liver was determined using the method described by Aebi [34]. Sera MDA and total antioxidant capacity (TAC) were measured as described by Habibi, Sadeghi [32].

2.6. Oxidation Stability of the Feed Rations

Lipid peroxidation of the dietary feed rations, including the different concentrations of ginger powder, was determined by measuring the peroxide value (PV) and the acid value (AV) of the feed for a period of 50 days, starting from day 10. Measurements were recorded every 10 days thereafter, until day 50. The PV and the AV were measured using the methods described in AOCS [35] and Rao, Xiang [36], respectively.

2.7. Hematological Measurements

The samples were analyzed using a Cell-Dyn 3500 Hematology system (Abbott Laboratories, Abbott Park, IL, USA) to measure total and differential WBC and blood quality parameters, including red blood cells (RBC), hemoglobin (HGB), hematocrit (HCT), mean corpuscular volume (MCV), mean corpuscular hemoglobin (MCH), mean corpuscular hemoglobin concentration (MCHC), red cell distribution width (RDW), and platelet count (PLT).

2.8. Plasma Cholesterol Content

Blood cholesterol, triglyceride, and high- and low-density lipoprotein contents were measured calorimetrically using a commercial kit from Bioassay Technology Laboratory, China. Serum was first collected by centrifugation at 2000–3000 rpm for 20 min. All reagents and standard solutions were used at room temperature. In total, 50 μL of standard solution were added to a standard well, and 40 μL of sample were added to the sample wells. Then, the corresponding LDL and HDL antibodies were added to the sample wells. Streptavidin-HRP was added to both the sample and standard wells. The plate was then incubated for 60 min at 37 °C, and 50 μL of substrate solutions A and B were added to all wells. After incubation for 10 min at 30 °C in the dark, 50 μL of stop solution were added to each well. The optical density (OD) of each well was then determined.

2.9. Statistical Analysis

Four experimental diets/treatments were used. For each treatment, 340 birds were randomly housed in four multi-floor batteries, each of which had five levels. Each level contained 17 birds, for 85 birds in the battery. Each level was considered as a replicate (5 replicates per treatment). Overall differences among the dietary treatments were evaluated using one-way ANOVAs via the general linear model procedure in Minitab. Differences among treatments were considered statistically different at $p \leq 0.05$. Data were arcsine transformed before analysis to improve normality. Where significant differences occurred, pairwise Tukey post-hoc comparisons were made to identify significant differences between groups.

3. Results

3.1. Growth Performance

All broilers appeared to be healthy, and no significant mortality occurred during the experiment. Table 2 shows the effects of different levels of ginger powder on body weight, feed consumption, feed efficiency, and weekly body weight gain of broiler chickens. The results showed that addition of ginger powder at all levels improved the body weight of the broiler chickens at 5 weeks of age ($p < 0.001$). Feed consumption of broilers fed 0, 5, 10, or 15 g/kg of ginger powder is shown in Table 2. The results showed that supplementing broiler chickens with ginger powder significantly reduced feed consumption ($p < 0.031$). Results in Table 2 showed that there was no significant effect of ginger powder on the feed efficiency of broilers. Results in Table 2 showed that the total body weight gain of birds supplemented with ginger was significantly higher than that of control birds ($p < 0.048$).

Table 2. The effects of different levels of ginger powder on body weight, feed consumption, feed efficiency, and weekly body weight gain of broiler chickens.

	Treatment (Ginger g/kg)					
Age	**0**	**5**	**10**	**15**	**SEM**	***p*-Value**
Bodyweight (g)						
1 day	43	42	42	43	0.000	–
1 week	99	108	107	97	11.21	<0.213
2 week	280	287.5	287.9	283	8.98	<0.342
3 week	640 [a]	715.5 [b]	679.9 [b]	665 [b]	4.13	<0.001
4 week	1090 [a]	1205.5 [b]	1110.9 [b]	1120 [b]	15.53	<0.001
5 week	1680 [a]	1792.5 [b]	1690.9 [b]	1730 [b]	9.85	<0.001
Feed Consumption (g)						
1 week	72.7	67.4	66.2	67.8	1.64	<0.060
2 week	210	215	230	211	2.56	<0.341
3 week	620.9 [a]	614.2 [b]	590.5 [b]	580.8 [b]	12.21	<0.001
4 week	950	953	920	934	15.31	<0.215
5 week	1410.5 [a]	1281.5 [b]	1280.6 [b]	1354.1 [b]	36.72	<0.050
Feed Efficiency						
1 week	1.9	2.3	2.1	2.4	0.14	<0.118
2 week	1.09	1.12	1.12	1.15	0.98	<0.213
3 week	1.4	1.5	1.3	1.4	0.05	<0.409
4 week	1.11	1.13	1.13	1.2	0.18	<0.134
5 week	1.2	1.2	1.1	1.2	0.13	<0.079
Weekly Body Weight Gain (g)						
1 week	56 [a]	66 [b]	65 [b]	54.0 [a]	2.65	<0.025
2 week	181	179.5	180.9	186	6.84	<0.642
3 week	360 [a]	428 [b]	392 [b]	382.0 [b]	11.75	<0.001
4 week	450	490	431	455	23.12	<0.216
5 week	590	587	580	610	18.88	<0.06
Total Body Weight Gain (g)	1637	1750.5	1649	1687	9.45	<0.048
Total Feed Consumption (g)	3264	3131	3087	3148	12.10	<0.031

Means within rows with different letters are statistically different at $p \leq 0.05$, SEM = standard error of the mean, calculated by one-way analysis of variance (ANOVA) and the general linear model procedure of Minitab.

3.2. Apparent Digestibility Coefficient

The effects of different levels of ginger powder on nutrient digestibility are shown in Table 3. The results showed that there were significant ($p \leq 0.05$) differences among the dietary treatment groups for DMD, CPU, CFU, and EEU.

Table 3. The effects of different levels of ginger powder on nutrient utilization coefficients in broilers.

	Treatment (Ginger g/kg)					
	Control	**5**	**10**	**15**	**SEM**	***p*-Value**
DMD	58.6 ^{a}	73.9 ^{b}	71.1 ^{b}	77.9 ^{c}	2.00	<0.001
CPU	60.5 ^{a}	74.3 ^{b}	76.1 ^{b}	73.9 ^{b}	5.00	<0.001
CFU	64.8 ^{a}	73.9 ^{b}	74.2 ^{b}	73.5 ^{c}	3.80	<0.001
EEU	58.2 ^{a}	65.6 ^{b}	65.8 ^{b}	64.9 ^{b}	4.62	<0.050

All treatment groups received a soybean basal diet, means within rows with different letters are statistically different at $p \leq 0.05$, $n = 5$ Dry matter digestibility (DMD), crude protein utilization (CPU), crude fiber utilization (CFU), and ether extract utilization (EEU).

3.3. Antioxidative Indices

Table 4 shows the effect of ginger on serum and liver antioxidant indices in broiler chickens. The results in Table 4 show that broiler chickens fed diets rich in ginger powder had increased serum TAC ($p < 0.021$). However, MDA concentrations were decreased with ginger supplementation ($p < 0.038$). For the liver parameters, the results of Table 4 show that broiler chickens fed a diet supplemented with 5 g/kg ginger did not have changes in liver TSOD. However, broilers fed a diet supplemented with 10 or 15 g/kg of ginger had increased liver TSOD compared to the control group and the 5 g/kg ginger-supplemented group ($p < 0.049$). Consumption of feed rations supplemented with ginger did not affect the liver GPX level ($p < 0.949$). The concentration of CAT was increased by ginger supplementation, compared to the control group ($p < 0.050$). In contrast, broiler chickens that consumed different concentrations of ginger had decreased liver MDA ($p < 0.020$) compared to the control group.

Table 4. Antioxidant parameters in serum and livers of broiler chickens fed various concentrations of ginger powder.

	Treatment (Ginger g/kg)					
Parameter	**0**	**5**	**10**	**15**	**SEM**	***p*-Value**
Serum						
TAC (mmol/L)	0.6 ^{a}	0.9 ^{b}	1.1 ^{b}	0.9 ^{b}	0.03	<0.021
MDA (nmol/mL)	4.1 ^{a}	2.9	2.8 ^{b}	3.0 ^{b}	0.05	<0.038
Liver						
TSOD (U/mg pro.)	4.0 ^{a}	3.9 ^{a}	4.5 ^{b}	4.9 ^{b}	0.95	<0.049
Gpx (U/mg pro.)	0.6	0.5	0.7	0.6	0.01	<0.949
CAT (k/mg pro.)	0.2 ^{a}	0.6 ^{b}	0.7 ^{b}	0.7 ^{b}	0.11	<0.050
MDA (nmol/mg pro.)	6.1 ^{a}	5.0 ^{b}	4.2 ^{b}	4.0 ^{b}	0.20	<0.020

Means within rows with different letters are statistically different at $p \leq 0.05$, $n = 5$. SEM = standard error of the mean, calculated by one-way analysis of variance (ANOVA) and the general linear model procedure of Minitab. TAC = antioxidant capacity; TSOD = total superoxide dismutase; Gpx = glutathione peroxidase; CAT = catalase; MDA = malondialdehyde.

3.4. Antioxidative Capacity of Feed Rations

Table 5 shows the PV of lipids extracted from stored ginger-supplemented dietary feed rations over 50 days of oxidation. From Table 5, it is evident that there was an interaction between PV value and the storage duration of ginger-supplemented feed rations. The control diet remained stable over the 50 days of storage. However, the diet supplemented with ginger at a concentration of 5 g/kg showed stability in PV value until 40 days but approached the control value by day 50. The PV value of the diet supplemented with ginger at a concentration of 10 g/kg of diet increased until day 30 but decreased to the initial value by day 40. In addition, the PV value of the feed ration supplemented with ginger at a concentration of 15 g/kg increased until day 30, started to decrease at day 40, and reached the control level at day 50 (Table 5).

Table 5. Peroxide value (PV) of lipid extracted from stored ginger-supplemented dietary feed rations over 50 days of oxidation.

	Peroxide Value (mEq/kg)					
	Treatment (Ginger g/kg)					
Time (Days)	**0**	**5**	**10**	**15**	**SEM**	***p*-Value**
10	16.11 [a]	20.02 [b]	22.09 [b]	17.01 [a]	2.02	<0.001
20	20.00 [a]	28.01 [b]	30.41 [b]	22.51 [a]	2.99	<0.001
30	19.52. [a]	30.89 [b]	40.74 [c]	40.19 [d]	2.84	<0.001
40	18.63. [a]	25.22 [b]	30.32 [b]	30.55 [c]	2.01	<0.001
50	19.51 [a]	21.03 [a]	25.42 [b]	22.84 [a]	3.15	<0.001

Means within rows with different letters are statistically different at $p \leq 0.05$, $n = 5$. SEM = standard error of the mean, calculated by one-way analysis of variance (ANOVA) and the general linear model procedure of Minitab.

Table 6 shows the AV of lipids extracted from stored, ginger-supplemented dietary feed rations over 50 days of oxidation. The results of Table 6 also indicate an interaction between AV and storage time. The AV value increased with increasing storage duration. At all days of storage, the AV value of the control group was significantly higher than that of the groups supplemented with ginger at different concentrations.

Table 6. Acid value (AV) of lipid extracted from stored ginger-supplemented dietary feed rations over the time of oxidation.

	Acid Value (mg KOH/g)					
	Treatment (Ginger g/kg)					
Time (Days)	**0**	**5**	**10**	**15**	**SEM**	***p*-Value**
10	10.22 [a]	6.02 [b]	7.01 [b]	7.99 [b]	4.01	<0.001
20	18.04 [a]	10.65 [b]	11.21 [b]	10.45 [b]	3.59	<0.001
30	24.05 [a]	18.54 [b]	20.01 [b]	15.22 [c]	2.32	<0.001
40	30.14 [a]	22.01 [b]	19.87 [b]	19.54 [b]	3.91	<0.001
50	12.44 [a]	14.01 [b]	13.98 [b]	12.45 [b]	4.00	<0.001

Means within rows with different letters are statistically different at $p \leq 0.05$, $n = 5$. SEM = standard error of the mean, calculated by one-way analysis of variance (ANOVA) and the general linear model procedure of Minitab.

3.5. Hematological Measurements

The effects of ginger powder on the blood composition of broiler chickens are shown in Table 7. The results in Table 7 show that providing broiler chickens with feed rations supplemented with ginger at 5, 10, or 15 g/kg of diet enhanced WBCs of broilers, compared to the control ($p < 0.001$). In addition, the percentages of heterophils in birds from groups fed ginger powder at 10 or 15 g/kg of diet were significantly higher than that of the control group, and lower than that of the group supplemented with ginger powder at a concentration of 15 g/kg. The results in Table 6 show that ginger supplementation had no effect on any other blood parameter.

Table 7. Hematological and biochemical parameters of broilers fed ginger powder.

	Treatment (Ginger g/kg)					
Parameter	**0**	**5**	**10**	**15**	**SEM**	***p*-Value**
WBC (K/uL)	47.9 [a]	79.6 [b]	75.3 [b]	74.2 [b]	4.42	<0.001
Heterophils (%)	13.1 [a]	24.6 [b]	29.3 [b]	32.6 [c]	1.26	<0.001
Lymphocytes (%)	72.3	73.9	68.2	65.4	4.06	<0.474
Monocytes (%)	9.2	7.6	11.03	5.84	1.82	<0.256
Eosinophils (%)	0.03	0.02	0.04	0.01	0.01	<0.318
Basophils (%)	2.9	2.4	4.15	2.5	0.76	<0.384
RBC (M/uL)	2.9	2.2	2.41	2.00	0.27	<0.469
HGB (g/dL)	13.2	13.3	12.26	12.7	0.51	<0.266
HCT%	34.7	34.7	31.28	34.6	1.09	<0.104
MCV (fL)	127.8	130.4	129.8	128.0	1.87	<0.702
MCH (pg)	49.9	51.7	50.9	50.2	0.64	<0.238
MCHC (g/dL)	39.0	39.6	39.3	39.2	0.64	<0.918
RDW%	12.1	12.3	12.1	12.4	0.45	<0.975
Thrombocyte PLT (K/uL)	1.79	3.7	2.1	0.8	1.05	<0.306

Means within rows with different letters are statistically different at $p \leq 0.05$, $n = 5$. SEM = standard error of the mean, calculated by one-way analysis of variance (ANOVA) and the general linear model procedure of Minitab. WBC = white blood cells; RBC = red blood cells; HGB = hemoglobin; HCT = hematocrit; MCV = ean corpuscular volume; MCH = mean corpuscular hemoglobin; MCHC = mean corpuscular hemoglobin concentration; RDW = red cell distribution width; PLT = platelet count.

3.6. Plasma Cholesterol Content

The effects of different levels of ginger powder on blood cholesterol, triglyceride, HDL, and VLDL levels are shown in Table 8. A significant decrease in total cholesterol (TC) was observed in treatment groups supplemented with 5–15 g/kg ginger powder. A similar trend was also observed in total glycerides (TG). The HDL levels were found to be increased ($p \leq 0.05$) in treatment groups fed 5–15 g/kg ginger powder, compared to the control group. However, LDL levels in the supplemented birds were decreased ($p \leq 0.05$).

Table 8. The effects of different levels of ginger powder on blood cholesterol, triglyceride, high- and low-density lipoprotein content in broiler chickens.

	Treatment (Ginger g/kg)					
	Control	**5**	**10**	**15**	**SEM**	***p*-Value**
TC, mg/dL	160 [a]	145 [b]	140 [b]	121 [c]	8.00	<0.001
TG, mg/dL	40.0 [a]	28.5 [b]	25.2 [b]	22.8 [b]	5.5	<0.001
HDL, mg/dL	69.8 [a]	83.0 [b]	90.9 [b]	92.9 [b]	7.9	<0.001
LDL, mg/dL	60.7 [a]	29.9 [b]	27.8 [b]	30.1 [b]	5.6	<0.001

All treatment groups received a soybean basal diet. Means within rows with different letters are statistically different at $p \leq 0.05$, $n = 5$, total cholesterol (TC), triglyceride (TG), high-density lipoprotein (HDL), and low-density lipoprotein (LDL).

4. Discussion

The current study was conducted to investigate the effects of increased concentrations of dietary ginger powder on the antioxidation status, production performance, and hematological parameters of broiler chickens. In this study, both the control and the treatment diets were consumed equally by chickens, indicating that the inclusion of ginger in the diet did not adversely affect the palatability of the diet. The significant increase in the total body weight gain observed in this study is in agreement with some previous studies that investigated the same variable [23,37]. However, other studies have reported that the inclusion of ginger in the diet of broiler chickens did not improve weight gain [38–40]. Zhang, Yang [24] investigated the effect of dried ginger root on the growth performance of broilers and found that supplementation with ginger powder led to better production performance. The positive effect of phytobiotics on the production performance of broilers was also investigated by Hashemi and Davoodi [41]. Moorthy, Ravi [42] reported that dried ginger powder increased the body weight of broilers when it was included in the diet at a level of 2%. In addition, Tekeli, Zengin [43] reported that supplementing broiler feed rations with 120, 240, and 360 ppm of ginger significantly enhanced the broiler body weight gain. Similar enhancement results in the body weight gain were reported by Onu [23], Kausar, Rizvi [44], Javandel, Navidshad [39], Herawati [37], and Ademola, Farinu [38]; however, no significant effect of ginger was observed in the average broiler daily weight gain in broilers when using 5 g of ginger/kg of diet [24] or 1 g of ginger/kg of diet [45]. In contrast, a reduction in the starter broiler growth rate was reported by Al-Homidan [46] when ginger was supplemented at a level of 60 g/kg of diet. The authors explained this reduction as a result of the toxic action of ginger. Interestingly, Zhang, Yang [24] reported a better carcass yield ($p < 0.014$) of the ginger-supplemented broilers, compared to the control group and attributed this effect to the antioxidant effect of ginger that stimulates protein and fat metabolic pathways. Conversely, Moorthy, Ravi [42] and Onu [23] suggested that supplementing broiler feed rations with ginger does not affect carcass quality.

Results of the current study showed that supplementing the broiler chickens with ginger powder reduced total feed consumption ($p < 0.031$). This result is in line with that of Herawati [37] who reported that broilers fed a 2% ginger-supplemented diet had significantly lower feed consumption than the control group. These results are in contrast with the findings of Onu [23], who reported no significant differences in the feed consumption of birds provided with different ginger treatments versus the control. In addition, Ademola, Farinu [47] observed significantly higher feed consumption in broilers fed a ginger-supplemented diet compared to the control group.

The results of the analysis of feed efficiency showed that there were no significant differences across the treatments. This finding agrees with some studies that have investigated the same variable [25,39,48]. However, Onimisi, Dafwang [49], Moorthy, Ravi [42], Onu [23], and Onimisi, Dafwang [49] reported significantly lower feed efficiency in ginger-supplemented groups compared to the control group. These authors suggested that this outcome may be due to improved gut micro-flora, which inhibited microbial fermentation and improved feed efficiency. Conversely, Ademola, Farinu [38] reported a significant, 5% increase in feed efficiency in birds supplemented with ginger compared to control birds.

There was a significant effect of ginger powder supplementation on nutrient digestibility in the chickens. This result could be attributed to stimulation of digestive enzymes by bioactive compounds of ginger, and thus improvement of overall digestion. The active compound gingerol contributes to the secretion of digestive enzymes, which aids in the digestive process and helps provide nutrients. The active compounds in herbs stimulate the pancreas to produce digestive enzymes in larger amounts, which leads to increased nutrient digestibility and absorption to support growth [48].

Biochemical studies of the blood of ginger-fed broiler chicken have previously been undertaken by several authors. Interestingly, there is a debate regarding the effect of ginger on lipid profile and blood parameters among different studies. This could be attributed to differences in strain, age, ginger level, genetics, and experiment circumstances. For example, Rehman, Durrani [50] fed broiler chickens on 10 mL of therapeutic plants (garlic, mberberine, and aloe vera)/L of drinking water alongside ginger. The authors reported that serum glucose, alanine aminotransferase, aspartate aminotransferase, and alkaline phosphatase levels were significantly reduced and the serum protein increase was significantly enhanced in supplemented broilers. Results of the same study revealed a significant reduction in the total cholesterol, triglyceride, LDL, and VLDL and significant enhancement of the HDL level in the supplemented broilers. Similarly, Zhang, Yang [24] reported that ginger inclusion increased the broiler total protein concentration and reduced cholesterol concentration at 21 and 42 days of broilers. On the other hand, Kausar, Rizvi [44] reported no effect of ginger inclusion at dosages of 2 or 4 mL/L of drinking water on serum albumin, globulin, or total protein. Al-Homidan [46] revealed a reduction in the total protein and globulin in the plasma of broilers after supplementing ginger at 60 g/kg, which could have been due to the toxic effect of ginger at that dose. However, Ademola, Farinu [47] reported that ginger supplementation at concentrations of 5, 10, or 15 g/kg did not affect total protein or albumin in the serum of broilers.

The current effects of different levels of ginger powder on blood cholesterol, triglycerides, HDL, and LDL showed significant results. Our findings are in line with those of Shewita and Taha [17], who showed that lipid profile parameters such as total cholesterol, total triglycerides, HDL, LDL, and VLDL were found to be significantly modulated in the ginger-supplemented groups. These findings could be due to the anti-hypercholesterolemia and hypolipidemic activity of ginger. Dietary ginger acts on total serum cholesterol by inhibition of hydroxylmethyl-glutaryl-coenzyme-A reductase (HMG-CoA), or by increasing the excretion of bile acid and fecal cholesterol. However, Hayajneh [15] observed no significant differences in total protein, albumin, total cholesterol, or triglyceride levels after dietary treatment, but plasma cholesterol was found to be higher in broilers fed a diet supplemented with ginger powder. Ginger-supplemented diets administered over short terms and at low doses have been implicated in lower plasma cholesterol levels. Similarly, Barazesh, Boujar Pour [51] studied the effect of 0%, 0.5%, 1%, and 1.5% ginger powder on the blood parameters of broiler chickens. They found significant effects of ginger on blood parameters, and cholesterol and triglyceride levels. Zomrawi, Abdel Atti [52] studied the effects of 0%, 0.5%, 1%, and 1.5% ginger root powder on the blood and serum constituents of broiler chickens. They observed significant differences in serum triglyceride and cholesterol levels. In addition, inclusion of ginger root powder in the diet at levels of 0.5% and 1% lowered the cholesterol level.

Notably, inclusion of ginger in the diet did not cause anemia in the broilers, as evidenced by the lack of a significant effect on RBC counts and hemoglobin concentration. White blood cells and their sophisticated interactions are essential for developing and stimulating immune responses [53]. Tan and Vanitha [54] concluded that essential oil constituents from the rhizomes of *Z. officinale* exert immune-stimulating effects by enhancing the phagocytic activity of heterophils. Furthermore, Vattem, Lester [55] found that dietary supplementation of *Zingiberaceae* spices significantly increased the number and viability of coelomcytes, in addition to promoting differentiation into neutrophil-like cells, thus increasing phagocytic activity. Ginger consumed at a level of 100 mg/kg of diet was found to be effective for stimulating innate immunity by increasing the phagocytic capacity of heterophils, and for humoral immunity by increasing the production of antibodies; consequently improving the immunological profile of broiler chickens [56]. Additionally, Azhir, Zakeri [57] found that adding ginger rhizome powder at a concentration of 10 g/kg improved the humoral immunity of broilers at 35 days of age. Nidaullah, Durrani [58] observed that an aqueous extract of ginger rhizome mixed with water acted as an immune stimulant against Newcastle disease and coccidiosis. Ademola, Farinu [38] observed that ginger provided to chickens at a concentration of 1.0% caused a significant decrease in the total number of WBCs. The authors also found that ginger failed to affect the RBCs of broiler chickens. According to Nasiroleslami and Torki [59], the differential count of WBCs was similarly not affected by the dietary inclusion of ginger essential oil. In the current study, the inclusion of ginger in the diet of broiler chickens did not affect the hematological parameters of the birds, except for the total WBC and percentage of heterophils. The total WBC count increased significantly with an increased level of ginger in the diet. This indicates an enhanced immune response of cells involved in the innate and/or specific immune system.

Oxidative stress in broilers is associated with a high concentration of MDA and fatty acid peroxidation, due to increased free radicals [60]. Ginger has been shown to enhance antioxidative status. The results of the current study are in agreement with those reported in the literature. For example, Safiullah, Chand [61] reported that the inclusion of ginger powder and ginger essential oil in the feed rations of broiler chickens decreased MDA in liver and sera samples compared to birds fed a control diet. Wen, Gu [20] reported that the addition of ginger extract significantly increased the total antioxidant potential, decreased MDA content, and increased glutathione peroxidase activity in the serum and breast muscles. Interestingly, Mountzouris, Paraskeuas [62] showed that inclusion of a phytogenic premix including ginger and other natural herbs elevated the expression of cytoprotective genes against oxidation. In particular, the cytoprotective genes opposing oxidation were upregulated generally in the duodenum and ceca, and secondarily in the jejunum [62].

5. Conclusions

Ginger inclusion in the broiler diet can be safely used to enhance the production performance, immune response, and antioxidative status of broiler chickens.

Author Contributions: H.A.-K. handled conceptualization, methodology, validation, formal analysis, investigation, resources, data curation, writing, review and editing, visualization, funding acquisition; A.A.-N. investigated, reviewed and supervised the manuscript; T.A.-S. handled methodology and reviewed the manuscript; D.A.-A. and R.A.-K. contributed to the methodology and data collection; H.S., F.D., A.A.-H. and H.A.-S. carried out the statistical analysis. All authors have read and agreed to the published version of the manuscript.

Funding: This research received no external funding.

Institutional Review Board Statement: This research study was approved by the department committee of the Environment and Life Sciences Research Center in Kuwait Institute for Scientific Research under project No. FA157K (2017). These procedures and protocols followed the official animal welfare guidelines and regulations encoded with reference No. PMO/PV/RP/032/2017. This protocol recommends humane treatment of experimental animals with no pain, stress, or harm.

Data Availability Statement: No new data were created or analyzed in this study. Data sharing is not applicable to this article.

Acknowledgments: The authors would like to extend her sincere thanks to the managements of Kuwait Institute for Scientific Research (KISR), Kuwait United Poultry Company (KUPCO), and Kuwait Foundation for the Advancement of Sciences (KFAS) for their continuous financial and technical support.

Conflicts of Interest: The authors declare no conflict of interest.

References

1. Tanweer, A.; Chand, N.; Khan, S.; Qureshi, M.; Akhtar, A.; Niamatullah, M. Impact of Methanolic Extract of Peganum Harmala on The Weight Gain, Feed Conversion Ratio, Feed Cost and Gross Return of Broiler Chicks. *J. Anim. Plant Sci.* **2012**, *22*, 264–267.
2. Al-Khalaifah, H. Modulatory Effect of Dietary Polyunsaturated Fatty Acids on Immunity, Represented by Phagocytic Activity. *Front. Vet. Sci.* **2020**, *7*, 672. [CrossRef]
3. Al-Khalaifah, H.; Al-Nasser, A. Dietary Supplementation with Various Fat Oils Affect Phytohemagglutinin Skin Test in Broiler Chickens. *Front. Immunol.* **2020**, *11*, 1735. [CrossRef] [PubMed]
4. Al-Khalaifah, H.; Al-Nasser, A.; Abdulmalek, N.; Al-Mansour, H.; Ahmed, A.; Ragheb, G. Impact of SARS-Con-V2 on the poultry industry in Kuwait: A case study. *Front. Vet. Sci.* **2020**, *7*, 656. [CrossRef]
5. Hafez, H.M.; Attia, Y.A. Challenges to the poultry industry: Current perspectives and strategic future after the COVID-19 outbreak. *Front. Vet. Sci.* **2020**, *7*, 516. [CrossRef] [PubMed]
6. Al-Khalaifah, H.; Al-Nasser, A. Dietary source of polyunsaturated fatty acids influences cell cytotoxicity in broiler chickens. *Sci. Rep.* **2021**, *11*, 10113. [CrossRef]
7. Attia, Y.A.; Al-Khalaifah, H.; El-Hamid, A.; Hatem, S.; Al-Harthi, M.A.; Alyileili, S.R.; El-Shafey, A.A. Antioxidant Status, Blood Constituents and Immune Response of Broiler Chickens Fed Two Types of Diets with or without Different Concentrations of Active Yeast. *Animals* **2022**, *12*, 453. [CrossRef] [PubMed]
8. Al-Khalaifah, H. Benefits of probiotics and/or prebiotics for antibiotic-reduced poultry. *Poult. Sci.* **2018**, *97*, 3807–3815. [CrossRef]
9. Ke, J.; Delin, W.; Larsen, K. Zingiberaceae. *Flora China* **2000**, *24*, 322–377.
10. Dieumou, F.; Teguia, A.; Kuiate, J.; Tamokou, J.; Fonge, N.; Dongmo, M. Effects of ginger (*Zingiber officinale*) and garlic (*Allium sativum*) essential oils on growth performance and gut microbial population of broiler chickens. *Livest. Res. Rural. Dev.* **2009**, *21*, 23–32.
11. Al-khalaifah, H.S.; Badawi, M.E.; Abd El-Aziz, R.M.; Ali, M.A.; Omar, A.E. Effect of Egyptian Leek Leaf Extract Supplementation on Productive and Economic Performance of Broilers. *Front. Vet. Sci.* **2020**, *7*, 890. [CrossRef] [PubMed]

12. Al-Khalaifah, H.S.; Khalil, A.A.; Amer, S.A.; Shalaby, S.I.; Badr, H.A.; Farag, M.F.M.; Altohamy, D.E.; Rahman, A.N.A. Effects of Dietary Doum Palm Fruit Powder on Growth, Antioxidant Capacity, Immune Response, and Disease Resistance of African Catfish, *Clarias gariepinus* (B.). *Animals* **2020**, *10*, 1407. [CrossRef] [PubMed]
13. Al-Nasser, A.; Al-Khalaifah, H.; Al-Mansour, H.; Ahmad, A.; Ragheb, G. Evaluating farm size and technology use in poultry production in Kuwait. *World's Poult. Sci. J.* **2020**, *76*, 365–380. [CrossRef]
14. Umatiya, R.; Srivastava, A.; Pawar, M.; Chauhan, H.; Jain, A. Efficacy of ginger (*Zingiber officinale*) and garlic (*Allium sativum*) powder as phytogenic feed additives in diet of broiler chickens. *J. Pharmacogn. Phytochem.* **2018**, *7*, 1136–1140.
15. Hayajneh, F. Natural feed additives for broiler chickens. *S. Afr. J. Anim. Sci.* **2019**, *49*, 869–875. [CrossRef]
16. Irivboje, O.A.; Olufayo, O.; Irivboje, Y.I. Phytogenic Compounds: A Review of Ginger and Garlic as an Alternative Feed Additive in Poultry Nutrition. 2020. Available online: http://eprints.federalpolyilaro.edu.ng/1254/1/pp%20299-302%20Irivboje%20et%20al%20for%20Abuja%202020.pdf (accessed on 24 January 2022).
17. Shewita, R.; Taha, A. Influence of dietary supplementation of ginger powder at different levels on growth performance, haematological profiles, slaughter traits and gut morphometry of broiler chickens. *S. Afr. J. Anim. Sci.* **2018**, *48*, 997–1008. [CrossRef]
18. An, S.; Liu, G.; Guo, X.; An, Y.; Wang, R. Ginger extract enhances antioxidant ability and immunity of layers. *Anim. Nutr.* **2019**, *5*, 407–409. [CrossRef]
19. Sahoo, N.; Mishra, S.; Swain, R.; Acharya, A.; Pattnaik, S.; Sethy, K.; Sahoo, L. Effect of turmeric and ginger supplementation on immunity, antioxidant, liver enzyme activity, gut bacterial load and histopathology of broilers. *Indian J. Anim. Sci.* **2019**, *9*, 774–779.
20. Wen, C.; Gu, Y.; Tao, Z.; Cheng, Z.; Wang, T.; Zhou, Y. Effects of Ginger Extract on Laying Performance, Egg Quality, and Antioxidant Status of Laying Hens. *Animals* **2019**, *9*, 857. [CrossRef]
21. Zhao, X.; Yang, Z.; Yang, W.; Wang, Y.; Jiang, S.; Zhang, G. Effects of ginger root (*Zingiber officinale*) on laying performance and antioxidant status of laying hens and on dietary oxidation stability. *Poult. Sci.* **2011**, *90*, 1720–1727. [CrossRef]
22. Yang, C.; Ding, X.; Zhao, X.; Guo, Y.; Mu, A.; Yang, Z. Effects of star anise (*Illicium verum* Hook. f.), salvia miltiorrhiza (*Salvia miltiorrhiza* Bge) and ginger root (*Zingiber officinale Roscoe*) on laying performance, antioxidant status and egg quality of laying hens. *Eur. Poult. Sci.* **2017**, *81*, 183.
23. Onu, P. Evaluation of two herbal spices as feed additives for finisher broilers. *Biotechnol. Anim. Husb.* **2010**, *26*, 383–392. [CrossRef]
24. Zhang, G.; Yang, Z.; Wang, Y.; Yang, W.; Jiang, S.; Gai, G. Effects of ginger root (*Zingiber officinale*) processed to different particle sizes on growth performance, antioxidant status, and serum metabolites of broiler chickens. *Poult. Sci.* **2009**, *88*, 2159–2166. [CrossRef] [PubMed]
25. Sa'aci, Z.; Alabi, O.; Brown, D.; Ng'ambi, J. Effect of aqueous ginger (*Zingiber officinale*) extract on growth performance, nutrient digestibility and economy of feed conversion of broiler chickens. *Anim. Nutr. Feed. Technol.* **2018**, *18*, 225–231. [CrossRef]
26. Zidan, D.E.; Kahilo, K.A.; El-Far, A.; Sadek, K.M. Ginger (*Zingiber officinale*) and thymol dietary supplementation improve the growth performance, immunity and antioxidant status in broilers. *Glob. Vet.* **2016**, *16*, 530–538.
27. *Cobb-500*; Cobb Broiler Performance and Nutrient Supplement Guide. Cobb-Vantress Inc.: Siloam Springs, AR, USA, 2013.
28. Al-Khalaifa, H.; Al-Nasser, A.; Al-Surayee, T.; Al-Kandari, S.; Al-Enzi, N.; Al-Sharrah, T.; Ragheb, G.; Al-Qalaf, S.; Mohammed, A. Effect of dietary probiotics and prebiotics on the performance of broiler chickens. *Poult. Sci.* **2019**, *98*, 4465–4479. [CrossRef]
29. Horwitz, W. Official Methods of Analysis of AOAC International. In *Agricultural Chemicals, Contaminants, Drugs*; Horwitz, W., Ed.; AOAC International, 1997: Gaithersburg, MD, USA, 2010; Volume I.
30. Attia, Y.; Al-Khalaifah, H.; El-Hamid, A.; Al-Harthi, M.; El-Shafey, A. Effect of different levels of multienzymes on immune response, blood hematology and biochemistry, antioxidants status and organs histology of broiler chicks fed standard and low-density diets. *Front. Vet. Sci.* **2020**, *6*, 510. [CrossRef]
31. Kishawy, A.T.Y.; Al-Khalaifah, H.S.; Nada, H.S.; Roushdy, E.M.; Zaglool, A.W.; Ismail, T.A.; Ibrahim, S.M.; Ibrahim, D. Black Pepper or Radish Seed Oils in a New Combination of Essential Oils Modulated Broiler Chickens' Performance and Expression of Digestive Enzymes, Lipogenesis, Immunity, and Autophagy-Related Genes. *Vet. Sci.* **2022**, *9*, 43. [CrossRef]
32. Habibi, R.; Sadeghi, G.; Karimi, A. Effect of different concentrations of ginger root powder and its essential oil on growth performance, serum metabolites and antioxidant status in broiler chicks under heat stress. *Br. Poult. Sci.* **2014**, *55*, 228–237. [CrossRef]
33. Paglia, D.E.; Valentine, W.N. Studies on the quantitative and qualitative characterization of erythrocyte glutathione peroxidase. *J. Lab. Clin. Med.* **1967**, *70*, 158–169.
34. Aebi, H. Catalase In Vitro. In *Methods in Enzymology*; Academic Press: Cambridge, MA, USA, 1984; Volume 105, pp. 121–126.
35. AOCS. *Methods and Recommended Practices of the American Oil Chemists' Society*; AOCS: Champaign, IL, USA, 1998.
36. Rao, Y.; Xiang, B.; Zhou, X.; Wang, Z.; Xie, S.; Xu, J. Quantitative and qualitative determination of acid value of peanut oil using near-infrared spectrometry. *J. Food Eng.* **2009**, *93*, 249–252. [CrossRef]
37. Herawati, O. The effect of red ginger as phytobiotic on body weight gain, feed conversion and internal organs condition of broiler. *Int. J. Poult. Sci.* **2010**, *9*, 963–967.
38. Ademola, S.; Farinu, G.; Babatunde, G. Serum lipid, growth and haematological parameters of broilers fed garlic, ginger and their mixtures. *World J. Agric. Sci.* **2009**, *5*, 99–104.

39. Javandel, F.; Navidshad, B.; Seifdavati, J.; Pourrahimi, G.; Baniyaghoub, S. The favorite dosage of garlic meal as a feed additive in broiler chickens ratios. *Pak. J. Biol. Sci.* **2008**, *11*, 1746–1749. [CrossRef] [PubMed]
40. Omage, J.; Onimisi, P.; Adegbite, E.; Agunbiade, M. The Effect of Ginger (*Zingiber officinale Roscoe*) Waste Meal on Growth Performance, Carcass characteristics, serum lipid and serum cholesterol profiles of rabbit. *Pak. J. Nutr.* **2007**, *6*, 359–362. [CrossRef]
41. Hashemi, S.; Davoodi, H. Phytogenics as new class of feed additive in poultry industry. *J. Anim. Vet. Adv.* **2010**, *9*, 2295–2304. [CrossRef]
42. Moorthy, M.; Ravi, S.; Ravikumar, M.; Viswanathan, K.; Edwin, S. Ginger, pepper and curry leaf powder as feed additives in broiler diet. *Int. J. Poult. Sci.* **2009**, *8*, 779–782. [CrossRef]
43. Tekeli, Y.; Zengin, G.; Aktumsek, A.; Sezgin, M.; Torlak, E. Antibacterial activities of extracts from twelve Centaurea species from Turkey. *Arch. Biol. Sci.* **2011**, *63*, 685–690. [CrossRef]
44. Kausar, R.; Rizvi, F.; Anjum, A. Effect of Carminative mixture on health of broiler chicks. *Pak. J. Biol. Sci.* **1999**, *2*, 1074–1077. [CrossRef]
45. El-Deek, A.; Attia, Y.; Maysa, M.; Hannfy, M. Effect of anise (*Pimpinella anisum*), ginger(*Zingiber officinale Roscoe*) and fennel (*Foeniculum vulgare*) and their mixture on performance of broilers. *Arch. fur Geflugelkd.* **2002**, *67*, 92–96.
46. Al-Homidan, A. Efficacy of using different sources and levels of *Allium sativum* and *Zingiber officinale* on broiler chicks performance. *Saudi J. Biol. Sci.* **2005**, *12*, 96–102.
47. Ademola, S.; Farinu, G.; Obe, A.A.; Babatunde, G. Growth, haematological and biochemical studies on garlic-and ginger-fed broiler chickens. *Moor J. Agric. Res.* **2006**, *5*, 122–128. [CrossRef]
48. Risdianto, D.; Suthama, N.; Suprijatna, E.; Sunarso, S. Inclusion effect of ginger and turmeric mixture combined with Lactobacillus spp. isolated from rumen fluid of cattle on health status and growth of broiler. *J. Indones. Trop. Anim. Agric.* **2019**, *44*, 423–433. [CrossRef]
49. Onimisi, P.; Dafwang, I.; Omage, J. Growth performance and water consumption pattern of broiler chicks fed graded levels of ginger waste meal. *J. Agric. For. Soc. Sci.* **2007**, *3*, 113–119. [CrossRef]
50. Rehman, S.; Durrani, F.; Chand, N.; Khan, R.U.; ur Rehman, F. Comparative efficacy of different schedules of administration of medicinal plants infusion on hematology and serum biochemistry of broiler chicks. *Res. Opin. Anim. Vet. Sci.* **2011**, *1*, 8–14.
51. Barazesh, H.; Boujar Pour, M.; Salari, S.; Mohammad Abadi, T. The effect of ginger powder on performance, carcass characteristics and blood parameters of broilers. *Int. J. Adv. Biol. Biomed. Res.* **2013**, *1*, 1645–1651.
52. Zomrawi, W.; Abdel Atti, K.A.; Dousa, B.; Mahala, A. The effect of ginger root powder (*Zingiber officinale*) supplementation on broiler chicks performance, blood and serum constituents. *Online J. Anim. Feed Res.* **2012**, *2*, 457–460.
53. Al-Kassie, G.A. Influence of two plant extracts derived from thyme and cinnamon on broiler performance. *Pak. Vet. J.* **2009**, *29*, 169–173.
54. Tan, B.K.; Vanitha, J. Immunomodulatory and antimicrobial effects of some traditional Chinese medicinal herbs: A review. *Curr. Med. Chem.* **2004**, *11*, 1423–1430. [CrossRef]
55. Vattem, D.A.; Lester, C.E.; DeLeon, R.C.; Jamison, B.Y.; Maitin, V. Dietary supplementation with two Zingiberaceae spices ginger and turmeric modulates innate immunity parameters in *Lumbricus terrestris*. *J. Nat. Pharm.* **2012**, *3*, 37. [CrossRef]
56. Saleh, N.; Allam, T.; El-latif, A.A.; Ghazy, E. The Effects of Dietary Supplementation of Different Levels of Thyme (*Thymus vulgaris*) and Ginger (*Zingiber officinale*) Essential Oils on Performance, Hematological, Biochemical and Immunological Parameters of Broiler Chickens. *Glob. Vet.* **2014**, *12*, 736–744.
57. Azhir, D.; Zakeri, A.; Kargare-Rezapour, A. Effect of ginger powder rhizome on humoral immunity of broiler chickens. *Eur. J. Exp. Biol.* **2012**, *2*, 2090–2092.
58. Nidaullah, H.; Durrani, F.; Ahmad, S.; Jan, I.; Gul, S. Aqueous extract from different medicinal plants as anticoccidial, growth promotive and immunostimulant in broilers. *J. Agric. Biol. Sci.* **2010**, *5*, 53–58.
59. Nasiroleslami, M.; Torki, M. Including essential oils of fennel (*Foeniculum vulgare*) and Ginger (*Zingiber officinale*) to diet and evaluating performance of laying hens, white blood cell count and egg quality characteristics. *Adv. Environ. Biol.* **2010**, *4*, 341–345.
60. Büyükkılıç Beyzi, S.; Konca, Y.; Kaliber, M.; Sarıözkan, S.; Kocaoğlu Güçlü, B.; Aktuğ, E.; Şentürk, M. Effects of thyme essential oil and A, C, and E vitamin combinations to diets on performance, egg quality, MDA, and 8-OHdG of laying hens under heat stress. *J. Appl. Anim. Res.* **2020**, *48*, 126–132. [CrossRef]
61. Safiullah; Chand, N.; Khan, R.U.; Naz, S.; Ahmad, M.; Gul, S. Effect of ginger (*Zingiber officinale Roscoe*) and organic selenium on growth dynamics, blood melanodialdehyde and paraoxonase in broilers exposed to heat stress. *J. Appl. Anim. Res.* **2019**, *47*, 212–216. [CrossRef]
62. Mountzouris, K.C.; Paraskeuas, V.V.; Fegeros, K. Prming of intestinal cytoprotective genes and antioxidant capacity by dietary phytogenic inclusion in broilers. *Anim. Nutr.* **2020**, *6*, 305–312. [CrossRef]

animals

Article

Dietary Soluble Non-Starch Polysaccharide Level Influences Performance, Nutrient Utilisation and Disappearance of Non-Starch Polysaccharides in Broiler Chickens

Hong T. Nguyen [1], Michael R. Bedford [2], Shu-Biao Wu [1] and Natalie K. Morgan [1,3,*]

[1] School of Environmental and Rural Science, University of New England, Armidale, NSW 2350, Australia; hnguye43@myune.edu.au (H.T.N.); swu3@une.edu.au (S.-B.W.)
[2] AB Vista Woodstock Court, Blenheim Road, Marlborough SN8 4AN, UK; mike.bedford@abvista.com
[3] School of Molecular and Life Sciences, Curtin University, Bentley, WA 6102, Australia
* Correspondence: natalie.morgan@curtin.edu.au; Tel.: +61-8-9266-4816

Simple Summary: Non-starch polysaccharides (NSP) constitute a major part of the dietary fibre component in plant-based feed ingredients, accounting for approximately 10% of the nutrients in a poultry diet. However, NSP are generally not considered during formulation of commercial broiler diets. The functions of dietary NSP in poultry diets, including both the soluble and insoluble fraction, has been extensively researched and discussed. The soluble fraction is of particular interest to poultry nutritionists and producers, as it increases digesta viscosity, affecting nutrient digestion and absorption and thus litter quality. Soluble NSP (sNSP) also provides fuel for beneficial microbiota species. The extent of impact of dietary sNSP level on broiler performance and nutrient utilisation is poorly understood. Consequently, in this study, broilers were fed commercial-type diets with varying sNSP levels, and the effects of the sNSP level on ileal and total tract nutrient digestibility and productive performance were evaluated. The results revealed that even a small variation in dietary sNSP content induces an impact in broilers, particularly in young birds. Thus, sNSP level and composition should be considered during formulation of commercial poultry diets.

Abstract: This study evaluated the effect of dietary soluble non-starch polysaccharides (sNSP) on performance and nutrient utilisation in broilers from d 0 to 35. Cobb 500 broilers (n = 480, 80 birds per treatment) were fed either wheat- or corn-soybean meal-based diets formulated to contain either a high, medium, or low sNSP content, in a 2 × 3 factorial arrangement, fed as Starter (d 0–14) and Grower (d 14–35). Birds fed the low sNSP level presented greater BWG at d 0–14 and lower feed intake at d 14–35 compared to birds fed the medium sNSP level ($p < 0.005$). At d 14, birds fed the high sNSP level presented greater ileal and total tract starch digestibility and total tract sNSP degradability and insoluble NSP degradability, compared to feeding the low sNSP level. At d 35, total tract DM and metabolisability of gross energy was greater in birds fed the medium sNSP level compared to those fed the high or low sNSP level ($p < 0.005$). Generally, bird performance and nutrient utilisation was greater in birds fed the corn-based diets compared to the wheat-based diets. These results illustrate that dietary sNSP level and composition influences bird performance and nutrient digestibility.

Keywords: broiler; non-starch polysaccharide degradation; growth performance; nutrient digestibility

Citation: Nguyen, H.T.; Bedford, M.R.; Wu, S.-B.; Morgan, N.K. Dietary Soluble Non-Starch Polysaccharide Level Influences Performance, Nutrient Utilisation and Disappearance of Non-Starch Polysaccharides in Broiler Chickens. *Animals* **2022**, *12*, 547. https://doi.org/10.3390/ani12050547

Academic Editor: Velmurugu Ravindran

Received: 16 January 2022
Accepted: 20 February 2022
Published: 22 February 2022

1. Introduction

The successful and sustainable development of broiler production, particularly during an era with reduced reliance on in-feed antibiotics and escalated ingredient prices, relies on minimising feed costs and managing bird gastrointestinal health. The majority of a poultry ration is derived from plant-based ingredients, which contain approximately 10–30% non-starch polysaccharides (NSP) [1]. Dietary NSP directly and indirectly affects nutritive value of the diet, as well as digestive function and metabolic processes [2], through its impact on development and morphology of the gastrointestinal tract [3,4] and on microbiota population and composition [5,6]. Consequently, there is increasing interest in the positive effects of feeding dietary NSP to poultry, including supplementing diets directly with sources of fibre [7,8]. However, crude, acid detergent, and neutral detergent fibre values are currently used as the indicator of dietary fibre availability for the bird, despite it being well-established that these values do not fully describe the actual fibre level or their physico-chemical properties in feedstuffs [1,9]. Total dietary fibre is defined as the sum of NSP and lignin [1,10]. This suggests that NSP values should be applied to poultry diet formulations in order to achieve the most from the fibre available in feed ingredients, and from NSP-degrading enzyme application.

The influence of dietary NSP in the gastrointestinal tract is dependent on its quantity and physio-chemical composition and structure [11,12]. Of particular interest is its solubility in water, with soluble and insoluble NSP presenting very different properties [13]. Commercial broiler diets contain around 10–12% total NSP, of which 1–2.5% is water soluble [14]. Soluble NSP (sNSP) with high molecular weight has a high water-holding capacity, resulting in increased viscosity and thus heightened water consumption, increased excreta moisture, and poorer litter quality, alongside increased endogenous secretions and thus nutritional losses [14–17]. In contrast, sNSP can also provide a source of fermentable substrates for beneficial intestinal bacteria through selective fermentation [18], resulting in production of short chain fatty acids, which provide a source of energy, and reduced ability for pathogenic bacteria to proliferate [19].

Generally, fibre content and composition is considered to be an issue only in diets containing viscous grains with high sNSP level, such as wheat and barley, which contain approximately 10–12% and 17–20% NSP, respectively (DM basis) [11]. Approximately 19% and 29% of the total NSP is water soluble in wheat and barley, respectively [20]. However, positive effects of supplementing NSP-degrading enzymes are also observed when feeding grains with low sNSP level, namely corn which has a total NSP content of approximately 8% (DM basis), less than 2% of which is soluble [21–25]. This suggests that the sNSP content should be accounted for when feeding both viscous and non-viscous grains, as even low levels of NSP can induce a negative impact on bird nutrient utilisation and performance. This was illustrated by Morgan et al. [15], who observed significant differences in feed conversion ratio and ileal digesta and excreta dry matter content when feeding diets with sNSP content ranging from 5.29–14.63 g/kg.

NSP-degrading enzymes are commonly applied to poultry diets, primarily as a tool to eliminate viscosity effects, but the sNSP substrates are not currently accounted for, which may partially explain the inconsistent results when applying these enzymes. This suggests that there is potential to manipulate dietary sNSP level to enhance gut health, and that it is beneficial to account for sNSP concentration and composition in feed ingredients when formulating diets. This was shown by Musigwa et al. [26], who reported an interaction between dietary sNSP to total NSP ratio and multi-carbohydrase supplementation on energy utilisation in broilers. This reaffirms the need to accurately quantify and account for NSP level and composition in diets fed to poultry. Currently, there is a deficit of data available about the levels of NSP to formulate to, and data that is available has been derived from studies of feeding diets specifically enriched with sources of NSP [27]. The NSP used in these diets is usually in purified form, which provides insight into the effects and mechanisms of NSP, but lacks industry relevance. Consequently, there is a lack of understanding about how sNSP level influences broiler performance in common commercial-type diets.

In this study, birds were fed commercial-type diets formulated to have similar crude fibre, energy, and protein contents but varying sNSP levels (low, medium, and high). The aim of this design was to both determine how much influence sNSP has on bird performance and nutrient utilisation and assess how reliable sNSP values are compared to crude fibre values during feed formulation. Wheat- and corn-based diets were used due to their differing NSP compositions, and to represent diets from different regions. The diets with 'medium' sNSP level represented the average sNSP level for commercial-type wheat- and corn- based diets in Australia, and the relative increase or decrease in sNSP level provided the 'high' and 'low' treatments, respectively. The hypothesis of this study was that sNSP level would have a direct impact on bird performance and nutrient utilisation, with a low level of sNSP having a detrimental impact, primarily due to lack of fuel for beneficial microbiota species. It was also hypothesised that outputs from this study would prove that crude fibre values are meaningless during feed formulation, and that NSP levels need to be considered.

2. Materials and Methods

2.1. Birds and Housing

Cobb 500-day-old mixed sex broilers (n = 480) were obtained from a local commercial hatchery at day of hatch. Upon arrival, birds were weighed and randomly distributed into 48 floor pens (120 cm length × 77 cm width), with ten birds per pen, bedded on clean wood shavings. The room was thermostatically controlled to produce an initial temperature of 32 °C, gradually reduced to 22 °C by d 21, and maintained until d 35. The lighting regimen used was 23 h of light at approximately 40 lux on d 1, with darkness increasing 1 h a day until 6 h of darkness was reached, and then 18 h of light per day at 10 lux was maintained for the remainder of the study. Feed and water were provided ab libitum. Starter diet was fed as crumble from d 0–7 and as pellets from d 8–14, and grower diet was fed as pellets from d 15–35. All experimental procedures were approved by the University of New England Animal Ethics Committee, Australia (Approval number: AEC18-058).

2.2. Diets and Experimental Design

All diets were formulated to meet or exceed the nutritional requirements for Cobb 500 broilers [28]. Prior to feed formulation, ingredients were ground through a 0.5 mm screen and the energy, protein, amino acid, mineral content, and crude fibre was analysed by near-infrared spectroscopy (NIRS, Evonik AminoProx, Essen, Germany). Soluble and insoluble NSP content was also analysed in the feed ingredients (Table 1) by measuring the constituent sugar components as alditol acetates using gas-liquid chromatography (Model CP3800, Varian Inc., Palo Alto, CA, USA), following the procedure of Englyst et al. [29] with some modifications as described by Theander et al. [30] and Morgan et al. [31]. Briefly, the sample was fat extracted using hexane and then free oligosaccharides were extracted by heating the sample at 80 °C with 80% ethanol. The starch in the resulting residue was gelatinised using acetate buffer (pH 5) and α-amylase and amyloglucosidase was added, at 95 °C and 55 °C, respectively, to remove the starch. The prepared sample was then incubated and centrifuged at 2000× g for 10 min and the resulting supernatant and residue were used for the analysis of soluble and insoluble NSP, respectively. For the soluble NSP analysis, the sugars released by the enzymes were removed using ethanol at 4 °C, the residue was dried and then 2 M trifluoroacetic acid added and heated at 125 °C. For the insoluble NSP analysis, the glucose released from starch digestion was removed with water and acetone, and the resulting supernatant was removed and the residue was dried. Following this, 12 M H_2SO_4 was added and the sample was heated to 35 °C, and then water was added and the sample was heated to 100 °C, cooled and then centrifuged at 3000× g for 15 min to sediment the insoluble materials. For the free sugar analysis, the extracted sample was dried, hydrolysed with 1 M H_2SO_4 at 100 °C, and centrifuged to sediment the insoluble material. Ammonium (28%) was added to an aliquot of the resulting supernatant from the insoluble NSP and free oligosaccharide samples. For all the resulting samples, an internal standard was added (allose, 4 mg/mL) and the sample was evaporated to dryness, and then re-dissolved in water with slight alkalinity. Freshly prepared $NaBH_4$ was then added, the sample was incubated, and any excess $NaBH_4$ was decomposed with glacial acetic acid. 1-methylimidazole and acetic anhydride were added followed by water, and then dichloromethane was added, the sample was centrifuged, and the bottom layer was collected and dried. Finally, ethyl acetate and water were added, the sample was centrifuged, and the supernatant was analysed by gas chromatography (Model CP3800, Varian Inc., Palo Alto, CA, USA).

Table 1. Content of dry matter (g/100 g, as is basis) and non-starch polysaccharides of the plant-based ingredients used in the study diets (g/kg, as is basis).

Ingredients	DM [1]	sNSP [2]	iNSP [3]	tNSP [4]
Wheat	90.53	14.23	83.08	97.31
Corn	88.97	2.86	64.56	67.42
Barley	90.96	42.36	137.4	179.7
Sorghum	88.45	1.65	53.54	55.2
Soybean meal	90.05	11.22	132.2	143.4
Canola meal	92.16	15.35	146.8	162.1
Wheat bran	91.21	23.16	385.1	408.2
Oat bran	92.40	52.24	65.02	117.3
Soy protein concentrate	93.05	14.69	157.4	172.1

[1] Dry matter; [2] Soluble non-starch polysaccharides; [3] Insoluble non-starch polysaccharides; [4] Total non-starch polysaccharides.

A 2 × 3 factorial arrangement of treatments was applied. The factors were: (1) grain type, wheat, or corn; and (2) sNSP level, high (8.47 and 11.69 g/kg for starter, and 8.74 and 12.75 g/kg for grower, in corn and wheat diets, respectively), medium (6.22 and 9.95 g/kg for starter and 8.01 and 11.45 g/kg for grower, in corn and wheat diets, respectively) or low (5.78 and 9.31 g/kg for starter and 6.32 and 9.95 g/kg for grower, in corn and wheat diets, respectively). This resulted in 6 dietary treatments, 8 replicate pens per treatment. The sNSP levels differed between the wheat- and corn-based diets due to the differing NSP compositions between these two primary grains, but need to ensure both diets met the nutrient requirements of the birds. Thus, the three sNSP levels represent the relative change in sNSP content within each diet, as opposed to the absolute sNSP level. The ingredient composition of the diets is presented in Tables 2 and 3. Tables 4 and 5 present the analysed nutrient composition of the diets. Titanium dioxide (TiO_2) was added at a rate of 0.5% as an inert marker for nutrient digestibility assessment.

2.3. Performance Data Collection

The body weight and feed intake of the birds were recorded on an individual pen basis on arrival and on bird age d 14 and d 35, for determination of individual feed intake (FI), body weight gain (BWG), and feed conversion ratio (FCR), corrected for mortality, at d 0–14, d 14–35 and d 0–35. The number of male birds per pen was determined, and used to calculate the percentage male birds per pen.

2.4. Sample Collection

2.4.1. Excreta Collection

On d 14 and d 35, excreta samples were collected from each pen by placing the birds onto a clean metal tray and collecting the fresh excreta as soon as possible post-defecation. Samples were then homogenised, weighed, frozen at −20 °C, and then freeze-dried to constant weight. The freeze-dried samples were then ground and homogenized. Approximately 1.5 g of ground samples was weighed into a crucible, oven-dried at 105 °C to constant weight, and reweighed to determine the dry matter content.

Table 2. Composition of starter diets (g/100 g, as fed basis).

Ingredient	Corn-Based Diet			Wheat-Based Diet		
	High sNSP	Medium sNSP	Low sNSP	High sNSP	Medium sNSP	Low sNSP
Corn	51.25	52.00	51.40	-	-	-
Wheat	-	-	-	52.14	50.02	50.00
Sorghum	-	2.34	8.00	0.10	5.00	8.00
Barley	8.00	5.57	-	8.00	5.13	-
Wheat bran	0.13	2.24	2.73	1.54	2.03	2.96
Oat bran	3.47	-	-	-	-	-
Soybean meal	19.26	21.03	23.89	19.46	20.15	27.66
Soy protein concentrate	6.00	6.00	6.00	6.00	6.00	3.14
Canola meal solvent	5.00	4.00	1.39	5.00	4.00	0.01
Canola oil	2.39	2.35	2.09	3.47	3.31	3.63
Limestone	1.08	1.12	1.14	1.23	1.20	1.30
Dicalcium phosphate 18P/21Ca	1.69	1.63	1.67	1.36	1.45	1.50
Salt	0.32	0.32	0.33	0.30	0.30	0.37
TiO_2	0.50	0.50	0.50	0.50	0.50	0.50
Vitamin Premix [1]	0.050	0.050	0.050	0.050	0.050	0.050
Mineral Premix [2]	0.075	0.075	0.075	0.075	0.075	0.075
Choline	0.12	0.12	0.12	0.10	0.10	0.10
L-Lysine HCl	0.24	0.23	0.30	0.22	0.22	0.21
DL-Methionine	0.30	0.31	0.32	0.31	0.32	0.35
L-Threonine	0.10	0.10	0.09	0.13	0.13	0.13
Calculated Values (as-is)						
Crude fibre, %	3.30	3.30	3.20	3.50	3.50	3.48
AME_n, kcal/kg	3008	3008	3008	3008	3008	3008
d Arg, %	1.280	1.279	1.305	1.315	1.307	1.316
d Lys, %	1.180	1.180	1.180	1.180	1.180	1.180
d Met, %	0.597	0.604	0.614	0.581	0.589	0.608
d Met + Cys, %	0.880	0.880	0.880	0.880	0.880	0.880
d Ileu, %	0.785	0.771	0.770	0.709	0.758	0.719
d Threo, %	0.690	0.690	0.690	0.690	0.690	0.690
d Val, %	0.906	0.893	0.890	0.800	0.846	0.800
Fat, %	5.38	5.32	4.83	5.87	5.67	5.58
Calcium, %	0.90	0.90	0.90	0.90	0.90	0.93
Available P, %	0.45	0.45	0.45	0.45	0.46	0.46

[1] Vitamin premix per kg diet: vitamin A, 12 MIU; vitamin D, 5 MIU; vitamin E, 75 mg; vitamin K, 3 mg; nicotinic acid, 55 mg; pantothenic acid, 13 mg; folic acid, 2 mg; riboflavin, 8 mg; cyanocobalamin, 0.016 mg; biotin, 0.25 mg; pyridoxine, 5 mg; thiamine, 3 mg; antioxidant, 50 mg; [2] Trace mineral concentrate supplied per kg diet: Cu (sulfate), 16 mg; Fe (sulfate), 40 mg; I (iodide), 1.25 mg; Se (selenate), 0.3 mg; Mn (sulfate and oxide), 120 mg; Zn (sulfate and oxide), 100 mg; cereal-based carrier, 128 mg; mineral oil, 3.75 mg.

2.4.2. Ileal Digesta Collection

Four (d 14) and two (d 35) birds per pen were randomly selected, respectively, for ileum digesta sample collection. Birds were weighed individually before being euthanised by cervical dislocation. Ileum digesta samples were then pooled per pen, homogenised, frozen at −20 °C, and then freeze dried for further chemical analyses.

2.4.3. Measurements and Chemical Analyses

The diet and freeze-dried excreta and ileum samples were ground through a 0.5 mm screen. Total dry matter content was determined in the diets and freeze-dried excreta and ileum samples by weighing approximately 1.5 g of sample into a crucible and drying it to constant weight in a 105 °C oven. Gross energy was determined in the diets, excreta, and ileum samples using an adiabatic bomb calorimeter (Model 6400, Parr Instruments, Moline, IL, USA), using benzoic acid as a calibration standard. Nitrogen content of these samples was determined using a combustion analyser (Leco model FP-2000N analyser, Leco Corp.,

St. Joseph, MI, USA) using EDTA as a calibration standard, and the value was multiplied by 6.25 to calculate the protein content. Total starch was measured using the Megazyme total starch assay (Megazyme International Ireland Ltd., Wicklow, Ireland). The calcium content of the diets was measured using the microwave digestion technique (Milestone UltraWAVE, Milestone Srl, Sorisole (BG), Italy) and the total phosphorus concentration was determined using an inductively coupled plasma optical emission spectrometer (ICP-OES, Agilent, Melbourne, Australia). The NSP content and composition of the diets, excreta, and ileum samples was analysed by gas-liquid chromatography, as described above, and titanium dioxide (TiO_2) content was quantified using UV-spectroscopy, according to the method described by Short et al. [32].

Table 3. Composition of the grower diets (g/100 g, as fed basis).

Ingredient	Corn-Based Diet			Wheat-Based Diet		
	High sNSP	Medium sNSP	Low sNSP	High sNSP	Medium sNSP	Low sNSP
Corn	49.72	54.62	58.28	-	-	-
Wheat	-	-	-	59.65	57.08	55.01
Sorghum	-	-	-	-	4.00	7.83
Barley	8.00	4.03	-	8.00	3.34	-
Wheat bran	1.93	2.82	3.60	-	-	-
Oat bran	2.00	1.03	-	-	-	
Soybean meal	18.76	18.25	19.82	14.11	16.68	22.68
Soy Protein Concentrate	6.00	6.00	5.17	6.00	6.00	0.53
Canola meal solvent	6.00	6.00	6.00	4.79	5.51	6.00
Canola oil	3.80	3.41	3.28	3.51	3.61	4.03
Limestone	1.04	1.05	1.06	1.17	1.15	1.15
Dicalcium phosphate 18P/21Ca	1.47	1.46	1.44	1.25	1.25	1.27
Salt	0.32	0.32	0.33	0.30	0.30	0.30
TiO_2	0.50	0.50	0.50	0.50	0.50	0.50
Vitamin Premix [1]	0.050	0.050	0.050	0.050	0.050	0.050
Mineral Premix [2]	0.075	0.075	0.075	0.075	0.075	0.075
L-lysine HCl	0.08	0.11	0.11	0.21	0.13	0.199
DL-Methionine	0.22	0.23	0.23	0.26	0.24	0.27
L-Threonine	0.03	0.04	0.04	0.11	0.08	0.11
Calculated Values (as-is)						
Crude fibre, %	3.60	3.60	3.60	3.00	3.00	3.00
AME_n, kcal/kg	3086	3086	3086	3086	3086	3086
d Arg, %	1.263	1.232	1.224	1.172	1.245	1.171
d Lys, %	1.050	1.050	1.050	1.050	1.050	1.050
d Met, %	0.518	0.523	0.525	0.512	0.504	0.517
d Met + Cys, %	0.800	0.800	0.800	0.800	0.800	0.800
d Ileu, %	0.785	0.771	0.770	0.709	0.758	0.719
d Threo, %	0.690	0.690	0.690	0.690	0.690	0.690
d Val, %	0.906	0.893	0.890	0.800	0.846	0.800
Fat, %	6.90	6.59	6.51	5.91	6.11	6.64
Calcium, %	0.84	0.84	0.84	0.84	0.84	0.84
Available P, %	0.42	0.42	0.42	0.42	0.42	0.42

[1] Vitamin premix per kg diet: vitamin A, 12 MIU; vitamin D, 5 MIU; vitamin E, 75 mg; vitamin K, 3 mg; nicotinic acid, 55 mg; pantothenic acid, 13 mg; folic acid, 2 mg; riboflavin, 8 mg; cyanocobalamin, 0.016 mg; biotin, 0.25 mg; pyridoxine, 5 mg; thiamine, 3 mg; antioxidant, 50 mg; [2] Trace mineral concentrate supplied per kg diet: Cu (sulfate), 16 mg; Fe (sulfate), 40 mg; I (iodide), 1.25 mg; Se (selenate), 0.3 mg; Mn (sulfate and oxide), 120 mg; Zn (sulfate and oxide), 100 mg; cereal-based carrier, 128 mg; mineral oil, 3.75 mg.

Table 4. Analysed nutrient composition of starter diets (as-fed basis).

Item	Corn-Based Diet			Wheat-Based Diet		
	High sNSP	Medium sNSP	Low sNSP	High sNSP	Medium sNSP	Low sNSP
DM, %	88.67	87.19	87.74	88.57	88.43	88.28
Gross energy, kcal/kg	16.77	16.45	16.50	16.82	16.75	16.68
Crude fibre, %	2.45	2.40	2.40	2.51	2.50	2.57
Crude protein, %	20.93	21.36	21.91	23.18	23.40	23.46
Starch, %	38.06	37.73	37.45	36.45	35.03	35.24
Calcium, %	1.04	0.99	0.96	1.00	1.01	1.03
Total phosphorus, %	0.72	0.72	0.66	0.75	0.74	0.75
Total NSP, g/kg	75.31	76.25	76.43	84.73	85.69	83.83
Soluble NSP, g/kg	8.47	6.22	5.78	11.69	9.95	9.32
Rhamnose	0.04	0.05	0.04	0.05	0.03	0.04
Fucose	0.07	0.06	0.07	0.06	0.06	0.06
Ribose	0.39	0.39	0.39	0.37	0.41	0.42
Arabinose	1.62	1.40	1.40	2.91	2.75	2.58
Xylose	0.72	0.66	0.73	2.82	2.65	2.73
Mannose	1.34	1.40	1.38	1.84	1.34	1.52
Galactose	1.87	1.77	1.74	2.28	2.19	2.18
Glucose	3.44	1.22	0.73	2.79	1.74	0.95
Insoluble NSP, g/kg	66.84	70.03	70.65	73.03	75.75	74.51
Rhamnose	0.59	0.60	0.62	0.57	0.57	0.61
Fucose	0.92	0.97	1.03	0.97	0.93	1.04
Ribose	0.27	0.24	0.26	0.27	0.25	0.24
Arabinose	16.99	17.90	18.37	20.56	20.46	20.82
Xylose	14.85	15.42	15.47	18.90	20.49	19.11
Mannose	2.05	2.06	2.05	1.64	2.08	1.87
Galactose	15.36	16.73	17.14	13.75	14.37	15.49
Glucose	23.97	24.65	24.34	25.38	25.95	24.52
sNSP:iNSP	0.13	0.09	0.08	0.16	0.13	0.13
Free oligosaccharides, g/kg	34.76	34.04	35.93	36.22	37.50	40.08

2.5. Calculation of Nutrient Digestibility and Flow of NSP Sugar Residues

The following equation was used for calculation of NSP degradability at the total tract level and for calculation of apparent ileal or total tract digestibility of nutrients and metabolisability of crude protein and gross energy:

$$\text{Digestibility (\%)} = 1 - [1 \times (\text{TiO}_{2,\%\ \text{diet}}/\text{TiO}_{2,\%\ \text{digesta/excreta}}) \times (\text{nutrient}_{\%\ \text{digesta/excreta}}/\text{nutrient}_{\%\ \text{diet}})] \times 100,$$

Nutrient refers to either dry matter, gross energy, crude protein, starch (assuming all glucose measured were derived from starch), or NSP.

Total tract flow of NSP sugar residues was calculated as g/kg DM intake using the equation:

$$\text{NSP sugar residue flow}_{\text{ excreta}} = \text{NSP}_{\text{excreta}} \times (\text{TiO}_{2\text{diet}}/\text{TiO}_{2\text{excreta}})$$

2.6. Statistical Analysis

All data were analysed using Minitab (Minitab 19. 2020.1, Minitab Inc., State College, PA, USA). Data were tested for normality using the Anderson—Darling test, and then analysed as a 2 × 3 factorial arrangement using General Linear Model (GLM). Percentage of males was used as a co-variate when analyzing the performance data, after first checking the model assumption. Tukey's multiple range test was used to determine the differences between individual treatment means when interactions were observed. Pen served as the experimental unit, and there were eight replicate pens per treatment. Significant differences between mean values were declared at $p < 0.05$.

Table 5. Analysed nutrient composition of grower diets (as-fed basis).

Item	Corn-Based Diet			Wheat-Based Diet		
	High sNSP	Medium sNSP	Low sNSP	High sNSP	Medium sNSP	Low sNSP
DM, %	90.92	87.62	89.72	90.21	89.90	90.58
Gross energy, kcal/kg	17.46	16.88	17.15	16.97	16.84	17.05
Crude fibre, %	2.50	2.63	2.48	2.74	2.82	2.97
Crude protein, %	21.61	20.81	21.04	20.99	20.71	21.14
Starch, %	36.44	36.04	36.32	39.15	39.91	37.03
Calcium, %	1.04	0.99	0.96	1.00	1.01	1.03
Total phosphorus, %	0.72	0.72	0.66	0.75	0.74	0.75
Total NSP, g/kg	80.14	84.10	83.56	87.45	79.62	78.63
Soluble NSP, g/kg	8.74	8.01	6.32	12.75	11.45	9.95
Rhamnose	0.05	0.05	0.05	0.05	0.04	0.03
Fucose	0.06	0.06	0.06	0.05	0.05	0.05
Ribose	0.42	0.41	0.40	0.40	0.39	0.41
Arabinose	1.71	1.65	1.55	3.38	3.09	3.18
Xylose	0.84	0.76	0.69	3.68	3.64	3.30
Mannose	1.61	1.78	1.62	1.63	1.57	1.28
Galactose	1.74	1.80	1.72	2.30	2.26	2.02
Glucose	3.34	2.46	0.99	2.84	1.84	0.94
Insoluble NSP, g/kg	71.40	76.09	77.23	74.69	68.17	68.67
Rhamnose	0.72	0.65	0.63	0.49	0.42	0.49
Fucose	1.01	1.01	1.00	0.81	0.77	0.85
Ribose	0.27	0.29	0.28	0.23	0.19	0.24
Arabinose	18.99	20.82	20.87	20.07	18.69	19.52
Xylose	16.31	18.95	18.20	21.63	19.12	17.94
Mannose	1.92	1.97	2.27	1.94	1.82	1.88
Galactose	15.72	16.08	16.77	11.96	12.11	12.34
Glucose	25.21	25.67	26.68	26.80	23.49	23.88
sNSP:iNSP	0.12	0.11	0.08	0.17	0.17	0.14
Free oligosaccharides, g/kg	36.81	35.32	34.04	35.35	42.70	36.03

3. Results

3.1. Growth Performance

Growth performance at d 0–14, 14–35 and 0–35 are presented in Table 6. At d 0–14, birds fed the corn-based diet presented higher body weight gain and feed intake ($p = 0.003$ and 0.034, respectively), and showed a tendency for improved FCR ($p = 0.058$), compared to those fed the wheat-based diets. Furthermore, at this age BWG was greater in birds fed the low sNSP level compared to those fed the medium sNSP level ($p = 0.014$). At d 14–35, birds fed the diets with medium sNSP level presented greater FI compared to those fed the low sNSP level ($p = 0.026$), but sNSP level had no impact on BWG or FCR, and grain source had no impact on any of the performance parameters measured in this period. Over the whole trial period (d 0–35), birds fed the corn-based diet presented a lower FCR ($p = 0.049$), and tendency for greater BWG ($p = 0.053$), compared to those fed the wheat-based diets. Dietary sNSP level had no impact on d 0–35 performance. No interactions between grain type and sNSP level were observed on productive performance in this study.

3.2. Nutrient Utilisation

The impact of the dietary treatments on nutrient utilisation, measured at ileal and total tract level, at both d 14 and d 35, are presented in Tables 7 and 8, respectively. As illustrated in Table 7, birds fed the corn-based diet presented higher apparent ileal digestibility of energy ($p = 0.010$) and total tract digestibility of DM ($p < 0.001$), starch ($p < 0.001$), and apparent total tract protein ($p = 0.001$) and energy ($p < 0.001$) metabolism at d 14 compared to those fed the wheat-based diet. However, apparent ileal protein digestibility was greater in birds fed wheat compared to those fed corn ($p < 0.001$). Apparent ileal starch digestibility at d 14 was greater in birds fed the high sNSP level compared to those fed the medium or low sNSP level ($p = 0.010$). Similarly, apparent total tract starch digestibility at d 14 was greater in birds fed the high sNSP level compared to those fed the low sNSP level ($p = 0.019$).

Table 6. Effect of dietary soluble non-starch polysaccharide (sNSP) level and grain source on body weight gain (BWG, g/bird), feed intake (FI, g/bird), and feed conversion ratio corrected to mortality (cFCR) in broilers during d 0–14, 14–35 and 0–35.

Grain	sNSP Level	Day 0–14			Day 14–35			Day 0–35		
		BWG	FI	cFCR	BWG [c]	FI	cFCR [c]	BWG [c]	FI	cFCR
Corn	High	482	583	1.227	1914	2950	1.547	2399	3533	1.478
	Medium	465	584	1.225	1966	3095	1.579	2432	3679	1.517
	Low	487	601	1.234	1912	2907	1.515	2396	3508	1.459
Wheat	High	448	569	1.271	1880	2934	1.570	2331	3504	1.510
	Medium	446	557	1.248	1910	3063	1.596	2352	3619	1.532
	Low	475	587	1.235	1881	2946	1.571	2358	3533	1.502
SEM		10	12	0.016	42	69	0.025	44	73	0.021
Grain										
Corn		478 [a]	589 [a]	1.229	1931	2984	1.547	2409	3573	1.485 [b]
Wheat		457 [b]	571 [b]	1.251	1890	2981	1.579	2347	3552	1.515 [a]
sNSP level										
High		465 [ab]	576	1.249	1897	2942 [ab]	1.558	2365	3518	1.494
Medium		456 [b]	570	1.237	1938	3079 [a]	1.587	2392	3649	1.524
Low		481 [a]	594	1.235	1896	2927 [b]	1.543	2377	3521	1.481
p-value										
Grain		0.003	0.034	0.058	0.180	0.953	0.081	0.053	0.682	0.049
sNSP level		0.014	0.068	0.568	0.431	0.026	0.137	0.783	0.071	0.063
Grain × sNSP level		0.411	0.756	0.325	0.933	0.821	0.644	0.855	0.795	0.755

Male percentage was used as covariate when analysing the d 35 performance data; [a,b] Means in the same column with no common superscripts are significantly different ($p < 0.05$). [c] Means in the same column were adjusted to 41.9% male birds as a covariate, if the covariate effect was significant.

Table 7. Effect of dietary soluble non-starch polysaccharide (sNSP) level and grain source on apparent ileal and total tract digestibility of dry matter (DM) and starch (%) and apparent total tract metabolisability of crude protein (CP) and gross energy (%) in broilers at d 14.

Items	Ileum				Total Tract			
	DM	CP	Energy	Starch	DM	CP	Energy	Starch
Grain								
Corn	66.62	81.34 [b]	74.16 [a]	97.71	73.26 [a]	71.06 [a]	75.78 [a]	98.36 [a]
Wheat	65.43	82.97 [a]	72.58 [b]	97.62	70.70 [b]	66.18 [b]	73.20 [b]	97.81 [b]
sNSP level								
High	66.27	81.77	73.33	98.16 [a]	71.42	68.32	74.23	98.35 [a]
Medium	65.91	82.44	73.60	97.49 [b]	71.85	68.61	74.53	98.07 [ab]
Low	65.89	82.25	73.17	97.34 [b]	71.72	68.94	74.71	97.83 [b]
SEM	0.94	0.55	0.83	0.31	0.72	1.84	0.81	0.20
p-value								
Grain	0.084	<0.001	0.010	0.700	<0.001	0.001	<0.001	<0.001
sNSP level	0.869	0.391	0.832	0.010	0.782	0.928	0.790	0.019
Grain × sNSP level	0.862	0.444	0.915	0.489	0.956	0.566	0.949	0.395

[a,b] Means in the same column within main effects or interactive effect with no common superscripts are significantly different ($p < 0.05$).

Table 8. Effect of dietary soluble non-starch polysaccharide (sNSP) level and grain source on apparent ileal and total tract digestibility of dry matter (DM) and starch (%) and metabolisability of gross energy (%) and crude protein (%) in broilers at d 35.

Items	Ileum				Total Tract			
	DM	Protein	Energy	Starch	DM	Protein	Energy	Starch
Grain								
Corn	66.12	82.30	72.81 [a]	96.74 [a]	70.68 [a]	61.07	75.42 [a]	97.60 [a]
Wheat	64.85	82.31	70.81 [b]	94.50 [b]	69.51 [b]	59.58	73.79 [b]	94.95 [b]
sNSP level								
High	66.08	82.33	71.88	96.44	69.41 [b]	60.65	74.06 [b]	96.68
Medium	65.97	82.95	72.56	95.87	71.52 [a]	61.12	75.73 [a]	96.56
Low	64.41	81.63	70.99	94.56	69.36 [b]	59.21	74.03 [b]	95.59
SEM	1.17	0.61	1.10	1.01	0.73	1.78	0.81	0.80
p-value								
Grain	0.132	0.978	0.015	0.003	0.028	0.243	0.007	<0.001
sNSP level	0.194	0.059	0.267	0.100	0.001	0.443	0.031	0.251
Grain × sNSP level	0.953	0.834	0.656	0.294	0.925	0.519	0.748	0.220

[a,b] Means in the same column within main effects or interactive effect with no common superscripts are significantly different ($p < 0.05$).

Table 8 presents that at d 35, birds fed the corn-based diets presented higher apparent ileal energy ($p = 0.015$) and starch ($p = 0.003$) digestibility and higher apparent total tract DM ($p = 0.028$), and starch ($p < 0.001$) digestibility and gross energy metabolisability ($p = 0.007$) compared to those fed the wheat-based diets. Birds fed the medium sNSP level also presented higher total tract DM ($p = 0.001$) and energy ($p = 0.031$) digestibility at d 35 compared to those fed the high or low sNSP levels. No interactions between grain type and iNSP level were observed for any of the measurements, at both bird ages.

3.3. Utilisation of NSP and Their Constituent Sugars

Table 9 presents the influence of the dietary treatments on the apparent total tract disappearance of soluble, insoluble, total NSP and free oligosaccharides (FO), as well as the ratio of soluble to insoluble NSP (sNSP:iNSP) in the excreta, at both d 14 and d 35. At d 35, an interaction between grain source and sNSP was observed for apparent total tract sNSP degradability, showing that when the medium sNSP level was fed, sNSP degradation was greater in birds fed the corn-based diet compared to those fed the wheat-based diet ($p = 0.034$). Degradability of sNSP at d 14 ($p < 0.001$), total NSP at d 14 ($p = 0.017$), and digestibility of FO at d 35 ($p = 0.001$) was higher in birds fed corn compared to those fed wheat. At d 14, birds fed the diets with high sNSP level presented greater apparent total tract sNSP degradability ($p = 0.030$), and lower apparent insoluble NSP degradation ($p = 0.021$), compared to those fed the diets with low sNSP level. At both d 14 and d 35, the excreta sNSP:iNSP ratio was lower in birds fed corn compared to those fed wheat ($p < 0.001$ for both), and at d 14 this ratio value was higher in birds fed the high sNSP level compared to the medium or low sNSP level ($p = 0.014$).

Table 10 presents the apparent total tract flow of the constituent soluble and insoluble sugars at d 14. Birds fed the wheat-based diets presented greater apparent total tract flow of insoluble ribose ($p < 0.001$), arabinose ($p = 0.006$), xylose ($p < 0.001$), mannose ($p = 0.036$) and glucose ($p < 0.001$), and soluble arabinose ($p < 0.001$), xylose ($p < 0.001$), galactose ($p = 0.016$), and glucose ($p < 0.001$), but lower insoluble galactose flow ($p < 0.001$) compared to those fed the corn-based diets. Soluble arabinose and glucose flow was greater in birds fed the high sNSP level compared to those fed the medium or low sNSP level ($p = 0.013$ and $p < 0.001$, respectively). The opposite was true for insoluble galactose and fucose flow, with greater levels seen in birds fed the low sNSP level compared to those fed the medium or high sNSP level ($p = 0.011$ and $p = 0$ 001, respectively). Apparent insoluble mannose and soluble xylose flow were greater in birds fed the high sNSP level compared to those fed the low sNSP level ($p = 0.029$ and $p = 0.026$, respectively). No interactions between grain type and sNSP level were observed for any measurements of total tract flow of NSP sugar residue at d 14.

Table 9. Effect of dietary soluble non-starch polysaccharide (sNSP) level and grain source on apparent disappearance of soluble (sNSP), insoluble (iNSP), and total NSP (tNSP) and free oligosaccharides (FO) (%) and ratio of soluble to insoluble non-starch polysaccharide (sNSP:iNSP) in the excreta in broilers at d 14 and 35.

Grain	sNSP Level	Day 14					Day 35				
		sNSP	iNSP	tNSP	FO	sNSP:iNSP	sNSP	iNSP	tNSP	FO	sNSP:iNSP
Corn	High	35.18	13.13	15.61	98.73	0.09	41.72 ab	14.56	18.07	98.83	0.07
	Medium	26.31	18.96	19.56	98.64	0.08	49.14 a	22.30	24.78	99.06	0.07
	Low	21.95	20.54	20.65	98.66	0.08	43.09 a	18.26	20.28	99.06	0.06
Wheat	High	13.25	13.10	13.12	98.69	0.16	29.14 ab	19.05	20.66	98.70	0.16
	Medium	9.46	16.92	16.05	98.52	0.14	20.92 b	14.64	15.73	98.80	0.15
	Low	5.08	16.03	15.17	98.42	0.14	23.29 ab	21.42	21.71	98.32	0.14
SEM		4.42	2.30	2.17	0.13	0.01	6.72	3.61	3.77	0.15	0.01
Grain											
Corn		27.81 a	17.54	18.60 a	98.68	0.09 b	44.65	18.37	21.04	98.98 a	0.07 b
Wheat		9.27 b	15.35	14.78 b	98.54	0.15 a	24.56	18.37	19.37	98.61 b	0.15 a
sNSP level											
High		24.22 a	13.11 b	14.36	98.71	0.13 a	35.60	16.80	19.37	98.77	0.12
Medium		17.89 ab	17.94 ab	17.80	98.58	0.11 b	35.03	18.47	20.25	98.93	0.11
Low		13.51 b	18.28 a	17.91	98.54	0.11 b	33.19	19.84	21.00	98.69	0.10
p-value											
Grain		<0.001	0.184	0.017	0.144	<0.001	0.002	0.999	0.533	0.001	<0.001
sNSP level		0.030	0.021	0.113	0.313	0.014	0.902	0.626	0.883	0.190	0.060
Grain × sNSP level		0.751	0.535	0.725	0.690	0.735	0.034	0.115	0.158	0.070	0.997

a,b Means in the same column within main effects or interactive effect with no common superscripts are significantly different ($p < 0.05$).

Table 10. Effect of dietary soluble NSP (sNSP) level and grain source on apparent total tract flow (g/kg DM intake) of soluble and insoluble non-starch polysaccharide sugar residues at d 14.

Items		Grain		sNSP Level			SEM	*p*-Values		
		Corn	Wheat	High	Medium	Low		Grain	sNSP Level	Grain × sNSP Level
Rhamnose	Soluble	0.06	0.07	0.06	0.07	0.07	0.01	0.124	0.287	0.106
	Insoluble	0.68	0.59	0.60	0.63	0.68	0.13	0.335	0.795	0.794
Fucose	Soluble	0.39	0.42	0.40	0.42	0.40	0.03	0.183	0.778	0.263
	Insoluble	1.24	1.20	1.19 b	1.19 b	1.28 a	0.03	0.086	0.001	0.360
Ribose	Soluble	0.07	0.08	0.07	0.07	0.08	0.01	0.150	0.824	0.411
	Insoluble	0.12 b	0.17 a	0.14	0.15	0.14	0.01	<0.001	0.419	0.356
Arabinose	Soluble	1.22 b	2.65 a	2.14 a	1.83 b	1.83 b	0.13	<0.001	0.013	0.294
	Insoluble	21.40 b	22.71 a	21.76	21.75	22.65	0.64	0.006	0.190	0.556
Xylose	Soluble	0.75 b	3.71 a	2.44 a	2.22 ab	2.04 b	0.17	<0.001	0.026	0.456
	Insoluble	21.74 b	25.62 a	24.12	23.66	23.25	0.65	<0.001	0.309	0.987
Mannose	Soluble	0.60	0.69	0.64	0.64	0.65	0.07	0.104	0.968	0.995
	Insoluble	1.76 b	1.95 a	1.98 a	1.90 ab	1.68 b	0.12	0.036	0.029	0.283
Galactose	Soluble	2.26 b	2.54 a	2.44	2.34	2.40	0.16	0.016	0.758	0.674
	Insoluble	17.82 a	16.01 b	16.44 b	16.65 b	17.66 a	0.47	<0.001	0.011	0.310
Glucose	Soluble	0.84 b	1.62 a	1.71 a	1.09 b	0.89 b	0.13	<0.001	< 0.001	0.257
	Insoluble	8.52 b	11.95 a	11.06	10.83	8.81	1.04	<0.001	0.303	0.571

a,b Means in the same row within main effects or interactive effect with no common superscripts are significantly different ($p < 0.05$).

Table 11 shows the impact of the dietary treatments on apparent soluble constituent sugar flow at d 35. An interaction between grain source and sNSP level was observed for apparent soluble rhamnose total tract flow, showing that feeding the medium sNSP level increased the apparent flow of total tract soluble rhamnose in corn-fed birds compared to feeding the low or high sNSP level ($p = 0.040$), but in wheat-fed birds. Birds fed the wheat-based diets presented greater apparent total tract flow of soluble fucose ($p = 0.001$), arabinose ($p < 0.001$), xylose ($p < 0.001$), and glucose ($p < 0.001$) compared to those fed the corn-based diets. Apparent total tract flow of soluble glucose was greater in birds fed the high sNSP level compared to those fed the low sNSP level ($p = 0.012$).

Table 11. Effect of dietary soluble NSP (sNSP) level and grain source on apparent total tract flow (g/kg DM intake) of soluble non-starch polysaccharide sugar residues at d 35.

Grain	sNSP Level	Rhamnose	Fucose	Ribose	Arabinose	Xylose	Mannose	Galactose	Glucose
Corn	High	0.04 ^{b}	0.26	0.09	1.29	0.82	0.51	1.40	1.06
	Medium	0.10 ^{a}	0.21	0.10	1.09	0.68	0.62	1.73	0.69
	Low	0.06 ^{b}	0.21	0.11	1.04	0.61	0.58	1.49	0.58
Wheat	High	0.05 ^{b}	0.33	0.12	2.78	4.00	0.46	1.71	1.68
	Medium	0.06 ^{ab}	0.28	0.11	2.53	3.97	0.52	2.01	1.63
	Low	0.05 ^{b}	0.29	0.12	2.50	3.41	0.53	1.58	0.96
SEM		0.01	0.03	0.02	0.25	0.38	0.07	0.18	0.22
Grain									
Corn		0.07	0.23 ^{b}	0.10	1.14 ^{b}	0.70 ^{b}	0.57	1.54	0.78 ^{b}
Wheat		0.06	0.30 ^{a}	0.12	2.61 ^{a}	3.79 ^{a}	0.50	1.77	1.42 ^{a}
sNSP									
High		0.05	0.30	0.10	2.03	2.41	0.49	1.56	1.37 ^{a}
Medium		0.08	0.24	0.11	1.81	2.32	0.57	1.87	1.16 ^{ab}
Low		0.06	0.25	0.11	1.77	2.01	0.55	1.54	0.77 ^{b}
p-value									
Grain		0.060	0.001	0.127	<0.001	<0.001	0.191	0.074	<0.001
sNSP level		<0.001	0.066	0.765	0.422	0.433	0.317	0.061	0.012
Grain × sNSP level		0.040	0.961	0.559	0.991	0.730	0.874	0.743	0.374

^{a,b} Means in the same column within a main effect or interactive effect with no common superscripts are significantly different ($p < 0.05$).

Table 12 illustrates the impact of the dietary treatments on apparent insoluble constituent sugar flow at d 35. An interaction between grain source and sNSP level was observed for apparent flow of insoluble xylose, showing that flow of insoluble xylose was greater in birds fed the high sNSP level compared to the low sNSP level, but only when the wheat-based diets were fed ($p = 0.004$). Birds fed the corn-based diet presented higher apparent total tract flow of rhamnose ($p < 0.001$), fucose ($p < 0.001$), arabinose ($p = 0.002$) and galactose ($p < 0.001$), and lower apparent flow of mannose ($p = 0.026$), compared to those fed the wheat-based diets. Apparent insoluble glucose, mannose, and ribose flow was greater in birds fed the diets with medium sNSP level compared to those fed the high or low sNSP level ($p < 0.001$ for all). Similarly, apparent insoluble rhamnose flow was greater in birds fed the medium sNSP level diets compared to those fed the high sNSP level ($p = 0.008$). In contrast, apparent insoluble arabinose flow was greater in birds fed the diets with high sNSP level compared to those fed the medium sNSP level ($p = 0.019$).

Table 12. Effect of dietary soluble NSP (sNSP) level and grain source on apparent total tract flow (g/kg DM intake) of insoluble non-starch polysaccharide sugar residues at d 35.

Grain	sNSP Level	Rhamnose	Fucose	Ribose	Arabinose	Xylose	Mannose	Galactose	Glucose
Corn	High	0.76	1.20	0.13	24.87	25.86 ^{ab}	1.07	17.47	4.31
	Medium	0.97	1.18	0.18	22.39	23.34 ^{b}	1.54	16.25	10.15
	Low	0.82	1.28	0.15	25.97	26.20 ^{ab}	1.11	17.76	5.12
Wheat	High	0.59	0.99	0.15	23.39	28.73 ^{a}	1.37	13.43	6.99
	Medium	0.66	1.04	0.20	20.92	25.02 ^{ab}	1.71	13.58	9.83
	Low	0.60	1.05	0.14	21.54	22.38 ^{b}	1.27	13.66	5.88
SEM		0.05	0.04	0.012	1.05	1.15	0.129	0.72	1.05
Grain									
Corn		0.85 ^{a}	1.22 ^{a}	0.15	24.13 ^{a}	25.14	1.24 ^{b}	17.16 ^{a}	6.53
Wheat		0.62 ^{b}	1.03 ^{b}	0.16	21.95 ^{b}	25.38	1.45 ^{a}	13.56 ^{b}	7.57
sNSP									
High		0.68 ^{b}	1.10	0.14 ^{b}	24.13 ^{a}	27.30	1.22 ^{b}	15.45	5.65 ^{b}
Medium		0.81 ^{a}	1.11	0.19 ^{a}	21.65 ^{b}	24.18	1.62 ^{a}	14.92	9.99 ^{a}
Low		0.71 ^{ab}	1.17	0.14 ^{b}	23.76 ^{ab}	24.29	1.19 ^{b}	15.71	5.50 ^{b}
p-value									
Grain		<0.001	<0.001	0.350	0.002	0.768	0.026	<0.001	0.185
sNSP level		0.008	0.118	<0.001	0.019	0.004	<0.001	0.438	<0.001
Grain × sNSP level		0.282	0.388	0.236	0.184	0.004	0.781	0.444	0.293

^{a,b} Means in the same column within a main effect or interactive effect with no common superscripts are significantly different ($p < 0.05$).

4. Discussion

It is well-established that dietary NSP regulates digestion and modulates gastrointestinal health in poultry, directly influencing bird performance and litter quality [33,34]. Traditionally, only the anti-nutritive effects of sNSP have been considered, but a recent report has highlighted the positive impact of sNSP on broiler gastrointestinal health and performance, namely through its role as a source of substrates for probiotic bacteria species [35]. In the current study, the extent to which sNSP level and composition influences bird performance and nutrient utilisation in common commercial wheat and corn-based diets was investigated. Although significant responses to sNSP were observed, it must be noted that formulating the diets to keep crude fibre, protein, and energy levels similar meant that the three sNSP levels were not extremely different from each other within each grain source. Thus, further research is warranted using more extreme differences in sNSP to determine the optimum sNSP level to formulate to, to derive more meaningful performance data. Removing the constraint of crude fibre values during feed formulation would enable for a greater range of sNSP levels. The contribution of the other grain sources on performance also warrants consideration; regression analysis including the grain sources revealed that barley had a positive impact on growth performance at d 14, without considering the impact of its sNSP content.

As expected, performance was better in birds fed the corn-based diets, based on d 0–35 cFCR, compared to those fed the wheat-based diets, which is in agreement with a number of studies [36–38]. This is likely because the sNSP content of the wheat-based diets was approximately 50% higher than that in the corn-based diets, owing to much lower levels of sNSP in corn. The corn used in this study contained around 0.3% sNSP, whilst the wheat contained approximately 1.4% sNSP, which is indeed much lower than the values reported previously by Back Knudsen [13], who indicated that the sNSP content ranged from 0.2–1.6% and 1.9–2.9% (DM basis) for corn and wheat, respectively. The wheat-based diet also contained a higher concentration of high molecular weight sNSP, such as arabinoxylans [39,40]; the estimated soluble arabinoxylan content of the corn- and wheat-based diets in the current study were approximately 0.20% and 0.54%, respectively. The presence of these polymers results in increased viscosity in the gastrointestinal tract [41], inducing a barrier between enzymes and perspective substrates, and between nutrients and the gastrointestinal lining. The consequence of this could impair nutrient utilisation, resulting in increasing undigested nutrient presence in the gastrointestinal tract. This, in turn, can alter the microbiota profile by providing fuel for pathogenic bacterial species, which then compete with the host for nutrients [18,42]. This may partly explain why birds fed the wheat-based diet, which had higher total protein content in the starter phase, presented lower performance. It is interesting to note that BWG was higher, and FI was numerically highest, in birds fed the low sNSP level compared to the medium sNSP level at d 0–14. This could again be related to viscosity, whereby sNSP induces signals of satiety due to prolonged digesta transit rates and heightened water retention in the gastrointestinal tract, suggesting birds fed the low sNSP diets in the current study probably had comparatively lower feeling of satiation and thus could consume more feed. However, this heightened feed intake was coupled with increased weight gain, suggesting that the NSP in this diet was at a low enough level to not induce anti-nutritional effects on viscosity. At d 35, total tract DM and metabolisability of gross energy was greater in birds fed the medium sNSP level compared to the high or low sNSP level, possibly indicating that the sNSP adequately stimulated the microbiota environment towards proficient manufacture of SCFA [41], providing energy. However, this did not translate into improvements in performance, suggesting that nutrient partitioning between the host and the microbiota was not improved with this treatment, likely due to an alteration in the intestinal microbial profile [43]. Additionally, the wheat-based diets with medium sNSP level contained the highest free oligosaccharide level, which may have fuelled bacteria earlier in the gastrointestinal tract. This might have caused undesirable proliferation of bacterial population in the upper gut, potentially resulting in nutrient competition with the host, as opposed to the desirable fermentation in caeca which is the main site for bacterial fermentation.

Results from this study suggest that considering the quantity of sNSP alone during feed formulation does not provide an accurate indication of the impact of dietary NSP on gastrointestinal status and bird performance, due to the variable ability of the constituent sugars to modulate microbiota composition, and influence of the sNSP on digesta viscosity. There are a lack of studies examining NSP content in commercial-type diets, with most studies investigating effects of fibre addition in purified or semi-purified diets. Thus, further research is needed into the influence of the composition and physiochemical properties of dietary fibre, as opposed to the total level, in common diets, with particular focus on the quantity of fermentable and non-fermentable fractions, and ratio between the soluble and insoluble components in the sNSP [13]. This was illustrated by Musigwa et al. [26], who observed that the proportion of total NSP that was soluble, as opposed to total level, directly influenced the efficacy of a multi-carbohydrase at enhancing energy utilisation in broilers. For this reason, it is likely that further refinements in the definition of sNSP to take into account sugar composition, molecular weight, and degree of substitution will further benefit the predictability of bird performance and response to exogenously added enzymes.

The lower insoluble NSP digestibility observed at d 14 in birds fed the high sNSP level, compared to the low sNSP level, may be a consequence of young birds possessing microbiota that is not yet adapted to digesting complex, insoluble NSP molecules. The suggestion is that the microbiota will preferentially ferment sNSP over iNSP, and as a result, overwhelming a naïve microbiota with sNSP can compromise its ability, and indeed need, to ferment insoluble NSP. However, exposure to these dietary polymers, or ideally oligosaccharides derived from them, in young birds can prime the microbiota to become more adept at utilising more complex NSP later in life [44]. This is elucidated by the significantly greater impact of sNSP level on the flow of soluble constituent sugars at d 14 than at d 35, and conversely, on the flow of the insoluble constituent sugars at d 35 than d 14. Regardless of the initial sNSP:iNSP ratio in diets, the ratio of soluble to insoluble NSP in the excreta was lowest in diets with the medium and low sNSP levels at d 14, and tended to be lower with the low sNSP level at d 35, revealing a more efficient breakdown of sNSP in birds fed the low sNSP diets. This reconfirms that young birds have a requirement for a minimum sNSP level in order to optimise early sNSP fermentation, which then influences the fermentation of total NSP at a later growing stage. This could also partially be explained by the observed greater sNSP degradability at d 35 in birds fed the corn-based diet with medium or low sNSP, compared to those fed the wheat-based diet with medium sNSP level. This inherent requirement for dietary fibre has also been demonstrated previously by Sadeghi et al. [45], in that broilers actively selected extra fibre sources to boost their immunity, regardless of impact on performance. Thus, it appears that it is beneficial to supplement diets containing poor fermentable NSP substrates, such as corn, with a fermentable NSP substrate, to enhance intestinal fermentation activities. Arabinoxylans (AX) are the predominant NSP in both wheat and corn, but the structure of AX in corn is more complex than that in wheat, with a higher degree of substitution and arabinose: xylose ratio [46], meaning corn AX is less readily fermentable than wheat AX [47]. In addition, the poor ileal and total tract starch digestibility observed at d 14 in birds fed the low sNSP diets could be attributed to production of high molecular weight viscous AX, from insoluble NSP degradation, exceeding their hydrolysis rate in the gastrointestinal tract in this treatment group. The NSP digestibility values in this study were lower than those presented by Jamroz et al. [48] and Zhang et al. [49], but higher than those seen by Meng et al. [50], probably due to differences in diet composition, such as source of fat and quality of wheat, and bird age.

5. Conclusions

In conclusion, outputs from this study indicated that considering sNSP level in commercial-type broiler diets provides a more meaningful indication of bird response to dietary fibre, in terms of performance and nutrient utilisation, compared to using conventional crude fibre values. Outputs from this study also demonstrated a need to supply birds with fermentable fibre substrates, indicating that the industry should consider supplementing sNSP-poor diets with a source of fermentable NSP. In conclusion, the current results suggest that it may be advantageous to include sNSP levels when formulating broiler diets, although further research is required to define what levels to formulate to, depending on the composition of the NSP in feed ingredients.

Author Contributions: Conceptualization, methodology and formal analysis, H.T.N. and N.K.M.; investigation, laboratory analysis, writing—original draft preparation, H.T.N.; supervision, editing and revising critically, N.K.M.; supervision, editing and providing comments on the manuscripts, M.R.B. and S.-B.W.; funding acquisition, N.K.M. All authors have read and agreed to the published version of the manuscript.

Funding: This research received no external funding.

Institutional Review Board Statement: The experimental procedures were approved by the University of New England Animal Ethics Committee, Australia (Approval number: AEC18-058).

Informed Consent Statement: Not applicable.

Data Availability Statement: All available data are incorporated in the manuscript.

Acknowledgments: The authors would like to acknowledge Shuyu Song, Craig Johnson and the CART team, University of New England for their technical assistance.

Conflicts of Interest: The authors declare no conflict of interest.

References

1. Choct, M. Feed non-starch polysaccharides for monogastric animals: Classification and function. *Anim. Prod. Sci.* **2015**, *55*, 1360–1366. [CrossRef]
2. Bedford, M.R. Interaction between ingested feed and the digestive system in poultry. *J. Appl. Poult. Res* **1996**, *5*, 86–95. [CrossRef]
3. Iji, P.A. The impact of cereal non-starch polysaccharides on intestinal development and function in broiler chickens. *Worlds Poult. Sci. J.* **1999**, *55*, 375–387. [CrossRef]

4. Kheravii, S.K.; Morgan, N.K.; Swick, R.A.; Choct, M.; Wu, S.-B. Roles of dietary fibre and ingredient particle size in broiler nutrition. *Worlds Poult. Sci. J.* **2018**, *74*, 301–316. [CrossRef]
5. Choct, M. Managing gut health through nutrition. *Br. Poult. Sci.* **2009**, *50*, 9–15. [CrossRef]
6. Mahmood, T.; Guo, Y. Dietary fiber and chicken microbiome interaction: Where will it lead to? *Anim. Nutr.* **2020**, *6*, 1–8. [CrossRef]
7. Branton, S.L.; Lott, B.D.; Deaton, J.W.; Maslin, W.R.; Austin, F.W.; Pote, L.M.; Keirs, R.W.; Latour, M.A.; Day, E.J. The effect of added complex carbohydrates or added dietary fiber on necrotic enteritis lesions in broiler chickens. *Poult. Sci.* **1997**, *76*, 24–28. [CrossRef]
8. Adibmoradi, M.; Navidshad, B.; Jahromi, M.F. The effect of moderate levels of finely ground insoluble fibre on small intestine morphology, nutrient digestibility and performance of broiler chickens. *Ital. J. Anim. Sci.* **2016**, *15*, 310–317. [CrossRef]
9. Bach Knudsen, K.E. The nutritional significance of "dietary fibre" analysis. *Anim. Feed Sci. Tech.* **2001**, *90*, 3–20. [CrossRef]
10. Theander, O.; Åman, P.; Westerlund, E.; Graham, H. Enzymatic/chemical analysis of dietary fiber. *J. AOAC Int.* **1994**, *77*, 703–709. [CrossRef]
11. Jha, R.; Fouhse, J.M.; Tiwari, U.P.; Li, L.; Willing, B.P. Dietary Fiber and Intestinal Health of Monogastric Animals. *Front. Vet. Sci.* **2019**, *6*, 48. [CrossRef] [PubMed]
12. Raninen, K.; Lappi, J.; Mykkänen, H.; Poutanen, K. Dietary fiber type reflects physiological functionality: Comparison of grain fiber, inulin, and polydextrose. *Nut. Rev.* **2019**, *69*, 9–21. [CrossRef] [PubMed]
13. Jiménez-Moreno, E.; Frikha, M.; de Coca-Sinova, D.; García, J.; Mateos, G.G. Oat hulls and sugar beet pulp in diets for broilers 1. Effects on growth performance and nutrient digestibility. *Anim. Feed Sci. Tech.* **2013**, *182*, 33–43. [CrossRef]
14. Bach Knudsen, K.E. Carbohydrate and lignin contents of plant materials used in animal feeding. *Anim. Feed Sci. Tech.* **1997**, *67*, 319–338. [CrossRef]
15. Morgan, N.K.; Choct, M.; Toghyani, M.; Wu, S.-B. Effects of dietary insoluble and soluble non-starch polysaccharides on performance and ieal and exreta moisture contents in broilers. In Proceedings of the 29th Annual Australian Poultry Science Symposium, Sydney, Australia, 4–7 February 2018.
16. Owusu-Asiedu, A.; Patience, J.F.; Laarveld, B.; Van Kessel, A.G.; Simmins, P.H.; Zijlstra, R.T. Effects of guar gum and cellulose on digesta passage rate, ileal microbial populations, energy and protein digestibility, and performance of grower pigs. *J. Anim. Sci.* **2006**, *84*, 843–852. [CrossRef]
17. Konieczka, P.; Smulikowska, S. Viscosity negatively affects the nutritional value of blue lupin seeds for broilers. *Animal* **2018**, *12*, 1144–1153. [CrossRef]
18. Bedford, M.R.; Cowieson, A.J. Exogenous enzymes and their effects on intestinal microbiology. *Anim. Feed Sci. Tech.* **2012**, *173*, 76–85. [CrossRef]
19. Bao, Y.M.; Choct, M. Dietary NSP nutrition and intestinal immune system for broiler chickens. *Worlds Poult. Sci. J.* **2010**, *66*, 511–518. [CrossRef]
20. Rodehutscord, M.; Rückert, C.; Maurer, H.P.; Schenkel, H.; Schipprack, W.; Bach Knudsen, K.E.; Schollenberger, M.; Laux, M.; Eklund, M.; Siegert, W. Variation in chemical composition and physical characteristics of cereal grains from different genotypes. *Arch. Anim. Nutr.* **2016**, *70*, 87–107. [CrossRef]
21. Choct, M. Feed non-starch polysaccharides: Chemical structures and nutritional significance. *Feed Milling Int.* **1997**, *191*, 13–26.
22. Kiarie, E.; Romero, L.F.; Ravindran, V. Growth performance, nutrient utilization, and digesta characteristics in broiler chickens fed corn or wheat diets without or with supplemental xylanase. *Poult. Sci.* **2014**, *93*, 1186–1196. [CrossRef] [PubMed]
23. Munyaka, P.M.; Nandha, N.K.; Kiarie, E.; Nyachoti, C.M.; Khafipour, E. Impact of combined β-glucanase and xylanase enzymes on growth performance, nutrients utilization and gut microbiota in broiler chickens fed corn or wheat-based diets. *Poult. Sci.* **2016**, *95*, 528–540. [CrossRef] [PubMed]
24. Saleh, A.A.; Kirrella, A.A.; Abdo, S.E.; Mousa, M.M.; Badwi, N.A.; Ebeid, T.A.; Nada, A.L.; Mohamed, M.A. Effects of dietary xylanase and arabinofuranosidase combination on the growth performance, lipid peroxidation, blood constituents, and immune response of broilers fed low-energy diets. *Animals* **2019**, *9*, 467. [CrossRef] [PubMed]
25. Gracia, M.I.; Aranibar, M.J.; Lazaro, R.; Medel, P.; Mateos, G.G. Alpha-amylase supplementation of broiler diets based on corn. *Poult. Sci.* **2003**, *82*, 436–442. [CrossRef] [PubMed]
26. Musigwa, S.; Cozannet, P.; Morgan, N.; Swick, R.A.; Wu, S.-B. Multi-carbohydrase effects on energy utilization depend on soluble non-starch polysaccharides-to-total non-starch polysaccharides in broiler diets. *Poult. Sci.* **2020**, *100*, 788–796. [CrossRef]
27. Maharjan, P.; Mayorga, M.; Hilton, K.; Weil, J.; Beitia, A.; Caldas, J.; England, J.; Coon, C. Non-cellulosic polysaccharide content in feed ingredients and ileal and total tract non-cellulosic polysaccharide digestibility in 21-and 42-day-old broilers fed diets with and without added composite enzymes. *Poult. Sci.* **2019**, *98*, 4048–4057. [CrossRef]
28. Cobb-Vantress. Cobb 500 Broiler: Performance and Nutrition Supplement, 14, Cobb-Vantress.com. Available online: https://www.cobb-vantress.com/assets/Cobb-Files/product-guides/bdc20a5443/70dec630-0abf-11e9-9c88-c51e407c53ab.pdf (accessed on 7 April 2021).
29. Englyst, H.N.; Quigley, M.E.; Hudson, G.J. Determination of dietary fibre as non-starch polysaccharides with gas–liquid chromatographic, high-performance liquid chromatographic or spectrophotometric measurement of constituent sugars. *Analyst* **1994**, *119*, 1497–1509. [CrossRef]
30. Theander, O.; Åman, P.; Westerlund, E.; Andersson, R.; Pettersson, D. Total dietary fiber determined as neutral sugar residues, uronic acid residues, and Klason lignin (the Uppsala method): Collaborative study. *J. AOAC Int.* **1995**, *78*, 1030–1044. [CrossRef]

31. Morgan, N.K.; Keerqin, C.; Wallace, A.; Wu, S.-B.; Choct, M. Effect of arabinoxylo-oligosaccharides and arabinoxylans on net energy and nutrient utilization in broilers. *Anim. Nutr.* **2019**, *5*, 56–62. [CrossRef]
32. Short, F.J.; Gorton, P.; Wiseman, J.; Boorman, K.N. Determination of titanium dioxide added as an inert marker in chicken digestibility studies. *Anim. Feed Sci. Tech.* **1996**, *59*, 215–221. [CrossRef]
33. Williams, P.E.V.; Geraert, P.A.; Uzu, G.; Annison, G. Factors affecting non-starch polysaccharide digestibility in poultry. In *Feed Manufactoring in Southern Europe: New Challages*; Morand-Fehr, P., Ed.; CIHEAM: Zaragoza, Spain, 1997; pp. 125–134.
34. Ward, N.E. Debranching enzymes in corn/soybean meal-based poultry feeds: A review. *Poult. Sci.* **2020**, *100*, 765–775. [CrossRef] [PubMed]
35. Singh, A.K.; Kim, W.K. Effects of dietary fiber on nutrients utilization and gut health of poultry: A review of challenges and opportunities. *Animals* **2021**, *11*, 181. [CrossRef]
36. Rodríguez, M.L.; Rebolé, A.; Velasco, S.; Ortiz, L.T.; Treviño, J.; Alzueta, C. Wheat-and barley-based diets with or without additives influence broiler chicken performance, nutrient digestibility and intestinal microflora. *J. Sci. Food Agric.* **2012**, *92*, 184–190. [CrossRef]
37. Yaghobfar, A.; Kalantar, M. Effect of Non-Starch Polysaccharide (NSP) of Wheat and Barley Supplemented with Exogenous Enzyme Blend on Growth Performance, Gut Microbial, Pancreatic Enzyme Activities, Expression of Glucose Transporter (SGLT1) and Mucin Producer (MUC2) Genes of Broiler Chickens. *Braz. J. Poult. Sci.* **2017**, *19*, 629–638.
38. Ghiasvand, A.R.; Khatibjoo, A.; Mohammadi, Y.; Gharaei, M.A.; Shirzadi, H. Effect of fennel essential oil on performance, serum biochemistry, immunity, ileum morphology and microbial population, and meat quality of broiler chickens fed corn or wheat-based diet. *Br. Poult. Sci.* **2021**, *62*, 562–572. [CrossRef] [PubMed]
39. Annison, G.; Choct, M. Anti-nutritive activities of cereal non-starch polysaccharides in broiler diets and strategies minimizing their effects. *Worlds Poult. Sci. J.* **1991**, *47*, 232–242. [CrossRef]
40. Bedford, M.R.; Schulze, H. Exogenous enzymes for pigs and poultry. *Nutr. Res. Rev.* **1998**, *11*, 91–114. [CrossRef]
41. Nguyen, H.T.; Bedford, M.R.; Wu, S.-B.; Morgan, N.K. Soluble non-starch polysaccharide modulates broiler gastrointestinal tract environment. *Poult. Sci.* **2021**, *100*, 101183. [CrossRef]
42. Saki, A.A.; Matin, H.R.H.; Tabatabai, M.M.; Zamani, P.; Harsini, R.N. Microflora population, intestinal condition and performance of broilers in response to various rates of pectin and cellulose in the diet. *Eur. Poult. Sci.* **2010**, *74*, 183–188.
43. Rinttilä, T.; Apajalahti, J. Intestinal microbiota and metabolites—Implications for broiler chicken health and performance1. *J. Appl. Poult. Res.* **2013**, *22*, 647–658. [CrossRef]
44. Bautil, A.; Verspreet, J.; Buyse, J.; Goos, P.; Bedford, M.R.; Courtin, C.M. Age-related arabinoxylan hydrolysis and fermentation in the gastrointestinal tract of broilers fed wheat-based diets. *Poult. Sci.* **2019**, *98*, 4606–4621. [CrossRef] [PubMed]
45. Sadeghi, A.; Toghyani, M.; Gheisari, A. Effect of various fiber types and choice feeding of fiber on performance, gut development, humoral immunity, and fiber preference in broiler chicks. *Poult. Sci.* **2015**, *94*, 2734–2743. [CrossRef] [PubMed]
46. Vangsøe, C.T.; Bonnin, E.; Joseph-Aime, M.; Saulnier, L.; Neugnot-Roux, V.; Bach Knudsen, K.E. Improving the digestibility of cereal fractions of wheat, maize, and rice by a carbohydrase complex rich in xylanases and arabinofuranosidases: An in vitro digestion study. *J. Sci. Food Agric.* **2021**, *101*, 1910–1919. [CrossRef] [PubMed]
47. Malunga, L.N.; Beta, T. Isolation and identification of feruloylated arabinoxylan mono-and oligosaccharides from undigested and digested maize and wheat. *Heliyon* **2016**, *2*, e00106. [CrossRef]
48. Jamroz, D.; Jakobsen, K.; Bach Knudsen, K.E.; Wiliczkiewicz, A.; Orda, J. Digestibility and energy value of non-starch polysaccharides in young chickens, ducks and geese, fed diets containing high amounts of barley. *Comp. Biochem. Physiol. Part A Mol. Integr. Physiol.* **2002**, *131*, 657–668. [CrossRef]
49. Zhang, L.; Xu, J.; Lei, L.; Jiang, Y.; Gao, F.; Zhou, G.H. Effects of xylanase supplementation on growth performance, nutrient digestibility and non-starch polysaccharide degradation in different sections of the gastrointestinal tract of broilers fed wheat-based diets. *Asian-Australas. J. Anim. Sci.* **2014**, *27*, 855–861. [CrossRef] [PubMed]
50. Meng, X.; Slominski, B.A.; Nyachoti, C.M.; Campbell, L.D.; Guenter, W. Degradation of cell wall polysaccharides by combinations of carbohydrase enzymes and their effect on nutrient utilization and broiler chicken performance. *Poult. Sci.* **2005**, *84*, 37–47. [CrossRef]

animals

MDPI

Article

Influences of L-Arginine In Ovo Feeding on the Hatchability, Growth Performance, Antioxidant Capacity, and Meat Quality of Slow-Growing Chickens

Panpan Lu, Thanidtha Morawong, Amonrat Molee and Wittawat Molee *

School of Animal Technology and Innovation, Institute of Agricultural Technology, Suranaree University of Technology, Nakhon Ratchasima 30000, Thailand; PanpanLu12@hotmail.com (P.L.); Thanidtha93@gmail.com (T.M.); amonrat@sut.ac.th (A.M.)
* Correspondence: wittawat@sut.ac.th; Tel.: +66-8-9920-4970

Simple Summary: The nutrition and health status of the embryo in the hatching process directly influence the hatchability and chicken performance post-hatch in poultry production. The in ovo feeding (IOF) technique provides a viable way to improve the embryonic development and chicken performance post-hatch. Thus, the hypothesis of this study was that supplementing L-arginine (Arg) into embryos could positively affect the hatchability, growth performance, antioxidant capacity, and meat quality of slow-growing chickens. The results of this study demonstrate that IOF of Arg positively affected the antioxidant capacity of the breast muscle in the starter period, and there was no effect on the hatchability, growth performance, carcass traits, and meat quality. Overall, our findings suggest that IOF of Arg may have beneficial effects on chicken health without compromising the hatchability, subsequent growth, and meat quality.

Abstract: The aim of this study was to evaluate the effects of in ovo feeding (IOF) of L-arginine (Arg) on the hatchability, growth performance, antioxidant capacity, and meat quality of slow-growing chickens. A total of 480 eggs were randomly divided into a non-injected control group (NC group) and a 1% Arg-injected group (Arg group). On day 18 of incubation, 0.5 mL of Arg solution was injected into the embryonic amnion in the Arg group. Upon hatching, 160 mixed-sex chickens were randomly assigned to two groups, with four replicates per group. This experiment lasted for 63 days. The results showed that the hatchability, growth performance, carcass traits, and meat quality were not significantly different ($p > 0.05$) between the two groups. However, the malondialdehyde (MDA) content was lower ($p < 0.05$), and the glutathione (GSH) level was higher ($p < 0.05$) on day of hatching in the Arg group. The total antioxidant capacity (T-AOC) activity was increased ($p < 0.05$) on day 21 post-hatch in the Arg group compared to that in the NC group. In conclusion, IOF of Arg increased the antioxidant capacity of the breast muscle in the starter period, which may have a positive effect on health status of slow-growing chickens post-hatch.

Keywords: L-arginine; in ovo feeding; growth performance; antioxidant capacity; meat quality; slow-growing chicken

check for updates

Citation: Lu, P.; Morawong, T.; Molee, A.; Molee, W. Influences of L-Arginine In Ovo Feeding on the Hatchability, Growth Performance, Antioxidant Capacity, and Meat Quality of Slow-Growing Chickens. *Animals* **2022**, *12*, 392. https://doi.org/10.3390/ani12030392

Academic Editors: Natalie Morgan and Youssef A. Attia

Received: 5 January 2022
Accepted: 3 February 2022
Published: 7 February 2022

1. Introduction

With the developing research on poultry nutrition, the nutrition and health status of the embryo during the hatching process—which have impacts on economic profits—have become the focus of research. It is known that avian embryonic development depends on the nutrients deposited in the fertile egg. A sufficient supply of nutrients is a good start for hatchability and subsequent growth. However, under commercial poultry production conditions, the nutrition of eggs may be insufficient to fulfill the requirement for reaching the maximum development of the embryo [1]. This is due to the variety of physiological activities of the embryo, which consume large amounts of energy that come from the nutrients deposited in the eggs, limiting the embryonic development and chicken growth post-hatch [2,3]. Meanwhile, the oxygen consumption and metabolic rates rise during and after hatching to meet the energy demand of the embryo's physiological activities [4], which tends to produce reactive oxygen species (ROS) [5]. Particularly, yolk lipid contains an abundant amount of polyunsaturated fatty acids that can be easily attacked by ROS [6], and the excess of which causes oxidative stress. This results in the oxidative damage of biological molecules [7], ultimately compromising the embryonic growth and chicken performance post-hatch [8].

An in ovo feeding (IOF) technique may be an effective way to achieve the full genetic potential of chick growth post-hatch; that is, IOF of additional nutrients to the embryonic amnion during the late period of incubation [9]. Growing evidence has reported that IOF of different types of exogenous nutrients is beneficial for embryonic development and subsequent growth in poultry [9–12].

Arginine (Arg) is known as an essential amino acid for poultry, wherein a high amount of Arg is required in the starter period and plays multiple roles in the biological and physiological activities [13]. In addition, Arg can be converted into glucose for regulating the energy metabolism [14]. Thereby adding Arg into embryo could positively affect the hatchability and subsequent performance post-hatch. Previous studies indicated that the IOF of Arg affected the breast muscle growth and energy metabolism of chickens in the starter period [15–17]. Additionally, the IOF of Arg positively affected the hatchability and subsequent growth of Japanese quails and pigeons [18,19]. Importantly, Arg could reduce oxidative damage and improve antioxidant capacity [20,21]. Some studies reported that dietary Arg enhanced the antioxidant capacity and chicken growth [22,23]. However, no study has reported about the effect of IOF of Arg on the hatchability, growth performance, antioxidant capacity, and meat quality of slow-growing chickens.

The slow-growing Korat chicken (KRC) is a crossbreed between a male of the Thai Leung Hang Khao line and a female of the Suranaree University of Technology (SUT) line in Thailand, which is characterized by superior meat quality with low fat, rich collagen, and good texture [24,25]. They are sent to the market at 1.2–1.5 kg bodyweight at about 10 weeks of age. Rearing of this breed is encouraged by the agriculture sector of Thailand which advocates the small-scale farmers to rear indigenous chickens to develop the rural economy. Thus, in order to increase the productivity of KRC, we aimed to assess the effects of the IOF of Arg on the hatchability, growth performance, antioxidant capacity, and meat quality of slow-growing chickens. We hypothesized that the IOF of Arg into the amnion may benefit the hatchability and performance of market age chickens.

2. Materials and Methods

The experimental protocols applied in this study were approved by the Ethics Committee on Animal Use of the SUT, Nakhon Ratchasima, Thailand (user application ID: U1-02633-2559). The experiment was conducted at SUT farm.

2.1. Eggs and Incubation

Fertile eggs (SUT female and Leung Hang Khao male) were collected from the SUT farm (Nakhon Ratchasima, Thailand). These eggs (57.0 ± 3.0 g) were randomly transferred into an automatic incubator (Model 192, Petersime Incubation Equipment Co., Ltd., Zulte, Belgium) with optimal conditions (37.8 °C and 60% relative humidity), and the eggs were turned automatically every hour. On day 14 of embryonic development, the eggs were candled by electric torch, and the unfertilized and nonviable eggs were discarded. A total of 480 viable embryos (59.0 ± 1.0 g) were randomly assigned to two treatment groups with four replicates of 60 eggs each, wherein two trays were used for each treatment group, and four incubator trays were used in this experiment.

2.2. IOF Procedure

On day 18 of embryonic development, the IOF procedure was performed. Before injection, the Arg solutions was freshly prepared with 0.9% saline (A. N. B. Laboratories Co., Ltd., Bangkok, Thailand). The concentration of the Arg solution was 1%, which was selected on the basis of a previous study [16], with minor modification. Specifically, 1.5 g of Arg (Sigma-Aldrich, St. Louis, MO, USA) was dissolved in 150 mL of 0.9% saline, which was equivalent to 5 mg of Arg per egg. The Arg solution was autoclaved at 120 °C for 15 min prior to injection.

After preparing the Arg solution, all the eggs were taken from the four incubator trays. The eggs from the two trays served as the non-injected group (NC group), and the other two trays were the 1% Arg-injected group (Arg group). The eggs were rechecked to make sure that the embryos were alive. The location (large end surface) of the eggs in the Arg group were disinfected with 75% alcohol, a hole was created by sterile needle, and then 0.5 mL solution was injected into the amniotic sac with a 21-gauge needle based on the method described by Uni et al. [26]. The holes were sealed using paraffin immediately after injection, and the eggs were sent back to the hatching baskets. The temperature and humidity of incubator were 37.2 °C and 60%, respectively. This IOF process was finished within 2 h. The eggs of the NC group were kept in the same environmental condition (outside of incubator) as that of the Arg group. The eggs from each of the two treatments were randomly allocated into four replicates with 60 eggs each, and each basket was regarded as a replicate. All eggs continued to perform the hatchery program.

2.3. Hatchability Rate

On day of hatching (DOH), the number of birds hatched was counted, and the hatchability rate was calculated as the numbers of chicks hatched divided by the fertilized eggs per replicate.

2.4. Animals, Experimental Design, and Management

Upon hatching, all chicks of each treatment group were pooled and weighed. In total, 160 mix-sexed chicks per treatment group were randomly divided into four replicates of 40 chicks, and the selected chicks had similar weights, which were close to the average body weight (BW) of each treatment group. Eight floor pens were provided for the two treatment groups, and each pen was a replicate. The housing conditions were monitored to make sure the similar environment condition in each pen followed the guidelines of the SUT farm. All chicks were allowed commercial feed (Charoen Pokphand Co., Ltd., Nakhon Ratchasima, Thailand) and fresh water ad libitum. This experiment lasted for 63 days. The nutrient composition of the basal diet for the starter (DOH–21), grower (D22–42), and finisher (D43–63) periods which analyzed by AOAC method [27] are shown in Table 1.

Table 1. Nutrient composition of the basal diets.

	Starter (DOH–21)	**Grower (D22–42)**	**Finisher (D43–63)**
	Analyzed nutrient composition, (g/kg)		
Dry matter	938.3	935.1	942.1
Gross energy (MJ/Kg)	125.4	129.6	133.8
Crude protein	227.2	204.6	186.5
Crude fat	52.0	67.4	66.6
Crude fiber	34.4	34.5	35.5
Crude ash	47.0	45.8	41.9
Lysine	17.8	14.3	9.2
Methionine	3.4	2.5	2.8
Threonine	10.1	8.5	7.3
Arginine	15.8	11.3	5.5

2.5. Growth Performance and Carcass Traits Indices

The BW and feed intake (FI) were recorded by a replicate weekly. Then, the body weight gain (BWG) and feed conversion ratio (FCR) were calculated by a replicate. No chickens died during the experiment. On day 63 (D63) post-hatch, two male chickens from each replicate with BW close to the average BW of their replicate were chosen following a 12 h fast, and killed after the electrical stunning, then were bled and defeathered. Then, carcasses (with the giblets, head, neck, and hocks removed) were chilled at 4 °C for 24 h. After chilling, the carcasses were weighed to determine the eviscerated yield percentage based on the live BW. The percentages of liver, heart, and gizzard were calculated based on the live BW. The entire right breast muscle was measured for meat quality.

2.6. Assay of the Malondialdehyde Level and Antioxidant Capacity in the Breast Muscle

On DOH, day 21 (D21), day 42 (D42), and D63 post-hatch, two male chickens per replicate with BW close to the average BW of their replicate were chosen, weighed, and killed after using chloroform after 12 h fasting. The left muscle tissue was stored at −80 °C for antioxidant capacity.

Malondialdehyde (MDA) is a marker for monitoring oxidative stress. The supernatant of the breast muscle was used to measure the MDA concentration by thiobarbituric acid (TBA) method using the Lipid Peroxidation (MDA) Assay Kit (Catalog Number MAK085, Sigma-Aldrich, St. Louis, MO, USA), which was scanned at 532 nm (A532). The details of measurements followed the manufacturer's instructions. The results obtained were expressed as nmole of MDA per mg muscle.

The level of glutathione (GSH) was measured through reaction with 5,5′-dithio-bis-nitrobenzoic acid at 412 nm using the Glutathione Assay Kit (Catalog Number CS0260, Sigma-Aldrich, St. Louis, MO, USA), following the manufactures' instructions. The results obtained were expressed as nmoles GSH per mg muscle.

The total antioxidant capacity (T-AOC) activity was determined by the reduction of Cu^{2+} to Cu^{+} that was scanned at 570 nm using the Total Antioxidant Capacity Assay Kit (Catalog Number MAK187, Sigma-Aldrich, St. Louis, MO, USA). The procedures were performed based on the manufacturer's instructions. The T-AOC values were expressed as nmole per mg of protein.

2.7. Determination of Meat Quality

The meat quality was measured using the following parameters: meat pH, color, shear force, drip loss, and cooking loss. The pH was determined using a hand-held digital pH meter (Ultra Basic pH meter, Model UB10A, Denver Instrument, Bohemia, NY, USA) on the breast meat at 45 min and 24 h postmortem. The color values of lightness (L*), redness (a*), and yellowness (b*) were measured by a chroma meter (Model CR 300, Minolta, Osaka, Japan) on the breast meat at 24 h postmortem. Drip loss was determined as described by Zhang et al. [28], with some modifications. Briefly, samples with a size of 3 × 2 × 1 cm were cut from the breast meat, weighed, and placed in a plastic bag, and left freely hanging at 4 °C. After 24 h, the samples were wiped and reweighed. The drip loss percentage was calculated as follows: (initial weight − final weight)/initial weight × 100. The cooking loss and shear force were determined following the method of Cong et al. [29], with some modifications. The samples were weighed at 24 h postmortem, packaged in a sealed plastic bag, and cooked in a digital water bath at 85 °C until the internal temperature reached 77 °C. Then the samples were taken out and cooled to room temperature and reweighed to calculate the cooking loss. The formula was as follows: (initial weight − final weight)/initial weight × 100. The cooked samples were used for the shear force determination. After the cooking loss determination, the samples were cut to small strips of 1 × 1 × 3 cm in size, and the values were measured using the Instron texture system (Model 5565, Instron Corporation, Burlington, ON, Canada).

2.8. Statistical Analysis

A completely randomized design (CRD) was applied in this study. The data were analyzed by an independent t-test using SPSS software (IBM Corp. 1989, 2013. New York, NY, USA), and the statistical significances between the two groups were denoted at $p < 0.05$. The results were expressed as the mean and standard error of the mean (SEM). Pearson correlation coefficients were evaluated to determine the relationship between the antioxidant capacity and meat quality.

3. Results

3.1. Hatchability

As shown in Table 2, the IOF of Arg did not significantly increase ($p > 0.05$) the hatchability as compared to that of the NC group.

Table 2. Effects of in ovo feeding of L-arginine on the hatchability of slow-growing chickens.

Treatment	Hatchability
NC	86.25
Arg	87.09
SEM	1.910
p-value	0.768

NC = non-injected control group. Arg = 1% L-arginine-injected group. SEM = standard error of the mean. Values are means with $n = 4$ per treatment.

3.2. Growth Performance and Carcass Traits

As presented in Table 3, there were no significant differences ($p > 0.05$) in the BWG, FI, and FCR between the NC and Arg groups. In Table 4, the IOF of Arg did not improve ($p > 0.05$) the carcass traits (eviscerated yield, heart, liver, and gizzard) as compared to that in the NC group.

Table 3. Effects of in ovo feeding of L-arginine on growth performance of slow-growing chickens.

Items	Treatments		SEM	*p*-Value
	NC	Arg		
BWG (g)				
DOH–21	280.21	273.40	2.394	0.122
D22–42	464.02	457.61	5.271	0.591
D43–63	488.37	452.69	16.085	0.218
FI (g)				
DOH–21	569.78	573.37	25.680	0.924
D22–42	1026.70	1024.82	21.011	0.955
D43–63	1261.64	1233.98	37.722	0.652
FCR (g/g)				
DOH–21	2.03	2.10	0.092	0.581
D22–42	2.21	2.24	0.052	0.794
D43–63	2.58	2.73	0.045	0.084

BWG = body weight gain; FI = feed intake; FCR = feed conversion ratio (FI: BWG). DOH = day of hatching; D21 = day 21; D42 = day 42; D63 = day 63. NC = non-injected control group. Arg = 1% L-arginine-injected group. SEM = standard error of the mean. Values are means with $n = 4$ per treatment.

3.3. MDA Level and Antioxidant Capacity

The results of MDA content and antioxidant capacity are shown in Table 5. Compared with that of the NC group, a decrease in MDA contents and an increase in GSH levels were found ($p < 0.05$) in the Arg group on DOH, but no significant differences were found on D21, D42, and D63 post-hatch, respectively. A significant improvement of T-AOC activities was found ($p < 0.05$) on D21 post-hatch in the Arg group compared to that in the NC group, but it had no effect on DOH, D42, and D63 post-hatch, respectively.

Table 4. Effects of in ovo feeding of L-arginine on carcass traits of slow-growing chickens on day 63.

Items	Treatments		SEM	*p*-Value
	NC	Arg		
Eviscerated yield (%)	65.36	66.89	0.910	0.320
Heart (%)	1.06	1.09	0.213	0.931
Liver (%)	2.00	1.83	0.100	0.418
Gizzard (%)	2.30	2.23	0.150	0.851

Eviscerated yield (%) = Eviscerated carcass weight/live body weight ∗ 100; Heart (%) = Heart weight/live body weight ∗ 100; Liver (%) = Liver weight/live body weight ∗ 100; Gizzard (%) = Gizzard weight/live body weight ∗ 100. NC = non-injected control group. Arg = 1% L-arginine-injected group. SEM = standard error of the mean. Values are means with $n = 8$ per treatment.

Table 5. Effects of in ovo feeding of L-arginine on antioxidant capacity in the breast muscle of slow-growing chickens.

Items	Treatments		SEM	*p*-Value
	NC	Arg		
MDA (nmol/mg muscle)				
DOH	0.18 ^{a}	0.11 ^{b}	0.016	0.044
D21	0.35	0.28	0.032	0.195
D42	0.32	0.29	0.035	0.680
D63	0.38	0.34	0.027	0.345
T-AOC (nmol/mg protein)				
DOH	12.15	14.41	0.667	0.077
D21	3.67 ^{b}	4.36 ^{a}	0.114	0.011
D42	4.79	4.83	0.192	0.881
D63	4.16	4.33	0.168	0.492
GSH (nmol/mg muscle)				
DOH	4.58 ^{b}	5.88 ^{a}	0.181	0.012
D21	1.93	2.80	0.312	0.127
D42	1.84	2.13	0.199	0.387
D63	1.36	1.61	0.178	0.448

MDA = malondialdehyde; T-AOC = total antioxidant capacity; GSH = glutathione. DOH = day of hatching; D21 = day 21; D42 = day 42; D63 = day 63. NC = non-injected control group. Arg = 1% L-arginine-injected group. SEM = standard error of the mean. Values are means with $n = 8$ per treatment. Means with different superscripts in the same row differ significantly at $p < 0.05$.

3.4. Meat Quality and Correlation between the Meat Quality and Antioxidant Capacity

The meat quality results are shown in Table 6. The meat quality ($pH^{45\ min}$, $pH^{24\ h}$, color, drip loss, cooking loss, and shear force) did not differ ($p > 0.05$) by the IOF of Arg compared to that in the NC group. No significant correlation was found ($p > 0.05$) between the meat quality and antioxidant capacity of slow-growing chickens (Table 7).

Table 6. The effects of in ovo feeding of L-arginine on meat quality of slow-growing chickens on day 63.

Items	Treatments		SEM	*p*-Value
	NC	Arg		
$pH^{45\ min}$	5.99	5.91	0.082	0.640
$pH^{24\ h}$	5.84	5.83	0.062	0.849
Color				
L*	52.24	51.81	1.274	0.819
a*	2.33	2.23	0.297	0.832
b*	1.95	1.26	0.339	0.255
Drip loss (%)	11.07	10.49	0.854	0.673
Cooking loss (%)	22.94	24.68	1.137	0.462
Shear force (kg/cm^2)	1.99	2.28	0.092	0.125

Color: L* = lightness; a* = redness; b* = yellow. NC = non-injected control group. Arg = 1% L-arginine-injected group. SEM = standard error of the mean. Values are means with $n = 8$ per treatment.

Table 7. Correlation coefficients between the meat quality and antioxidant capacity of slow-growing chickens on day 63.

	$pH^{45\ min}$	$pH^{24\ h}$	a*	b*	L*	Drip Loss	Cooking Loss	Shear Force
MDA	0.682	0.479	−0.079	0.047	−0.372	−0.177	−0.796	−0.107
T-AOC	0.099	0.166	−0.385	−0.342	0.690	0.341	0.690	0.087
GSH	−0.592	0.842	−0.865	−0.838	−0.017	−0.381	−0.108	−0.589

Color: L* = lightness; a* = redness; b* = yellow. MDA = malondialdehyde; T-AOC = total antioxidant capacity; GSH = glutathione.

4. Discussion

The hatchability is one of the main indices for determining the success of the IOF technique. In the current study, the hatchability was similar between the two groups. This result is similar to that in previous studies in poultry [14,30,31]. On the contrary, two studies reported that the hatchability was increased in poultry [18,19], while Tahmasebi and Toghyani [32] reported that the hatchability of broiler chickens decreased after the IOF of Arg. The hatching process is related to the energy metabolism activity because the reserved glycogen of the fertilized egg would be consumed by the embryo to fuel the energy demand needed for hatching activities [2]. The energy supply may be insufficient to meet the needs of maximum hatching activities [33], which in turn forces muscle to break down protein and then produce glucose by gluconeogenesis, which negatively influences the embryonic development [34]. Thus, high glycogen storage is necessary to improve the hatchability [33]. External nutrients have the ability to improve the energy status to meet the high demand of glucose for hatching activities [26]. Arginine has a vital role in regulating energy metabolisms that can convert glucose by gluconeogenesis [35]. It has been reported that the IOF of 1% Arg increased the glycogen and glucose concentrations of the liver and pectoral muscle for regulating energy metabolism in broiler chickens [16]. Combining the results of the current study with those of previous studies, it is speculated that the IOF of 1% Arg may not improve glucose deposition and may limit the energy utilization for the hatching process. On the other hand, the unaffected hatchability indicates that the IOF technique is a safe method for the current study. However, further study should be undertaken to explore the energy metabolism by the IOF of Arg.

In this study, the growth performance and carcass traits did not respond to the in ovo administration of Arg. These results are inconsistent with those of previous studies. Gao et al. [31] demonstrated that the FI and BWG were increased during 1 to 21 and 1 to 42 d post-hatch by the IOF of Arg in broiler chickens. Toghyani et al. [36] observed that the IOF of Arg caused a significant increase in BWG and FI from 1 to 42 d post-hatch in broiler chickens. Growth performance is associated with the gastrointestinal tract development that is controlled by gastrointestinal hormones and intestinal enzyme activity [37]. It has been reported that the IOF of Arg into the amnion promoted the release of gastrointestinal hormones and intestinal enzyme activity, and then improved the gastrointestinal tract development, finally increasing FI and BWG [31,38,39]. According to the current results, it is speculated that the IOF of Arg may not affect gastrointestinal tract development. In other words, the gizzard growth of chickens cannot be affected by the addition of Arg solution, and chickens are unable to store, digest, and absorb more feed. Similar to the results obtained for the carcass traits in this study, no significant differences were found between the two groups on market day. The current result is in line with the report of Tahmasebi and Toghyani [32], who found that the carcass, liver, and heart were not affected by the IOF of Arg in broiler chickens on market day. Conversely, Al-Daraji et al. [19] obtained the expected results (carcass, liver, heart, and gizzard) after in ovo injection of Arg in Japanese quails on market day. However, further study is necessary to reveal the gastrointestinal tract development, such as the release of gastrointestinal hormones and the digestive and absorptive capacity of the gastrointestinal tract.

The incubation in birds is associated with the production of oxidative stress. Malondialdehyde is known as a biomarker that monitors the degree of oxidative stress [40].

Our study revealed that the MDA content in the breast muscle was decreased on DOH post-hatch by the IOF of Arg. In agreement with our report, Duan et al. [23] found that supplementing Arg in the diet of late-laying hens significantly reduced the MDA contents in the serum and egg yolk of broiler breeders as well as the tissues of broilers on D1 post-hatch. These results indicate that the Arg deposited in the egg could be transferred to their offspring and exhibit the function of eliminating oxygen free radicals. Moreover, Atakisi et al. [41] and Ruan et al. [22] reported that dietary Arg decreased the MDA content in Japanase quails and yellow-feathered chickens. Our observation suggests that a certain amount of Arg is needed to scavenge free radicals produced by physiological metabolic activities in embryonic development, which may benefit the chick quality post-hatch.

The antioxidant defense system plays an important role in the maintenance of prooxidant–antioxidant balance of normal physiological metabolic activity in animals. The GSH is a biomarker of cellular antioxidant defense capacity [42], which can act against ROS generation and decrease the oxidative stress of cells because it is related to the enzymatic processes that reduce H_2O_2 into oxidized glutathione and other hybrid disulfides by GSH metabolism [43]. Our results showed that the GSH level was increased on DOH by the IOF of Arg. This result is similar to those in the reports of Liang et al. [21] and Xiao et al. [44], wherein it was stated that supplemental Arg in rats increased the GSH levels in the liver, plasma, and jejunum. The GSH level depends on the nutritional status of their body. Arginine is a substrate of glutamate synthesis that may contribute to GSH synthesis and is responsible for the antioxidant system [45,46]. The T-AOC is used as an integrative indicator of total antioxidant capacity in animal bodies [47]. Duan et al. [23] indicated that dietary supplementation with Arg increased the T-AOC activities in the serum and egg yolk of laying hens as well as tissue of broilers on D1 post-hatch. Ruan et al. [22] found that dietary Arg improved the T-AOC capacity of the small intestine in yellow-feathered chickens. Atakisi et al. [41] reported that dietary Arg in Japanese quails improved the T-AOC activity. In agreement with earlier studies, our data showed that T-AOC activity was significantly increased by the IOF of Arg on D21 post-hatch. These results for the GSH and T-AOC suggest that increased Arg in the breast muscle enhanced the antioxidant capacity against lipid peroxidation of slow-growing chickens during the starter period.

Meat quality is closely associated with the purchasing desire of consumers. The pH value is an important index to monitor the rate of muscle anaerobic glycolysis after slaughter [48]. The pH of meat is highly related to color [49]. The drip loss, cooking loss, and shear force are also important indicators of meat quality for detecting sensory characteristics (tenderness, juiciness, and flavor) [50]. Previous studies reported that dietary Arg did not have any effect on the pH, color, drip loss, and cooking loss in broiler chickens [51,52]. These results are consistent with that of the current study, wherein no significant differences in pH, color, drip loss, cooking loss, and shear force were found between the two groups. Moreover, the pH values observed in our study were within the acceptable range (5.7 to 6.1) for chicken breast meat [53]. Conversely, in pigs, dietary Arg decreased the drip loss and cooking loss and maintained the meat quality [54,55]. The different results of these studies may be due to the difference in species. In addition, the correlation between the antioxidant capacity and meat quality was further tested in this study. A previous study reported that dietary Arg enhanced meat quality, while increasing the antioxidant capacity and attenuating oxidative stress in pigs [54]. However, we did not find any correlation between the antioxidant capacity and meat quality in our study. It is suggested that the IOF of Arg may not cause any improvement in meat quality. Due to the limited information about the effects of Arg by in ovo administration in chicken meat, the differences in the results may be due to the long duration between the IOF and market age.

5. Conclusions

In conclusion, the IOF of 1% Arg did not influence their performance nor meat quality on market day, and the antioxidant capacity was time-limited and limited to the starter period only. Thus, these results suggest that the IOF of Arg serving as an early nutrition strategy may have a beneficial effect on chicken health without compromising the hatchability, subsequent growth, and meat quality.

Author Contributions: Conceptualization, W.M. and A.M.; methodology, W.M. and A.M.; investigation, P.L. and T.M.; formal analysis, P.L. and W.M.; writing—original draft, P.L.; writing—review and editing, W.M. and P.L.; supervision, W.M. and A.M.; project administration, W.M.; funding acquisition, W.M. and A.M. All authors have read and agreed to the published version of the manuscript.

Funding: This research was funded by the Center of Excellence on Technology and Innovation of Korat chicken Business Development, which was granted by the Suranaree University of Technology (grant number CoE3-303-62-60-02).

Institutional Review Board Statement: All experimental protocols were approved by the Ethics Committee on Animal Use of the Suranaree University of Technology (SUT), Nakhon Ratchasima, Thailand (user application ID: U1-02633-2559).

Informed Consent Statement: Not applicable.

Data Availability Statement: The data presented in this study are available on request from the corresponding author.

Acknowledgments: The first author gratefully acknowledge the Suranaree University of Technology for their financial support through the Potential Graduate Students Scholarship.

Conflicts of Interest: The authors declare no conflict of interest.

References

1. Uni, Z.; Yadgary, L.; Yair, R. Nutritional limitations during poultry embryonic development. *J. Appl. Poult. Res.* **2012**, *21*, 175–184. [CrossRef]
2. Christensen, V.L.; Wineland, M.J.; Fasenko, G.M.; Donaldson, W.E. Egg storage effects on plasma glucose and supply and demand tissue glycogen concentrations of broiler embryos. *Poult. Sci.* **2001**, *80*, 1729–1735. [CrossRef] [PubMed]
3. Ohta, Y.; Tsushima, N.; Koide, K.; Kidd, M.; Ishibashi, T. Effect of amino acid injection in broiler breeder eggs on embryonic growth and hatchability of chicks. *Poult. Sci.* **1999**, *78*, 1493–1498. [CrossRef] [PubMed]
4. Hohtola, E. Facultative and obligatory thermogenesis in young birds: A cautionary note. *Comp. Biochem. Physiol. A Mol. Integr. Physiol.* **2002**, *131*, 733–739. [CrossRef]
5. Surai, P.F. Tissue-specific changes in the activities of antioxidant enzymes during the development of the chicken embryo. *Br. Poult. Sci.* **1999**, *40*, 397–405. [CrossRef]
6. Speake, B.K.; Noble, R.C.; Murray, A.M.B. The utilization of yolk lipids by the chick embryo. *World Poult. Sci. J.* **1998**, *54*, 319–334. [CrossRef]
7. Surai, P.F. *Natural Antioxidants in Avian Nutrition and Reproduction*; Nottingham University Press: Nottingham, UK, 2002.
8. Surai, P.F.; Fisinin, V.I.; Karadas, F. Antioxidant systems in chick embryo development. Part 1. Vitamin E, carotenoids and selenium. *Anim. Nutr.* **2016**, *2*, 1–11. [CrossRef]
9. Kadam, M.M.; Barekatain, M.R.; Bhanja, S.K.; Iji, P.A. Prospects of in ovo feeding and nutrient supplementation for poultry: The science and commercial applications—A review. *J. Sci. Food Agric.* **2013**, *93*, 3654–3661. [CrossRef]
10. Araújo, I.C.S.; Café, M.B.; Noleto, R.A.; Martins, J.M.S.; Ulhoa, C.J.; Guareshi, G.C.; Reis, M.M.; Leandro, N.S.M. Effect of vitamin E in ovo feeding to broiler embryos on hatchability, chick quality, oxidative state, and performance. *Poult. Sci.* **2019**, *98*, 3652–3661. [CrossRef]
11. Elwan, H.A.M.; Elnesr, S.S.; Xu, Q.; Xie, C.; Dong, X.; Zou, X. Effects of In Ovo Methionine-Cysteine Injection on Embryonic Development, Antioxidant Status, IGF-I and TLR4 Gene Expression, and Jejunum Histomorphometry in Newly Hatched Broiler Chicks Exposed to Heat Stress during Incubation. *Animals* **2019**, *9*, 25. [CrossRef]
12. Zhang, H.; Elliott, K.E.C.; Durojaye, O.A.; Fatemi, S.A.; Schilling, M.W.; Peebles, E.D. Effects of in ovo injection of L-ascorbic acid on growth performance, carcass composition, plasma antioxidant capacity, and meat quality in broiler chickens1,2,3. *Poult. Sci.* **2019**, *98*, 3617–3625. [CrossRef]
13. Wu, G.; Bazer, F.W.; Davis, T.A.; Kim, S.W.; Li, P.; Rhoads, J.M.; Satterfield, M.C.; Smith, S.B.; Spencer, T.E.; Yin, Y. Arginine metabolism and nutrition in growth, health and disease. *Amino Acids* **2009**, *37*, 153–168. [CrossRef]
14. Tangara, M.; Chen, W.; Xu, J.; Huang, F.R.; Peng, J. Effects of in ovo feeding of carbohydrates and arginine on hatchability, body weight, energy metabolism and perinatal growth in duck embryos and neonates. *Br. Poult. Sci.* **2010**, *51*, 602–608. [CrossRef] [PubMed]
15. Yu, L.L.; Gao, T.; Zhao, M.M.; Lv, P.A.; Zhang, L.; Li, J.L.; Jiang, Y.; Gao, F.; Zhou, G.H. Effects of in ovo feeding of l-arginine on breast muscle growth and protein deposition in post-hatch broilers. *Animal* **2018**, *12*, 2256–2263. [CrossRef] [PubMed]

16. Yu, L.L.; Gao, T.; Zhao, M.M.; Lv, P.A.; Zhang, L.; Li, J.L.; Jiang, Y.; Gao, F.; Zhou, G.H. In ovo feeding of L-arginine alters energy metabolism in post-hatch broilers. *Poult. Sci.* **2018**, *97*, 140–148. [CrossRef]
17. Li, Y.; Wang, Y.; Willems, E.; Willemsen, H.; Franssens, L.; Buyse, J.; Decuypere, E.; Everaert, N. In ovo L-arginine supplementation stimulates myoblast differentiation but negatively affects muscle development of broiler chicken after hatching. *J. Anim. Physiol. Anim. Nutr.* **2016**, *100*, 167–177. [CrossRef] [PubMed]
18. Zhang, X.Y.; Wan, X.P.; Miao, L.P.; Zou, X.T.; Dong, X.Y. Effects of in ovo feeding of l-arginine on hatchability, hatching time, early posthatch development, and carcass traits in domestic pigeons (Columba livia). *J. Anim. Sci.* **2017**, *95*, 4462–4471. [CrossRef]
19. Al-Daraji, H.J.; Al-Mashadani, A.A.; Al-Mashadani, W.K.; Al-Hassani, A.S.; Mirza, H.A. Effect of in ovo injection with L-arginine on productive and physiological traits of Japanese quail. *S. Afr. J. Anim. Sci.* **2012**, *42*, 139–145. [CrossRef]
20. Lin, W.T.; Yang, S.C.; Tsai, S.C.; Huang, C.C.; Lee, N.Y. L-Arginine attenuates xanthine oxidase and myeloperoxidase activities in hearts of rats during exhaustive exercise. *Br. J. Nutr.* **2006**, *95*, 67–75. [CrossRef]
21. Liang, M.; Wang, Z.; Li, H.; Cai, L.; Pan, J.; He, H.; Wu, Q.; Tang, Y.; Ma, J.; Yang, L. l-Arginine induces antioxidant response to prevent oxidative stress via stimulation of glutathione synthesis and activation of Nrf2 pathway. *Food Chem. Toxicol.* **2018**, *115*, 315–328. [CrossRef]
22. Ruan, D.; Fouad, A.M.; Fan, Q.L.; Huo, X.H.; Kuang, Z.X.; Wang, H.; Guo, C.Y.; Deng, Y.F.; Zhang, C.; Zhang, J.H.; et al. Dietary L-arginine supplementation enhances growth performance, intestinal antioxidative capacity, immunity and modulates gut microbiota in yellow-feathered chickens. *Poult. Sci.* **2020**, *99*, 6935–6945. [CrossRef] [PubMed]
23. Duan, X.; Li, F.; Mou, S.; Feng, J.; Liu, P.; Xu, L. Effects of dietary L-arginine on laying performance and anti-oxidant capacity of broiler breeder hens, eggs, and offspring during the late laying period. *Poult. Sci.* **2015**, *94*, 2938–2943. [CrossRef] [PubMed]
24. Katemala, S.; Molee, A.; Thumanu, K.; Yongsawatdigul, J. Meat quality and Raman spectroscopic characterization of Korat hybrid chicken obtained from various rearing periods. *Poult. Sci.* **2021**, *100*, 1248–1261. [CrossRef] [PubMed]
25. Sangsawad, P.; Kiatsongchai, R.; Chitsomboon, B.; Yongsawatdigul, J. Chemical and Cellular Antioxidant Activities of Chicken Breast Muscle Subjected to Various Thermal Treatments Followed by Simulated Gastrointestinal Digestion. *J. Food Sci.* **2016**, *81*, C2431–C2438. [CrossRef]
26. Uni, Z.; Ferket, P.R.; Tako, E.; Kedar, O. In ovo feeding improves energy status of late-term chicken embryos. *Poult. Sci.* **2005**, *84*, 764–770. [CrossRef]
27. Association of Official Analytical Chemists (AOAC). *International Official Methods of Analysis of AOAC International*, 17th ed.; AOAC International: Arlington, VA, USA, 2000.
28. Zhang, C.; Wang, L.; Zhao, X.H.; Chen, X.Y.; Yang, L.; Geng, Z.Y. Dietary resveratrol supplementation prevents transport-stress-impaired meat quality of broilers through maintaining muscle energy metabolism and antioxidant status. *Poult. Sci.* **2017**, *96*, 2219–2225. [CrossRef] [PubMed]
29. Cong, J.; Zhang, L.; Li, J.; Wang, S.; Gao, F.; Zhou, G. Effects of dietary supplementation with carnosine on growth performance, meat quality, antioxidant capacity and muscle fiber characteristics in broiler chickens. *J. Sci. Food. Agric.* **2017**, *97*, 3733–3741. [CrossRef]
30. Foye, O.T.; Ferket, P.R.; Uni, Z. The effects of in ovo feeding arginine, β-hydroxy-β-methyl-butyrate, and protein on jejunal digestive and absorptive activity in embryonic and neonatal turkey poults. *Poult. Sci.* **2007**, *86*, 2343–2349. [CrossRef]
31. Gao, T.; Zhao, M.; Zhang, L.; Li, J.; Yu, L.; Gao, F.; Zhou, G. In ovo feeding of l-arginine regulates intestinal barrier functions of posthatch broilers by activating the mTOR signaling pathway. *J. Sci. Food Agric.* **2018**, *98*, 1416–1425. [CrossRef]
32. Tahmasebi, S.; Toghyani, M. Effect of arginine and threonine administered in ovo on digestive organ developments and subsequent growth performance of broiler chickens. *J. Anim. Physiol. Anim. Nutr.* **2016**, *100*, 947–956. [CrossRef]
33. Moran, E.T., Jr. Nutrition of the developing embryo and hatchling. *Poult. Sci.* **2007**, *86*, 1043–1049. [CrossRef] [PubMed]
34. John, T.M.; George, J.C.; Moran, E.T., Jr. Metabolic changes in pectoral muscle and liver of turkey embryos in relation to hatching: Influence of glucose and antibiotic-treatment of eggs. *Poult. Sci.* **1988**, *67*, 463–469. [CrossRef] [PubMed]
35. Foye, O.T.; Uni, Z.; McMurtry, J.P.; Ferket, P.R. The effects of amniotic nutrient administration, "in-ovo feeding" of arginine and/or ß-hydroxy-ß-methyl butyrate (HMB) on insulin-like growth factors, energy metabolism and growth in turkey poults. *Int. J. Poul. Sci.* **2006**, *5*, 309–317. [CrossRef]
36. Toghyani, M.; Tahmasebi, S.; Modaresi, M.; Fosoul, S.S.A.S. Effect of arginine and threonine in ovo supplementation on immune responses and some serum biochemical attributes in broiler chickens. *Ital. J. Anim. Sci.* **2018**, *18*, 342–349. [CrossRef]
37. Hakim, Y.; Uni, Z.; Hulata, G.; Harpaz, S. Relationship between intestinal brush border enzymatic activity and growth rate in tilapias fed diets containing 30% or 48% protein. *Aquaculture* **2006**, *257*, 420–428. [CrossRef]
38. Gao, T.; Zhao, M.M.; Li, Y.J.; Zhang, L.; Li, J.L.; Yu, L.L.; Gao, F.; Zhou, G.H. Effects of in ovo feeding of L-arginine on the development of digestive organs, intestinal function and post-hatch performance of broiler embryos and hatchlings. *J. Anim. Physiol. Anim. Nutr.* **2018**, *102*, e166–e175. [CrossRef]
39. Gao, T.; Zhao, M.; Zhang, L.; Li, J.; Yu, L.; Lv, P.; Gao, F.; Zhou, G. Effect of in ovo feeding of L-arginine on the hatchability, growth performance, gastrointestinal hormones, and jejunal digestive and absorptive capacity of posthatch broilers. *J. Anim. Sci.* **2017**, *95*, 3079–3092. [CrossRef]
40. Cherian, D.A.; Peter, T.; Narayanan, A.; Madhavan, S.S.; Achammada, S.; Vynat, G.P. Malondialdehyde as a Marker of Oxidative Stress in Periodontitis Patients. *J. Pharm. Bioallied Sci.* **2019**, *11*, S297–S300. [CrossRef]

41. Atakisi, O.; Atakisi, E.; Kart, A. Effects of dietary zinc and l-arginine supplementation on total antioxidants capacity, lipid peroxidation, nitric oxide, egg weight, and blood biochemical values in Japanase quails. *Biol. Trace Elem. Res.* **2009**, *132*, 136–143. [CrossRef]
42. Garcia-Ruiz, C.; Fernandez-Checa, J.C. Mitochondrial glutathione: Hepatocellular survival–death switch. *J. Gastroenterol. Hepatol.* **2006**, *21*, S3–S6. [CrossRef]
43. Lu, S.C. Regulation of glutathione synthesis. *Mol. Aspects Med.* **2009**, *30*, 42–59. [CrossRef]
44. Xiao, L.; Cao, W.; Liu, G.; Fang, T.; Wu, X.; Jia, G.; Chen, X.; Zhao, H.; Wang, J.; Wu, C.; et al. Arginine, *N*-carbamylglutamate, and glutamine exert protective effects against oxidative stress in rat intestine. *Anim. Nutr.* **2016**, *2*, 242–248. [CrossRef]
45. Wu, G.; Morris, S.M., Jr. Arginine metabolism: Nitric oxide and beyond. *Biochem. J.* **1998**, *336*, 1–17. [CrossRef]
46. Lu, S.C. Glutathione synthesis. *Biochim. Biophys. Acta* **2013**, *1830*, 3143–3153. [CrossRef]
47. Ren, W.; Yin, Y.; Liu, G.; Yu, X.; Li, Y.; Yang, G.; Li, T.; Wu, G. Effect of dietary arginine supplementation on reproductive performance of mice with porcine circovirus type 2 infection. *Amino Acids* **2012**, *42*, 2089–2094. [CrossRef] [PubMed]
48. Van Laack, R.L.J.M. Determinants of ultimate pH of meat and poultry. In Proceedings of the 53rd Annual Reciprocal Meat of the Conference, Columbus, OH, USA, 18–21 June 2000; pp. 74–75.
49. Fletcher, D.L.; Qiao, M.; Smith, D.P. The relationship of raw broiler breast meat color and pH to cooked meat color and pH. *Poult. Sci.* **2000**, *79*, 784–788. [CrossRef] [PubMed]
50. Aaslyng, M.D.; Oksama, M.; Olsen, E.V.; Bejerholm, C.; Baltzer, M.; Andersen, G.; Bredie, W.L.P.; Byrne, D.V.; Gabrielsen, G. The impact of sensory quality of pork on consumer preference. *Meat Sci.* **2007**, *76*, 61–73. [CrossRef] [PubMed]
51. Zampiga, M.; Laghi, L.; Petracci, M.; Zhu, C.; Meluzzi, A.; Dridi, S.; Sirri, F. Effect of dietary arginine to lysine ratios on productive performance, meat quality, plasma and muscle metabolomics profile in fast-growing broiler chickens. *J. Anim. Sci. Biotechnol.* **2018**, *9*, 79–92. [CrossRef] [PubMed]
52. Zampiga, M.; Soglia, F.; Petracci, M.; Meluzzi, A.; Sirri, F. Effect of different arginine-to-lysine ratios in broiler chicken diets on the occurrence of breast myopathies and meat quality attributes. *Poult. Sci.* **2019**, *98*, 2691–2697. [CrossRef]
53. Zhang, L.; Barbut, S. Rheological characteristics of fresh and frozen PSE, normal and DFD chicken breast meat. *Br. Poult. Sci.* **2005**, *46*, 687–693. [CrossRef] [PubMed]
54. Ma, X.; Lin, Y.; Jiang, Z.; Zheng, C.; Zhou, G.; Yu, D.; Cao, T.; Wang, J.; Chen, F. Dietary arginine supplementation enhances antioxidative capacity and improves meat quality of finishing pigs. *Amino Acids* **2010**, *38*, 95–102. [CrossRef] [PubMed]
55. Shi, H.; Kim, J.K.; Kim, I.H. Effects of dietary l-arginine on growth performance, nutrient digestibility, gas emission, and meat quality in finishing pigs. *Anim. Feed Sci. Tech.* **2019**, *253*, 93–100. [CrossRef]

animals

MDPI

Article

Dietary Cinnamon Bark Affects Growth Performance, Carcass Characteristics, and Breast Meat Quality in Broiler Infected with *Eimeria tenella* Oocysts

Mohammed M. Qaid *, Saud I. Al-Mufarrej, Mahmoud M. Azzam, Maged A. Al-Garadi, Abdulmohsen H. Alqhtani *, Abdulaziz A. Al-abdullatif, Elsayed O. Hussein and Gamaleldin M. Suliman

Animal Production Department, College of Food and Agriculture Sciences, King Saud University, Riyadh 11451, Saudi Arabia; salmufarrej@ksu.edu.sa (S.I.A.-M.); mazzam@ksu.edu.sa (M.M.A.); malgaradi@ksu.edu.sa (M.A.A.-G.); aalabdullatif@ksu.edu.sa (A.A.A.-a.); shessin@ksu.edu.sa (E.O.H.); gsuliman@ksu.edu.sa (G.M.S.)

* Correspondence: mqaid@ksu.edu.sa (M.M.Q.); ahalqahtani@ksu.edu.sa (A.H.A.); Tel.: +966-509-844-024 (M.M.Q.); Tel: +966-540-000-232 (A.H.A.)

Simple Summary: Antimicrobial tolerance problems have culminated in an increased focus on raising broiler chickens without using any antibiotics, and an increasing interest has developed in non-antibiotic feed supplements with potential productivity and health benefits. Previous studies have shown beneficial results linked to the addition of cinnamon to broiler diets under health conditions without induced coccidiosis. In this study, different amounts of dietary cinnamon, a phytogenic of the Lauraceae family, were evaluated for their effects on carcass characteristics and meat quality in broilers infected with oocysts of *Eimeria tenella*. Overall, emeriosis negatively affects bird slaughter weight, carcass yield, and most carcass traits. Compared with the positive control, cinnamon increased slaughter weight, carcass yield, and the percentage weights of heart, proventriculus, gizzard, breast, and pancreas. In addition, cinnamon at 2 g/kg diet improved performance, cocking loss, and meat tenderness among cinnamon groups. The ionophore salinomycin (Sacox®) group had the highest slaughter yield, myofibril fragmentation index (MFI), and texture profile analysis (TPA) of meat. The current research offers equivalent and unbiased findings from a study of substitutes for commercial coccidiostats in a consistent experimental paradigm that applies well to commercial conditions.

Abstract: A total of 150 broiler chicks were used to determine the impact of dietary cinnamon bark powder (CBP; *Cinnamomum verum*) on breast meat quality, growth performance, and carcass characteristics of birds under coccidiosis, as one of the protozoan parasitic diseases. A total of 5 replicates of birds received 1 of the following 6 groups for 34 days: control groups (1 and 2) received a basal diet without the addition of CBP or salinomycin; group 3 received a basal diet with 0.066 g salinomycin; groups 4–6 were given a basal diet supplemented with 2, 4, and 6 g CBP/kg feed, respectively. On day 21, $4 \times 10^4/100\ \mu L$ of *Eimeria tenella* oocysts/bird were challenged, except for the negative control (NC). At the end of the experimental trial, five birds/group were sampled for carcass characteristics and breast attributes. Overall, emeriosis negatively affects slaughter body mass, carcass yield, and the majority of carcass characteristics in birds, and cinnamon can mitigate these effects. Cinnamon groups, particularly at the 2 g level, alleviated the negative effect on performance caused by coccidia infestation to the same or greater extent as the negative control and salinomycin treatment groups. Furthermore, when compared with the other experimental groups, the addition of cinnamon improved some physicochemical properties with some affecting meat quality, such as decreasing MFI and increasing toughness in cinnamon-treated groups. In summary, it can be concluded that CBP can enhance the shelf life, carcass, and quality of birds' meat by maximizing the productive performance efficiency and breast meat productivity of birds under coccidiosis infestation. Further research is required to investigate the use of cinnamon to optimize the quality of meat and the productivity of both healthy and diseased broilers.

Citation: Qaid, M.M.; Al-Mufarrej, S.I.; Azzam, M.M.; Al-Garadi, M.A.; Alqhtani, A.H.; Al-abdullatif, A.A.; Hussein, E.O.; Suliman, G.M. Dietary Cinnamon Bark Affects Growth Performance, Carcass Characteristics, and Breast Meat Quality in Broiler Infected with *Eimeria tenella* Oocysts. *Animals* **2022**, *12*, 166. https://doi.org/10.3390/ani12020166

Academic Editors: Natalie Morgan and Youssef A. Attia

Received: 27 October 2021
Accepted: 5 January 2022
Published: 11 January 2022

Keywords: breast; broiler; carcass traits; cinnamon bark; meat quality

1. Introduction

Total imports of Saudi Arabian chicken meat were reduced from 618,000 to 520,000 tons in 2020 and 2021 as local production of Saudi Arabian chicken meat increased. Globally, increases in broiler meat production were also expected, with ~100.827 million tons produced in 2020 and projections for 2022 of ~100.9 million tons [1]. These increases in production are related to the high demand for poultry meat, which exceeds that of other animals, reflecting its rich nutritional value and favorable profit margin. As a result of the sharp increase in the production and popularity of broiler meat, the demand for improving the palatability and quality of meat has increased [2]. Important criteria for assessing the slaughter value of poultry carcasses include the percentage of dressing, the percentage of culinary, the tenderness of the meat, and the muscle content of the carcass. According to [3], broilers have a greater dressing percentage, reaching ~75%. This indicator is considered to be essential for evaluating the quantity of produce. Likewise, the composition of the carcass is essential because a high muscle content, particularly of the breast muscle, and a low fat content stimulate the consumer's appetite. While the quality of the meat comprises a sequence of attributes, the consumer is most concerned about the tenderness, color, juiciness, and palatability of the meat [4–6]. Meat color is a simple and quick criterion for evaluation and seems to be an important indicator to consumers. Fresh chicken meat should be light red in color. Thus, at first glance by a consumer, any change in color indicates declining nutritional value and deterioration in quality. The capacity of bird meat to retain its natural and added moisture during processing, storage, and when subjected to an external force is termed its water-holding capacity (WHC). Measuring WHC is one of the easiest ways to detect the quality of meat, and it has a direct impact on yield and appearance [7].

The larger body and breast weights of birds are related to larger areas and diameters of muscle fibers (myofiber), and older birds have lower myofiber density than younger birds [8]. Consumers expect tenderness; therefore, measuring tenderness is an appropriate way of measuring consumer satisfaction in eating meat products, and it is subjectively judged to be a sense of the hardness or elasticity of tough or tender meat [9]. Texture profile analysis (TPA) is a constructive technique that mimics the bite action of the mouth through a popular double compression test to determine the textural properties of foods, and it does not require a large number of investigators to assess the texture of poultry meat [10].

Humans that consume poultry meat could be directly affected by antibiotic growth promoters (AGPs) residues in the meat, or indirectly affected by the development of antibiotic-resistant pathogens in the meat [11]. Some researchers have investigated the growth and meat quality of birds fed diets containing natural herbs or extracts of natural herbs [12–16]. Coccidiosis is a parasitic disease caused by a protozoan that causes enteritis, hemorrhagic cecal lesions, and bloody diarrhea, with significant economic losses worldwide to the poultry industry [17].

Cinnamon spice is obtained from the inner bark of *Cinnamomum verum*, which is a vigorous, evergreen, annual, and aromatic ethnomedicinal plant belonging to the Lauraceae family. Various herb extracts, including cinnamon plant oils and their bioactive constituents, such as cinnamaldehyde and eugenol, are used as dietary supplements in poultry production. These extracts have a variety of uses, including antibacterial activity against many pathogens and acceleration of the growth of good bacteria, such as lactic acid bacteria and bifidobacterial, in the poultry intestine [18,19]. Furthermore, cinnamon oil has potent hypercholesterolemia, anticandidal, antioxidant, analgesic, and antiulcer activities [20], and the health-promoting and performance-enhancing effects of dietary aromatic herbs and extracts have been shown in both healthy and diseased farm animals [21]. Dietary cinnamon supplementation improves the quality and shelf life of broiler meat and maximizes meat productivity by lowering abdominal fat, increasing the dressing percentage and redness, and reducing drip loss and antioxidant activity of the breast meat of stressed broilers [22]. The leaves and bark of the cinnamon herb and its metabolites are receiving more attention as phytogenic feed additive substitutes for AGPs because they are natural antibiotics—readily available, non-toxic, and residue-free [20]—as well as enhancing poultry growth and improving carcass characteristics and the quality of broiler meat, and enhancing immunity and microbiological factors. However, limited studies are available on the use of cinnamon powder as a phytogenic alternative antibiotic and potential nutrition enhancer in the diets of broilers facing coccidiosis challenges.

This study hypothesized that the cinnamon herb could be used to enhance growth performance, carcass traits, meat quality, and physico-chemical properties of the breast muscle of broiler chickens experiencing coccidiosis infestation. Various levels of cinnamon bark plant were evaluated to validate this hypothesis. Therefore, this study aimed to test the ability of the cinnamon powder to be used as a dietary AGP substitute for enhancing the breast quality, carcass characteristics, and growth performance of Ross broiler chickens challenged by *Eimeria tenella*.

2. Materials and Methods

2.1. Ethical Approval

The research was carried out in accordance with the Kingdom of Saudi Arabia's ethical standards for animal use (Ethic committee King Saud University, Ethical approval number: KSU-SE-20-44).

2.2. Infection with Eimeria tenella

Our previous study [23] described the source of *Eimeria tenella* (*E. tenella*) oocysts, the sporulation of unsporulated oocysts, the identification, passage, and propagation of sporulated oocysts, and the inoculation procedure. All birds, with the exception of NC, were administered 1 mL of double distilled water containing 4×10^4/100 µL/bird of live sporulated oocysts of *E. tenella* at 21 days of age, according to previous research [24–26]. The authors choose the 7 dpi for anticoccidial indices (data published in [23]). Thus, this study is an extension of the same broiler batch experiment in which the anticoccidial indicators of CBP evaluation, namely the number of fecal oocysts, survival rate, bloody diarrhea, and lesion scores, were included. As a result, this investigation does not address the assessment of anticoccidial indicators. We found that CBP was effective on *E. tenella*. Furthermore, salinomycin, commercially called (Sacox®; Huvepharma NV, Belgium), is a standard product that protects birds from coccidiosis. For growth performance sampling, we chose 7 and 14 dpi. Following [27], meat quality and carcass variables were measured in all treatments on the last day of the trial, at 34 days of age (14 dpi here), in order to mimic the carcass traits of commercial poultry at marketing weights and to identify meat quality.

2.3. Birds and Husbandry

The study was carried out in a controlled environment—a heated battery room at the Animal Production Department's experimental poultry research unit at King Saud University, Riyadh, Saudi Arabia. The trial operated during spring month "March–April 2019" with average temperature varying between 20.4 °C and 33.4 °C and the average relative humidity in Riyadh in April was 28% during the experimental period. A total of 150, 1-day-old, mixed-sex, commercial Ross 308 broiler birds were collected from the national commercial hatchery (Al Wadi Poultry Company "Al Khomasia," Riyadh, Saudi Arabia). At the hatchery, the birds were immunized against Newcastle and infectious bronchitis diseases. The chicks were randomly divided into 30 experimental cages with 5 replicates of 6 treatments with 5 chicks per replicate. At 1 day of age, the temperature was set at 35 °C and gradually decreased by 1 °C every 2 days until a permanent temperature of 22 °C was reached. Then, it was maintained until the end of the trial. Relative humidity ranged from 65–85%. Feed and water were supplied *ad libitum*, and birds were kept on a "23 h on and 1 h off" light schedule.

2.4. Experimental Dietary Treatments

Poultry diets were obtained from the Arabian Agricultural Services Company. The ingredients and chemical properties of the commercial starter (1–21 days) and finisher (22–34 days) broiler chicks' diets were analyzed, formulated, and mixed in a mashed form, based on Ross 308′s recommendation guidelines (Supplementary Table S1). On arrival, the birds were randomly allocated to one of the following 6 treatments:

T1—negative control group given an unmedicated diet, these were unchallenged coccidiosis chicks (NC); T2—positive control group given an unmedicated diet + coccidial challenge (PC); T3—medicated diet with salinomycin sodium (66 mg salinomycin/kg diet) + coccidial challenge; T4–T6—2, 4, and 6 g CBP/kg diet, respectively, + coccidial challenge. The supplemented levels of purchased CBP or salinomycin powder were mixed with the basal broiler diet.

2.5. Preparation and Compositions: Cinnamon Bark Powder

Cinnamomum verum bark was acquired from a nearby store in Riyadh, Saudi Arabia, for use in this study. The dried bark used was ground to a fine powder. Moreover, as described by [23], high-performance liquid chromatography (HPLC) and gas chromatography–mass spectrometry (GC–MS) were used to detect the biologically relevant compounds in the CBP extract mixture. A total of 26 different active compounds with the highest quality were detected by GC–MS in the CNB extract, particularly Cinnamaldehyde, 3-phenyl-, hexadecanoic acid, (E)-2-propenal, methyl ester, 14-methyl-,methyl ester, pentadecanoic acid, oxime-, methoxy-phenyl-, and 2-methyl-benzofuran, as previously accounted for in [28].

2.6. Performance Measurements and Production Efficiency

Body weight and feed intake (FI) of birds were recorded at 1, 7, and 14 dpi per replicate to evaluate performance. Then, for each pen, the body weight gain (BWG) and FI were recorded. The average feed conversion efficiency (FCR) was calculated by dividing FI (g) by BWG (g). The European production efficiency factor (PEF) was calculated as follows:

$$\text{PEF} = ((\text{Live weight (kg)} \times \text{Livability})/(\text{Age in days} \times \text{FCR})) \times 100 \quad (1)$$

2.7. Carcass Relative Weights

A total of 30 34-day-old birds (Ross 308) of each treatment (n = 5, 1 male bird per cage, per treatment) were randomly selected for slaughter and subjected to feed withdrawal for 10 h. Before slaughtering, the birds were weighed. After bleeding, birds were defeathered and eviscerated (about two minutes after jugular vein incision). The feathers of the birds were plucked, and the viscera was eviscerated. The carcasses were dissected after the head, feathers, and shanks were removed. Live weight at slaughter (SW, kg) and carcass weight (CW, kg) were registered to calculate the yield of (CY%) (dressing percentage) = (CW/SW) × 100). Then, the CW were dissected into commercial parts, including the offal (heart, liver, and gizzard), neck, abdominal fat, heart, liver, proventriculus, gizzard, lymphoid organs, breast, leg, and pancreas, and weighed. Then, the relative weights of these weights were expressed in relation to the SW.

2.8. Meat Quality Indices

In addition to carcass parameters, the left and right parts of the breast meat (pectoralis major) of each selected bird (one breast per cage/treatment) were used for qualitative analyses. For the pH and color measurements, breast samples were kept at 4 °C; meanwhile, for the other quality measurements, they were frozen at −20 °C until further measurements were conducted for other meat quality assessments, and then thawed overnight in the fridge at 4 °C before analysis. The initial and ultimate pH and color components of dissected breast muscle were determined at 1 and 24 h.

2.8.1. Breast Meat Physicochemical Characteristics (pH, Temperature, and Color Indicators)

The breast pH was measured after processing with a microprocessor Hanna Instruments pH meter, and incisions were made in the cranial left side of the pectoral muscle. At 1 and 24 h, an average of 3 pH measurements for each sample had been taken.

A thermocouple thermometer probe from Eutech Instruments was placed deep in the center of the muscle 1 h postmortem to monitor internal core temperature values of the pectoral muscle.

Breast flesh color measurements, developed by Dr. Richard Hunter as the Hunter values—lightness (L*), redness (a*), and yellowness (b*)—were set on CIELAB scales and assessed with a Chroma meter, 1 and 24 h after the slaughter on 2 different areas of the inner side of the cranial position of the breast muscles. Values for L*, a*, and b* were converted to estimate the total color change (ΔE), Chroma meter (saturation index), hue angle, browning index (BI), and whiteness index (WI), as described by [29–31]. According to [32], these measurements obtain a much more accurate assessment of how consumers perceive the color of meat. The averages of the two readings of the color components were taken.

2.8.2. Water-Holding Capacity

The breast meat water-holding capacity (WHC) of the frozen meat samples was measured immediately after thawing overnight at 4 °C using the compression method outlined by [33].

2.8.3. Cooking Loss

The cooking loss (CL) was measured as follows [33]: CL was determined by weighing muscle samples with a semi-analytical balance, placing them in a commercial indoor table-top grill, cooking them until they reached an internal temperature of 75 °C, and reweighing them after cooking. The CL percentage = [(Initial weight − Cooked weight)/Initial weight] × 100.

2.8.4. Drip Loss

The parts of the breasts used for the drop test were weighed separately, packed, and kept at 4 ± 1 °C for 24 h. Then, the difference in weights before (Wi) and after (Wu) storage was calculated and expressed as a drop loss proportion [34]. Drip loss (DL) (%) = [(Wi − Wu)/Wi] × 100.

2.8.5. Myofibril Fragmentation Index

The procedures mentioned by [33] were used to evaluate myofibril fragmentation (MFI) as an indirect measure of calpain intracellular proteases. An amount of 4 g of muscle, minced with scissors, was homogenized in a blender for 30 s with 40 mL of cold MFI buffer at 2 °C. After many washes, the suspension aliquots were diluted in MFI buffer to a final concentration of 0.5 mg/mL and poured into a cuvette for immediate measurement of absorbance at 540 nm using a spectrophotometer. Each sample's MFI was calculated to be $A_{540\ nm} \times 200$.

2.8.6. Meat Texture Analysis

The shear force (SF) and TPA of the samples were determined using a texture analyzer (TA.HD. Stable Micro Systems, Surrey, UK) in 2 parts per replicate (1 breast/replicate/treatment). After the cooked samples had cooled at 22 °C, 5 round core meat slices (1.27 and 2.5 cm diameter for SF and TPA, respectively) were cut from each sample, parallel to the longitudinal direction of the muscle fibers, using a handheld coring tool. During the SF test, the maximum force (kgf) was applied vertically to the fibers using a TA.HD. Texture Analyzer, designed for a Warner-Bratzler shear blade, with a triple-slotted cutting edge. The speed of the crosshead was set at 200 mm/min. The SF values were estimated from the maximum point of the generated curve. A cylindrical piston (75 mm diameter) was used to compress the TPA sample to within 80% of its original height over two test cycles in 5 s. The texture analyzers conditions were used to generate force–time curves of deformation. The hardness, springiness, chewiness, and cohesion parameters were measured following [35].

2.9. Statistical Analysis

In the statistical analysis system [36], a general linear model (GLM) was used to analyze slaughter characteristics and meat quality data. Six groups were arranged in five replicates in a completely randomized design. Each replicate cage represented an experimental unit. On the 34th day of age, male birds were sampled (n = 5 birds per treatment, 1 bird from each replicate).

All data were analyzed using one-way ANOVA and expressed as a statistical mean ± standard error of the mean (SE) using the following models:

$$Yij = \mu + Ti + \varepsilon ij$$

where Yij is the observed j parameters in the ith treatment, μ is the overall mean of the measurements, Ti is the effect of the ith treatment, and Ɛij is the random residual error. To assess significant differences between means for measurements using Duncan's multiple range test, a statistical significance level of $p < 0.05$ was used.

3. Results

According to our previous research [23], clinical coccidiosis symptoms were observed in birds after infection with oocysts of *E. tenella*. It was found that 6 g of CBP had moderate anti-coccidial activity and could be used to treat poultry emeriosis in the field. Consequently, CBP decreased the severity of lesions and reduced oocyst excretion per gram in chickens' droppings. Moreover, based on the HPLC and GC–MS results, we found that cinnamaldehyde and other important bioactive compounds are present in the cinnamon bark extract.

3.1. Performance Measurements and Production Efficiency

Table 1 shows that the challenge of coccidiosis had adverse effects on BWG, FCR, and PEF of the birds at 1st and 2nd week post-infection and suffered from it over the entire period compared with an unchallenged control group. During the second week after infection, the BWG, FCR, and PEF of the birds were dose-dependent, increasing as the cinnamon level decreased. Birds receiving a 2g CBP/kg diet had higher BWG, FCR, and PEF over the entire post-infection period (0–14 dpi) than those which received 4g CBP, 6g CBP, and PC ($p < 0.05$), but was similar to those receiving 66 mg salinomycin and NC. This means that birds given 2g CBP/kg gained more and converted feed more efficiently. The statistical models of FI and FCR did not differ significantly ($p > 0.05$) during the first or second weeks or during the entire period after the *Eimeria tenella* oocyst challenge.

3.2. Carcass Characteristic Variables

The effects of the CBP on carcass variables at 34 d of age (14 days post-inoculation) are shown in Table 2. Except for liver, leg, fat, and lymphoid organs (bursa, thymus, and spleen) values, there were significant differences in slaughter variables between treatments. The live weight, CW, and the carcass yield of slaughtered birds were statistically different ($p < 0.05$); this result indicates that the adverse effects of *Eimeria tenella* infection in birds were clearly observed in the PC group and were compensated for in the salinomycin and cinnamon groups. Thus, the heart, proventriculus, gizzard, breast, and pancreas % CW values did differ ($p < 0.05$) between treatments. Birds receiving cinnamon at levels of 4 g with their basal diet had higher percentage heart, proventriculus, gizzard, and pancreas weights, respectively, compared with those in other dietary groups. The lowest abdominal fat yield exhibited an insignificant decrease ($p < 0.05$) with increased CBP compared with the NC group.

Table 1. Live body weight (BW), average live body gain (BWG), average feed intake (FI), feed conversion ratio (FCR), and production efficiency factor (PEF) of broiler chickens given experimental diets (cinnamon), post coccidial challenge period (0–14 dpi).

				Cinnamon (g/kg)			
Treatment [1]	NC	PC	Salinomycin	2	4	6	*p*-Value
1 to 7 dpi							
BW (kg)	1.273 ± 0.01 [2a]	1.122 ± 0.01 [c]	1.238 ± 0.03 [ab]	1.178 ± 0.03 [abc]	1.161 ± 0.02 [bc]	1.228 ± 0.02 [abc]	0.043
BWG (g)	82.91 ± 1.77 [a]	56.45 ± 1.86 [b]	57.75 ± 0.78 [b]	54.28 ± 1.38 [b]	61.03 ± 1.96 [b]	64.45 ± 1.96 [b]	0.001
FI (g)	117.2 ± 2.00	99.6 ± 2.06	104.4 ± 1.82	99.5 ± 1.88	111.2 ± 8.28	102.3 ± 5.52	0.131
FCR (g:g)	1.41 ± 0.02 [b]	1.77 ± 0.06 [a]	1.81 ± 0.04 [a]	1.84 ± 0.05 [a]	1.83 ± 0.04 [a]	1.59 ± 0.05 [ab]	0.055
PEF	336.2 ± 11.4 [a]	230.6 ± 12.2[b]	254.2 ± 7.1 [b]	239.3 ± 11.0 [b]	237.2 ± 15.4 [b]	289.5 ± 14.2 [ab]	0.008
8 to 14 dpi							
BW (kg)	1.870 ± 0.03	1.649 ± 0.02	1.850 ± 0.03	1.858 ± 0.01	1.752 ± 0.02	1.737 ± 0.03	0.067
BWG (g)	85.23 ± 3.34[b]	75.58 ± 2.68 [c]	87.25 ± 2.26 [b]	97.08 ± 0.81 [a]	84.36 ± 4.11 [b]	72.76 ± 3.08[c]	0.007
FI (g)	134.6 ± 2.78	130.2 ± 2.27	130.7 ± 3.85	126.6 ± 1.75	133.7 ± 2.01	125.3 ± 1.91	0.095
FCR (g:g)	1.58 ± 0.05 [ab]	1.73 ± 0.05 [a]	1.51 ± 0.04 [ab]	1.31 ± 0.02 [b]	1.60 ± 0.06 [ab]	1.73 ± 0.06 [a]	0.043
PEF	438.8 ± 14.0 [ab]	354.2 ± 12.6 [b]	456.8 ± 16.5 [ab]	537.1 ± 13.3 [a]	411.4 ± 17.7 [b]	375.6 ± 15.5 [b]	0.038
1 to 14 dpi							
BWG (g)	84.07 ± 1.81 [a]	66.02 ± 1.89 [d]	72.50 ± 0.86 [bc]	75.68 ± 0.86 [b]	72.70 ± 1.62 [bc]	68.60 ± 1.50 [cd]	<0.001
FI (g)	125.88 ± 2.16	114.9 ± 1.88	117.52 ± 2.20	113.07 ± 1.13	122.43 ± 0.70	113.78 ± 1.46	0.148
FCR (g:g)	1.50 ± 0.02 [c]	1.74 ± 0.05 [a]	1.62 ± 0.03 [b]	1.50 ± 0.02 [c]	1.69 ± 0.03 [ab]	1.66 ± 0.04 [ab]	0.001
PEF	387.5 ± 9.1 [a]	292.4 ± 12.1 [c]	355.5 ± 8.0 [ab]	388.2 ± 8.5 [a]	324.3 ± 16.0 [bc]	332.6 ± 12.1 [b]	<0.001

[1] Treatments: NC—negative control, unsupplemented, unchallenged; PC—positive control, unsupplemented, challenged; Salinomycin—basal diet supplemented with coccidiostat salinomycin, challenged; Cinnamon—groups whose basal diet was supplemented with 2, 4, and 6 g cinnamon powder/kg diet, respectively, challenged. [a–d] Different letters indicate statistically significant differences ($p < 0.05$). [2] Values are presented in means ± standard error (SE) ($n = 5$).

Table 2. Carcass variables of broiler chickens supplemented with cinnamon powder, 14 days post-infection.

				Cinnamon (g/kg)			
Treatment [1]	NC	PC	Salinomycin	2	4	6	Probability
Live wt (kg)	1.779 ± 32.3 [2a]	1.505 ± 37.4 [c]	1.736 ± 24.2 [a]	1.872 ± 41.7 [a]	1.810 ± 29.5 [a]	1.791 ± 32.0 [a]	<0.0001
Carcass wt (kg)	1.194 ± 19.6 [a]	0.974 ± 24.9 [b]	1.179 ± 18.9 [a]	1.245 ± 28.0 [a]	1.222 ± 24.7 [a]	1.204 ± 27.6 [a]	<0.0001
CY% [3]	67.14 ± 0.33 [a]	64.72 ± 0.15 [b]	67.88 ± 0.39 [a]	66.53 ± 0.73 [a]	67.78 ± 0.63 [a]	67.19 ± 0.42 [a]	0.003
Heart	0.44 ± 0.01 [b]	0.45 ± 0.01 [b]	0.46 ± 0.01 [b]	0.54 ± 0.01 [a]	0.54 ± 0.02 [a]	0.46 ± 0.002 [b]	<0.0001
Liver	1.82 ± 0.08	1.95 ± 0.06	2.00 ± 0.08	1.84 ± 0.02	1.75 ± 0.12	1.96 ± 0.01	0.133
Proventriculus	0.37 ± 0.02 [bc]	0.39 ± 0.02 [bc]	0.34 ± 0.04[c]	0.41 ± 0.03 [bc]	0.53 ± 0.03 [a]	0.44 ± 0.03 [b]	0.003
Gizzard	1.96 ± 0.08 [c]	2.35 ± 0.09 [b]	2.00 ± 0.15[c]	2.55 ± 0.11 [ab]	2.69 ± 0.04 [a]	2.63 ± 0.08 [ab]	<0.0001
Bursa	0.19 ± 0.03	0.22 ± 0.03	0.15 ± 0.03	0.21 ± 0.01	0.25 ± 0.01	0.21 ± 0.01	0.115
Spleen	0.09 ± 0.002	0.10 ± 0.001	0.11 ± 0.01	0.07 ± 0.002	0.10 ± 0.01	0.09 ± 0.01	0.267
Thymus	0.27 ± 0.02	0.32 ± 0.02	0.33 ± 0.02	0.35 ± 0.05	0.40 ± 0.06	0.43 ± 0.03	0.094
Breast	26.87 ± 0.53 [ab]	25.64 ± 0.14 [b]	27.71 ± 0.60 [ab]	26.14 ± 0.19 [b]	26.77 ± 0.44 [ab]	28.69 ± 0.74 [a]	0.051
Leg	19.14 ± 0.48	20.17 ± 0.35	19.40 ± 0. 0.36	20.21 ± 0.53	19.92 ± 0.22	19.53 ± 0.21	0.275
Fat	0.80 ± 0.14	0.60 ± 0.08	0.85 ± 0.17	0.78 ± 0.03	0.57 ± 0.16	0.55 ± 0.08	0.274
Pancreas	0.307 ± 0.03 [c]	0.389 ± 0.02 [b]	0.418 ± 0.03 [ab]	0.417 ± 0.02 [ab]	0.474 ± 0.03 [a]	0.403 ± 0.02 [ab]	0.006

[1] Treatments: NC—negative control, unsupplemented, unchallenged; PC—positive control, unsupplemented, challenged; Salinomycin—basal diet supplemented with coccidiostat salinomycin, challenged; Cinnamon—groups whose basal diet supplemented with 2, 4, and 6 g of cinnamon powder/kg of diet, respectively, challenged. [2] The data is presented as means with standard errors ($n = 5$). [a–c] Means in the rows with different superscripts differ significantly ($p < 0.05$), [3] Dressing percentage or carcass yield (CY%) = (carcass weight/Live weight) * 100.

3.3. *Breast Meat Physicochemical Characteristics*

Tables 3 and 4 display the influence of CBP on the physicochemical characteristics of broiler breast samples at 1 and 24 h postmortem, respectively. The core temperature, initial and ultimate lightness, total color change, and WI of samples were different ($p < 0.05$). Otherwise, the initial and ultimate values for pH, redness, yellowness, total color change (Delta E: ΔE), hue angle, BI, and saturation index (Chroma) did not differ among experimental groups ($p > 0.05$). The core temperature values of breast muscle were significantly different ($p < 0.05$), with the broilers from the 2 g of CBP/kg treatment having the highest (26.55 °C) temperature values and the broilers fed the control diet and exposed to challenge with coccidia (PC) having the lowest values (24.89 °C). Although initial and ultimate pH values did not differ significantly ($p > 0.05$) between cinnamon-treated groups, pH decreased 24 h postmortem with increasing CBP content. Birds receiving 4 g of CBP and NC had higher initial color lightness than those receiving 66 mg of salinomycin and NC ($p = 0.038$) but were similar to those receiving 2 g of CBP, 6 g of CBP, and in the PC group. Birds that received 66 mg of salinomycin had lower final color lightness than those in the NC group and 6 g of CBP ($p < 0.01$) but were similar to those receiving 2 g of CBP, 4 g of CBP, and in the PC group.

Table 3. Core temperature, pH, and color of the pectoralis major at 34 d of age in broilers fed diets containing varying amounts of cinnamon bark powder (CBP) were measured 1 h postmortem.

				Cinnamon (g/kg)			Probability
Treatment [1]	**NC**	**PC**	**Salinomycin**	**2**	**4**	**6**	
Core Temperature	25.5 ± 0.15 [2cd]	24.89 ± 0.23 [d]	25.78 ± 0.23 [bc]	26.55 ± 0.10 [a]	25.6 ± 0.19 [c]	26.41 ± 0.05 [ab]	<0.0001
Initial pH	6.08 ± 0.07	6.16 ± 0.08	6.23 ± 0.02	6.25 ± 0.10	6.16 ± 0.09	6.36 ± 0.04	0.104
L1	42.12 ± 0.94 [b]	45.29 ± 0.74 [a]	42.12 ± 0.29 [b]	44.10 ± 0.59 [ab]	45.35 ± 0.60 [a]	44.48 ± 0.47 [ab]	0.002
a1	6.59 ± 1.04	5.43 ± 0.81	5.38 ± 0.52	5.88 ± 0.57	6.48 ± 0.38	6.25 ± 0.45	0.695
b1	7.7 ± 0.34	7.36 ± 0.27	5.96 ± 0.39	6.32 ± 0.25	7.5 ± 0.87	6.91 ± 0.48	0.104
ΔE	52.71 ± 1.05 [a]	49.4 ± 0.73 [b]	52.44 ± 0.31 [a]	50.55 ± 0.61 [ab]	49.49 ± 0.54 [b]	50.25 ± 0.50 [ab]	0.004
Hue angle (○)	50.86 ± 4.54	54.39 ± 4.25	48.31 ± 1.31	47.62 ± 2.00	48.28 ± 2.60	47.86 ± 2.04	0.589
Saturation index	10.28 ± 0.75	9.27 ± 0.52	8.03 ± 0.62	8.65 ± 0.55	9.95 ± 0.86	9.34 ± 0.58	0.205
Browning index	31.5 ± 2.91	26.11 ± 1.29	24.26 ± 1.96	24.88 ± 1.65	28.11 ± 2.56	26.88 ± 1.98	0.223
Whiteness index	41.20 ± 1.04 [b]	44.49 ± 0.71 [a]	41.55 ± 0.32 [b]	43.43 ± 0.61 [ab]	44.41 ± 0.50 [a]	43.69 ± 0.52 [ab]	0.004

[1] Treatments: NC—negative control, unsupplemented, unchallenged; PC—positive control, unsupplemented, challenged; Salinomycin—basal diet supplemented with coccidiostat salinomycin, challenged; Cinnamon—groups whose basal diet supplemented with 2, 4, and 6 g cinnamon powder/kg diet, respectively, challenged. [2] Each mean is based on measurements from 5 birds per treatment. [a–d] Means in the same rows with different superscripts differ significantly at $p < 0.05$; L1, a1, and b1—initial lightness, redness, and yellowness, respectively; ΔE—total color change.

Table 4. pH, and color of the pectoralis major at 34 d of age in broilers fed diets containing varying amounts of cinnamon bark powder (CBP) were measured 24 h postmortem.

				Cinnamon (g/kg)			Probability
Treatment [1]	**NC**	**PC**	**Salinomycin**	**2**	**4**	**6**	
Ultimate pH	5.78 ± 0.02 [2]	5.79 ± 0.02	5.76 ± 0.02	5.83 ± 0.04	5.77 ± 0.02	5.71 ± 0.03	0.090
pH decline	0.31 ± 0.08	0.37 ± 0.08	0.46 ± 0.03	0.42 ± 0.07	0.39 ± 0.05	0.65 ± 0.06	0.067
L2	49.91 ± 0.26 [a]	44.29 ± 1.01 [bc]	43.18 ± 0.77 [c]	45.46 ± 0.32 [bc]	46.26 ± 0.88 [bc]	47.47 ± 1.31 [ab]	0.0001
a2	7.25 ± 0.89	7.94 ± 0.53	7.55 ± 0.87	7.61 ± 0.53	8.03 ± 0.16	8.54 ± 1.14	0.873
b2	10.48 ± 0.70 [bc]	10.79 ± 0.85 [bc]	9.95 ± 0.51 [c]	13.22 ± 0.80 [a]	11.97 ± 0.63 [abc]	12.17 ± 0.48 [ab]	0.021
ΔE	45.48 ± 0.30 [c]	51.09 ± 1.10 [ab]	52.01 ± 0.85 [a]	50.28 ± 0.36 [ab]	49.33 ± 0.90 [ab]	48.38 ± 1.11 [bc]	0.004
Hue angle (○)	55.68 ± 2.73	53.49 ± 0.73	53.44 ± 2.36	60.12 ± 0.92	55.94 ± 1.00	55.4 ± 4.34	0.589
Saturation index	12.81 ± 0.95	13.4 ± 0.99	12.54 ± 0.87	15.26 ± 0.93	14.42 ± 0.61	15.09 ± 0.41	0.205
Browning index	33.82 ± 2.60	41.13 ± 3.96	38.71 ± 3.16	46.25 ± 3.13	42.44 ± 2.52	42.52 ± 1.10	0.223
Whiteness index	48.25 ± 0.29 [a]	42.66 ± 1.12 [b]	41.79 ± 0.84 [b]	43.33 ± 0.38 [b]	44.34 ± 0.92 [b]	45.31 ± 1.17 [ab]	0.004

Ultimate pH and color values of the breast muscle were measured. [1] Treatments: NC—negative control, un challenged, not supplemented; PC—positive control, challenged, not supplemented; Salinomycin—challenged, basic diet supplemented with salinomycin; Cinnamon—groups whose basal diet supplemented with 2, 4, and 6 g CBP/kg diet, respectively, challenged. [2] Each mean is calculated using data from 5 birds per group with standard errors. [abc] Means differ significantly in the same rows with different superscripts at $p < 0.05$; abbreviations: L2, a2, and b2, final lightness; redness; and yellowness, respectively; ΔE—total color change.

The birds that received 66 mg of salinomycin had lower final color lightness than those in the NC group and fed 6 g of CBP ($p < 0.01$) but were similar to those receiving 2 g of CBP, 4 g of CBP, and in the PC group. Generally, the pH and color measurements of the samples were similar across all treatment groups tested, with the exception of the lightness and WI.

The higher initial lightness results in the control challenged group (PC) resulted in a lower color change and a higher WI compared with the control unchallenged group (NC), and vice versa at the ultimate lightness estimate. These findings may be attributed to the PC group having a higher pH decline than the NC group, indicating that the emeriosis challenge might alter the appearance of the lightness of breast samples. The birds that received cinnamon had lower final yellowness than those in the controls and salinomycin group ($p < 0.021$).

3.4. Water-Holding Capacity, Dripping Loss, Cooking Loss, Myofibril Fragmentation Index, and Shear Force

The broiler breast quality, WHC, DL, CL, MFI, and SF, of the samples at 34 d of age (14 days post-inoculation) are presented in Table 5. The CL, MFI, and SF values of the meat samples differed ($p < 0.05$) between groups, while the WHC and DL values did not differ ($p > 0.05$). The most favorable numerical values ($p > 0.05$) for WHC were obtained for the broilers fed 2 g of CBP/kg. Broilers fed 4 g of CBP/kg had the highest and worst CL value (36.64%; $p < 0.05$). In contrast, those fed 2 g of CBP/kg, the NC group, and the PC group had the lowest and the most convenient CL value (26.64%, 23.36%, and 22.46%, respectively) but were similar to those receiving 66 mg of salinomycin/kg and 6 g of CBP/kg (35.75% and 32.97%, respectively). The CL results demonstrate that breasts from chickens that received the anticoccidial (salinomycin) and 4 and 6 g of CBP shrank more when cooked than those given 2 g of CBP/kg and the control treatments ($p < 0.001$).

Table 5. Water-holding capacity (WHC), dripping loss (DL), cooking loss (CL), myofibril fragmentation index (MFI), and shear force (SF) in broilers fed diets containing varying amounts of cinnamon bark powder (CBP).

Parameter Treatment [1]	NC	PC	Salinomycin	Cinnamon (g/kg) 2	4	6	p-Value
WHC	31.85 ± 1.23 [2]	32.33 ± 0.70	30.38 ± 0.48	28.8 ± 1.10	31.92 ± 0.65	27.90 ± 0.82	0.0715
DL	1.55 ± 0.06	0.99 ± 0.07	1.01 ± 0.05	1.31 ± 0.10	1.45 ± 0.07	1.25 ± 0.09	0.1026
CL	23.35 ± 0.86 ^{b}	22.45 ± 0.95 ^{b}	35.75 ± 1.18 ^{a}	26.65 ± 1.19 ^{b}	36.63 ± 0.85 ^{a}	32.98 ± 1.06 ^{a}	<0.0001
MFI	112.85 ± 5.51 ^{a}	104.80 ± 6.30 ^{a}	123.33 ± 6.12 ^{a}	64.88 ± 3.28 ^{b}	73.38 ± 5.59 ^{b}	67.35 ± 4.78 ^{b}	<0.0001
SF (kgf)	1.33 ± 0.05 ^{b}	1.40 ± 0.05 ^{b}	1.47 ± 0.06 ^{b}	1.69 ± 0.04 ^{ab}	2.10 ± 0.09 ^{a}	1.96 ± 0.08 ^{a}	0.0002

[1] Treatments: NC—negative control, unsupplemented, unchallenged; PC—positive control, unsupplemented, challenged; Salinomycin—basal diet supplemented with coccidiostat salinomycin, challenged); Cinnamon—groups whose basal diet supplemented with 2, 4, and 6 g of cinnamon powder/kg of diet, respectively, challenged. [2] The data is presented as means of 5 birds per treatment with standard errors of each mean ($n = 5$). ^{abc} Means in the same rows with different superscripts differ significantly at $p < 0.05$.

The broilers fed CBP had the lowest MFI value ($p < 0.05$), with 2 g of CBP/kg being the most convenient compared with the PC, NC, and salinomycin groups. The CBP treatments had higher SF values ($p < 0.05$) than the salinomycin and control groups, indicating tough meat. However, the 2 g of CBP/kg group had the lowest SF (1.69 kgf) among the cinnamon groups, indicating that it was the CBP group with the most tenderness.

3.5. Texture Analysis

TPA quality of breast samples from broilers fed with CBP inoculation and exposed to *E. tenella* challenge is revealed in Table 6. There were significant differences between the experimental groups in the values for hardness, springiness, and chewiness ($p < 0.05$). However, there was a tendency for the cohesiveness value to differ ($p = 0.04$) among treatments. TPA values were found to be higher ($p < 0.05$) in the breast meat of the broiler chickens receiving 66 mg salinomycin/kg. Higher hardness values ($p < 0.05$) of 2 g CBP/kg compared with 4 g and 6 g of CBP/kg groups indicated that hardness improved in broilers fed diets containing 2 g CBP/kg. Springiness and chewiness did not differ significantly among groups with different levels of CBP ($p > 0.05$). Springiness values were lower in CBP groups compared with controls and salinomycin groups ($p < 0.0001$).

Table 6. Texture profile in broilers fed diets containing varying amounts of cinnamon bark powder (CBP).

Treatments [1]	NC	PC	Salinomycin	Cinnamon (g/kg) 2	4	6	*p*-Value
Hardness (kg)	0.51 ± 0.02 [2b]	0.62 ± 0.03 [b]	0.94 ± 0.02 [a]	0.73 ± 0.02 [ab]	0.67 ± 0.02 [b]	0.67 ± 0.02 [b]	<0.0001
Springiness (mm)	0.91 ± 0.01 [a]	0.88 ± 0.02 [a]	0.90 ± 0.01 [a]	0.74 ± 0.01 [b]	0.75 ± 0.01 [b]	0.79 ± 0.02 [b]	<0.0001
Cohesiveness (dimensionless)	0.43 ± 0.01 [ab]	0.44 ± 0.01 [ab]	0.45 ± 0.01 [a]	0.40 ± 0.01 [b]	0.41 ± 0.01 [ab]	0.41 ± 0.01 [b]	<0.0001
Chewiness (g mm)	2.07 ± 0.05 [b]	2.46 ± 0.05 [b]	3.67 ± 0.03 [a]	2.10 ± 0.07 [b]	2.11 ± 0.07 [b]	2.06 ± 0.06 [b]	<0.0001

[1] Treatments: NC—negative control, unsupplemented, unchallenged; PC—positive control, unsupplemented, challenged; Salinomycin—basal diet supplemented with *coccidiostat* salinomycin, challenged); Cinnamon—groups whose basal diet supplemented with 2, 4, and 6 g of cinnamon powder/kg of diet, respectively, challenged). [2] Each mean is based on measurements of 5 birds per treatment with standard errors of each mean (mean ± SE) (n = 5) at the end of the trial. [a] and [b] Means in the same row with different superscripts differ significantly at $p < 0.05$.

In this trial, texture profiles differed significantly between treatments, with the CBP groups having lower texture scores compared with the salinomycin treatment, but not from the control groups. However, the CBP groups had lower springiness than the control groups. The NC and 6 g CBP/kg treatments exhibited the lowest values for hardness (0.51) and for chewiness (2.06), while the salinomycin treatment demonstrated the highest values for hardness (0.94) and for chewiness (3.07).

4. Discussion

This study assessed the success of different levels of cinnamon as a natural herb compared with the standard synthetic anticoccidial product (salinomycin) in broilers exposed to experimentally induced challenge of coccidiosis. Few or no studies have reported the effect on the performance, carcass traits, and breast quality of broilers supplemented with CBP under coccidial challenge. The null hypothesis states that the effects of CBP on growth performance, carcass traits, and meat quality of broiler breasts are the same as the effects of the control groups (non-CBP), which is based on the p-value is calculated using a probability level of $\alpha = 0.05$. The alternative hypothesis is that CBP inhibits the weight loss in the breast and carcass caused by *E. tenella*, which is consistent with the results of [23] who found that CBP extract inhibits the weight loss caused by *Eimeria* infection.

Several studies have evaluated cinnamon powder supplementation at different levels (g/kg) in feed: 2 [37], 10 and 20 [38], 10, 30, and 50 [39], and 30, 50, and 70 [40]. The above studies have had significant effects of the levels of cinnamon powder evaluated in broiler chickens, such as BW, FI, and FCR. The authors in [41] reported that cinnamon at different levels (2.5, 5.0, or 7.5 g) did not show any positive impact on the performance or performance index, as opposed to the results of the current study.

As expected, the results of this study showed that BW, weight gain, and feed efficiency were most adversely affected by the positive control (coccidiosis-exposed group, not treated with any natural or synthetic drug). The effect of salinomycin sodium was similar to that of the cinnamon groups as a natural herb in the 1st week after the challenge. Luckily, cinnamon outcomes, particularly cinnamon at level 2 g per kilogram of diet, have been the best during the 2nd week and entire period after challenge since BWG, FCR, and PEF were improved compared with infected groups. The exploration of alternatives to synthetic antibacterial and anti-coccidial drugs is an interesting field of research for poultry scientists [42]. The performance of birds was recovered closer to the negative control and drug-treated groups at the 2nd week following a coccidiosis challenge, which is encouraging; fortunately, the effect was exceeded in the 2g CBP group and was equal in other CBP groups to salinomycin as the synthetic drug. The positive effects of the anticoccidial cinnamon herb have been linked to the presence of active compounds present in plants that reduce the parasitic oocyst, modulate intestinal microflora, improve immunity and antioxidant status, and reduce intestinal inflammation [13,43–45]. Herbal anticoccidial agents maintain the growth of the broilers by reducing the destructive effects of coccidiosis [46]. The harmful effect of coccidiosis is clear in the infected group (PC), which has been improved in cinnamon powder-treated birds. Confirming our document, prior literature have also reported improved performance in response to anticoccidials and the ability of natural products to decrease the induced weight loss due to infection of birds [24,42,47]. Cinnamon bark oil up to 0.05% has a better preservative impact on the quality of lamb meat during storage it is supplemented at levels 0.01, 0.25, 0.5, and 5 g in the diet [48]. Cinnamon and/or citral supplementation in feed improved growth performance of chicken vaccinated or not vaccinated against coccidiosis to the level comparable with bacitracin and alter cecal microbiota composition [13]. The addition of cinnamon powder up to 5% appears to have no noticeable impact on growth and carcass characteristics except for the spleen and heart weight percentage [39]. However, they have discovered that it can be used to enhance cellular immune responses in broiler chicks.

Conventional coccidiostats have been associated with undesirable residues in meat in some areas since the late 1990s [49]. However, there is no credible scientific proof to boost the allegation that residues exist or have caused issues for consumers. There is no proof of residual effects from salinomycin or other coccidiostats, with the exception of diclazuril [50]. According to [50], any coccidiostat residue in poultry meat poses a low direct hazard to human health (<1%). Moreover, none of the natural ingredients have already been studied to determine whether they also produce residues. Furthermore, the impacts of CBP on meat quality, carcass characteristics, and marketing growth of broilers infected with *E. tenella* oocysts were studied at different doses and with different results. As a follow-up to a previously published study on the effects of CBP as a coccidiosis prevention product, the effects of CBP on meat quality, growth performance, and carcass characteristics of broiler chickens infected with *E. tenella* were investigated here. Thus, this study examined the efficacy of different quantities of CBP as a natural herb compared with the current synthetic anticoccidial product (salinomycin) in birds exposed to an experimentally induced coccidial challenge. Previously, there was limited research on the effects of CBP on the carcass traits and breast quality of birds facing a coccidial challenge.

The bird became infected with *E. tenella* after ingesting sporulated oocysts which penetrated and damaged the intestinal epithelium of the caecum due to endogenous and exogenous multiplication of the *E. tenella* stage, whereupon the developed oocysts were excreted in the feces [51]. In order to reduce the cost of eliminating the damage caused by *E. tenella* in poultry farms, the authors are actively developing various prevention methods for monitoring *E. tenella* [24,52]. The effects of *E. tenella* infection range from localized intestinal tissue degeneration to death in the most severe instances [53]. To minimize detrimental impacts on high-quality broiler meat production, new agents with cheap costs and minimum adverse effects against *E. tenella* are needed.

Here, the positive impact of cinnamon additives, particularly at level 4 g, was obvious on most carcass traits such as the CW, carcass yield, as well as breast, heart, proventriculus, gizzard, and pancreas percentage relative to CW, increased as compared with the PC. This result may be attributed to that cinnamon can improve the digestive system due to the presence of cinnamon's active components, such as cinnamaldehyde, which can stimulate appetite and digestion [54]. Therefore, birds who received a cinnamon diet, particularly at level 6 g, had a higher breast weight than those in the PC and similar to those received a diet supplemented with salinomycin. Our data showed that experimental treatments did not influence some of the carcass characteristics of meat, such as the relative weights of liver, leg, and fat. These findings are in agreement with [55], who found that the carcass characteristics did not change with the addition of various cinnamon oil levels and sodium butyrate except that the cholesterol level of the muscles was lowered in broilers. In part, this agrees with [56], who found that including CBP in broiler meal had no substantial effect ($p > 0.05$) on heart, breast, gizzard, cholesterol, abdominal fat, or triglycerides. However, [57] found that broilers fed CBP had a higher dressing percentage ($p < 0.05$). The findings of [58] corroborated our findings that dietary treatment with CBP had no significant effect on lymphoid organ relative weights ($p > 0.05$). Immune organ weight was significantly higher in the 5.0% CBP group, according to [40].

As expected, the results of this study showed that slaughter weight, carcass weight, and then dressing or carcass yield were adversely affected by the positive control compared with medicated or NC groups. Our findings support those of [59,60], who were orally challenged with Eimeria-populated oocysts harmed performance indices and carcass yield. Where [59] observed *Rumex nervosus* leaves can mitigate *E. tenella* suffering and improve dressing percentage. In addition, [60] observed that the dressing percentage was significantly higher in the group receiving a mixture of all 3 herbs (*Aloe barbadensis*, *Ferulafoetida regal*, and *Tamarindus indica*) at 2 mL/L mixed with citric acid and lowest in the control group (without medicinal herbs supplementation and challenged with Eimeria-populated oocysts).

Instead of glucose under aerobic conditions, muscle glycogen is the primary metabolic fuel for anaerobic glycolysis after slaughter. When anaerobic glycolysis occurs, pyruvate is reduced to lactate. The accumulation of lactate causes a decrease in pH whenever the muscle is converted to meat [61]. When the pH reaches acidic conditions, glycolytic enzymes are probably inactivated. Thus, [62] mention that pH is one of the most important alterations that occur during rigor mortis and that it has a direct influence on the quality characteristics of the meat, such as juiciness, texture (tenderness), WHC, color, and shelf life. The meat of birds with a high pH has a higher WHC than meat with a lower pH. The pH of meat is easily determined by its color. The pH of meat is high when it is very dark and low when it is very light. The lower pH in bird meat groups with herbs may be responsible for inhibiting the integration of the deterioration of the growth of microorganisms [63]. The meat quality (PSE, DFD) of birds can be measured quickly and precisely to determine the pHi value of meat samples. The threshold pH value categories of the breast meat of broilers are 5.8 (reddish, soft, and exudative), 5.9–6.2 (standard meat quality), and 6.3 (pale, firm, and non-exudative or dark, firm, and dry) [64]. A duration of 15 min postmortem, the pH parameters were a good predictor of meat traits [65] and ranged from 5.78 to 6.59 [66]. The authors of [61] reviewed the literature and found that the highest quality commercial poultry meat products are more likely to fall within the pH range of 5.7–6.0. In this study, the initial pH ranged from 6.08 to 6.36, and the ultimate pH ranged from 5.71 to 5.83. There was acceptable color and increased wateriness (reddish, soft, and exudative) as well as pale color and good juice retention (pale, firm, and non-exudative). Although there were no significant differences between the experimental groups, the pH of chicken breast meat in the present study decreased mathematically with a rise in CBP doses. The variation in the pH could be attributed to the high antioxidant of cinnamon and its other hydroxyl derivatives attributed to the action of hydroxyl radicals ($^{\bullet}$OH) present in the phenyl ring of phenolic compounds acting as hydrogen donors [67]. Therefore, increasing CBP nutritional levels in diets increase the donation of hydroxyl groups, effectively reducing the pH value noticed in broiler breast muscle.

Meat color is influenced by many influences, such as pre-slaughter factors, stunning methods, cooling regimes, moisture content, heme pigments, protein physical status, strain, stress, and sex [4,5]. In [68], it was pointed out that raising the L* value was desirable in terms of consumer acceptance. Here, the 4 g of CBP/kg group had a higher initial L* value and then a higher WI and lower color change, while the ultimate L* value and WI were increased with increasing doses of CBP. Metmyoglobin cumulating on the exterior part of the storage meat contributes to the discoloration of the meat [69], which eliminates the a* value discrepancies. The formation of metmyoglobin and an elevation of lipid oxidation are the key components responsible for distinctions in the b* value [70]. Differences in initial and ultimate of L* values, total color change (ΔE), and WI were found by [31] when testing CBP. Few researchers have described ΔE and BI parameters, which could assist in elucidating the L*, a*, and b* behavior in bird breast meat, as impacted by CBP intake in the feed. These findings revealed that partial alterations in color variables could be due to the effects of experimental groups on the ΔE and WI. However, other authors, such as those of [55], did not discover any effect of CBP supplemented diets on broiler meat color measurements. Thus, decreasing water retention tends to lead to less reflective surface light that reduces L* values [71], and is associated with a decrease in the nutritional value of the meat due to the loss of some nutrients, and as a result, the breast meat becomes less tender. The experimental treatment had no effect on both WHC and DL ($p > 0.05$). WHC is a phrase used to refer to a muscle's ability to bind water under a particular set of circumstances. Commonly, the increase in muscle fat content results in higher WHC and a reduction in the percentage CL [72]. After death, oxygen deprivation causes lactic acid production, leading to a decline in pH, which causes protein denaturation, loss of protein solubility, and an overall reduction in the number of reactive groups available for water binding on muscle protein [62]. CL is a measurement of how much water is lost during cooking because of shrinkage. The degree of shrinkage that occurs during cooking is proportional to the loss of juiciness on the palate. CL was considerably lower in the study when birds were fed a diet containing 2 g of CBP/kg. In contrast, [73] did not observe any influence of 0.5 or 1 mL of cinnamon oil in broiler diets on CL.

The CBP treatment affected the MFI of the breast muscle. Myofibril fragmentation refers to the degree to which homogenization causes myofibrils to be destroyed. The authors of [74] have shown that the values of MFI are strongly correlated with other muscle measurements, such as tenderness and SF. Therefore, cinnamon supplementation could cause less fragmentation of myofibrils. On another hand, the SF in the breast muscle of birds ranged from 5.5 to 5.8 kgf/g [75] and between 2.71 and 3.31 kgf/g [76]. Therefore, the CBP groups in this trial had no effect on meat tenderness as the SF values were between 1.69 and 2.10 kgf/g, and they were almost 67% and 37% lower than the values reported by [75], respectively. However, the CBP treatments had higher SF values than the controls, and they were similar to the salinomycin group. These findings are in contrast with [55], who found that different levels of diets supplemented with cinnamon oil and sodium butyrate did not influence the meat SF value (kg force/cm^2) of the broilers.

In comparison with the salinomycin group, the CBP groups had lower levels of texture profiles. Additionally, the CBP groups had lower springiness than the control groups. Meat texture was evaluated using TPA and SF as having a myofibril structure. Recent investigations have been performed on the quality of meat or carcass characters of birds fed diets containing either powder or plant extracts [27,30,77–80]. However, little or no research has been conducted on the effect of CBP on TPA. The authors in [81] found that springiness and cohesiveness decreased as more cinnamon extract was added to sourdough bread. However, hardness, chewiness, and gumminess were reversed. It was thought that adding the cinnamon extract to a bread recipe would be beneficial. In our experiment, the hardness, springiness, cohesiveness, and chewiness decreased in CBP treatments compared with the salinomycin treatment, but there was no difference between the control groups.

Despite the fact that the supplemented CBP resulted in conflicting patterns in terms of SF and TPA, the treatment without CBP (NC) produced the best SF and hardness values. This was also the case for cohesiveness and springiness. The addition of CBP to broiler diets resulted in an increase in meat toughness in general. As a result, adding natural antioxidant compounds to meat can improve its quality, and cinnamon has the highest antioxidant capacity due to its high phenolic content [20,82]. Natural antioxidants, on the other hand, have been found to have little or no effect on the sensory characteristics of meat by some researchers. Supplementation with cinnamon oil, for example, had no effect on the quality of chicken meat, according to [55].

5. Conclusions

In summary, *Eimeria tenella* infection has a negative impact on growth performance, slaughter weight, carcass yield, and most carcass characteristics of broiler chicken; on the other hand, the use of cinnamon as alternatives to anticoccidials and ionophore coccidiostats, can mitigate these effects. Moreover, the addition of cinnamon was able to improve some physicochemical properties without affecting the meat's quality. However, in the breasts of birds given cinnamon, MFI decreased and toughness increased, when compared with the other experimental groups. Although this is a consistent experimental paradigm that is highly applicable to commercial conditions, more research into the use of cinnamon to improve the meat quality and productivity of broiler chickens in both healthy and sick conditions is needed.

Supplementary Materials: The following supporting information can be downloaded at: https://www.mdpi.com/article/10.3390/ani12020166/s1, Table S1: Ingredients and calculated nutrients of broilers starter and finisher diets.

Author Contributions: Conceptualization, M.M.Q. and G.M.S.; data curation, M.M.Q. and E.O.H.; formal analysis, M.M.Q. and A.H.A.; funding acquisition, M.A.A.-G. and A.H.A.; investigation, M.M.Q. and M.A.A.-G.; methodology, M.M.Q., A.A.A.-a. and G.M.S.; project administration, S.I.A.-M., M.A.A.-G., and A.H.A.; supervision, S.I.A.-M. and M.M.A.; validation, M.A.A.-G.; visualization, M.M.Q.; writing—original draft preparation, M.M.Q.; writing—review and editing, M.M.Q. and S.I.A.-M. All authors have read and agreed to the published version of the manuscript.

Funding: This study was supported by the Research Supporting Project (RSP-2022R439), King Saud University (Riyadh, Saudi Arabia).

Institutional Review Board Statement: The study complied with the King Saudi Arabia standards on animal use (KSU-SE-20-44) and was approved by the local animal care and welfare committee of King Saud University.

Informed Consent Statement: Not applicable.

Data Availability Statement: All data sets collected and analyzed during the current study are available from the corresponding author on fair request.

Acknowledgments: This work was supported by the Research Supporting Project (RSP-2022R439), King Saud University (Riyadh, Saudi Arabia). The authors extend their thanks to RSSU at King Saud University for their technical support.

Conflicts of Interest: The authors declare no conflict of interest.

References

1. USDA (United States Department of Agriculture). Livestock and Poultry: World Markets and Trade. 2021. Available online: https://usda.library.cornell.edu/concern/publications/73666448x?locale=en (accessed on 12 October 2021).
2. Hussein, E.; Suliman, G.; Al-Owaimer, A.; Ahmed, S.; Abudabos, A.; Abd El-Hack, M.; Taha, A.; Saadeldin, I.M.; Swelum, A. Effects of stock, sex, and muscle type on carcass characteristics and meat quality attributes of parent broiler breeders and broiler chickens. *Poult. Sci.* **2019**, *98*, 6586–6592. [CrossRef]
3. Shewita, R.; Taha, A. Influence of dietary supplementation of ginger powder at different levels on growth performance, haematological profiles, slaughter traits and gut morphometry of broiler chickens. *S. Afr. J. Anim. Sci.* **2018**, *48*. [CrossRef]
4. Attia, Y.; El-Hanoun, A.; Bovera, F.; Monastra, G.; El-Tahawy, W.; Habiba, H. Growth performance, carcass quality, biochemical and haematological traits and immune response of growing rabbits as affected by different growth promoters. *J. Anim. Physiol. Anim. Nutr.* **2014**, *98*, 128–139. [CrossRef]
5. Attia, Y.A.; Al-Harthi, M.A.; Korish, M.A.; Shiboob, M.M. Evaluation of the broiler meat quality in the retail market: Effects of type and source of carcasses. *Rev. Mex. Cienc. Pecu.* **2016**, *7*, 321–339. [CrossRef]
6. Ashour, E.A.; Abd El-Hack, M.E.; Swelum, A.A.; Osman, A.O.; Taha, A.E.; Alhimaidi, A.R.; Ismail, I.E. Does the dietary graded levels of herbal mixture powder impact growth, carcase traits, blood indices and meat quality of the broilers? *Ital. J. Anim. Sci.* **2020**, *19*, 1228–1237. [CrossRef]
7. Bowker, B. Developments in our understanding of water-holding capacity. In *Poultry Quality Evaluation*; Elsevier: Amsterdam, The Netherlands, 2017; pp. 77–113.
8. Li, J.; Yang, C.; Peng, H.; Yin, H.; Wang, Y.; Hu, Y.; Yu, C.; Jiang, X.; Du, H.; Li, Q. Effects of Slaughter Age on Muscle Characteristics and Meat Quality Traits of Da-Heng Meat Type Birds. *Animals* **2020**, *10*, 69. [CrossRef]
9. Zdanowska-Sąsiadek, Ż.; Lipińska-Palka, P.; Damaziak, K.; Michalczuk, M.; Grzybek, W.; Kruzińska, B.; Jasińska, K.; Róg, D.; Kordos, K.; Ząbek, K. Antioxidant effects of phytogenic herbal-vegetable mixtures additives used in chicken feed on breast meat quality. *Anim. Sci. Pap. Rep.* **2019**, *37*, 393–408.
10. Xiong, R.; Cavitt, L.; Meullenet, J.F.; Owens, C. Comparison of Allo–Kramer, Warner–Bratzler and razor blade shears for predicting sensory tenderness of broiler breast meat. *J. Texture Stud.* **2006**, *37*, 179–199. [CrossRef]

11. Attia, Y.A.; Al-Harthi, M.A.; Hassan, S.S. Turmeric (*Curcuma longa* Linn.) as a phytogenic growth promoter alternative for antibiotic and comparable to mannan oligosaccharides for broiler chicks. *Rev. Mex. Cienc. Pecu.* **2017**, *8*, 11–21. [CrossRef]
12. Namdeo, S.; Baghel, R.; Nayak, S.; Khare, A.; Prakash, R.; Pal, A.C.; Thakur, S.; Reddy, B. Essential oils: An potential substitute to antibiotics growth promoter in broiler diet. *J. Entomol. Zool. Stud.* **2020**, *8*, 1643–1649.
13. Yang, C.; Kennes, Y.M.; Lepp, D.; Yin, X.; Wang, Q.; Yu, H.; Yang, C.; Gong, J.; Diarra, M.S. Effects of encapsulated cinnamaldehyde and citral on the performance and cecal microbiota of broilers vaccinated or not vaccinated against coccidiosis. *Poult. Sci.* **2020**, *99*, 936–948. [CrossRef]
14. Granstad, S.; Kristoffersen, A.B.; Benestad, S.L.; Sjurseth, S.K.; David, B.; Sørensen, L.; Fjermedal, A.; Edvardsen, D.H.; Sanson, G.; Løvland, A. Effect of Feed Additives as Alternatives to In-feed Antimicrobials on Production Performance and Intestinal Clostridium perfringens Counts in Broiler Chickens. *Animals* **2020**, *10*, 240. [CrossRef]
15. Hussein, E.O.; Ahmed, S.H.; Abudabos, A.M.; Suliman, G.M.; El-Hack, A.; Mohamed, E.; Swelum, A.A.; N Alowaimer, A. Ameliorative Effects of Antibiotic-, Probiotic-and Phytobiotic-Supplemented Diets on the Performance, Intestinal Health, Carcass Traits, and Meat Quality of Clostridium perfringens-Infected Broilers. *Animals* **2020**, *10*, 669. [CrossRef]
16. El-Deek, A. Effect of anise (*Pimpinella anisum*), fennel (*Foeniculum vulgare*) and ginger (*Zingiber officinale* Roscoe) on growth performance, carcass criteria and meat quality of broilers. *Arch. Geflugelkd.* **2003**, *67*, 92–96.
17. Abouelenien, F.A.; Elshahawy, I.S.; Hamad, M.E.; Elsaidy, N.R. Biosecurity assessment in relation to the occurrence of some coccidian parasites in poultry farms, with in vitro evaluation of Psidium gujava as coccidia sporulation inhibitor. *Emir. J. Food Agric.* **2021**, *33*, 532–543.
18. Adarsh, A.; Chettiyar, B.; Kanthesh, B.; Raghu, N. Phytochemical Screening and Antimicrobial Activity of "*Cinnamon zeylanicum*". *Int. J. Pharm. Res. Innov.* **2020**, *13*, 22–33.
19. Bedier, M.M. Comparative Evaluation of the Antibacterial Activity of Cinnamon zylanicum (True Cinnamon) and Azadirachta indica (Neem) Herbal Extracts Versus Calcium Hydroxide Intracanal Medication against Enterococcus faecalis in Single Rooted Premolars Teeth: A Randomized In Vitro Study. *Acta Sci. Dent. Sci.* **2020**, *4*, 210. [CrossRef]
20. El-Hack, M.E.A.; Alagawany, M.; Abdel-Moneim, A.-M.E.; Mohammed, N.G.; Khafaga, A.F.; Bin-Jumah, M.; Othman, S.I.; Allam, A.A.; Elnesr, S.S. Cinnamon (*Cinnamomum zeylanicum*) Oil as a Potential Alternative to Antibiotics in Poultry. *Antibiotics* **2020**, *9*, 210. [CrossRef]
21. Skoufos, I.; Bonos, E.; Anastasiou, I.; Tsinas, A.; Tzora, A. Effects of phytobiotics in healthy or disease challenged animals. In *Feed Additives*; Elsevier: Amsterdam, The Netherlands, 2020; pp. 311–337.
22. Ciftci, M.; Simsek, U.G.; Yuce, A.; Yilmaz, O.; Dalkilic, B. Effects of dietary antibiotic and cinnamon oil supplementation on antioxidant enzyme activities, cholesterol levels and fatty acid compositions of serum and meat in broiler chickens. *Acta Vet. Brno* **2010**, *79*, 33–40. [CrossRef]
23. Qaid, M.M.; Al-Mufarrej, S.I.; Azzam, M.M.; Al-Garadi, M.A. Anticoccidial effectivity of a traditional medicinal plant, Cinnamomum verum, in broiler chickens infected with Eimeria tenella. *Poult. Sci.* **2021**, *100*, 100902. [CrossRef] [PubMed]
24. Al-Quraishy, S.; Qasem, M.A.; Al-Shaebi, E.M.; Murshed, M.; Mares, M.M.; Dkhil, M.A. Rumex nervosus changed the oxidative status of chicken caecum infected with Eimeria tenella. *J. King Saud Univ. -Sci.* **2020**, *32*, 2207–2211. [CrossRef]
25. Soutter, F.; Werling, D.; Kim, S.; Pastor-Fernández, I.; Marugán-Hernández, V.; Tomley, F.M.; Blake, D.P. Impact of Eimeria tenella Oocyst Dose on Parasite Replication, Lesion Score and Cytokine Transcription in the Caeca in Three Breeds of Commercial Layer Chickens. *Front. Vet. Sci.* **2021**, *8*, 70. [CrossRef] [PubMed]
26. Mengistu, B.M.; Bitsue, H.K.; Huang, K. The Effects of selenium-enriched probiotics on growth performance, oocysts shedding, intestinal cecal lesion scores, antioxidant capacity, and mRNA gene expression in chickens infected with *Eimeria tenella*. *Biol. Trace Elem. Res.* **2021**, *199*, 278–291. [CrossRef] [PubMed]
27. Suliman, G.M.; Alowaimer, A.N.; Al-Mufarrej, S.I.; Hussein, E.O.; Fazea, E.H.; Naiel, M.A.; Alhotan, R.A.; Swelum, A.A. The effects of clove seeds (*Syzygium aromaticum*) dietary administration on carcass characteristics, meat quality and sensory attributes of broiler chickens. *Poult. Sci.* **2020**, *100*, 100904. [CrossRef] [PubMed]
28. Qaid, M.M.; Al-Mufarrej, S.I.; Azzam, M.M.; Al-Garadi, M.A.; Albaadani, H.H.; Alhidary, I.A.; Aljumaah, R.S. Growth Performance, Serum Biochemical Indices, Duodenal Histomorphology, and Cecal Microbiota of Broiler Chickens Fed on Diets Supplemented with Cinnamon Bark Powder at Prestarter and Starter Phases. *Animals* **2021**, *11*, 94. [CrossRef]
29. Korley Kortei, N.; Tawia Odamtten, G.; Obodai, M.; Appiah, V.; Toah Akonor, P. Determination of color parameters of gamma irradiated fresh and dried mushrooms during storage. *Hrvat. Časopis Prehrambenu Tehnol. Biotehnol. Nutr.* **2015**, *10*, 66–71.
30. Cázares-Gallegos, R.; Silva-Vázquez, R.; Hernández-Martínez, C.; Gutiérrez-Soto, J.; Kawas-Garza, J.; Hume, M.; Méndez-Zamora, G. Performance, Carcass Variables, and Meat Quality of Broilers Supplemented with Dietary Mexican Oregano Oil. *Braz. J. Poult. Sci.* **2019**, *21*. [CrossRef]
31. Valizadeh, S.; Naseri, M.; Babaei, S.; Hosseini, S.M.H.; Imani, A. Development of bioactive composite films from chitosan and carboxymethyl cellulose using glutaraldehyde, cinnamon essential oil and oleic acid. *Int. J. Biol. Macromol.* **2019**, *134*, 604–612. [CrossRef]
32. Ledesma, E.; Laca, A.; Rendueles, M.; Díaz, M. Texture, colour and optical characteristics of a meat product depending on smoking time and casing type. *LWT-Food Sci. Technol.* **2016**, *65*, 164–172. [CrossRef]
33. Wilhelm, A.E.; Maganhini, M.B.; Hernández-Blazquez, F.J.; Ida, E.I.; Shimokomaki, M. Protease activity and the ultrastructure of broiler chicken PSE (pale, soft, exudative) meat. *Food Chem.* **2010**, *119*, 1201–1204. [CrossRef]

34. Malila, Y.; Petracci, M.; Benjakul, S.; Visessanguan, W. Effect of tumbling marination on marinade uptake of chicken carcass and parts quality. *Braz. J. Poult. Sci.* **2017**, *19*, 61–68.
35. Novaković, S.; Tomašević, I. A comparison between Warner-Bratzler shear force measurement and texture profile analysis of meat and meat products: A review. In Proceedings of the 59th International Meat Industry Conference, IOP Conf. Series: Earth and Environmental Science, Zlatibor, Serbia, 1–4 October 2017.
36. SAS. *SAS Institute/OR 9.3 User's Guide: Mathematical Programming Examples*; SAS Institute: Cary, NC, USA, 2012.
37. Mustafa, M.; Ibrahim, S.; Hussein, M. Effect of the Use of Black Seed Powder, Cinnamon, Yeast and their Mixtures on the Productive Performance of the Ross Broiler Chicken. *J. Univ. Babylon Pure Appl. Sci.* **2019**, *27*, 196–204.
38. Tiwari, G.; Rahal, A.; Kumar, A.; Kumar, A. Influence of Supplementing Giloe and Cinnamon on Production Performance in Commercial Broiler Chicken. *Indian J. Anim. Nutr.* **2019**, *36*, 94–98. [CrossRef]
39. Ali, M.; Ismail, Z.; Ali, A.; Sultan, S. Physiological responses and productive performance of broiler chicks fed diets supplemented with different levels of cinnamon powder. *Egypt. Poult. Sci. J.* **2018**, *38*, 1171–1184.
40. Sang-Oh, P.; Chae-Min, R.; Byung-Sung, P.; Jong, H. The meat quality and growth performance in broiler chickens fed diet with cinnamon powder. *J. Environ. Biol.* **2013**, *34*, 127. [PubMed]
41. Chowlu, H.; Vidyarthi, V.; Zuyie, R.; Maiti, C. Effect of Dietary Supplementation of Cinnamon on the Performance of Broiler Chicken. *Livest. Res. Int.* **2019**, *7*, 83–87.
42. Alhotan, R.A.; Abudabos, A. Anticoccidial and antioxidant effects of plants derived polyphenol in broilers exposed to induced coccidiosis. *Environ. Sci. Pollut. Res.* **2019**, *26*, 14194–14199. [CrossRef]
43. Khater, H.F.; Ziam, H.; Abbas, A.; Abbas, R.Z.; Raza, M.A.; Selim, A. Avian Coccidiosis: Recent Advances in Alternative Control Strategies and Vaccine Development. *Agrobiol. Rec.* **2020**, *1*, 11–25. [CrossRef]
44. Abbasi, R.; Abdi-Hachesoo, B.; Razavi, S.M.; Namazi, F.; Nazifi, S. In vitro and in vivo activity of cinnamaldehyde against Eimeria kofoidi in chukar partridge (*Alectoris chukar*). *Exp. Parasitol.* **2020**, *218*, 107978. [CrossRef]
45. Ali, S.A.-F.; Ismail, A.; Abdel-Hafez, S.A.; El-Genaidy, H.M.A. The Protective Effect of Cinnamon against Thermally Oxidized Palm Oil in Broiler Chickens. *Egypt. Acad. J. Biol. Sci. C Physiol. Mol. Biol.* **2020**, *12*, 23–39. [CrossRef]
46. Song, X.; Li, Y.; Chen, S.; Jia, R.; Huang, Y.; Zou, Y.; Li, L.; Zhao, X.; Yin, Z. Anticoccidial Effect of Herbal Powder "Shi Ying Zi" in Chickens Infected with Eimeria tenella. *Animals* **2020**, *10*, 1484. [CrossRef]
47. Abudabos, A.M.; Alyemni, A.H.; Swilam, E.O.; Al-Ghadi, M.a.Q. Comparative Anticoccidial Effect of some Natural Products against Eimeria spp. Infection on Performance Traits, Intestinal Lesion and Occyte Number in Broiler. *Pak. J. Zool.* **2017**, *49*, 989–1995. [CrossRef]
48. Hussain, Z.; Li, X.; Zhang, D.; Hou, C.; Ijaz, M.; Bai, Y.; Xiao, X.; Zheng, X. Influence of adding cinnamon bark oil on meat quality of ground lamb during storage at 4 °C. *Meat Sci.* **2020**, *171*, 108269. [CrossRef] [PubMed]
49. Clarke, L.; Fodey, T.L.; Crooks, S.R.; Moloney, M.; O'Mahony, J.; Delahaut, P.; O'Kennedy, R.; Danaher, M. A review of coccidiostats and the analysis of their residues in meat and other food. *Meat Sci.* **2014**, *97*, 358–374. [CrossRef]
50. McDougald, L.R.; Seibert, B.P. Residual activity of anticoccidial drugs in chickens after withdrawal of medicated feeds. *Vet. Parasitol.* **1998**, *74*, 91–99. [CrossRef]
51. El-Shazly, K.A.; El-Latif, A.A.; Abdo, W.; El-Morsey, A.; El-Aziz, M.I.A.; El-Mogazy, H. The anticoccidial activity of the fluoroquinolone lomefloxacin against experimental Eimeria tenella infection in broiler chickens. *Parasitol. Res.* **2020**, *119*, 1955–1968. [CrossRef]
52. Choi, J.; Kim, W.K. Dietary Application of Tannins as a Potential Mitigation Strategy for Current Challenges in Poultry Production: A Review. *Animals* **2020**, *10*, 2389. [CrossRef]
53. Qasem, M.A.; Dkhil, M.A.; Al-Shaebi, E.M.; Murshed, M.; Mares, M.; Al-Quraishy, S. Rumex nervosus leaf extracts enhance the regulation of goblet cells and the inflammatory response during infection of chickens with Eimeria tenella. *J. King Saud Univ.-Sci.* **2020**, *32*, 1818–1823. [CrossRef]
54. Kumar, M.; Kumar, V.; Roy, D.; Kushwaha, R.; Vaiswani, S. Application of Herbal Feed Additives in Animal Nutrition—A Review. *Int. J. Livest. Res.* **2014**, *4*, 1–8. [CrossRef]
55. Gomathi, G.; Senthilkumar, S.; Natarajan, A.; Amutha, R.; Purushothaman, M.R. Effect of dietary supplementation of cinnamon oil and sodium butyrate on carcass characteristics and meat quality of broiler chicken. *Vet. World* **2018**, *11*, 959. [CrossRef]
56. Shirzadegan, K. Reactions of modern broiler chickens to administration of cinnamon powder in the diet. *Iran. J. Appl. Anim. Sci.* **2014**, *4*, 367–371.
57. Eltazi, S.M. Effect of using cinnamon powder as natural feed additive on performance and carcass quality of broiler chicks. *Int. J. Innov. Agric. Biol. Res.* **2014**, *2*, 1–8.
58. Najafi, S.; Taherpour, K. Effects of Dietary Ginger, Cinnamon, Synbiotic and Antibiotic Supplementation on Performance of Broilers. *J. Anim. Sci. Adv.* **2014**, *4*, 658–667.
59. Qaid, M.M.; Al-Mufarrej, S.I.; Azzam, M.M.; Al-Garadi, M.A.; Alqhtani, A.H.; Fazea, E.H.; Suliman, G.M.; Alhidary, I.A. Effect of Rumex nervosus Leaf Powder on the Breast Meat Quality, Carcass Traits, and Performance Indices of Eimeria tenella Oocyst-Infected Broiler Chickens. *Animals* **2021**, *11*, 1551. [CrossRef]
60. Moryani, A.; Rajput, N.; Naeem, M.; Shah, A.; Soomro, H. Effect of different herbs on the histomorphology and performance of coccidiosis challenged broiler chickens. *Sarhad J. Agric.* **2021**, *37*, 1280–1290. [CrossRef]

61. El Rammouz, R.; Berri, C.; Le Bihan-Duval, E.; Babile, R.; Fernandez, X. Breed differences in the biochemical determinism of ultimate pH in breast muscles of broiler chickens–a key role of AMP deaminase? *Poult. Sci.* **2004**, *83*, 1445–1451. [CrossRef]
62. Mir, N.A.; Rafiq, A.; Kumar, F.; Singh, V.; Shukla, V. Determinants of broiler chicken meat quality and factors affecting them: A review. *J. Food Sci. Technol.* **2017**, *54*, 2997–3009. [CrossRef]
63. Alexandre, A.C.S.; Albergaria, F.C.; Fernandes, L.A.C.; de Sousa Gomes, M.E.; Pimenta, C.J. Effect of natural and synthetic antioxidants on oxidation and storage stability of mechanically separated tilapia meat. *LWT* **2022**, *154*, 112679. [CrossRef]
64. Ristic, M.; Damme, K. Significance of pH-value for meat quality of broilers: Influence of breed lines. *Vet. Glas.* **2013**, *67*, 67–73. [CrossRef]
65. Fletcher, D. Poultry meat quality. *World's Poult. Sci. J.* **2002**, *58*, 131–145. [CrossRef]
66. Dodge, J.; Peters, F. Temperature and pH changes in poultry breast muscles at slaughter. *Poult. Sci.* **1960**, *39*, 765–768. [CrossRef]
67. Godlewska-Żyłkiewicz, B.; Świsłocka, R.; Kalinowska, M.; Golonko, A.; Świderski, G.; Arciszewska, Ż.; Nalewajko-Sieliwoniuk, E.; Naumowicz, M.; Lewandowski, W. Biologically Active Compounds of Plants: Structure-Related Antioxidant, Microbiological and Cytotoxic Activity of Selected Carboxylic Acids. *Materials* **2020**, *13*, 4454. [CrossRef] [PubMed]
68. Yetişir, R.; Karakaya, M.; İlhan, F.; Yılmaz, M.T.; Özalp, B. Effects of different lighting programs and sex on some broiler meat quality properties affecting consumer preference. *Anim Prod.* **2008**, *49*, 20–28.
69. Bekhit, A.; Cassidy, L.; Hurst, R.; Farouk, M. Post-mortem metmyoglobin reduction in fresh venison. *Meat Sci.* **2007**, *75*, 53–60. [CrossRef] [PubMed]
70. Xiong, Y. Protein Oxidation and Implications for Muscle Food Quality. In *Antioxidants in Muscle Foods: Nutritional Strategies to Improve Quality*; Decker, E., Faustman, C., Lopez-Bote, C.J., Eds.; Wiley: Hoboken, NJ, USA, 2000.
71. Hughes, J.; Oiseth, S.; Purslow, P.; Warner, R. A structural approach to understanding the interactions between colour, water-holding capacity and tenderness. *Meat Sci.* **2014**, *98*, 520–532. [CrossRef] [PubMed]
72. Corzo, A.; Schilling, M.; Loar, R.; Jackson, V.; Kin, S.; Radhakrishnan, V. The effects of feeding distillers dried grains with solubles on broiler meat quality. *Poult. Sci.* **2009**, *88*, 432–439. [CrossRef]
73. Symeon, G.K.; Athanasiou, A.; Lykos, N.; Charismiadou, M.A.; Goliomytis, M.; Demiris, N.; Ayoutanti, A.; Simitzis, P.E.; Deligeorgis, S.G. The Effects of Dietary Cinnamon (*Cinnamomum zeylanicum*) Oil Supplementation on Broiler Feeding Behaviour, Growth Performance, Carcass Traits and Meat Quality Characteristics. *Ann. Anim. Sci.* **2014**, *14*, 883–895. [CrossRef]
74. Olson, D.G.; Stromer, M. Myofibril fragmentation and shear resistance of three bovine muscles during postmortem storage. *J. Food Sci.* **1976**, *41*, 1036–1041. [CrossRef]
75. Pelicano, E.R.L.; De Souza, P.; De Souza, H.; Oba, A.; Norkus, E.; Kodawara, L.; De Lima, T. Effect of different probiotics on broiler carcass and meat quality. *Braz. J. Poult. Sci.* **2003**, *5*, 207–214. [CrossRef]
76. Al-Owaimer, A.N.; Suliman, G.M.; Alyemni, A.H.; Abudabos, A.M. Effect of different probiotics on breast quality characteristics of broilers under Salmonella challenge. *Ital. J. Anim. Sci.* **2014**, *13*, 3189. [CrossRef]
77. Ashour, E.A.; El-Hack, M.E.A.; Shafi, M.E.; Alghamdi, W.Y.; Taha, A.E.; Swelum, A.A.; Tufarelli, V.; Mulla, Z.S.; El-Ghareeb, W.R.; El-Saadony, M.T. Impacts of Green Coffee Powder Supplementation on Growth Performance, Carcass Characteristics, Blood Indices, Meat Quality and Gut Microbial Load in Broilers. *Agriculture* **2020**, *10*, 457. [CrossRef]
78. Singh, J.; Kaur, P.; Sharma, M.; Mehta, N.; Singh, N.; Sethi, A.; Sikka, S. Effect of combination of garlic powder with black pepper, cinnamon and aloe vera powder on the growth performance, blood profile, and meat sensory qualities of broiler chickens. *Indian J. Anim. Sci.* **2019**, *89*, 1370–1376.
79. Ashour, E.A.; Bin-Jumah, M.; Abou Sayed-Ahmed, E.T.; Osman, A.O.; Taha, A.E.; Momenah, M.A.; Allam, A.A.; Swelum, A.A.; Abd El-Hack, M.E. Effects of dried okra fruit (*Abelmoschus esculentus* L.) powder on growth, carcass characteristics, blood indices, and meat quality of stored broiler meat. *Poult. Sci.* **2020**, *99*, 3060–3069. [CrossRef]
80. Wang, L.; Mandell, I.; Bohrer, B. Effects of feeding essential oils and benzoic acid to replace antibiotics on finishing beef cattle growth, carcass characteristics, and sensory attributes. *Appl. Anim. Sci.* **2020**, *36*, 145–156. [CrossRef]
81. Huh, C.K.; Shim, K.H. Quality characteristics of sourdough bread added with lactic acid bacteria culture solution and cinnamon extract. *Korean J. Food Preserv.* **2017**, *24*, 764–770. [CrossRef]
82. Shah, M.A.; Bosco, S.J.D.; Mir, S.A. Plant extracts as natural antioxidants in meat and meat products. *Meat Sci.* **2014**, *98*, 21–33. [CrossRef]

animals

Article

Chicken Immune Cell Assay to Model Adaptive Immune Responses In Vitro

Filip Larsberg [1,*], Maximilian Sprechert [1], Deike Hesse [1], Gudrun A. Brockmann [1] and Susanne Kreuzer-Redmer [2,*]

[1] Albrecht Daniel Thaer-Institute, Breeding Biology and Molecular Genetics, Humboldt University of Berlin, Unter den Linden 6, 10099 Berlin, Germany; max.sprechert@web.de (M.S.); deike.hesse-wilting@hu-berlin.de (D.H.); gudrun.brockmann@hu-berlin.de (G.A.B.)

[2] Institute of Animal Nutrition and Functional Plant Compounds, University of Veterinary Medicine Vienna, Veterinärplatz 1, 1210 Vienna, Austria

* Correspondence: filip.larsberg.1@hu-berlin.de (F.L.); susanne.kreuzer-redmer@vetmeduni.ac.at (S.K.-R.)

Simple Summary: Knowledge about the modes of action of immunomodulating compounds such as pathogens, drugs, or feed additives, e.g., probiotics, will allow the development of targeted nutrition strategies, prevent infectious diseases and the usage of antimicrobials, and promote the health of animals. To investigate the mechanisms of action of immunomodulating compounds, controlled in vitro systems using freshly isolated immune cells from blood represent a promising alternative to animal experiments. Immune cell isolation from the blood of chickens is a complex and difficult process since the immune cell fractions are significantly contaminated with red blood cells and platelets. To our knowledge, a robust protocol for immune cell isolation from chicken blood and the subsequent cultivation of immune cells is not available. Here, we established a protocol for blood sampling and immune cell isolation and cultivation from chicken blood, which could be applied for the investigation of direct effects of immunomodulating compounds. This protocol, combining different techniques of immune cell isolation, cultivation, and differentiation of distinct immune cell populations, will serve as a potential alternative to animal testing in vivo. By gaining knowledge about the mechanisms of action of immunomodulating compounds, this in vitro model will contribute to promote health and welfare in chicken farming.

Abstract: Knowledge about the modes of action of immunomodulating compounds such as pathogens, drugs, or feed additives, e.g., probiotics, gained through controlled but animal-related in vitro systems using primary cultured peripheral blood mononuclear cells (PBMCs) will allow the development of targeted nutrition strategies. Moreover, it could contribute to the prevention of infectious diseases and the usage of antimicrobials, and further promote the health of the animals. However, to our knowledge, a protocol for the isolation of PBMCs with reduced thrombocyte count from chicken blood and subsequent cell culture over several days to assess the effects of immunomodulating compounds is not available. Therefore, we established an optimized protocol for blood sampling and immune cell isolation, culture, and phenotyping for chicken PBMCs. For blood sampling commercial Na–citrate tubes revealed the highest count of vital cells compared to commercial Li–heparin ($p < 0.01$) and K3EDTA ($p < 0.05$) tubes. Using combined dextran and ficoll density gradient separation, the thrombocyte count was significantly reduced ($p < 0.01$) compared to slow-speed centrifugation with subsequent ficoll. For cell culture, the supplementation of RPMI-1640 medium with 10% chicken serum resulted in the lowest relative cell count of thrombocytes compared to fetal calf serum (FCS) ($p < 0.05$). To validate the ability of the cell culture system to respond to stimuli, concanavalin A (conA) was used as a positive control. The optimized protocol allows the isolation and cultivation of vital PBMCs with reduced thrombocyte count from chicken blood for subsequent investigation of the modes of action of immunomodulating compounds.

Keywords: chicken; PBMCs; primary cell culture; immunomodulating compounds

check for updates

Citation: Larsberg, F.; Sprechert, M.; Hesse, D.; Brockmann, G.A.; Kreuzer-Redmer, S. Chicken Immune Cell Assay to Model Adaptive Immune Responses In Vitro. *Animals* **2021**, *11*, 3600. https://doi.org/10.3390/ani11123600

Academic Editors: Natalie Morgan and Youssef A. Attia

Received: 17 October 2021
Accepted: 15 December 2021
Published: 19 December 2021

1. Introduction

Today, there is growing concern about the resistance of pathogenic bacteria against antibiotics, the residual effects of antibiotics in meat products [1], and the public health risk from zoonotic pathogens like *Salmonella* and *Campylobacter*. With the ban of the subtherapeutic usage of antibiotics in farming (Council Regulation EC 70/524/ EEC), immunomodulating alternatives have arisen to improve animal and human health. The immune response to a given stimulus varies between different species [2]. The knowledge of immunomodulatory properties is of high importance, particularly regarding those immunomodulatory compounds with a high potential to improve animal health via improved defence against infection [3]. Therefore, knowledge about the specific modes of action of immunomodulating compounds is needed in order to develop specific diets as alternatives to widely used antimicrobials [4] on farms and to improve the health and welfare of animals, and thereby also humans [5–7]. Peripheral blood mononuclear cells (PBMCs) are often used as a cell model to investigate the direct effects of immunomodulating compounds. Methods are well established to isolate PBMCs from the blood of humans [8,9] and animals [5,10,11]. However, in avian species, although it is often performed [5,10,11], it is a complex and difficult process to isolate a population of PBMCs for subsequent cultivation and in vitro assays using common isolation methods, without contamination of nucleated thrombocytes and erythrocytes in the immune cell fractions [11–14]. Furthermore, erythrocytes, which can be eliminated by red blood cell lysis in mammals, are nucleated in avian species and show a high degree of resistance to common lysis procedures [15]. The majority of erythrocytes are usually eliminated using density gradient centrifugation with ficoll. Thrombocytes can be excluded in further analysis, e.g., by flow cytometry. In contrast to lymphocytes, they express lower levels of the pan-leukocyte marker CD45 [16–18]. Moreover, they lack the T-cell marker CD3, the B-cell marker Bu-1a, and the monocyte/macrophage marker Kul-01 [16,19,20], and appear positive for the fibrinogen receptor CD41/CD61 [21] and the CD51/CD61 integrin [17]. Therefore, a dual-labeling approach was established to eliminate thrombocytes [19]. However, as thrombocytes interfere with the quantification of avian leukocytes and can result in shifting percentages of the latter, further markers are needed for the exclusion of thrombocytes [19]. In cell culture, thrombocytes were shown to suffer apoptotic cell death when cultured together with monocytes. The addition of lymphocytes or their soluble factors could delay apoptosis [22]. However, functional ex vivo analyses of, for example, T-cell responses, which are used to determine responses to infections and vaccination in chickens, require an efficient and pure isolation of PBMCs without contaminating thrombocytes [12,20,23]. Furthermore, thrombocytes have been shown to be a major cytokine producer in chickens [24], which would distort the responses of the lymphocytes. The aim of this work was the establishment of a cell culture system with chicken PBMCs to assess the direct effects of potentially immunomodulating compounds on chicken immune cells in vitro, which requires the isolation of a PBMC population with reduced thrombocyte count. Recently, a combined dextran and ficoll density gradient separation was reported to yield large populations of chicken PBMCs without contaminating thrombocytes [25]. Furthermore, a few slow-speed centrifugation approaches have been described previously [12,14,17,26].

However, the isolation of adaptive immune cells from peripheral blood with subsequent primary cell culture over a period of several days for the assessment of the effects of immunomodulating compounds has not yet been described. In this study, different optimization steps were performed to establish an innovative in vitro approach to assess the properties of immunomodulatory compounds. The established in vitro model will help to investigate the modes of action of immunomodulatory compounds such as feed additives and other immune cell stimuli used to improve health in chicken farming and prevent infectious diseases and the usage of antimicrobials.

2. Materials and Methods

2.1. Animals

Five- to 6-week-old broiler chickens of the commercial layer variety Cobb500 (Cobb Germany Avimex GmbH, Wiedemar, Germany) were used for the establishment of an in vitro cell culture model with chicken PBMCs. The birds were stunned and decapitated. The blood was sampled in tubes containing an anticoagulant. All chickens were fed a starter diet from day 1 to day 14 post hatch, and a grower diet afterwards (H. Wilhelm Schaumann GmbH, Pinneberg, Germany). The ration was fed on an ad libitum basis and water was always available. The light duration was 24 h on days 1 and 2, followed by 16 h/day until sampling. The chickens were kept in groups of approximately 20 chickens in suitable 4 m^2 pens.

The study was approved by the local State Office for Health and Social Affairs, Landesamt für Gesundheit und Soziales Berlin (LaGeSo, T 0151/19).

2.2. Blood Sampling Methods

Different anticoagulants were tested for blood sampling. Commercial 9 mL tripotassium ethylenediaminetetraacetic acid (K3EDTA), lithium heparin (Li–heparin), and sodium citrate (Na–citrate) pre-filled polystyrene tubes (VACUETTE®) were used (all from Greiner Bio-One, Kremsmünster, Austria). Additionally, different volumes of 0.5 M EDTA (Carl Roth, Karlsruhe, Germany) in dH_2O (200 µL, 1 mL, and 8 mL) were tested in 50 mL tubes. Per biological replicate, 30 mL of blood was used for investigation of the best anticoagulant. To test anticoagulants, subsequent cell isolation was performed via combined slow-speed centrifugation and density gradient centrifugation. Isolated PBMCs were resuspended in 5 to 10 mL RPMI-1640 medium (Gibco™, ThermoFisher Scientific, Waltham, MA, USA).

2.3. PBMC Isolation Methods

2.3.1. Isolation of PBMCs Using Combined Slow-Speed Centrifugation and Density Gradient Centrifugation

Blood samples were diluted 1:2 with PBS (Gibco™, ThermoFisher Scientific, Waltham, MA, USA) containing 2 mM EDTA (Carl Roth, Karlsruhe, Germany). The samples were mixed and centrifuged for 15 min at 60× *g*. Upon centrifugation, three layers could be observed. The lymphocytes, laid on top of the erythrocyte layer, were swirled up using a dropper and transferred into a new 50 mL tube. The cells were washed once with PBS/EDTA (the tube was filled up with PBS/EDTA to 50 mL), after which the pellet was resuspended in 10 mL PBS/EDTA upon centrifugation, layered 1:2 onto ficoll (Histopaque-1077, Sigma-Aldrich, St. Louis, MO, USA), and centrifuged without a break for 30 min at 400× *g*. The buffer layer was collected at the interface of the plasma and ficoll, transferred to a new 50 mL tube, and washed once by centrifugation at 350× *g* for 10 min. After centrifugation, cells were resuspended in 10 mL RPMI-1640 medium.

2.3.2. Isolation of PBMCs Using Combined Dextran–Ficoll Separation

Blood samples were diluted 1:2 with PBS/EDTA. The diluted blood samples were mixed with 3% dextran in a ratio of 1:0.4 and centrifuged for 20 min at 50× *g*. The upper layer containing the PBMCs was carefully collected and layered onto ficoll in a ratio of 1:2 in a 50 mL tube. After centrifugation without a break for 30 min at 900× *g*, the buffer layer containing the PBMCs was collected, washed twice, and centrifuged for 10 min at 400× *g*. After centrifugation, cells were resuspended in 10 mL RPMI-1640 medium.

2.4. PBMC Culture

2.4.1. Cell Counting

The number of isolated vital cells was counted using a Tali® Image-Based Cytometer (Invitrogen™, ThermoFisher Scientific, Waltham, MA, USA). Therefore, 25 µL of cell suspension was mixed with 1 µL of 1 mg/mL propidium iodide (PI) as a viability marker and transferred onto a Tali® Cellular Analysis Slide for measurement.

2.4.2. Cell Seeding

For immune cell culture, cells were seeded in Nunc™ Non-treated T25 EasYFlasks™ (ThermoFisher Scientific, Waltham, MA, USA) at a density of 5×10^6 cells/mL and cultured for several days. For co-culture experiments, cells were seeded in nontreated, flat-bottomed 24-well plates (Eppendorf, Hamburg, Germany) at a density of 1×10^6 cells/mL and cultured from 24 h up to 72 h. All cells were cultured at 41 °C with 5% CO_2 in RPMI-1640 medium with 2 g/L glucose, 100 U/mL penicillin, 100 µg/mL streptomycin.

2.4.3. Serum Supplementation of Culture Medium

For cultivation of chicken PBMCs over several days, different sera were tested as cell culture supplements. Porcine serum (Sigma-Aldrich, St. Louis, MO, USA), chicken serum (Gibco™, ThermoFisher Scientific, Waltham, MA, USA), or fetal calf serum (FCS, Gibco™, ThermoFisher Scientific, Waltham, MA, USA) was added to the cells cultured in T25 flasks in a concentration of 10% to the RPMI-1640 medium. Immune cells were cultured for 3 days. After 24 h and 72 h, cells were subjected to flow cytometric measurement.

2.4.4. Supplementation of Culture Medium with Additional L-Glutamine

The effect of additional supplementation of 2 mM L-glutamine (Gibco™, ThermoFisher Scientific, Waltham, MA, USA) to the cell culture medium was tested. Therefore, PBMCs with or without additional L-glutamine were cultured in RPMI-1640 medium supplemented with 10% chicken serum for 24 h and subjected to flow cytometry.

2.4.5. Response Capacity of PBMCs towards Immune Cell Stimulants

To validate the cell culture system's responses to stimuli, concanavalin A (conA, Vector Laboratories, Burlingame, CA, USA) was used as a positive control. Therefore, 5 µg/mL or 10 µg/mL conA was added to the cells cultured in 24-well plates. PBMCs were cultured in RPMI-1640 medium supplemented with 10% chicken serum for 24 h and subjected to flow cytometry.

2.5. Immunophenotyping

For immunophenotyping, 1×10^6 cells per antibody staining set were harvested, centrifuged for 10 min at $400\times g$, washed once with cold PBS/EDTA containing 0.05% bovine serum albumin (BSA, Sigma-Aldrich, St. Louis, MO, USA), and stained with different panels of monoclonal antibodies. Therefore, 25 µL of the antibody mix, containing the diluted antibodies in the cold staining buffer PBS/EDTA, was added to the samples. After labeling, the samples were stored on ice in the dark for 30 min. The samples were washed with 600 µL PBS/EDTA/BSA and centrifuged for 10 min at $400\times g$. Subsequently, the supernatant was discarded, the samples were resuspended in 200 µL PBS/EDTA, and analyzed on a flow cytometer. Initially, the most suitable antibody concentration was determined by titration of the respective antibody from 1:25 to 1:400. In this study, immune cells were stained with mouse anti-chicken CD3-Allophycocyanin (APC) (CT-3, SouthernBiotech, Birmingham, AL, USA), CD4-Spectral Red (SPRD) (CT-4, SouthernBiotech, Birmingham, AL, USA), CD28-Phycoerythrin (PE) (AV7, SouthernBiotech, Birmingham, AL, USA), CD8-APC (CT-8, ThermoFisher Scientific, Waltham, MA, USA), CD45-Fluorescein isothiocyanate (FITC) (LT40, ThermoFisher Scientific, Waltham, MA, USA), CD41/CD61-(R)PE (11C3, ThermoFisher Scientific, Waltham, MA, USA), and human anti-chicken CD25-FITC (AbD13504, Bio-Rad Laboratories, Hercules, CA, USA) antibodies. Lymphocytes were gated using forward and sideward scatter by exclusion of debris, erythrocytes, and granulocytes. Dead cells were excluded using 1 µL 4′,6-diamidino-2-phenylindole (DAPI, Sigma-Aldrich, St. Louis, MO, USA) (1 mg/mL). Thereafter, doublets were excluded and at least 20,000 cells in the vital lymphocyte region were acquired on a Canto II (Becton Dickinson (BD), Franklin Lakes, NJ, USA) flow cytometer.

2.6. Statistical Analysis

The relative cell count of antibody-positive cells in the flow cytometer was calculated relative to the number of vital lymphocytes. Statistical analysis for blood sampling, PBMC isolation, and PBMC culture was performed using an unpaired Student's *t*-test. All tests were executed using GraphPad Prism 8.0.2 (GraphPad Software, San Diego, CA, USA). Differences between groups were considered statistically significant at $p < 0.05$.

3. Results

3.1. Blood Sampling

Blood sampling in commercial Na–citrate tubes revealed the highest count of vital cells (1.11×10^8) compared to commercial K3EDTA (5.37×10^7) ($p < 0.05$) and heparin (2.42×10^7) ($p < 0.01$) tubes (Figure 1). We found no significant difference between blood sampling in commercial Na–citrate tubes and 200 µL 0.5 M EDTA (6.81×10^7).

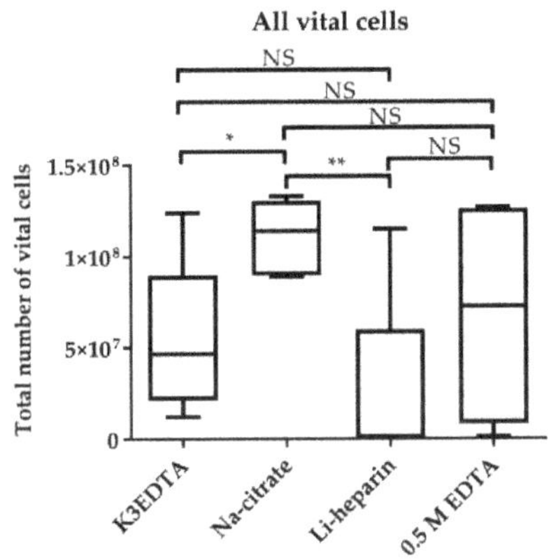

Figure 1. Influence of the anticoagulants K3EDTA, Na–citrate, and Li–heparin as well as 200 µL 0.5 M EDTA on cell survival. After PBMC isolation, chicken immune cells were counted with a Tali image-based cytometer and the viability was assessed using propidium iodide (PI). Five biological replicates are displayed. All bars represent one experiment. A box-and-whisker plot is displayed. Significance is shown as **; $p < 0.01$; *, $p < 0.05$. NS: not significant. Significance was analyzed using an unpaired Student's *t*-test.

However, flow cytometric analysis of the relative cell count of CD45-high leukocytes and CD45-low and integrin CD41/CD61+ thrombocytes within the vital lymphocyte population revealed the highest thrombocyte count in blood sampled in commercial Na–citrate tubes after the cell isolation (Figure S1a,b). Since Na–citrate is often used in immunological studies and we also detected the highest total number of live cells after measurement with a Tali image-based cytometer (Figure 1) and BD FACS Canto II (Data not shown), we decided to find a more suitable isolation method to reduce the high thrombocyte count.

3.2. PBMC Isolation Method

After blood sampling, the optimal PBMC isolation method was evaluated. In this step, we wanted to decrease the number of thrombocytes and increase the number of leukocytes in blood sampled in commercial Na–citrate tubes. Therefore, the relative cell counts of CD45-high leukocytes and CD45-low and integrin CD41/CD61+ thrombocytes in the vital lymphocyte population were assessed after the isolation of PBMCs using either a combined slow-speed and ficoll, or a combined dextran andficoll separation protocol (Figure 2).

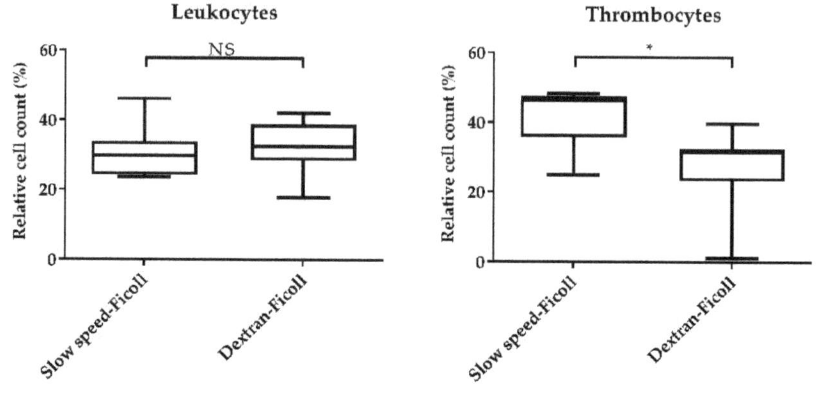

(**a**) (**b**)

Figure 2. Yield of leukocytes and thrombocytes using two immune cell isolation methods, combined slow-speed-ficoll and dextran–ficoll separation. After PBMC isolation, immune cells were stained with the pan-leukocyte marker CD45 and the thrombocyte marker CD41/CD61 and subjected to flow cytometry. (**a**) Leukocytes and (**b**) thrombocytes relative to the total live cell count. A total of 20,000 vital lymphocytes were recorded on a BD Canto II flow cytometer. DAPI was used as a viability marker. Data represent seven biological replicates and two technical replicates, each in two independent experiments. A box-and-whisker plot is displayed. Significance is shown as *, $p < 0.05$. NS: not significant. Significance was analyzed using an unpaired Student's *t*-test.

The dextran–ficoll separation did not change the relative cell count of vital leukocytes compared to the slow-speed–ficoll separation method (Figure 2a). Moreover, the mean of the relative cell count of thrombocytes was significantly lower after dextran–ficoll separation (27.23%) compared to the combined slow-speed–ficoll separation (41.19%) ($p < 0.05$) (Figure 2b). The latter was in line with the decrease of cells in the lymphocyte gate ($p < 0.05$, data not shown).

3.3. *PBMC Culture Conditions*

3.3.1. Medium Supplementation with Serum of Different Species

Chicken PBMCs were cultivated in RPMI-1640 medium supplemented with 10% porcine, chicken, or the standard fetal calf serum. The addition of chicken serum was associated with the highest mean of the relative cell count of leukocytes (84.21%) compared to the cultivation with the often-used FCS (54.55%) ($p < 0.05$); interestingly, there was no clear difference to porcine serum (77.04%) (Figure 3a). The thrombocyte count was low in cells cultivated in RPMI-1640 supplemented with chicken serum (13.44%) compared to those supplemented with FCS (43.01%) ($p < 0.05$), but not different to supplementation with porcine serum (21.17%) after 1 day of cultivation (Figure 3b).

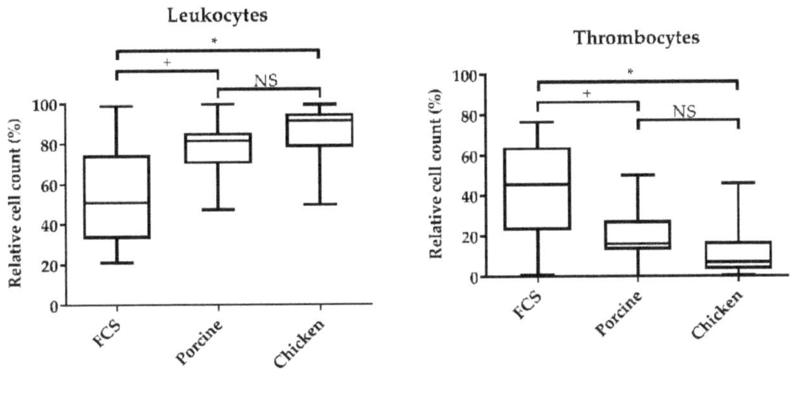

Figure 3. Influence of the addition of different sera to the cell culture on the survival of immune cells. Isolated PBMCs were cultured in RPMI-1640 medium with glucose, 100 U/mL penicillin, 100 µg/mL streptomycin, and either 10% chicken, 10% porcine, or 10% FCS. The relative cell counts of leukocytes and thrombocytes were assessed after 24 h of cultivation. (**a**) CD45-high leukocytes, relative to the vital lymphocyte population; (**b**) CD45-low thrombocytes relative to the vital lymphocyte population. A total of 20,000 cells were recorded on a BD Canto II flow cytometer. DAPI was used as a viability marker. Data represent seven biological replicates and two technical replicates each. A box-and-whisker plot is displayed. Significance is shown as +, $p < 0.1$; *, $p < 0.05$. NS: not significant. Significance was analyzed using an unpaired Student's *t*-test.

After one day of cultivation, the highest number of lymphocytes (Figure S2a) and vital cells (Figure S2b) was found in PBMCs cultured in RPMI-1640 medium with FCS (60.96%). The relative lymphocyte count was higher compared to culture with porcine (43.16%) ($p < 0.01$) or chicken serum (44.58%) ($p < 0.05$). The live cell count of cells cultivated in FCS (91.16%) was higher compared to those with porcine serum (86.74%) ($p < 0.05$), but did not differ between FCS and chicken serum (89.64%). The relative cell count of leukocytes cultured in RPMI-1640 supplemented with porcine serum (77.04%) was higher compared to cells cultured in medium with FCS (54.55%) ($p < 0.1$). The thrombocyte count was lower in cells cultured in medium with porcine serum (21.17%) compared to FCS (43.01%) ($p < 0.1$). The supplementation of porcine and chicken serum did not differ significantly for leukocytes and thrombocytes.

3.3.2. Medium Supplementation with Additional L-Glutamine

Supplementation with an additional 2 mM L-glutamine to the RPMI-1640 medium with glucose did not affect the viability of cultured PBMCs (Figure 4a). Furthermore, the relative cell count of leukocytes did not change after the addition of L-glutamine, compared to the control without additional L-glutamine (Figure 4b).

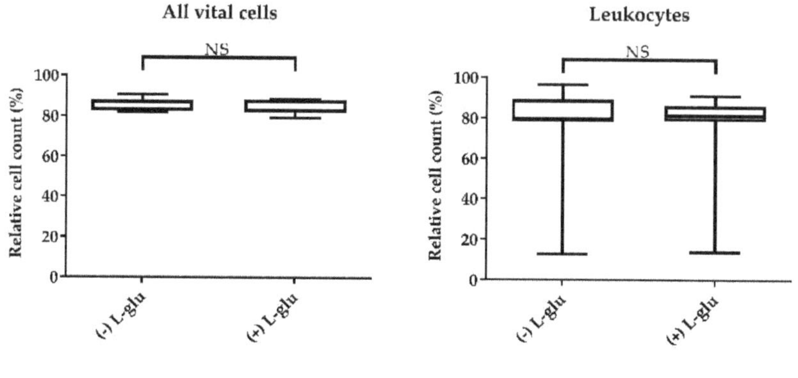

(**a**) (**b**)

Figure 4. The effect of additional L-glutamine on cell viability. Isolated chicken PBMCs were cultured in RPMI-1640 medium with glucose, 100 U/mL penicillin, 100 µg/mL streptomycin, and 10% chicken serum with or without additional L-glutamine (2 mM) for 24 h. (**a**) Vital cells relative to the lymphocytes; (**b**) CD45-high leukocytes, relative to the vital lymphocyte population. A total of 20,000 cells were recorded on a BD Canto II flow cytometer. DAPI was used as a viability marker. Data represent seven biological replicates and two technical replicates each. A box-and-whisker plot is displayed. NS: not significant. Significance was analyzed using an unpaired Student's *t*-test.

3.3.3. ConA as a Positive Control for the Validation of the Response Capacity of the Cell Culture System

To test if the cell culture system was a valid system to examine the direct effects of potentially immunomodulating compounds, the effect of conA, a well-known lymphocyte mitogen which stimulates mainly T-cells, was examined via measurement of CD8+ cytotoxic T-cells (Figure 5a,b) and CD4+ T-helper cells (Figure 5c,d).

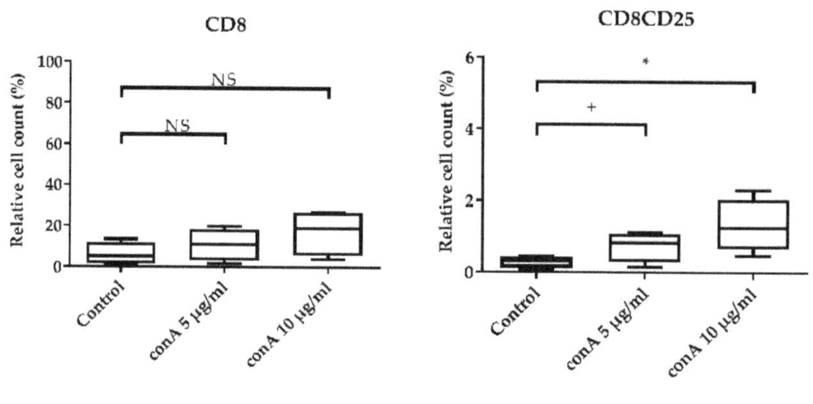

(**a**) (**b**)

Figure 5. *Cont.*

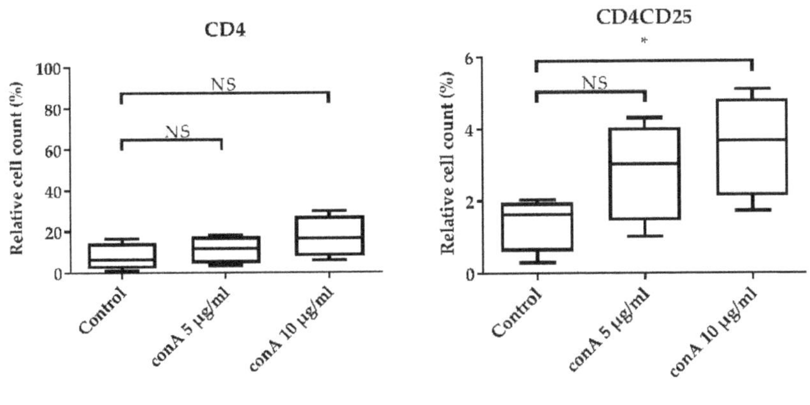

(c) (d)

Figure 5. Effects of conA on CD8+ cytotoxic T-cells and CD4+ T-helper cells. The effects of two concentrations (5 µg/mL and 10 µg/mL) of conA on the activation and proliferation of cytotoxic T-cells and T-helper cells relative to the vital lymphocyte population were assessed after 24 h of cultivation. (**a**) CD8+ cytotoxic T-cells; (**b**) CD8+CD25+ activated cytotoxic T-cells; (**c**) CD4+ T-helper cells; (**d**) CD4+CD25+ activated T-helper cells. A total of 20,000 cells were recorded on a BD Canto II flow cytometer. DAPI was used as a viability marker. Data represent four biological replicates and two technical replicates each. A box-and-whisker plot is displayed. Significance is shown as +, $p < 0.1$; *, $p < 0.05$. NS: not significant. Significance was analyzed using an unpaired Student's *t*-test.

As expected, conA treatment increased the relative cell count of CD8+ cytotoxic T-cells (Figure 5a), although the result was not significant. However, at a concentration of 10 µg/mL, conA increased the mean of the relative cell count of CD8+CD25+ activated cytotoxic T-cells from 0.29% to 1.32% ($p < 0.05$) (Figure 5b). For T-helper cells, conA stimulation increased the relative cell count of CD4+ T-helper cells numerically (Figure 5a), but this also did not reach significance. However, looking at CD4+CD25+ activated T-helper cells, conA treatment in a concentration of 10 µg/mL increased the mean relative cell count from 1.40% to 3.54% ($p < 0.05$) (Figure 5d). These effects were visible for two different organs, blood (Figure 5) and spleen (data not shown).

For validation, the effect of 10 µg/mL conA was tested on a higher number of biological replicates (Figure 6). Therefore, the relative cell count of conA-treated PBMCs was assessed by measurement of T-helper cells (Figure 6a), activated T-helper cells (Figure 6b), and all T-cells, except $\gamma\delta$ T-cells, via an additional marker, CD28-PE (Figure 6c).

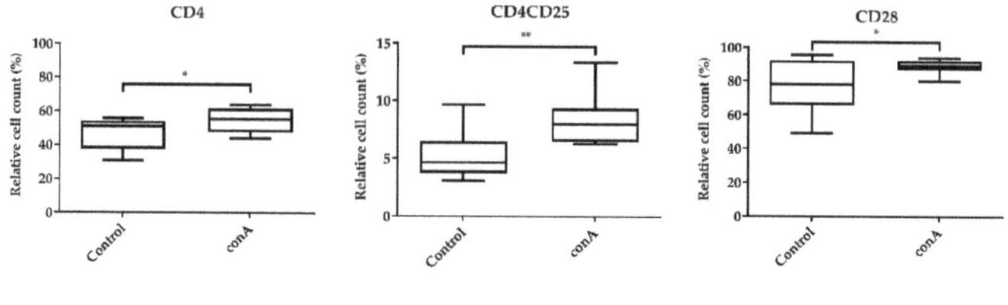

Figure 6. Effects of conA on activation and proliferation of CD4+ T-helper cells and CD28+ T-cells. The effect of 10 μg/mL conA on the activation and proliferation of T-helper cells relative to the vital lymphocyte population was assessed after 24 h of cultivation. (**a**) CD4+ T-helper cells; (**b**) CD4+CD25+ activated T-helper cells; (**c**) CD28+ T-cells. A total of 20,000 cells were recorded on a BD Canto II flow cytometer. DAPI was used as a viability marker. Data represent eight biological replicates and two technical replicates each. A box-and-whisker plot is displayed. Significance is shown as **, $p < 0.01$; *, $p < 0.05$. NS: not significant. Significance was analyzed using an unpaired Student's *t*-test.

As shown by testing two different conA treatment concentrations on PBMCs (Figure 5), the validation of the effect of 10 μg/mL conA on PBMCs by testing more biological replicates produced similar results (Figures 6 and S3). ConA treatment increased the mean relative cell count of CD4+ T-helper cells significantly from 46.93% to 54.81% ($p < 0.05$) (Figures 6a and S3a,d). Furthermore, conA increased the mean relative cell count of CD4+ CD25+ activated T-helper cells from 5.33% to 8.44% ($p < 0.01$) (Figures 6b and S3b,d). Moreover, the mean relative cell count of CD28+ T-cells increased significantly after conA treatment from 76.79% to 88.54% (Figures 6d and S3c,e), indicating T-cell proliferation.

In summary, conA induced T-cell activation and proliferation and can be used as a positive control.

3.4. Immune Cell Phenotyping Using Flow Cytometry

To assess the percentage of immune cell subsets, the following gating strategy was used (Figure 7a–d). First, lymphocytes were gated with a SSC/FSC plot (Figure 7a). From the lymphocyte population, only live cells were considered for further analysis (Figure 7b). Doublets were gated out (Figure 7c). Immune cell subsets were analyzed with antibodies labeling leukocytes (CD45-FITC), thrombocytes (CD41/CD61-(R)PE), T-helper cells (CD4-SPRD), cytotoxic T-cells (CD8-APC), T-cells (CD28-PE and CD3-APC), and activated T-cells (CD25-FITC) (example CD45-FITC; Figure 7d and Table 1).

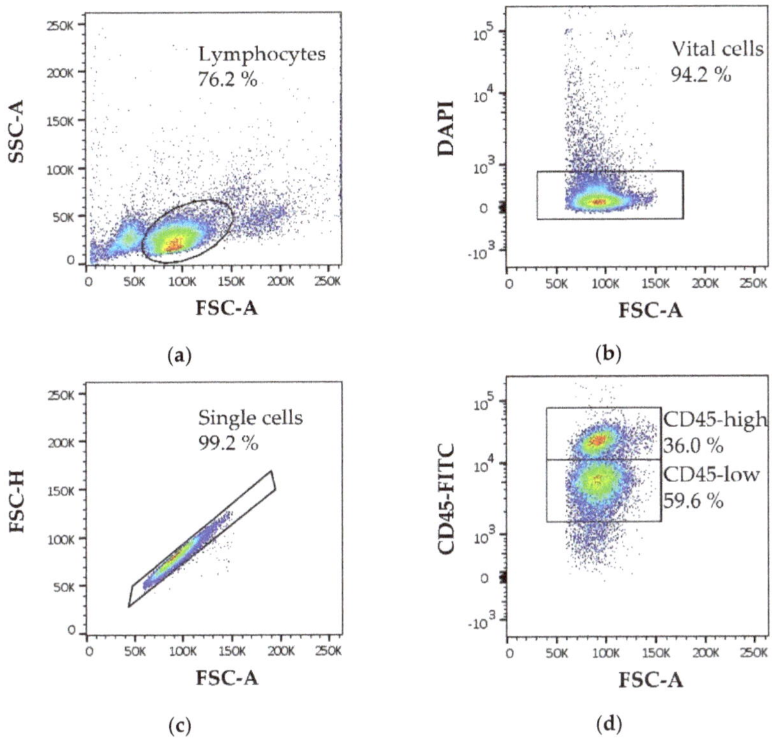

Figure 7. Example of the gating strategy for chicken PBMCs (**a**) *y* axes: SSC-A, *x* axes: FSC-A, the egg-shaped gate represents the lymphocyte gate; (**b**) *y* axes: DAPI, *x* axes: FSC-A, DAPI was used as a live/dead marker, the rectangle gate represents the vital lymphocyte population; (**c**) *y* axes: FSC-H, *x* axes: FSC-A, the rectangle gate represents the single cell population out of the vital lymphocyte population in B; (**d**) *y* axes: CD45-FITC, *x* axes: FSC-A, the two rectangle gates represent the CD45-high leukocyte and CD45-low thrombocyte populations out of the single cell and vital lymphocyte populations in B and C. Data represent one biological replicate. A total of 20,000 cells were recorded on a BD Canto II flow cytometer.

Table 1. Antibodies with the conjugated fluorophore, its isotype, and the concentration used after antibody titration.

Antibody	Isotype	Final Concentration
CD45-FITC	LT40	1:50
CD41/CD61-(R)PE	11C3	1:50
CD4-SPRD	CT-4	1:50
CD8-APC	CT-8	1:50
CD25-FITC	AbD13504	1:25
CD28-PE	AV7	1:50
CD3-APC	CT-3	1:25

All antibodies used in this study were titrated and evaluated by mean fluorescence intensity (MFI) to obtain optimal concentrations for the staining assays (Table 1 and Figure S4). The following antibody sets were used: CD45-FITC, CD41/CD61-(R)PE, CD3-APC; CD4-SPRD, CD28-PE, CD25-FITC; CD8-APC, CD28-PE, CD25-FITC.

4. Discussion

To investigate the mode of action of immunomodulating compounds such as pathogens, drugs, or feed additives, e.g., probiotics, primary cell culture systems are inevitable and represent a good alternative to in vivo models. Our aim was the establishment of an in vitro cell culture model with chicken PBMCs, without contaminating thrombocytes, to assess the properties of immunomodulating compounds, especially feed additives. The study will help to increase knowledge about precise mechanisms of action, which will allow the development of targeted nutrition strategies, prevent infectious diseases and the usage of antimicrobials, and further improve the health in poultry production. Chicken PBMCs are often isolated using common methods such as density gradient centrifugation using ficoll, which result in the isolation of immune cells contaminated by nucleated thrombocytes and erythrocytes [11–14]. Therefore, the isolation of PBMCs without either nucleated thrombocytes or erythrocytes is complex and difficult in avian species. Concerning this matter, a dextran–ficoll separation method was recently published [25]. However, PBMCs were not cultivated for a longer period after the isolation. Here, we present a robust protocol for longer cultivation of chicken PBMCs, which will enable functional in vitro studies in chicken PBMCs.

4.1. Blood Sampling

Our data suggest that, despite a higher relative thrombocyte and a lower relative leukocyte count, the cell number and viability in Na–citrate tubes was highest compared to commercial K3EDTA and Li–heparin tubes, as well as to 200 µL 0.5 M EDTA. In other species, it has been shown that a slightly purer population of PBMCs is obtained by using EDTA as an anticoagulant compared to heparin [27]. Furthermore, for the purification of mononuclear cells from other sources than peripheral blood, heparin was shown to promote clumping and pre-activation of unstimulated control cells [28]. Citrate as an anticoagulant may result in better quality of RNA and DNA compared with other anticoagulants and furthermore produce a higher yield of mononuclear cells, which is consistent with the results of the current study. Li–heparin, which revealed the highest leukocyte and the lowest thrombocyte count in this study, is reported to affect T-cell proliferation and to bind to many proteins. RNA yields from EDTA-treated blood have been shown to be higher compared to heparin-sampled blood [29]. Furthermore, EDTA was shown to affect PBMCs by a progressive and irreversible loss of antigen-specific lymphoprolerative responses when PBMCs were exposed to EDTA for a longer time period [30]. Therefore, ethyleneglycol-bis-(beta-aminoethylether)tetraacetate (EGTA) was suggested [30]. However, in the current study, K3EDTA revealed lower relative lymphocyte and leukocyte counts and a lower count of total isolated vital cells. Taking these findings from the literature and our results into account, we choose Na–citrate as the best choice for anticoagulation.

4.2. PBMC Isolation Method

The reduction or elimination of thrombocytes, which are a large part of chicken PBMCs, is of high importance in immunomodulatory studies, since chicken thrombocytes have been shown to play roles in inflammation and antimicrobial defence [24,31,32]. Initially, we used a combined slow-speed centrifugation with subsequent density gradient centrifugation using ficoll, which was modified according to Viertlboeck and Göbel (2007) [17], Lavoie et al. (2005) [14], Sundaresan et al. (2005) [26], and Gogal et al. (1997) [12]. In comparison to those studies, we were not able to reduce the thrombocyte and erythrocyte counts in our samples (Figure 2). Other studies used only density gradient centrifugation to isolate PBMCs from chickens, without considering the thrombocyte count [10,27,33,34]. However, in this study, a method by Jergović et al. (2017) [25] was additionally tested. The method included the use of dextran, a slow-speed centrifugation step and a density gradient centrifugation step afterwards. Dextran, a high-molecular polysaccharide, is often used for the purification of neutrophils [35]. Other studies used 1% methylcellulose instead of dextran [36,37]. Our results showed a significant decrease of the relative cell count of thrombocytes by about

14% when using the dextran–ficoll separation method (Figure 2b). However, compared to the study of Jergović et al. (2017), where the thrombocyte count was about 2.85% compared to 96.9% of PBMCs, we found a thrombocyte count of 27.23%. However, the thrombocyte count was lower compared to the slow-speed centrifugation and ficoll separation method, in which we obtained 41.19% thrombocytes. However, the leukocyte count remained unaffected in our study (Figure 2a). One possible explanation is the dextran used in the different studies, which was provided by different companies. Furthermore, we centrifuged for 20 min after the addition of dextran, because we could not discriminate the phases after 10 min of centrifugation. In the study of Jergović et al. (2017), the centrifugation step was 10 min. All in all, we could reduce the thrombocyte count in our samples and start the cell culture with a low number of thrombocytes when we used the combined separation method based on the addition of dextran, a slow-speed centrifugation step, and a density gradient centrifugation.

4.3. PBMC Culture Conditions

We tested sera from different species as cell culture supplementation. Fetal calf serum is very often used even for nonbovine species such as mice, humans, and chickens [38–41]. Since we detected an increased number of thrombocytes in the cell culture with RPMI-1640 medium supplemented with FCS compared to supplementation with chicken serum, we do not recommend the addition of FCS. Thrombocyte cytokine expression was reported up to 18 h in culture [42], which could interfere with assays that are carried out to investigate immunomodulatory functions. DaMatta et al. [22] and Lam [43] showed that thrombocytes cultured in DMEM with 10% fetal bovine serum displayed cytoplasm and chromatin condensation and were suggested to suffer an apoptotic cell death in culture of about 70% after 24 h and 85% after 48 h of cultivation. Apoptosis of thrombocytes was additionally reported by Kaspers and Kaiser (2014), who showed that cells die by apoptosis after 48–72 h [44]. Despite the fact that thrombocytes are active for a short period, our aim was to reduce the thrombocyte count as much as possible to ensure the least interference with subsequent immunomodulatory assays. However, in our study, thrombocytes were not cultured under agitation to prevent adherence. Interestingly, there seemed to be a difference between the adherence in FCS-supplemented medium to the adherence in porcine- and chicken-serum supplemented media. Whether the serum influences the adhesion of thrombocytes still needs to be elucidated. In addition to the reduced thrombocyte count in cultures supplemented with chicken serum, we found that supplementation with chicken serum resulted in higher counts of vital leukocytes. Therefore, we showed that chicken serum is the most suitable supplement for long-term chicken PBMC cultures with reduced thrombocyte counts.

As L-glutamine is very unstable in medium, the supplementation of additional L-glutamine has been reported in different studies for PBMC culture in RPMI-1640 medium [10,45–47] in different concentrations. In our study, we added L-glutamine in a concentration of 2 mM to the cell cultures. We could not detect differences for cell viability or cell counts between media with and without additional supplementation of L-glutamine. Hence, we did not include the supplementation of additional L-glutamine in our protocol.

It is of high importance to validate the response capacity of cultured chicken immune cells to a stimulus. We chose conA, since it is a well-reported T-cell mitogen [12,13,48–50]. We observed a clear effect on the proliferation and activation of CD8+ cytotoxic T-cells (Figure 5a,b) and CD4+ T-helper cells (Figure 5c,d) in the presence of conA. Furthermore, we validated the effect on T-cell activation and proliferation (Figure 6a–c and Figure S3a–f). In a study by Alvarez et al. (2020), it was shown that chicken splenocytes have a low viability after conA treatment in a concentration of 1 μg/mL for 3 days [51]. In fact, the cell viability decreased also in our study after treatment with 10 μg/mL conA, but not with 5 μg/mL. However, we aimed initially to validate the responsiveness of the chicken PBMCs in culture as a positive control. Therefore, we were able to validate the presented protocol for a system to culture chicken immune cells.

Summarized brief protocol:

1. Sample blood in Na–citrate VACUETTE® tubes
2. Dilute sampled blood 1:2 with PBS containing 2 mM EDTA
3. Add 3% dextran solution in a ratio of 1:0.4
4. Centrifuge 50× *g* 20 min
5. Collect upper phase
6. Overlay Histopaque-1077 1:2 with upper phase from step 5
7. Centrifuge 900× *g* 30 min
8. Collect PBMCs at interphase btw. serum and Histopaque-1077 and transfer into a new collection tube
9. Wash twice with PBS/EDTA, centrifuge 400× *g* 10 min
10. Resuspend PBMCs in RPMI-1640 supplemented with 10% chicken serum, 100 U/mL penicillin and 100 µg/mL streptomycin
11. Count cells and adjust to 1×10^6/mL up to 5×10^6/mL
12. Culture cells at 41 °C and 5% CO_2

5. Conclusions

After several optimization steps, we established a valid in vitro cell culture system to assess the direct effects of potentially immunomodulating compounds.

Here, we tested and optimized blood sampling, PBMC isolation, PBMC culture, and immune cell phenotyping of chicken PBMCs using monoclonal antibodies. This cell culture system will help to evaluate and understand the underlying mechanisms of the immunomodulatory properties of potentially immunomodulating compounds, e.g., feed additives, which could serve as potential alternatives to antibiotics, and may further serve as an alternative to animal testing in vivo. Besides testing feed additives, immunomodulation by challenges with pathogenic bacteria like *Salmonella* or *Campylobacter*, viruses, or particular drugs could be tested in our established chicken immune cell assay to model adaptive immune responses in vitro.

Supplementary Materials: The following are available online at https://www.mdpi.com/article/10.3390/ani11123600/s1, Figure S1: Influence of the anticoagulants K3EDTA, Na-citrate, and Li-heparin as well as 200 µL 0.5 M EDTA on leukocyte and thrombocyte count in isolated PBMCs. After cell isolation, cells were resuspended in PBS/EDTA and counted with Tali image-based cytometer and the viability was assessed using propidium iodide (PI). 1×10^6 cells/mL in duplicates per anticoagulant were labelled with the thrombocyte marker CD41/CD61-(R)PE and the leukocyte marker CD45-FITC. 20,000 cells were recorded on a BD Canto II flow cytometer. DAPI was used as a live/dead marker. Data represent 5 biological replicates. A box-and-whisker plot is displayed. Significance is shown as + $p < 0.1$; ** $p < 0.01$. NS: not significant. Significance was analyzed using an unpaired Student's *t*-test, Figure S2: Comparison of sera in cell culture of chicken PBMCs. Isolated PBMCs were cultured in RPMI-1640 medium with glucose and 100 U/mL penicillin and 100 µg/mL streptomycin and either 10% chicken, 10% porcine, or 10% fetal calf serum. The relative cell count of lymphocytes and vital lymphocytes was assessed after 24 h of cultivation (a) Gated lymphocytes relative to the measured cells; (b) Vital cells relative to the lymphocyte population. 20,000 cells were recorded on a BD Canto II flow cytometer. Data represent 7 biological replicates and two technical replicates each in two independent experiments. A box-and-whisker plot is displayed. Significance is shown as * $p < 0.05$; ** $p < 0.01$; *** $p < 0.001$. NS: not significant. Significance was analyzed using an unpaired Student's *t*-test, Figure S3: Effect of conA on PBMCs. (a–c) the upper row shows the relative cell count of CD4+ (a), CD4+CD25+ (b), and CD28+ (c) cells in untreated PBMCs as negative controls; (d–f) the low-er row displays the relative cell count of CD4+ (d), CD4+CD25+ (e), and CD28+ cells in conA-treated PBMCs as positive controls for activation and proliferation. After 24 h of cultivation, the relative cell count was investigated on a BD Canto II flow cytometer. The populations of interest out of the vital lymphocyte population, are represented in rectangle gates. Data represent 1 bio-logical replicate. 20,000 cells were recorded, Figure S4: Example of a serial antibody dilution. 1×10^6 PBMCs were labelled with CD4-SPRD in different concentrations. The y-axes of each plot represent the CD4-SPRD fluorescence signal and the x-axes show FSC-A. (a) unlabelled negative control; (b–f) cells labelled with CD4-SPRD in a concentration 1:25 (b), 1:50 (c), 1:100 (d), 1:200 (e), and 1:400 (f). CD4-SPRD in a concentration of 1:25 will be used in further experiments since this concentration enabled clear discrimination between the positive and negative population. The relative cell count of CD4-SPRD labelled cells was investigated on a BD Canto II flow cytometer. The populations of interest out of the vital lymphocyte population, are represented in rectangle gates. Data represent 1 biological replicate. 20,000 cells were recorded.

Author Contributions: Conceptualization, G.A.B. and S.K.-R.; Data curation, F.L.; Formal analysis, F.L.; Investigation, F.L. and M.S.; Supervision, D.H., G.A.B. and S.K.-R.; Writing—original draft, F.L.; Writing—review and editing, D.H., G.A.B. and S.K.-R. All authors will be informed about each step of manuscript processing including submission, revision, revision reminder, etc. via emails from our system or assigned Assistant Editor. All authors have read and agreed to the published version of the manuscript.

Funding: This research was funded by the Federal Ministry of Education and Research, Bundesministerium für Bildung und Forschung (BMBF, FKZ 031B0743).

Institutional Review Board Statement: The study was approved by the local State Office for Health and Social Affairs, Landesamt für Gesundheit und Soziales Berlin (LaGeSo, T 0151/19).

Informed Consent Statement: Not applicable.

Data Availability Statement: Data supporting reported results will be provided upon request.

Acknowledgments: We thank K. Männer (Freie Universität Berlin, Berlin, Germany) for providing Cobb500 broiler chicken. We acknowledge support by the German Research Foundation (DFG) and the Open Access Publication Fund of Humboldt-Universität zu Berlin.

Conflicts of Interest: The authors declare no conflict of interest. The funders had no role in the design of the study; in the collection, analyses, or interpretation of data; in the writing of the manuscript, or in the decision to publish the results.

References

1. Sato, Y.; Kuroki, Y.; Oka, K.; Takahashi, M.; Rao, S.; Sukegawa, S.; Fujimura, T. Effects of Dietary Supplementation with *Enterococcus faecium* and *Clostridium butyricum*, Either Alone Or in Combination, on Growth and Fecal Microbiota Composition of Post-Weaning Pigs at a Commercial Farm. *Front. Vet. Sci.* **2019**, *6*, 26. [CrossRef]
2. Hartung, T.; Sauer, A.; Wendel, A. Testing of Immunomodulatory Properties In Vitro. *Dev. Biol. Stand.* **1996**, *86*, 85–96. [PubMed]
3. Kiczorowska, B.; Samolińska, W.; Al-Yasiry, A.; Kiczorowski, P.; Winiarska-Mieczan, A. The Natural Feed Additives as Immunostimulants in Monogastric Animal Nutrition—A Review. *Ann. Anim. Sci.* **2017**, *17*, 605–625. [CrossRef]
4. Jørgensen, P.S.; Wernli, D.; Carroll, S.P.; Dunn, R.R.; Harbarth, S.; Levin, S.A.; So, A.D.; Schlüter, M.; Laxminarayan, R. Use Antimicrobials Wisely. *Nature* **2016**, *537*, 159–161. [CrossRef] [PubMed]
5. Swaggerty, C.L.; He, H.; Genovese, K.J.; Callaway, T.R.; Kogut, M.H.; Piva, A.; Grilli, E. A Microencapsulated Feed Additive Containing Organic Acids, Thymol, and Vanillin Increases In Vitro Functional Activity of Peripheral Blood Leukocytes from Broiler Chicks. *Poult. Sci.* **2020**, *99*, 3428–3436. [CrossRef] [PubMed]
6. Mata-Haro, V.; Reséndiz-Sandoval, M.; Hernández, J. In Vitro Differential Modulation of Immune Response by Probiotics in Porcine Peripheral Blood Mononuclear Cells. *Food Agric. Immunol.* **2014**, *25*, 209–219. [CrossRef]
7. Dhama, K.; Saminathan, M.; Jacob, S.S.; Singh, M.; Karthik, K.; Amarpal; Tiwari, R.; Sunkara, L.T.; Malik, Y.S.; Singh, R.K. Effect of Immunomodulation and Immunomodulatory Agents on Health with Some Bioactive Principles, Modes of Action and Potent Biomedical Applications. *Int. J. Pharmacol.* **2015**, *11*, 253–290. [CrossRef]
8. Lan, K.; Verma, S.C.; Murakami, M.; Bajaj, B.; Robertson, E.S. Isolation of Human Peripheral Blood Mononuclear Cells (PBMCs). *Curr. Proc. Microbiol.* **2007**, *6*, A.4C.1–A.4C.9. [CrossRef]
9. Grievink, H.W.; Luisman, T.; Kluft, C.; Moerland, M.; Malone, K.E. Comparison of Three Isolation Techniques for Human Peripheral Blood Mononuclear Cells: Cell Recovery and Viability, Population Composition, and Cell Functionality. *Biopreserv. Biobank.* **2016**, *14*, 410–415. [CrossRef]
10. Kaiser, M.G.; Cheeseman, J.H.; Kaiser, P.; Lamont, S.J. Cytokine Expression in Chicken Peripheral Blood Mononuclear Cells after In Vitro Exposure to Salmonella Enterica Serovar Enteritidis. *Poult. Sci.* **2006**, *85*, 1907–1911. [CrossRef]
11. Désert, C.; Merlot, E.; Zerjal, T.; Bed'hom, B.; Härtle, S.; Le Cam, A.; Roux, P.F.; Baeza, E.; Gondret, F.; Duclos, M.J.; et al. Transcriptomes of Whole Blood and PBMC in Chickens. *Comp. Biochem. Physiol. Part D Genom. Proteom.* **2016**, *20*, 1–9. [CrossRef]
12. Gogal, R.M., Jr.; Ahmed, S.A.; Larsen, C.T. Analysis of Avian Lymphocyte Proliferation by a New, Simple, Nonradioactive Assay (Lympho-Pro). *Avian. Dis.* **1997**, *41*, 714–725. [CrossRef] [PubMed]
13. Grasman, K.A. Assessing Immunological Function in Toxicological Studies of Avian Wildlife. *Integr. Comp. Biol.* **2002**, *42*, 34–42. [CrossRef] [PubMed]

14. Lavoie, E.T.; Grasman, K.A. Isolation, Cryopreservation, and Mitogenesis of Peripheral Blood Lymphocytes from Chickens (Gallus Domesticus) and Wild Herring Gulls (Larus Argentatus). *Arch. Environ. Contam. Toxicol.* **2005**, *48*, 552–558. [CrossRef]
15. Coles, E.H. *Veterinary Clinical Pathology*, 4th ed.; W. B. Saunders Company: Philadelphia, PA, USA, 1986; p. 297.
16. Horiuchi, H.; Tanaka, K.; Shigeta, A.; Yoshida, K.; Kushima, K.; Ohta, H.; Furusawa, S.; Matsuda, H. A Monoclonal Antibody against Chicken Thrombocytes Reacts with the Cells of Thrombocyte Lineage. *J. Vet. Med. Sci.* **2004**, *66*, 243–250. [CrossRef] [PubMed]
17. Viertlboeck, B.C.; Göbel, T.W. Chicken Thrombocytes Express the CD51/CD61 Integrin. *Vet. Immunol. Immunopathol.* **2007**, *119*, 137–141. [CrossRef]
18. De Boever, S.; Croubels, S.; Demeyere, K.; Lambrecht, B.; De Backer, P.; Meyer, E. Flow Cytometric Differentiation of Avian Leukocytes and Analysis of Their Intracellular Cytokine Expression. *Avian Pathol.* **2010**, *39*, 41–46. [CrossRef] [PubMed]
19. Bohls, R.L.; Smith, R.; Ferro, P.J.; Silvy, N.J.; Li, Z.; Collisson, E.W. The Use of Flow Cytometry to Discriminate Avian Lymphocytes from Contaminating Thrombocytes. *Dev. Comp. Immunol.* **2006**, *30*, 843–850. [CrossRef] [PubMed]
20. Seliger, C.; Schaerer, B.; Kohn, M.; Pendl, H.; Weigend, S.; Kaspers, B.; Härtle, S. A Rapid High-Precision Flow Cytometry Based Technique for Total White Blood Cell Counting in Chickens. *Vet. Immunol. Immunopathol.* **2012**, *145*, 86–99. [CrossRef] [PubMed]
21. Lacoste-Eleaume, A.S.; Bleux, C.; Quéré, P.; Coudert, F.; Corbel, C.; Kanellopoulos-Langevin, C. Biochemical and Functional Characterization of an Avian Homolog of the Integrin GPIIb-IIIa Present on Chicken Thrombocytes. *Exp. Cell Res.* **1994**, *213*, 198–209. [CrossRef] [PubMed]
22. DaMatta, R.A.; Manhães, L.; Lassounskaia, E.; de Souza, W. Chicken Thrombocytes in Culture: Lymphocyte-Conditioned Medium Delays Apoptosis. *Tissue Cell* **1999**, *31*, 255–263. [CrossRef]
23. Dalgaard, T.S.; Norup, L.R.; Rubbenstroth, D.; Wattrang, E.; Juul-Madsen, H.R. Flow Cytometric Assessment of Antigen-Specific Proliferation in Peripheral Chicken T Cells by CFSE Dilution. *Vet. Immunol. Immunopathol.* **2010**, *138*, 85–94. [CrossRef] [PubMed]
24. Scott, T.; Owens, M.D. Thrombocytes Respond to Lipopolysaccharide through Toll-Like Receptor-4, and MAP Kinase and NF-kappaB Pathways Leading to Expression of Interleukin-6 and Cyclooxygenase-2 with Production of Prostaglandin E2. *Mol. Immunol.* **2008**, *45*, 1001–1008. [CrossRef] [PubMed]
25. Jergović, M.; Nedeljković, G.; Cvetić, Ž.; Gottstein, Ž.; Bendelja, K. Combined Dextran and Ficoll Separation Yields Pure Populations of Chicken Peripheral Blood Mononuclear Cells—Short Communication. *Vet. Arh.* **2017**, *87*, 623–630. [CrossRef]
26. Sundaresan, N.R.; Ahmed, K.A.; Saxena, V.K.; Sastry, K.V.; Saxena, M.; Pramod, A.B.; Nath, M.; Singh, K.B.; Rasool, T.J.; DevRoy, A.K.; et al. Differential Expression of Inducible Nitric Oxide Synthase and Cytokine mRNA in Chicken Lines Divergent for Cutaneous Hypersensitivity Response. *Vet. Immunol. Immunopathol.* **2005**, *108*, 373–385. [CrossRef] [PubMed]
27. Bøyum, A. Isolation of Lymphocytes, Granulocytes and Macrophages. *Scand. J. Immunol.* **1976**, *5*, 9–15. [CrossRef] [PubMed]
28. Baien, S.H.; Langer, M.N.; Heppelmann, M.; von Köckritz-Blickwede, M.; de Buhr, N. Comparison between K_3EDTA and Lithium Heparin as Anticoagulant to Isolate Bovine Granulocytes From Blood. *Front. Immunol.* **2018**, *9*, 1570. [CrossRef] [PubMed]
29. Marteau, J.B.; Mohr, S.; Pfister, M.; Visvikis-Siest, S. Collection and Storage of Human Blood Cells for mRNA Expression Profiling: A 15-Month Stability Study. *Clin. Chem.* **2005**, *51*, 1250–1252. [CrossRef]
30. Kumar, P.; Satchidanandam, V. Ethyleneglycol-Bis-(Beta-Aminoethylether)Tetraacetate as a Blood Anticoagulant: Preservation of Antigen-Presenting Cell Function and Antigen-Specific Proliferative Response of Peripheral Blood Mononuclear Cells from Stored Blood. *Clin. Diagn. Lab. Immunol.* **2000**, *7*, 578–583. [CrossRef] [PubMed]
31. Ferdous, F.; Saski, C.; Bridges, W.; Burns, M.; Dunn, H.; Elliott, K.; Scott, T.R. Transcriptome Profile of the Chicken Thrombocyte: New Implications as an Advanced Immune Effector Cell. *PLoS ONE* **2016**, *11*, e0163890.
32. Ferdous, F.; Scott, T. Bacterial and Viral Induction of Chicken Thrombocyte Inflammatory Responses. *Dev. Comp. Immunol.* **2015**, *49*, 225–230. [CrossRef]
33. Slawinska, A.; Dunislawska, A.; Plowiec, A.; Gonçalves, J.; Siwek, M. TLR-Mediated Cytokine Gene Expression in Chicken Peripheral Blood Mononuclear Cells as a Measure to Characterize Immunobiotics. *Genes* **2021**, *12*, 195. [CrossRef] [PubMed]
34. Mirzaie, S. Optimization of Enhanced Expression of Interferon-γ and Interleukin-2 in Chicken Peripheral Blood Mononuclear Cells by Polysaccharide Immunomodulators Using Historical-Data Response Surface Methodology. *Gene Cell Tissue* **2019**, *6*, e97089. [CrossRef]
35. Quach, A.; Ferrante, A. The Application of Dextran Sedimentation as an Initial Step in Neutrophil Purification Promotes Their Stimulation, Due to the Presence of Monocytes. *J. Immunol. Res.* **2017**, *2017*, 1254792. [CrossRef] [PubMed]
36. Musa, H.; Wu, S.; Zhu, C.; Zhu, J.; Zhu, G. Immune Response of Peripheral Blood Mononuclear Cells to Avian Pathogenic Escherichia coli. *Ann. Microbiol.* **2009**, *59*, 587–592. [CrossRef]
37. Kogut, M.H.; Genovese, K.J.; Lowry, V.K. Differential Activation of Signal Transduction Pathways Mediating Phagocytosis, Oxidative Burst, and Degranulation by Chicken Heterophils in Response to Stimulation with Opsonized Salmonella enteritidis. *Inflammation* **2001**, *25*, 7–15. [CrossRef]
38. Peng, L.; van den Biggelaar, R.H.G.A.; Jansen, C.A.; Haagsman, H.P.; Veldhuizen, E.J.A. A Method to Differentiate Chicken Monocytes into Macrophages with Proinflammatory Properties. *Immunobiology* **2020**, *225*, 152004. [CrossRef] [PubMed]
39. Grönberg, A.; Halapi, E.; Ferm, M.; Petersson, M.; Patarroyo, M. Regulation of Lymphocyte Aggregation and Proliferation through Adhesion Molecule CD54 (ICAM-1). *Cell Immunol.* **1993**, *147*, 12–24. [CrossRef]
40. Matsuda, Y.; Imamura, R.; Takahara, S. Evaluation of Antigen-Specific IgM and IgG Production during an In Vitro Peripheral Blood Mononuclear Cell Culture Assay. *Front. Immunol.* **2017**, *8*, 794. [CrossRef]

41. Watarai, H.; Nakagawa, R.; Omori-Miyake, M.; Dashtsoodol, N.; Taniguchi, M. Methods for Detection, Isolation and Culture of Mouse and Human Invariant NKT Cells. *Nat. Protoc.* **2008**, *3*, 70–78. [CrossRef]
42. St Paul, M.; Paolucci, S.; Barjesteh, N.; Wood, R.D.; Schat, K.A.; Sharif, S. Characterization of Chicken Thrombocyte Responses to Toll-Like Receptor Ligands. *PLoS ONE* **2012**, *7*, e43381. [CrossRef] [PubMed]
43. Lam, K.M. Activation, Adhesion, Migration and Death of Chicken Thrombocytes. *Comp. Haematol. Int.* **2007**, *7*, 81–87. [CrossRef]
44. Kaspers, B.; Kaiser, P. Avian Antigen-Presenting Cells. In *Avian Immunology*, 2nd ed.; Schat, K.A., Kaspers, B., Kaiser, P., Eds.; Academic Press: Cambridge, MA, USA, 2014; Chapter 9; pp. 169–188.
45. Suklek, A.; Kayan, A.; Rattanasrisomporn, J.; Boonkaewwan, C. Isolation of Peripheral Blood Mononuclear Cells and the Expression of Toll-Like Receptors in Betong Chickens. *Vet. World* **2020**, *13*, 1372–1375. [CrossRef] [PubMed]
46. Song, L.; Xiong, D.; Hu, M.; Kang, X.; Pan, Z.; Jiao, X. Enhanced Humoural and Cellular Immune Responses to Influenza H7N9 Antigen HA1-2 Fused with Flagellin in Chickens. *BMC Vet. Res.* **2017**, *13*, 190. [CrossRef] [PubMed]
47. Tantikositruj, C.; Buadkhunthod, A.; Rattanasrisomporn, J.; Kitpipit, W.; Boonkaewwan, C. Assessment of Chicken Peripheral Blood Mononuclear Cells Isolated from Freshly Drawn Blood versus 24 H Refrigerated Blood. *Vet. World* **2021**, *14*, 2549–2553. [CrossRef] [PubMed]
48. Palacios, R. Concanavalin A Triggers T Lymphocytes by Directly Interacting with Their Receptors for Activation. *J. Immunol.* **1982**, *128*, 337–342.
49. Ando, Y.; Yasuoka, C.; Mishima, T.; Ikematsu, T.; Uede, T.; Matsunaga, T.; Inobe, M. Concanavalin A-mediated T Cell Proliferation Is Regulated by Herpes Virus Entry Mediator Costimulatory Molecule. *In Vitro Cell Dev. Biol. Anim.* **2013**, *50*, 313–320. [CrossRef]
50. Al-Khalifa, H. Immunological Techniques in Avian Studies. *Worlds Poult. Sci. J.* **2015**, *72*, 573–584. [CrossRef]
51. Alvarez, K.; Poma-Acevedo, A.; Fernández-Sánchez, M.; Fernández-Díaz, M. An EdU-Based Flow Cytometry Assay to Evaluate Chicken T Lymphocyte Proliferation. *BMC Vet. Res.* **2020**, *16*, 230. [CrossRef] [PubMed]

animals

MDPI

Article

Influence of Age on the Standardized Ileal Amino Acid Digestibility of Corn and Barley in Broilers

Mukti Barua [1,2,*], Mohammad Reza Abdollahi [1], Faegheh Zaefarian [1], Timothy J. Wester [1], Channarayapatna Krishnegowda Girish [3], Peter V. Chrystal [4] and Velmurugu Ravindran [1]

[1] Monogastric Research Center, School of Agriculture and Environment, Massey University, Palmerston North 4442, New Zealand; M.Abdollahi@massey.ac.nz (M.R.A.); F.Zaefarian@massey.ac.nz (F.Z.); T.J.Wester@massey.ac.nz (T.J.W.); V.Ravindran@massey.ac.nz (V.R.)

[2] Department of Animal Science and Nutrition, Faculty of Veterinary Medicine, Chattogram Veterinary and Animal Sciences University, Khulshi, Chattogram 4225, Bangladesh

[3] Nutrition and Care, Animal Nutrition, Evonik (SEA) Pte. Ltd., Singapore 609927, Singapore; Girish.channarayapatna@evonik.com

[4] Complete Feed Solutions, Howick, Auckland 2145, New Zealand; Peter@completefeeds.co.nz

* Correspondence: M.Barua@massey.ac.nz; Tel.: +64-2-75364926

Simple Summary: Efficient amino acid (AA) utilization in broilers is crucial concerning the accuracy of feed formulation, economy of diet and minimizing nitrogenous pollution in the environment. A range of factors (ingredient type, age, sex, feed form and bird type) can affect the AA digestion in poultry. The first week of broiler life is considered the most critical period while chicks source nutrition from residual yolk. Dynamic changes are noted in the digestive tract development, secretion and activity of protein digestive enzymes during the first few weeks in broiler. Limited data exist on the age effect on AA digestibility in broilers, and the results are paradoxical. The aim of the present study was to investigate the influence of age on the standardized ileal digestibility coefficients (SIDCs) of AAs in corn and barley from hatching to the end of the broiler growth cycle (day 42). Based on the results, the age influence on AA digestibility is grain- and AA-dependent. The pattern of the age effect on the SIDC AA in corn was not consistent. In the case of barley, the SIDC AA increased with advancing age.

Abstract: The aim of this study was to determine the standardized ileal digestibility coefficients (SIDCs) of nitrogen (N) and amino acids (AAs) in corn and barley at six different ages (days 7, 14, 21, 28, 35 and 42) of broilers using the direct method. The apparent AA digestibility coefficients were corrected using age-appropriate basal endogenous AA losses. No age effect ($p > 0.05$) was noted for the SIDC of N in corn. The average SIDC of indispensable AAs (IAAs) and total AAs (TAAs) was influenced in a quadratic manner ($p < 0.05$) with the values being higher at day 7 that decreased at day 14, increased and plateaued between days 21 and 35 and dropped again at day 42. The average SIDC of dispensable AAs (DAAs) was influenced linearly ($p < 0.05$). In barley, the SIDC of N and average IAAs, DAAs and TAAs was affected (quadratic; $p < 0.001$) by age. The digestibility increased from day 7 to 21 and then plateaued up to day 42. The present findings confirm that the SIDC of AA in corn and barley are influenced by broiler age and that the age effect on AA digestibility may need to be considered for precise feed formulation.

Keywords: age; amino acid; barley; broilers; corn; digestibility

Citation: Barua, M.; Abdollahi, M.R.; Zaefarian, F.; Wester, T.J.; Girish, C.K.; Chrystal, P.V.; Ravindran, V. Influence of Age on the Standardized Ileal Amino Acid Digestibility of Corn and Barley in Broilers. *Animals* **2021**, *11*, 3575. https://doi.org/10.3390/ani11123575

Academic Editors: Natalie Morgan and Youssef A. Attia

Received: 17 November 2021
Accepted: 14 December 2021
Published: 16 December 2021

1. Introduction

To improve broiler performance and production economics, accurate feed formulation that closely matches nutrient requirements is crucial. Determination of amino acid (AA) digestibility in feed ingredients is an important way to reach the goal of meeting the AA requirements. Poultry feed formulations based on digestible AA are superior to those based on total AAs because they are reflective of the actual amounts utilized for maintenance and production [1]. A better understanding of the factors influencing AA digestibility is vital to supply available AAs at optimum levels for accurate feed formulation and reduction of diet cost and nitrogen pollution into the environment [2].

The AA digestibility in poultry was measured at the excreta level in the past [3,4] but is currently measured at the terminal ileal level. The excreta digestibility assay has several shortcomings including possible modifying action of hindgut microflora on excreta AA profile and contamination with nitrogen (N) and AAs from urine [5]. Ileal AA digestibility can be categorized as either apparent or standardized/true. For the calculation of standardized ileal digestibility coefficients (SIDCs), the apparent ileal digestibility coefficients (AIDCs) are corrected for basal endogenous AA (EAA) losses originating from various digestive, pancreatic and enzymatic secretions [1]. The SIDC is more additive than AIDC in broiler feed formulations [2].

Grains are the major energy sources in broiler diets. However, they also supply about 40% of the total dietary protein and contribute significantly to the provision of some indispensable AAs (IAAs). Corn (*Zea mays* L.) is used extensively worldwide in poultry diets because of its high palatability, low fiber, high energy and essential fatty acids. Despite the low protein content in corn, owing to its higher inclusion levels (50–70%), it may contribute approximately up to one-third of the protein requirement of broilers. Barley (*Hordeum vulgare* L.) is another grain used in the European Union, Western Canada, New Zealand and Australia. Nevertheless, the inclusion of barley in poultry diets remains limited because of relatively low metabolizable energy, high content of fiber (220 g/kg) and high soluble (45 g/kg) and insoluble (122 g/kg) non-starch polysaccharide (NSP) contents [6]. Because of their viscous nature, the water-soluble fractions of barley exert a negative impact on the digestion and absorption of nutrients, including AAs [7].

Several datasets are available on the AIDC AA in feed ingredients for broilers [8,9]. However, the AA digestibility varies depending on ingredient type [9], class of bird (rooster, broiler, layer) [10] and feed form (mash vs. pellet) [11,12]. Despite the potential effects of age [3,13,14], only sporadic and inconsistent data exist on the age influence on the AA digestibility of ingredients in broilers [3,4,13,15]. Although a number of studies [1,9,12,16] have reported the SIDC AA in a range of feed ingredients, only a few [15,17–20] exist on the age-related standardized ileal digestibility (SID) of AAs, and the data are limited to two or three specific broiler ages. To the authors' knowledge, no studies to date have investigated the SIDC AA in grains from hatching to the end of the growth cycle of broilers. The current study was designed to determine the SIDC AA in corn and barley at six different ages (days 7, 14, 21, 28, 35 and 42 posthatch) of broilers.

2. Materials and Methods

The experimental procedure complied with the New Zealand Revised Code of Ethical Conduct for the use of live animals for research, testing and teaching and was approved by the Massey University Animal Ethics Committee.

2.1. Diets and Experimental Design

Corn and barley were obtained from a commercial supplier and ground in a hammer mill to pass through a screen size of 3.0 mm. Two experimental diets were developed with similar inclusions (938 g/kg) of either corn or barley as the only source of AAs in the diet (Table 1). Titanium dioxide (5 g/kg; Merck KGaA, Darmstadt, Germany) was incorporated in both diets as an indigestible marker. The diets were steam-conditioned at 70 °C for 30 s and pelleted using a pellet mill (Model Orbit 15; Richard Size Limited Engineers, Kingston upon Hull, UK) capable of manufacturing 180 kg of feed/h and equipped with a die ring with 3 mm holes and 35 mm thickness. Pellets were crumbled for the feeding of young chicks during the first two weeks of the experiment.

Representative grain samples were analyzed, in duplicate, for dry matter (DM), N, starch, crude fat, crude fiber, neutral detergent fiber (NDF), gross energy (GE), AAs, calcium

(Ca), phosphorus (P) and ash. The AIDC of N and AAs in each grain was determined using the direct method. The AIDCs were then standardized using the age-dependent (days 7, 14, 21, 28, 35 and 42) basal endogenous N and AA losses measured in a previous experiment [21].

Table 1. Composition of the experimental diets (g/kg, as-fed basis).

Ingredient	Corn	Barley
Corn	938	-
Barley	-	938
Soybean oil	20	20
Dicalcium phosphate	18	18
Limestone	13	13
Titanium dioxide [1]	5.0	5.0
Sodium chloride	2.0	2.0
Sodium bicarbonate	2.0	2.0
Trace mineral premix [2]	1.0	1.0
Vitamin premix [2]	1.0	1.0

[1] Merck KGaA, Darmstadt, Germany. [2] Supplied per kilogram of diet: antioxidant (ethoxyquin), 100 mg; biotin, 0.2 mg; calcium pantothenate, 12.8 mg; cholecalciferol, 0.06 mg; cyanocobalamin, 0.017 mg; folic acid, 5.2 mg; menadione, 4 mg; niacin, 35 mg; pyridoxine, 10 mg; trans-retinol, 3.33 mg; riboflavin, 12 mg; thiamine, 3.0 mg; dl-α-tocopheryl acetate, 60 mg; choline chloride, 638 mg; Co, 0.3 mg; Cu, 3.0 mg; Fe, 25 mg; I, 1 mg; Mn, 125 mg; Mo, 0.5 mg; Se, 0.2 mg; Zn, 60 mg.

2.2. Birds and Housing

A total of 696, one-day-old male broilers (Ross 308), obtained from a commercial hatchery, were used in this study. The birds were raised in floor pens and fed a commercial broiler starter diet (12.14 MJ/kg metabolizable energy; 225 g/kg crude protein; CP) from day 1 to 21 and a commercial broiler finisher diet (12.69 MJ/kg metabolizable energy; 190 g/kg CP) from day 22 until day 42 in pelleted form (Table 2).

On day 1, 168 chicks were individually weighed and allotted to 12 cages (14 chicks per cage) in such a way that group mean body weight (BW) per replicate was identical. The remaining chicks were allotted to 12 cages at 5 different ages, namely day 7 (12 birds per cage), day 14 (10 birds per cage), day 21 (8 birds per cage), day 28 (8 birds per cage) and day 35 (6 birds per cage). The test diets were offered for 4 days (days 3–7 and 10–14 (crumbled); days 17–21, 24–28, 31–35 and 38–42 (pelleted)) before collecting ileal digesta on days 7, 14, 21, 28, 35 and 42 posthatch, respectively.

The birds were offered ad libitum feed, and water was freely available throughout the whole experimental period. The room temperature was 32 ± 1 °C in the first week that was gradually reduced to 23 °C by the end of the third week. The floor pens, battery brooders and grower cages were housed in an environmentally controlled room with 20 h of fluorescent illumination per day.

2.3. Growth Performance Data

During the 4-day study period, feed intake and BW were recorded on a cage basis each week.

2.4. Determination of the Coefficient of Apparent Ileal Digestibility

At the end of each experimental period (days 7, 14, 21, 28, 35 and 42), all birds were euthanized by intravenous injection (0.5 mL per kg BW) of sodium pentobarbitone solution (Provet NZ Pty. Ltd., Auckland, New Zealand). The digesta were collected from the lower half of the ileum and processed as described by Ravindran et al. [8]. The ileum was marked as that portion of the small intestine extending from the Meckel's diverticulum to a point ~40 mm proximal to the ileocecal junction. In brief, the ileum was excised and divided into halves (proximal and distal ileum), and the digesta samples were collected from the lower half toward the ileocecal junction after gently flushing with distilled water into plastic containers. The ileal digesta from birds within a cage were pooled after collection, frozen immediately and then lyophilized (Model 0610, Cuddon Engineering, Blenheim, New Zealand). Diet and lyophilized digesta samples were ground to pass through a 0.5 mm sieve and stored in airtight plastic containers at 4 °C pending analysis.

Table 2. Composition and calculated analysis (g/kg, as-fed basis) of broiler starter and finisher diets.

Ingredient	Starter Diet (0–21 days)	Finisher Diet (22–42 days)
Corn	574.2	660
Soybean meal, 460 g/kg	381.4	295.6
Soybean oil	8.8	13.6
Limestone	11.3	9.9
Dicalcium phosphate	10.7	8.2
DL-methionine	3.3	3.0
L-lysine HCl	2.0	1.9
L-threonine	1.0	0.7
Sodium bicarbonate	2.7	2.5
Sodium chloride	2.5	2.5
Trace mineral premix [1]	1.0	1.0
Vitamin premix [1]	1.0	1.0
Phytase	0.1	0.1
	Calculated analysis	
Apparent metabolizable energy (MJ/kg)	12.14	12.69
Crude protein	225	190
Digestible lysine	11.0	9.2
Digestible methionine	6.2	5.6
Digestible methionine + cysteine	9.2	8.3
Digestible threonine	7.2	6.0
Crude fat	32	39
Crude fiber	29.3	27.5
Calcium	9.8	8.5
Available phosphorus	4.9	4.2
Sodium	2.2	2.1
Chloride	2.3	2.3
Potassium	11.5	9.7

[1] Supplied per kilogram of diet: antioxidant (ethoxyquin), 100 mg; biotin, 0.2 mg; calcium pantothenate, 12.8 mg; cholecalciferol, 0.06 mg; cyanocobalamin, 0.017 mg; folic acid, 5.2 mg; menadione, 4 mg; niacin, 35 mg; pyridoxine, 10 mg; trans-retinol, 3.33 mg; riboflavin, 12 mg; thiamine, 3.0 mg; dl-α-tocopheryl acetate, 60 mg; choline chloride, 638 mg; Co, 0.3 mg; Cu, 3.0 mg; Fe, 25 mg; I, 1 mg; Mn, 125 mg; Mo, 0.5 mg; Se, 0.2 mg; Zn, 60 mg.

2.5. Gizzard pH and Jejunal Digesta Viscosity

From the birds euthanized for ileal digesta collection, two birds from each replicate cage were used for the measurement of gizzard pH by a digital pH meter (pH spear, Oakton Instruments, Vernon Hill, IL, USA). The glass probe was inserted through an opening made in the gizzard and was placed directly in the digesta. Three values were taken from the proximal, middle and distal regions, and the average value was considered as the final pH value. The jejunal digesta viscosity was also determined from these birds. The digesta was collected from the distal jejunum, followed by centrifugation at 3000 × *g* at 20 °C for 15 min. A 0.5 mL aliquot of the supernatant was used in a viscometer (Brookfield digital viscometer, Model DV2TLV; Brookfield Engineering Laboratories Inc., Stoughton, MA, USA) fitted with CP-40 cone spindle with shear rates of 5–500/s to measure the digesta viscosity.

2.6. Chemical Analysis

Dry matter was measured using the standard procedure (Method 930.15) [22]. Titanium was analyzed on a UV spectrophotometer (Berthold Technologies GmbH and Co. KG, Bad Wildbad, Germany) following the method described by Short et al. [23]. Gross energy was determined by an adiabatic bomb calorimeter (Gallenkamp autobomb, Weiss Gallenkamp Ltd., Loughborough, UK) standardized with benzoic acid. Starch was analyzed using the Megazyme Total Starch Assay kit (Megazyme International Ireland Ltd., Wicklow, Ireland) based on thermostable α-amylase and amyloglucosidase [22,24]. Nitrogen was determined by combustion (Method 968.06) [22] using a carbon nanosphere-200 carbon, N and sulfur autoanalyzer (LECO Corporation, St. Joseph, MI, USA). The CP content was calculated as N × 6.25. Fat was determined using the Soxhlet extraction procedure (Method 2003.06) [22]. Neutral detergent fiber was determined (Method 2002.04) [22] using Tecator Fibertec™ (FOSS Analytical AB, Höganäs, Sweden). Ash was measured by ashing in a muffle furnace at 550 °C for 16 h (Method 942.05) [22]. Calcium and phosphorus concentrations were measured by Inductively Coupled Plasma-Optical Emission Spectroscopy (ICP-OES) using a Thermo Jarrell Ash IRIS instrument (Thermo Jarrell Ash Corporation, Franklin, MA, USA).

Amino acids were analyzed following standard procedures (Method 994.12) [25]. Briefly, the samples were hydrolyzed with 6 N HCl containing phenol for 24 h at 110 ± 2 °C in glass tubes in an oven. Amino acids were measured using AA analyzer (ion exchange) with ninhydrin post-column derivatization. The chromatograms detected at 570 and 440 nm were integrated using dedicated software (Agilent Open Lab software, Waldbronn, Baden-Württemberg, Germany). Cys and Met were analyzed as cysteic acid and methionine sulphone, respectively, by oxidation with performic acid–phenol for 16 h at 0 °C prior to hydrolysis. For the measurement of Trp, the samples were saponified under alkaline conditions with barium hydroxide solution in the absence of air at 110 °C for 20 h in an autoclave. The internal standard α-methyl Trp was added to the mixture following hydrolysis. After adjusting the hydrolysate to pH 3.0 and diluting with 30% methanol, Trp and the internal standard were separated by reverse phase chromatography (RP-18) on an HPLC column (CORTECS C18 Column; 2.7 µm, Waters Corporation, Dublin, Ireland). Finally, detection was selectively performed by means of a fluorescence detector to prevent interference by other AAs and constituents.

2.7. Calculations

Data were expressed on a DM basis. The AIDCs of AAs were calculated from the dietary ratio of AA to Ti relative to the corresponding ratio in the ileal digesta using the following formula.

$$\text{AIDC of AA} = [(\text{AA}/\text{Ti})_\text{d} - (\text{AA}/\text{Ti})_\text{i}]/(\text{AA}/\text{Ti})_\text{d}$$

where $(\text{AA}/\text{Ti})_\text{d}$ = ratio of AA to Ti in the diet, and $(\text{AA}/\text{Ti})_\text{i}$ = ratio of AA to Ti in the ileal digesta.

Apparent digestibility values for N and AAs were then standardized using the age-appropriate basal endogenous N and AA estimates (EAA; grams per kilogram of DM intake (DMI)) analyzed at different ages (days 7, 14, 21, 28, 35 and 42) in a previous experiment [21].

$$\text{SIDC} = \text{AIDC} + [\text{Basal EAA (g/kg DMI)}/\text{Ing. AA (g/kg DM)}]$$

where SIDC = standardized ileal digestibility coefficient of the AA, AIDC = apparent ileal digestibility coefficient of the AA, Basal EAA = basal endogenous AA loss, and Ing. AA = concentration of the AA in the ingredient.

2.8. Data Analysis

Cage was considered as the experimental unit. Data were analyzed by the GLM procedure of SAS (version 9.4; 2015; SAS Institute, Cary, NC, USA) for each grain. Differences were considered significant at $p < 0.05$. Orthogonal polynomial contrasts were performed to determine the linear and quadratic effects of age. The relationships between SIDC AA and other parameters were analyzed by Pearson correlation.

3. Results

3.1. Proximate and Nutrient Composition

The proximate and nutrient composition of the grains is summarized in Table 3. The results are presented on an "as-received" basis.

Table 3. Proximate, carbohydrate and amino acid composition of grains (g/kg, as-received basis).

Item	Corn	Barley
Dry matter	859	875
Starch	590	541
Nitrogen (N)	10.8	18.4
Crude protein (N × 6.25)	67.8	115
Fat	32.4	21.0
Neutral detergent fiber	83.1	110
Gross energy (MJ/kg)	16.3	16.3
Ash	20.5	18.6
Calcium	0.17	0.14
Phosphorus	2.47	3.02
	Indispensable amino acids (IAAs)	
Arg	3.43	5.28
His	2.07	2.37
Ile	2.49	3.98
Leu	8.14	7.53
Lys	2.33	3.72
Met	1.31	1.77
Thr	2.59	3.54
Trp	0.58	1.35
Val	3.44	5.52
Total IAA	26.4	35.1
	Dispensable amino acids (DAAs)	
Ala	5.14	4.54
Asp	4.77	6.78
Cys [1]	1.37	2.23
Glu	12.4	24.3
Gly [1]	2.85	4.44
Pro	5.60	10.8
Ser	3.29	4.41
Total DAA	35.4	57.5
Total AA [2]	61.8	92.5

[1] Semi-indispensable amino acids for poultry. [2] Total AA = IAA + DAA.

In both grains, starch was the main component followed by NDF in corn and CP in barley. The starch content was higher in corn than in barley, and the opposite was observed for the CP content. The NDF in corn was determined to be 83.1 g/kg which was lower than that in barley (110 g/kg). The contents of Ca in both corn (0.17 g/kg) and barley (0.14 g/kg) were negligible.

Among the IAAs, the content of Leu was the highest followed by Val, Arg, Ile and Thr in both grains, whereas lower contents were determined for Trp and Met. The Glu was the major dispensable AA (DAA) followed by Pro in both corn and barley. The variations in CP contents between the two grains were reflected in total AA (TAA) contents with a higher value in barley (92.5 g/kg) compared to corn (61.8 g/kg).

3.2. Growth Performance, Gizzard pH and Jejunal Digesta Viscosity

Weekly data on the performance, gizzard pH and jejunal digesta viscosity of birds fed corn- or barley-based diets are presented in Table 4.

Table 4. Daily feed intake (DFI; g/bird/d), daily weight gain (DWG; g/bird/d), gizzard pH and viscosity (cP) in jejunal digesta of broilers fed corn- and barley-based diets at different ages [1].

	Age (Days)							**Orthogonal Polynomial Contrasts**	
Parameter	7	14	21	28	35	42	Pooled SEM	Linear	Quadratic
					Corn				
DFI [2]	12.1	34.8	77.9	123	135	165	1.249	0.001	0.001
DWG [2]	10.5	28.4	35.8	54.3	58.9	59.7	0.476	0.001	0.001
Gizzard pH [3]	2.53	2.07	2.24	2.59	3.19	3.61	0.108	0.001	0.001
Viscosity [3]	2.03	2.33	2.23	2.25	2.11	2.34	0.079	0.168	0.480
					Barley				
DFI [2]	13.6	37.3	84.8	129	142	170	1.153	0.001	0.001
DWG [2]	12.5	31.4	38.8	56.2	61.1	62.9	0.855	0.001	0.001
Gizzard pH [3]	2.14	2.08	2.75	3.24	3.38	3.05	0.105	0.001	0.001
Viscosity [3]	2.94	2.70	2.69	2.80	2.75	2.94	0.093	0.745	0.024

[1] Each value represents the mean of six replicates (14, 12 and 10 birds per replicate for 7-, 14- and 21-days old birds, respectively; eight birds per replicate for 28- and 35-days old birds; and six birds per replicate for 42-days old birds). [2] Measured during 4-day feeding of experimental diets. [3] Calculated as the mean of six replicates (two birds per replicate).

Mortality during the experiment was negligible. Out of the 696 birds, only four died, and the deaths were not related to any specific treatment. The daily feed intake (DFI) and daily weight gain (DWG) increased (quadratic; $p < 0.001$) in both corn- and barley-based diets as birds grew older. Gizzard pH increased in a quadratic manner ($p < 0.001$) with advancing age in both grains. A decline in gizzard pH was observed from day 7 to day 14 but increased beyond day 21. The jejunal viscosity in corn was unaffected ($p > 0.05$) by age. In the case of barley, however, the jejunal digesta viscosity was influenced quadratically ($p < 0.05$) by age. Higher viscosity was observed on days 7 and 42 (2.94 cP). After day 7, a reduction in viscosity was observed at day 14 that plateaued until day 35. A further increase was observed at day 42.

3.3. Ileal Digestibility Coefficients of N and AAs in Corn

The influence of broiler age on the AIDC, SIDC and SID content of N and AAs in corn is presented in Tables 5–7, respectively.

Table 5. Apparent ileal digestibility coefficients [1] of nitrogen (N) and amino acids of corn at different ages of broilers [1].

	Age (Days)							**Orthogonal Polynomial Contrasts**	
Parameter	7	14	21	28	35	42	Pooled SEM	Linear	Quadratic
N	0.651	0.669	0.779	0.766	0.764	0.766	0.0089	0.001	0.001
					Indispensable amino acids				
Arg	0.745	0.752	0.864	0.849	0.850	0.859	0.0056	0.001	0.001
His	0.718	0.744	0.805	0.795	0.792	0.802	0.0080	0.001	0.001
Ile	0.628	0.645	0.805	0.779	0.771	0.789	0.0100	0.001	0.001
Leu	0.799	0.817	0.901	0.882	0.872	0.882	0.0059	0.001	0.001
Lys	0.467	0.473	0.753	0.709	0.701	0.712	0.0142	0.001	0.001
Met	0.724	0.721	0.882	0.856	0.850	0.859	0.0098	0.001	0.001
Thr	0.464	0.435	0.636	0.635	0.635	0.648	0.0138	0.001	0.001
Trp	0.464	0.469	0.648	0.635	0.654	0.662	0.0143	0.001	0.001
Val	0.604	0.616	0.794	0.779	0.768	0.786	0.0100	0.001	0.001
IAA	0.624	0.630	0.788	0.769	0.766	0.778	0.0093	0.001	0.001

Table 5. *Cont.*

	Age (Days)							Orthogonal Polynomial Contrasts	
Parameter	7	14	21	28	35	42	Pooled SEM	Linear	Quadratic
				Dispensable amino acids					
Ala	0.771	0.779	0.874	0.854	0.839	0.846	0.0072	0.001	0.001
Asp	0.609	0.601	0.773	0.753	0.739	0.753	0.0100	0.001	0.001
Cys [2]	0.709	0.737	0.747	0.738	0.730	0.747	0.0127	0.077	0.209
Glu	0.794	0.798	0.893	0.869	0.862	0.866	0.0067	0.001	0.001
Gly [2]	0.551	0.534	0.724	0.708	0.703	0.721	0.0121	0.001	0.001
Pro	0.706	0.725	0.811	0.801	0.798	0.813	0.0079	0.001	0.001
Ser	0.625	0.589	0.769	0.757	0.761	0.772	0.0119	0.001	0.001
DAA	0.681	0.680	0.799	0.783	0.776	0.788	0.0089	0.001	0.001
TAA	0.649	0.652	0.792	0.775	0.770	0.782	0.0087	0.001	0.001

[1] Each value represents the mean of six replicates (14, 12 and 10 birds per replicate for 7-, 14- and 21-days old birds, respectively; eight birds per replicate for 28- and 35-days old birds; and six birds per replicate for 42-days old birds). [2] Semi-indispensable amino acids for poultry. DAA = Average digestibility of dispensable amino acids; IAA = Average digestibility of indispensable amino acids; TAA = Average digestibility of all amino acids.

Table 6. Standardized ileal digestibility coefficients [1] of nitrogen (N) and amino acids of corn at different ages of broilers [2].

	Age (Days)							Orthogonal Polynomial Contrasts	
Parameter	7	14	21	28	35	42	Pooled SEM	Linear	Quadratic
N	0.936	0.817	0.922	0.911	0.907	0.868	0.0089	0.295	0.684
				Indispensable amino acids					
Arg	0.914	0.828	0.943	0.939	0.930	0.909	0.0056	0.001	0.003
His	0.839	0.810	0.866	0.862	0.854	0.844	0.0080	0.034	0.039
Ile	0.844	0.757	0.917	0.898	0.875	0.859	0.0100	0.001	0.001
Leu	0.902	0.868	0.952	0.939	0.923	0.915	0.0059	0.001	0.001
Lys	0.704	0.583	0.858	0.827	0.805	0.776	0.0142	0.001	0.001
Met	0.899	0.801	0.963	0.948	0.927	0.904	0.0097	0.001	0.001
Thr	0.912	0.678	0.872	0.856	0.871	0.821	0.0138	0.348	0.119
Trp	0.769	0.640	0.828	0.823	0.834	0.792	0.0143	0.001	0.048
Val	0.807	0.723	0.902	0.892	0.873	0.861	0.0100	0.001	0.001
IAA	0.843	0.743	0.900	0.887	0.877	0.854	0.0091	0.001	0.002
				Dispensable amino acids					
Ala	0.896	0.841	0.934	0.923	0.901	0.887	0.0072	0.043	0.001
Asp	0.863	0.733	0.902	0.887	0.868	0.843	0.0100	0.002	0.017
Cys [3]	0.968	0.906	0.908	0.894	0.895	0.877	0.0089	0.001	0.013
Glu	0.912	0.853	0.948	0.931	0.918	0.903	0.0067	0.026	0.002
Gly [3]	0.784	0.654	0.845	0.836	0.823	0.807	0.0121	0.001	0.034
Pro	0.845	0.799	0.885	0.874	0.871	0.866	0.0079	0.001	0.048
Ser	0.902	0.733	0.915	0.897	0.907	0.876	0.0119	0.001	0.996
DAA	0.881	0.788	0.905	0.892	0.883	0.865	0.0083	0.010	0.109
TAA	0.860	0.763	0.902	0.889	0.879	0.859	0.0087	0.001	0.011

[1] Apparent digestibility values were standardized using the following basal ileal endogenous flow values (g/kg DM intake), determined by feeding nitrogen-free diet at different ages [21]: Day 7: N, 3.59; Arg, 0.68; His, 0.29; Ile, 0.63; Leu, 0.97; Lys, 0.64; Met, 0.27; Thr, 1.35; Trp, 0.21; Val, 0.81; Ala, 0.75; Asp, 1.41; Cys, 0.47; Glu, 1.71; Gly, 0.78; Pro, 0.91; and Ser, 1.06. Day 14: N, 1.87; Arg, 0.30; His, 0.16; Ile, 0.33; Leu, 0.49; Lys, 0.30; Met, 0.12; Thr, 0.73; Trp, 0.12; Val, 0.43; Ala, 0.37; Asp, 0.73; Cys, 0.27; Glu, 0.80; Gly, 0.39; Pro, 0.48; and Ser, 0.55. Day 21: N, 1.79; Arg, 0.31; His, 0.15; Ile, 0.32; Leu, 0.49; Lys, 0.28; Met, 0.12; Thr, 0.71; Trp, 0.12; Val, 0.43; Ala, 0.36; Asp, 0.72; Cys, 0.26; Glu, 0.80; Gly, 0.40; Pro, 0.48; and Ser, 0.56. Day 28: N, 1.82; Arg, 0.36; His, 0.16; Ile, 0.34; Leu, 0.55; Lys, 0.32; Met, 0.14; Thr, 0.67; Trp, 0.13; Val, 0.45; Ala, 0.41; Asp, 0.74; Cys, 0.25; Glu, 0.89; Gly, 0.43; Pro, 0.48; and Ser, 0.54. Day 35: N, 1.81; Arg, 0.32; His, 0.15; Ile, 0.30; Leu, 0.49; Lys, 0.28; Met, 0.12; Thr, 0.71; Trp, 0.12; Val, 0.42; Ala, 0.37; Asp, 0.72; Cys, 0.26; Glu, 0.81; Gly, 0.40; Pro, 0.47; and Ser, 0.56. Day 42: N, 1.29; Arg, 0.20; His, 0.10; Ile, 0.20; Leu, 0.31; Lys, 0.18; Met, 0.07; Thr, 0.52; Trp, 0.09; Val, 0.29; Ala, 0.25; Asp, 0.49; Cys, 0.21; Glu, 0.53; Gly, 0.28; Pro, 0.34; and Ser, 0.40. [2] Each value represents the mean of six replicates (14, 12 and 10 birds per replicate for 7-, 14- and 21-days old birds, respectively; eight birds per replicate for 28- and 35-days old birds; and six birds per replicate for 42-days old birds). [3] Semi-indispensable amino acids for poultry. DAA = Average digestibility of dispensable amino acids; IAA = Average digestibility of indispensable amino acids; TAA = Average digestibility of all amino acids.

Table 7. Influence of age (days) on standardized digestible protein (CP) and amino acid contents [1] (g/kg) of corn, as-received basis.

	Age (Days)							Orthogonal Polynomial Contrasts	
Parameter	7	14	21	28	35	42	Pooled SEM	Linear	Quadratic
CP	63.4	55.4	62.5	61.7	61.5	58.9	0.607	0.295	0.685
Indispensable amino acids									
Arg	3.14	2.84	3.23	3.22	3.19	3.12	0.019	0.001	0.003
His	1.74	1.68	1.79	1.78	1.77	1.75	0.017	0.034	0.039
Ile	2.10	1.88	2.28	2.24	2.18	2.14	0.025	0.001	0.001
Leu	7.34	7.06	7.75	7.65	7.51	7.45	0.048	0.001	0.001
Lys	1.64	1.36	1.99	1.93	1.88	1.81	0.033	0.001	0.001
Met	1.18	1.05	1.26	1.24	1.21	1.18	0.013	0.001	0.001
Thr	2.36	1.75	2.26	2.22	2.26	2.13	0.036	0.348	0.119
Trp	0.45	0.37	0.48	0.48	0.48	0.46	0.008	0.001	0.048
Val	2.78	2.49	3.10	3.07	3.00	2.96	0.034	0.001	0.001
IAA	22.7	20.5	24.2	23.8	23.5	22.9	0.22	0.001	0.001
Dispensable amino acids									
Ala	4.60	4.32	4.80	4.74	4.63	4.56	0.037	0.043	0.001
Asp	4.11	3.49	4.30	4.23	4.14	4.02	0.048	0.002	0.017
Cys [2]	1.33	1.24	1.24	1.22	1.23	1.20	0.012	0.001	0.013
Glu	11.3	10.6	11.8	11.5	11.4	11.2	0.083	0.026	0.002
Gly [2]	2.24	1.86	2.41	2.38	2.35	2.29	0.034	0.001	0.034
Pro	4.73	4.47	4.96	4.89	4.88	4.85	0.044	0.001	0.047
Ser	2.97	2.41	3.01	2.95	2.99	2.88	0.039	0.001	0.996
DAA	31.2	28.4	32.5	31.9	31.6	31.0	0.28	0.002	0.018
TAA	54.1	48.9	56.6	55.8	55.1	53.9	0.49	0.001	0.005

[1] Each value represents the mean of six replicates (14, 12 and 10 birds per replicate for 7-, 14- and 21-days old birds, respectively; eight birds per replicate for 28- and 35-days old birds; and six birds per replicate for 42-days old birds). [2] Semi-indispensable amino acids for poultry. DAA = Average digestibility of dispensable amino acids; IAA = Average digestibility of indispensable amino acids; TAA = Average digestibility of all amino acids.

A quadratic increase ($p < 0.001$) was observed for the AIDC of N and average digestibility of IAAs, DAAs and TAAs of corn with the advancing age of broilers (Table 5). The AIDC of N and average AIDC of IAAs, DAAs and TAAs increased from day 7 to 21 then plateaued up to day 42. The AIDC of all individual IAAs increased in a quadratic manner ($p < 0.001$) with age. With the exception of Cys ($p > 0.05$), an increase (quadratic; $p < 0.001$) was observed for the AIDC of all individual DAAs.

The SIDC of N in corn was unaffected ($p > 0.05$; Table 6) by age. Bird age, however, quadratically influenced the average SIDC of IAAs ($p < 0.002$) and TAAs ($p < 0.05$). The higher values were recorded on day 7 than day 14, and the SIDC values increased until day 21 and plateaued until day 35, followed by a decrease on day 42. The SIDC of average DAAs was influenced in a linear ($p < 0.05$) manner with a higher value on day 7 (0.881) than day 14 (0.788). Afterward, an increase in the SIDC was observed on day 21 which plateaued until day 35, followed by a decrease on day 42. Except for Thr ($p > 0.05$), the SIDC of all individual IAAs and DAAs was influenced (linear or quadratic; $p < 0.05$ to < 0.001) by broiler age.

No age influence ($p > 0.05$) was recorded for the SID protein content of corn ($p > 0.05$; Table 7). The SID contents of total IAAs, DAAs and total AAs was influenced by age (quadratic; $p < 0.05$ to < 0.001). The SID content of total AAs was higher at day 7 (54.1 g/kg), increased from day 14 to 21 and plateaued until day 42. The SID contents of all individual AAs, except Thr and Ser, were influenced quadratically ($p < 0.05$ to < 0.001) by bird age. A linear pattern ($p < 0.001$) was observed for age effect on the SID content of Ser.

3.4. Ileal Digestibility Coefficients of N and AAs in Barley

The impact of broiler age on the AIDC, SIDC and SID content of N and AAs in barley is presented in Tables 8–10, respectively.

Table 8. Apparent ileal digestibility coefficients [1] of nitrogen (N) and amino acids of barley at different ages of broilers [1].

	Age (Days)							Orthogonal Polynomial Contrasts	
Parameter	7	14	21	28	35	42	Pooled SEM	Linear	Quadratic
N	0.504	0.670	0.731	0.713	0.730	0.720	0.0142	0.001	0.001
Indispensable amino acids									
Arg	0.611	0.705	0.803	0.772	0.802	0.791	0.0135	0.001	0.001
His	0.544	0.703	0.726	0.702	0.729	0.719	0.0142	0.001	0.001
Ile	0.436	0.660	0.749	0.725	0.747	0.749	0.0155	0.001	0.001
Leu	0.537	0.717	0.794	0.766	0.788	0.783	0.0133	0.001	0.001
Lys	0.230	0.545	0.739	0.682	0.717	0.717	0.0211	0.001	0.001
Met	0.479	0.692	0.812	0.777	0.796	0.789	0.0186	0.001	0.001
Thr	0.261	0.524	0.646	0.633	0.669	0.655	0.0196	0.001	0.001
Trp	0.430	0.646	0.709	0.694	0.724	0.713	0.0163	0.001	0.001
Val	0.449	0.652	0.756	0.737	0.753	0.752	0.0144	0.001	0.001
IAA	0.442	0.649	0.748	0.721	0.747	0.741	0.0155	0.001	0.001
Dispensable amino acids									
Ala	0.423	0.633	0.734	0.699	0.722	0.717	0.0158	0.001	0.001
Asp	0.328	0.569	0.703	0.680	0.707	0.699	0.0179	0.001	0.001
Cys [2]	0.620	0.764	0.743	0.731	0.739	0.728	0.0171	0.004	0.001
Glu	0.709	0.812	0.845	0.822	0.832	0.825	0.0107	0.001	0.001
Gly [2]	0.360	0.561	0.675	0.658	0.676	0.673	0.0158	0.001	0.001
Pro	0.681	0.782	0.805	0.793	0.803	0.799	0.0112	0.001	0.001
Ser	0.413	0.622	0.727	0.708	0.741	0.728	0.0159	0.001	0.001
DAA	0.505	0.678	0.747	0.727	0.746	0.738	0.0140	0.001	0.001
TAA	0.469	0.662	0.748	0.724	0.747	0.739	0.0148	0.001	0.001

[1] Each value represents the mean of six replicates (14, 12 and 10 birds per replicate for 7-, 14- and 21-days old birds, respectively; eight birds per replicate for 28- and 35-days old birds; and six birds per replicate for 42-days old birds). [2] Semi-indispensable amino acids for poultry. DAA = Average digestibility of dispensable amino acids; IAA = Average digestibility of indispensable amino acids; TAA = Average digestibility of all amino acids.

The AIDC of N and average digestibility of IAAs, DAAs and TAAs increased (quadratic; $p < 0.001$) from days 7 to 21 and then plateaued from days 21 to 42 (Table 8). The AIDC of all individual AAs in barley increased in a quadratic manner ($p < 0.001$) as the birds grew older.

The SIDC of N and average SIDC of IAAs, DAAs and TAAs increased quadratically ($p < 0.001$) with the advancing age of broilers (Table 9). The values increased from days 7 to 21 and then plateaued until day 42. The lower SIDC of TAAs was recorded at day 7 (0.617), followed by day 14 (0.738). The SIDC of every single AA in barley increased (quadratic; $p < 0.05$ to 0.001) with bird age with lower values on day 7.

Table 9. Standardized ileal digestibility coefficients [1] of nitrogen (N) and amino acids of barley at different ages of broilers [2].

	Age (Days)							Orthogonal Polynomial Contrasts	
Parameter	7	14	21	28	35	42	Pooled SEM	Linear	Quadratic
N	0.674	0.759	0.816	0.799	0.816	0.782	0.0143	0.001	0.001
Indispensable amino acids									
Arg	0.724	0.755	0.855	0.831	0.854	0.824	0.0135	0.001	0.001
His	0.652	0.762	0.780	0.762	0.783	0.756	0.0143	0.001	0.001
Ile	0.574	0.732	0.821	0.801	0.813	0.793	0.0155	0.001	0.001
Leu	0.650	0.773	0.851	0.829	0.844	0.819	0.0133	0.001	0.001
Lys	0.382	0.616	0.806	0.757	0.783	0.758	0.0211	0.001	0.001
Met	0.611	0.752	0.874	0.847	0.854	0.823	0.0186	0.001	0.001
Thr	0.595	0.705	0.822	0.798	0.845	0.784	0.0196	0.001	0.001
Trp	0.564	0.721	0.787	0.776	0.802	0.769	0.0163	0.001	0.001
Val	0.578	0.719	0.825	0.809	0.819	0.799	0.0144	0.001	0.001
IAA	0.592	0.726	0.825	0.801	0.822	0.792	0.0155	0.001	0.001

Table 9. *Cont.*

Parameter	Age (Days) 7	14	21	28	35	42	Pooled SEM	Orthogonal Polynomial Contrasts Linear	Quadratic
				Dispensable amino acids					
Ala	0.566	0.703	0.803	0.778	0.793	0.764	0.0158	0.001	0.001
Asp	0.509	0.664	0.796	0.776	0.800	0.763	0.0179	0.001	0.001
Cys [3]	0.806	0.870	0.844	0.829	0.843	0.809	0.0171	0.553	0.045
Glu	0.771	0.841	0.874	0.854	0.861	0.844	0.0107	0.001	0.001
Gly [3]	0.513	0.639	0.753	0.742	0.755	0.729	0.0158	0.001	0.001
Pro	0.755	0.821	0.844	0.831	0.842	0.827	0.0112	0.001	0.001
Ser	0.623	0.732	0.838	0.815	0.852	0.808	0.0159	0.001	0.001
DAA	0.649	0.753	0.822	0.804	0.821	0.792	0.0140	0.001	0.001
TAA	0.617	0.738	0.823	0.802	0.822	0.792	0.0148	0.001	0.001

[1] Apparent digestibility values were standardized using the following basal ileal endogenous flow values (g/kg DM intake), determined by feeding nitrogen-free diet at different ages [21]; see Table 6. [2] Each value represents the mean of six replicates (14, 12 and 10 birds per replicate for 7-, 14- and 21-days old birds, respectively; eight birds per replicate for 28- and 35-days old birds; and six birds per replicate for 42-days old birds). [3] Semi-indispensable amino acids for poultry. DAA = Average digestibility of dispensable amino acids; IAA = Average digestibility of indispensable amino acids; TAA = Average digestibility of all amino acids.

Table 10. Influence of age (days) on standardized digestible protein (CP) and amino acid contents [1] (g/kg) of barley, as-received basis.

Parameter	Age (Days) 7	14	21	28	35	42	Pooled SEM	Orthogonal Polynomial Contrasts Linear	Quadratic
CP	77.7	87.5	94.1	92.1	92.9	90.1	1.57	0.001	0.001
				Indispensable amino acids					
Arg	3.82	3.99	4.51	4.39	4.51	4.35	0.071	0.001	0.001
His	1.55	1.81	1.85	1.81	1.86	1.79	0.034	0.001	0.001
Ile	2.28	2.91	3.27	3.19	3.24	3.16	0.062	0.001	0.001
Leu	4.89	5.82	6.41	6.25	6.36	6.17	0.100	0.001	0.001
Lys	1.42	2.29	2.99	2.82	2.91	2.82	0.079	0.001	0.001
Met	1.08	1.33	1.55	1.49	1.51	1.46	0.033	0.001	0.001
Thr	2.10	2.49	2.91	2.82	2.99	2.78	0.069	0.001	0.001
Trp	0.76	0.97	1.06	1.05	1.08	1.04	0.021	0.001	0.001
Val	3.19	3.97	4.55	4.46	4.53	4.41	0.079	0.001	0.001
IAA	21.1	25.6	29.1	28.3	28.9	27.9	0.53	0.001	0.001
				Dispensable amino acids					
Ala	2.57	3.19	3.65	3.53	3.60	3.47	0.072	0.001	0.001
Asp	3.45	4.50	5.39	5.26	5.43	5.17	0.122	0.001	0.001
Cys [2]	1.79	1.94	1.88	1.85	1.88	1.80	0.038	0.553	0.044
Glu	18.8	20.5	21.2	20.8	20.9	20.5	0.259	0.001	0.001
Gly [2]	2.28	2.84	3.34	3.29	3.35	3.24	0.070	0.001	0.001
Pro	8.12	8.84	9.08	8.94	9.06	8.89	0.121	0.001	0.001
Ser	2.75	3.23	3.69	3.59	3.76	3.56	0.069	0.001	0.001
DAA	38.1	43.3	46.6	45.6	48.0	45.0	0.70	0.001	0.001
TAA	60.8	70.6	77.4	75.5	76.9	74.6	1.25	0.001	0.001

[1] Each value represents the mean of six replicates (14, 12 and 10 birds per replicate for 7-, 14- and 21-days old birds, respectively; eight birds per replicate for 28- and 35-days old birds; and six birds per replicate for 42-days old birds). [2] Semi-indispensable amino acids for poultry. DAA = Average digestibility of dispensable amino acids; IAA = Average digestibility of indispensable amino acids; TAA = Average digestibility of all amino acids.

The SID protein content of barley increased (quadratic; $p < 0.001$) from 77.7 g/kg on day 7 to 94.1 g/kg on day 21 and then plateaued from days 21 to 42 (Table 10). The SID contents of total IAAs, DAAs and total AAs increased (quadratic; $p < 0.001$) from day 7 to 21 and then plateaued until day 42. The SID contents of all individual AAs increased in a quadratic manner ($p < 0.05$ to 0.001) with lower values on day 7, which increased either at day 14 or 21 and then plateaued until day 42.

3.5. Uplift in Digestibility Coefficients Due to Correction for Age-Appropriate Endogenous Amino Acid Losses

The percentage increase in the digestibility coefficients of N and AAs after standardization of apparent values for basal endogenous N and AA losses is shown in Table 11.

Table 11. Percentage increase in digestibility coefficients of nitrogen (N) and amino acids in corn and barley after correction of apparent ileal digestibility coefficients for age-appropriate endogenous amino acid losses of broilers.

	Corn						Barley					
	Age (Days)						Age (Days)					
Parameter	7	14	21	28	35	42	7	14	21	28	35	42
N	43.8	22.1	18.4	18.9	18.7	13.3	33.7	13.3	11.6	12.1	11.8	8.61
Indispensable amino acids												
Arg	22.7	10.1	9.14	10.6	9.41	5.82	18.5	7.09	6.48	7.64	6.48	4.17
His	16.9	8.87	7.58	8.43	7.83	5.24	19.9	8.39	7.44	8.55	7.41	5.15
Ile	34.4	17.4	13.9	15.3	13.5	8.87	31.7	10.9	9.61	10.5	8.84	5.87
Leu	12.9	6.24	5.66	6.46	5.85	3.74	21.0	7.81	7.18	8.22	7.11	4.60
Lys	50.7	23.3	13.9	16.6	14.8	8.99	66.1	13.0	9.07	11.0	9.21	5.72
Met	24.2	11.1	9.18	10.8	9.06	5.24	27.6	8.67	7.64	9.01	7.29	4.31
Thr	96.6	55.9	37.1	34.8	37.2	26.7	128	34.5	27.2	26.1	26.3	19.7
Trp	65.7	36.5	27.8	29.6	27.5	19.6	31.2	11.6	11.0	11.8	10.8	7.85
Val	33.6	17.4	13.6	14.5	13.7	9.54	28.7	10.3	9.13	9.77	8.76	6.25
IAA	35.1	17.9	14.2	14.0	14.5	9.77	33.9	11.9	10.3	11.1	10.0	6.88
Dispensable amino acids												
Ala	16.2	7.96	6.90	8.08	7.39	4.85	33.8	11.1	9.40	11.3	9.83	6.56
Asp	41.7	21.9	16.7	17.8	17.5	11.9	55.2	16.7	13.2	14.1	13.2	9.16
Cys [1]	36.5	22.9	21.6	21.1	22.6	17.4	30.0	13.9	13.6	13.4	14.1	11.1
Glu	14.9	6.89	6.16	7.13	6.50	4.27	8.74	3.57	3.43	3.89	3.49	2.30
Gly [1]	42.3	22.5	16.7	18.1	17.1	11.9	42.5	13.9	11.6	12.8	11.7	8.32
Pro	19.7	10.2	9.12	9.11	9.15	6.52	10.9	4.99	4.84	4.79	4.86	3.50
Ser	44.3	24.5	18.9	18.5	19.2	13.5	50.8	17.7	15.3	15.1	14.9	10.9
DAA	29.4	15.9	13.3	13.9	13.8	9.77	28.5	11.1	10.0	10.6	10.1	7.32
TAA	32.5	17.0	13.9	14.7	14.2	9.85	31.6	11.5	10.0	10.8	10.0	7.17

[1] Semi-indispensable amino acids for poultry. DAA = Average digestibility of dispensable amino acids; IAA = Average digestibility of indispensable amino acids; TAA = Average digestibility of all amino acids.

The correction of AIDC for age-appropriate endogenous N and AA losses resulted in an increase in the SIDC regardless of age, though the extent of increase reduced as the birds grew older. After standardization of the AIDC estimates, the average TAA digestibility coefficients increased in corn by 32.5% (day 7), 17% (day 14), 13.9% (day 21), 14.7% (day 28), 14.2% (day 35) and 9.85% (day 42). In the case of barley, the corresponding increases were recorded as 31.6% (day 7), 11.5% (day 14), 10.0% (day 21), 10.8% (day 28), 10.0% (day 35) and 7.17% (day 42).

4. Discussion

Most available data on the AIDC and SIDC AA of ingredients have been determined using older broilers (22 to 35 days of age), and the estimates are applied in feed formulations regardless of broiler age. Several reports exist on age-related AA digestibility in ingredients for poultry, but the results are contradictory and inconclusive. Some studies have documented reductions in protein or AA digestibility [3,4] with advancing age, while others reported increases in digestibility estimates [13,26]. The present experiment aimed at identifying whether the broiler age has an impact on the SIDC of AAs in corn and barley.

Although it may be intuitively expected that the AA digestibility in broilers will vary depending on broiler age, studies comparing the SID of AAs corrected using age-appropriate EAAs are limited [18–20]. A previous study in our laboratory [14] reported the SIDC AA in wheat and sorghum at six different ages (days 7, 14, 21, 28, 35 and 42) of broilers. In the current research, the AIDC and SIDC AA in corn and barley were determined from hatching to the end of the growth cycle of broilers, and the AIDC values were standardized using age-appropriate basal EAA losses.

4.1. Nutrient Composition

The proximate nutrient contents of corn and barley were comparable to the values reported previously [12,16,27]. The higher starch content in corn (590 g/kg) than barley (541 g/kg) was expected. Due to high starch (620 to 720 g/kg) and crude fat (34 to 52 g/kg) contents in corn, it contains higher energy than any other grain. The CP content of corn (67.8 g/kg) was lower than the range (71 to 94 g/kg) reported by Cowieson [28].

The CP content (115 g/kg) in barley was marginally higher than the value (101 g/kg) reported by Perera et al. [27] and lower than the range (121 to 180 g/kg) of Bandegan et al. [16]. The higher CP content in barley compared to corn was in agreement with previous studies [9]. The AA contents of corn and barley were identical or close to those reported in previous studies [9,12,16,27].

4.2. Performance, Gizzard pH and Jejunal Digesta Viscosity

As anticipated, regardless of the grain type, an increase in both the DFI and DWG was observed with advancing age. The gizzard pH on day 7 of birds fed corn (2.53) and barley diets (2.14) was close to the values of 2.39 and 2.33 observed by David et al. [29] and Morgan et al. [30], respectively, for broilers of similar age. The reduction of gizzard pH at day 14 compared to day 7 was in line with the findings of David et al. [29] when feeding a corn-based diet. Based on the review of 15 published studies, Angel et al. [31] also reported a reduction in gizzard pH in broilers at day 14. According to Rynsburger [32], the secretion of gastric acid in the proventriculus increased from day 2 to 15, causing a decrease in pH.

As observed by Nitsan et al. [33], the secretion and activity of digestive enzymes and hydrochloric acid (HCl) secretion from proventriculus increase with broiler age. However, the observed increase in gizzard pH with bird age after day 14 in both diets could be, at least in part, explained by the increasing intake of feed with neutral pH with advancing age [34]. An increased feed load can dilute the HCl secreted and consequently increase the gizzard pH. The secretion of HCl is fundamental to sustain an acidic environment and to convert pepsinogen to pepsin, the first step in protein digestion [32]. Thus, the implication is that a high gizzard pH would compromise the protein digestion. In this experiment, however, gizzard pH was not correlated ($r = 0.242$; $p > 0.05$) with the average SIDC of TAAs in corn. On the contrary, in the case of barley, there was a positive correlation ($r = 0.644$; $p < 0.001$) between the gizzard pH and the average SIDC of TAAs. Therefore, it could be speculated that, though the gizzard pH was elevated by age in barley, contrary to expectations, AA digestibility also increased. Besides the pH, there are several other factors, to be discussed later, that potentially influence the AA digestibility in broilers.

The jejunal digesta viscosity in corn was not influenced by age, and no correlation ($p > 0.05$) existed between the digesta viscosity and the average SIDC of TAAs. In barley, the range (2.69–2.94) of jejunal digesta viscosity at different ages was notably higher than that of corn (2.03–2.34), with viscosity being higher on days 7 and 42. Yu et al. [35] replaced corn with increasing barley inclusions (0, 125, 250, 500 and 1000 g/kg) in 3-week-old broilers and reported an increase in the duodenal digesta viscosity from 1.46 cP (0 g/kg) to 2.40 (125 g/kg), 2.15 (250 g/kg), 2.71 (500 g/kg) and 2.81 cP (1000 g/kg). A positive relationship exists between the soluble NSP content and digesta viscosity [7]. Corn contains negligible amounts of soluble NSP compared to wheat and barley [28]. The NSP in plant cell walls such as β-glucans, the major NSP in barley, and the pentosans of rye and wheat exhibit significant antinutritive effects in poultry [36]. A portion of NSP of high molecular weight dissolves in the gastrointestinal tract, increasing the viscosity of gut contents that impedes the digestion and absorption of nutrients [37]. In the case of barley, however, the jejunal digesta viscosity tended ($r = -0.292$; $p = 0.084$) to be negatively correlated with the average SIDC of TAAs.

Besides the soluble NSP content, a myriad of factors such as growing location, storage time, ingredient inclusion level, age of bird, heat processing and pelleting temperature have been shown to influence the digesta viscosity in barley-containing diets [6]. It is difficult to explain the high viscosity at day 42 observed in the current work since previous studies [38] have reported decreased intestinal viscosity in older birds fed barley-based diets. During the evaluation of a high viscosity hull-less barley, Salih et al. [38] recorded a drop in the digesta viscosity in broilers from two weeks (2.59 cP) to eight weeks of age (1.74 cP).

4.3. Ileal Digestibility Coefficients of Nitrogen and Amino Acids

The first week is the most critical period in a bird's life when they consume only small amounts of feed and depend mostly on the nutrients from residual yolk [39]. Notable changes occur in the morphology and development of the gastrointestinal tract during the first few weeks of life that contribute to improved digestion and absorption of nutrients. During the first few weeks, there is high protein demand for the development of organs and muscle [39]. The secretion and activity of different proteolytic enzymes such as trypsin, chymotrypsin, intestinal peptidase and dipeptidase also generally increase with age [33].

From previous studies, it is evident that the AIDC of AAs is variable depending on broiler age [3,13,40,41]. In the current experiment, with advancing broiler age, an increase in the AIDC of N, all individual AAs (except Cys) and average of TAAs was observed for corn. Compared to day 7, the average AIDC of all AAs increased by 0.46, 22.0, 19.4, 18.6 and 20.5% at days 14, 21, 28, 35 and 42, respectively. The increased AIDC AA with age is in agreement with previous findings [13,19,26]. An increase in ileal N digestibility from 78% at day 4 to about 90% at day 21 has been reported in broilers fed diets based on corn–soybean meal by Noy and Sklan [40]. They concluded that the hydrolysis of exogenous and endogenous proteins was not optimum due to insufficient proteolytic activity at the early posthatch period. An increase in apparent AA digestibility from 1 to 10 days of age was reported with a corn–soybean meal diet by Batal and Parsons [41]. Wallis and Balnave [26] recorded an increase in AIDC AA from day 30 to 50 posthatch (0.732 vs. 0.814) feeding a finisher diet containing a wide range of feed ingredients (wheat, sorghum, soybean meal, cottonseed meal, meat and bone meal, poultry tallow, poultry offal meal, feather meal). Huang et al. [13] determined the AIDC AA of eight ingredients (corn, wheat, sorghum, soybean meal, canola meal, meat and bone meal, cottonseed meal and millrun) at three broiler ages (days 14, 28 and 42). Combining all the results, it was concluded that the digestibility increased with advancing age. The trends, however, were variable depending on the AA and ingredient type. In their study, higher AIDC AA was recorded at days 28 and 42 compared to day 14 in the case of corn, soybean meal, canola meal and meat and bone meal. The AIDC in millrun at day 42 was higher than those at 14 and 28 days. Whilst there was no age influence on the AIDC of most AAs in cottonseed meal, Lys and Arg digestibility increased with age. On the contrary, in wheat, the AIDC of most individual AAs was recorded to be higher at day 14 than at days 28 and 42. In sorghum, the AIDC AA was higher at day 42 compared to day 28 but similar to those at day 14 with the exception of His, Lys, Ser and Gly which were higher at day 42.

In the present study, with the exception of Thr, linear or quadratic responses to broiler age were observed for the SIDC of all individual AAs in corn. Unlike the AIDC, the pattern of increase in SIDC AA in corn was not gradual with increasing age. Rather, a decline was observed from day 7 to 14 followed by an increase from day 14 to 21, a plateau between days 21 and 35 and a decline at day 42.

Differing age-related trends between the AIDC and SIDC estimates have also been observed by Adedokun et al. [19]. These researchers, comparing AA digestibility between days 5 and 21 of broilers for five ingredients (corn, light and dark distiller's dried grains with solubles, canola meal and soybean meal), reported an increased AID AA with age in all test ingredients. However, increasing broiler age elevated the SID AA only in corn and distiller's dried grains with solubles and had no influence on the SID AA in soybean meal and canola meal.

Apart from the age effect, another notable observation was made on the AA digestibility in corn in the current study. Though maize contained lower CP and TAA contents compared to barley (Table 3), the average SIDC in maize was 39.4% (day 7), 3.4% (day 14), 9.6% (day 21), 10.8% (day 28), 6.9% (day 35) and 8.5% (day 42) higher than barley. Compared to other grains, maize contains low levels of soluble NSP (1 g/kg) and highly digestible nutrients for broilers. Barley contains a high amount of soluble NSP (45 g/kg) compared to maize [28]. According to Andriotis et al. [42], the barley endosperm is composed of β-glucans (70%) and arabinoxylans (20%). The high content of soluble β-glucan is the major antinutritional factor in barley, and, to assess the potential feed value of barley for poultry, determination of the content and properties of β-glucan is crucial [27]. When birds are fed diets with high inclusions of barley, the NSP impedes nutrient digestion and absorption by two mechanisms. First, the soluble NSP forms gel-like viscous matter, which impairs the interaction between nutrient substrates and endogenous enzymes. Second, the insoluble NSP fraction exerts a "cage effect" by encapsulating the nutrients (starch, protein) in endosperm cells, impeding the contact between nutrients and digestive enzymes. High levels of β-glucans result in thicker cell walls than low levels of β-glucans [43,44]. Though barley has a potential to be included in broiler diets without compromising growth performance [45,46], its inclusion is not recommended for young birds (<7 days) in this experiment due to very low SIDC AA at day 7.

In the present study, an increase in the AIDC of N and average AIDC of IAAs, DAAs and TAAs with age was observed in barley. The AIDC of all AAs increased from day 7 to 21 and then plateaued until day 42. Similar patterns of increase were observed in the SIDC of N, SIDC of all individual AAs and average SIDC of IAAs, DAAs and TAAs with lower values on day 7, then an increase until day 21, which plateaued from day 21 to 42. The average SIDC of TAAs in barley at day 7 was 16.4% lower than day 14, 25.0% lower than day 21 and 23.4% lower than the average of days 28, 35 and 42. Szczurek et al. [20] determined the AA digestibility in three grain sources (barley, triticale and wheat) in broilers at day 14 vs. 28 posthatch. In barley and triticale, the average AIDC of IAAs was notably higher at day 28 than 14. The average AIDC of IAAs in barley was reported to be 0.730 on day 14 and 0.780 on day 28. The SIDC of all AAs in barley and most AAs in triticale was also higher at day 28. In the case of wheat, however, no difference was recorded for the average AIDC of IAAs between the two ages. The effect of age on the AA digestibility in wheat and sorghum at six different ages (days 7, 14, 21, 28, 35 and 42) of broilers has already been documented [14]. The AIDC of all individual AAs in wheat increased as the birds grew older. Although the average SIDC of TAAs was unaffected by age, the digestibility of some individual AAs (Met, Trp, Asp and Cys) was higher in the older birds. It is possible that the GIT of older birds is developed sufficiently enough to counteract any adverse effects of digesta viscosity induced by β-glucan. Similar observations were made in other studies [38,47], indicating that the digestibility of nutrients in barley diets improves with age and that viscosity is not a limiting factor in older birds.

Several factors may explain the increased AIDC AA with advancing age for corn and barley. Lower production and activities of pancreatic enzymes are a constraint for nutrient digestion in young chicks [39]. Nitsan et al. [33] measured the activities (units/kg BW) of trypsin and chymotrypsin in the pancreas and small intestine from hatching to 23 days of age. The activity of trypsin was low at days 3 to 6 posthatch and increased on day 14. A gradual increase in chymotrypsin activity was also observed from hatch to day 14, and this remained constant afterward. Maximum activities of trypsin and chymotrypsin were observed on day 11. In the intestinal contents, the activity of trypsin increased 10-fold from hatching to day 14, and that of chymotrypsin increased 3-fold by day 20. As stated by Tarvid [48], the total intestinal peptidase (aminopeptidase and dipeptidase) activity increases with the advancing age of broilers.

Another possible factor contributing to the increasing apparent AA digestibility with age may be the greater absorptive area [39]. According to Nitsan et al. [33], the weight of the small intestine increased 10-fold at day 8 and 20-fold at day 23. Insoluble NSP is known to promote the development of gizzard [49]. As barley contains high insoluble NSP (122–142 g/kg) [6,27], it is likely that a more developed gizzard in barley will facilitate better nutrient digestion by mechanical breakdown of digesta. Well-developed GIT in older birds can overcome the negative viscosity effects of β-glucans [47].

Digesta retention time in the GIT plays a vital role in nutrient digestion because it determines the actual contact time among nutrients, digestive enzymes and microbiota. An increase in digesta retention enhances the absorption of nutrients by increasing the exposure time between the digesta and intestinal absorptive surface. Feed passage time through the GIT is reduced with bird age [50]. Increased digesta retention, in combination with a well-developed digestive tract, especially foregut, in older birds may also increase protein and AA digestibility by enhancing intestinal refluxes, subsequently re-exposing the digesta to pepsin [51].

Higher AA digestibility in barley with advancing age might be due to better tolerance of older birds to high NSP and viscosity owing to the increased stability of intestinal microbiota with advancing age [7]. Gut microbiota, a major consumer of AAs, may potentially influence digestion by competing for nutrients [52]. The GIT of the newly hatched chick is sterile. The primary source of initial microbiota is the farm environment where the hatchlings are reared. With increasing age, a cascade of changes occurs in microbiota including species diversity, followed by the complexity of population structure and finally maturation and stabilization. This process continues with age in commercial broilers, and the microbiota profile is stabilized by week 3 [53]. According to Choct [54], dietary NSP can promote microbial growth and fermentation. Increased degradation of soluble NSP and the resultant reduction in the viscosity with age may limit bacterial growth and partly mitigate the adverse effects caused by the excessive population of microbiota.

In contrast to the present findings, earlier age-related studies measuring the excreta or total tract AA digestibility in poultry [3,15,55] indicated higher protein and AA digestibility in younger birds. Fonolla et al. [3] reported a decline in the excreta digestibility of protein in broilers with advancing age (day 21 vs. 52), which was attributed to an increased excretion of metabolic N. A drop in true digestibilities of protein and AAs from 3 to 6 weeks of age was reported by Zuprizal et al. [15]. According to Carré et al. [55], the apparent protein digestibility in pea was lower in adult roosters than in 3-week-old broilers. Following feeding of a mixed diet with a wide variety of ingredients, ten Doeschate et al. [4] observed a reduction in protein digestibility coefficients with broiler age. The values were reported to be 0.849 (days 13 to 15), 0.830 (days 27 to 29) and 0.840 (days 41 to 43), respectively. However, these studies are based on total tract digestibility and are not comparable with current findings due to possible contamination of AAs from urine, microbial fermentation in the hindgut and, consequently, modifications in the AA constituents [5].

With advancing broiler age, the reduction in ileal digestibility of AAs along with other major nutrients in feed ingredients has also been reported. Adedokun et al. [18] reported 18.5% higher AIDC for TAAs in meat and bone meal in 5-day-old broilers (0.705) compared to that of 21-day broilers (0.595). This pattern remained unchanged even after the standardization of AIDC, with SIDC values of 0.760 and 0.632 at days 5 and 21, respectively. In a previous experiment [14], though there was no age influence on the average AIDC of TAAs in sorghum, the average SIDC of TAAs on day 7 (0.903) was 6.9% higher than on day 14 (0.844) and 8.5% higher than the average from day 21 to day 42 (0.819–0.854). Comparisons among published age-related SID digestibility estimates are not straightforward because of differences in ingredient type [1,13], assumed age-appropriate endogenous losses [21], secretion and activities of digestive enzymes [33], development and activity of gastrointestinal microbiota [53], environment, chemical nature of nutrients and methodology.

In the SIDC calculations, the AIDCs were corrected for the basal endogenous AA losses from various digestive, pancreatic and enzymatic secretions [56], and unsurprisingly, these two values were different with SIDC being higher. In practical feed formulations, the SIDC AA of ingredients is preferred because it is more additive especially when mixed into a complete diet. Moreover, standardization eliminates the underestimation of some nutritionally critical and less additive AAs in poultry diets [2]. Though there are some constraints in AIDC estimates, the AIDC is reported in this study to better understand the magnitude of age impact on the SIDC due to correction for age-dependent EAA losses. The differences between the AIDC and SIDC illustrated in Figures 1 and 2 were due to the correction of apparent values by age-appropriate EAA losses determined in a previous study in our laboratory [21]. Although data [1,9,16] exist on the SID AA of different poultry feed ingredients, only a few studies [14,15,17–20] have determined the SID AA at different broiler ages. Except for some [14,18–20], all previous studies have used a single EAA flow value, derived from older birds, to standardize the AIDC values at different ages. From our previous findings [21], the basal EAA losses of TAAs at day 7 (12.93 g/kg DMI) were twice that of the average of days 14 to 35 (6.61 g/kg DMI) and almost three times higher than day 42 (4.48 g/kg DMI). These resulted in substantial differences in the percentage of increase in SIDC over AIDC (Table 11) after correction of apparent digestibility data with age-appropriate EAA losses. The current findings suggest that correcting AIDC AA using a single EAA flow value across ages underestimates the SIDC AA of feed ingredients in young birds and overestimates in older birds.

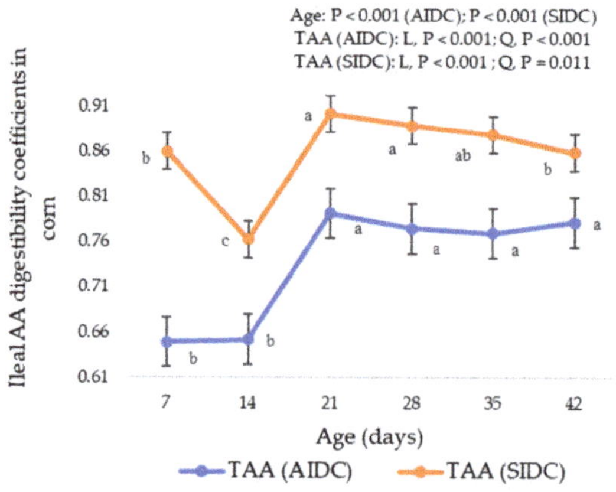

Figure 1. Apparent and standardized ileal digestibility coefficients of total amino acids (TAAs) in corn (bars represent means ± SE) as influenced by broiler age. [a, b, c] Values with different superscripts differ significantly ($p < 0.05$). L, Linear; Q, Quadratic.

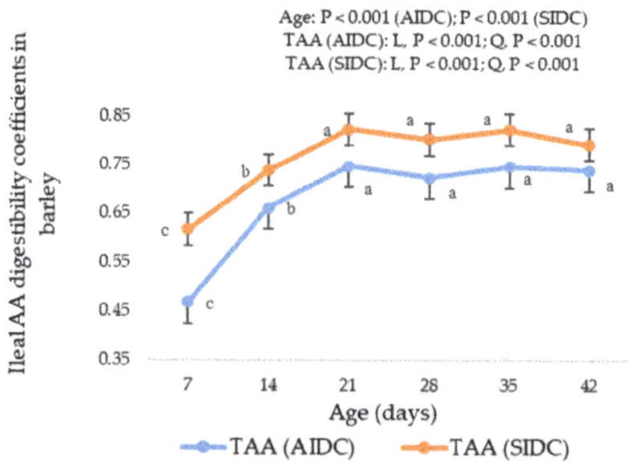

Figure 2. Apparent and standardized ileal digestibility coefficients of total amino acids in barley (bars represent means ± SE) as influenced by broiler age. [a, b, c] Values with different superscripts differ significantly ($p < 0.05$). L, Linear; Q, Quadratic.

In general, the magnitude of increase in the SIDC AA of corn and barley after correction for age-appropriate EAAs decreased with advancing broiler age. After correcting for age-appropriate EAAs, an increase of 32.5% in the average SIDC of TAAs at day 7 was observed in corn that was almost two times higher compared to that from day 14 to 35 (13.9–17.0%) and three times higher than the increase at day 42 (9.85%). Similar to corn, the average SIDC of TAAs in barley at day 7 was 31.5% higher than the average AIDC of TAAs, which decreased to 7.17% at day 42.

To summarize, several factors such as the composition of grains, antinutritive factors, digestive tract development, secretion of enzymes, feed intake, nutrient load, digesta retention time and gut microbiota contribute to the age effect on the AA digestibility in broilers. The endogenous AA losses are a key factor with the greatest impact on the SIDC being observed in young broilers.

5. Conclusions

The present study provides information on the SIDC of AAs in corn and barley from hatching to the end of the broiler growth cycle. The findings suggest that the age influence on AA digestibility is dependent on the grain type and AA. The AIDC AA in corn increased with advancing age. The SIDC AA was higher at day 7, decreased at day 14 and increased and plateaued between days 21 and 35. A further decrease was observed at day 42. In the case of barley, both the AIDC and SIDC of AAs increased as the birds grew older. Standardization of AIDC AA with age-appropriate EAA flows resulted in marked differences in the SIDC of both grains. Application of a single EAA value for correction of the AIDC for broilers of different ages can result in the underestimation of the SIDC AA in young birds and overestimation in older birds. It is concluded that the precision of feed formulations can be improved by using age-specific EAA values for the standardization of AIDC AA values.

Author Contributions: Conceptualization and designing, M.B. and M.R.A.; methodology, M.B. and M.R.A.; investigation, writing, original draft preparation, M.B.; data curation, M.B. and F.Z.; supervision, editing and revising critically, V.R. and M.R.A.; supervision, providing comments on the manuscript, T.J.W., C.K.G. and P.V.C. All authors have read and agreed to the published version of the manuscript.

Funding: This research was funded by the "AgriFutures Australian Chicken Meat Program (Project No: PRJ-011835)".

Institutional Review Board Statement: The experimental procedure complied with the New Zealand Revised Code of Ethical Conduct for the use of live animals for research, testing and teaching and was approved by the Massey University Animal Ethics Committee (New Zealand) (MUAEC, 20/32, May 2020).

Informed Consent Statement: Not applicable.

Data Availability Statement: All available data are incorporated in the manuscript.

Acknowledgments: The authors would like to sincerely thank the "AgriFutures Australian Chicken Meat Program" for funding the project and Preethi Ramesh, AMINOLab EMSEA, Singapore, for amino acid analyses.

Conflicts of Interest: The authors declare no conflict of interest.

References

1. Lemme, A.; Ravindran, V.; Bryden, W.L. Ileal digestibility of amino acids in feed ingredients for broilers. *World's Poult. Sci. J.* **2004**, *60*, 423–438. [CrossRef]
2. Cowieson, A.; Sorbara, J.O.; Pappenberger, G.; Abdollahi, M.R.; Roos, F.F.; Ravindran, V. Additivity of apparent and standardized ileal amino acid digestibility of corn and soybean meal in broiler diets. *Poult. Sci.* **2019**, *98*, 3722–3728. [CrossRef]
3. Fonolla, J.; Prirto, C.; Sanz, R. Influence of age on nutrient utilization of diets for broilers. *Anim. Feed Sci. Technol.* **1981**, *6*, 405–411. [CrossRef]

4. ten Doeschate, R.A.H.M.; Scheele, C.W.; Schreurs, V.V.A.M.; van der Klis, J.D. Digestibility studies in broiler chickens: Influence of genotype, age, sex and method of determination. *Br. Poult. Sci.* **1993**, *34*, 131–146. [CrossRef]
5. Ravindran, V.; Bryden, W.L. Amino acid availability in poultry—In vitro and in vivo measurements. *Aust. J. Agric. Res.* **1999**, *50*, 889–908. [CrossRef]
6. Jacob, J.P.; Pescatore, A.J. Using barley in poultry diets-a review. *J. Appl. Poult. Res.* **2012**, *21*, 915–940. [CrossRef]
7. Choct, M.; Annison, G. The inhibition of nutrient digestion by wheat pentosans. *Br. J. Nutr.* **1992**, *67*, 123–132. [CrossRef]
8. Ravindran, V.; Hew, L.I.; Ravindran, G.; Bryden, L. Apparent ileal digestibility of amino acids in dietary ingredients for broiler chickens. *Anim. Sci.* **2005**, *81*, 85–97. [CrossRef]
9. Bryden, W.L.; Li, X.; Ravindran, G.; Hew, L.I.; Ravindran, V. *Ileal Digestible Amino Acid Values in Feedstuffs for Poultry*; RIRDC Publication No: 09/077; Rural Industries Research and Development Corporation: Canberra, Australia, 2009; ISBN 1-741151-870-9.
10. Huang, K.H.; Li, X.; Ravindran, V.; Bryden, W.L. Comparison of apparent ileal amino acid digestibility of feed ingredients measured with broilers, layers, and roosters. *Poult. Sci.* **2006**, *85*, 625–634. [CrossRef]
11. Barua, M.; Abdollahi, M.R.; Zaefarian, F.; Wester, T.J.; Girish, C.K.; Ravindran, V. Standardized ileal amino acid digestibility of protein sources for broiler chickens is influenced by the feed form. *Poult. Sci.* **2020**, *99*, 6925–6934. [CrossRef]
12. Barua, M.; Abdollahi, M.R.; Zaefarian, F.; Wester, T.J.; Girish, C.K.; Ravindran, V. Influence of feed form on the standardized ileal amino acid digestibility of common grains for broiler chickens. *Anim. Feed Sci. Technol.* **2021**, *272*, 114743. [CrossRef]
13. Huang, K.H.; Ravindran, V.; Li, X.; Bryden, W.L. Influence of age on the apparent ileal amino acid digestibility of feed ingredients for broiler chickens. *Br. Poult. Sci.* **2005**, *46*, 236–245. [CrossRef] [PubMed]
14. Barua, M.; Abdollahi, M.R.; Zaefarian, F.; Wester, T.J.; Girish, C.K.; Chrystal, P.V.; Ravindran, V. An investigation into the influence of age on the standardized amino acid digestibility of wheat and sorghum in broilers. *Poult. Sci.* **2021**, *100*, 101466. [CrossRef] [PubMed]
15. Larbier, M.; Chagneau, A.M. Effect of age and sex on true digestibility of amino acids of rapeseed and soybean meals in growing broilers. *Poult. Sci.* **1992**, *71*, 1486–1492.
16. Bandegan, A.; Golian, A.; Kiarie, E.; Payne, R.L.; Crow, G.H.; Guenter, W.; Nyachoti, C.M. Standardized ileal amino acid digestibility in wheat, barley, pea and flaxseed for broiler chickens. *Can. J. Anim. Sci.* **2011**, *91*, 103–111. [CrossRef]
17. Garcia, A.R.; Batal, A.B.; Dale, N.M. A comparison of methods to determine amino acid digestibility of feed ingredients for chickens. *Poult. Sci.* **2007**, *86*, 94–101. [CrossRef] [PubMed]
18. Adedokun, S.A.; Parsons, C.M.; Lilburn, M.S.; Adeola, O.; Applegate, T.J. Standardized ileal amino acid digestibility of meat and bone meal from different sources in broiler chicks and turkey poults with a nitrogen-free or casein diet. *Poult. Sci.* **2007**, *86*, 2598–2607. [CrossRef]
19. Adedokun, S.A.; Adeola, O.; Parsons, C.M.; Lilburn, M.S.; Applegate, T.J. Standardized ileal amino acid digestibility of plant feedstuffs in broiler chickens and turkey poults using a nitrogen-free or casein diet. *Poult. Sci.* **2008**, *87*, 2535–2548. [CrossRef] [PubMed]
20. Szczurek, W.; Szymczyk, B.; Arczewska-Włosek, A.; Świątkiewicz, S. Apparent and standardized ileal digestibility of amino acids in wheat, triticale and barley for broiler chickens at two different ages. *Br. Poult. Sci.* **2020**, *61*, 63–69. [CrossRef] [PubMed]
21. Barua, M.; Abdollahi, M.R.; Zaefarian, F.; Wester, T.J.; Girish, C.K.; Chrystal, P.V.; Ravindran, V. Basal ileal endogenous amino acid flow in broiler chickens as influenced by age. *Poult. Sci.* **2021**, *100*, 101269. [CrossRef] [PubMed]
22. AOAC International. *Official Methods of Analysis*, 20th ed.; Association of Official Analytical Chemists: Washington, DC, USA, 2016.
23. Short, F.J.; Gorton, P.; Wiseman, J.; Boorman, K.N. Determination of titanium dioxide added as an inert marker in chicken digestibility studies. *Anim. Feed Sci. Technol.* **1996**, *59*, 215–221. [CrossRef]
24. McCleary, B.V.; Gibson, T.S.; Mugford, D.C. Measurement of total starch in cereals products by amyloglucosidase vs amylase method: Collaborative study. *J. AOAC Int.* **1997**, *80*, 571–579. [CrossRef]
25. AOAC International. *Official Methods of Analysis*, 18th ed.; Association of Official Analytical Chemists: Washington, DC, USA, 2011.
26. Wallis, I.R.; Balnave, D. The influence of environmental temperature, age and sex on the digestibility of amino acids in growing broiler chickens. *Br. Poult. Sci.* **1984**, *25*, 401–407. [CrossRef]
27. Perera, W.N.U.; Abdollahi, M.R.; Ravindran, V.; Zaefarian, F.; Wester, T.J.; Ravindran, G. Nutritional evaluation of two barley cultivars, without and with carbohydrase supplementation, for broilers: Metabolizable energy and standardized amino acid digestibility. *Br. Poult. Sci.* **2019**, *60*, 404–413. [CrossRef] [PubMed]
28. Cowieson, A.J. Factors that affect the nutritional value of maize for broilers. *Anim. Feed Sci. Technol.* **2005**, *119*, 293–305. [CrossRef]
29. David, L.S.; Abdollahi, M.R.; Bedford, M.R.; Ravindran, V. Effect of age and dietary crude protein content on the apparent ileal calcium digestibility of limestone in broiler chickens. *Anim. Feed Sci. Technol.* **2020**, *263*, 114468. [CrossRef]
30. Morgan, N.K.; Walk, C.L.; Bedford, M.R.; Burton, E.J. The effect of dietary calcium inclusion on broiler gastrointestinal pH: Quantification and method optimization. *Poult. Sci.* **2014**, *93*, 354–363. [CrossRef] [PubMed]
31. Angel, R.; Humphreys, B.; Saylor, W. Age changes in gastrointestinal pH in broilers. *Poult. Sci.* **2010**, *89*, 650.
32. Rynsburger, J.M. Physiological and Nutritional Factors Affecting Protein Digestion in Broiler Chickens. Master's Thesis, University of Saskatchewan, Saskatoon, SK, Canada, 2009.

33. Nitsan, Z.; Ben-Avraham, G.; Zoref, Z.; Nir, I. Growth and development of the digestive organs and some enzymes in broiler chicks after hatching. *Br. Poult. Sci.* **1991**, *32*, 515–523. [CrossRef]
34. Ravindran, V. Feed enzymes: The science, practice, and metabolic realities. *J. Appl. Poult. Res.* **2013**, *22*, 628–636. [CrossRef]
35. Yu, B.; Hsu, J.C.; Chiou, P.W.S. Effects of β-glucanase supplementation of barley diets on growth performance of broilers. *Anim. Feed Sci. Technol.* **1998**, *70*, 353–361. [CrossRef]
36. Antoniou, T.; Marquardt, R.R.; Cansfield, E. Isolation, partial characterization, and antinutritional activity of a factor (pentosans) in rye grain. *J. Agric. Food Chem.* **1981**, *29*, 1240–1247. [CrossRef]
37. Fengler, A.I.; Marquardt, R.R. Water-soluble pentosans from rye. 1. Isolation, partial-purification, and characterization. *Cereal Chem.* **1988**, *65*, 291–297.
38. Salih, M.E.; Classen, H.L.; Campbell, G.L. Response of chickens fed on hull-less barley to dietary β-glucanase at different ages. *Anim. Feed Sci. Technol.* **1991**, *33*, 139–149. [CrossRef]
39. Ravindran, V.; Abdollahi, M.R. Nutrition and digestive physiology of the broiler chicks: State of the art and outlook. *Animals* **2021**, *11*, 2795. [CrossRef]
40. Noy, Y.; Sklan, D. Digestion and absorption in the young chick. *Poult. Sci.* **1995**, *74*, 366–373. [CrossRef]
41. Batal, A.B.; Parsons, C.M. Effects of age on nutrient digestibility in chicks fed different diets. *Poult. Sci.* **2002**, *81*, 400–407. [CrossRef] [PubMed]
42. Andriotis, V.M.; Rejzek, M.; Barclay, E.; Rugen, M.D.; Field, R.A.; Smith, A.M. Cell wall degradation is required for normal starch mobilization in barley endosperm. *Sci. Rep.* **2016**, *6*, 33215. [CrossRef] [PubMed]
43. Åman, P.; Graham, H. Mixed Linked β-(1→ 3), (1→ 4)-D Glucans in the cell walls of barley and oats-chemistry and nutrition. *Scand. J. Gastroenterol.* **1987**, *22*, 42–51. [CrossRef]
44. Izydorczyk, M.S.; Dexter, J.E. Barley β-glucans and arabinoxylans: Molecular structure, physicochemical properties, and uses in food products—A Review. *Food Res. Int.* **2008**, *41*, 850–868. [CrossRef]
45. Perera, W.N.U.; Abdollahi, M.R.; Zaefarian, F.; Wester, T.J.; Ravindran, G.; Ravindran, V. Influence of inclusion level of barley in wheat-based diets and supplementation of carbohydrase on growth performance, nutrient utilization and gut morphometry in broiler starters. *Br. Poult. Sci.* **2019**, *60*, 736–748. [CrossRef]
46. Perera, W.N.U.; Abdollahi, M.R.; Ravindran, V.; Zaefarian, F.; Wester, T.J.; Ravindran, G. The effect of graded inclusions of waxy starch hull-less barley and a multicomponent exogenous carbohydrase on the growth performance, nutrient digestibility and intestinal morphometry of broiler chickens. *Br. Poult. Sci.* **2020**, *61*, 442–453. [CrossRef]
47. Almirall, M.; Esteve-Garcia, E. Rate of passage of barley diets with chromium oxide: Influence of age and poultry strain and effect of β-glucanase supplementation. *Poult. Sci.* **1994**, *73*, 1433–1440. [CrossRef]
48. Tarvid, I. The development of protein digestion in poultry. *Avian Poult. Biol. Rev.* **1995**, *6*, 35–54.
49. Svihus, B. The gizzard: Function, influence of diet structure and effects on nutrient availability. *World's Poult. Sci. J.* **2011**, *67*, 207–224. [CrossRef]
50. Wilson, E.K.; Pierson, F.W.; Hester, P.Y.; Adams, R.L.; Stadelman, W.J. The effects of high environmental temperature on feed passage time and performance traits of White Pekin Ducks. *Poult. Sci.* **1980**, *59*, 2322–2330. [CrossRef]
51. Abdollahi, M.R.; Ravindran, V. Influence of pellet length on pellet quality and performance of broiler starters. *J. Appl. Poult. Res.* **2013**, *22*, 516–522. [CrossRef]
52. Stanley, D.; Geier, M.S.; Denman, S.E.; Haring, V.R.; Crowley, T.M.; Hughes, R.J.; Moore, R.J. Identification of chicken intestinal microbiota correlated with the efficiency of energy extraction from feed. *Vet. Microbiol.* **2013**, *164*, 85–92. [CrossRef]
53. Oakley, B.B.; Lillehoj, H.S.; Kogut, M.H.; Kim, W.K.; Maurer, J.J.; Pedroso, A.; Lee, M.D.; Collett, S.R.; Johnson, T.J.; Cox, N.A. The chicken gastrointestinal microbiome. *FEMS Microbiol. Lett.* **2014**, *360*, 100–112. [CrossRef]
54. Choct, M. Feed non-starch polysaccharides: Chemical structures and nutritional significance. *J. Feed Mill. Int.* **1997**, *191*, 13–26.
55. Carré, B.; Beaufils, E.; Melcion, J. Evaluation of protein and starch digestibility and energy value of pelleted or unpelleted pea seed from winter or spring cultivars in adult and young chicks. *J. Agric. Food Chem.* **1991**, *39*, 468–472. [CrossRef]
56. Ravindran, V. Progress in ileal endogenous amino acid flow research in poultry. *J. Anim. Sci. Biotechnol.* **2021**, *12*, 5. [CrossRef] [PubMed]

animals

MDPI

Article

Performance and Egg Quality of Laying Hens Fed Diets Containing Raw, Hydrobarothermally-Treated and Fermented Rapeseed Cake

Magdalena Kopacz [1], Aleksandra Alicja Drażbo [1,*], Katarzyna Śmiecińska [2] and Katarzyna Ognik [3]

[1] Department of Poultry Science and Apiculture, University of Warmia and Mazury in Olsztyn, Oczapowskiego 5, 10-719 Olsztyn, Poland; magdalena.kopacz@uwm.edu.pl

[2] Department of Commodity Science and Animal Raw Material Processing, University of Warmia and Mazury in Olsztyn, Oczapowskiego 5, 10-719 Olsztyn, Poland; katarzyna.smiecinska@uwm.edu.pl

[3] Department of Biochemistry and Toxicology, Faculty of Animal Sciences and Bioeconomy, University of Life Sciences in Lublin, Akademicka 13, 20-950 Lublin, Poland; katarzyna.ognik@up.lublin.pl

* Correspondence: aleksandra.drazbo@uwm.edu.pl

Simple Summary: Rapeseed cake (RC) has recently gained increasing interest as a dietary protein source alternative to soybean meal (SBM). However, its wider use in poultry diets (including laying hen diets) is limited due to the high concentrations of antinutritional compounds. Technological processes such as thermal treatment and fermentation may improve the nutritional value of RC by reducing the content of non-starch polysaccharides (NSPs), glucosinolates (GLS) and phytate phosphorus (PP). The present study revealed that the inclusion of 20% RC in layer diets improves the fatty acid (FA) profile of egg yolks while maintaining a desirable redox status. Hydrobarothermally-treated RC (HRC) and fermented RC (FRC) exerted a greater beneficial influence on the laying performance of hens than raw RC (RRC). FCR appears to be the optimal substitute for SBM because it contributed to the highest albumen quality and the highest sensory quality of eggs.

check for updates

Citation: Kopacz, M.; Drażbo, A.A.; Śmiecińska, K.; Ognik, K. Performance and Egg Quality of Laying Hens Fed Diets Containing Raw, Hydrobarothermally-Treated and Fermented Rapeseed Cake. *Animals* **2021**, *11*, 3083. https://doi.org/10.3390/ani11113083

Academic Editors: Natalie Morgan and Youssef A. Attia

Received: 22 September 2021
Accepted: 27 October 2021
Published: 28 October 2021

Abstract: The present study was conducted to investigate how raw rapeseed cake (RRC), hydrobaro thermally-treated rapeseed cake (HRC) and fermented rapeseed cake (FRC) fed to laying hens over a period of 12 weeks affected their performance, and the quality, fatty acid (FA) profile and oxidative stability of eggs. A total of 304 Hy-Line Brown laying hens at 36 weeks of age were distributed in a completely randomized design to four treatment groups with 38 replicates per treatment and two hens per replicate. The birds had ad libitum access to feed and water throughout the study. During the experiment, the birds were fed isonitrogenous and isocaloric diets in mash form, with various protein sources. In the control group (C), soybean meal (SBM) was the main source of dietary protein, whereas the experimental groups were fed diets containing 20% of RRC, HRC or FRC. Hydrobarothermal treatment and fermentation decreased the glucosinolate (GLS) content of RC, and fermentation reduced the concentration of phytate phosphorus (PP). In comparison with the RRC group, layers from the HRC and FRC groups were characterized by higher laying performance, comparable with that in group C. Irrespective of its physical form, RC added to layer diets adversely affected eggshell quality in all experimental groups, whereas albumen quality was highest in the FRC group. In comparison with group C, diets containing RRC, HRC and HRC led to a significant decrease in the content of saturated fatty acids (SFAs), an increase in the proportion of n-3 and n-6 polyunsaturated fatty acids (PUFAs) in the total FA pool in egg yolks, and a decrease in the n-6/n-3 PUFA ratio. The inclusion of RRC, HRC and FRC in layer diets decreased the activity of superoxide dismutase (SOD) in egg yolks, relative to group C. Group FCR eggs were characterized by the highest activity of catalase (CAT) and the lowest lipid peroxides LOOH concentration, compared with the remaining groups. The addition of RC to layer diets did not compromise the sensory quality of eggs, and eggs produced in group FRC received the highest overall score. It can be concluded that the inclusion of 20% RRC, HRC and FRC in layer diets does not compromise the sensory quality of eggs and has a beneficial influence on the FA profile and antioxidant potential of egg yolks. The use of

FRC is recommended because it contributes to the highest laying performance, superior albumen quality and the highest sensory quality of eggs, relative to RRC and HRC.

Keywords: laying hen; nutrition; egg quality; rapeseed cake; fermentation; hydrobarothermal treatment

1. Introduction

Soybean meal (SBM) is presently the main source of protein in animal diets, including laying hen diets. However, the global demand for feed protein is met by SBM imports from a limited number of countries. Due to the high prices and variable supply of SBM, as well as concerns over genetically modified soybeans, alternative domestic sources of vegetable protein, such as rapeseed meal (RSM) or rapeseed cake (RC), are being sought.

According to Food and Agriculture Organization (FAO) [1], in 2019/2020, global oilseed production reached 584.3 million tons, including 69.2 million tons of rapeseed. Rapeseed is widely used in the production of both foodstuffs and feedstuffs. Rapeseed oil by-products are utilized in the feed industry. These by-products include RC which can be a valuable component in layer diets [2]. The production technology of RC is less expensive and more environmentally friendly than organic solvent extraction [3]. RC is also a richer source of oil and metabolizable energy for poultry than RSM. Rapeseed oil is abundant in omega-3 (approx. 8–9%) and omega-6 (approx. 20–24%) polyunsaturated fatty acids (PUFAs), and it can increase the content of these fatty acids in the egg yolk [4,5]. Rapeseed oil also contains bioactive compounds, including vitamin E, phenolic compounds, flavonoids, phytosterols and other antioxidants that deliver health benefits for humans and animals [6].

The high content of non-starch polysaccharides (NSPs), glucosinolates (GLS) and phytic acid (PA) limits the use of RC in layer diets. Indigestible NSPs compromise the utilization of nutrients, in particular protein, energy and phosphorus [7]. Phytic acid is poorly available to poultry, therefore poultry diets have to be supplemented with phytase enzymes or inorganic phosphorus [8,9]. Glucosinolates, the major antinutritional compounds in rapeseed, increase mortality and decrease egg production and egg weight [10]. Some GLS degradation products contribute to the production of fish-tainted eggs whose yolks contain high levels of trimethylamine (TMA), a product of bacterial fermentation of choline in the lower gut [11]. Goitrin, an antinutritional factor (ANF) formed by the action of myrosinase on GLS, inhibits the oxidation of TMA to odorless compounds, which produces a fishy taint in eggs laid by genetically susceptible birds, specifically brown layers [12,13]. In the literature, there is no consensus on the optimal proportions of rapeseed products in layer diets. According to some authors, the rapeseed content of layer diets should not exceed 10%, whereas in other studies, diets containing 20% of rapeseed had no negative effect on laying performance or egg quality [14–18].

The nutritional value of rapeseed can be improved by reducing the content of some ANFs, which can be accomplished through physical, chemical and biological treatment or by modifying crop breeding practices [19–21]. Numerous studies have demonstrated that thermal processing effectively deactivates myrosinase and decreases GLS levels in rapeseed feeds, thus improving their nutritional value for monogastric animals, including poultry [3,22,23]. According to Zentek and Goodarzii-Boorojeni [24], hydrothermal processes can reduce the content of many ANFs in poultry feed components, but their effectiveness is determined by the type and intensity of the applied treatment. Lichovníková et al. [25] and Sasyte et al. [26] observed that extruded rapeseed can be incorporated into layer diets without compromising egg production or egg quality, but little is known about other thermal processing methods for improving the nutritional value of layer diets.

Fermentation treatments have also been long used to improve the nutritional value of feed components, including rapeseed, in animal diets [27–30]. According to Gao et al. [31], fermentation can reduce the levels of aliphatic compounds (by 57.7%), indole GLS (by 97.3%), oligosaccharides (by 73%), lignin and NDF (by 25%), and PA (by 86%) in RSM, depending on processing conditions and inoculum type. However, the efficacy of fermented rapeseed feeds in laying hen nutrition has not been investigated to date. In our previous experiment performed on female turkeys, the dietary inclusion of fermented RC (FRC) at 15% did not compromise the growth performance of birds and had a greater beneficial influence on breast muscle quality than SBM used as the sole protein source [32]. Engberg et al. [33] demonstrated that fermented feeds fed to laying hens positively affected egg weight and eggshell quality.

In light of previous findings, the research hypothesis postulates that the partial replacement of SBM with RC in laying hen diets can improve productivity, and the antioxidant status and quality of eggs. Therefore, the aim of this study was to determine the effect of raw RC (RRC), hydrobarothermally-treated RC (HRC) and FRC added at 20% to layer diets on laying performance, egg traits, and the FA profile and antioxidant status of egg yolks.

2. Materials and Methods

The trial with laying hens was conducted at the Animal Research Laboratory (Department of Poultry Science and Apiculture, University of Warmia and Mazury in Olsztyn, Poland) in accordance with EU Directive 2010/63/EU on the protection of animals used for scientific purposes [34].

2.1. Preparation of Hydrobarothermally-Treated Rapeseed Cake (HRC) and Fermented Rapeseed Cake (FRC)

Locally produced RRC was purchased from "Agrolok" *Agrolok* in Golub Dobrzyn, Poland. Purified RRC was ground in a roller mill to pass through a 3 mm sieve, and then conditioned at a temperature of 90–95 °C and humidity of around 18%. Next RRC was processed in a hydrothermal reactor in the presence of steam at a temperature of 95–100 °C for 15–20 min. The raw material was expanded at a temperature of 105–110 °C under pressure of 40–60 atm, dried and cooled to the optimal storage temperature and humidity.

Raw RC was fermented as described by Drażbo et al. [35]. RRC was ground and thoroughly mixed with water in a ratio of 1:2. The fermentation process was carried out with the use of commercial 6-phytase enzyme preparation expressed by Pichia pastoris, which was added to RRC in a weight ratio of 1:1000, and thoroughly mixed. Solid-state fermentation was conducted for 24 h at a temperature of 30 °C under anaerobic conditions. Then the enzyme was deactivated at a temperature of 70 °C for 15 min, and the fermented biomass was dried at a temperature of 55 °C. The fermentation process was carried out under patent pending procedure No. 422849.

Rapeseed samples were analyzed in duplicate for the content of dry matter (DM), crude protein (CP), ether extract (EE), crude fat (CF) and crude ash (CA) according to Association of Official Agricultural Chemists AOAC [36] methods 934.01, 976.05, 920.39, 978.10, 942.05 and 984.27, respectively. Gross energy (GE) was determined with an adiabatic bomb calorimeter (KL 12 Mn, Precyzja-Bit PPHU, Bydgoszcz, Poland) standardized with benzoic acid. NSPs were determined by gas—liquid chromatography (constituent neutral sugars) using an SP-2340 column and a Varian CP3380 gas chromatograph (Varian Inc., Palo Alto, CA, USA) and by colorimetry (uronic acids) using a Biochrom Ultrospec 50 (Biochrom Ltd., Cambridge, UK), according to the procedure described by Englyst and Cummings [37,38] with modifications [39]. Uronic acids were determined as described by Scott [40]. Glucosinolates were determined by gas—liquid chromatography as described by Slominski and Campbell [41].

2.2. Birds and Housing

The experimental materials comprised 304 Hy-Line Brown laying hens aged 36 weeks, obtained from a local commercial flock. Before the experiment, all birds were weighed individually and were placed in three-tier battery cages (40 × 35 × 60) cm, with a floor slope of 12°) with two hens per cage. The birds were distributed in a completely randomized design to four treatment groups with 38 replicates per treatment and two hens per replicate. The replicates were equally distributed among three cage levels to minimize the cage level effect. Room temperature was maintained at 20 to 22 °C, and the light cycle was set at 16 h of continuous light and 8 h of darkness. The birds had ad libitum access to feed and water throughout the study. The trial started when the hens were 36 weeks old, and lasted for 12 weeks (until 48 weeks of age).

During the experiment, the birds were fed isonitrogenous and isocaloric diets in mash form, with various protein sources. In the control group (C), SBM was the main source of dietary protein, whereas the experimental groups were fed diets containing 20% of RRC, HRC or FRC. All diets were formulated to meet the nutrient requirements of laying hens [42]. The composition of control and experimental diets is presented in Table 1.

Table 1. Ingredient composition and nutritional value of experimental diets (%).

Ingredient	Diet			
	C	RRC	HRC	FRC
Wheat	49.756	36.972	36.972	36.972
Maize	20.000	20.000	20.000	20.000
Soybean meal	17.273	7.700	7.700	7.700
Rapeseed cake	-	20	-	-
Hydrobarothermally-treated rapeseed cake	-	-	20	-
Fermented rapeseed cake	-	-	-	20
Soybean oil	1.144	3.842	3.842	3.842
Sodium chloride	0.370	0.372	0.372	0.372
Limestone	9.537	9.319	9.319	9.319
Monocalcium phosphate	1.153	1.123	1.123	1.123
L-Lysine HCL [1]	0.096	0.079	0.079	0.079
DL-Methionine	0.172	0.092	0.092	0.092
Vitamin-mineral premix [2]	0.500	0.500	0.500	0.500
Analyzed nutrients				
Crude protein	17.0	16.9	17.1	17.2
Crude fat	3.1	7.1	7.0	7.1
Calculated nutritional value [3]				
AME, kcal/kg	2700	2700	2700	2700
Crude fiber	2.358	4.765	4.765	4.765
Lysine	0.790	0.790	0.790	0.790
Arginine	0.961	0.936	0.936	0.936
Methionine	0.412	0.371	0.371	0.371
Methionine + Cysteine	0.720	0.720	0.720	0.720
Threonine	0.555	0.601	0.601	0.601
Tryptophan	0.189	0.189	0.189	0.189
Calcium	3.900	3.900	3.900	3.900
Available phosphorus	0.400	0.400	0.400	0.400

[1] Lysine hydrochloride,[2] Supplied the following per kilogram of feed: 8000 IU vit A, 2500 IU vit D3, 20 mg vit E, 1.0 mg vit K_3, 1.5 mg vit B_1, 4 mg vit B_2, 1.0 mg vit B_6, 0.02 mg vit B_{12}, 0.1 mg biotin, 6.0 mg pantothenic acid, 65.0 mg Mn from manganese oxide, 52 mg zinc from zinc oxide, 45.0 mg I from ethylene diamine dihydroiodide, 0.15 mg Se from sodium selenite, 6 mg Cu. [3] Calculated according to Polish Feedstuff Analysis Tables [42].

2.3. Laying Performance

The body weights (BW) of hens were determined at the beginning (36 weeks of age) and at the end (48 weeks of age) of the experiment. Daily records of egg production were kept, and the egg laying rate was expressed as an average hen-day production. Eggs were collected daily, and egg production was expressed based on the number of days (% working days) at 4-week intervals. Individual egg weight was recorded by individual egg weighing per cage every 2 weeks, and it was used to calculate average egg weight. Total egg weight was calculated by multiplying average egg weight by egg production. Feed intake was controlled every 4 weeks, at 40, 44 and 48 weeks of age. Average daily feed intake (ADFI) per bird was calculated based on total feed intake and the number of days in the analyzed period. The feed conversion ratio—FCR (kg feed/kg eggs laid) was estimated based on egg weight and feed intake.

2.4. Egg Quality

At the end of the experiment (at 48 weeks of hens' age), 12 eggs from each group were picked randomly to determine their physicochemical properties. All quality analyses were completed within 24 h of egg collection. The eggs were weighed individually and were broken on the EQM plate measurement stand (Egg Quality Microprocessor, Technical Services & Supplies Ltd., Dunnington, York, UK) to determine albumen height. The average of two measurements of albumen height and egg weight were used to compute the Haugh unit score (HU) [43]. Yolk color intensity was evaluated and scored according to the yolk color fan (DSM, Heerlen, Netherlands) (1, light yellow; 15, orange). Then the yolk was separated from the albumen using a teflon spoon (Tefal, Rumilly, France). The yolk was rolled on a blotting paper towel. Albumen weight was calculated by subtracting the weights of yolk and eggshell from whole egg weight. To determine eggshell weight, eggshells were cleaned of any adhering albumen and membranes, and they were dried at room temperature. The proportions of the yolk, albumen and eggshell were expressed as a percentage of whole egg weight. Eggshell thickness was measured at three different locations (middle, broad and narrow ends) with the use of a digital micrometer gauge (Mitutoyo QuantuMike, Poland Ltd., Wrocław, Poland), and it was expressed as the mean value. Eggshell breaking strength was measured using an egg force reader (Orka Food Technology, Herzliya, Israel).

2.5. Cholesterol Concentration and Fatty Acid Composition of the Egg Yolk

The concentration of cholesterol and FA profile were determined in fat extracted from the yolk, according to the method proposed by Folch et al. [44]. Cholesterol was separated from fat after saponification with potassium hyroxide KOH and extraction with ethyl ether, by the modified method of the International Dairy Federation [45]. The samples were analyzed on a PU4600 chromatograph (Unicam, Cambridge, UK) with a flame ionization detector (FID), under the following conditions: glass column length—1 m, inner diameter—4 mm, film thickness—0.25 m, temperature: detector—300 °C, injector—290 °C, column—260 °C, carrier gas—argon, flow rate—50 cm^3/min, and internal standard—dotriacontane (Sigma, St. Louis, MO, USA). The extracted fat was esterified with a chloroform, methanol, and sulfuric acid mixture, as described by Peisker [46]. The resulting FA methyl esters (FAMEs) were analyzed on a 7890A gas chromatograph (Agilent Technologies Inc., Palo Alto, CA, USA) with a FID and a Supelcowax 10 capillary column (column length—30 m, internal diameter—0.32 mm, film thickness—0.25 m, carrier gas—helium, and temperature: detector—250 °C, injector—230 °C, and column—195 °C). FA peaks were identified by comparing their relative retention times with those of individual FAME reference standards (Supelco) diluted in hexane (1:1, 1:2, 1:3, and 1:4 v/v).

2.6. Antioxidant Status

Homogenates from egg yolks were also analyzed for lipid peroxidation products, i.e., lipid peroxide content according to Gay and Gebicki [47] and malondialdehyde (MDA) content according to Botsoglou et al. [48]. Malondialdehyde was determined based on the reaction of thiobarbituric acid (TBA) with lipid peroxidation end products in an acidic environment and increased temperature to generate a colored adduct. To eliminate the quantities of a complex series of adducts from TBA, the assay was run in the presence of inhibitors, e.g., butylated hydroxytoluene (BHT). The activity of an antioxidant enzyme, superoxide dismutase (SOD), was determined spectrophotometrically in extracted eggs using the adrenaline method [49] at a modified wavelength of 320 nm to achieve greater selectivity of intermediate reaction products. SOD activity was determined by measuring the rate of auto-oxidation of adrenaline at 30 °C based on an increase in absorbance at 320 nm (which corresponds to the increase in the concentrations of various products of adrenaline oxidation). The activity of catalase (CAT) was determined according to Claiborne pp. 283–284 in [49]. The assay consisted of measuring the rate of substrate (hydrogen peroxide) decomposition catalyzed by this enzyme.

2.7. Sensory Evaluation

Hard-boiled eggs were subjected to a sensory evaluation. Before the analysis, the eggs were immersed in a water bath and held at a temperature of 100 °C for 15 min. At the completion of the thermal treatment, the eggs were cooled under a stream of cold running water for 15 min. Immediately after cooling, the eggs were evaluated on a five-point hedonic scale where: 5—very high quality, 4—high quality, 3—satisfactory quality, 2—unsatisfactory quality, 1—poor quality. The following sensory quality attributes were evaluated: appearance (including the appearance of the whole egg and its longitudinal cross-section), aroma, albumen texture, yolk texture, taste. The analysis was performed by six trained panelists selected for their sensory sensitivity [50] (The panelists assessed samples in individual compartments. Fluorescent white lights (500 lux) that simulated daylight were installed at a height of approximately 1 m to evenly illuminate the table. Relative air humidity of minimum 60% and temperature of 21 °C were maintained in the panel room.

2.8. Statistical Analysis

For a statistical analysis of performance parameters, a cage was considered as a replicate experimental unit. For analyses of egg quality, FA profile, and the antioxidant parameters of yolks, individual eggs were considered as experimental units. One-way ANOVA was performed with the use of Statistica 13.1 software (StatSoft Corp., Cracow, Poland). When a significant treatment effect was noted, Tukey's post hoc test was applied to determine differences between treatment groups. Data are presented as means with pooled standard error of the mean (SEM) estimates, and the value of $p \leq 0.05$ was considered statistically significant.

3. Results

The effect of hydrobarothermal treatment and fermentation on the concentrations of nutrients and ANFs in RC is presented in Table 2. In comparison with RRC, HRC was characterized by lower concentrations of CF (9.91 vs. 10.80%) and GLS (13.32 vs. 15.35 µmol/g), whereas the content of the remaining components did not change. Fermentation had a more significant influence on the nutritional value of RC; FRC had a higher content of DM (94.57 vs. 88.87%), CP (39.44 vs. 37.74%) and GE (21.86 vs. 21.19 MJ/kg) than RRC, and the lowest concentration of CF (9.29%) of all analyzed rapeseed products. The greatest changes were observed in the content of ANFs in FRC—the concentrations of PP and GLS decreased two-fold (0.833 vs. 1.607%) and nearly 17-fold (0.92 vs. 15.35 µmol/g), respectively, compared with RRC. A minor increase in NSP content was also noted (21.64 vs. 20.48%).

Table 2. Nutrient composition of rapeseed products and the presence of antinutritional factors (%, dry matter basis).

Content	RRC	HRC	FRC
Dry matter	88.87	88.45	94.57
Crude protein	37.74	37.02	39.44
Crude fat	10.80	9.91	9.29
Gross energy, MJ/kg	21.19	21.41	21.86
Phytate phosphorus	1.607	1.684	0.833
Non-starch polysaccharides [1]	20.48	20.10	21.64
Glucosinolates (μmol/g) [2]	15.35	13.32	0.92

[1] Including rhamnose, arabinose, xylose, mannose, galactose, glucose, and uronic acids. [2] Including gluconapin, glucobrassicinapin, progoitrin, glucobrassicin, hydroxyglucobrassicin, and neoglucobrassicin.

At the end of the experiment, an increase in the BW of hens was noted only in groups HRC and FRC, whereas hens fed HRC had the significantly highest final BW, relative to group C (p = 0.037, Table 3). The applied dietary treatments had no effect on egg weight, but the inclusion of 20% RRC in layer diets significantly decreased laying rate and, consequently, egg mass during the entire experiment, compared with group C (p = 0.001 and p = 0.014, respectively). Although ADFI was comparable in all groups, the FCR was highest in group RRC, and a significant difference was noted relative to group C (p = 0.026). The inclusion of HRC and FRC in layer diets had a greater beneficial influence on the above parameters, and the best results were noted in group FRC where the laying rate was comparable with that in group C and significantly higher than in group RRC. The numerical values of egg mass and the FCR were more favorable in groups HRC and FRC, compared with group RRC, but less satisfactory than in group C.

Table 3. Performance of laying hens fed raw (RRC), hydrobarothermally-treated (HRC) and fermented (FRC) rapeseed cake.

Item	Diet				SEM	p-Value
	C	RRC	HRC	FRC		
Initial body weight, kg	2.079	2.040	2.040	2.052	0.014	0.763
Final body weight, kg	1.997 ^{b}	2.019 ^{ab}	2.096 ^{a}	2.054 ^{ab}	0.013	0.037
Laying rate, %	96.2 ^{a}	92.9 ^{b}	94.5 ^{ab}	95.4 ^{a}	0.285	0.001
Egg weight (g)	61.9	61.5	60.7	61.3	0.243	0.325
Egg mass (kg/hen)	4.95 ^{a}	4.76 ^{b}	4.81 ^{ab}	4.84 ^{ab}	0.022	0.014
Average daily feed intake (g/hen/day)	112.6	113.2	112.2	113.3	0.252	0.374
Feed conversion ratio (kg/kg eggs)	1.899 ^{b}	1.980 ^{a}	1.959 ^{ab}	1.952 ^{ab}	0.009	0.026

^{a,b} Means within the same column with different superscripts differ significantly ($p < 0.05$).

An analysis of the physicochemical properties of eggs revealed that eggshell quality was highest in group C (Table 4). The inclusion of HRC and FRC in layer diets decreased eggshell thickness and strength. Eggshell thickness was significantly lower in group HRC than in group C, and in group FRC than in group RRC (p = 0.001). Eggshell strength was lowest in group HRC (p = 0.018). Fermentation had a greater beneficial influence on the parameters of albumen quality than hydrobarothermal treatment. Both albumen height and HU values were highest in eggs produced in group FRC, and significant differences were found relative to group HRC (p = 0.001). The addition of RC to layer diets, regardless of its form, improved yolk color, and the highest values of yolk color intensity (DSM yolk color fan) were noted in group HRC (p = 0.005). The applied dietary treatments had no effect on the percentage content of yolk, albumen and shell in eggs.

Table 4. Effect of raw (RRC), hydrobarothermally-treated (HRC) and fermented (FRC) rapeseed cake on the physicochemical properties of eggs [1].

Item	Diet				SEM	*p*-Value
	C	RRC	HRC	FRC		
Eggshell strength (kg)	4.05 [a]	3.78 [ab]	3.30 [b]	3.70 [ab]	0.086	0.018
Eggshell thickness, mm	0.314 [a]	0.306 [ab]	0.288 [bc]	0.279 [c]	0.003	0.001
Yolk color	2.42 [b]	2.92 [ab]	3.50 [a]	3.17 [ab]	0.115	0.005
Albumen height, mm	7.39 [ab]	7.29 [bc]	6.43 [c]	8.22 [a]	0.146	0.001
Haugh units	84.0 [a]	83.7 [ab]	77.6 [b]	89.4 [a]	0.985	0.001
Yolk content, %	25.8	25.9	25.3	25.4	0.758	0.229
Albumen content, %	64.9	65.0	65.8	65.7	0.265	0.505
Shell content, %	9.3	9.1	8.9	8.9	0.087	0.247

[1] Data representing mean values of 12 eggs per treatment. [a,b,c] Means within the same column with different superscripts differ significantly ($p < 0.05$).

Dietary protein sources had a significant effect on the FA profile of yolk lipids (Table 5). The inclusion of RRC, HRC, and FRC in layer diets contributed to an increase ($p = 0.001$) in the concentrations of linoleic acid (C18:2, n-6), linolenic acid (C18:3, n-3) and docosahexaenoic acid (DHA, C22:6, n-3). The addition of HRC led to a decrease in the content of *arachidonic acid* (AA, C20:4, n-6) in egg yolks, compared with the remaining groups ($p = 0.002$). The concentrations of SFAs were highest in egg yolks in group C ($p = 0.001$). The eggs laid by hens receiving RRC, HRC, and FRC were characterized by significantly highest proportions of PUFAs, n-6 PUFAs, and n-3 PUFAs in the total FA pool, compared with group C ($p = 0.001$), and a much more favorable n-6/n-3 PUFA ratio ($p = 0.001$). In all experimental groups fed 20% RC, eggs had a significantly higher content of hypocholesterolemic FAs and a lower content of hypercholesterolemic FAs, relative to group C ($p = 0.001$). Cholesterol concentration in egg yolks was lowest in group HRC ($p = 0.001$).

Table 5. Effect of raw (RRC), hydrobarothermally-treated (HRC) and fermented (FRC) rapeseed cake on the fatty acid composition (% of total fatty acid content) and cholesterol content of yolk lipids (%) [1].

Item	Diet				SEM	*p*-Value
	C	RRC	HRC	FRC		
C14:0	0.290 [a]	0.230 [b]	0.233 [b]	0.218 [b]	0.006	0.001
C14:1	0.060 [a]	0.024 [b]	0.022 [b]	0.021 [b]	0.003	0.001
C15:0	0.063 [ab]	0.061 [ab]	0.057 [b]	0.069 [a]	0.002	0.040
C16:0	25.4 [a]	21.9 [b]	21.5 [b]	20.8 [b]	0.340	0.001
C16:1	3.17 [a]	1.73 [b]	1.75 [b]	1.60 [b]	0.125	0.001
C17:0	0.167 [c]	0.211 [b]	0.216 [b]	0.245 [a]	0.005	0.001
C17:1	0.129	0.136	0.128	0.145	0.002	0.468
C18:0	8.13	8.54	8.05	8.45	0.086	0.120
C18:1	42.1	41.4	41.8	41.7	0.208	0.706
C18:2 (n-6)	16.2 [b]	20.5 [a]	21.2 [a]	21.3 [a]	0.451	0.001
C18:3 (n-6)	0.221 [b]	0.252 [a]	0.232 [ab]	0.255 [a]	0.004	0.006
C18:3 (n-3)	0.782 [b]	1.366 [a]	1.488 [a]	1.465 [a]	0.060	0.001
C20:1	0.223 [a]	0.190 [b]	0.189 [b]	0.210 [ab]	0.004	0.006
C20:2 (n-6)	0.134	0.157	0.167	0.168	0.005	0.096
C20:3 (n-6)	0.141	0.145	0.142	0.142	0.003	0.945
C20:4 (n-6)	1.79 [a]	1.81 [a]	1.63 [b]	1.83 [a]	0.022	0.002
C22:6 (n-3)	0.912 [c]	1.326 [ab]	1.166 [b]	1.410 [a]	0.038	0.001
SFAs	34.1 [a]	30.9 [b]	30.0 [b]	29.7 [b]	0.435	0.001
MUFAs	45.7 [a]	43.5 [b]	43.9 [ab]	43.7 [b]	0.271	0.010
PUFAs	20.2 [b]	25.6 [a]	26.0 [a]	26.6 [a]	0.539	0.001
n-3 PUFAs	1.69 [b]	2.69 [a]	2.65 [a]	2.87 [a]	0.091	0.001
n-6 PUFAs	18.5 [b]	22.9 [a]	23.4 [a]	23.7 [a]	0.455	0.001
n-6/n-3 PUFA ratio	11.52 [a]	8.52 [b]	8.81 [b]	8.26 [b]	0.267	0.001
Hypercholesterolemic FAs	26.0 [a]	22.4 [b]	22.0 [b]	21.3 [b]	0.340	0.001
Hypocholesterolemic FAs	74.0 [b]	77.6 [a]	78.0 [a]	78.7 [a]	0.340	0.001
Cholesterol	21.0 [a]	21.4 [a]	17.0 [b]	21.8 [a]	0.522	0.001

[1] Data representing mean values of 10 eggs per treatment. [a,b,c] Means within the same column with different superscripts differ significantly ($p < 0.05$); SFAs—saturated fatty acids, MUFAs—monounsaturated fatty acids, PUFAs—polyunsaturated fatty acids.

Indicators of the redox status of egg yolks are presented in Table 6. Irrespective of its form, RC had no effect on the concentrations of MDA and total glutathione (GSH/GSSG) in egg yolks. A significant decrease in SOD activity ($p = 0.001$) and an increase in CAT activity in eggs were observed in groups RRC, HRC, and FRC, and CAT activity was higher in group FRC ($p = 0.002$) than in group C. An analysis of LOOH concentrations revealed that the content of lipid oxidation products increased significantly in the yolks of eggs laid by hens fed RRC ($p = 0.001$). Both hydrobarothermal treatment and fermentation of RC decreased LOOH concentrations in egg yolks, compared with RRC and significant differences were noted in group FRC where LOOH concentrations were comparable with those in group C.

Table 6. Redox parameters of egg yolks sampled from layers fed a control diet (C), and diets containing raw (RRC), hydrobarothermally-treated (HRC) or fermented rapeseed cake (FRC) [1].

Item	Diet				SEM	*p*-Value
	C	RRC	HRC	FRC		
SOD, U/g	123.7 ^{a}	99.8 ^{b}	98.5 ^{b}	104.7 ^{b}	2.252	0.001
CAT, U/g	29.7 ^{b}	32.9 ^{b}	37.0 ^{ab}	42.3 ^{a}	1.295	0.002
MDA, µmol/kg	7.38	7.67	7.61	7.66	0.148	0.900
LOOH, µmol/L	5.14 ^{b}	6.51 ^{a}	5.61 ^{ab}	5.12 ^{b}	0.151	0.001
GSH + GSSG, µmol/kg	4.01	4.06	4.05	4.59	0.142	0.443

[1] Data representing mean values of 10 eggs per treatment. ^{a,b} Means within the same column with different superscripts differ significantly ($p < 0.05$); SOD, superoxide dismutase; CAT, catalase; MDA, malondialdehyde; LOOH, lipid peroxides; GSH + GSSG, total glutathione.

Regardless of its form, RC added to layer diets had no negative effect on the major sensory attributes of eggs, such as general appearance, albumen texture, yolk texture, and, in particular, taste and aroma (Table 7). A statistical analysis of the results of sensory evaluation demonstrated that eggs produced in group FRC received the highest overall score.

Table 7. Average intensity values of egg sensory attributes [1].

Item	Diet				SEM	*p*-Value
	C	RRC	HRC	FRC		
General appearance	3.65 ^{b}	3.80 ^{b}	3.70 ^{b}	4.25 ^{a}	0.065	0.002
Aroma	3.75 ^{b}	3.75 ^{b}	3.75 ^{b}	4.15 ^{a}	0.057	0.020
Albumen texture	3.80 ^{ab}	3.80 ^{ab}	3.65 ^{b}	4.25 ^{a}	0.069	0.008
Yolk texture	3.80 ^{b}	3.70 ^{b}	3.80 ^{b}	4.30 ^{a}	0.067	0.004
Taste	3.60 ^{b}	3.65 ^{b}	3.65 ^{b}	4.35 ^{a}	0.071	0.001

[1] Data representing mean values of 10 eggs per treatment. ^{a,b} Means within the same column with different superscripts differ significantly ($p < 0.05$).

4. Discussion

Previous research has demonstrated that hydrothermal and fermentation processes can improve the quality of poultry diets by modifying their chemical composition and reducing the content of ANFs in selected feed components [24,51,52]. In the present study, the content of essential nutrients in RC was not affected by hydrobarothermal treatment or fermentation. The small increase in the crude protein content of FRC can be probably attributed to changes in DM content, rather than an actual increase in protein content [28]. The above observation could also explain the small increase in NSP levels in FRC. In turn, significant changes in ANF content were noted in the components of experimental diets. Glucosinolate levels decreased by 2% in HRC, but they were nearly 17 times lower in FRC than in RRC. Vig and Walia [27] and Chiang et al. [28] observed that the decrease in the GLS content of RMS was proportional to fermentation time, and that isothiocyanate concentration decreased by as much as 83% after 30 days of fermentation. The decrease in the content of GLS and their degradation products during fermentation could be attributed to the breakdown of glucose and sulfur molecules by microbial enzymes [10]. In turn, high-pressure thermal processing inactivates myrosinase that contributes to GLS hydrolysis, leading to the formation of ANFs [53]. The content of PP was two times lower in FRC than in RRC and HRC. According to Fazhi et al. [54], microbial phytases can be used to reduce the PA content of rapeseed during fermentation. In the literature, various microorganisms have been analyzed for their ability to produce phytase and reduce the PA levels in RSM during fermentation [55,56].

Diets with increased content of rapeseed products can compromise feed intake in poultry because GLS degradation products have a bitter taste [12]. Raw RC is also high in fiber, which increases satiety and can additionally decrease feed intake [57]. Numerous studies have demonstrated that diets containing more than 100 g of RSM/kg can reduce feed intake and decrease egg weight [58–60]. Interestingly, the above relationship was not observed in the current study, where ADFI and egg weight were comparable in all groups. However, the inclusion of RRC in hen diets significantly decreased the laying rate and egg mass, and it compromised the FCR during the entire experiment. Both rapeseed treatments improved the laying rate, and in groups HRC and FRC, all parameters (that deteriorated in group RRC) were comparable to those noted in group C. Similar results were reported by Jeroch et al. [61,62] who observed that the rapeseed content of layer diets can even be increased to 30% without compromising performance when ANF levels are reduced through combined chemical and hydrothermal treatment.

Eggshell quality plays a very important role in the egg industry [63]. In the present study, layer diets containing 20% of raw or processed RC had a negative influence on eggshell thickness and strength. In the experiments conducted by Riyazi et al. [64], Świątkiewicz et al. [65] and Zhu et al. [66], the incorporation of rapeseed products into layer diets did not affect eggshell quality, whereas Sasyte et al. [26] reported that even small amounts of extruded rapeseed decreased eggshell thickness. Interestingly, in the present study, RRC produced more desirable effects than HRC or FRC, since despite lower numerical values of the analyzed parameters, the results noted in group RRC were statistically comparable to those noted in group C. This was an unexpected outcome because the application of FRC, which was far less abundant in PP than other experimental feed components, should have improved eggshell quality. According to Skřivan et al. [67], high dietary levels of PP decrease phosphorus utilization by laying hens, which can compromise eggshell quality. In turn, FRC improved the parameters of albumen quality. The eggs produced in group FRC were characterized by the highest albumen quality, determined based on albumen height and HU values, and the noted differences were significant relative to group HRC. The above can probably be attributed to a much lower content of GLS in FRC, compared with RRC and HRC. According to Zhu et al. [66], GLS metabolites can inhibit ovalbumin synthesis, decrease albumen height and HU values. However, studies examining the influence of rapeseed products on albumen quality produced ambiguous results. In the work of Lichovníková et al. [68] and Riyazi et al. [64], the incorporation of RSM into layer diets did not affect HU values, whereas in a study by Najib and Al.-Khateeb [59], HU values increased with a rise in the proportion of canola meal in layer diets. However, the effect of treated rapeseed feeds on the above parameters has not been investigated to date. In the present study, the inclusion of rapeseed products in layer diets increased yolk color intensity, which corroborates the findings of other authors [59,60,66]. Treated RC, in particular HRC, delivered the most satisfactory results, but further research is needed to confirm this relationship.

The FA profile of eggs is an important quality attribute for consumers, and efforts are being made to modify layer diets and further improve the FA profile of eggs [69]. Egg yolk PUFAs that deliver the greatest health benefits include linoleic acid (C18:2 n-6, a substrate for the biosynthesis of other long-chain FAs), linolenic acid (C18:3 n-3), EPA (C20:5 n-3), and DHA (C22:6 n-3) [70]. Most of these FAs belong to the family of n-3 PUFAs that contribute to cardiovascular health, brain development and cognitive functioning, and reduce the risk of cancer, autoimmune diseases and diabetes [71,72]. These PUFAs are not synthesized by the body and must be supplied in the diet. In the current study, none of the applied treatments influenced the FA profile of egg yolks relative to group C, whereas all diets containing RRC, HRC, or FRC improved the FA profile of egg yolks relative to group C by increasing the content of linoleic acid, linolenic acid, and DHA, and decreasing the n-6/n-3 PUFA ratio. Buckiuniene et al. [73] reported that an increase in the concentrations of n-3 PUFAs decreased the n-6/n-3 PUFA ratio and enhanced the FA profile of egg yolks. According to the World Health Organization (WHO) recommendations for dietary fat intake, the optimal n-6/n-3 PUFA ratio in the human diet is 4:1 [74]. Kaczmarek et al. [75] observed that RC, a rich source of n-3 PUFAs (including linoleic and linoleic acids) with an n-6/n-3 PUFA ratio of around 2, delivers greater health benefits for monogastric animals than other feeds, which undoubtedly affected the results of the present study. The beneficial effects of rapeseed oil on the FA profile of chicken eggs have been confirmed by many researchers [5,76,77]. However, the influence of fermented or hydrobarothermally-treated RC on the FA profile of chicken eggs has never been examined in the literature. It should be noted that despite the absence of significant differences, egg yolks in group FRC were characterized by the numerically highest content of n-3 PUFAs and the most desirable n-6/n-3 PUFA ratio.

An analysis investigating the effects of various RRC treatments on the cholesterol content of eggs produced rather surprising results. Panaite et al. [78] demonstrated that diets rich in PUFAs can effectively reduce egg cholesterol levels. In the present study, the eggs produced in groups RRC, HRC, and FRC were characterized by the significantly lowest levels of hypercholesterolemic FAs and the highest levels of hypocholesterolemic Fas, which is why egg cholesterol levels were expected to decrease in all experimental groups. Despite the above, egg cholesterol levels decreased significantly only in group HRC, which is difficult to explain. It can only be speculated that lower egg cholesterol levels in group HRC were associated with the metabolism of carotenoids that were easily absorbed, transferred to the egg yolk and significantly improved yolk color. According to Yeum and Russel [79], diets rich in carotenoids decrease cholesterol levels in the blood serum and, probably, in egg yolks.

As previously noted, the inclusion of RC in layer diets increased the content of n-3 PUFAs in egg yolks. However, egg yolks are abundant in lipids, and their susceptibility to peroxidation may increase when layers are fed diets rich in PUFAs [80]. This process leads to the production of MDA, the end product of lipid peroxidation, and the MDA content of tissues indirectly reflects the degree of lipid peroxidation by reactive oxygen species (ROS) [81]. In the current study, the MDA content of egg yolks was comparable in all experimental groups, whereas SOD activity decreased significantly in all groups fed RC, and an increase was observed in CAT activity, in particular in group FRC. Superoxide dismutase and CAT are the main antioxidant enzymes that scavenge ROS via a chain reaction [82]. The MDA content of eggs remained stable, which suggests that dietary modifications did not induce oxidative processes. The observed changes in the activity of antioxidant enzymes could also indicate that RC, in particular FRC, enhances the protective potential of antioxidant enzymes, as demonstrated by the lowest MDA levels and increased CAT activity. According to Ognik and Krauze [83], SOD and CAT activity can be increased or decreased through the addition of feed additives and components with antioxidant properties, but the direction of changes in enzyme activity can be determined by the initial redox potential.

Numerous researchers have demonstrated that the addition of rapeseed products to layer diets can compromise the sensory attributes of eggs, in particular by imparting a fishy taint to eggs [84–86]. In the present study, the inclusion of 20% RRC and HRC in layer diets did not affect the taste, aroma, or texture of eggs. Interestingly, FRC significantly improved the above attributes in comparison with control group eggs. In the work of Lichovníková et al. [68] and Świątkiewicz et al. [87], these attributes were already compromised after the addition of 8% untreated rapeseed and rapeseed expeller cake to layer diets, which was attributed to the presence of TMA in the eggs laid by brown hens. The above observation was not confirmed in the current study, and the inclusion of FRC in layer diets led to the production of eggs with the highest sensory quality. These differences could be explained by lower GLS content and, consequently, lower levels of goitrin which inhibits the oxidation of TMA to odorless compounds.

5. Conclusions

Hydrobarothermal treatment and fermentation decreased the GLS content of RC, and fermentation reduced PP concentration. As a result, HRC and FRC exerted a greater beneficial influence on the laying performance of hens than raw RRC. RRC, HRC, and FRC can be included at 20% in layer diets because they do not compromise the sensory quality of eggs and have a beneficial influence on the FA profile and antioxidant potential of egg yolks. The use of FRC is recommended because it contributes to the highest laying performance, superior albumen quality, and the highest sensory quality of eggs, relative to RRC and HRC.

Author Contributions: Conceptualization, A.A.D.; data curation, A.A.D. and M.K.; formal analysis, M.K., A.A.D., K.Ś. and K.O.; investigation, K.O. and A.A.D.; methodology, A.A.D., K.Ś. and K.O.; writing—original draft, M.K. and A.A.D.; writing—review and editing, A.A.D., M.K. and K.O. All authors have read and agreed to the published version of the manuscript.

Funding: Project financially supported by the Minister of Education and Science under the program entitled "Regional Initiative of Excellence" for the years 2019-2022, Project No. 010/RID/2018/19, amount of funding 12.000.000 PLN. under the program entitled "Regional Initiative of Excellence" for the years 2019–2022, Project No. 010/RID/2018/19, amount of funding PLN 12,000,000.

Institutional Review Board Statement: The trial with laying hens was conducted at the Animal Research Laboratory (De-partment of Poultry Science and Apiculture, University of Warmia and Mazury in Olsztyn, Poland) in accordance with EU Directive 2010/63/EU on the protection of animals used for scientific purposes.

Acknowledgments: The authors would like to thank Anna Rogiewicz and Bogdan A. Slonimski from the Department of Animal Science at the University of Manitoba, Winnipeg, MB, Canada, for technical support.

Conflicts of Interest: The authors declare that they have no competing interest.

References

1. FAO. *Food Outlook—Biannual Report on Global Food Markets: June 2020*; FAO: Rome, Italy, 2020. Available online: https://reliefweb.int/report/world/food-outlook-biannual-report-global-food-markets-june-2020 (accessed on 11 June 2020).
2. Angelovičová, M.; Angelovič, M. Rapeseed Cakes as an Important Feed Raw Material for Laying Hens. *Sci. Pap. Anim. Sci. Biotechnol.* **2013**, *46*, 335–338.
3. Smulikowska, S.; Czerwiński, J.; Mieczkowska, A. Nutritional value of rapeseed expeller cake for broilers: Effect of dry extrusion. *J. Anim. Feed Sci.* **2006**, *15*, 447–455. [CrossRef]
4. Bedekovic, D.; Pintar, J.; Janjecic, Z.; Mužic, S.; Kos, I. Using Rapeseed Cake to Improve Fatty Acid Composition of Turkey Meat. *J. Agr. Sci. Technol.* **2014**, *16*, 1279–1287.
5. Faitarone, A.B.G.; Garcia, E.A.; Roca, R.O.; Andrade, E.N.; Vercese, F.; Pelicia, K. Yolk Color and Lipid Oxidation of the Eggs of Commercial White Layers Fed Diets Supplemented with Vegetable Oils. *Rev. Bras. Cienc. Avic.* **2016**, *18*, 9–16. [CrossRef]
6. Szydłowska-Czerniak, A. Rapeseed and its Products—Sources of Bioactive Compounds: A Review of their Characteristics and Analysis. *Crit. Rev. Food Sci. Nutr.* **2013**, *53*, 307–330. [CrossRef]
7. Knudsen, K.E. Carbohydrate and lignin contents of plant materials used in animal feeding. *Anim. Feed Sci. Technol.* **1997**, *67*, 319–338. [CrossRef]
8. Lim, H.S.; Namkung, H.; Paik, I.K. Effects of phytase supplementation on the performance, egg quality, and phosphorous excretion of laying hens fed different levels of dietary calcium and nonphytate phosphorous. *Poult. Sci.* **2003**, *82*, 92–99. [CrossRef]
9. Kozłowski, K.; Jankowski, J.; Jeroch, H. Efficacy of different levels of Escherichia coli phytase in hens fed maize-soyabean meal based diets with a decreased non-phytate phosphorus content. *J. Anim. Feed Sci.* **2011**, *20*, 224–235. [CrossRef]
10. Tripathi, M.K.; Mishra, A.S. Glucosinolates in animal nutrition: A review. *Anim. Feed Sci. Technol.* **2007**, *132*, 1–27. [CrossRef]

11. Ward, A.K.; Classen, H.L.; Buchanan, F.C. Fishy-egg tainting is recessively inherited when brown-shelled layers are fed canola meal. *Poult. Sci.* **2009**, *88*, 714–721. [CrossRef]
12. Fenwick, G.R.; Curl, C.L.; Butler, E.J.; Greenwood, N.M.; Pearson, A.W. Rapeseed meal and egg taint: Effects of low glucosinolate Brassica napus meal, dehulled meal and hulls, and of neomycin. *J. Sci. Food Agric.* **1984**, *35*, 749–756. [CrossRef]
13. Mawson, R.; Heaney, R.K.; Zduńczyk, Z.; Kozłowska, H. Rapeseed meal-glucosinolates and their antinutritional effects. Part 6. Taint in end-products. *Food/Nahrung* **1995**, *39*, 21–31. [CrossRef]
14. Thomas, D.; Robblee, A.R.; Clandinin, D.R. Effects of low and high glucosinolate rapeseed meals on productive performance, egg quality, composition of liver and incidence of haemorrhagic liver syndrome in laying birds. *Br. Poult. Sci.* **1978**, *19*, 449–454. [CrossRef]
15. Mawson, R.; Heaney, R.; Zduńczyk, Z.; Kozłowska, H. Rapeseed meal-glucosinolates and their antinutritional effects Part 5. Animal reproduction. *Nahrung* **1994**, *38*, 588–598. [CrossRef] [PubMed]
16. Jia, W.; Slominski, B.A.; Guenter, W.; Humphreys, A.; Jones, O. The effect of enzyme supplementation on egg production parameters and omega-3 fatty acid deposition in laying hens fed flaxseed and canola seed. *Poult. Sci.* **2008**, *87*, 2005–2014. [CrossRef] [PubMed]
17. Khajali, F.; Slominski, A. Factors that affect the nutritive value of canola meal for poultry. *Poult. Sci.* **2012**, *91*, 2564–2575. [CrossRef] [PubMed]
18. Savary, R.K.; MacIsaac, J.; Rathgeber, B.M.; McLean, N.L.; Anderson, D.M. Evaluating Brassica napus and Brassica juncea meals with supplemental enzymes for use in brown-egg laying hen diets: Production performance and egg quality factors. *Can. J. Anim. Sci.* **2019**, *99*, 820–832. [CrossRef]
19. Dänicke, S.; Kracht, W.; Jeroch, H.; Zachmann, R.; Heidenreich, E.; Löwe, R. Effect of different technical treatments of rapeseed on the feed value for broilers and laying hens. *Arch. Anim. Nutr.* **1998**, *51*, 53–62. [CrossRef] [PubMed]
20. Feng, D.; Zuo, J. Nutritional and anti-nutritional composition of rapeseed meal and its utilization as a feed ingredient foranimal. In Proceedings of the 12th IRC Rapeseed Conference, Wuhan, China, 26–30 March 2007; pp. 265–270.
21. Kozłowski, K.; Jeroch, H. Enhancing the Nutritional Value of Poultry Feedstuffs Using the Example of Rapeseed Products—A Review. *Ann. Anim. Sci.* **2014**, *14*, 245–256. [CrossRef]
22. Newkirk, R.W.; Classen, H.L. The effects of toasting canola meal on body weight, feed conversion efficiency, and mortality in broiler chickens. *Poult. Sci.* **2002**, *81*, 815–825. [CrossRef]
23. Milošević, N.; Stanaćev, V.; Perić, L.; Žikić, D.; Dukić-Stojčić, M.; Milić, D. Use extruded rapeseed meal in the feed of broiler chickens. *Biotechnol. Anim. Husb.* **2011**, *27*, 1681–1687. [CrossRef]
24. Zentek, J.; Goodarzi-Boroojeni, F. (Bio)Technological processing of poultry and pig feed: Impact on the composition, digestibility, anti-nutritional factors and hygiene. *Anim. Feed Sci. Technol.* **2020**, *268*, 114576. [CrossRef]
25. Lichovníková, M.; Klecker, D.; Zeman, L. The effect of extruded rapeseed feed on laying performance and quality of eggs. *Czech J. Anim. Sci.* **2000**, *45*, 493–499.
26. Sasyte, V.; Grashorn, M.A.; Klementaviciute, J.; Viliene, V.; Raceviciute-Stupeliene, A.; Gruzauskas, R.; Dauksiene, A.; Alijosius, S. Effect of extruded full-fat rapeseed on egg quality in laying hens. *Eur. Poult. Sci.* **2017**, *81*, 201. [CrossRef]
27. Vig, A.P.; Walia, A. Beneficial effects of Rhizopus oligosporus fermentation on reduction of glucosinolates, fibre and phytic acid in rapeseed (*Brassica napus*) meal. *Bioresour. Technol.* **2001**, *78*, 309–312. [CrossRef] [PubMed]
28. Chiang, L.; Lu, W.Q.; Piao, X.S.; Hu, J.K.; Gong, L.M.; Thacker, P.A. Effects of feeding solid-state fermented rapeseed meal on performance, nutrient digestibility, intestinal ecology and intestinal morphology of broiler chickens. *Asian-Aust. J. Anim. Sci.* **2010**, *23*, 263–271. [CrossRef]
29. Zaworska, A.; Kasprowicz-Potocka, M.; Frankiewicz, A.; Zduńczyk, Z.; Jaśkiewicz, J. Effects of fermentation of narrow-leafed lupine (*L. angustifolius*) seeds on their chemical composition and physiological parameters in rats. *J. Anim. Feed Sci.* **2016**, *25*, 326–334. [CrossRef]
30. Drażbo, A.; Mikulski, D.; Jankowski, J.; Zduńczyk, Z. The effect of diets containing raw and fermented faba beans on gut functioning and growth performance in young turkeys. *J. Anim. Feed Sci.* **2018**, *27*, 65–73. [CrossRef]
31. Gao, M.; Cieślak, A.; Kierończyk, B.; Huang, H.; Yanza, Y.R.; Zaworska-Zakrzewska, A.; Józefiak, D.; Szumacher-Strabel, M. Effects of Raw and Fermented Rapeseed Cake on Growth Performance, Methane Production, and Breast Meat Fatty Acid Composition in Broiler Chickens. *Animals* **2020**, *10*, 2250. [CrossRef] [PubMed]
32. Drażbo, A.; Kozłowski, K.; Ognik, K.; Zaworska, A.; Jankowski, J. The effect of raw and fermented rapeseed cake on growth performance, carcass traits, and breast meat quality in turkey. *Poult. Sci.* **2019**, *98*, 6161–6169. [CrossRef] [PubMed]
33. Engberg, R.M.; Hammershoj, M.; Johansen, N.F.; Abousekken, M.S.; Steenfeldt, S.; Jensen, B.B. Fermented feed for laying hens: Effects on egg production, egg quality, plumage condition and composition and activity of the intestinal microflora. *Br. Poult. Sci.* **2009**, *50*, 228–239. [CrossRef]
34. OJEU. Directive 2010/63/EU of the European parliament and of the council on the protection of animals used for scientific purposes. *Off. J. Eur. Union.* **2010**, *276*, 33–79.
35. Drażbo, A.; Ognik, K.; Zaworska, A.; Ferenc, K.; Jankowski, J. The effect of raw and fermented rapeseed cake on the metabolic parameters, immune status, and intestinal morphology of turkeys. *Poult. Sci.* **2018**, *97*, 3910–3920. [CrossRef] [PubMed]
36. AOAC. *Official Methods of Analysis of AOAC International*, 18th ed.; Association of Official Analytical Chemists: Washington, DC, USA, 2007.

37. Englyst, H.N.; Cummings, J.H. Simplified method for the measurement of total non-starch polysaccharides by gas-liquid chromatography of constituent sugars as alditol acetates. *Analyst* **1984**, *109*, 937–942. [CrossRef]
38. Englyst, H.N.; Cummings, J.H. Improved method for the determination of dietary fiber as non-starch polysaccharidesin plant foods. *J. AOAC Int.* **1988**, *71*, 808–814. [CrossRef]
39. Slominski, B.A.; Campbell, L.D. Non-starch polysaccharides of canola meal: Quantification, digestibility in poultry and potential benefit of dietary enzyme supplementation. *J. Sci. Food Agric.* **1990**, *53*, 175–184. [CrossRef]
40. Scott, R.W. Colorimetric determination of hexuronic acids in plant materials. *Anal. Chem.* **1979**, *51*, 936–941. [CrossRef]
41. Slominski, B.A.; Campbell, L.D. Gas chromatographic determination of indole glucosinolates-a re-examination. *J. Sci. Food Agric.* **1987**, *40*, 131–143. [CrossRef]
42. Smulikowska, S.; Rutkowski, A. (Eds.) Recommended Allowances and Nutritive Value of Feedstuffs. In *Poultry Feeding Standards*, 4th ed.; The Kielanowski Institute of Animal Physiology and Nutrition, PAS: Jablonna, Poland, 2005. (in Polish)
43. Haugh, R.R. The Haugh unit for measuring egg quality. *US Egg Poult. Mag.* **1937**, *43*, 522–555.
44. Folch, J.; Lees, M.; Sloane-Stanley, G.H. A simple method for isolation and purification of total lipides from animal tissues. *J. Biol. Chem.* **1957**, *226*, 497–509. [CrossRef]
45. International Dairy Federation. *Provisional Standard (159): Milk and Milk Fat Product-Determination of Cholesterol Content*; IDF: Brussels, Belgium, 1992.
46. Peisker, K.V. A rapid semi-micro method for preparation of methyl esters from triglycerides using chloroform, methanol, sulphuric acid. *J. Am. Oil. Chem. Soc.* **1964**, *41*, 87–88. [CrossRef]
47. Gay, C.; Gebicki, J.M. Perchloric acid enhances sensitivity and reproducibility of the ferric-xylenol orange peroxide assay. *Anal. Biochem.* **2002**, *304*, 42–46. [CrossRef]
48. Botsoglou, N.A.; Fletouris, D.J.; Papageorgiou, G.E.; Vassilopoulos, V.N.; Mantis, A.J.; Trakatellis, A.G. Rapid sensitive and specific thiobarbituric acid method for measuring lipid peroxidation in animal tissue, food and feed-stuff samples. *J. Agr. Food Chem.* **1994**, *42*, 1931–1937. [CrossRef]
49. Greenwald, R.A. *CRC Handbook of Methods for Oxygen Radical Research*; CRC Press: Boca Raton, FL, USA, 1985.
50. ISO 2012. *Sensory Analysis—General Guidelines for the Selection, Training and Monitoring of Selected Assessors and Expert Sensory Assessors*; International Organization for Standardization: Geneva, Switzerland, 2012.
51. Fenwick, G.R.; Spinks, E.A.; Wilkinson, A.P.; Heany, R.K.; Legoy, M.A. Effect of processing on the antinutrient content of rape seed. *J. Sci. Food Agric.* **1986**, *37*, 735–741. [CrossRef]
52. Goodarzi-Boroojeni, F.; Kozłowski, K.; Jankowski, J.; Senz, M.; Wiśniewska, M.; Boros, D.; Drażbo, A.; Zentek, J. Fermentation and enzymatic treatment of pea for turkey nutrition. *Anim. Feed Sci. Technol.* **2018**, *237*, 78–88. [CrossRef]
53. Huang, S.; Liang, M.; Lardy, G.; Huff, H.E.; Kerley, M.S.; Hsieh, F. Extrusion processing of rapeseed meal for reducing glucosinolates. *Anim. Feed Sci. Technol.* **1995**, *56*, 1–9. [CrossRef]
54. Fazhi, X.; Lvmu, L.; Jiaping, X.; Kun, Q.; Zhide, Z.; Zhangyi, L. Effects of Fermented Rapeseed Meal on Growth Performance and Serum Parameters in Ducks. *Asian-Aust. J. Anim. Sci.* **2011**, *24*, 678–684. [CrossRef]
55. Al-Asheh, S.; Duvnjak, Z. Phytase production and decrease of phytic acid content in canola meal by Aspergillus carbonarius in solid state fermentation. *World J. Microbiol. Biotechnol.* **1995**, *11*, 228–231. [CrossRef]
56. Ebune, A.; Al-Asheh, S.; Duvnjak, Z. Production of phytase during solid state fermentation using Aspergillus ficuum NRRL 3135 in canola meal. *Bioresour. Technol.* **1995**, *53*, 7–12. [CrossRef]
57. Razdan, A.; Pettersson, D.; Pettersson, J. Broiler chickenbody weights, feed intakes, plasma lipid and small-intestinal bileacid concentrations in response to feeding of chitosan and pectin. *Br. J. Nutr.* **1997**, *78*, 283–291. [CrossRef]
58. Leslie, A.; Summers, J. Feeding value of rapeseed for laying hens. *Can. J. Anim.* **1972**, *52*, 563–566. [CrossRef]
59. Najib, H.; Al-Khateeb, S.A. The effect of incorporating different levels of locallyproduced canola seeds (*Brassica napus* L.) in the diet of laying hen. *Int. J. Poult. Sci.* **2004**, *3*, 490–496. [CrossRef]
60. Zhu, L.P.; Wang, J.P.; Ding, X.M.; Bai, S.P.; Zeng, Q.F.; Su, Z.W.; Xuan, Y.; Zhang, K.Y. Effects of dietary rapeseed meal on laying performance, egg quality, apparent metabolic energy, and nutrient digestibility in laying hens. *Livest. Sci.* **2018**, *214*, 265–271. [CrossRef]
61. Jeroch, H.; Dänicke, S.; Zachmann, R. Zum Futterwert und zur Eignung von Rapsexpellern in der Legehennenfütterung. *Agribiol. Res.* **1995**, *48*, 248–256.
62. Jeroch, H.; Jankowski, J.; Schöne, F. Rapsfuttermittel in der Broiler- und Legehennenfütterung. *Arch. Geflügelk.* **2008**, *72*, 49–55.
63. Roberts, J.R. Factors Affecting Egg Internal Quality and Egg Shell Quality in Laying Hens. *J. Poult. Sci.* **2004**, *41*, 161–177. [CrossRef]
64. Riyazi, S.R.; Ebrahimnezhad, Y.; Nazeradl, K.; Maheri-Sis, N.; Salamatdust, R.; Vahdatpour, T. The effects of replacing soybean meal with different levels of rapeseed meal on egg quality characteristics of commercial laying hens. *Asian J. Anim. Vet. Adv.* **2009**, *4*, 337–341. [CrossRef]
65. Świątkiewicz, S.; Koreleski, J.; Arczewska, A. Laying performance and eggshell quality in laying hens fed diets supplemented with prebiotics and organic acids. *Czech J. Anim. Sci.* **2010**, *55*, 294–306. [CrossRef]
66. Zhu, L.P.; Wang, J.P.; Ding, X.M.; Bai, S.P.; Zeng, Q.F.; Su, Z.W.; Xuan, Y.; Applegate, T.J.; Zhang, K.Y. The effects of varieties and levels of rapeseed expeller cake on egg production performance, egg quality, nutrient digestibility, and duodenum morphology in laying hens. *Poult. Sci.* **2019**, *98*, 4942–4953. [CrossRef]

67. Skřivan, M.; Englmaierová, M.; Skřivanová, V. Negative Effect of Phytase Superdosing in Laying Hens. *Czech J. Anim. Sci.* **2018**, *63*, 182–187. [CrossRef]
68. Lichovníková, M.; Zeman, L.; Jandásek, J. The effect of feeding untreated rapeseed and iodine supplement on egg quality. *Czech J. Anim. Sci.* **2008**, *53*, 77–82. [CrossRef]
69. Favé, G.; Coste, T.C.; Armand, M. Physicochemical properties of lipids: New strategies to manage fatty acid bioavailability. *Cell. Mol. Biol.* **2004**, *50*, 815–831. [CrossRef]
70. Tomaszewska, E.; Muszyński, S.; Arczewska-Włosek, A.; Domaradzki, P.; Pyz-Łukasik, R.; Donaldson, J.; Świątkiewicz, S. Cholesterol Content, Fatty Acid Profile and Health Lipid Indices in the Egg Yolk of Eggs from Hens at the End of the Laying Cycle, Following Alpha-Ketoglutarate Supplementation. *Foods* **2021**, *10*, 596. [CrossRef]
71. MacLean, C.H.; Newberry, S.J.; Mojica, W.A.; Khanna, P.; Issa, A.M.; Suttorp, M.J.; Lim, Y.W.; Traina, S.B.; Hilton, L.; Garland, R.; et al. Effects of omega-3 fatty acids on cancer risk: A systematic review. *JAMA* **2006**, *295*, 403–415. [CrossRef]
72. Koppenol, A.; Delezie, E.; Aerts, J.; Willems, E.; Wang, Y.; Franssens, L.; Everaert, N.; Buyse, J. Effect of the ratio of dietary n-3 fatty acids eicosapentaenoic acid and docosahexaenoic acid on broiler breeder performance, egg quality, and yolk fatty acid composition at different breeder ages. *Poult. Sci.* **2014**, *93*, 564–573. [CrossRef] [PubMed]
73. Buckiuniene, V.; Alencikiene, G.; Miezeliene, A.; Raceviciute-Stupeliene, A.; Bliznikas, S.; Gruzauskas, R. Effect of sunflower and rapeseed oil, organic and inorganic selenium and vitamin e in the diet on yolk fatty acids profile, malondialdehydes concentration and sensory quality of laying hens eggs. *Vet. Med. Zoot.* **2018**, *76*, 21–28.
74. Osek, J.; Gallien, P. Molecular analysis of Escherichia coli O157 strains isolated from cattle and pigs by the use of PCR and pulsed-field gel electrophoresis methods. *Vet. Med.—Czech* **2002**, *47*, 149–158. [CrossRef]
75. Kaczmarek, P.; Korniewicz, D.; Lipiński, K.; Mazur, M. Chemical composition of rapeseed products and their use in pig nutrition. *Pol. J. Natur. Sc.* **2016**, *31*, 545–562.
76. Halle, I.; Schöne, F. Influence of rapeseed cake, linseed cake and hemp seed cake on laying performance of hens and fatty acid composition of egg yolk. *J. Verbrauch. Lebensm.* **2013**, *8*, 185–193. [CrossRef]
77. Wang, H.; Zhang, H.J.; Wang, X.C.; Wu, S.G.; Wang, J.; Xu, L.; Qi, G.H. Dietary choline and phospholipid supplementation enhanced docosahexaenoic acid enrichment in egg yolk of laying hens fed a 2% Schizochytrium powder-added diet. *Poult. Sci.* **2017**, *96*, 2786–2794. [CrossRef]
78. Panaite, T.D.; Turcu, R.P.; Soica, C.; Visinescu, P. Nutritional parameters of eggs from laying hens fed with flaxseed meal or mixture with rapeseed meal or rice bran. *J. Appl. Anim. Res.* **2020**, *48*, 566–574. [CrossRef]
79. Yeum, K.J.; Russell, R.M. Carotenoid bioavailability and bioconversion. *Annu. Rev. Nutr.* **2002**, *22*, 483–504. [CrossRef]
80. An, S.Y.; Guo, M.Y.; Ma, S.D.; Yuan, M.J.; Liu, G.Z. Effect of different oil sources and vitamin E in breeder diet on egg quality, hatchability and development of the neonatal offspring. *Asian-Aust. J. Anim. Sci.* **2010**, *23*, 234–239. [CrossRef]
81. Yang, R.L.; Li, W.; Shi, Y.H.; Le, G.W. Lipoic acid prevents high-fat diet-induced dyslipidemia and oxidative stress: A microarray analysis. *Nutrition* **2008**, *24*, 582–588. [CrossRef]
82. Rajput, I.R.; Li, Y.L.; Xu, X.; Huang, Y.; Zhi, W.C.; Yu, D.Y.; Li, W. Supplementary Effects of Saccharomyces boulardii and Bacillus subtilis B10 on Digestive Enzyme Activities, Antioxidation Capacity and Blood Homeostasis in Broiler. *Int. J. Agric. Biol.* **2013**, *15*, 231–237.
83. Ognik, K.; Krauze, M. The potential for using enzymatic assays to assess the health of turkeys. *World's Poul. Sci. J.* **2016**, *72*, 535–550. [CrossRef]
84. Elangovan, A.V.; Verma, S.V.S.; Sastry, V.R.B.; Singh, S.D. Effect of Feeding High Glucosinolate Rapeseed Meal to Laying Japanese Quail. *Asian-Aust, J. Anim. Sci.* **2001**, *14*, 1304–1307. [CrossRef]
85. Goldberg, E.M.; Ryland, D.; Alianj, M.; House, J.D. Interactions between canola meal and flaxseed oil in the diets of White Lohmann hens on fatty acid profile and sensory characteristics of table eggs. *Poult. Sci.* **2016**, *94*, 1805–1812. [CrossRef]
86. Grigorova, S.; Surdjiiska, S.; Gjorgovska, N. Effects of the use of high-protein rapeseed feed on laying productivity and egg quality in japanese quails. *Scientific Pap. Ser. D Anim. Sci.* **2017**, *60*, 85–90.
87. Świątkiewicz, S.; Koreleski, J. Effect of crude glycerin level in the diet of laying hens on egg performance and nutrient utilization. *Poult. Sci.* **2009**, *88*, 615–619. [CrossRef]

animals

MDPI

Article

Fermented Corn–Soybean Meal Mixed Feed Modulates Intestinal Morphology, Barrier Functions and Cecal Microbiota in Laying Hens

Yinglu Liu †, Jia Feng †, Yamin Wang, Jing Lv, Jinghe Li, Lijuan Guo and Yuna Min *

College of Animal Science and Technology, Northwest A&F University, Yangling 712100, China; liuyinglu1996@163.com (Y.L.); fengjiacaas@163.com (J.F.); nwafuyummy@163.com (Y.W.); lvjing96@163.com (J.L.); z13389192603@126.com (J.L.); glj1430151801@163.com (L.G.)

* Correspondence: minyuna2010@nwsuaf.edu.cn; Tel.: +86-29-87092102

† These authors contributed equally to this work.

Simple Summary: Fermented feed has been of wide concern in livestock and poultry production because of its many advantages. In this study, the nutritional quality of the feed before and after fermentation was assessed, and four supplemental levels of fermented feed were used to replace unfermented feed to study the influence of fermented feed on the gut health of the laying hens during the laying peak period. The results suggest that fermented feed can improve the intestinal morphology and barrier functions of laying hens, possibly by altering the cecal microbiome.

Abstract: This study aimed to evaluate the effects of fermented corn–soybean meal mixed feed on intestinal barrier function and cecal microbiota in laying hens. A total of 360 Jingfen No.6 laying hens (22 wk-old) were assigned to 4 dietary treatments, which were offered basal diets (without antibiotics) containing 0, 4, 6 and 8% of fermented mixed feed respectively. The results showed that the pH value and anti-nutritional factor concentrations in fermented mixed feed were lower than those in unfermented feed ($p < 0.05$). Moreover, fermentation in the feed significantly increased the crude protein content ($p < 0.05$). Supplementation with fermented feed significantly reduced the crypt depth and increased the villi height:crypt depth ratio of duodenum and jejunum ($p < 0.05$). Meanwhile, fermented feed increased the secretory immunoglobulin A content and MUC2 mRNA expression of jejunum ($p < 0.05$). These beneficial effects were exhibited at the addition level $\geq$6% and microbial composition of caeca in the control, and so 6% fermented feed groups were analyzed. The structure of the gut microbiota was remarkably altered by additions, characterized by increased abundances of some health-promoting bacteria, such as *Parasutterella*, *Butyricicoccus* and *Erysipelotrichaceae* ($p < 0.05$). In summary, fermented mixed feed modulated cecal flora, subsequently contributing to improvements in intestinal morphology and barrier functions in laying hens.

Keywords: fermented feed; laying hen; cecal microbiota; gut health

Citation: Liu, Y.; Feng, J.; Wang, Y.; Lv, J.; Li, J.; Guo, L.; Min, Y. Fermented Corn–Soybean Meal Mixed Feed Modulates Intestinal Morphology, Barrier Functions and Cecal Microbiota in Laying Hens. *Animals* **2021**, *11*, 3059. https://doi.org/10.3390/ani11113059

Academic Editors: Natalie Morgan and Youssef A. Attia

Received: 11 October 2021
Accepted: 20 October 2021
Published: 26 October 2021

1. Introduction

To achieve resistance-free breeding, a growing number of studies has been conducted to evaluate non-antibiotic alternatives and their effects on the health and production performance of animals [1]. Fermented feed has attracted wide attention in livestock and poultry breeding because of their potential to improve the nutritional quality of feedstuffs by increasing nutrient bioavailability and reducing feed costs [2,3]. Previous research has demonstrated that fermentation may enhance the bioavailability of feed nutrients by the following processes: (1) increasing crude protein content [4], (2) decreasing fiber content, (3) enhancing the utilization of vitamins, (4) improving amino acid patterns and protein solubility [5] and (5) degrading anti-nutritional factors with enzymes, such as phytase, xylanase, cellulase, and glucanase enzymes [6]. Besides this, the probiotics and metabolites produced during the fermentation process could reduce the colonization of gut pathogenic microorganisms via competitive exclusion and the secretion of antibacterial substances (like bacteriocin), and thus exert beneficial effects on animals [7].

Gut integrity and function are essential factors in maintaining animal health and production performance. Intestinal morphology is an important criterion for evaluating intestinal health of animals. Villus height and crypt depth were related to nutrient digestion and poultry performance [8]. The composition and diversity of intestinal microorganisms also have a profound impact on the gut health of poultry. Colonization of harmful bacteria will disrupt the intestinal microbial balance, and then affect the host's physiology, immunity, and nutritional metabolism [9]. Probiotic fermentation has been regarded as an effective method to enhance gut health [10] and it has been widely used in pig rations for several years [11]. Recently, there have been more studies on the utilization of fermented products in poultry industry, particularly focusing on gut health and production parameters of birds. It was reported that the use of fermented feed had a positive impact on the production performance and egg quality of 16 wk Babcock pullets [12]. Semjon et al. [13] observed that fermented wheat bran supplementation could improve broiler performance and meat quality. Fermented feed has been demonstrated to be beneficial to the maintenance of gut microbial ecosystems and intestinal morphology, possibly due to low pH, elevated numbers of probiotics, high short-chain fatty acid concentrations and reduced pathogens [14,15]. In addition, fermented feed could modulate gut microbiota by providing energy and nutrients to probiotics in the microbial community [16]. Therefore, fermented feed was speculated to exert beneficial effects on gut health by altering intestinal microbial composition and subsequently contributed to improvements in laying hen performance. However, the effects of fermented feed on intestinal community and gut health of laying hens remain unclear.

In China, fermented feed mainly refers to the fermented single feed material, such as soybean meal [17], rapeseed meal [15] and cottonseed meal [18]. However, there were few studies evaluating the effectiveness of corn–soybean meal mixed feed in laying hens. Based on the results of previous studies, we hypothesized that a certain concentration of fermented corn–soybean mixed feed could improve the gut health of laying hens. Therefore, the present study was carried out to assess the physicochemical characteristics of mixed feed following fermentation and then to investigate the effects of graded levels of fermented feed on gut morphology and mechanical and immunological barriers, as well as the cecal microbiota of laying hens.

2. Materials and Methods

2.1. Preparation of Fermented Mixed Feed

The probiotic purchased from Baide Biotechnology Co., Ltd. (Shandong, China), was a lyophilized powder containing *Bacillus* 2×10^9 CFU/g, *Lactobacillus* 3×10^9 CFU/g, *Saccharomyces cerevisiae* 5×10^8 CFU/g. The probiotic powder was dissolved in 1 L sterile water at 37 °C, and stirred evenly to make a probiotic solution. The basal substrate (12 kg) included 60% corn, 20% soybean meal and 20% wheat bran, which was mixed and inoculated in probiotics with probiotic solution. The fermentation process according to the method of Shi et al. [19], with appropriate adjustments made to adapt to the actual situation of the experimental farm. Sterile water was added to the mixed substrate to reach a moisture content of 30%, and aerobic fermentation was carried out in a fermenter vessel at 37 °C for 24 h. After the first stage, a mixture of aerobic fermented mixture was transferred to a plastic drum equipped with a gas-pressure opening valve for 37 °C anaerobic fermentation, then fermented under anaerobic conditions at 37 °C for 5 days (the second stage of fermentation). The same proportion of 37 °C sterile water was added to the unfermented mixture and sealed for 5 days as control. Fermented feed was produced every week, and the feed intake of laying hens was measured once a week to adjust the amount of fermented feed.

2.2. Chemical Analysis of Fermented Mixed Feed

Fermented and unfermented feed were collected and dried at 65 °C for 48 h. Feed samples (n = 3) were tested to determine their crude fiber (CF; AOAC #978.10), crude protein (CP; AOAC #984.13) and ether extract (EE; AOAC #2003.05) contents according to AOAC International guidelines [20]. Phytic acid levels were determined as in a previous study [21]. Trypsin inhibitor and β-glucan in fermented and unfermented mixture were tested using a commercial kit (Jianglai Bio Company, Shanghai, China). To determine pH, 5 g of fermented and unfermented mixed feed were dissolved in 50 mL distilled water. After centrifuging at 4000× g for 5 min, the supernatant pH was tested using a probe style-pH meter (H170 Hach pH meter, Hach, Loveland, CO, USA). Protein extraction and SDS-polyacrylamide gel electrophoresis (PAGE) of the feed samples were performed according to the method described in reference [22].

2.3. Birds, Housing and Dietary Treatments

The trial was carried out in the non-antibiotic breeding demonstration plant of Chunmanyuan Farm (Tongchuan, Shaanxi, China). A total of 360 Jingfen No.6 layers aged 22 weeks were randomly assigned to four numerically equal groups, with 6 replicates per treatment and 15 birds per replicate. The layers had a similar body weight and good health. The feeding period lasted 10 weeks, commencing when the layers were 25 weeks of age and ending when they were 35 weeks of age, with an addition of 3 weeks for feed adaptation. During the experiment, three laying hens were kept in each cage and had ad libitum access to feed and water. Proper indoor temperature (15~22 °C) and humidity (30%~50%) were maintained.

According to the NRC (1994) layer feeding standard, the corn–soybean meal basal diet (antibiotic-free) was designed based on the actual situation of layer feeding in the experimental chicken farm. Unfermented corn, soybean meal, and wheat bran were replaced in the basic diet with increasing levels of fermented mixed feed (0, 4%, 6% and 8%; F0, F4, F6 and F8, respectively). The ingredients were mixed in a mixer for 15 min. Feed composition and nutrient content of the experimental diets are shown in Table 1.

Table 1. Composition and nutrient levels of experimental diets (air-dry basis, %).

Items	Group			
	0	4%	6%	8%
Ingredients				
Corn	56.00	56.00	56.00	56.00
Soybean meal	20.60	20.60	20.60	20.60
Wheat bran	2.40	2.40	2.40	2.40
Fermented feed	0.00	4.00	6.00	8.00
Unfermented feed	8.00	4.00	2.00	0.00
Stone powder	7.50	7.50	7.50	7.50
Soybean oil	0.50	0.50	0.50	0.50
Premix [1]	5.00	5.00	5.00	5.00
Nutrient levels [2]				
ME(MJ/kg)	12.09	12.09	12.09	12.09
Crude protein	15.99	16.00	16.01	16.01
Calcium	3.60	3.60	3.60	3.60
Total phosphorus	0.42	0.42	0.42	0.42
Lys	0.94	0.94	0.94	0.94
Met	0.44	0.44	0.44	0.44
Thr	0.70	0.70	0.70	0.70

[1] The premix composed of: Vitamin A 10,000 IU, Vitamin D 31, 800 IU, Vitamin E 10 IU, Vitamin K 10 mg, Vitamin B 125 ug, Vitamin B1 1 mg, Vitamin B24.5 mg, calcium pantothenate 50 mg, niacin 24.5 mg, pyridoxine 5 mg, biotin 1 mg, folic acid 1 mg, choline 500 mg, iodine 0.4 mg, ferrum 80 mg, copper 8 mg, selenium 0.3 mg. [2] Crude protein of the 0 group was measured, and the rest was calculated.

2.4. Sample Collection

At 35 wk, 24 chickens (one with an average body weight from each replicate pen) were selected to collect samples. All birds were killed by cervical dislocation and small intestinal (duodenum, jejunum and ileum) mucosa was scraped off at the forepart of individual small intestinal segments with a glass microscope slide on ice and frozen and stored immediately at −80 °C liquid nitrogen tanks for further analysis. About 3 cm of the mid-portion of the small intestines were excised carefully and washed in saline solution before storing in formalin (10%). The cecal contents were gathered using sterile spatulas into sterile plastic tubes, immediately refrigerated (maximum 2 h) and stored at −80 °C refrigeration until the DNA extraction.

2.5. Morphological Investigations

Three-centimeter lengths from the medial portions of duodenum, jejunum and ileum from chickens were fixed with 10% formalin for 24 h. Tissues were later embedded in paraffin wax blocks, mounted onto glass slides, and then stained with Haematoxylin & Eosin (H&E). The fixed segments were sectioned and observed under the light microscope, and the villus height VD, crypt depth CD and the ratio of VD and CD evaluated.

2.6. Enzyme-Linked Immunosorbent Assay (ELISA) and Real-Time Polymerase Chain Reaction (PCR)

Secretory immunoglobulin A (sIgA) content of mucous membrane samples of the duodenum, jejunum, and ileum were measured by ELISA kits (Cloud-Clone Crop Biological Technology Co., Ltd., Wuhan, China) according to the kit instructions. Total RNA was extracted from the snap-frozen jejunal tissue samples with a RNeasy mini kit (Qiagen, Germantown, MD, USA), following the instructions. Complementary DNA (cDNA) was synthesized from 1 µg of total RNA using MMLV reverse transcriptase (TaKaRa, Dalian, China). Then, the transcriptional changes were identified by quantitative PCR using Premix Ex TaqTM with SYBR Green (TaKaRa, Dalian, China) and the Bio-Rad CFX 96™ Real-Time Detection System (Bio-Rad Laboratories, Richmond, CA, USA). The thermocycle protocol lasted for 30 s at 94 °C, followed by 40 cycles of 5 s denaturation at 94 °C, 34 s annealing/extension at 60 °C, and then a final melting curve analysis to monitor the purity of the PCR product. The $2-\Delta\Delta Ct$ method was used to estimate mRNA abundance (Livak et al., 2002). Relative gene expression levels were normalized using β-actin as an internal control. The primers were synthesized by the Xi'an Qingke Biological Company, and the sequences are shown in Table 2.

Table 2. Gene name, primer sequences.

Gene	Primers Sequence (5′–3′)
β-actin	F: ACACCCACACCCCTGTGATGAA R: TGCTGCTGACACCTTCACCATTC
ZO-1	F: TATAGAAGATCGTGCCGCCTCC R: GAGGTCTGCCATCGTAGCTC
Occludin	F: ACAGCCCTCAATACCAGGATGTG R: ACCATGCGCTTGATGTGGAA
MUC2	F: TTCATGATGCCTGCTCTTGTG R: CCTGAGCCTTGGTACATTCTTGT

F = forward primer; R = reverse primer. .

2.7. 16S rRNA Gene Sequencing

Total genomic DNA was extracted from samples using the CTAB/SDS method and then stored at −80 °C until sequencing analysis. The V3 + V4 fragments of the 16S rRNA gene were amplified using the Phusion® High-Fidelity PCR Master Mix (New England Biolabs, Cambridge, MA, USA) and primers F341 and R806. The 30 µL reaction system was used in PCR reactions with 15 µL Mix, 0.2 M of forward and reverse primers, and about 10 ng template DNA. Thermal cycling consisted of initial denaturation at 98 °C for 1 min, followed by 30 cycles of denaturation at 98 °C for 10 s, annealing at 50 °C for 30 s, elongation at 72 °C for 30 s and a final elongation at 72 °C for 5 min. The amplicons were examined using 2% agarose gel electrophoresis, and then the GeneJET Gel Extraction Kit was used (Thermo Fisher Scientific, Waltham, MA, USA) to purify the excised target fragments. Finally, 16S rRNA gene sequencing was performed using the Illumina NovaSeq 6000 PE250 (Illumina, Santa Clara, CA, USA) with the MiSeq Reagent Kit from Novogene Bioinformatics Technology Co., Ltd., Beijing, China.

2.8. Statistical Analysis

Alpha diversity (Shannon and Simpson indices) and abundance (Chao1 and ACE indices) were analyzed withQiime v.1.7.1. (http://qiime.org/index.html). Data regarding alpha diversity indices, intestinal morphology, total sIgA concentration and relative mRNA expression levels among all treatments in this study were analyzed by one-way ANOVA with SPSS 22.0. Differences among groups means were determined by Duncan's multiple comparison test. The significance of differentiations in microbial structure among groups was assessed by ANOSIM using R package "vegan". *t*-test was used to compare the chemical composition of feed and cecal microflora at the phylum and genus level. The correlations between the cecal microbial composition and gut health were assessed by Spearman's correlation analysis using GraphPad Prism version 8.00. $p < 0.05$ was considered statistically significant and $0.05 < p \leq 0.10$ a trend.

3. Results

3.1. Chemical Composition of Fermented Mixed Feed

The chemical composition of unfermented mixed feed and fermented mixed feed are summarized in Table 3. The pH value and phytic acid, trypsin inhibitor and β-glucan concentrations in fermented mixed feed were lower than in unfermented mixed feed. Moreover, the fermentation of feed significantly increased the crude protein content. Furthermore, there were no significant differences of the total ether extract and crude fiber among the two different kinds of feed. As shown in Figure 1, unfermented mixed feed contained greater amounts of large- (>60 kDa) and medium-size (20–60 kDa) peptide than fermented mixed feed.

Table 3. Chemical composition of unfermented mixture and fermented mixture.

	Unfermented Mixed Feed	Fermented Mixed Feed	SEM	*p*-Value
pH	6.23 ^{a}	4.49 ^{b}	0.40	<0.01
Crude protein, %	15.90 ^{b}	16.18 ^{a}	0.07	0.016
Crude fiber, %	3.33	3.20	2.80	0.680
Ether extract, %	0.57	1.09	0.15	0.060
Phytic acid, %	0.65 ^{a}	0.37 ^{b}	0.07	0.025
Trypsin inhibitor, µg/g	359.29 ^{a}	216.39 ^{b}	32.04	<0.01
β-glucan, µg/g	1588.89 ^{a}	1204.32 ^{b}	86.35	<0.01

Composition of fermented mixture: corn 60%, soybean meal 20%, wheat bran 20%;. ^{a,b} There are statistically significant differences in the mean values per line for different superscripts ($p < 0.05$).

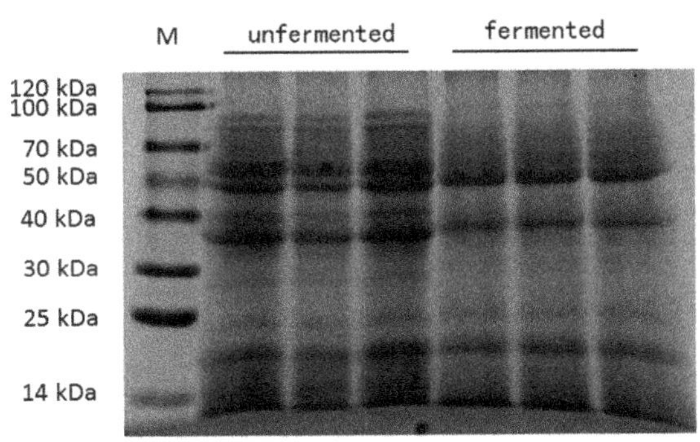

Figure 1. SDS-PAGE map of unfermented mixed feed and fermented mixed feed. M = protein marker.

3.2. Intestinal Morphology

The measurement of intestinal morphology is a common way to judge the integrity and function of the intestinal barrier. The results on villi height (VH), crypt depth (CD) and the villi height:crypt depth ratio (VH/CD) are displayed in Table 4. In the duodenum, three levels of fermented mixed feed significantly decreased the CD and increased the VH/CD ratio. In the jejunum, dietary supplementation with fermented mixed feed significantly increased the VH and VH/CD ratio compared to the control group. However, the 4% fermented mixed feed group significantly increased CD compared with all other treatments. In the ileum, no significant effect of dietary fermented mixture supplementation was observed on the VH, CD and VH/CD ratio.

Table 4. Effect of fermented mixed feed on intestinal morphology in laying hens.

	Fermented Mixed Feed				SEM	p-Value
	0%	4%	6%	8%		
Duodenum						
VH [1], μm	1767.04	1731.68	1774.26	1753.69	21.83	0.889
CD [2], μm	294.56 [a]	256.96 [b]	254.26 [b]	221.31 [c]	4.24	<0.01
VH/CD [3], μm/μm	6.18 [bc]	6.86 [b]	7.16 [b]	8.02 [a]	0.13	<0.01
Jejunum						
VH, μm	922.89 [c]	1315.79 [a]	1079.89 [b]	1121.96 [b]	19.41	<0.01
CD, μm	177.47 [b]	222.91 [a]	180.14 [b]	169.85 [b]	3.94	<0.01
VH/CD, μm/μm	5.28 [b]	6.23 [a]	6.28 [a]	6.88 [a]	0.12	<0.01
Ileum						
VH, μm	882.98	936.10	968.38	961.42	15.84	0.208
CD, μm	136.40	145.14	145.96	138.04	1.97	0.201
VH/CD, μm/μm	6.57	6.64	6.63	7.13	0.11	0.241

[1] VH: villi height. [2] CD: crypt depth. [3] VH/CD: villi height: crypt depth ratio. [a–c] There are statistically significant differences in the mean values per line for different superscripts ($p < 0.05$).

3.3. Total sIgA Concentration and Physical Barrier mRNA Abundance in the Intestinal Mucosa

Secretory IgA (sIgA) acts as the first immune defense for intestinal epithelium and maintains the homeostasis of the gut. Therefore, we assessed the intestinal immune function by measuring the intestinal sIgA contents. As shown in Figure 2a, no statistical differences in sIgA content were observed in the duodenum and ileum. However, the jejunal sIgA content was significantly higher in the 6% and 8% fermented mixed feed group compared to the other treatments. In the presence of intact epithelial cell layers, intercellular paracellular pathways must be closed. This function is effectuated through physical barriers—especially by tight junctions [23]. In addition, mucin 2 (MUC2) is the most abundant mucin, which creates the first defense line against invading microorganisms [24]. Thus, we focused on the mRNA expression levels of zonula occludens 1 (ZO-1), occludin (OCLN) and MUC2 in jejunal mucosa to further explore the physical barriers of laying hens. Figure 2b showed that fermented mixed feed at three levels all significantly increased MUC2 gene expression compared with the control group, whereas 4% fermented mixed feed significantly decreased the mRNA expression of ZO-1.

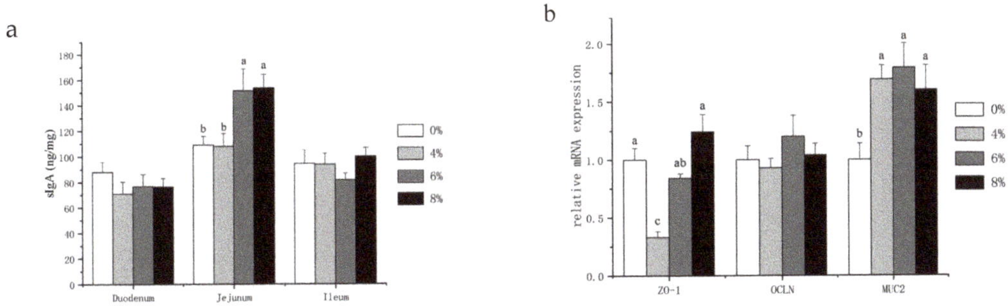

Figure 2. Effects of fermented mixture supplementation on the intestinal sIgA content (**a**) and the physical barrier in the jejunum (**b**) of laying hens. Data are expressed as mean ± SEM of 6 replicates per treatment; $^{a–c}$ There are statistically significant differences in the mean values per line for different superscripts ($p < 0.05$).

3.4. Cecal Microbial Diversity and Community

As the microbial community plays an important role in the intestinal health and barrier function of laying hens, the composition of the cecal microbial community was analyzed using the 16S rRNA gene amplicon sequence. After filtering, an average of 60,788 ± 6481 reads were obtained for each sample. In the 16S amplicon data analysis, the Shannon and Simpson indices were used to assess the community diversity, and the ACE and Chao1 indices reflected the community richness. Therefore, alpha diversity was measured using these four indices, and there were no significant differences in cecal feces microbiome taxon abundance and diversity ($p > 0.05$; Table 5) among the groups, indicating that fermented mixed feed did not change the alpha diversity of the microbiota of the chicken cecum. Beta diversity was illustrated via principal component analysis (PCA) in Figure 3, showing fermented mixed feed treatment significantly affected cecal microbiota composition. The composition of the microbiota was similar between the 6% and 8% fermented mixed feed group. ANOSIM results, shown in Table 6, indicated that the variations of the inter-group microbiota composition of the 6% and 8% group were considered significant ($p < 0.05$) to the control group and were larger than those of the inner-group (R > 0.05). Hence, to illuminate how fermented mixed feed served to improve the gut health of laying hens, the 6% group was selected as the representative group among the 3 treated groups to identify the roles of fermented mixed feed in regulating the cecal microbiota.

Since pairwise comparisons (0 and 6%) and the data were in accordance with normal distribution, *t*-test analysis was used to evaluate the differential bacteria (relative abundance > 0.1%) on the phylum and genus (Figure 4). Compared with the basal diet, at the phylum level, supplementation with 6% fermented mixed feed significantly improved the *Tenericutes* abundance but reduced the abundance of *Actinobacteria*; at the genus levels, the abundance of *Parasutterella, Butyricicoccus, unidentified_Erysipelotrichaceae* and *Mailhella* were found to be significantly increased and *Alloprevotella, Gallibacterium, Romboutsia* ($p < 0.05$) and *Enterococcus* were significantly decreased in the 6% fermentation group when compared with the 0% control group ($p < 0.05$).

Table 5. Effect of fermented mixed feed on α-diversity of cecal microflora in layers.

	Fermented Mixed Feed				SEM	*p*-Value
	0%	4%	6%	8%		
Chao1	693.32	682.26	712.44	731.14	10.61	0.398
Ace	702.49	694.70	723.97	740.77	10.52	0.419
Shannon	6.42	5.90	6.41	6.68	0.12	0.137
Simpson	0.97	0.94	0.96	0.97	0.01	0.287

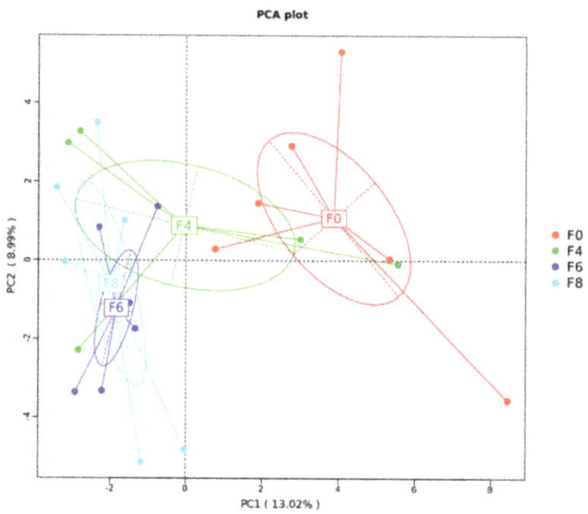

Figure 3. Comparison of the compositions of the cecal microbiota by principal component analysis (PCA). F0, basal diet; F4, 4% fermented mixed feed; F6, 6% fermented mixed feed; F8, 8% fermented mixed feed.

Table 6. Comparison of similarities in microbiota composition between the three treatments by ANOSIM analysis.

Treatment	R-Value	*p*-Value
F0–F4	0.07407	0.4
F0–F8	0.6	0.016
F4–F8	0.4769	0.025
F0–F6	0.3457	0.045
F4–F6	0.2099	0.214
F6–F8	−0.02667	0.563

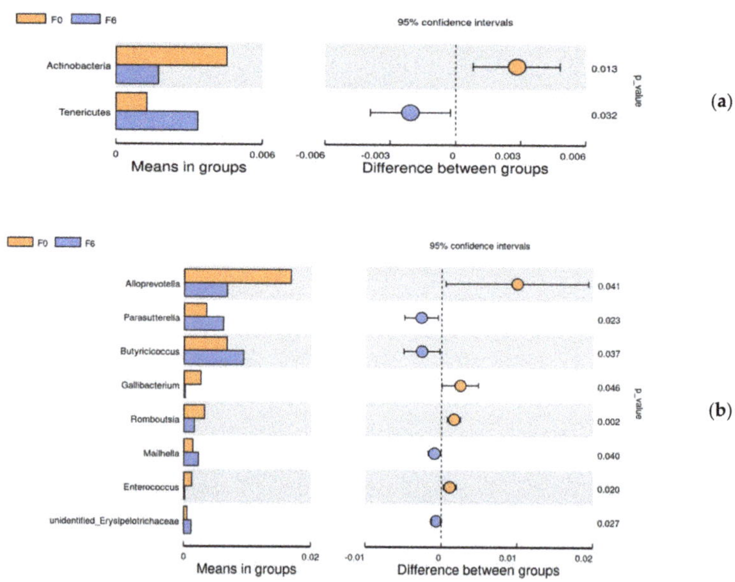

Figure 4. The phylum (**a**) and genera (**b**) differentially abundant between F0 and F6 in the cecal by *t*-test analysis. F0, basal diet; F6, 6% fermented mixed feed.

3.5. Correlations between Microbiota and Gut Health

In order to explore the specific bacteria related to gut health, the Spearman correlation coefficient (Figure 5) was used to analyze the correlation between the abundance of the cecal microbiota and the intestinal barrier. This analysis can identify species that are significantly correlated with certain environmental factors. The heatmap revealed significant positive correlations between intestinal morphology (higher villi height VH, shorter crypt depth CD and lager villi height:crypt depth VH/CD indicate superior development) and *unidentified_Lachnospiraceae*, *unidentified_Spirochaetaceae*, *Barnesiella Helicobacter*, *Parasutterella* and *Synergistes*. In contrast, the intestinal morphology was negatively correlated with *Bacteroides*, *Megamonas*, *Desulfovibrio*, *Alistipes*, *Tyzzerella*, *Fournierella Succinatimonas*, *Gallibacterium* and *Elusimicrobium* significantly. In addition, the abundance of genus *Faecalibacterium* and *Desulfovibrio* were negatively correlated with the jejunal sIgA contents, and the abundance of genus *Sutterella* and *Lachnoclostridium* were negatively correlated with the duodenal sIgA content. Moreover, *Fournierella* was positively correlated with ileal sIgA contents.

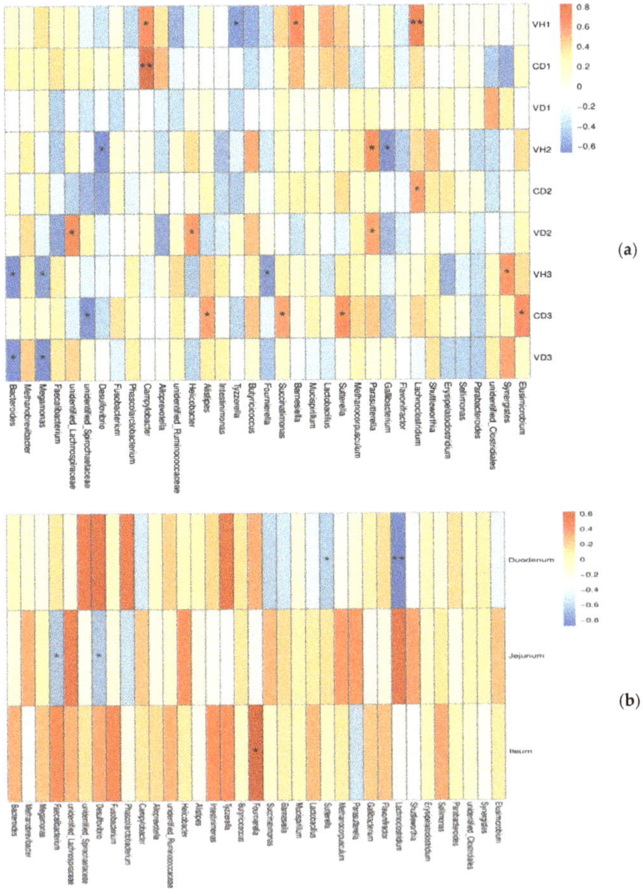

Figure 5. Heatmap of spearman's correlation between the microbiota from chicken cecum and gut health. (**a**) cecal microbiota and intestinal morphology; (**b**) cecal microbiota and sIgA contents of intestine. * $p < 0.05$, ** $p < 0.01$. Red and blue cells indicate positive and negative correlations, respectively. Color intensity is in proportion to magnitude.

4. Discussion

The corn–soybean meal diet is the most widely applied in poultry production in China. However, ordinary corn–soybean meal diets contain some anti-nutritional factors, for instance, phytic acid, soybean antigenic protein and soy oligosaccharides, which may compromise the nutrient bioavailability and exhibit negative effects on animal health [25]. Solid-state fermentation was proposed to improve the nutritional properties of coarse plant materials and improve their application in animal feeds [26].

Among a variety of parameters, pH is an important reference to estimate the quality of fermented feed, because lower pH may favor the digestion of intestinal nutrients and inhibit the growth of pathogenic microorganisms in feed [27]. Our results showed that the pH of mixed feed following fermentation decreased from 6.23 to 4.49, indicating that the fermented feed has reached a low pH requirement. In addition, compared to the control, the fermented feed contained higher concentrations of crude protein, which was similar to previous research [28]. The increase in CP concentration might be related to the synthesis of microbial protein or the loss of dry matter during fermentation. Furthermore, the decreased contents of phytic acid, trypsin inhibitor and β-glucan in mixed feed were observed after fermentation, which have been considered as the major anti-nutritional factors in the current feed industry negatively affecting nutrient digestion and absorption of laying hens. A decreased amount of phytic acid in fermented feed may be due to the production of phytase by microorganisms [29]. The reduction of β-glucan could be a possible reason for the soluble fraction of the polymer [30]. Probiotics in fermented feed could decompose proteins, including trypsin inhibitors, through the secretion of proteases, which may have a beneficial impact on intestinal health [31]. In addition, SDS-PAGE analysis showed that large-sized protein contents in fermented mixed feed were reduced, consistent with the study of Shi et al. [19]. This indicated that large-sized proteins can be degraded into small-sized proteins or peptides during fermentation, presumptively assisting intestinal protein digestion in laying hens [32].

This study found that fermented mixed feed could partially improve the intestinal morphology of laying hens. This was consistent with the past study of Missotten et al. [33], who noticed higher villi height, lower crypt depth and increased villus crypt ratios in broilers fed fermented moist feed. Similar results were also shown by Li et al. [34] in the duodenum and jejunum of broilers fed 10% fermented soybean meal. Gut morphology parameters, including villus height VH, crypt depth CD, and villus crypt VH/CD ratios, are regarded as gold standards for assessing intestinal health status [8]. The increase of villi height suggested a greater area addressed to the absorption of available nutrients [35], which was conducive to enhanced intestinal function. On the contrary, shortening of villi and deepening of crypts could cause malabsorption of nutrients and consequently compromise the production performance of laying hens [36]. The positive impact of fermented feed on gut morphology is likely attributed to its regulatory roles in gut microbiota balance and microbial metabolites, which can promote enterocyte differentiation and proliferation [37]. It might be also related to the reduction of anti-nutrient factors in the fermented feed, as supported by the negative correlation between trypsin inhibitor in soybean meal and villi height [38]. Surprisingly, fermented mixed feed addition at 4% has opposite effects on jejunum morphology (i.e., the increased villi height and crypt depth) and the exact reasons for this phenomenon need further exploration.

The gut barrier can prevent the intestinal tract from the colonization of pathogens, and sIgA is the critical component of the immune barrier, limiting epithelial contact with pathogens and other antigens [39]. In this study, we observed a significant improvement in jejunum sIgA concentration with the addition of 6% and 8% fermented feed, indicating positive effects of fermented feed on the barrier function of gut mucosa to reduce the adverse effects on gut health. This was in combination with the improvements in jejunum morphology, suggesting that the fermented feed played a role in the intestinal mucosa health in the jejunum of laying hens. Jejunum, the longest part of the small intestine with the longest retention time of nutrients [40], is considered the best segment for exerting favorable effects of fermented feed on gut health. Therefore, the jejunum was chosen for further study. OCLN and ZO are unique proteins that form the extracellular barrier of the gut [41]. This barrier is famous for tight junctions, which make up a wall against invading pathogens in the intestine [42]. MUC2, the main mucin of the intestinal mucosa, is involved in providing nutrients and attachment sites for host bacteria, and it can contribute to selecting species-specific gut microbiota [43]. Little information is available on physical barrier gene expression in birds fed with fermented feed. This study indicated that fermented feed can apparently increase the gene expression of MUC2, suggesting that fermented feed may prevent intestinal epithelial cells from pathogen invasion by modulating MUC2 expression in the jejunum. Fermentation products, including probiotics [44] and organic acids [45], have been proven to significantly improve the expression of poultry gut barrier-related genes, but the mechanism of their action is complex and needs to be studied in depth. However, this study showed no significant effects on the expression of OCLN. In particular, the addition of 4% fermented mixed feed significantly downregulated ZO-1 gene expression, in accordance with the observations that 4% fermented feed significantly increased crypt depth. It is necessary to research an in vivo pathogenic attack model as a further study to confirm the influences of fermented feed on gut barrier functions.

To further explore the underlying mechanism of its modulation on gut health, we have used Illumina MiSeq sequencing to analyze the cecal microbiota. The present study revealed no significant effect on the α-diversity of the cecal microbiota, whereas the β-diversity analysis showed significant clustering between the 6% and 8% groups and controls, indicating that the cecal microorganism community profiles in the 6% and 8% fermented groups could be altered following fermented feed addition. This was in line with the results on the intestinal morphology and the barrier-related gene expression, indicating that the 6% and 8% fermented feed could significantly improve intestinal health with no remarkable difference between the two groups. It was possible that fermented mixed feed at levels above a certain threshold (which was 6% in this study) could result in changes of the gut microflora and subsequently improve intestinal morphology and barrier functions. Therefore, control and 6% groups were selected for further analysis. At the genus level, *Parasutterella*, *Butyricicoccus*, *unidentified_Erysipelotrichaceae* and *Mailhella* were identified as the main microbes with increased abundances in response to the addition of fermented feed. The increased proportion of *Parasutterella* has been reported to be beneficial to intestinal mucosal homeostasis [46]. Similarly, our results also confirmed a positive correlation between *Parasutterella* abundance and villus height VH and VH/CD ratio. *Butyricicoccus*, as butyrate producers, are assumed to improve growth performance, inhibit the proliferation of pathogens and relieve intestinal inflammation in broilers [47]. *Erysipelotrichaceae* may be associated with the degradation of feed ingredients and the production of short-chain fatty acids [48]. On the other hand, the decreased abundances of *Alloprevotella*, *Gallibacterium*, *Romboutsia* and *Enterococcus* were observed in 6% fermented feed group compared to the control. *Gallibacterium* has been recognized as a main cause of peritonitis and salpingitis in laying hens [49], which leads to decreased productive performance. *Alloprevotella* is known as an opportunistic pathogen; however, previous studies have noted that the increased abundance of *Alloprevotella* genera was linked to better intestinal health [50]. Besides this, the genus *Romboutsia* is a valuable intestinal biomarker maintaining host health, and *Enterococcus* with natural antimicrobial probiotic properties could prevent diarrhea in animal production [51]. However, their roles in gut health and functions of birds need to be explored. *Gallibacterium* is directly associated with poultry intestinal disease, which was also confirmed in this study by a negative relationship between abundance and villus height [52]. Therefore, the ameliorated gut morphology and enhanced epithelial barrier functions might be mainly attributed to the increased abundances of some health-promoting bacteria in the fermented feed supplementation group. Further investigations based on metabolomics should be done to detect changes in metabolites caused by fermentation and their effects on gut health in a future study.

5. Conclusions

In conclusion, fermented mixed feed could improve the morphology and barrier functions of the intestine, and alter the cecal microflora. This study demonstrated that fermented mixed feed could be used as a novel feed ingredient for laying hens, and that their favorable effects could be exhibited at addition levels of ≥6%. The specific mechanism of fermented feed-changing of cecal microflora of laying hens needs further study.

Author Contributions: Conceptualization, Y.L.; methodology, Y.L. and J.L. (Jing Lv); formal analysis, Y.L.; investigation, Y.L., Y.W. and J.L. (Jinghe Li); resources, Y.M.; data curation, J.F., and Y.M.; writing—original draft preparation, J.F., Y.L. and L.G.; writing—review and editing, J.F. and Y.L.; supervision, J.F. and Y.M.; project administration, Y.M.; funding acquisition, Y.M. All authors have read and agreed to the published version of the manuscript.

Funding: This research was funded by the China Agriculture Research System of MOF and MARA (CARS-40-S20) and the Province Key Research and Development Program of Shaanxi (2021NY-005).

Institutional Review Board Statement: The animal feeding and experimental protocols in this study were approved by the Institutional Animal Care and Use Committee of Northwest A&F University (protocol number NWAFAC1378).

Data Availability Statement: Informed consent was obtained from all subjects involved in the study. The datasets analyzed in the present study are available from the corresponding author on reasonable request.

Conflicts of Interest: The authors declare no conflict of interest.

References

1. Khalique, A.; Zeng, D.; Shoaib, M.; Wang, H.; Qing, X.; Rajput, D.S.; Pan, K.; Ni, X. Probiotics mitigating subclinical necrotic enteritis (SNE) as potential alternatives to antibiotics in poultry. *AMB Express* **2020**, *10*, 50. [CrossRef] [PubMed]
2. Wang, J.; Han, Y.; Zhao, J.-z.; Zhou, Z.-j.; Fan, H. Consuming fermented distillers' dried grains with solubles (DDGS) feed reveals a shift in the faecal microbiota of growing and fattening pigs using 454 pyrosequencing. *J. Integr. Agric.* **2017**, *16*, 900–910. [CrossRef]
3. Plumed-Ferrer, C.; von Wright, A. Fermented pig liquid feed: Nutritional, safety and regulatory aspects. *J. Appl. Microbiol.* **2009**, *106*, 351–368. [CrossRef] [PubMed]
4. Wang, C.; Shi, C.; Su, W.; Jin, M.; Xu, B.; Hao, L.; Zhang, Y.; Lu, Z.; Wang, F.; Wang, Y.; et al. Dynamics of the Physicochemical Characteristics, Microbiota, and Metabolic Functions of Soybean Meal and Corn Mixed Substrates during Two-Stage Solid-State Fermentation. *mSystems* **2020**, *5*, 501–519. [CrossRef] [PubMed]
5. Wang, T.Y.; Wu, Y.H.; Jiang, C.Y.; Liu, Y. Solid state fermented potato pulp can be used as poultry feed. *Br. Poult. Sci.* **2010**, *51*, 229–234. [CrossRef] [PubMed]
6. Taheri, H.R.; Moravej, H.; Tabandeh, F.; Zaghari, M.; Shivazad, M. Screening of lactic acid bacteria toward their selection as a source of chicken probiotic. *Poult. Sci.* **2009**, *88*, 1586–1593. [CrossRef] [PubMed]
7. Jin, W.; Zhang, Z.; Zhu, K.; Xue, Y.; Xie, F.; Mao, S. Comprehensive Understanding of the Bacterial Populations and Metabolites Profile of Fermented Feed by 16S rRNA Gene Sequencing and Liquid Chromatography-Mass Spectrometry. *Metabolites* **2019**, *9*, 239. [CrossRef]
8. Ducatelle, R.; Goossens, E.; De Meyer, F.; Eeckhaut, V.; Antonissen, G.; Haesebrouck, F.; Van Immerseel, F. Biomarkers for monitoring intestinal health in poultry: Present status and future perspectives. *Vet. Res.* **2018**, *49*, 43. [CrossRef]
9. Hiippala, K.; Jouhten, H.; Ronkainen, A.; Hartikainen, A.; Kainulainen, V.; Jalanka, J.; Satokari, R. The Potential of Gut Commensals in Reinforcing Intestinal Barrier Function and Alleviating Inflammation. *Nutrients* **2018**, *10*, 988. [CrossRef]
10. Sugiharto, S.; Ranjitkar, S. Recent advances in fermented feeds towards improved broiler chicken performance, gastrointestinal tract microecology and immune responses: A review. *Anim. Nutr.* **2019**, *5*, 1–10. [CrossRef] [PubMed]
11. Canibe, N.; Jensen, B.B. Fermented liquid feed-Microbial and nutritional aspects and impact on enteric diseases in pigs. *Anim. Feed Sci. Technol.* **2012**, *173*, 17–40. [CrossRef]
12. Engberg, R.M.; Hammershoj, M.; Johansen, N.F.; Abousekken, M.S.; Steenfeldt, S.; Jensen, B.B. Fermented feed for laying hens: Effects on egg production, egg quality, plumage condition and composition and activity of the intestinal microflora. *Br. Poult Sci.* **2009**, *50*, 228–239. [CrossRef]
13. Semjon, B.; Bartkovsky, M.; Marcincakova, D.; Klempova, T.; Bujnak, L.; Hudak, M.; Jaduttova, I.; Certik, M.; Marcincak, S. Effect of Solid-State Fermented Wheat Bran Supplemented with Agrimony Extract on Growth Performance, Fatty Acid Profile, and Meat Quality of Broiler Chickens. *Animals* **2020**, *10*, 942. [CrossRef] [PubMed]

14. Xie, Y.; Liu, J.; Wang, H.; Luo, J.; Chen, T.; Xi, Q.; Zhang, Y.; Sun, J. Effects of fermented feeds and ginseng polysaccharides on the intestinal morphology and microbiota composition of Xuefeng black-bone chicken. *PLoS ONE* **2020**, *15*, e0237357. [CrossRef] [PubMed]
15. Chiang, G.; Lu, W.Q.; Piao, X.S.; Hu, J.K.; Gong, L.M.; Thacker, P.A. Effects of Feeding Solid-state Fermented Rapeseed Meal on Performance, Nutrient Digestibility, Intestinal Ecology and Intestinal Morphology of Broiler Chickens. *Asian Australas. J. Anim. Sci.* **2009**, *23*, 263–271. [CrossRef]
16. Wang, C.; Shi, C.; Zhang, Y.; Song, D.; Lu, Z.; Wang, Y. Microbiota in fermented feed and swine gut. *Appl. Microbiol. Biotechnol.* **2018**, *102*, 2941–2948. [CrossRef] [PubMed]
17. Mathivanan, R.; Selvaraj, P.; Nanjappan, K. Feeding of Fermented Soybean Meal on Broiler Performance. *Int. J. Poult. Sci.* **2006**, *5*, 868–872. [CrossRef]
18. Zhang, W.-J.; Xu, Z.-R.; Zhao, S.-H.; Sun, J.-Y.; Yang, X. Development of a microbial fermentation process for detoxification of gossypol in cottonseed meal. *Anim. Feed Sci. Technol.* **2007**, *135*, 176–186. [CrossRef]
19. Shi, C.; Zhang, Y.; Yin, Y.; Wang, C.; Lu, Z.; Wang, F.; Feng, J.; Wang, Y. Amino acid and phosphorus digestibility of fermented corn-soybean meal mixed feed with *Bacillus subtilis* and *Enterococcus faecium* fed to pigs. *J. Anim. Sci.* **2017**, *95*, 3996–4004. [CrossRef]
20. AOAC. *Official Methods of Analysis*, 18th ed.; Association of Official Analytical Chemists: Arlington, VA, USA, 2006.
21. Buddrick, O.; Jones, O.A.H.; Cornell, H.J.; Small, D.M. The influence of fermentation processes and cereal grains in wholegrain bread on reducing phytate content. *J. Cereal. Sci.* **2014**, *59*, 3–8. [CrossRef]
22. Hong, K.J.; Lee, C.H.; Kim, S.W. *Aspergillus oryzae* GB-107 fermentation improves nutritional quality of food soybeans and feed soybean meals. *J. Med. Food* **2004**, *7*, 430–435. [CrossRef]
23. Anderson, J.M.; Van Itallie, C.M.; Fanning, A.S. Setting up a selective barrier at the apical junction complex. *Curr. Opin. Cell Biol.* **2004**, *16*, 140–145. [CrossRef] [PubMed]
24. Peterson, L.W.; Artis, D. Intestinal epithelial cells: Regulators of barrier function and immune homeostasis. *Nat. Rev. Immunol.* **2014**, *14*, 141–153. [CrossRef] [PubMed]
25. Gu, C.; Pan, H.; Sun, Z.; Qin, G. Effect of soybean variety on anti-nutritional factors content, and growth performance and nutrients metabolism in rat. *Int. J. Mol. Sci.* **2010**, *11*, 1048–1056. [CrossRef] [PubMed]
26. Chi, C.-H.; Cho, S.-J. Improvement of bioactivity of soybean meal by solid-state fermentation with *Bacillus amyloliquefaciens* versus *Lactobacillus* spp. and *Saccharomyces cerevisiae*. *LWT Food Sci. Technol.* **2016**, *68*, 619–625. [CrossRef]
27. Varsha, K.K.; Priya, S.; Devendra, L.; Nampoothiri, K.M. Control of spoilage fungi by protective lactic acid bacteria displaying probiotic properties. *Appl. Biochem. Biotechnol.* **2014**, *172*, 3402–3413. [CrossRef]
28. Shi, C.; Zhang, Y.; Lu, Z.; Wang, Y. Solid-state fermentation of corn-soybean meal mixed feed with *Bacillus subtilis* and *Enterococcus faecium* for degrading antinutritional factors and enhancing nutritional value. *J. Anim. Sci. Biotechnol.* **2017**, *8*, 50. [CrossRef] [PubMed]
29. Sokrab, A.M.; Mohamed Ahmed, I.A.; Babiker, E.E. Effect of fermentation on antinutrients, and total and extractable minerals of high and low phytate corn genotypes. *J. Food Sci. Technol.* **2014**, *51*, 2608–2615. [CrossRef] [PubMed]
30. Livak, K.J.; Schmittgen, T.D. Analysis of relative gene expression data using real-time quantitative PCR and the 2(-Delta Delta C(T)) Method. *Methods* **2001**, *25*, 402–408. [CrossRef] [PubMed]
31. Hu, J.; Lu, W.; Wang, C.; Zhu, R.; Qiao, J. Characteristics of Solid-state Fermented Feed and its Effects on Performance and Nutrient Digestibility in Growing-finishing Pigs. *Asian Australs. J. Anim. Sci.* **2008**, *21*, 1635–1641. [CrossRef]
32. Silk, D.B.; Chung, Y.C.; Berger, K.L.; Conley, K.; Beigler, M.; Sleisenger, M.H.; Spiller, G.A.; Kim, Y.S. Comparison of oral feeding of peptide and amino acid meals to normal human subjects. *Gut* **1979**, *20*, 291–299. [CrossRef]
33. Missotten, J.A.; Michiels, J.; Dierick, N.; Ovyn, A.; Akbarian, A.; De Smet, S. Effect of fermented moist feed on performance, gut bacteria and gut histo-morphology in broilers. *Br. Poult. Sci.* **2013**, *54*, 627–634. [CrossRef]
34. Li, L.; Li, W.F.; Liu, S.Z.; Wang, H.H. Probiotic fermented feed improved the production, health and nutrient utilisation of yellow-feathered broilers reared in high altitude in Tibet. *Br. Poult. Sci.* **2020**, *61*, 746–753. [CrossRef]
35. Caspary, W.F. Physiology and pathophysiology of intestinal absorption. *Am. J. Clin. Nutr.* **1992**, *55*, 299S–308S. [CrossRef] [PubMed]
36. Xu, Z.R.; Hu, C.H.; Xia, M.S.; Zhan, X.A.; Wang, M.Q. Effects of dietary fructooligosaccharide on digestive enzyme activities, intestinal microflora and morphology of male broilers. *Poult. Sci.* **2003**, *82*, 1030–1036. [CrossRef]
37. Jazi, V.; Foroozandeh, A.D.; Toghyani, M.; Dastar, B.; Rezaie Koochaksaraie, R.; Toghyani, M. Effects of *Pediococcus acidilactici*, mannan-oligosaccharide, butyric acid and their combination on growth performance and intestinal health in young broiler chickens challenged with Salmonella Typhimurium. *Poult. Sci.* **2018**, *97*, 2034–2043. [CrossRef]
38. Zarkadas, L.N.; Wiseman, J. Influence of processing of full fat soya beans included in diets for piglets. *Anim. Feed. Sci. Technol.* **2005**, *118*, 121–137. [CrossRef]
39. Kumar, N.; Arthur, C.P.; Ciferri, C.; Matsumoto, M.L. Structure of the secretory immunoglobulin A core. *Science* **2020**, *367*, 1008–1014. [CrossRef] [PubMed]
40. Mitjans, M.; Barniol, G.; Ferrer, R. Mucosal surface area in chicken small intestine during development. *Cell Tissue Res.* **1997**, *290*, 71–78. [CrossRef] [PubMed]

41. Schneeberger, E.E.; Lynch, R.D. The tight junction: A multifunctional complex. *Am. J. Physiol. Cell Physiol.* **2004**, *286*, C1213–C1228. [CrossRef]
42. Awad, W.A.; Hess, C.; Hess, M. Enteric Pathogens and Their Toxin-Induced Disruption of the Intestinal Barrier through Alteration of Tight Junctions in Chickens. *Toxins* **2017**, *9*, 60. [CrossRef] [PubMed]
43. Johansson, M.E.; Larsson, J.M.; Hansson, G.C. The two mucus layers of colon are organized by the MUC2 mucin, whereas the outer layer is a legislator of host-microbial interactions. *Proc. Natl. Acad. Sci. USA* **2011**, *108* (Suppl. S1), 4659–4665. [CrossRef] [PubMed]
44. Gharib-Naseri, K.; de Paula Dorigam, J.C.; Doranalli, K.; Kheravii, S.; Swick, R.A.; Choct, M.; Wu, S.B. Modulations of genes related to gut integrity, apoptosis, and immunity underlie the beneficial effects of *Bacillus amyloliquefaciens* CECT 5940 in broilers fed diets with different protein levels in a necrotic enteritis challenge model. *J. Anim. Sci. Biotechnol.* **2020**, *11*, 104. [CrossRef]
45. Pham, V.H.; Kan, L.; Huang, J.; Geng, Y.; Zhen, W.; Guo, Y.; Abbas, W.; Wang, Z. Dietary encapsulated essential oils and organic acids mixture improves gut health in broiler chickens challenged with necrotic enteritis. *J. Anim. Sci. Biotechnol.* **2020**, *11*, 18. [CrossRef]
46. Ju, T.; Kong, J.Y.; Stothard, P.; Willing, B.P. Defining the role of Parasutterella, a previously uncharacterized member of the core gut microbiota. *ISME J.* **2019**, *13*, 1520–1534. [CrossRef] [PubMed]
47. Eeckhaut, V.; Wang, J.; Van Parys, A.; Haesebrouck, F.; Joossens, M.; Falony, G.; Raes, J.; Ducatelle, R.; Van Immerseel, F. The Probiotic *Butyricicoccus pullicaecorum* Reduces Feed Conversion and Protects from Potentially Harmful Intestinal Microorganisms and Necrotic Enteritis in Broilers. *Front. Microbiol.* **2016**, *7*, 1416. [CrossRef]
48. Stanley, D.; Denman, S.E.; Hughes, R.J.; Geier, M.S.; Crowley, T.M.; Chen, H.; Haring, V.R.; Moore, R.J. Intestinal microbiota associated with differential feed conversion efficiency in chickens. *Appl. Microbiol. Biotechnol.* **2012**, *96*, 1361–1369. [CrossRef] [PubMed]
49. Neubauer, C.; De Souza-Pilz, M.; Bojesen, A.M.; Bisgaard, M.; Hess, M. Tissue distribution of haemolytic *Gallibacterium anatis* isolates in laying birds with reproductive disorders. *Avian Pathol.* **2009**, *38*, 1–7. [CrossRef] [PubMed]
50. Downes, J.; Dewhirst, F.E.; Tanner, A.C.R.; Wade, W.G. Description of *Alloprevotella rava* gen. nov., sp. nov., isolated from the human oral cavity, and reclassification of *Prevotella tannerae* Moore et al. 1994 as *Alloprevotella tannerae* gen. nov., comb. nov. *Int. J. Syst. Evol. Microbiol.* **2013**, *63*, 1214–1218. [CrossRef] [PubMed]
51. Feng, J.; Lu, M.; Wang, J.; Zhang, H.; Qiu, K.; Qi, G.; Wu, S. Dietary oregano essential oil supplementation improves intestinal functions and alters gut microbiota in late-phase laying hens. *J. Anim. Sci. Biotechnol.* **2021**, *12*, 72. [CrossRef]
52. Wiersema, M.L.; Koester, L.R.; Schmitz-Esser, S.; Koltes, D.A. Comparison of intestinal permeability, morphology, and ileal microbial communities of commercial hens housed in conventional cages and cage-free housing systems. *Poult. Sci.* **2021**, *100*, 1178–1191. [CrossRef] [PubMed]

animals

MDPI

Article

Limitation and Potential Effects of Different Levels of Aging Corn on Performance, Antioxidative Capacity, Intestinal Health, and Microbiota in Broiler Chickens

Ahsan Mustafa, Shiping Bai, Qiufeng Zeng, Xuemei Ding, Jianping Wang, Yue Xuan, Zhuowei Su and Keying Zhang *

Institute of Animal Nutrition, Key Laboratory for Animal Disease-Resistance Nutrition of China, Ministry of Education, Sichuan Agricultural University, Chengdu 611130, China; dr.ahsan.mustafa@gmail.com (A.M.); shipingbai@sicau.edu.cn (S.B.); zqf@sicau.edu.cn (Q.Z.); dingxuemei0306@163.com (X.D.); wangjianping@sicau.edu.cn (J.W.); xuanyuede1007@hotmail.com (Y.X.); wzs698@126.com (Z.S.)

* Correspondence: zkeying@sicau.edu.cn; Tel./Fax: +86-28-8629-0922

Simple Summary: Corn is an important ingredient and staple food in China; thus, corn storage has a certain importance to ensure domestic food resources. Normally, corn has been stored for 3 or more years under the proper storage conditions in national barns before it is used as a feed ingredient. This study aimed to investigate the effect of different levels of aging corn (AC) on performance, antioxidative capacity, intestinal health, and microbiota in broilers. In the present study, AC grains were stored for 4 years under the proper storage conditions at the national storage facility. The results indicated that a lower level of AC diet showed improved performance and overall bird health than a higher level of AC, and comparable with a normal corn diet. However, antioxidative capacity is reduced by AC diets.

Abstract: Three-hundred and sixty-day-old male broilers underwent three treatments with six replicates of 20 birds per treatment. The experimental diets included NC: normal corn diet; ACL: lower level (39.6–41.24%) of AC; and ACH: a higher level (56.99–59.12%) of AC. During phase 1 (0–21 d), broilers fed on AC showed lower ($p < 0.05$) body weight (BW), body weight gain (BWG), and feed conversion ratio (FCR) as compared with the NC group. During phase 2 (22–42 d), the NC group and ACL group showed better ($p < 0.05$) BW, BWG, and FCR than the ACH group. The footpad lesion score ($p = 0.05$) and litter moisture percentage ($p < 0.05$) were found to be higher in the ACH group. During phase 1, the ACL group showed a lower level of malondialdehyde (MDA) contents ($p < 0.05$) in serum; moreover, catalase (CAT) ($p < 0.05$) and glutathione peroxidase (GSH-Px) activities ($p < 0.05$) were found lower in both AC-containing groups. During phase 2, CAT activity in serum was found higher ($p < 0.05$) in the ACH group. During phase 1, the NC group showed higher CAT ($p = 0.05$), GSH-Px ($p < 0.05$), and superoxide dismutase (SOD) activity ($p = 0.03$); however, it showed lower MDA ($p < 0.05$) and total-antioxidative capability (T-AOC) ($p < 0.05$) in the liver. During phase 1, in breast muscle, CAT, SOD, and T-AOC were higher ($p < 0.05$) in the NC group. During phase 1, total cholesterol and high-density lipoprotein were found to be lower ($p < 0.05$) in the ACL group. Similarly, triglyceride and low-density lipoprotein were found to be lower ($p < 0.05$) in the ACL group than the ACH group. During phase 1, villus height was found to be higher ($p < 0.05$) in the ACH group. Moreover, the goblet cell (GC) was found to be higher ($p < 0.05$) in the NC group than the ACL group. During phase 2, GC was found to be higher ($p < 0.05$) in the ACL group. In ileal digesta, during phase 1, acetic acid, propionic acid, and butyric acid (BA) levels were found to be higher ($p < 0.05$) in the ACL group. In cecal digesta, BA was significantly lower ($p < 0.05$) in the NC group.

Keywords: aging corn; oxidative stress; broiler performance; intestinal health; microbiota

check for updates

Citation: Mustafa, A.; Bai, S.; Zeng, Q.; Ding, X.; Wang, J.; Xuan, Y.; Su, Z.; Zhang, K. Limitation and Potential Effects of Different Levels of Aging Corn on Performance, Antioxidative Capacity, Intestinal Health, and Microbiota in Broiler Chickens. *Animals* **2021**, *11*, 2832. https://doi.org/10.3390/ani11102832

Academic Editors: Natalie Morgan and Youssef A. Attia

Received: 14 September 2021
Accepted: 21 September 2021
Published: 28 September 2021

1. Introduction

Corn contains rich nutrients, including carbohydrates, water, vitamins, fiber, and minerals [1], and plays an essential role in China's food crops. China is a major corn producer with more than 300 million acres of corn and is the second largest corn producer in the world [2]. A large number of corn has been stored as the country's grain reserves to deal with unexpected natural disasters, and therefore the storage corn may be used as an animal feed when it was not suitable to store further [3]. It is expected that currently there are 200 million tons of stored corn, which could be stored for 3 to 4 years in China. During the storage time, some physiological and biochemical reactions happened in corn that may result in the loss of the nutritional value and accumulation of oxidation products in corn [4–6]. Ensuring the efficient use of stored corn in livestock and poultry, feed production is of great significance to the timely absorption of stock corn [7].

During storage, alpha- (α-) and beta- (β-) amylase enzyme activities could be decreased; however, lipoxygenase, lipases, and proteases enzyme activities of stored rice are increased [8,9]. Therefore, storage could reduce the digestibility and solubility of protein [9,10]. Moreover, Galliard [11] reported higher contents of free fatty acids (FFA) and oxidation of lipids in whole flour during storage, and Salman and Copeland [12] found an inversely proportional relationship between increased acidity of fat contents with decreased iodine binding capacity in wheat. The production of FFA in corn was found to be interlinked with the higher H_2O_2 production, which ultimately affects the activities of peroxidase (POD) and catalase (CAT) [13]. The lipid peroxidation of the cell membrane in stored grains could also alter the POC and CAT activities, which have been used as an indicator to evaluate the corn quality after storage [14].

The oxidation process and production of peroxides in aging corn, due to storage, can cause oxidative stress, which might be a cause of extreme oxidation inside the body tissues and create extremely reactive molecules, i.e., reactive species of nitrogen and radical of oxygen [15]. The free radicals can cause oxidative stress by the mean of a sequence of synthetic reactions comprising immunity, biochemistry, and physiology that ultimately affect the overall health and performance of livestock [16]. Likewise, the aging corn can take part in the production of resistant starch due to the degradation of starch during storage [17]. In the corn storage process, due to improper storage conditions, there are chances of mycotoxin contents, besides the alterations in corn's chemical and physical properties, which are considered as a main cause of nutrients damage and limit the usage of aged corn [18]. Nutrients in corn expose to oxidation (such as oxidation of fatty acids) and corresponding activity (such as antioxidant enzymes) appears lower when these changes to a certain extent led to the generation of aging corn (AC).

It is reported that AC used as a feed ingredient in animals shows a negative impact on their body and health. These negative effects can be addressed and improved. Rational use and animal nutrition research is increasingly focusing on these issues. Studies have found that feed peroxidation destroys the redox balance of animals and leads to oxidative stress. Oxidative stress affects animal health, performance, reduced the FI and FCR, causes animal destruction of the intestinal integrity of the mucosa and barrier function of the intestine [19,20]. Lipid oxidation and FFA contents of AC are a major reason for poor performance and decrease serum oxidative capacity in broilers [3,21], laying hens [22], ducks [23], and weaned piglets [24], whereas Zhou et al. reported no adverse effect on the productive performance of laying hens [25]. The data found above suggest that the storage of corn under suitable conditions could not modify the nutritive value of corn. Contrary, the storage length had a harmful effect on the content of energy due to the increased FFA and decreased fat content [26]. However, to our best knowledge, there are no reports about the effect of AC levels on broilers' health. Therefore, the present study was designed to investigate the effect of the higher and lower level of 4-year-old stored corn on the performance, antioxidative capacity, intestinal health, and microbiota in broilers.

2. Materials and Methods

2.1. Birds, Diets, and Management

The present study was accomplished by following guidelines of the standard recommendations of the National Institutes of Health for the Care and Use of Laboratory Animals. The current study protocol was approved by the Animal Care and Use Committee of Sichuan Agricultural University, China (Ethic Approval Code: SICAUAC201710-7).

Three-hundred and sixty-day-old male broiler chicks (Ross 308) were obtained from a local commercial hatchery (Yuguan Co. Ltd., Chengdu, China) and reared in a broiler house, which was environmentally controlled, on floor pens (dimension of 6.56 × 3.28 ft.), with a litter of rice husk at broiler experimental farms, Sichuan Agricultural University, Ya'an, China. This experiment was a completely randomized block study and comprises of 3 treatments with 6 replicates of 20 birds per treatment. Broilers were randomly assigned to the treatments with two phase-feeding programs (Phase 1: Day 1–21, Phase 2: Day 22–42). The basal diet was formulated as a corn–wheat–soybean-based diet according to the NRC [27] nutrient recommendations (Table 1). Experimental diet comprises as follows; NC: control diet with normal corn; ACL: control diet with a lower level (39.6–41.24%) of aging corn; ACH: control diet with a higher level (56.99–59.12%) of aging corn. All the diets were processed in pellet (3 mm) form. Feed and water were provided to broilers ad libitum. Bird management was conducted as described in the Ross 308 Broiler Commercial Management Guide.

Table 1. Formulation of basal diets [1] fed to broilers during Phases 1 and 2.

Ingredients (%)	Phase 1 (1 to 21 d)		Phase 2 (21 to 42 d)	
	NC/ACL	ACH	NC/ACL	ACH
Corn (Normal or Aging)	39.6	56.99	41.24	59.12
Wheat	20.1	0	21	0
Soybean Meal (43%)	33.76	36.7	30.74	34
Soybean Oil	2.82	2.6	3.53	3.4
Limestone ($CaCO_3$)	1.05	1.1	0.86	0.93
Calcium hydrogen phosphate	1.4	1.4	1.33	1.33
Salt (NaCl)	0.4	0.4	0.4	0.4
Choline chloride	0.15	0.15	0.15	0.15
Multi-vitamins [2]	0.03	0.03	0.03	0.03
Mineral premix [3]	0.2	0.2	0.2	0.2
Lys HCl (99%)	0.24	0.18	0.27	0.2
DL-methionine (99%)	0.25	0.25	0.25	0.24
Total	100	100	100	100
Nutrient level (calculated)				
ME (kcal/kg)	2950	2950	3020	3020
CP (%)	21	21.03	20	20.07
Calcium (%)	1	1	0.9	0.9
Available phosphorus (%)	0.45	0.45	0.43	0.43
Lysine (%)	1.15	1.15	1.1	1.1
Methionine (%)	0.5	0.5	0.48	0.48
Methionine + Cysteine (%)	0.86	0.85	0.83	0.81

[1] Normal corn diet (NC); low aging corn (ACL); high aging corn (ACH). [2] Vitamin premix per kilogram feed provided: Vitamin A, 16,000 IU (trans retinol); Vitamin D3, 4000 IU; Vitamin E, 1IU (dl-α-tocopheryl acetate); Vitamin B1, 0.8 mg; Vitamin B2, 6.4 mg; Vitamin B12, 0.012 mg; Vitamin B6, 2.4 mg; calcium pantothenate, 10 mg; niacin acid, 14 mg; biotin, 0.1 mg; folic acid, 0.2 mg; Vitamin K3, 2 mg. [3] Mineral premix per kilogram feed provided: Fe ($FeSO_4 \cdot H_2O$), 100 mg; Cu ($CuSO_4 \cdot 5H_2O$), 12.5 mg; Mn ($MnSO_4 \cdot H_2O$), 88 mg; Zn ($ZnSO_4 \cdot H_2O$), 95 mg; I (KI), 0.9 mg; Se (Na_2SeO_3), 0.3 mg.

2.2. Aging Corn

Normal corn and AC originated from the national barns in Yinchuan, China, and Changchun, China, which had been stored for 6 months and 4 years, respectively. All corn samples were stored in brick structures, and some phytochemical properties were determined (Table 2). According to the methods described by AOAC International [28], the dry matter (DM), crude protein (CP), and crude fiber (CF) were analyzed by oven drying (Method No. 934.01), Kjeldahl (Method No. 990.3), and Soxhlet fat analysis (Method No. 920.39), respectively. Gross energy was analyzed by adiabatic bomb calorimetry (Parr Instrument Company, Moline, IL, USA). The quantity of potassium hydroxide (KOH) required for acid neutralization in a 100 g sample was used for the analysis of titratable acidity (GB/T 20570-2015). The POD and CAT activity, as well as malondialdehyde (MDA) contents, were measured by using precise detection kits from Nanjing Jiancheng Bioengineering Institute (Nanjing, China) with a Multiskan Spectrum Reader (Model 1500; Thermo Scientific, Nyon, Switzerland). Deoxynivalenol, zearalenone, and aflatoxin were measured by using the national standard methods, i.e., SN/T 1571-2005, GB/T 28716-2012, and GB/T 30955-2014, respectively. Similarly, the standard method for fatty acids (FAs) was used for its quantification (GB 5009.168-2016).

Table 2. Phytochemical properties of normal and aging corn (air-dried basis).

Items	Content	
	Normal Corn	Aging Corn
Gross energy (kcal/g)	3.80	3.85
Crude protein (%)	7.56	7.73
Moisture (%)	14.31	13.06
Crude fat (%)	3.42	3.31
ADF (%)	9.80	10.1
aNDF (%)	10.7	12.4
Xanthophylls (mg/kg)	15.5	4.9
Fatty acids acidity (KOH mg/100 g)	64	126
MDA (nmol/mL)	96.03	40.30
POD (U/mg)	64.93	34.26
CAT (U/mg)	28.49	17.00
Aflatoxin (μg/kg)	1.9	†
Vomitoxin (μg/kg)	†	240.9
Zeranol (μg/kg)	63.4	87.4
Fatty acids (mg/g)		
Palmitic (C16:0)	3.156	2.448
Stearic (C18:0)	0.314	0.256
Oleic (C18:1)	5.727	3.872
Linoleic (C18:2)	11.391	9.232
Linolenic (C18:3)	0.318	0.012

† Non-detected element.

2.3. Growth Performance

Individual broilers, after 12 h fasting, body weight (BW), and the amount of feed intake (FI; offered–refused) by pen were measured at the end of each phase (1 and 2) at 21 and 42 d. Then, average body weight (ABW), average body weight gain (ABWG), average feed intake (AFI), and feed conversion ratio (FCR) by pen were calculated. Birds were monitored to account for their morbidity (health status) and mortality on a daily basis. During the experiment, the BW of dead birds was included in growth performance calculations.

2.4. Footpad Dermatitis Score

Footpad dermatitis scoring was conducted at the end of phase 1 and phase 2 by visual inspection of the footpad of both feet of all live birds in each pen. A seven-point scoring system was followed, according to the procedure of Mayne et al. [29]. The better quality of the footpad is associated with a lower footpad dermatitis score. The total footpad score (*TFPS*) for each pen was calculated as follows.

$$TFPS = \frac{SUM\ OF\ AVERAGE\ OF\ 2\ FEET\ FOOTPAD\ SCORE}{TOTAL\ NUMBER\ OF\ BIRDS\ EXAMINED\ IN\ EACH\ PEN} \quad (1)$$

2.5. Litter Moisture Percentage

Litter samples were taken from each pen at the end of phase 1 and phase 2. Four sub-samples were taken from each pen from the surface to the full depth of the accumulation at locations shown in Figure 1. Sub-samples were taken from the middle of each side except the side with the door, avoiding the drinker and feeders. Side sub-samples were taken approximately 30 cm from the wall at the side of the pen. The four sub-samples were mixed per pen for analysis for moisture. Litter moisture was determined by placing litter samples in a forced-air drying oven at 60 °C for more than 24 h.

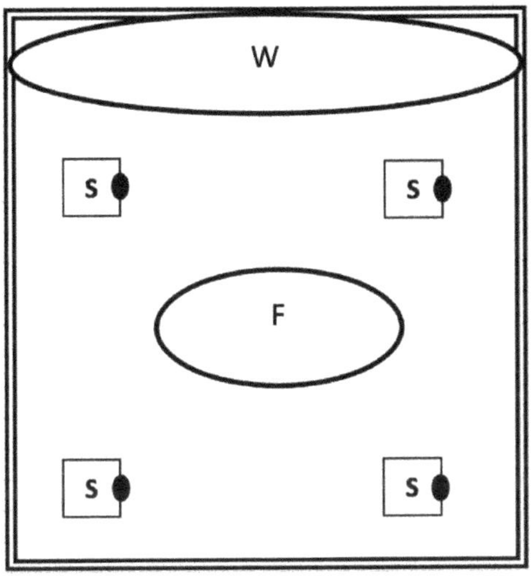

Figure 1. Litter sampling in each pen. Waterer (W), feeder (F), and sampling location (S).

2.6. Sample Collections

After phases 1 and 2, two birds from each replicate (12 broilers from each treatment), with BW near to the average BW of the pen, were selected and slaughtered by severing their jugular vein. Before slaughtering, the samples of blood were collected through the jugular vein, and afterward, for serum collection, these blood samples were centrifuged (2000× *g*, 10 min, 4 °C). The collected serum was then stored at −20 °C for further biochemical parameter analysis. After serum was collected from 12 birds, from them, 6 birds were slaughtered by severing their jugular vein. Liver tissue, breast muscle, and gastrointestinal tract (GIT) samples were collected. Serum, liver, and breast muscle samples were stored at −20 °C for antioxidant activity analysis. The GIT was immediately dissected after euthanization, and a half portion of jejunum was stored in 4% paraformaldehyde solution of histological analysis. However, another half portion of jejunum was used for jejunum mucosa flash-frozen in liquid nitrogen −80 °C until gene expression analysis. Moreover, from the same birds, ileal and cecal digesta samples were collected by squeezing them gently and stored at −20 °C for short-chain fatty acid (SCFA) analysis. On days 21 and 42, the remaining six birds were slaughtered to collect cecal digesta and this was stored at −80 °C for microbial community analysis.

2.7. Antioxidant Activity

Six serum, liver, and breast muscle samples from each treatment were used to analyze MDA content, the enzymatic activity of CAT, superoxide dismutase (SOD), glutathione peroxidase (GSH-Px), and the total antioxidative capability (T-AOC) using commercial assay kits (Nanjing Jiancheng Institute, Nanjing, China) and procedure described previously by Che et al. [30].

2.8. Serum Profile

Six serum samples from each treatment were used to analyze triglyceride (TG), total cholesterol (TC), high-density lipoprotein (HDL), and low-density lipoprotein (LDL) concentrations, as well as alanine aminotransferase (ALT) and aspartate aminotransferase (AST) activities, which were important indices to evaluate lipid metabolism and liver function, using an automatic biochemistry analyzer Hitachi 7020 (Hitachi High Technologies Inc., Tokyo, Japan) with kits being purchased from the Nanjing Jiancheng Bioengineering Research Institute, Nanjing, China.

2.9. Intestinal Characteristics

2.9.1. Jejunum Morphology and Counting of Goblet Cell

Fixed segments in 4% neutral buffered paraformaldehyde solution were rinsed in ethyl alcohol and embedded in paraffin wax. The samples were cut (5 μm) using a propeller slicer (Leica-2016, Leica Inc., Bensheim, Germany), with 3 slices per treatment, and stained using the hematoxylin and eosin method. The micrographs were taken using a microscope (BA400Digital, Motic China Group Co. Ltd., Xiamen, China) was used to take micrograph, and measurements were performed for villus height (VH), crypt depth (CD), and calculated ratio of villus height to crypt depth (VH:CD), as well as count the goblet cells (GC) and calculate the number of goblet cells per unit area by using Image-Pro Plus 6.0 (Media Cybernetics, Rockville, MD, USA) for each structure per slice. The tip of the villus to the villus-crypt junction was defined as VH, whereas the CD was measured from the depth of the invagination to adjacent villi [31].

2.9.2. Jejunum Mucosal mRNA Gene Expression

After phase I and II, from jejunal mucosa, total RNA was extracted by using a TRIzol reagent kit (Takara, Dalian, China), and synthesis of cDNA was completed by using the reagent kit (PrimeScript RT, Takara, Kusatsu, Japan). Primers for Tight Junction Protein (TJP), 3 genes in association with the intestinal barrier including Claudin1 (CLDN1), Zonula Occludens-1 (ZO-1), Occludin (OCLN), and β-actin (housekeeping gene) were designed using Primer Express 3.0 (Applied Biosystems, Waltham, MA, USA; Table 3). Real-time quantitative PCR was performed according to Livak and Schmittgen, [32].

Table 3. Primers used for the quantitative RT-PCR of the target genes.

Target Gene	Forward/Reverse Sequence (5′ to 3′)	Gen Bank Accession No.	Reference
β-actin	F: TTGGTTTGTCAAGCAAGCGG R: CCCCCACATACTGGCACTTT	NM_205518.1	[33]
Zonula Occludens-1	F: TGTAGCCACAGCAAGAGGTG R: CTGGAATGGCTCCTTGTGGT	XM_413773	
Claudin-1	F: TGGAGGATGACCAGGTGAAGA R: CGAGCCACTCTGTTGCCATA	NM_001013611.2	[34]
Occludin	F: TCATCGCCTCCATCGTCTAC R: TCTTACTGCGCGTCTTCTGG	NM_205128.1	

2.10. Ileal and Cecal Digesta Analysis

2.10.1. Volatile Fatty Acid Analysis

At 21 and 42 d, proportions of the volatile fatty acid (VFA), i.e., acetic acid (AA), propionic acid (PA), and butyric acid (BA), were determined from ileal and cecal digesta by using the HPLC system following the method Qin et al. [35] with some modifications. Approximately 0.5 g ileal and cecal contents were gently transferred into a micro-centrifuge tube containing 2 ml of ultrapure water. The solution was thoroughly mixed using a vortex mixer and centrifuged at 5000 rpm for 10 min at 4 °C. Taking 1 mL supernatant added 0.2 mL ice-cold 25% (w/v) meta-phosphoric acid solution, incubated at 4 °C for 30 min, and again centrifuged @12,000 rpm for 10 min. The supernatant was filtered through 0.22 μm syringe filters. The VFA contents of this filtrate were measured using a gas chromatograph (CP-3800, Varian, Palo Alto, CA, USA).

2.10.2. 16S rDNA Gene Amplicons Analysis

At d 21 and 42, cecal digesta samples were subjected for extraction of DNA by using QIAamp PowerFecal DNA Kit (Qiagen, Hilden, Germany) according to the manufacturer's instructions. The DNA concentration and quality were checked using a NanoDrop Spectrophotometer. The Novo gene platform (Illumina Hiseq, Novogene Bioinformation Technology, Beijing, China) was used to performed 16S rDNA gene amplicons analysis. All methods including extraction of DNA, 16S rRNA sequencing, processing of sequences, and analysis of data were performed according to Qin et al. [36]. Concisely, sterile water was used to dilute DNA up to 10 ng/µL. The 16S rRNA genes of distinct regions (16S V4) were amplified using a specific primer (515F GTGCCAGCMGCCGCGGTAA; 806R GGACTACHVGGGTWTCTAAT) with the unique barcodes. The Phusion High-Fidelity PCR Master Mix (New England Biolabs) was used for the accomplishment of all PCR reactions. The end products of PCR were mixed in equal density ratios. Then, Qiagen Gel Extraction Kit (Qiagen) was used for purification of these PCR mixture products. Sequencing libraries were generated using TruSeq DNA PCR-Free Sample Preparation Kit (Illumina, San Diego, CA, USA) following the manufacturer's recommendations, and index codes were added. The Agilent Bioanalyser 2100 system and the Qubit 2.0 Fluorometer (Thermo Scientific) were used to assess the quality of these libraries. Lastly, the sequencing of eligible libraries was completed on an Illumina HiSeq 2500 platform and generated the 250 bp paired-end reads. QIIME quality filters were used for the filtration of selected reads. Sequences with ≥97% similarity were assigned to the same optimal taxonomic units (OTUs). The relative abundance of each OTU was examined at different taxonomic levels. Diversity within communities (Alpha diversity) calculations and taxonomic community assessments were performed by Qiime 1.7.0, and Beta diversity included both unweighted and weighted Unifrac distances calculated with 10 times subsampling; distances were visualized by principal component analysis (PCA; Lozupone and Knight) [37], and the separation was tested using R in Anosim.

2.11. Statistical Analysis

The experiment was a completely randomized design with a pen as the experimental unit. For serum parameter, antioxidative parameters, intestinal characteristics, short-chain fatty acids, and diversity and structure of the cecal microbiota analysis, the randomly selected birds were the experimental unit, and results were analyzed by a one-way analysis of variance (ANOVA) using the GLM procedure of SAS (SAS Institute Inc., Cary, NC, USA). Differences among means were tested with Duncan's multiple range tests. $p \leq 0.05$ is considered significant.

3. Results

3.1. Growth Performance

The growth performance responses of the broiler chickens under different experimental groups are shown in Table 4. During phase 1, the ACL group has lowered ($p < 0.05$) ABW and AWBG from NC and ACH groups, whereas ACH groups have lowered ($p < 0.05$) ABW and AWBG from the NC group, and similarly FCR was poor ($p < 0.05$) in ACL group from NC diet, and numerically poor from the ACH group. On 42 d, ABW was significantly ($p < 0.05$) and numerically lower in the ACH group from the NC group and ACL group, respectively. Similarly, during phase 2 and the whole experiment (1–42 d), AWBG and FCR were found to be poor ($p < 0.05$) in the ACH group among all dietary treatments. The FI and mortality were not significantly different among treatments during phase 1, phase 2, and during the whole experiment (1–42 d). Overall, during the whole experimental period, a lower level (39.6–41.24%) of AC has shown comparable performance with normal corn, whereas a high level (56.99–59.12%) of AC has shown lower performance than other dietary treatments.

Table 4. Effect of different levels of aging corn on performance parameters under the three experimental diets [1] in broilers.

Measurements	NC	ACL	ACH	SEM	p-Value
Average Body Weight (g)					
1 d	45.02	45.07	45.08	0.07	0.78
21 d	1000.2 ^{a}	935.7 ^{c}	972.1 ^{b}	9.0	<0.0001
42 d	2732.0 ^{a}	2692.0 ^{ab}	2589.0 ^{b}	37.4	0.032
Average Body Weight Gain (g)					
1–21 d	955.2 ^{a}	890.6 ^{c}	927.0 ^{b}	9.0	<0.0001
22–42 d	1738.0 ^{a}	1757.0 ^{a}	1617.0 ^{b}	34.2	0.028
1–42 d	2687.0 ^{a}	2647.0 ^{ab}	2543.0 ^{b}	37.4	0.032
Average Feed Intake (g)					
1–21 d	1299.0	1261.0	1279.0	11.9	0.08
22–42 d	3680.0	3571.0	3644.0	64.3	0.46
1–42 d	4849.0	4741.0	4821.2	64.0	0.47
Feed Conversion Ratio (g/g)					
1–21 d	1.36 ^{b}	1.42 ^{a}	1.38 ^{ab}	0.01	0.009
22–42 d	2.13 ^{ab}	2.03 ^{b}	2.24 ^{a}	0.04	0.014
1–42 d	1.81 ^{b}	1.79 ^{b}	1.88 ^{a}	0.02	0.028
Mortality (%)					
1–21 d	3.33	2.50	0.83	1.14	0.28
22–42 d	7.77	10.58	3.82	2.42	0.24
1–42 d	10.00	11.66	4.17	2.54	0.16

[1] Normal corn diet (NC); low aging corn (ACL); high aging corn (ACH). The data are presented as the mean of $n = 6$ for each group. $^{a–c}$ Mean within each column with no common superscript differ significantly ($p < 0.05$).

3.2. Footpad Dermatitis Score and Litter Quality

Both footpads were gross examined on the d 21 and d 42 for allotting the proper score according to their pathological condition (Table 5). A similar trend was observed in TFPS between experimental groups after phase 1 and phase 2. After 21 d, TFPS was significantly lower ($p = 0.05$) in the ACL group from the ACH group and numerically lower from the NC group. Similarly, after 42 d, TFPS in the ACL group was numerically lower and higher from ACH and NC groups, respectively. Litter moisture percentage (%) was also measured on 21 d and 42 d (Table 5) and found similar results as TFPS. On d 21, litter moisture was numerically lower in the ACL from NC and ACH groups. After 42, the litter moisture percentage was significantly higher ($p < 0.05$) in the ACH group than in the NC and ACL groups. Convincingly, a higher level (56.99–59.12%) of AC in the diet caused adverse effects on footpad dermatitis score and litter moisture as compared with normal corn and lower levels (39.6–41.24%) of AC.

Table 5. Effect of different levels of aging corn on litter moisture and foot-pad dermatitis score in the broiler.

Treatments [1]	Footpad Score		Litter Moisture%	
	Phase I	Phase II	Phase I	Phase II
NC	1.69 ^{ab}	4.17	34.60	51.10 ^{b}
ACL	1.46 ^{b}	4.63	31.50	51.50 ^{b}
ACH	2.44 ^{a}	5.35	33.93	55.63 ^{a}
SEM	0.26	0.53	1.57	1.28
p-value	0.050	0.271	0.345	0.044

[1] Normal corn diet (NC); low aging corn (ACL); high aging corn (ACH). The data are presented as the mean of $n = 6$ for each group. ^{a,b} Mean within each column with no common superscript differ significantly ($p < 0.05$).

3.3. Antioxidant Activity

The enzymatic activity of CAT, SOD, GSH-Px, T-AOC, and MDA contents were measured in serum, breast muscle, and liver on d 21 and 42 (Table 6). On d 21, CAT ($p < 0.05$) and GSH-Px ($p < 0.05$) were found higher in the NC group than ACL and ACH groups, and MDA was found lower ($p < 0.05$) in the ACL group than NC and ACH groups in serum; however, SOD and T-AOC were not significantly different among all treatment groups. For the liver, the NC group shows higher CAT ($p = 0.05$) than the ACH group, and higher GSH-Px ($p < 0.05$) than ACL and ACH groups, similarly SOD was found higher ($p < 0.05$) in the NC group than ACL group; however, MDA found lower ($p < 0.05$) in the NC group from all other treatment, and TAOC also found lower ($p = 0.05$) in the NC group than the ACL group. For breast muscles, CAT was higher ($p < 0.05$) in the NC group than the ACH group, and SOD and TAOC were found higher ($p < 0.05$) in the NC group than all other treatment groups. However, the lowest TAOC was found in the ACH group. Moreover, GSH-Px and MDA were found to be similar among all the treatment groups.

Table 6. Effect of different levels of aging corn on the oxidative ability of serum, liver, and breast muscle in broilers.

	Measurements [2]	Treatments [1]				
		NC	ACL	ACH	SEM	*p*-Value
		Phase I (21 d)				
Serum	CAT (U/mL)	0.80 [a]	0.63 [b]	0.56 [b]	0.05	0.005
	GSH-Px (U/mL)	260.4 [a]	200.9 [b]	191.4 [b]	17.5	0.019
	SOD (U/mL)	71.23	72.68	75.70	4.67	0.788
	MDA (nmol/mL)	8.41 [a]	6.99 [b]	8.66 [a]	0.31	0.004
	TAOC (nmol/mL)	1.51	1.43	1.28	0.09	0.211
Liver	CAT (U/mgprot)	5.60 [a]	4.20 [ab]	3.85 [b]	0.54	0.05
	GSH-Px (U/mgprot)	104.8 [a]	89.12 [b]	81.05 [b]	4.73	0.006
	SOD (U/mgprot)	383.6 [a]	327.2 [b]	341.0 [ab]	15.2	0.036
	MDA (nmol/mgprot)	0.35 [b]	0.65 [a]	0.73 [a]	0.05	0.0002
	TAOC(nmol/mgprot)	0.080 [b]	0.100 [a]	0.095 [ab]	0.005	0.05
Breast Muscle	CAT (U/mgprot)	0.76 [a]	0.69 [ab]	0.54 [b]	0.05	0.03
	GSH-Px (U/mgprot)	16.44	14.01	14.54	0.75	0.071
	SOD (U/mgprot)	104.50 [a]	78.16 [b]	84.89 [b]	3.83	0.0003
	MDA (nmol/mgprot)	2.17	1.76	1.82	0.15	0.128
	TAOC (nmol/mgprot)	0.160 [a]	0.130 [b]	0.080 [c]	0.009	<0.0001
		Phase II (42 d)				
Serum	CAT (U/mL)	0.60 [b]	0.58 [b]	0.81 [a]	0.04	0.001
	GSH-Px (U/mL)	275.0	315.6	276.6	18.9	0.266
	SOD (U/mL)	80.5	85.3	75.4	4.7	0.381
	MDA (nmol/mL)	8.09	7.91	8.72	0.48	0.488
	TAOC (nmol/mL)	1.47 [a]	1.03 [b]	1.23 [ab]	0.09	0.011
Liver	CAT (U/mgprot)	3.45	3.37	3.07	0.29	0.619
	GSH-Px (U/mgprot)	83.18	73.47	73.39	3.75	0.114
	SOD (U/mgprot)	314.0	308.6	350.7	16.4	0.19
	MDA (nmol/mgprot)	0.63	0.80	0.72	0.06	0.151
	TAOC (nmol/mgprot)	0.103	0.090	0.111	0.007	0.138
Breast Muscle	CAT (U/mgprot)	0.85 [a]	0.60 [b]	0.62 [b]	0.06	0.018
	GSH-Px (U/mgprot)	10.58	10.53	8.30	0.74	0.078
	SOD (U/mgprot)	84.2 [b]	85.3 [b]	98.0 [a]	3.0	0.009
	MDA (nmol/mgprot)	1.54	1.35	1.39	0.09	0.277
	TAOC (nmol/mgprot)	0.076	0.077	0.076	0.002	0.967

[1] Normal corn diet (NC); low aging corn (ACL); high aging corn (ACH). [2] Catalase (CAT), superoxide dismutase (SOD), glutathione peroxidase (GSH-Px), malondialdehyde (MDA), antioxidative capability (T-AOC). The data are presented as the mean of $n = 6$ for each group. [a–c] Mean within each row with different superscripts differ significantly ($p < 0.05$).

On d 42, CAT was found to be higher ($p < 0.05$) in the ACH group from other treatment groups; however, T-AOC was found higher ($p < 0.05$) in the NC group than ACL group in serum, whereas GSH-Px, SOD, and MDA were found similar among all treatment groups in serum. However, for the liver, all antioxidant indices were found to be similar among all dietary treatments. For breast muscle, CAT was found to be higher ($p < 0.05$) in the NC group than other treatment groups, and SOD was found to be higher ($p < 0.05$) in the ACH group than other treatment groups. Other indices including GSH-Px, MDA, and TAOC were found to be similar among all treatment groups.

3.4. Serum Profile

Results of important indices to evaluate liver function (AST and ALT) and lipid metabolism (TC, TG, HDL-C, and LDL-C) through serum analysis during both phases 1 and 2 are shown in Table 7. During phase 1, both AST and ALT were found to be similar among all dietary treatments. However, TC and HDL-C were found to be lower ($p < 0.05$) in the ACL group than NC and ACH groups. Similarly, TG and LDL-C were found to be lower ($p < 0.05$) in the ACL group from the ACH group. Although during phase 2, serum levels of AST, ALT, TC, TG, HDL-C, and LDL-C were found to be non-significantly different from each other, the ACL group has a lower value for all serum indices.

Table 7. Effect of different levels of aging corn on blood lipid and liver enzymes of broilers fed three different experimental diets [1].

Measurements [2]	NC	ACL	ACH	SEM	p-Value
Phase 1					
ALT (U/L)	6.38	6.33	5.50	0.86	0.732
AST (U/L)	239.1	213.8	276.7	22.0	0.181
TC (nmol/L)	3.50 ^{a}	2.42 ^{b}	3.61 ^{a}	0.22	0.003
TG (nmol/L)	0.43 ^{ab}	0.37 ^{b}	0.47 ^{a}	0.02	0.033
HDL-C (nmol/L)	2.02 ^{a}	1.35 ^{b}	2.10 ^{a}	0.14	0.004
LDL-C (nmol/L)	0.44 ^{ab}	0.35 ^{b}	0.53 ^{a}	0.04	0.031
Phase 2					
ALT (U/L)	14.88	11.17	17.33	3.53	0.507
AST (U/L)	713.5	403.8	664.5	122.5	0.193
TC (nmol/L)	3.29	2.77	3.06	0.17	0.113
TG (nmol/L)	0.41	0.41	0.39	0.04	0.94
HDL-C (nmol/L)	1.71	1.51	1.53	0.08	0.17
LDL-C (nmol/L)	0.59	0.40	0.56	0.07	0.147

[1] Normal corn diet (NC); low aging corn (ACL); high aging corn (ACH). [2] Level of serum alanine aminotransferase (ALT); level of serum aspartate aminotransferase (AST); Level of serum total cholesterol (TC); level of serum total triglyceride (TG); level of serum high-density lipoprotein (HDL-C); and level of serum low-density lipoprotein (LDL-C). The data are presented as the mean of $n = 6$ for each group. ^{a,b} Mean within each row with different superscripts differ significantly ($p < 0.05$).

3.5. Intestinal Morphology and mRNA Gene Expression

On day 21, jejunal morphological results were observed to be higher VH in the ACH group from NC and ACL group ($p < 0.05$) shown in Table 8. Similarly, VH:CD was also found to be higher ($p < 0.05$) in the ACH group from the NC group, whereas CD was found to be similar among the dietary treatments. Moreover, the GC was found to be higher in the NC group ($p < 0.05$; Table 8; Figure 2) than in the ACL group. On day 42, VH:CD was found higher in the NC group than ACL and ACH groups ($p < 0.05$), and GC was found higher in the ACL group than NC and ACH groups ($p < 0.05$), whereas VH and CD were found to be similar among all the treatments. Jejunal mRNA gene expression for TJP results (Table 8) indicated that CLDN1, ZO-1, and OCLN were not significant among all dietary treatments during both phases 1 and 2, whereas the ACL group has shown more improved results than the ACH group, and these results are comparable with the NC group.

Table 8. Effect of different levels of aging corn on jejunum morphology and mRNA gene expression in broilers fed three different experimental diets [1].

Measurment [2]	NC	ACL	ACH	SEM	p-value
			Phase 1		
Jejunal Morphology					
VH (μm)	943.3 ^{b}	1086.5 ^{b}	1406.1 ^{a}	58.9	<0.0001
CD (μm)	182.7	173.5	202.7	16.6	0.462
VH/CD	5.34 ^{b}	6.55 ^{ab}	7.20 ^{a}	0.50	0.048
GC ($\times 10^{-3}$)/μm^2	3.56 ^{a}	2.13 ^{b}	2.69 ^{ab}	0.33	0.017
Jejunal Tight Junction Proteins					
CLDN1	1.00	0.88	0.69	0.29	0.754
ZO-1	0.99	0.81	0.75	0.14	0.392
OCLN	1.00	0.53	0.49	0.25	0.266
			Phase 2		
Jejunal Morphology					
VH (μm)	1216.8	1273.5	1211.6	83.4	0.845
CD (μm)	220.9	272.7	271.1	25.9	0.297
VH/CD	6.12 ^{a}	4.76 ^{b}	4.55 ^{b}	0.43	0.037
GC ($\times 10^{-3}$)/μm^2	1.76 ^{b}	2.74 ^{a}	1.95 ^{b}	0.21	0.009
Jejunal Tight Junction Proteins					
CLDN1	1.00	1.27	1.28	0.26	0.654
ZO-1	0.99	1.30	1.18	0.12	0.199
OCLN	1.00	1.12	1.25	0.11	0.319

[1] Normal corn diet (NC); low aging corn (ACL); high aging corn (ACH). [2] VH = villus height; CD = crypt depth; GC = goblet cell; CLDN1 = Claudin1; ZO-1 = Zona Occludin-1; OCLN = Occludin. The data are presented as the mean of n = 6 for each group. [a,b] Mean within each row with no common superscript differ significantly (p < 0.05).

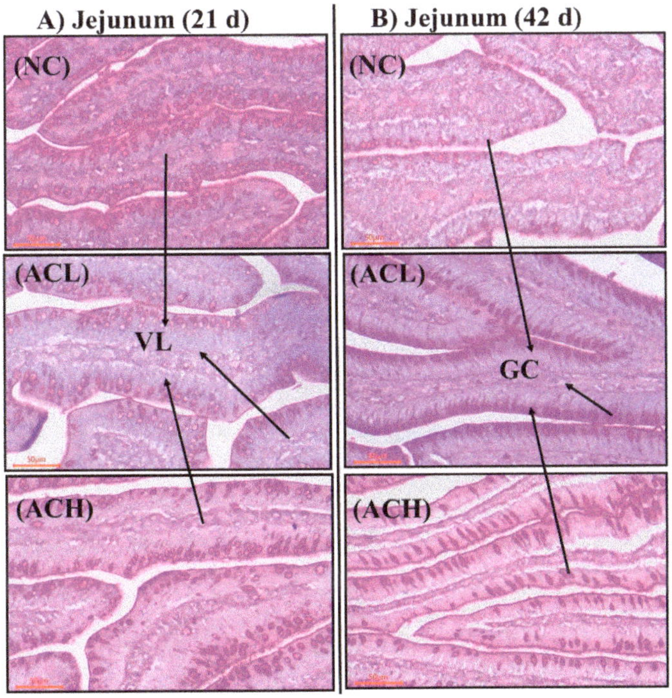

Figure 2. Jejunum villus morphology. Intestinal specimens (n = 6) were collected at 21 d (**A**) and 42 d (**B**); normal corn diet (NC); low aging corn (ACL); high aging corn (ACH). VH = villus height; GC = goblet cell.

3.6. Ileal and Cecal VFA Contents

In ileal digesta, after phase I, AA level was higher ($p < 0.05$) in the ACL group than NC group and similar with the ACH group; moreover, PA and BA levels were found significantly ($p < 0.05$) higher in the ACL group from all other treatment groups (Table 9). After phase II, AA and PA were found similar among all the experimental groups; however, BA was significantly lower in the NC group than all other treatment groups ($p < 0.05$). In cecal digesta, AA and PA were found to be similar among all the experimental groups; however, BA was significantly lower in the NC group than the ACH treatment group ($p < 0.05$) after phase I, whereas after phase II, PA and BA were found to be lower in the NC group than the ACL group ($p < 0.05$) and similar to the ACH group. The level of AA was found to be similar among all experimental groups.

Table 9. Effect of different levels of aging corn on volatile fatty acids in ileal and cecal chyme of broilers.

Treatments [1]	NC	ACL	ACH	SEM	*p*-Value
Ileal VFA [2] (μmol/g)		Phase I			
AA	23.82 ^{b}	100.10 ^{a}	65.14 ^{ab}	16.50	0.01
PA	14.79 ^{b}	36.31 ^{a}	18.22 ^{b}	4.39	0.006
BA	15.58 ^{b}	73.99 ^{a}	49.05 ^{a}	8.47	0.0004
		Phase II			
AA	45.56	48.88	68.29	7.63	0.094
PA	27.46	36.76	32.18	4.54	0.355
BA	45.08 ^{b}	105.39 ^{a}	93.89 ^{a}	14.00	0.013
Cecal VFA [2] (μmol/g)		Phase I			
AA	63.58	98.33	98.05	16.30	0.22
PA	25.58	37.74	28.33	6.24	0.375
BA	41.00 ^{b}	70.51 ^{ab}	86.16 ^{a}	11.50	0.031
		Phase II			
AA	59.25	94.76	70.99	12.60	0.154
PA	22.61 ^{b}	38.69 ^{a}	27.61 ^{ab}	3.85	0.025
BA	48.90 ^{b}	85.86 ^{a}	60.47 ^{ab}	9.12	0.023

[1] Normal corn diet (NC); low aging corn (ACL); high aging corn (ACH). [2] VFA = volatile fatty acid; AA = acetic acid; PA = propionic acid; BA = butyric acid. The data are presented as the mean of $n = 6$ for each group. ^{a,b} Mean within each row with no common superscript differ significantly ($p < 0.05$).

3.7. Cecal Microbial Community

The microbial communities were compared in the cecum among five dietary groups, using Illumina Hiseq high-throughput sequencing. On day 21, a total of 1,083,368 sequencing reads were obtained from the cecal digesta samples, and through cutting and filtering of reads, an average of 83,336 reads was measured per sample, and an average of 78,125 valid data was obtained after quality control. The effective rate of quality control was 93.8%. The sequences were clustered into OTUs (operational taxonomic units) with 97% identity. At the phylum level, the most dominant species were *Firmicutes*, *Proteobacteria*, and *Bacteroidetes*, and their abundance was not significantly different among all experimental groups (Figure 3A). At the genus level, dominant but significantly similar, the species present were *Enterococcus*, *Faecalibacterium*, and *Parabacteroides* (Figure 3B). Similarly, at the species level, the dominant species were *Parabacteroides distasonis*, *Bacteroides uniformis*, and *Lactobacillus salivarius*, and were found to be similar among the all-treatment groups. On day 42, a total of 1,546,555 sequencing reads were obtained from the cecal digesta samples, and through cutting and filtering of reads, an average of 85,919 reads was measured per sample, and an average of 81,499 valid data was obtained after quality control. The effective rate of quality control was 94.8%. The sequences were clustered into OTUs (operational taxonomic units) with 97% identity, and a total of 3110 OTUs were obtained. At the phylum level, the most dominant species were *Firmicutes*, *Bacteroidetes*, and *Proteobacteria*, and their abundance was not significantly different among all experimental groups (Figure 3D), whereas *Fusobacteria* ($p < 0.05$) was significantly higher in the ACL group as compared with all other treatments. At the genus level, the dominant species were *Bacteroides*, *Faecalibacterium*, and *Phyllobacterium*, and Bacteroides and *Phyllobacterium* were found to be statistically similar, whereas *Faecalibacterium* was significantly higher ($p < 0.05$) in the NC group than other dietary treatments (Figure 3E). At the species level, the dominant species were *Bacteroides uniformis*, *Bacteroides plebeius*, and *E. Coli*, and found to be similar among all treatment groups. Moreover, all other species were found to be non-significantly different from each other (Figure 3F).

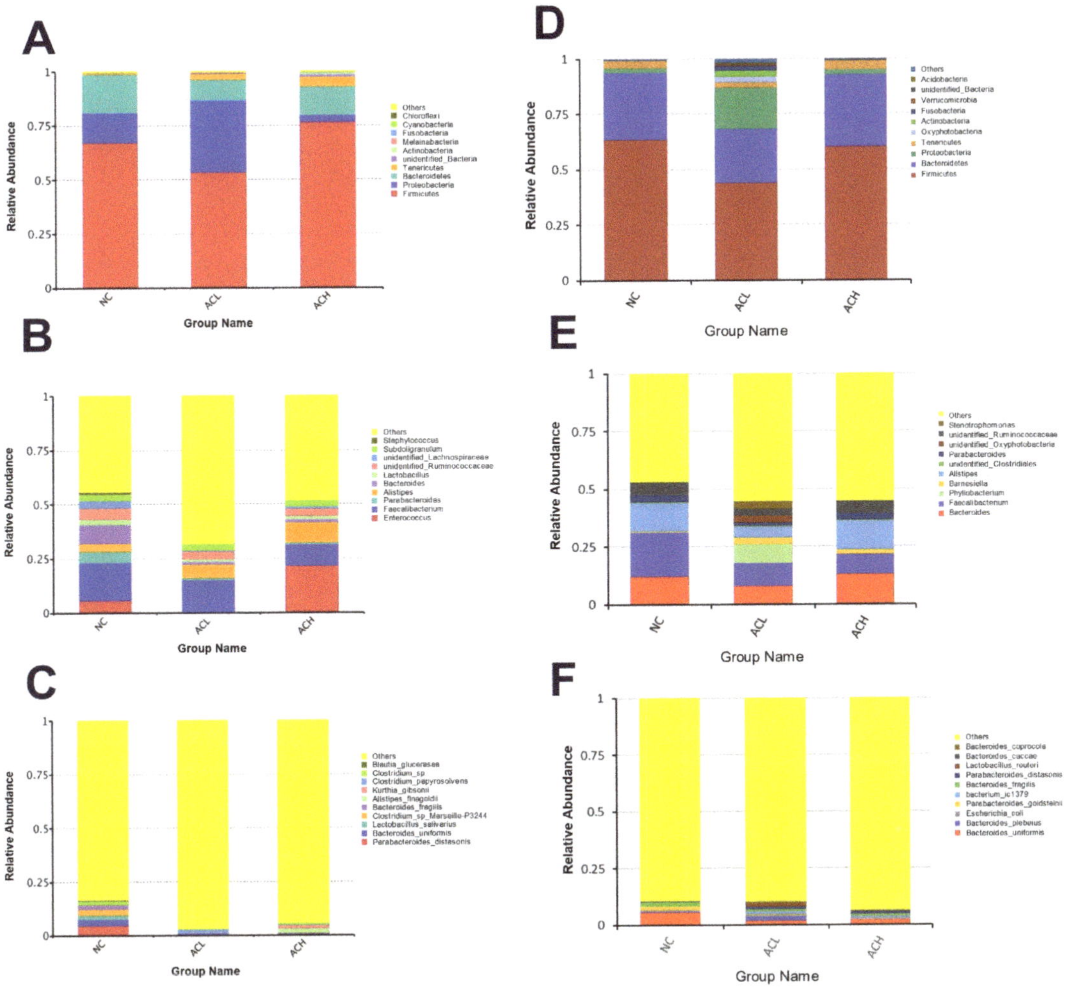

Figure 3. Taxonomic differences in the microbial community of the caeca in broilers ($n = 6$) on 21 d with relative abundance levels of the bacterial (**A**) phyla, (**B**) genu, and their (**C**) species, and on 42 d with relative abundance levels of the bacterial (**D**) phyla, (**E**) genu, and their (**F**) species.

After phase 1, the result of alpha (α) diversity including observed species (OS), Shannon (SH), Simpson (SI), Chao1 (CH), abundance-based coverage estimator metric (ace), good coverage (gc), and phylogenetic distance (PD) were found to be similar among all dietary treatments (Table 10). However, after phase 2, the result of all parameters of α-diversity except PD was found similar among all experimental groups (Table 10). PD was observed to be higher ($p < 0.05$) in the ACL group (Table 10). Additionally, the relationships between communities of various bacteria belonging to different treatments were characterized by PCA, and the results exhibited that there was no significant difference among microbial communities of cecal digesta from different experimental groups during both phase 1 and phase 2.

Table 10. Effect of different levels of aging corn on alpha (α) diversity in cecal chyme of broilers.

Treatment [1]	NC	ACL	ACH	SEM	p-Value
Cecal Digesta [2]			Phase I		
OS	246.2	230.3	288.0	54.2	0.763
SH	4.66	4.36	5.33	0.88	0.753
SI	0.88	0.85	0.89	0.07	0.925
CH	284.3	264.3	314.2	55.4	0.838
ACE	282.8	271.9	314.0	50.9	0.846
GC	0.998	0.998	0.999	0.0002	0.317
PD	16.98	15.72	19.53	2.85	0.671
			Phase II		
OS	699.5	996.8	729.5	93.1	0.075
SH	6.39	6.43	6.73	0.20	0.447
SI	0.96	0.95	0.97	0.01	0.422
CH	755.7	1073.5	794.4	94.6	0.062
ACE	779.9	1078.8	789.2	95.4	0.071
GC	0.998	0.998	0.998	0.0003	0.214
PD	40.72^{b}	71.07^{a}	38.04^{b}	8.65	0.029

[1] Normal corn diet (NC); low aging corn (ACL); high aging corn (ACH). [2] OS = observed species; SH = Shannon; SI = Simpson; CH = Chao1; ACE = abundance-based coverage estimator metric; GC = good coverage; PD = phylogenetic distance. The data are presented as the mean of $n = 6$ for each group. ^{a,b} Mean within each row with no common superscript differ significantly ($p < 0.05$).

4. Discussion

Various studies have designated storage conditions and time had adverse effects on cereal grains, specifically their chemical composition [6,38,39], as storage time is directly proportional to the acidity value of stored corn [40]. The results of the present study corroborated earlier findings that AC fat acidity was found to be higher when compared with corn that had been stored for a few months. These findings indicated that prolonged storage is not suitable for AC. Moreover, others scientists reported that lipid present in AC can be oxidized and formed hydroperoxides (H_2O_2) [41], and CAT and POD were decreased during grain storage [3]. These findings can be used as an indicator of cereal grain deterioration during storage, i.e., rice [42]. The AC quality evaluation mainly depends on the POD activity, and FAs' value includes the important parameters of the quality evaluation of corn storage. In the present study, after chemical evaluation of AC, it was observed that the FAs' value and POD activity of AC were found to be higher and lower, respectively, than normal corn. The AC came from the grain national depot, therefore the storage conditions were suitable, and the content of mycotoxins was not considerable, thus the adverse effects of the AC-containing diet on broilers were caused by changes in the corn nutritional components, such as FA oxidation.

In the present study, the performance was affected during both the starter phase and growing phase in broilers fed on the lower and higher level of AC, respectively. The BW and BWG were found to be highest in the normal corn diet, followed by a higher AC level diet, and the lowest was found in a lower AC level diet. The higher AC level diet showed higher BW and BWG than the lower AC level diet during the starter period because it contains a higher level of soybean, as this diet is based on corn–soybean, whereas other diets based on corn–wheat–soybean diets, and soybean meal is a major source of protein due to its higher digestibility and amino acid availability [43]. However, during the growing phase, a higher AC diet attained the lowest BW and BWG, as well as having poorer

FCR than the other groups. Interestingly, a lower level of AC diet shows improved FCR even from the normal corn diet. That indicated the positive effect of the lower level of AC as compared with a higher level of AC. A similar result was reported in ducks as a higher level of AC produced a more adverse effect on duck health as compared with a lower level of AC [23]. Another report in layers [25] as AC produced no negative effect on production performance. The performance results are interlinked with intestinal morphology as it was found to be balanced during the starter phase in broilers fed on a higher level of AC, whereas during the growing phase, it was found improved in broilers fed on the lower level of AC. As improved intestinal morphology can enhance digestibility, this ultimately improves performance in birds [44]. Therefore, it can be assumed that a lower level of AC could enhance digestibility and reduce the water content in excreta, as high moisture in excreta is a major cause of footpad lesions in broilers [45]. Improved intestinal health and digestibility could be a reason for lower litter moisture and footpad score in broilers fed on lower level of AC as compared with a higher level of AC.

The antioxidant capacity was found to be lower in serum, liver, and breast muscle by the inclusion of both higher and lower quantities of AC during the starter and growing period. These results indicating that the storage of corn can decrease the activity of antioxidant enzymes due to the presence of FFA, because FFA oxidized easily to produce H_2O_2, and these can affect the activities of enzymes such as POD and CAT in maize [13]. Similar results were reported as old corn lowers the serum antioxidative capacity in broilers [46]. Under the condition of oxidative stress, the activities of T-AOC, GSH-Px, CAT, and SOD in the serum were decreased; however, the MDA contents were increased [43]. Long-term corn storage could cause a negative impact on birds' health. Ducks fed AC diets were more likely to obtain oxidative damage, which resulted in reduced growth performance [23,47]. Moreover, MDA in serum and liver was found higher with a higher level of AC in the diet. MDA is a lipid peroxidation (LPO) and the main product of degradation, which reflects the lipid peroxidation (damage) in the body, and cells are attacked by free radicals. The MDA content was also found to be higher in stored corn, which can cause lipid peroxidation in broilers [21]. Liu et al. [46] also reported higher serum MDA content in broilers fed on the old corn.

In the growing animals, nutritional status is mainly reflected by their serum biochemical parameters [48]. Our results showed that a lower level of AC influences lipid metabolism during the starter phase. The lower TG in AC groups may be attained by lowering the FAs synthesis inside the liver and the activation of peroxisome proliferator activating receptor α, which enhances β-oxidation of FAs [49,50]. In this study, a lower level of AC significantly reduced various forms or types of cholesterol including TC, HDL-C, and LDL-C in serum. In another study, rats were fed on oxidized oil containing products of lipid peroxidation, and resulted in reducing cholesterol and triacylglycerols concentrations in the plasma and liver [51]. Likewise, Koch et al. [52] reported that oxidized oil significantly decreased the contents of VLDL-C, HDL-C, and TC in the liver and plasma. A similar effect of AC on serum lipid profile was reported in layer [22].

The villus inside the small intestine, i.e., the jejunum, is a major site for digestion of food and absorption of nutrients in monogastric animals. Consequently, the proper functioning of gut microbiota, immune system, and nutrition, could be achieved by a healthy mucosa of the small intestine. During the growing phase, a higher level of AC had adverse effects on intestinal morphology as compared with a lower level of AC and NC groups. Similar results have been reported by various scientists [47,53]. It can be assumed that lipid peroxidation in a higher level of AC might stimulate the jejunal mucosa and causes oxidative stress that ultimately increases the energy demand of birds as the free radicals damage the mucosa of the small intestine. The VH:CD is considered to play a key role during digestion and absorption. In the present study, a higher level of AC shows lower VH:CD in the jejunum indicated that AC in diets may damage the intestinal mucous to some extent and causes the reduction in the absorptive capacity. Moreover, intestinal morphology, tight junction proteins have also played key roles in maintaining the integrity of the intestinal barrier and regulating intestinal permeability [54]. In the present study, although no significant effect was observed between all dietary treatments, the ACL group has shown improve TJP than the ACH group.

In broilers, the major concentration of analyzed SCFAs of ileal and cecal chyme can be varied mainly due to the ingredients of chicken feed [55]. The present study provides novel information regarding ileal and cecal SCFAs in broilers fed on the higher and lower level of AC as it has never been tested previously. During both dietary phases, all SCFAs in ileum and cecum were found to be higher in broilers fed on a diet containing AC aging corn as compared with boilers fed on the diet containing normal corn. Mainly SCFAs are produced in the GIT due to the fermentation of complex carbohydrates [56]. Therefore, it could be assumed that AC might contain the complex carbohydrates that could be contributed to the production of SCFAs in the GIT of broilers. No study has been reported regarding the effect of AC on the intestinal microbial community in the broiler. Although microbiota including taxonomy, alpha diversity, and beta diversity were found to be similar among all the treatment and control groups, the SCFAs improved the intestinal health in broilers fed AC, especially the lower level of AC. Various researchers reported the positive effect of SCFAs on intestinal health in broilers [57] and pigs [58].

5. Conclusions

After phase 1, the higher level (56.9%) of AC had improved the performance compared with the lower level (39.6%); however, after phase 2, the higher level (59.12%) had more negative effects on performance and broilers' health than lower level (41.24%). As the higher level (59.12%) showed lower performance, antioxidant capacity, and lipid profile, which interlinked with decreased intestinal health that causes more water in dropping and resulting in higher footpad necrosis. However, a lower level (41.24%) of AC diet showed a more improved performance than a higher level (59.12%) of AC and was comparable with a normal corn diet. However, antioxidative capacity was found to be lower in both AC groups. Better performance and overall bird health could be obtained from lower levels (39.6–41.24%) of AC in the broiler diet by using some antioxidants as a supplement in the broiler diet.

Author Contributions: Conceptualization, A.M., S.B. and K.Z.; data curation, A.M.; formal analysis, A.M.; funding acquisition, K.Z.; methodology, A.M., S.B., Q.Z., X.D., J.W., Y.X., Z.S. and K.Z.; project administration, A.M., S.B. and K.Z.; writing—original draft preparation, A.M.; writing—review and editing, S.B. and K.Z. All authors have read and agreed to the published version of the manuscript.

Funding: This work was supported financially by Sichuan Provincial Science and Technology Project (2016NZ0003) and Nutreco R&D. project.

Institutional Review Board Statement: The study was approved by the Animal Care and Use committee, Sichuan Agricultural University (Ethic Approval Code: SICAUAC201710-7; Chengdu, China).

Informed Consent Statement: Not applicable.

Data Availability Statement: Data are available upon request.

Conflicts of Interest: The authors declare no conflict of interest. The funders had no role in the design of the study; in the collection, analyses, or interpretation of data; in the writing of the manuscript, or in the decision to publish the results.

References

1. Siyuan, S.; Tong, L.; Liu, R. Corn phytochemicals and their health benefits. *Food Sci. Hum. Wellness* **2018**, *7*, 185–195. [CrossRef]
2. Trostle, R. *Global Agricultural Supply and Demand: Factors Contributing to the Recent Increase in Food Commodity Prices*; Diane Publishing: Collingdale, PA, USA; United States Department of Agriculture (USDA): Washington, DC, USA, 2010; ISBN 1-4379-2109-4.
3. Yin, D.; Yuan, J.; Guo, Y.; Chiba, L.I. Effect of storage time on the characteristics of corn and efficiency of its utilization in broiler chickens. *Anim. Nutr.* **2017**, *3*, 252–257. [CrossRef] [PubMed]
4. Garbaba, C.A.; Denboba, L.G.; Ocho, F.L.; Hensel, O. Nutritional deterioration of stored Zea mays L. along supply chain in southwestern Ethiopia: Implication for unseen dietary hunger. *J. Stored Prod. Res.* **2017**, *70*, 7–17. [CrossRef]

5. Tian, G.; Wu, X.; Chen, D.; Yu, B.; He, J. Adaptation of gut microbiome to different dietary Nonstarch polysaccharide fractions in a porcine model. *Mol. Nutr. Food Res.* **2017**, *61*, 1700012. [CrossRef] [PubMed]
6. Rehman, Z.U. Storage effects on nutritional quality of commonly consumed sereals. *Food Chem.* **2006**, *95*, 53–57. [CrossRef]
7. Subedi, K.D.; Ma, B.L. *Corn Crop Production: Growth, Fertilization and Yield (Agriculture Issues and Policies)*; Nova Science Publishers, Inc.: Hauppauge, NY, USA, 2009.
8. Dhaliwal, Y.S.; Sekhon, K.S.; Nagi, H.P.S. Enzymatic activities and rheological properties of stored rice. *Cereal Chem.* **1991**, *68*, 18–21.
9. Chrastil, J. Protein-starch interactions in rice grains. Influence of storage on oryzenin and starch. *J. Agric. Food Chem.* **1990**, *38*, 1804–1809. [CrossRef]
10. Chrastil, J. chemical and physicochemical changes of rice during storage at different temperatures. *J. Cereal Sci.* **1990**, *11*, 71–85. [CrossRef]
11. Galliard, T. Hydrolytic and oxidative degradation of lipids during storage of wholemeal flour: Effects of bran and germ components. *J. Cereal Sci.* **1986**, *4*, 179–192. [CrossRef]
12. Salman, H.; Copeland, L. Effect of storage on fat acidity and pasting characteristics of wheat flour. *Cereal Chem.* **2007**, *84*, 600–606. [CrossRef]
13. Bailly, C.; Bogatek-Leszczynska, R.; Côme, D.; Corbineau, F. Changes in activities of antioxidant enzymes and lipoxygenase during growth of sunflower seedlings from seeds of different vigour. *Seed Sci. Res.* **2002**, *12*, 47–55. [CrossRef]
14. Zhou, X.Q.; Zhang, Y.; Zhang, Y.R. Studies on membrane lipid peroxidatic indexes of stored-maize. *J. Maize Sci.* **2007**, *15*, 80–83.
15. Setiawan, S.; Widjaja, H.; Rakphongphairoj, V.; Jane, J. Effects of drying conditions of corn kernels and storage at an elevated humidity on starch structures and properties. *J. Agric. Food Chem.* **2010**, *58*, 12260–12267. [CrossRef] [PubMed]
16. Yuan, S.; Chen, D.; Zhang, K.; Yu, B. Effects of oxidative stress on growth performance, nutrient digestibilities and activities of antioxidative enzymes of weanling pigs. *Asian Australas. J. Anim. Sci.* **2007**, *20*, 1600–1605. [CrossRef]
17. García-Rosas, M.; Bello-Pérez, A.; Yee-Madeira, H.; Ramos, G.; Flores-Morales, A.; Mora-Escobedo, R. Resistant starch content and structural changes in maize (*Zea mays*) tortillas during storage. *Starch Stärke* **2009**, *61*, 414–421. [CrossRef]
18. Gregori, R.; Meriggi, P.; Pietri, A.; Formenti, S.; Baccarini, G.; Battilani, P. Dynamics of fungi and related mycotoxins during cereal storage in silo bags. *Food Control* **2013**, *30*, 280–287. [CrossRef]
19. Liu, F.; Yin, J.; Du, M.; Yan, P.; Xu, J.; Zhu, X.; Yu, J. Heat-stress-induced damage to porcine small intestinal epithelium associated with downregulation of epithelial growth factor signaling. *J. Anim. Sci.* **2009**, *87*, 1941–1949. [CrossRef]
20. Santos, R.R.; Awati, A.; den Hil, P.J.R.; Tersteeg-Zijderveld, M.H.G.; Koolmees, P.A.; Fink-Gremmels, J. Quantitative histo-morphometric analysis of heat-stress-related damage in the small intestines of broiler chickens. *Avian Pathol.* **2015**, *44*, 19–22. [CrossRef]
21. Liu, B.; He, Y.; Yin, D.; Xia, Z.; Yuan, J. Corn at different storage periods affects serum antioxidant function of broilers. *Chin. J. Anim. Nutr.* **2013**, *25*, 1077–1084.
22. Mu, Y.; Zhang, K.; Bai, S.; Wang, J.P.; Zeng, Q.; Ding, X. Effects of vitamin E supplementation on performance, serum biochemical parameters and fatty acid composition of egg yolk in laying hens fed a diet containing ageing corn. *J. Anim. Physiol. Anim. Nutr.* **2019**, *103*, 135–145. [CrossRef]
23. Qamar, S.H.; Zeng, Q.; Ding, X.; Bai, S.; Wang, J.; Xuan, Y.; Zhou, Q.; Su, Z.; Zhang, K. Effect of oil supplementation on growth performance, meat quality and antioxidative ability in meat ducks fed a diet containing aging corn. *Int. J. Agric. Biol.* **2019**, *21*, 201–208.
24. Luo, B.; Chen, D.; Tian, G.; Zheng, P.; Yu, J.; He, J.; Mao, X.; Luo, Y.; Luo, J.; Huang, Z.; et al. Effects of dietary aged maize with oxidized fish oil on growth performance, antioxidant capacity and intestinal health in weaned piglets. *Animals* **2019**, *9*, 624. [CrossRef]
25. Zhou, L.; Ding, X.M.; Wang, J.P.; Bai, S.P.; Zeng, Q.F.; Su, Z.W.; Xuan, Y.; Zhang, K.Y. Ageing corn results in the pale-coloured egg yolks with decreased polyunsaturated fatty acids. *Anim. Feed Sci. Technol.* **2019**, *256*, 114242. [CrossRef]
26. Pomeranz, Y.; Zeleny, L. Biochemical and functional changes in stored cereal grains. *Crit. Rev. Food Technol.* **1971**, *2*, 45–80. [CrossRef]
27. NRC. *Nutrient Requirements of Poultry*; National Academy Press: Washington, DC, USA, 1994; pp. 57–63.
28. AOAC International. *Official Methods of Analysis of AOAC International*, 17th ed.; Association of Official Analytical Chemists: Gaithersburg, MD, USA, 2006.
29. Mayne, R.K.; Else, R.W.; Hocking, P.M. High litter moisture alone is sufficient to cause footpad dermatitis in growing Turkeys. *Br. Poult. Sci.* **2007**, *48*, 538–545. [CrossRef]
30. Che, L.; Xuan, Y.; Hu, L.; Liu, Y.; Xu, Q.; Fang, Z.; Lin, Y.; Xu, S.; Wu, D.; Zhang, K. Effect of postnatal nutrition restriction on the oxidative status of neonates with intrauterine growth restriction in a pig model. *Neonatology* **2015**, *107*, 93–99. [CrossRef] [PubMed]
31. Han, H.Y.; Zhang, K.Y.; Ding, X.M.; Bai, S.P.; Luo, Y.H.; Wang, J.P.; Zeng, Q.F. Effect of dietary fiber levels on performance, gizzard development, intestinal morphology, and nutrient utilization in meat ducks from 1 to 21 d of age. *Poult. Sci.* **2017**, *96*, 4333–4341. [CrossRef] [PubMed]
32. Livak, K.J.; Schmittgen, T.D. Analysis of relative gene expression data using real-time quantitative PCR and the $2^{-\Delta\Delta CT}$ method. *Methods* **2001**, *25*, 402–408. [CrossRef] [PubMed]

33. Li, J.; Cheng, Y.; Chen, Y.; Qu, H.; Zhao, Y.; Wen, C.; Zhou, Y. Dietary chitooligosaccharide inclusion as an alternative to antibiotics improves intestinal morphology, barrier function, antioxidant capacity, and immunity of broilers at early age. *Animals* **2019**, *9*, 493. [CrossRef]
34. Shao, Y.; Guo, Y.; Wang, Z. β-1,3/1,6-Glucan alleviated intestinal mucosal barrier impairment of broiler chickens challenged with salmonella enterica serovar Typhimurium. *Poult. Sci.* **2013**, *92*, 1764–1773. [CrossRef]
35. Qin, S.; Zhang, K.; Ding, X.; Bai, S.; Wang, J.; Zeng, Q. Effect of dietary graded resistant potato starch levels on growth performance, plasma cytokines concentration, and intestinal health in meat ducks. *Poult. Sci.* **2019**, *98*, 3523–3532. [CrossRef]
36. Qin, S.M.; Bai, W.Q.; Zhang, K.Y.; Ding, X.M.; Bai, S.P.; Wang, J.P.; Peng, H.W.; Yang, Y.F.; Chen, C.; Zeng, Q.F. Different microbiomes are found in healthy breeder ducks and those with foot pad dermatitis. *Poult. Sci.* **2019**, *98*, 6340–6348. [CrossRef]
37. Lozupone, C.; Knight, R. UniFrac: A new phylogenetic method for comparing microbial communities. *Appl. Environ. Microbiol.* **2005**, *71*, 8228–8235. [CrossRef] [PubMed]
38. Liu, R.; Cong, L.; Fu, H.; Liu, B.; Huang, W.; Qin, Q. Removal of methyl blue by magnetic CuFe2O4 nanoparticles prepared via the novel solution combustion process. *Adv. Sci. Eng. Med.* **2016**, *8*, 188–195. [CrossRef]
39. Labuschagne, M.; Phalafala, L.; Osthoff, G.; van Biljon, A. The influence of storage conditions on starch and amylose content of south African quality protein maize and normal maize hybrids. *J. Stored Prod. Res.* **2014**, *56*, 16–20. [CrossRef]
40. Zhou, Z.; Robards, K.; Helliwell, S.; Blanchard, C. Ageing of stored rice: Changes in chemical and physical attributes. *J. Cereal Sci.* **2002**, *35*, 65–78. [CrossRef]
41. Wang, F.; Wang, R.; Jing, W.; Zhang, W. Quantitative dissection of lipid degradation in rice seeds during accelerated aging. *Plant. Growth Regul.* **2012**, *66*, 49–58. [CrossRef]
42. Matsukura, U.; Kaneko, S.; Momma, M. Method for measuring the freshness of individual rice grains by means of a color reaction of catalase activity. *Nippon Shokuhin Kagaku Kogaku Kaishi* **2000**, *47*, 523–528. [CrossRef]
43. de Coca-Sinova, A.; Valencia, D.G.; Jiménez-Moreno, E.; Lázaro, R.; Mateos, G.G. Apparent ileal digestibility of energy, nitrogen, and amino acids of soybean meals of different origin in broilers. *Poult. Sci.* **2008**, *87*, 2613–2623. [CrossRef] [PubMed]
44. Oviedo-Rondón, E.O. Holistic view of intestinal health in poultry. *Anim. Feed Sci. Technol.* **2019**, *250*, 1–8. [CrossRef]
45. McIlroy, S.G.; Goodall, E.A.; McMurray, C.H. A Contact dermatitis of broilers—Epidemiological findings. *Avian Pathol.* **1987**, *16*, 93–105. [CrossRef]
46. Liu, J.B.; Yan, H.L.; Zhang, Y.; Hu, Y.D.; Zhang, H.F. Effects of stale maize on growth performance, immunity, intestinal morphology and antioxidant capacity in broilers. *Asian Australas. J. Anim. Sci.* **2020**, *33*, 605–614. [CrossRef]
47. McFarlane, J.M.; Curtis, S.E.; Shanks, R.D.; Carmer, S.G. Multiple concurrent stressors in chicks: 1. Effect on weight gain, feed intake, and behavior. *Poult. Sci.* **1989**, *68*, 501–509. [CrossRef]
48. Attia, Y.A.; Al-Harthi, M.A.; Elnaggar, A.S. Productive, Physiological and immunological responses of two broiler strains fed different dietary regimens and exposed to heat stress. *Ital. J. Anim. Sci.* **2018**, *17*, 686–697. [CrossRef]
49. Sülzle, A.; Hirche, F.; Eder, K. Thermally oxidized dietary fat upregulates the expression of target genes of PPARα in rat liver. *J. Nutr.* **2004**, *134*, 1375–1383. [CrossRef]
50. Chao, P.-M.; Chao, C.-Y.; Lin, F.-J.; Huang, C. Oxidized frying oil up-regulates hepatic Acyl-CoA oxidase and cytochrome P450 4 A1 genes in rats and activates PPARα. *J. Nutr.* **2001**, *131*, 3166–3174. [CrossRef]
51. Eder, K. The effects of a dietary oxidized oil on lipid metabolism in rats. *Lipids* **1999**, *34*, 717–725. [CrossRef] [PubMed]
52. Koch, A.; König, B.; Spielmann, J.; Leitner, A.; Stangl, G.I.; Eder, K. Thermally oxidized oil increases the expression of insulin-induced genes and inhibits activation of sterol regulatory element-binding protein-2 in rat liver. *J. Nutr.* **2007**, *137*, 2018–2023. [CrossRef] [PubMed]
53. Chen, Z.; Xie, J.; Wang, B.; Tang, J. Effect of γ-aminobutyric acid on digestive enzymes, absorption function, and immune function of intestinal mucosa in heat-stressed chicken1 1this work is supported by research grants from national natural science foundation of china (NSFC31060312; 31260555, Beijing). *Poult. Sci.* **2014**, *93*, 2490–2500. [CrossRef]
54. Li, X.; Mao, M.; Zhang, Y.; Yu, K.; Zhu, W. Succinate modulates intestinal barrier function and inflammation response in pigs. *Biomolecules* **2019**, *9*, 486. [CrossRef] [PubMed]
55. Dauksiene, A.; Ruzauskas, M.; Gruzauskas, R.; Zavistanaviciute, P.; Starkute, V.; Lele, V.; Klupsaite, D.; Klementaviciute, J.; Bartkiene, E. A Comparison study of the caecum microbial profiles, productivity and production quality of broiler chickens fed supplements based on medium chain fatty and organic acids. *Animals* **2021**, *11*, 610. [CrossRef] [PubMed]
56. Kimiaeitalab, M.V.; Cámara, L.; Mirzaie Goudarzi, S.; Jiménez-Moreno, E.; Mateos, G.G. Effects of the inclusion of sunflower hulls in the diet on growth performance and digestive tract traits of broilers and pullets fed a broiler diet from zero to 21 d of age. a comparative study1. *Poult. Sci.* **2017**, *96*, 581–592. [CrossRef] [PubMed]
57. Liao, X.; Shao, Y.; Sun, G.; Yang, Y.; Zhang, L.; Guo, Y.; Luo, X.; Lu, L. The relationship among gut microbiota, short-chain fatty acids, and intestinal morphology of growing and healthy broilers. *Poult. Sci.* **2020**, *99*, 5883–5895. [CrossRef]
58. Jha, R.; Fouhse, J.M.; Tiwari, U.P.; Li, L.; Willing, B.P. Dietary fiber and intestinal health of monogastric animals. *Front. Vet. Sci.* **2019**, *6*, 48. [CrossRef] [PubMed]

animals
MDPI

Article

Soybean Oil Replacement by Poultry Fat in Broiler Diets: Performance, Nutrient Digestibility, Plasma Lipid Profile and Muscle Fatty Acids Content

Ahmed A. Saleh [1], Abdulrahman S. Alharthi [2], Rashed A. Alhotan [2], Mustafa Shukry Atta [3,*] and Abdel-Moneim Eid Abdel-Moneim [4]

1 Department of Poultry Production, Faculty of Agriculture, Kafrelsheikh University, Kafrelsheikh 33516, Egypt; ahmed.saleh1@agr.kfs.edu.eg
2 Department of Animal Production, College of Food and Agriculture Sciences, King Saud University, Riyadh 11451, Saudi Arabia; Abalharthi@ksu.edu.sa (A.S.A.); ralhotan@ksu.edu.sa (R.A.A.)
3 Department of Physiology, Faculty of Veterinary Medicine, Kafrelsheikh University, Kafrelsheikh 33516, Egypt
4 Nuclear Research Center, Biological Applications Department, Egyptian Atomic Energy Authority, Abu-Zaabal 13759, Egypt; aeabdelmoneim@gmail.com
* Correspondence: mostafa.ataa@vet.kfs.edu.eg

Simple Summary: The effect of partial or complete substitution of soybean oil (SO) by poultry fat (PF) on growth, nutrient digestibility, plasma lipids, and the pectoral muscle content of fatty acids (FAs) was examined in this study. Dietary PF supplementation improved breast muscle FA profile but did not affect muscle vitamin E content and liver thiobarbituric acid reactive substances (TBARS). By adding PF to the diet, economic efficiency was greatly improved in a dose-dependent manner. Therefore, the results of this study revealed that PF could be used as a partial or total replacement of SO in broiler nutrition without affecting their performance or physiological response with a tendency to improve their meat products.

Abstract: Continuous genetic improvements of commercial broiler strains has led to the necessity of using fats in their rations to fulfill a large portion of the energetic requirements. Several fat sources have been introduced in poultry nutrition, such as rendering poultry fat (PF) an available and cheap lipid source compared to conventional sources such as soybean oil (SO). The present study investigated the effect of partial or full replacement of SO by PF on performance, nutrient digestibility, blood lipids, and fatty acids (FAs) content of pectoral muscle. Four hundred and eighty one-day-old male Ross-308 chicks were distributed into four experimental groups (12 replicates each): the first group (control) was fed a diet formulated with soybean oil as a fat source while the second to fourth groups (PF25, PF50, and PF100) were fed diets formulated with 25, 50 and 100% of PF as a fat source instead of SO. Results revealed no synergistic effect between SO and PF in any of the studied parameters. Replacing SO by PF did not alter birds' growth, carcass characteristics, and plasma indices of birds. Abdominal fat% was increased ($p < 0.01$) in PF50 and PF100. Dry matter digestibility was improved ($p < 0.05$) in PF50 and PF100, while crude fat and protein digestibility was not affected. Contents of palmitic and docosahexaenoic acids in the pectoral muscle of PF50 and PF100 were reduced ($p < 0.01$) while concentrations of oleic and linolenic acids, total unsaturated FAs, and polyunsaturated FAs/Saturated FAs ratio were elevated ($p < 0.05$) in the same groups. Liver thiobarbituric acid reactive substances (TBARS) and muscle vitamin E contents were not altered. The dietary addition of PF greatly improved economic parameters. In conclusion, PF can be used as a lipid source in broiler diets to produce inexpensive meat while maintaining its growth performance.

Keywords: poultry fat; growth performance; plasma lipid; muscle FAs content; broilers

Citation: Saleh, A.A.; Alharthi, A.S.; Alhotan, R.A.; Atta, M.S.; Abdel-Moneim, A.-M.E. Soybean Oil Replacement by Poultry Fat in Broiler Diets: Performance, Nutrient Digestibility, Plasma Lipid Profile and Muscle Fatty Acids Content. *Animals* **2021**, *11*, 2609. https://doi.org/10.3390/ani11092609

Academic Editors: Natalie Morgan and Youssef A. Attia

Received: 6 July 2021
Accepted: 1 September 2021
Published: 5 September 2021

1. Introduction

The metabolizable energy requirements of modern commercial broiler strains are increasing with continuous improvements in their performance and the swift evolution of the intensive poultry industry. Additionally, due to the limited amount and incessant rising in prices of conventional energy sources feedstuffs, it has necessitated the search for alternative materials to ensure the future profitability of poultry production [1]. Lipid sources (oils and fats) are commonly included in broiler feeds to fulfill their high energy requirements. Fats added to broiler diets can improve the absorption and digestion of fat-soluble vitamins and other nutrients, enhancing their growth performance under armor neutral and heat stress conditions [2,3]. However, the prices of different lipid sources vary, and their utilization also varies depending on their physical and chemical properties [4]. The chain length of fatty acids (FAs) and their saturation degree affect the digestibility and metabolizable energy of the fat source [5]. Although animal fats are cheaper than vegetable oils, it is generally believed that the nutritional value of the former is less than that of the latter. Furthermore, among animal fats, poultry fat (PF) can be utilized by poultry species at a higher rate than tallow and lard oil [6].

In the Middle East, soybean oil (SO) is the most common fat source used in poultry diet formulation, particularly after several hydrations, filtration, and degumming [7]. However, limited supply and consistently higher prices are anticipated for it. Rendered poultry fats, obtained from processed wastes of poultry slaughterhouses, can be sustainable alternatives to SO due to their wide availability and relatively low price. Moreover, dietary addition of PF has other benefits, including reducing dustiness, improving feed texture, increasing palatability, and enhancing nutrient absorption by reducing digesta rate of passage through the gut. However, rancidity and low utilization, particularly in young chicks, are disadvantages of using animal fats in poultry diets [8]. No differences in performance parameters were noticed in the literature when PF was used instead of SO [9,10]. However, it's worth noting that a synergistic effect was observed when vegetable oils were mixed with animal fats such as PF and tallow [6,11].

PF is commonly used in feed mills; however, the prohibitions and specific rules of the European Union regarding this practice have caused problems for these facilities with the EU countries [12–14]. Nevertheless, these by-products must be reintroduced into production and economy because of their high nutritional content, economic value, large production cost, and the high cost of alternative implementation. Therefore, the legal regulations and prohibitions put into action must be reconsidered to keep pace with the continuous improvement in production, use, and sales following current scientific developments.

The present study aimed to investigate the potential synergistic impact of mixing SO with PF in broiler diets as well as the effect of full replacement of SO by PF on performance parameters, nutrient digestibility, plasma lipids, and contents of fatty acids and α-tocopherol in pectoral muscle and TBARS in hepatic tissues.

2. Materials and Methods

2.1. Ethical Statement

The study was approved by the Ethics Committee of Local Experimental Animals Care Committee and conducted following the guidelines of Kafrelsheikh University, Egypt (Number 4/2016 EC). All precautions were followed to minimize suffering during the entire experimental period.

2.2. Chemical Analysis of Fat Sources

Poultry fat by-product and soybean oil were supplied by the Al-Sabeel Al-Gadidah Company (Tanta, Al-Gharbia, Egypt). Poultry fat was produced from the processed waste of poultry slaughterhouses (feathers, non-edible viscera, feet, head, blood, etc.) in a rendering unit with a batch-cooker and fat presser at the Fat Hanz Company (Tanta city, Egypt). Poultry fat was assessed for *Escherichia coli* and *Salmonella* spp., and it was incorporated into the diets only after confirming the absence of these pathogens. Metabolizable energy, peroxide value, and fatty acids profile of SO and PF were determined according to the procedures described by AOAC [15], Abd El-Moneim and Sabic [1], and Abd El-Moneim, et al. [16]. Values of the thrombogenic and atherogenic indexes were estimated following the equations of Ulbricht and Southgate [17]:

$$\textbf{Atherogenic index} = \frac{\textbf{C12:0 + 4} * \textbf{C14:0 + C16:0}}{\textbf{Sum of unsaturated FAs}} \quad (1)$$

$$\text{Thrombogenic index} = \frac{\text{C14}:0 + \text{C16}:0 + \text{C18}:0}{0.5 * \text{OA} + 0.5 * (\text{MUFA} - \text{OA}) + 0.5 * \text{n6PUFA} + 3 * \text{n3PUFA} + \left(\frac{\text{n3PUFA}}{\text{n6PUFA}}\right)} \quad (2)$$

where OA = oleic acid (C18:1), MUFA = monounsaturated FAs, PUFA = polyunsaturated FAs.

2.3. Experimental Design

A total of 480 one-day-old male Ross-308 broiler chicks (43 g) were allocated into 48 ground pens and distributed equally into four experimental groups (12 replicates each). Pens, the stocking density of which was ten birds/m^2, were equipped with an automatic nipple cup drinker and a chain feeder system. The starter (0–10 d), grower (11–24 d), and finisher (25–35 d) experimental diets (Table 1) were formulated to meet the recommendation of Aviagen [18] for male broilers.

Table 1. Composition of the experimental starter (1–11 d), grower (11–24 d), and finisher (25–35 d) diets.

Ingredient, g/kg	Control			PF25			PF50			PF100		
	Starter	Grower	Finisher	Starter	Grower	Finisher	Starter	Grower	Finisher	Starter	Grower	Finisher
Yellow corn	536	586	646	533	585	645	535	585	644	534	582	641
Soybean meal, 46%	358	302	225	364	307	231	369	313	237	377	323	250
Corn gluten meal, 62%	40	42	55	37	38	50	30	32	45	23	25	35
Soybean oil	24.00	29.00	32.00	18.00	21.75	24.00	12.00	14.50	16.00	0.00	0.00	0.00
Poultry Fat	0.00	0.00	0.00	6.00	7.25	8.00	12.00	14.50	16.00	24.00	29.00	32.00
Dicalcium phosphate	16	15	15	16	15	15	16	15	15	16	15	15
Dl Methionine, 99%	2.1	2.0	1.4	2.1	2.0	1.4	2.1	2.0	1.4	2.1	2.0	1.4
l-Lysine HCl, 98%	3.4	3.7	4.6	3.4	3.7	4.6	3.4	3.7	4.6	3.4	3.7	4.6
l-Threonine	0.5	0.3	0.1	0.5	0.3	0.1	0.5	0.3	0.1	0.5	0.3	0.1
CaCo3	12	12	11	12	12	11	12	12	11	12	12	11
NaCl	3.5	3.5	3.5	3.5	3.5	3.5	3.5	3.5	3.5	3.5	3.5	3.5
Premix *	3	3	3	3	3	3	3	3	3	3	3	3
NaCo3	1.5	1.5	1.6	1.5	1.5	1.6	1.5	1.5	1.6	1.5	1.5	1.6
K2Co3	0.0	0.0	1.8	0.0	0.0	1.8	0.0	0.0	1.8	0.0	0.0	1.8
Chemical Analysis												
Crude protein, %	23.22	21.14	19.00	23.23	21.14	19.01	23.21	21.15	19.02	23.21	21.15	19.02
AME kcal/kg	2967	3060	3165	2967	3061	3166	2967	3062	3167	2968	3062	3167
Ca, %	0.950	0.896	0.864	0.950	0.895	0.864	0.950	0.897	0.864	0.960	0.894	0.864
Available P, %	0.422	0.408	0.388	0.422	0.408	0.388	0.422	0.408	0.388	0.422	0.408	0.388
Crude fiber, %	3.444	3.547	3.333	3.786	3.992	3.909	4.311	4.327	4.334	5.444	5.131	5.509
Na, %	0.193	0.193	0.196	0.193	0.193	0.196	0.193	0.193	0.196	0.193	0.193	0.196
Cl, %	0.250	0.250	0.250	0.250	0.250	0.249	0.250	0.254	0.249	0.250	0.248	0.249

PF25, 25% of soybean oil replaced by PF; PF50, 50% of soybean oil replaced by PF; PF100, full replacement of soybean oil by PF. Apparent metabolizable energy (*AME*); Calcium (Ca); Available Phosphorus (Available P)l Sodium (Na); Chloride (Cl). * Composition of Hero mix® premix. (per 1 kg): Vitamin A 4,000,000 IU, vitamin E 3334 mg, vitamin D3 833,000 IU, vitamin B1 250 mg, vitamin K3 500 mg, vitamin B6 500 mg, vitamin B2 1250 mg, vitamin B12 3.33 mg, folic acid 333.4 mg, biotin 16.7 mg, niacin 10,000 mg, pantothenic acid 3334 mg, iron 10,000 mg, zinc 16,700 mg, manganese 20,000 mg, selenium 33.4 mg, iodine 100 mg, copper 1334 mg, and cobalt 33.4 mg).

The first diet (basal diet; control) was formulated using soybean oil as a fat source. In the second to fourth experimental diets, 25, 50, and 100% of the soybean oil were replaced by poultry fat. Diets and freshwater were offered ad libitum to the birds. The feed trial took place in a temperature-controlled chamber, started from 33 ± 1 °C and decreased by one degree per 3 days until reaching 24 ± 1 °C and kept till 35 days of age, with a proportional humidity between 50% and 70% and a 22:2 h light: dark cycle. Mortalities were recorded throughout the experimental phases.

2.4. Growth Performance and Organ Weights

Initial and final body weight and feed consumption were measured individually on a pen basis. Feed conversion ratio (FCR) was calculated as a g feed:g gain. The European production efficiency factor (EPEF) was calculated as liveability (%) × body weight (g)/FCR × age (d). At the end of the trial (35 days), 48 birds (one bird/replicate; 12 birds/treatment) were individually weighed, slaughtered by cutting the carotid artery, and dissected to evaluate the relative weights of the thigh and breast muscles, abdominal fat, and liver [19]. Blood samples were collected with heparinized test tubes for blood biochemical analyses.

2.5. Nutrient Digestibility

At the end of the experiment (35 d), the twelve birds per group were weighed and individually caged in metabolic pens for collecting their excreta for four days. Before the commencement of this period, an adaptation period of 24 h had elapsed. Fresh water and diets were offered ad libitum to all birds during the manure collection stage. The approximate analysis of dry matter (#930.15), crude protein (#954.01), and crude fat (#920.29) of dried excreta and diets were performed according to AOAC [15]. The trichloroacetic acid procedure was used to estimate fecal nitrogen [20].

2.6. Plasma Biochemical Analysis

At 35 days, to separate plasma, collected blood samples were centrifuged (2500× g for 15 min at 4 °C). Plasma samples were kept at −20 °C pending analysis. Plasma high-density lipoprotein (HDL), total cholesterol, aspartate aminotransferase (AST), total protein, and albumin were calorimetrically evaluated using commercial kits and following the manufacturer's instructions (Diamond Diagnostics, Cairo city, Egypt), using spectrophotometric analysis (Spectronic 1201; Milton Roy, Ivyland, PA, USA).

2.7. Muscle and Liver Biochemical Analysis

The analysis of superficial pectoral muscle FAs was conducted on 48 birds (1 bird per replicate; 12 birds per treatment) using gas-liquid chromatography (GLC) as described by [21,22]. The concentrations of muscle vitamin E [23] and liver thiobarbituric acid-reactive substances (TBARS) [24] were also determined.

2.8. Economic Efficiency

Average feed cost per bird was calculated as feed consumption per bird × cost of one kg diet (0.53, 0.52, 0.51 and 0.50 US $ for control, PF25, PF50, and PF100, respectively, considering the price of one kg of SO (1.47 US $) and PF (0.32 US $). Feed cost per kg gain was estimated by multiplying the cost of a one kg diet by the FCR. Total costs were measured by summing the feed cost/bird and all fixed costs, including housing labor, vaccines, drugs, day-old chick, disinfectant, veterinary supervision, etc. Subtracting total costs from the total return, considering the average price of the bird (1.72 US $ per one kg live body weight), was considered the net return. Benefit/cost ratio (B/C ratio) was estimated by the following equation: net return/total costs × 100 [25].

2.9. Statistical Analysis

Differences between the experimental groups were analyzed using one-way ANOVA was applied to determine the effects of replacing SO with PF, in which pens were the statistical units for performance parameters, birds for the carcass, organ weights, and samples for biochemical and other parameters, the General Linear Model package of SPSS (Version 19.0, Chicago, IL, USA). Tukey's multiple range test was used to identify the significant ($p < 0.05$) differences among means of experimental groups.

3. Results

3.1. Chemical Analysis of Fat Sources

The main difference between the FAs profile of SO and PF (Table 2) can be found in the ratio of unsaturated FAs over-saturated FAs (U/S ratio). The U/S ratio of SO (3.84) was 1.7 times that of PF (2.28). Concentrations of polyunsaturated (PU) FAs in SO were higher than that of PF, while the latter was richer in monounsaturated (MU) FAs. The PU/S ratio of SO was 2.4 times that of PF, while the MU/PU ratio of PF was 2.2 times that of SO.

Table 2. Fatty acid profile and energy values of soybean oil and poultry fat.

Item	Soybean Oil	Poultry Fat
Myristic acid (C14:0), %	0.01	0.47
Palmitic acid (C16:0), %	14.69	21.83
Palmitoleic acid (C16:1), %	-	3.14
Heptadecanoic acid (C17:0), %	-	0.20
Stearic acid (C18:0), %	5.50	7.59
Oleic acid (C18:1 n-9), %	26.80	32.06
Vaccenic acid (C18:1 n-7), %	-	1.79
Octadecanedioic acid (C18:2 n-4), %	-	0.24
Linoleic acid (C18:2 n-6), %	44.90	29.26
Linolenic acid (C18:3 n-3), %	6.95	1.69
Arachidic acid (C20:0), %	0.35	0.20
Gadoleic acid (C20:1 n-9),%	0.31	0.20
Eicosatrienoic acid (C20:3 n-3), %	-	0.20
Arachidonic acid (C20:4 n-6), %	-	0.55
Non identified fatty acids	0.49	0.58
Saturated fatty acids	20.55	30.29
Unsaturated fatty acids	78.96	69.13
Monounsaturated fatty acids	27.11	37.19
Polyunsaturated fatty acids	51.85	31.94
Unsaturated fatty acids/Saturated fatty acids	3.842	2.282
Polyunsaturated fatty acids/Saturated fatty acids	2.523	1.054
Monounsaturated fatty acids/Polyunsaturated fatty acids	0.523	1.164
Atherogenic index	0.187	0.343
Thrombogenic index	0.354	0.762
Peroxide value, meq/kg	1.85	4.26
Gross energy, kcal/kg	8400	9460

Milliequivalents per kilogram (mEq/kg); Kilocalorie per Kilogram (kcal/kg).

3.2. Growth Performance and Organ Weights

The partial or full replacement of SO by PF did not alter the final body weight, feed consumption, and FCR of broiler chickens at marketing age (Table 3). The EPEF of PF groups was slightly higher than that of the control. The highest value of EPEF was recorded in PF100. Additionally, the dietary inclusion of different lipid sources had no significant impact on carcass percentage and the relative weight of thigh and breast muscles and liver of 35-day-old broiler chicks. The abdominal fat percentage was increased ($p < 0.01$) in PF100 and PF50 compared to control and PF25.

Table 3. Effect of replacing soybean oil with poultry fat on growth performance and organ weights of broilers.

Item	Experimental Diets [1]				SEM [2]	p-Value
	Control	PF25	PF50	PF100		
Initial body weight, g	43.11	43.00	43.20	43.02	0.032	0.896
Body weight 35 d, g	2131.7	2142.1	2148.3	2157.5	10.33	0.851
Feed consumption 35 d, g	3340.4	3335.0	3330.8	3324.6	10.04	0.957
FCR, g feed:g gain	1.60	1.59	1.58	1.57	0.041	0.188
Mortality, %	0.833	0.833	0.833	0.833	0.089	0.894
EPEF	377.6	382.2	384.9	388.9	3.228	0.664
	Organ weights, %					
Carcass	67.22	67.41	67.72	67.91	0.172	0.497
Breast muscle	23.13	23.27	23.43	23.62	4.243	0.279
Thigh muscle	16.52	16.54	16.22	16.32	0.068	0.262
Liver	2.24	2.28	2.14	2.31	0.028	0.152
Abdominal fat	1.31 ^{b}	1.47 ^{b}	1.90 ^{a}	1.93 ^{a}	0.055	0.002

[1] PF25, 25% of soybean oil replaced by PF; PF50, 50% of soybean oil replaced by PF; PF100, full replacement of soybean oil by PF. [2] SEM, standard error of means. EPEF, European production efficiency factor; FCR, feed conversion ratio. ^{a,b} Means within the same row with different superscripts differ. Gram (g).

3.3. Nutrient Digestibility

As presented in Table 4, digestibility coefficients of dry matter were improved ($p < 0.01$) in PF50 and PF100 compared to the control. Digestibility coefficients of crude protein and fat were not influenced by dietary replacement of PF instead of SO.

Table 4. Effect of replacing soybean oil with poultry fat on nutrient digestibility of broilers.

Item	Experimental Diets [1]				SEM [2]	p-Value
	Control	PF25	PF50	PF100		
Dry matter, %	71.14 ^{b}	72.57 ^{ab}	74.27 ^{a}	74.80 ^{a}	0.491	0.022
Nitrogen, %	69.21	69.93	70.11	70.32	0.264	0.506
Fat, %	77.89	79.93	80.65	80.73	0.392	0.053

[1] PF25, 25% of soybean oil replaced by PF; PF50, 50% of soybean oil replaced by PF; PF100, full replacement of soybean oil by PF. [2] SEM, standard error of means. ^{a,b} Means within the same row with different superscripts differ.

3.4. Plasma Biochemical Analysis

The data presented in Table 5 shows the impact of partial or total replacement of SO by PF on plasma biochemical parameters of broilers at marketing age. Total protein, albumin, AST, HDL-cholesterol, and total cholesterol were not significantly affected by dietary inclusion of PF compared to the control.

Table 5. Effect of replacing soybean oil with poultry fat on plasma biochemical parameters of broilers.

Item	Experimental Diets [1]				SEM [2]	p-Value
	Control	PF25	PF50	PF100		
Aspartate aminotransferase, mg/dL	232.3	225.4	234.9	221.2	3.960	0.607
Total protein, mg/dL	3.72	3.58	3.57	3.77	0.053	0.457
Albumin, mg/dL	2.05	2.03	2.11	2.13	0.038	0.743
Total cholesterol, mg/dL	160.2	157.0	146.3	149.3	2.441	0.155
HDL-cholesterol, mg/dL	80.17	81.00	84.75	85.83	1.071	0.169

[1] PF25, 25% of soybean oil replaced by PF; PF50, 50% of soybean oil replaced by PF; PF100, full replacement of soybean oil by PF. [2] SEM, standard error of means. High Density lipoprotein (HDL-cholesterol); milligrams per decilitre (mg/dL).

3.5. Muscle and Liver Biochemical Analysis

The fatty acid profile of the pectoral muscle was influenced by the type of dietary fat source (Table 6). The main differences in the FAs profile of breast muscle can be found in the concentrations of certain FAs and the PU/S ratio. Concentrations of palmitic acid and docosahexaenoic acids were reduced ($p < 0.01$) in PF50 and PF100. However, levels of oleic acid, linolenic acid, total unsaturated FAs, and PU/S ratio were elevated ($p < 0.05$) in the pectoral muscle of birds fed 50% and 100% PF. Pectoral muscle concentration of α-tocopherol and content of TBARS in the liver were not significantly affected by replacing SO with PF (Table 6). However, numerical reduction in α-tocopherol and elevation in TBARS levels in PF treated groups were observed.

Table 6. Effect of replacing soybean oil with poultry fat on fatty acids profile of broilers pectoral muscle.

Item	Experimental Diets [1]				SEM [2]	p-Value
	Control	PF25	PF50	PF100		
Myristic acid (C14:0), %	1.39	1.38	1.38	1.37	0.035	0.981
Palmitic acid (C16:0), %	22.12 [a]	21.07 [ab]	20.05 [b]	20.02 [b]	0.270	0.008
Palmitoleic acid (C16:1), %	5.46	5.41	5.49	5.43	0.175	0.977
Stearic acid (C18:0), %	9.01	9.17	9.07	9.03	0.189	0.953
Oleic acid (C18:1 n-9c), %	41.55 [b]	43.23 [ab]	44.01 [a]	44.06 [a]	0.329	0.012
Vaccenic acid (C18:1 n-7), %	5.22	5.45	5.48	5.08	0.160	0.807
Linoleic acid (C18:2 n-6), %	9.37	9.26	9.48	9.89	0.132	0.363
Linolenic acid (ALA, C18:3 n-3), %	0.66 [b]	0.64 [b]	0.78 [a]	0.88 [a]	0.027	0.001
Arachidonic acid (AA, C20:4 n-6), %	2.32	2.89	2.56	2.62	0.104	0.304
Eicosapentaenoic acid (EPA, C20:5 n-3), %	0.064	0.068	0.062	0.064	0.002	0.811
Docosapentaenoic acid (DPA, C22:5n-3), %	0.32	0.319	0.317	0.318	0.009	0.916
Docosahexaenoic acid (DHA, C22:6n-3),%	0.979 [a]	0.990 [a]	0.809 [b]	0.872 [b]	0.022	0.002
Saturated fatty acids	32.53	31.62	30.51	30.43	0.356	0.109
Unsaturated fatty acids	65.94 [b]	68.24 [ab]	68.99 [a]	69.22 [a]	0.510	0.049
Monounsaturated fatty acids	52.23	54.08	54.99	54.57	0.438	0.114
Polyunsaturated fatty acids	13.71	14.16	14.00	14.65	0.154	0.182
Unsaturated fatty acids/Saturated fatty acids	2.032 [b]	2.160 [ab]	2.271 [a]	2.275 [a]	0.031	0.006
Polyunsaturated fatty acids/Saturated fatty acids	0.423 [b]	0.448 [ab]	0.461 [a]	0.482 [a]	0.007	0.027
Monounsaturated fatty acids/Polyunsaturated fatty acids	3.810	3.822	3.932	3.744	0.040	0.431
Atherogenic index	0.420 [a]	0.390 [b]	0.371 [b]	0.369 [b]	0.006	0.001
Thrombogenic index	0.853 [a]	0.804 [ab]	0.772 [b]	0.759 [b]	0.011	0.006
Vitamin E, mg/100 g muscle	0.327	0.312	0.288	0.285	0.067	0.224
Liver TBARS, nmol/g	18.33	18.67	19.50	21.50	0.478	0.126

[1] PF25, 25% of soybean oil replaced by PF; PF50, 50% of soybean oil replaced by PF; PF100, full replacement of soybean oil by PF. [2] SEM, standard error of means. [a,b] Means within the same row with different superscripts differ. ThioBarbituric Acid Reactive Substances (*TBARS*); milligrams per 100 g (mg/100 g).

3.6. Economic Efficiency

As presented in Table 7, economic parameters were greatly influenced by partial and full replacement of SO by PF in broiler diets. Feed cost/bird, feed cost/kg gain, and total cost/bird were decreased ($p < 0.001$) in PF25, PF50, and PF100 compared to the control. Net return and benefit/cost ratio were increased ($p < 0.001$) in PF25, PF50, and PF100, and the highest values were recorded in group PF100.

Table 7. Effect of replacing soybean oil with poultry fat on the economic parameters.

Item	Experimental Diets [1]				SEM [2]	p-Value
	Control	PF25	PF50	PF100		
Feed cost/bird, US $	1.770 [a]	1.734 [b]	1.699 [c]	1.662 [d]	0.01	<0.001
Feed cost/kg gain, US $	0.848 [a]	0.827 [b]	0.807 [c]	0.787 [d]	0.01	<0.001
Total cost, US $/bird	2.951 [a]	2.890 [b]	2.831 [c]	2.771 [d]	0.01	<0.001
Total return, US $/bird	3.592	3.610	3.621	3.637	0.02	0.850
Net return, US $/bird	0.642 [c]	0.720 [b]	0.790 [b]	0.866 [a]	0.03	<0.001
B/C ratio, %	21.73 [d]	24.89 [c]	27.90 [b]	31.21 [a]	0.61	<0.001

[1] PF25, 25% of soybean oil replaced by PF; PF50, 50% of soybean oil replaced by PF; PF100, full replacement of soybean oil by PF; B/C, benefit/cost. [2] SEM, standard error of means. [a–d] Means within the same row with different superscripts differ.

4. Discussion

The analyzed composition of the lipids sources in the present study revealed that PUFAs in SO were higher than that of PF while the MUFAs in PF were higher. The U/S and PU/S ratios of SO were 1.7 and 2.4 times that of PF, while the MU/PU ratio of PF was 2.2 times that of SO. These findings are almost similar to NRC [26] and previous findings [8,27]. The predominant FA in PF was oleic acid followed by linoleic acid, while in SO, it was linoleic acid followed by oleic acid. This revealed that the two fat sources were rich in the unsaturated FAs with the superiority of SO by 12.5%, which may explain the effects of these lipids on studied parameters.

Fats are a high-energy feedstuff commonly incorporated in the formulation of commercial poultry diets. Results of earlier studies investigating the impact of dietary addition of different lipid sources on poultry performance were equivocal. Several investigations have reported that supplementation of vegetable oils to poultry feed can improve their performance, carcass traits, and production efficiency by elevating the diet's energy level better than animal fat [28–30]. Others reported a synergistic effect between vegetable oils and animal fat [6,11]. Nevertheless, some studies revealed non-significant differences between animal fat and vegetable oils [9,10,27]. In the present study, no synergy effect between PF and SO was noticed. We also found that partial and total replacement of SO by PF in broiler diets had no significant impact on final body weight, FCR, EFEF, and carcass traits except abdominal fat, which increased PF50 and PF100. A similar trend was reported by Okur [31] and Sanz [32], who noticed an elevation in abdominal fat weight when animal fats were used in broiler diets. The growth performance of broilers is greatly influenced by dietary fat sources and their FA profiles, particularly essential FAs, such as α-linolenic acid and linoleic acid, as their deficiency may retard broilers growth [33]. As our results revealed, the differences between SO and PF in these FAs were insufficient to induce significant differences in birds' growth performance. The lack of differences in growth performance of birds fed diets with SO or PF could also be attributed to the equilibrium ratios of energy-to-protein and energy-to-amino acid in these diets [27,31]. Additionally, Pesti et al. [27] reported that feeding on fat sources with high metabolizable energy resulted in a high amount of fat being deposited. This might explain the increase of abdominal fat in birds fed diets with high levels of PF.

In our study, replacing SO by PF improved dry matter digestibility while digestibility of crude protein and crude fat was not affected. These results are in line with previous findings [34,35]. Fatty acids chain length and U/S ratio greatly affect nutrient digestibility. Fat sources such as tallow and palm oil characterized by a low U/S ratio showed drastic negative impacts on nutrients digestibility [36,37]. Tancharoenrat et al. [35] noticed a reduction in crude fat digestibility for broilers fed diets with fat sources with low U/S ratio such as palm oil (U/S 0.93) and tallow (U/S 0.80) compared to SO (U/S 5.07).

In the present study, the difference between the U/S ratio of PF (2.28) and SO (3.84) was lower than mentioned in the study of Tancharoenrat et al. [35], which might explain the insignificant changes in fat and protein digestibility. Nevertheless, they did not observe a significant difference for crude fat digestibility between PF- (U/S 2.07) and SO-based diets. The authors suggested that the changes in FAs composition of PF were not enough to exert a drastic effect on crude fat digestibility. Furthermore, dietary addition of PF could enhance nutrients' digestion and absorption by reducing digesta rate of passage through the gut, which might explain the improvement in dry matter digestibility [8].

To our knowledge, limited investigations have studied the impact of dietary addition of fat sources on the blood biochemistry of poultry species. The majority of these studies focused on the impact of fat types on the quality of animal products for human uses without studying their effect on birds' health status during production [38]. In the present study, we evaluated the effect of replacing SO by PF on broiler chickens' hepatic function and blood lipids. No significant alterations were observed in all studied parameters among experimental groups. Hu [9] noticed similar results who reported insignificant impact of dietary SO and PF on HDL levels- and LDL-cholesterol and total cholesterol in the serum of Cherry Valley ducks. Donaldson et al. [38] also noticed non-significant changes in serum levels of AST, total protein, albumin, and cholesterol of Japanese quail-fed diets with SO and lard. However, results of serum triglycerides as affected by various dietary fat sources and levels were somewhat contradictory. Some studies reported significant elevated serum triglycerides of humans and birds fed high dietary fat [39,40]. Others reported non-significant changes [9,41], while some observed a significant reduction in its level [38,42]. Donaldson et a. [38] attributed the decrease in serum triglycerides to a possible reduction in de novo synthesis of FAs in the liver as large amounts of FAs were being supplied to the birds via dietary fat sources. The lack of consistency between findings of these studies suggests some differences in lipid handling pathways and multiple potential mechanisms contribute to regulating serum concentrations of cholesterol and triglycerides between different avian species, including postabsorptive lipid metabolism and/or hepatic uptake of HDL-cholesterol.

Consumers have become more concerned about the nutritional aspects, including the lipid profile and FA contents. Chicken meat with its low-fat and high-protein contents has been characterized as the main source of PUFAs [43]. Functional and beneficial foods that contribute to preventing chronic diseases, such as coronary heart disease and metabolic disorders, are characterized by higher concentrations of PUFAs [44,45]. It has been documented that FAs and lipid profiles of chicken meat can be modified by changing broilers' feed composition [46,47]. Reducing SFAs and elevation of PUFAs contents in chicken meat would improve its nutritional value and quality [48]. Our results showed that pectoral muscle levels of palmitic acid and docosahexaenoic acids decreased while concentrations of oleic acid, linolenic acid, total UFAs, and PU/S ratio were elevated PF50 and PF100. These findings are considered positive as the reduction in PU/S ratio, thrombogenic index, and the atherogenic index, and the elevation of linolenic acid (n-3) are favorable in healthy and functional food for human consumption. The added value of n-3 PUFAs to human foods and their favorable impacts of on human health were investigated. Bostami et al. [47] reported the health benefits of long-chain n-3 FAs to animals and humans, such as reducing the risk of heart diseases and lowering the concentration of circulating cholesterol.

Moreover, Pinchasov and Nir [49] documented that PUFAs can inhibit the activity of the 9-desaturase enzyme complex, which responsible for converting SFAs to MUFAs, thereby downregulating the synthesis of MUFAs. Furthermore, the reduction in the atherogenic and thrombogenic indexes in PF50 and PF100 is considered favorable, as Ulbricht and Southgate [17] recommended the low values of these indexes in healthy human diets. Generally, supplementation of PF in broiler diets instead of SO tends to improve the lipid profile of breast meat.

Lipid oxidation, one of the major factors responsible for the deterioration of the quality of meat products, is primarily initiated in the UFAs of membrane phospholipids [17]. Vitamin E (α-tocopherol) plays a fundamental role in protecting these susceptible cellular structures against oxygen-containing free radicals and reduces their content of TBARS [50]. The primary location of vitamin E is within the biological membranes, such as mitochondria and microsomes, which allow its effective function compared to other antioxidants [17]. Therefore, increasing the muscle membrane content of α-tocopherol by dietary manipulation is required. Dietary fat sources generally contain high fat-soluble vitamins, including α-tocopherol, but they vary among themselves. Vegetable oils and most plant-origin feedstuffs are rich in vitamin E, while their content is lower than most animal products [51–54]. In the present study, levels of α-tocopherol in pectoral muscle and TBARS in the liver did not differ by replacing SO with PF. Numerical reduction in α-tocopherol and elevation in TBARS levels in PF treated groups were observed. These findings agree with those of Polycarpo et al. [55], who reported significant changes in hepatic contents of vitamins E and A (fat-soluble vitamins) of broilers fed on corn-based diets with SO or beef tallow as lipid sources. Lauridsen, et al. [56] also noticed that fat sources did not influence the concentration of vitamin E in muscle membranes. Contrarily, Dänicke, et al. [56] and Gatellier, et al. [57] observed higher hepatic vitamin A and muscular TBARS concentrations in birds fed diets with SO compared with tallow. The lack of significance in α-tocopherol concentrations in the present study might be attributed to the low incorporation levels of SO and PF in broilers diet, eliminating the added value of vitamin E to the feed.

Furthermore, the insignificant differences in fat digestibility observed in this study may be considered another explanation as absorption of fat-soluble vitamins depends on fat digestibility and the emulsification process. Knarreborg, et al. [58] documented that good conditions of micelle formation and emulsion increase the bioavailability of α-tocopherol. The numerical reduction in α-tocopherol and increase in TBARS concentrations in the breast and liver tissues of birds fed on PF may be due to the relatively high susceptibility of broiler meat to lipid oxidation when fed diets incorporated with PF [59,60].

As expected, the economic efficiency and benefit-to-cost ratio were significantly improved by dietary replacement of SO by PF in a dose-dependent manner. This effect is due to the large difference between the prices of SO and PF; under our study's condition, the price of SO was 4.6 times that of PF. The differences in the prices of SO and PF are reasonable; since SO is one of the most commonly used fat sources in poultry rations and its various industrial uses. While PF is a cheap by-product of poultry slaughterhouses, its utilization is affordable and reduces its adverse impacts on the environment. Several studies have reported the economic benefits of using PF instead of SO [4,9,34,38].

5. Conclusions

The present study investigated the effect of partial or full replacement of SO by PF on growth performance, nutrient digestibility, plasma lipids, and vitamin E and FAs contents of the pectoral muscle of broilers. Neither a synergistic effect between SO and PF nor effects of the dietary changes on broilers' growth, carcass parts, and blood biochemistry were noticed in the present study. Dietary supplementation of PF improved the FA profile of breast muscle but did not affect muscle content of vitamin E and liver TBARS levels. Economic efficiency was greatly improved in a dose-dependent manner by dietary addition of PF. Therefore, this study revealed that PF can be used as a partial or total replacement of SO in broilers' nutrition without affecting their performance or physiological response to improve their meat products.

Author Contributions: Conceptualization, A.A.S., A.S.A., R.A.A., A.-M.E.A.-M., M.S.A.; data curation, A.A.S., A.S.A., R.A.A., A.-M.E.A.-M., M.S.A.; formal analysis, A.A.S., A.S.A., R.A.A., A.-M.E.A.-M., M.S.A.; funding acquisition, A.A.S., A.S.A., R.A.A., A.-M.E.A.-M., M.S.A.; investigation, A.A.S., A.S.A., R.A.A., A.-M.E.A.-M., M.S.A.; methodology, A.A.S., A.S.A., R.A.A., A.-M.E.A.-M., M.S.A.; supervision, A.A.S., A.S.A., R.A.A., A.-M.E.A.-M., M.S.A.; writing—original draft, A.A.S., A.S.A., R.A.A., A.-M.E.A.-M., M.S.A.; All authors have read and agreed to the published version of the manuscript.

Funding: This research received no specific grant from any funding agency in public, commercial or not-for-profit sectors.

Institutional Review Board Statement: This study was approved by the Local Experimental Animals Care Committee's Ethics Committee and done according to the rules of Kafrelsheikh University, Egypt. (No. 4/2016EC).

Informed Consent Statement: Not applicable.

Data Availability Statement: Data are available upon request.

Conflicts of Interest: The authors declare no conflict of interest.

References

1. Abd El-Moneim, A.E.; Sabic, E.M. Beneficial effect of feeding olive pulp and *Aspergillus awamori* on productive performance, egg quality, serum/yolk cholesterol and oxidative status in laying Japanese quails. *J. Anim. Feed Sci.* **2019**, *28*, 52–61. [CrossRef]
2. Fébel, H.; Mezes, M.; Palfy, T.; Herman, A.; Gundel, J.; Lugasi, A.; Balogh, K.; Kocsis, I.; Blazovics, A. Effect of dietary fatty acid pattern on growth, body fat composition and antioxidant parameters in broilers. *J. Anim. Physiol. Anim. Nutr.* **2008**, *92*, 369–376. [CrossRef] [PubMed]
3. Abdel-Moneim, A.-M.E.; Shehata, A.M.; Khidr, R.E.; Paswan, V.K.; Ibrahim, N.S.; El-Ghoul, A.A.; Aldhumri, S.A.; Gabr, S.A.; Mesalam, N.M.; Elbaz, A.M. Nutritional manipulation to combat heat stress in poultry–A comprehensive review. *J. Therm. Biol.* **2021**, *98*, 102915. [CrossRef] [PubMed]
4. Liu, J.; Yan, H.; Zhang, Y.; Hu, Y.; Zhang, H. Effects of dietary energy and protein content and lipid source on growth performance and carcass traits in Pekin ducks. *Poult. Sci.* **2019**, *98*, 4829–4837. [CrossRef]
5. Jimenez-Moya, B.; Barroeta, A.C.; Tres, A.; Soler, M.D.; Sala, R. Soybean Oil Replacement by Palm Fatty Acid Distillate in Broiler Chicken Diets: Fat Digestibility and Lipid-Class Content along the Intestinal Tract. *Animals* **2021**, *11*, 1035. [CrossRef]
6. Zhang, B.; Haitao, L.; Zhao, D.; Guo, Y.; Barri, A. Effect of fat type and lysophosphatidylcholine addition to broiler diets on performance, apparent digestibility of fatty acids, and apparent metabolizable energy content. *Anim. Feed Sci. Technol.* **2011**, *163*, 177–184. [CrossRef]
7. Beauregard, L.; Moustafa, A.; Sampaio, J. Puntos críticos a considerar en la refinación de aceites para la producción de grasas y margarinas. *Soya Notícias* **1996**, *16*, 10–15.
8. Aardsma, M.; Mitchell, R.; Parsons, C. Relative metabolizable energy values for fats and oils in young broilers and adult roosters. *Poult. Sci.* **2017**, *96*, 2320–2329. [CrossRef] [PubMed]
9. Hu, X.; Wang, W.; Liu, L.; Wang, C.; Feng, W.; Luo, Q.; Han, R.; Wang, X. Effects of fat type and emulsifier in feed on growth performance, slaughter traits, and lipid metabolism of Cherry Valley ducks. *Poult. Sci.* **2019**, *98*, 5759–5766. [CrossRef]
10. Patra, A.K.; Samanta, G.; Pal, K. Effects of an emulsifier on the performances of Khaki Campbell ducks added with different sources of fats. *Front. Agric. China* **2011**, *5*, 605–611.
11. Tufarelli, V.; Bozzo, G.; Perillo, A.; Laudadio, V. Effects of feeding different lipid sources on hepatic histopathology features and growth traits of broiler chickens. *Acta Histochem.* **2015**, *117*, 780–783. [CrossRef] [PubMed]
12. Hygiene of foodstuffs. In *Regulation (EC) No 852/2004 of the European Parliament and of the Council*; European Commission: London, UK, 2004.
13. Laying down specific hygiene rules for food animal origin. In *Regulation (EC) No 853/2004 of the European Parliament and of the Council*; European Commission: London, UK, 2004.
14. Laying down specific rules for the organisation of official controls on products of animal origin intended for human consumption. In *Regulation (EC) No 854/2004 of the European Parliament and of the Council*; European Commission: London, UK, 2004.
15. AOAC. *Official Methods of Analysis of AOAC*, 17th ed.; AOAC: Gaithersburg, MD, USA, 2003.
16. Abd El-Moneim, A.E.; Sabic, E.M.; Abu-Taleb, A.M. Influence of dietary supplementation of irradiated or non-irradiated olive pulp on biochemical profile, antioxidant status and immune response of Japanese quails. *Biol. Rhythm Res.* **2019**, 1–16. [CrossRef]
17. Ulbricht, T.; Southgate, D. Coronary heart disease: Seven dietary factors. *Lancet* **1991**, *338*, 985–992. [CrossRef]
18. Aviagen, R. *Ross Broiler Management Handbook*; Aviagen Limited: Scotland, UK, 2014.
19. Abo Ghanima, M.M.; Abd El-Hack, M.E.; Othman, S.I.; Taha, A.E.; Allam, A.A.; Abdel-Moneim, A.-M.E. Impact of different rearing systems on growth, carcass traits, oxidative stress biomarkers and humoral immunity of broilers exposed to heat stress. *Poult. Sci.* **2020**, *99*, 3070–3078. [CrossRef] [PubMed]
20. Jacobsen, D.; Gertovey, S.; Nielson, H. Digestibility trials with poultry. 322 Bertning fra forsg slabooratoriel udgbet of statens. In *Husdyrbugsudvaly–Kobengaven*; Københavns Universitet: Copenhagen, Denmark, 1960.
21. Saleh, A.A. Effects of fish oil on the production performances, polyunsaturated fattyacids and cholesterol levels of yolk in hens. *Emir. J. Food Agric.* **2013**, *25*, 605–612. [CrossRef]
22. Saleh, A.A.; Paray, B.A.; Dawood, M.A. Olive cake meal and Bacillus licheniformis impacted the growth performance, muscle fatty acid content, and health status of broiler chickens. *Animals* **2020**, *10*, 695. [CrossRef]
23. Faustman, C.; Cassens, R.; Schaefer, D.; Buege, D.; Williams, S.; Scheller, K. Improvement of pigment and lipid stability in Holstein steer beef by dietary supplementation with vitamin E. *J. Food Sci.* **1989**, *54*, 858–862. [CrossRef]

24. Saleh, A.A.; Amber, K.A.; Mousa, M.M.; Nada, A.L.; Awad, W.; Dawood, M.A.; El-Moneim, A.E.-M.E.A.; Ebeid, T.A.; Abdel-Daim, M.M. A mixture of exogenous emulsifiers increased the acceptance of broilers to low energy diets: Growth performance, blood chemistry, and fatty acids traits. *Animals* **2020**, *10*, 437. [CrossRef]
25. Abo Ghanima, M.M.; Bin-Jumah, M.; Abdel-Moneim, A.-M.E.; Khafaga, A.F.; Abd El-Hack, M.E.; Allam, A.A.; El-Kasrawy, N.I. Impacts of Strain Variation on Response to Heat Stress and Boldo Extract Supplementation to Broiler Chickens. *Animals* **2020**, *10*, 24. [CrossRef] [PubMed]
26. NRC. *Nutrient Requirements of Swine*; Eleventh Revised Edition; The National Academies Press: Washington, DC, USA. [CrossRef]
27. Pesti, G.; Bakalli, R.; Qiao, M.; Sterling, K. A comparison of eight grades of fat as broiler feed ingredients. *Poult. Sci.* **2002**, *81*, 382–390. [CrossRef] [PubMed]
28. Nayebpor, M.; Hashemi, A.; Farhomand, P. Influence of soybean oil on growth performance, carcass properties, abdominal fat deposition and humoral immune response in male broiler chickens. *J. Anim. Vet. Adv.* **2007**, *6*, 1317–1322.
29. Zollitsch, W.; Knaus, W.; Aichinger, F.; Lettner, F. Effects of different dietary fat sources on performance and carcass characteristics of broilers. *Anim. Feed Sci. Technol.* **1997**, *66*, 63–73. [CrossRef]
30. Zampiga, M.; Meluzzi, A.; Sirri, F. Effect of dietary supplementation of lysophospholipids on productive performance, nutrient digestibility and carcass quality traits of broiler chickens. *Ital. J. Anim. Sci.* **2016**, *15*, 521–528. [CrossRef]
31. Okur, N. The effects of soy oil, poultry fat and tallow with fixed energy: Protein ratio on broiler performance. *Arch. Anim. Breed.* **2020**, *63*, 91–101. [CrossRef]
32. Sanz, M. Higher lipid accumulation in broilers fed on saturated fats than in those fed on unsaturated fats. *Br. Poult. Sci.* **1999**, *40*, 95–101. [CrossRef] [PubMed]
33. Józefiak, D.; Kierończyk, B.; Rawski, M.; Hejdysz, M.; Rutkowski, A.; Engberg, R.M.; Højberg, O. Clostridium perfringens challenge and dietary fat type affect broiler chicken performance and fermentation in the gastrointestinal tract. *Animal* **2014**, *8*, 912–922. [CrossRef]
34. Jansen, M.; Nuyens, F.; Buyse, J.; Leleu, S.; Van Campenhout, L. Interaction between fat type and lysolecithin supplementation in broiler feeds. *Poult. Sci.* **2015**, *94*, 2506–2515. [CrossRef] [PubMed]
35. Tancharoenrat, P.; Ravindran, V.; Zaefarian, F.; Ravindran, G. Influence of age on the apparent metabolisable energy and total tract apparent fat digestibility of different fat sources for broiler chickens. *Anim. Feed Sci. Technol.* **2013**, *186*, 186–192. [CrossRef]
36. Tancharoenrat, P.; Ravindran, V.; Zaefarian, F.; Ravindran, G. Digestion of fat and fatty acids along the gastrointestinal tract of broiler chickens. *Poult. Sci.* **2014**, *93*, 371–379. [CrossRef]
37. Viveros, A.; Ortiz, L.; Rodríguez, M.; Rebolé, A.; Alzueta, C.; Arija, I.; Centeno, C.; Brenes, A. Interaction of dietary high-oleic-acid sunflower hulls and different fat sources in broiler chickens. *Poult. Sci.* **2009**, *88*, 141–151. [CrossRef]
38. Donaldson, J.; Madziva, M.T.; Erlwanger, K.H. The effects of high-fat diets composed of different animal and vegetable fat sources on the health status and tissue lipid profiles of male Japanese quail (Coturnix coturnix japonica). *Asian-Australas. J. Anim. Sci.* **2017**, *30*, 700. [CrossRef] [PubMed]
39. Ayala, I.; Castillo, A.M.; Adanez, G.; Fernandez-Rufete, A.; Pérez, B.G.; Castells, M.T. Hyperlipidemic chicken as a model of non-alcoholic steatohepatitis. *Exp. Biol. Med.* **2009**, *234*, 10–16. [CrossRef] [PubMed]
40. Kim, H.-J.; Kim, J.H.; Noh, S.; Hur, H.J.; Sung, M.J.; Hwang, J.-T.; Park, J.H.; Yang, H.J.; Kim, M.-S.; Kwon, D.Y. Metabolomic analysis of livers and serum from high-fat diet induced obese mice. *J. Proteome Res.* **2011**, *10*, 722–731. [CrossRef] [PubMed]
41. Donaldson, J.; Dangarembizi, R.; Mtetwa, B.; Madziva, M.; Erlwanger, K. The progressive effects of a high-fat diet on erythrocyte osmotic fragility, growth performance and serum triglyceride and cholesterol levels in Guinea fowl (N umida meleagris) and M uscovy duck (C airina moschata). *J. Anim. Physiol. Anim. Nutr.* **2014**, *98*, 867–874. [CrossRef]
42. Peebles, E.D.; Cheaney, J.; Brake, J.; Boyle, C.R.; Latour, M.A.; McDaniel, C. Effects of added lard fed to broiler chickens during the starter phase. 2. Serum lipids. *Poult. Sci.* **1997**, *76*, 1648–1654. [CrossRef]
43. Howe, P.; Meyer, B.; Record, S.; Baghurst, K. Dietary intake of long-chain ω-3 polyunsaturated fatty acids: Contribution of meat sources. *Nutrition* **2006**, *22*, 47–53. [CrossRef]
44. Russo, G.L. Dietary n − 6 and n − 3 polyunsaturated fatty acids: From biochemistry to clinical implications in cardiovascular prevention. *Biochem. Pharmacol.* **2009**, *77*, 937–946. [CrossRef]
45. Calder, P.C. Functional roles of fatty acids and their effects on human health. *J. Parenter. Enter. Nutr.* **2015**, *39*, 18S–32S. [CrossRef]
46. Scaife, J.; Moyo, J.; Galbraith, H.; Michie, W.; Campbell, V. Effect of different dietary supplemental fats and oils on the tissue fatty acid composition and growth of female broilers. *Br. Poult. Sci.* **1994**, *35*, 107–118. [CrossRef]
47. Bostami, A.; Mun, H.; Yang, C. Breast and thigh meat chemical composition and fatty acid profile in broilers fed diet with dietary fat sources. *J. Food Process. Technol* **2017**, *8*, 672. [CrossRef]
48. Wood, J.; Richardson, R.; Nute, G.; Fisher, A.; Campo, M.; Kasapidou, E.; Sheard, P.; Enser, M. Effects of fatty acids on meat quality: A review. *Meat Sci.* **2004**, *66*, 21–32. [CrossRef]
49. Pinchasov, Y.; Nir, I. Effect of dietary polyunsaturated fatty acid concentration on performance, fat deposition, and carcass fatty acid composition in broiler chickens. *Poult. Sci.* **1992**, *71*, 1504–1512. [CrossRef]
50. Leonhardt, M.; Gebert, S.; Wenk, C. Vitamin E content of different animal products: Influence of animal nutrition. *Z. Ernährungswissenschaft* **1997**, *36*, 23–27. [CrossRef]
51. Meydani, M. Vitamin E. *Lancet* **1995**, *345*, 170–175. [CrossRef]

52. Saleh, A.A.; Kirrella, A.A.; Abdo, S.E.; Mousa, M.N.; Badwi, N.A.; Ebeid, T.A.; Nada, A.L.; Mohamed, M.A. Effects of Dietary Xylanase and Arabinofuranosidase Combination on the Growth Performance, Lipid Peroxidation, Blood Constituents, and Immune Response of Broilers Fed Low-Energy Diets. *Animals* **2019**, *9*, 467. [CrossRef] [PubMed]
53. Saleh, A.A.; Eid, Y.Z.; Ebeid, T.A.; Ohtsuka, A.; Yamamoto, M.; Hayash, K. Feeding Aspergillus awamori reduces skeletal muscle protein breakdown and stimulates growth in broilers. *Anim. Sci. J.* **2012**, *83*, 594–598. [CrossRef] [PubMed]
54. Saleh, A.; Alzawqari, M. Effects of Replacing Yellow Corn with Olive Cake Meal on Growth Performance, Plasma Lipid Profile, and Muscle Fatty Acid Content in Broilers. *Animals* **2021**, *11*, 2240. [CrossRef] [PubMed]
55. Lauridsen, C.; Buckley, D.; Morrissey, P. Influence of dietary fat and vitamin E supplementation on α-tocopherol levels and fatty acid profiles in chicken muscle membranal fractions and on susceptibility to lipid peroxidation. *Meat Sci.* **1997**, *46*, 9–22. [CrossRef]
56. Dänicke, S.; Simon, O.; Jeroch, H.; Bedford, M. Interactions between dietary fat type and xylanase supplementation when rye-based diets are fed to broiler chickens 2. Performance, nutrient digestibility and the fat-soluble vitamin status of livers. *Br. Poult. Sci.* **1997**, *38*, 546–556. [CrossRef]
57. Gatellier, P.; Mercier, Y.; Rock, E.; Renerre, M. Influence of dietary fat and vitamin E supplementation on free radical production and on lipid and protein oxidation in turkey muscle extracts. *J. Agric. Food Chem.* **2000**, *48*, 1427–1433. [CrossRef] [PubMed]
58. Knarreborg, A.; Lauridsen, C.; Engberg, R.M.; Jensen, S.K. Dietary antibiotic growth promoters enhance the bioavailability of α-tocopheryl acetate in broilers by altering lipid absorption. *J. Nutr.* **2004**, *134*, 1487–1492. [CrossRef] [PubMed]
59. Saleh, A.A.; Alhotan, R.A.; Alharthi, A.S.; Nassef, E.; Kassab, M.A.; Farrag, F.A.; Hendam, B.M.; Abumnadour, M.M.A.; Shukry, M. Insight View on the Role of in Ovo Feeding of Clenbuterol on Hatched Chicks: Hatchability, Growth Efficiency, Serum Metabolic Profile, Muscle, and Lipid-Related Markers. *Animals* **2021**, *11*, 2429. [CrossRef]
60. Racanicci, A.; Menten, J.; Regitano-d'Arce, M.; Torres, E.; Pino, L.; Pedroso, A. Dietary oxidized poultry offal fat: Broiler performance and oxidative stability of thigh meat during chilled storage. *Braz. J. Poult. Sci.* **2008**, *10*, 29–35. [CrossRef]

animals

MDPI

Article

Effects of Dietary Supplementation of *Salvia officinalis* L. in Organic Laying Hens on Egg Quality, Yolk Oxidative Stability and Eggshell Microbiological Counts

Dimitrios Galamatis [1,2,*], Georgios A. Papadopoulos [1], Diamanto Lazari [3], Dimitrios Fletouris [4], Evanthia Petridou [5], Georgios I. Arsenos [1] and Paschalis Fortomaris [1]

1 Laboratory of Animal Husbandry, Faculty of Veterinary Medicine, School of Health Sciences, Aristotle University of Thessaloniki, 54124 Thessaloniki, Greece; geopaps@vet.auth.gr (G.A.P.); arsenosg@vet.auth.gr (G.I.A.); fortomap@vet.auth.gr (P.F.)
2 Hellenic Agricultural Organization DIMITRA (ELGO DIMITRA), General Management Assurance of Agricultural Products Quality, Kifissias 33 Str., 54248 Thessaloniki, Greece
3 Laboratory of Pharmacognosy, Faculty of Pharmacy, School of Health Sciences, Aristotle University of Thessaloniki, 54124 Thessaloniki, Greece; dlazari@pharm.auth.gr
4 Laboratory of Milk Hygiene and Technology, Faculty of Veterinary Medicine, School of Health Sciences, Aristotle University of Thessaloniki, 54124 Thessaloniki, Greece; djflet@vet.auth.gr
5 Laboratory of Microbiology, Faculty of Veterinary Medicine, School of Health Sciences, Aristotle University of Thessaloniki, 54124 Thessaloniki, Greece; epetri@vet.auth.gr
* Correspondence: galamatis@elog.gr; Tel.: +30-639-227-5112

Citation: Galamatis, D.; Papadopoulos, G.A.; Lazari, D.; Fletouris, D.; Petridou, E.; Arsenos, G.I.; Fortomaris, P. Effects of Dietary Supplementation of *Salvia officinalis* L. in Organic Laying Hens on Egg Quality, Yolk Oxidative Stability and Eggshell Microbiological Counts. *Animals* **2021**, *11*, 2502. https://doi.org/10.3390/ani11092502

Academic Editors: Natalie Morgan and Youssef A. Attia

Received: 23 July 2021
Accepted: 23 August 2021
Published: 26 August 2021

Simple Summary: Organic laying hen systems are considered welfare-friendly, because hens are raised mainly outdoors under natural conditions. The notion is that eggs produced in those systems are better in terms of quality. Research has found that aromatic plants and their extracts can tackle many of the latter challenges when added to poultry diets due to their antioxidant and antimicrobial effects. The current study investigated the effects of a dietary supplementation of *Salvia officinalis* L. in organically raised laying hens during two experimental periods. The results showed improved oxidative stability of the eggs and reduced microbial counts in the eggshells. The egg quality parameters were partly affected, with the yolk weight showing the largest differences between treatments.

Abstract: Aromatic plants of Labiatae family are used in poultry diets because of their antimicrobial and antioxidant activity. The notion is that hens raised in organic systems face several health and environmental challenges. Hence, the objective here was to assess hens' performances and the quality of their eggs in such systems following a dietary supplementation of *Salvia officinalis* L. in powder form. The experiments were conducted over two successive years (1 and 2). They lasted 16 weeks each and involved 198 laying hens aged 40 weeks old randomly assigned to three groups: Con (control diet), Sal-0.5%, and Sal-1.0% (diets supplemented with *Salvia officinalis* L. at 0.5% or 1.0%, respectively). The malondialdehyde (MDA) levels in egg yolks in year 2 were lower in both Sal-0.5% and Sal-1.0% compared to the Con ($p < 0.05$). The total number of *Enterobacteriaceae* in eggshells were lower in Sal-1.0% compared to the Con ($p < 0.05$) in both years. The results suggest that a dietary supplementation of *Salvia officinalis* L. at 1.0% improves the antioxidant status and reduces the microbial load of eggs produced in organic systems.

Keywords: organic; aromatic plants; laying hens; *Salvia officinalis* L.; malondialdehyde (MDA); *Enterobacteriaceae*

1. Introduction

Egg production systems have evolved because of continuous efforts to ensure the health, balanced nutrition, and wellbeing of birds. Commercial egg production systems comprise confined systems, free range, and organic. The latter are considered as the ones providing the highest welfare status by both egg consumers and the public. Consumers' knowledge and perception of egg production systems prioritizes animal welfare and dictates their purchasing decisions [1]. The prevailing view is that laying hens in free-range systems enjoy improved welfare compared to those housed indoors [2]. Free-range systems provide birds' natural conditions to express their behaviors [3], and their eggs are considered as better in terms of quality [4]. However, it has been documented that a major challenge in these systems is the management of health-related issues [5], while in organic eggs, production welfare issues are under debate [6]. Furthermore, the evaluation of microbial contamination is also used to assess the egg quality [7]. Eggshells derived from alternative housing systems have been identified with a higher level of microorganisms than the ones collected from cage systems [7]. Thus, there is a necessity to produce eggs from organic systems with limited microbial contamination.

The ban of antibiotics as growth promoters (AGP) in animal feeds by the EU [8] forced research to seek alternative methods to cater the need of improving animal performances in livestock systems. The latter also became a priority in organic poultry production, seeking alternatives to control diseases and improve bird performances [9]. Hence, phytogenic feed additives, organic acids, and probiotics became the favorable nonantibiotic growth promoters considering their established use in animal nutrition [10]. Probiotic supplementation in organic laying hens' diets influenced positively in the gut of hens the counts of beneficial bacteria such as *Lactobacillus* spp. and *Bifidobacterium* spp. and reduced the counts of harmful bacteria such as *E. coli*, clostridia, and staphylococci [11]. In the latter study, however, the level of eggshell contamination was not evaluated.

Essential oils and botanicals could serve the same scope as feed ingredients [12]. There is abundant evidence in the literature that aromatic plants and their extracts have antioxidant [9,13] and antimicrobial effects [14–16] following their supplementation in poultry diets. Moreover, such dietary supplements have a positive impact on the gastrointestinal function and nutrient digestibility [17]. Aromatic plants of the Labiatae family have been used in poultry diets due to their antimicrobial and antioxidant activity [9]. According to Bozkurt et al. [18], *Salvia officinalis* L., thyme, and oregano are among the most promising plants of the Labiatae family. The supplementation of *Salvia officinalis* L. extracts in roosters' diets had positive effects on the quantity and quality of the sperm produced, while it increased the testosterone levels [19]. The dietary supplementation of such an extract reduced the *Salmonella* counts in the liver, spleen, and cecum of broilers infected with *Salmonella enteritidis* [20]. Levkut et al. [21] and Farhadi et al. [22] showed that the incorporation of *Salvia officinalis* L. extract in the broilers' diet significantly improved their average daily gain and other performance parameters, whereas, in another study, Ryzner et al. [23] showed that it reduced the oxidative stress parameters evaluated in the red blood cells and kidneys of broilers. Recently, the supplementation of *Salvia officinalis* L. aqueous leaf extract in broilers' diets enhanced the immunity response of broilers and significantly reduced the ileal counts of *E. coli* [24]. Elsewhere, the supplementation of *Salvia officinalis* essential oil in laying strain chicks' diets showed that, when supplemented at lower concentrations, the antioxidant defense mechanisms were improved by the induction of antioxidant enzymes [25].

Considering the available literature, the evidence for the use of *Salvia officinalis* L. in the diets of organic laying hens is limited. Hence, the objective of the present study was to assess the egg quality parameters, oxidative stability of egg yolks, and microbial counts of eggshells of organically raised laying hens following the dietary supplementation of *Salvia officinalis* L. in powder form.

2. Materials and Methods

2.1. Animals and Housing

The experiment was conducted on a commercial organic laying hen farm located in Vasilika Village of Thessaloniki Prefecture, Central Macedonia, Greece. The farm was established in 1994 and started its commercial operation in 1996. Hy-Line Brown laying hens were housed under a free-range system in multiple buildings of the farm covering an area of 1500 m^2. All hens had outdoor access to a field of 25,000 m^2 planted with olive trees. The farm also had 200,000 m^2 of cultivated land with cereals and other feeds following the organic agriculture guidelines used in the hens' diet. All the feeds were formulated and produced in the mill of the farm. The farm organic registration code is 0EL54035. Organic certification was provided and monitored by an independent organization, "Physiologike" (https://physiologike.gr/; (accessed on 20 February 2021), according to EN ISO/IEC 17065.

2.2. Experimental Facility

Laying hens were housed in a designated chamber with the dimensions 6 m × 12 m, covering a surface of 72 m^2. The indoor facility was split into 6 compartments (dimensions 2 m × 4.2 m), each one covering a surface of 8.4 m^2. The compartments were constructed and separated using a portable wooden structure together with a plastic net. Each compartment had an egg nest with 8 available places (1 place per 3 laying hens), a conic feeder with a capacity of 25 kg, and a bell type drinker with a diameter of 38 cm (Plasson Livestock, Maagan Michael, D.N. Menashe 3780500, Israel). To enable outdoor access for all laying hens, each compartment had a specially constructed square pop hole with the dimensions of 0.5 m × 0.5 m. Moreover, each compartment had a door to allow access to personnel for the everyday husbandry practices, e.g., egg collection and feed renewal (Figure 1). The designated chamber had 3 windows, with dimensions of 1.8 m length × 0.8 m height, to ensure proper air circulation. The chamber walls and all equipment used were pressure washed and disinfected before the start of each experiment. Sawdust was used for bedding. An 8-h darkness period was set, with the duration of physical light about 6 h, on average.

(a) (b) (c)

Figure 1. Indoor and outdoor parts of the experimental facility used in the study. (**a**) Corridor in front of the 6 experimental compartments. Experimental diets were kept in front of each respective compartment. (**b**) Inside view of an experimental compartment. Each compartment contained an egg nest with 8 available places, a conic feeder, and a bell-type drinker. A square pop hole enabled outdoor access of the laying hens. (**c**) Outside view of the field planted with olive trees in which laying hens had access. Bell-type drinkers were also placed in the field.

In each experiment, 198 Hy-Line brown laying hens approximately 40 weeks old were used. The laying hens were randomly assigned to 3 experimental groups (n = 66 per group; 2 pens of 33 laying hens per group) as follows: (i) group Con: hens were fed the standard organic diet and served as the control group, (ii) group Sal-0.5%: hens were fed the standard organic diet supplemented with dehydrated *Salvia officinalis* L. in powder form at the level of 0.5%, and (iii) group Sal-1.0%: hens were fed the standard organic diet supplemented with dehydrated *Salvia officinalis* L. in powder form at the level of 1.0%. The study was subdivided into two successive phases. The duration of each experimental phase was 16 weeks and started in July and ended in November of the same year. The weather data of the area where the farm is located were collected from the Hellenic National Meteorological Service (http://www.hnms.gr; accessed on 20 February 2021). The meteorological data analysis of the study period showed a higher average temperature and a lower average precipitation rate during the first experimental period (year 1).

2.3. Ethics Approval Statement and Experimental Design

The research protocol of the study was part of a PhD project of DG and was approved by the General Assembly of the Veterinary Faculty of Aristotle University of Thessaloniki. The General Assembly of the Veterinary Faculty of Aristotle University of Thessaloniki approved the specific PhD protocol in its decision: 55/27-5-2015.

2.4. Ingredient Sources, Diet Formulation, and Analysis

Salvia officinalis L., used in the present study, originated from plants that were cultivated in a field in Western Macedonia, Greece. The harvest was made each year at the beginning of June after blossom. The harvested plants were placed in designated buildings that allowed natural drying, a process that lasted approximately 15 days. Sprouts, leaves, and flowers were subsequently ground into a powder form and stored for future use in the experiment. Table 1 shows the main ingredients of the experimental diets and their nutrient analysis. The proximate analysis and chemical composition of the diets are provided in Table 2, whereas Table 3 shows the chemical composition (%) of the *Salvia officinalis* L. extract used in the experiment. The essential oil of *Salvia officinalis* L. was analyzed at the Laboratory of Pharmacognosy, School of Pharmacy, Faculty of Health Sciences, Aristotle University of Thessaloniki. The dried and powdered leaves of the plant, which were received by the company "DIOSKOURIDIS", were subjected to water distillation for two hours in a Clevenger apparatus according to Europaea Pharmacopeia and connected to a modified refrigerated container of essential oils. Additional cooling was used in order to reduce the byproducts of the heat treatment. After distillation, the essential oil was taken up in 2 mL of pentane (GC grade) and filtered through anhydrous sodium sulfate to dehydrate it. The obtained essential oil was kept at −4 °C until it was analyzed. The essential oil yield was expressed in mL.100^{-1} g d.w. Essential oil analyses were performed on a Shimadzu GC-2010-GCMS-QP2010 system operating at 70 eV. This was equipped with a split/spitless injector (230 °C) and a fused silica HP-5 MS capillary column (30 m × 0.25 mm i.d., film thickness 0.25 μm). The temperature program was from 50–290 °C at a rate of 4 °C/min. Helium was used as a carrier gas at a flow rate of 1.0 mL/min. The injection volume of each sample was 1 μL. The retention indices (RI) for all the compounds were determined according to Van den Dool and Kratz [26], using n-alkanes as the standards. The identification of the components was based on a comparison of their mass spectra with those of NIST21 and NIST107 [27] and by comparison of their retention indices with the literature data [28]. The essential oils were often subjected to co-chromatography with authentic compounds (Fluka, Sigma).

The cultivated *Salvia officinalis* L. had a 2.37% extract concentration with the following basic components:

- *cis*- or α-Thujone (33.80%),
- *trans*- or β-Thujone (6.97%),
- 1,8-Cineole (=Eucalyptol) (11.61%), and
- Camphor (24.54%).

Approximately 98.95% of the composition of the extract was identified, and of 36 active substances, the concentration was analyzed (Table 3).

Table 1. Gross ingredient composition (kg/1000 kg, as fed basis) of the experimental laying hen diets Con (control diet), Sal-0.5%, and Sal-1.0% (diets supplemented with *Salvia officinalis* L. at 0.5% or 1.0%, respectively).

Ingredients	Treatments		
	Con	Sal-0.5%	Sal-1.0%
Corn	518	510	505
Soybean-44%CP	238	240	242
Barley	100	98	93
Limestone	104.4	104.4	104.4
Monocalcium Phosphate	9.5	9.5	9.5
DL-Methionine	1.5	1.5	1.5
Choline	0.5	0.5	0.5
[1] Premix	2	2	2
Soybean oil	22	25	28
Salvia officinalis L.	0	5	10
Sodium carbonate	2.3	2.3	2.3
Salt	1.8	1.8	1.8

[1] per kg of diet: vitamin A, 10,000 I.U.; vitamin D3, 2500 I.U; vitamin E, 30 mg; vitamin K, 4 mg; vitamin B1, 1 mg; vitamin B2, 5 mg; vitamin B6, 3 mg; vitamin B12, 0.02 mg; vitamin B3, 30 mg; vitamin B5, 10 mg; folic acid, 1 mg; biotin, 0.05 mg; vitamin C, 10 mg; choline, 400 mg; cobalt, 0.20 mg; copper, 10 mg; iodine, 1 mg; iron, 40 mg; manganese, 120 mg; selenium, 0.30 mg, zinc, 100 mg.

Table 2. Proximate analysis and chemical composition of the experimental diets of experimental laying hen diets: the Con (control diet), Sal-0.5%, and Sal-1.0% (diets supplemented with *Salvia officinalis* L. at 0.5% or 1.0%, respectively).

Nutrients	Treatments		
	Con	Sal-0.5%	Sal-1.0%
Crude protein	15.6	15.6	15.6
Crude fat	4.69	4.97	5.24
Crude fiber	3.08	3.07	3.06
Total calcium	4.17	4.17	4.17
Total phosphorus	0.51	0.51	0.51
Available phosphorus	0.32	0.32	0.32
Metabolizable Energy (Kcal/kg)	2808	2805	2803
Lysine	0.83	0.83	0.83
Methionine	0.41	0.41	0.41
Methionine-Cystine	0.68	0.68	0.68
Threonine	0.6	0.6	0.6
Tryptophan	0.18	0.18	0.18
Sodium	0.15	0.15	0.15
Chloride	0.14	0.14	0.14

Table 3. Chemical composition (%) of *Salvia officinalis* L. essential oil used in the experiment.

Chemical Compounds	RI [a]	%	Method [b]
Tricyclene	920	tr	AI, MS
α-Thujene	926	tr	AI, MS
α-Pinene	932	2.47	AI, MS, Co-GC
Camphene	946	2.46	AI, MS
β-Pinene	974	1.04	AI, MS, Co-GC
β-Myrcene	992	0.64	AI, MS, Co-GC
α-Phellandrene	1004	0.06	AI, MS
α-Terpinene	1016	0.11	AI, MS
p-Cymene	1024	0.51	AI, MS, Co-GC
Limonene	1028	1.54	AI, MS
1,8-Cineole (=Eucalyptol)	1030	11.61	AI, MS, Co-GC

Table 3. *Cont.*

γ-Terpinene	1059	0.17	AI, MS
trans-Linalool oxide (furanoid)	1073	tr	AI, MS
Terpinolene	1088	0.19	AI, MS
Linalool	1100	0.34	AI, MS, Co-GC
α-Thujone	1104	33.80	AI, MS
β-Thujone	1116	6.97	AI, MS
α-Campholenal	1126	tr	AI, MS
Isothujol	1135	0.09	AI, MS
cis-Sabinol	1140	0.07	AI, MS
Camphor	1143	24.54	AI, MS
Neoisothujol	1150	0.07	AI, MS
trans-Pinocamphone	1161	0.10	AI, MS
Borneol	1165	2.93	AI, MS, Co-GC
Menthol	1173	0.08	AI, MS
Terpinen-4-ol	1177	0.56	AI, MS, Co-GC
p-Cymen-8-ol	1186	0.08	AI, MS
α-Terpineol	1191	0.28	AI, MS
Myrtenol	1197	0.14	AI, MS
Isobornyl acetate	1286	1.73	AI, MS
trans-Pinocarvyl acetate	1294	0.23	AI, MS
β-Caryophyllene	1421	0.39	AI, MS, Co-GC
α-Caryophyllene	1455	1.34	AI, MS, Co-GC
Caryophyllene oxide	1586	0.20	AI, MS, Co-GC
Viridiflorol	1594	3.07	AI, MS
Humulene epoxide	1612	0.96	AI, MS

[a] HP-5MS column. [b] Identification method: RI = Retention Index determined on a HP-5 MS capillary column using a homologous series of n-alkanes (C9-C25), MS = mass spectrum, Co-GC = co-injection with authentic compound, and tr = traces, concentrations <0.05.

2.5. Laying Hen Performance

During both experimental periods, the feed intake of laying hens was measured on a weekly basis, and the average daily feed intake was calculated. The feed refusals and egg weight were recorded on a weekly basis to have an estimation of the average daily feed intake and egg mass production, respectively.

The daily egg production was also recorded for each experimental group by collecting eggs every morning (between 9:00–10:00 a.m.). The weekly egg-laying rate was expressed in % on a treatment basis. The feed conversion ratio was calculated by dividing the feed intake with the average egg weight for the respective periods and was expressed as kg of feed per kg of eggs produced. The individual body weight of hens was measured with a digital balance (Supra: SS3242, precision 5 g, Dinaksa Pesaje Industrial, Arrigorriaga—Vizcaya, Spain) at the start, the middle, and at the end of each experimental period.

2.6. Egg Quality

At weeks 2, 4, 6, 8, 10, 12, 14, and 16 of each experimental period, 12 eggs per treatment were collected to assess the egg quality parameters. The eggs were collected randomly from each treatment and weighed to calculate the average egg weight. The average egg mass was calculated on a weekly and on a pen basis by multiplying the egg weekly laying performance (%) with the average egg weight and dividing by 100: Egg mass = (Hen week % egg production * Egg Weight)/100. The eggs were weighed with a digital balance with 0.1-g accuracy (Navigator TM, N2B110, OHAUS Corporation, Parsippany, NJ, USA). The length and width of the eggs were measured with a digital caliper (EMC, LTD, China) with 0.01-mm accuracy. The egg shape index was calculated using the formula: shape index = (width/length) × 100. The eggshell color was measured with a reflectometer (EQ Reflectometer, York Electronics Centre, York, UK), while the egg-specific gravity was calculated using the method based on the Archimedes principle. The eggshell was washed to remove the adhering albumen and air-dried. The thickness of the eggshell with the membranes was measured with a caliper (accuracy 0.001 in, AMES, Waltham, MA, USA), while its weight was measured using a digital balance (Navigator TM, N2B110, OHAUS Corporation, Parsippany, NJ, USA). The albumen weight was calculated by subtracting the weights of the egg yolk and shell from the weight of the egg. The yolk color was estimated using the Roche Colour Yolk Fan, while the Haugh units were measured using designated equipment by the EQM York Electronics Center (Egg Quality Microprocessor, Technical Services & Supplies Ltd., Dunnington, York, UK). The egg quality parameters were measured at the Laboratory of Animal Husbandry, Faculty of Veterinary Medicine, Aristotle University of Thessaloniki.

2.7. Oxidative Parameters Analysis

2.7.1. Iron-Induced Lipid Oxidation of Egg Yolk

The oxidative stability was assessed in 12 egg yolks from each treatment sampled at the end of each experimental period (36 eggs/experimental period in total) following iron-induced lipid oxidation. Twelve (12) yolks of each treatment were mixed to create four mixtures of three yolks each. Four 1-g samples from each of the four mixtures were weighed into 50-mL centrifuge tubes, and iron-induced lipid oxidation was carried out with a modification of the method of Galobart et al. [29]. According to that, 0.5 mL of 5-mM ferrous sulphate and 0.5 mL of 2-mM ascorbic acid were added to the yolk samples, and the contents of the tubes were vortex-mixed vigorously for 15 s. Following incubation at 37 °C for either 0, 50, 100, or 150 min, all yolk samples were immediately submitted to malondialdehyde (MDA) determination for evaluating the extent of lipid oxidation.

2.7.2. Evaluation of Lipid Oxidation in Egg Yolk

The evaluation of lipid oxidation was based on malondialdehyde (MDA) determination using the selective third-order derivative spectrophotometric method developed by Botsoglou et al. [30]. According to this method, the yolk sample (12 egg yolks per treatment) was mixed with 5 mL of 0.8% butylated hydroxytoluene (Sigma Chemical, Co., St. Louis, MO, USA) in hexane and 8 mL of 5% aqueous trichloroacetic acid (Sigma Chemical, Co., St. Louis, MO, USA). The mixture was homogenized (Ultra-Turrax, Janke & Kunkel-IKA-Labortechnik, Staufen, Germany) for 30 s and centrifuged (Hettich Universal-1200) for 3 min at 2000× *g*. The top hexane layer was discarded, and a 2.5-mL aliquot of the bottom aqueous layer was mixed with 1.5 mL of 0.8% aqueous 2-thiobarbituric acid (Sigma Chemical, Co., St. Louis, MO, USA). The mixture was incubated at 70 °C for 30 min and, following cooling under tap water, was submitted to third-order derivative spectrophotometry (Shimadzu, Model UV-160A, Tokyo, Japan). The height of the peak at 521.5 nm was used for calculation of the MDA concentration in the yolk extracts based on the slope and intercept data of the computed least-squares fit of a freshly prepared standard calibration curve.

2.8. Microbiology Analysis

All samples were aseptically collected, placed in sterile bags, and transferred in cool bags to the Laboratory of Microbiology and Infectious Diseases, School of Veterinary Medicine, Aristotle University of Thessaloniki for bacteriological analyses. The detection of *Campylobacter* and *Salmonella* was performed according to ISO 10272-1 [31] and ISO 6579-1 [32], respectively. The detection and enumeration of *Enterobacteriaceae* was performed according to ISO 21528-2 [33]. For *Campylobacter* detection, three eggshells were separated from their contents and were aseptically placed and weighted in a sterile stomacher bag. A tenfold dilution was obtained by adding an enrichment medium Bolton broth. After homogenization, the samples were incubated in a microaerophilic atmosphere at 37 °C for 4 h and then at 41.5 °C for 24 h. From the enrichment culture, 10 μL were transferred and spread in the selective mCCD agar (OXOID) and incubated at 41.5 °C for 44 h in the above atmosphere. All plates were examined for the growth of suspected *Campylobacter* colonies. For *Salmonella* detection, 25 g of both eggshells and contents (yolk and albumen) of 2 eggs were weighted and pre-enriched in 225 mL of buffered peptone water (BPW; Biolife, Italy) at 37 °C for 18 h. Then, 100 μL were plated in 3 drops equally spaced onto the surface of Modified Semi-solid Rappaport Vassiliadis agar (MSRV, OXOID) and incubated at 41.5 °C for 24 h. Plates with no growth were additionally incubated for 24 h. All plates were examined for the growth of white grey colonies with a turbid zone around the droplet. Moreover, 1 mL of the pre-enriched culture was transferred to 10-mL Muller Kauffman tetrathionate/novobiocin broth (MKTTn-Biolife, Italian S.r.L, Milano, Italy) and incubated at 37 °C for 24 h. Suspected colonies from the MSRV plates, as well as a loop from MKTTn, was spread to XLD (Merck, Germany) and RAMBACH (Merck, Germany) agar plates. After the incubation at 37 °C for 24 h, the growth of typical or atypical *Salmonella* colonies was evaluated. For the detection and enumeration of *Enterobacteriaceae*, 10 g of the shell sample was transferred aseptically to 90 mL of diluent buffered peptone water (BPW). After homogenization, 1 mL of the initial dilution was transferred to 9 mL of diluent in the tubes. This resulted in 10^{-2} and 10^{-3} dilutions. Double-row Petri dishes (Ø 90 mm) were inoculated with a sterile pipette in 1 mL of the initial dilution, as well as the next 2 decimal dilutions. Approximately 10 mL of the Violet Red Bile Glucose agar (VRBG) substrate (Biolife, Italian S.r.L, Milan, Italy) was added to each plate. After complete solidification of the material, approximately 10 mL of the VRBG agar covering layer was added and incubated at 37 °C for 24 h ± 2 h. Pink to red or purple colonies were counted as *Enterobacteriaceae*, while 5 colonies from each plate was subjected to further biochemical identification according to the ISO methodology mentioned above.

2.9. Statistical Analysis

The data were analyzed using the Statistical Package for Social Sciences software (SPSS 25.0 Version, Chicago, IL, USA). Statistical significance was considered at $p < 0.05$, while a statistical trend was considered for those values between $0.05 < p < 1.0$. The results were presented as the mean ± standard deviation (SD) where appropriate. Parameters were analyzed with one-way ANOVA, and post-hoc comparisons between treatments were made by Tukey's test. Treatment was included in the model as the fixed factor. In accordance with the recently published study in organic laying hens by van der Heide et al. [34], which consisted of a similar experimental layout as the current study, the laying hen performance data were not subjected to statistical analysis due to a limited number of repetitions ($n = 2$ per treatment), and thus, only descriptive data are presented for years 1 and 2. The combined laying hen performance data for years 1 and 2 were analyzed for differences between the treatments. The egg quality data, egg lipid oxidation data and eggshell microbiology data were analyzed statistically, considering the sample (egg) as the statistical unit. The egg quality data and egg lipid oxidation were also analyzed with one-way ANOVA. Data on the microbiology analysis of the eggshells and, specifically, of the enumeration of *Enterobacteriaceae* were analyzed with a nonparametric Mann–Whitney test.

3. Results

3.1. Laying Hen Performance

The performance parameters during the first experimental period (year 1) are summarized in Table 4. The laying% and egg mass were numerically higher in the Con and Sal-1.0% groups compared to Sal-0.5%. The average daily feed intake (ADFI) was numerically higher in Sal-1.0% than Sal-0.5% and intermediate in the Con group. The feed conversion ratio (FCR) was numerically lower in the Con and Sal-1.0% groups than Sal-0.5%. The body weight of laying hens did not differ either between treatments at the start or at the end of the experimental period.

Table 4. Effects of treatments on the performance parameters of laying hens during the 1st experimental period ($n = 2$ per treatment for the laying performance (%), ADFI, egg mass, and FCR; $n = 66$ per treatment for the body weight measurements).

Parameter	Treatments			
	Con	Sal-0.5%	Sal-1.0%	p-Value
Laying performance (%)	60.56 ± 6.780	51.95 ± 11.349	59.86 ± 7.197	na
ADFI (g/day)	104.17 ± 12.867	101.24 ± 11.791	105.17 ± 10.161	na
Egg mass (g)	38.25 ± 4.418	32.32 ± 6.657	38.45 ± 3.888	na
FCR	2.77 ± 0.277	3.25 ± 0.797	2.78 ± 0.368	na
Body weight start	1649.17 ± 166.24	1631.82 ± 159.65	1611.59 ± 167.90	0.424
Body weight 16th week	1766.78 ± 197.61	1725.77 ± 199.93	1755.59 ± 170.19	0.506

Con: hens fed the control diet, Sal-0.5%: hens were fed the control diet supplemented with 0.5% of *Salvia officinalis* L., Sal-1.0% hens were fed the control diet supplemented with 1.0% of *Salvia officinalis* L., and na: not applicable.

The results of the performance parameters during the second experimental period (year 2) are summarized in Table 5. The laying%, ADFI, and egg mass were numerically higher in the Sal-0.5% compared to the other two groups. The body weights of laying hens did not differ between treatments either at the start or at the end of the experimental period.

Table 5. Effects of treatments on the performance parameters of laying hens during the 2nd experimental period ($n = 2$ per treatment for the laying performance (%), ADFI, egg mass, and FCR; $n = 66$ per treatment for the body weight measurements).

Parameter	Treatments			
	Con	Sal-0.5%	Sal-1.0%	p-Value
Laying performance (%)	57.72 ± 4.366	62.89 ± 4.476	55.49 ± 5.441	na
ADFI (g/day)	111.62 ± 10.622	117.78 ± 7.294	110.46 ± 5.805	na
Egg mass (g)	37.75 ± 3.464	41.44 ± 3.691	36.51 ± 3.453	na
FCR	2.98 ± 0.364	2.86 ± 0.251	3.06 ± 0.402	na
Body weight start	1549.85 ± 191.46	1519.5 ± 169.11	1490.30 ± 138.80	0.128
Body weight 16th week	1832.69 ± 188.90	1840.00 ± 244.49	1789.83 ± 159.90	0.358

Con: hens fed the control diet, Sal-0.5%: hens were fed the control diet supplemented with 0.5% of *Salvia officinalis* L., Sal-1.0%: hens were fed the control diet supplemented with 1.0% of *Salvia officinalis* L., and na: not applicable.

When investigating the results for both the experimental periods, it was revealed that the egg mass was higher in the Sal-0.5% than the Con group, while, in the Sal-1.0% group, it was intermediate ($p = 0.008$) (Table 6). The data on the FCR showed a better feed conversion in the Sal-0.5%compared to the Con group ($p = 0.031$).

Table 6. Effects of treatments on the performance parameters of laying hens for both the 1st and 2nd experimental periods ($n = 4$ per treatment).

Parameter	Treatments			
	Con	Sal-0.5%	Sal-1.0%	p-Value
Laying performance (%)	58.67 ± 6.15	57.36 ± 10.08	58.25 ± 6.43	0.620
ADFI (g/day)	108.59 ± 11.59	109.21 ± 12.70	108.17 ± 7.36	0.928
Egg mass (g)	35.85 ± 6.37 ^{a}	39.60 ± 4.16 ^{b}	36.96 ± 3.58 ^{ab}	0.008
FCR	3.14 ± 0.80 ^{a}	2.77 ± 0.33 ^{b}	2.95 ± 0.38 ^{ab}	0.031

Con: hens fed the control diet, Sal-0.5%: hens were fed the control diet supplemented with 0.5% of *Salvia officinalis* L., and Sal-1.0% hens were fed the control diet supplemented with 1.0% of *Salvia officinalis* L. ^{a,b} Values sharing different superscripts differ between them significantly at $p < 0.05$.

3.2. Egg Quality Parameters

The treatment differences for the egg quality parameters were investigated separately for each year and for each sampling timepoint—namely, weeks 2, 4, 6, 8, 10, 12, 14, and 16. The results presented here to the overall average values obtained in each experimental period.

The results on the egg quality parameters during year 1 are summarized in Table 7. The egg weights tended to be greater in the Sal-1.0% group than the Con (p = 0.087). The yolk weight was higher in Sal-1.0% compared to Con, with the Sal-0.5% was intermediate (p = 0.043). The shell weight and egg width tended to greater in the Sal-1.0% than the other two groups (p = 0.060 and p = 0.067, respectively). The albumen and yolk pH were significantly higher in the Sal-1.0% compared to the Con (p = 0.001 and p = 0.002), while the yolk pH was also higher in the Sal-0.5% than the Con group.

Table 7. Effects of the treatments on the egg quality parameters during year 1 (n = 96 per treatment).

Parameter	Treatments			
	Con	Sal-0.5%	Sal-1.0%	*p*-Value
Egg weight (g)	63.52 ± 5.818 ^{x}	63.99 ± 6.202 ^{xy}	64.82 ± 6.431 ^{y}	0.087
Yolk weight (g)	15.20 ± 1.660 ^{a}	15.30 ± 1.675 ^{ab}	15.50 ± 1.712 ^{b}	0.043
Shell weight (g)	5.99 ± 0.724 ^{x}	5.97 ± 0.682 ^{x}	6.11 ± 0.725 ^{y}	0.060
Albumen weight (g)	42.33 ± 4.671	42.72 ± 5.381	43.15 ± 5.152	0.268
Shell thickness (mm)	0.44 ± 0.039	0.43 ± 0.038	0.44 ± 0.036	0.143
Egg length (mm)	58.06 ± 2.552	58.38 ± 2.443	58.52 ± 2.381	0.138
Egg width (mm)	43.93 ± 1.421 ^{x}	43.96 ± 1.488 ^{x}	44.24 ± 1.550 ^{y}	0.067
Shape index	0.76 ± 0.028	0.75 ± 0.027	0.76 ± 0.027	0.363
Shell color	29.07 ± 4.927	29.41 ± 4.348	29.46 ± 4.964	0.653
Yolk weight	6.94 ± 0.988	6.99 ± 0.827	7.04 ± 0.792	0.546
Haugh units	90.45 ± 9.077	90.62 ± 10.140	89.28 ± 10.368	0.327
Albumen pH	8.62 ± 0.274 ^{a}	8.59 ± 0.291 ^{a}	8.68 ± 0.214 ^{b}	0.001
Yolk pH	6.13 ± 0.158 ^{a}	6.09 ± 0.101 ^{b}	6.10 ± 0.068 ^{b}	0.002
Yolk color	6.94 ± 0.988	7.00 ± 0.889	7.04 ± 0.792	0.466
Specific gravity	1.08 ± 0.007	1.08 ± 0.006	1.08 ± 0.006	0.655

Con: hens fed the control-standard diet, Sal-0.5%: hens were fed the control diet supplemented with 0.5% of *Salvia officinalis* L., and Sal-1.0% hens were fed the control diet supplemented with 1.0% of *Salvia officinalis* L. ^{a,b} Values sharing different superscripts differ between them significantly at $p < 0.05$. ^{x,y} Values sharing different superscripts tend to differ between them at $1 > p > 0.05$.

The results on the egg quality parameters during year 2 are summarized in Table 8. The majority of the egg quality parameters evaluated were similar between the treatments, and significant differences appeared only for the albumen and yolk pH. Albumen pH was higher in the Sal-1.0% group compared to the other two groups and was higher in the Sal-0.5% than the Con group ($p < 0.001$). The yolk pH was higher in both groups supplemented with *Salvia officinalis* L. compared to the Con group ($p < 0.001$).

3.3. Oxidative Stability of Egg Yolk

The results of the extent of lipid oxidation (levels of malondialdehyde, MDA) for year 1 are presented in Table 9. No significant difference was detected between the treatments.

Table 8. Effects of treatments on the egg quality parameters during year 2 (n = 96 per treatment).

Parameter	Treatments			
	Con	Sal-0.5%	Sal-1.0%	p-Value
Egg weight (g)	65.37 ± 5.497	65.87 ± 6.411	65.84 ± 5.037	0.597
Yolk weight (g)	15.42 ± 1.226	15.58 ± 1.819	15.61 ± 1.468	0.362
Shell weight (g)	6.13 ± 0.649	6.16 ± 0.756	6.18 ± 0.594	0.695
Albumen weight (g)	43.82 ± 4.612	44.13 ± 5.089	44.04 ± 4.179	0.787
Shell thickness (mm)	0.44 ± 0.035	0.43 ± 0.034	0.44 ± 0.0316	0.428
Egg length (mm)	58.17 ± 1.958	58.42 ± 2.408	58.26 ± 2.087	0.493
Egg width (mm)	44.62 ± 1.409	44.70 ± 1.543	44.77 ± 1.218	0.586
Shape index	0.77 ± 0.025	0.77 ± 0.026	0.77 ± 0.025	0.471
Shell color	25.27 ± 5.384	25.80 ± 4.790	25.23 ± 3.987	0.349
Yolk weight	7.66 ± 0.979	7.77 ± 0.915	7.82 ± 0.852	0.231
Haugh units	91.77 ± 9.992	92.17 ± 8.034	90.41 ± 10.431	0.254
Albumen pH	8.40 ± 0.293 ^{a}	8.48 ± 0.232 ^{b}	8.52 ± 0.220 ^{c}	<0.001
Yolk pH	6.01 ± 0.127 ^{a}	6.02 ± 0.109 ^{a}	6.06 ± 0.208 ^{b}	0.001
Yolk color	7.66 ± 0.979	7.77 ± 0.915	7.82 ± 0.852	0.212
Specific gravity	1.09 ± 0.006	1.09 ± 0.006	1.09 ± 0.007	0.711

Con: hens fed the control-standard diet, Sal-0.5%: hens were fed the control diet supplemented with 0.5% of *Salvia officinalis* L., and Sal-1.0% hens were fed the control diet supplemented with 1.0% of *Salvia officinalis* L. $^{a–c}$: Values sharing different superscripts differ between them significantly at $p < 0.05$.

Table 9. Effects of the treatments on the oxidative stability of egg yolk during year 1 (n = 12 per treatment).

Time	Treatments			
	Con	Sal-0.5%	Sal-1.0%	p-Value
T = 0 min	43.59 ± 7.749	56.23 ± 12.571	41.62 ± 6.538	0.107
T = 50 min	859.71 ± 424.689	951.88 ± 336.77	688.70 ± 104.288	0.521
T = 100 min	772.75 ± 247.486	780.29 ± 103.399	756.52 ± 295.707	0.989
T = 150 min	721.16 ± 272.843	628.99 ± 102.299	561.16 ± 132.086	0.497

Con: hens fed the control-standard diet, Sal-0.5%: hens were fed the control diet supplemented with 0.5% of *Salvia officinalis* L., and Sal-1.0% hens were fed the control diet supplemented with 1.0% of *Salvia officinalis* L.

The results of the extent of lipid oxidation for year 2 are presented in Table 10. At 50 min and 100 min, the lipid oxidation was significantly lower in the Sal-1.0% group compared to the control one, while it was intermediate for the Sal-0.5% group (p = 0.034 and p = 0.038, respectively). At 150 min, the lipid oxidation was lower in both the groups supplemented with *Salvia officinalis* L. compared to the control one (p = 0.010).

Table 10. Effects of the treatments on the oxidative stability of egg yolk during year 2 (n = 12 per treatment).

Time	Treatments			
	Con	Sal-0.5%	Sal-1.0%	p-Value
T = 0 min	45.78 ± 18.961	55.65 ± 21.167	62.61 ± 30.821	0.630
T = 50 min	582.03 ± 53.818 ^{a}	533.33 ± 59.482 ^{ab}	448.67 ± 66.819 ^{b}	0.034
T = 100 min	585.51 ± 38.917 ^{a}	553.62 ± 113.305 ^{a}	430.15 ± 48.732 ^{b}	0.038
T = 150 min	731.01 ± 168.776 ^{a}	513.04 ± 101.913 ^{b}	413.33 ± 33.73 ^{b}	0.010

Con: hens fed the control-standard diet, Sal-0.5%: hens were fed the control diet supplemented with 0.5% of *Salvia officinalis* L., and Sal-1.0% hens were fed the control diet supplemented with 1.0% of *Salvia officinalis* L. ^{a,b} Values sharing different superscripts differ between them significantly at $p < 0.05$.

3.4. Microbiological Analysis of Eggshells

The testing of both egg yolk and eggshell revealed the absence of *Salmonella* spp. and *Campylobacter* spp. during the first experiment. For the colonies of *Enterobacteriaceae*, their number per experimental group is presented in Table 11. The total number of

Enterobacteriaceae was significantly lower in the Sal-1.0% group compared to the Con ($p = 0.032$).

Table 11. Effects of the treatments on the total number of *Enterobacteriaceae* of eggshells collected during year 1 ($n = 12$ per treatment).

Parameter	Treatments		
	Con	**Sal-0.5%**	**Sal-1.0%**
Enterobacteriaceae (N/g)	235.0 ± 147.16	179.0 ± 154.95	108.8 ± 130.71
p-Value	Con vs. Sal-0.5%	Con vs. Sal-1.0%	Sal-0.5% vs. Sal-1.0%
	0.282	0.032	0.342

Con: hens fed the control-standard diet, Sal-0.5%: hens were fed the control diet supplemented with 0.5% of *Salvia officinalis* L., and Sal-1.0% hens were fed the control diet supplemented with 1.0% of *Salvia officinalis* L.

Additionally, in year 2, the testing of both egg yolk and eggshell revealed the absence of *Salmonella* spp. and *Campylobacter* spp. during the second experiment. For the colonies of *Enterobacteriaceae*, their number per experimental group is presented in Table 12. The total number of Enterobacteriaceae was significantly lower in the Sal-1.0% group compared to the Con ($p = 0.041$) and tended to be lower in Sal-1.0% compared to Sal-0.5% ($p = 0.087$).

Table 12. Effects of the treatments on the total number of Enterobacteriaceae of the eggshells collected during year 2 ($n = 12$ per treatment).

Parameter	Treatments		
	Con	**Sal-0.5%**	**Sal-1.0%**
Enterobacteriaceae (N/g)	233.3 ± 149.74	231.0 ± 184.89	123.3 ± 87.00
p-Value	Con vs. Sal-0.5%	Con vs. Sal-1.0%	Sal-0.5% vs. Sal-1.0%
	0.863	0.041	0.087

Con: hens fed the control-standard diet, Sal-0.5%: hens were fed the control diet supplemented with 0.5% of *Salvia officinalis* L., and Sal-1.0% hens were fed the control diet supplemented with 1.0% of *Salvia officinalis* L.

4. Discussion

The development in poultry production systems from traditional cages to enriched cages, free-range systems, and organic systems resulted in multiple challenges for the health and welfare of laying hens. The new systems also increased the efforts to maintain the high performance and health of hens [35]. As asserted in the Introduction, the supplementation of aromatic plants in the diet of laying hens has been a promising practice recently due to their antimicrobial–bacteriostatic and antioxidant properties [10]. However, in organic systems, there is limited evidence regarding the benefits of dietary supplementation of aromatic plants in production output and the health of laying hens [9]. Hence, in the present study, we chose to study the role of a common and distinctive aromatic plant, *Salvia officinalis* L., on the performance and health of laying hens raised in a commercial organic farm. The results showed that *Salvia officinalis* L. can improve certain egg quality characteristics subject to several environmental parameters at the farm level that are difficult to control in commercial farms [36]. Moreover, our intention was to avoid major alterations in the daily practices of the farm.

The primary focus of the study was on the egg quality, yolk oxidative stability, and eggshell microbiological counts. Although assessing the effects on the performance parameters was not within the main objectives, we presented performance data to provide more conclusive information for the readers. The practical aspects of the experimental setup of the study obliged us to split the 66 laying hens into two pens that were replicates of each treatment and for each year, resulting in 33 hens per pen and with two pens per treatment. Likewise, in a recent study with organic laying hens, each treatment comprised of two pens with 35 hens in each pen [34]. Similarly, in a study investigating the effects of probiotic supplementation in organic laying hens, each treatment comprised three replicates with 20 hens in each replicate [11]. Due to the housing requirements for organic laying hens for outdoor access, such an experimental setup for organic laying hens' feed experiments is inevitable, as housing hens individually in experimental battery cages is not possible. Under these practical considerations, and in agreement with the study of van der Heide et al. [34], we decided not to proceed with the statistical evaluation of the performance data for years 1 and 2. Instead, we assessed the treatment differences for the performance data of both years. Still, this approach enabled the inclusion of four replicate performance data per treatment, which is not an optimum number of replicates for statistical comparisons. Nevertheless, the descriptive results in year 1 showed that the performances of the hens differed between the groups supplemented with *Salvia officinalis* L. at 0.5% compared to the controls, with the laying percentage lower and feed conversion ratio higher. This may be attributed to the increased feed intake in the specific group during the transition from the summer to the autumn period [35,37]. It is plausible that the external factors may have been more detrimental in this group compared to the other two groups. Mashaly et al. [38] and Bozkurt et al. [39] reported increased feed intakes in birds during periods of lower environmental temperatures following periods of higher temperatures. During the second experimental period (year 2), the performance parameters were similar between the experimental groups. However, the daily feed intake was higher for hens of all groups compared to the first year. This could be due to the higher levels of humidity noticed during year 2. Other factors could also be involved for the increased feed intake, especially in the groups supplemented with the aromatic plant, i.e., a positive influence on the feed palatability in the group supplemented with 0.5% of *Salvia officinalis* L. but not in that supplemented with 1% [13]. Bölükbası et al. [40] showed that the inclusion of a 200-mg/kg extract of *Salvia sclarea* L. reduced the feed consumption of laying hens. In the latter study, the feed conversion was also improved in the hens supplemented with *Salvia sclarea* L., but there was no effect on the body weight or laying performance. Moreover, the supplementation of *Salvia officinalis* L. leaves at 2.5% in the laying hens' diet did not improve any of the performance parameters [41]. The supplementation of the extract of *Salvia sclarea* L. in the diets of laying hens did not affect the body weight or laying percentage but improved the feed conversion ratio [40]. The data on the overall experiment revealed that the hens supplemented with 0.5% of *Salvia officinalis* L. showed an improved feed conversion compared to the Con group, which corroborated with the previous findings. Meanwhile, a favorable increase in the egg mass for the overall experiment was noted in the Sal-0.5% compared to Con group, which can be attributed to the improvement of feed conversion in this specific *Salvia officinalis* L.-supplemented group. According to Çabuk et al. [42], the supplementation of a product containing extracts from various aromatic plants, including *Salvia triloba* L., in laying hens' diets increased the weight of hens and improved the feed conversion ratio. Özek et al. [43] and Bozkurt et al. [39] used an extract also containing *Salvia triloba* L. in their experiment but did not find any effects on the performance of laying hens. It should be noted that the previous studies have been conducted in conventional systems and not in organic ones. Laying hens in organic production systems show greater feed conversion rates than those reared under conventional systems [44]. The latter difference is attributed to the greater level of activity and due to a greater variability of the environmental temperature in organic systems [44,45].

In our study, the yolk weight was increased in eggs from hens supplemented with 1.0% *Salvia officinalis* L. in both experiments compared to the other two groups. This effect could be attributed to the greater abundance of antioxidant substrates in a specific group, which helped hens to tolerate thermal stress during the production of the yolk [46,47]. The improved digestion and absorption of nutrients in the groups supplemented with *Salvia officinalis* L. could also have contributed to the higher yolk weight [48–50]. Similarly, Bozkurt et al. [39] reported that the yolk weight was improved when the diets of laying hens were supplemented with an extract containing, among others, *Salvia triloba* L. In the latter study, the hens were subjected to thermal stress conditions, and the outcome was an increase in yolk weight with a concurrent reduction in albumen weight. In the past, other have also investigated the effects of *Salvia officinalis* L. supplementation on the egg quality characteristics. Loetscher et al. [41] showed that the dietary supplementation of leaves of *Salvia officinalis* L. at 2.5% did not improve the egg quality parameters. On the other hand, the supplementation of an extract of *Salvia sclarea* L. increased the egg weight and Haugh units, while it reduced the yolk percentage [40]. Under thermal stress conditions, the supplementation of an extract containing *Salvia triloba* L. increased only the egg Haugh unit [42]. Elsewhere, aromatic plant supplementation of the family of Labiatae (oregano, thyme, and rosemary) increased the egg yolk weight [51].

In the present study, it is plausible that laying hens were subjected to chronic heat stress conditions. According to Akbarian et al. [52], chronic heat stress is a combination of high ambient temperatures during a prolonged period. This was probably the case in our study, with the average environmental temperatures being higher during the summer months and especially during the first experimental period. It is known that chronic heat stress induces a depletion of antioxidant reserves in poultry [52], and therefore, it is necessary to replace them by dietary means. Phytochemicals are among those dietary antioxidant ingredients beneficial in chronic heat-stressed poultry [52]. In our study, based on the results of MDA in the egg yolks during year 1, it can be hypothesized that chronic heat stress had a detrimental effect on the antioxidant mechanisms of laying hens supplemented with *Salvia officinalis* L. However, during year 2, the MDA levels in egg yolks were significantly lower in both groups supplemented with *Salvia officinalis* L. This finding suggests that chronic heat stress was less pronounced compared to year 1 and that, under these conditions, *Salvia officinalis* L. supplemented with either 0.5% or 1.0% was able to counteract the oxidative stress conditions. Loetscher et al. [41], showed that the supplementation of leaves of *Salvia officinalis* L. at a level of 2.5% improved the antioxidative properties of egg yolks. Elsewhere, it was shown that the supplementation of an extract also containing *Salvia triloba* L. resulted in a significant reduction of MDA in egg yolks and increased the levels of liver enzymes involved in the antioxidative pathways [53]. It is apparent that *Salvia officinalis* L. supplementation improves not only the antioxidative properties of the eggs but also protects laying hens from pro-oxidative conditions.

The results also showed that the dietary supplementation of *Salvia officinalis* L. at 1% significantly reduced the counts of Enterobacteriaceae in eggshells compared to the control group in both experimental periods. This effect could be attributed to the antibacterial properties of the *Salvia officinalis* L. components, such as α-thujone, which had a concentration of 40% in the plants used in our study. Previous studies showed that the supplementation of 200 mg/kg of extract of *Salvia sclarea* L. in laying hen diets reduced the Enterobacteriaceae counts in the feces [40]. These findings are important, as eggs collected from free-range systems were shown to be more contaminated with Enterobacteriaceae counts than those eggs collected from conventional ones [54–56].

5. Conclusions

The supplementation of *Salvia officinalis* L. in powder form, especially at a level of 1%, can improve the oxidative stability of eggs produced by laying hens raised in organic systems. The dietary treatments reduced significantly the counts of Enterobacteriaceae in the eggshells. The potential of the use of *Salvia officinalis* L. in a powder form in organic laying hens' diets is promising and requires further investigation.

Author Contributions: Conceptualization, D.G. and P.F.; methodology, D.G., D.L., D.F., E.P., and P.F.; formal analysis, D.G. and G.A.P.; investigation, D.G.; data curation, D.G. and G.A.P.; writing—original draft preparation, D.G.; writing—review and editing, D.G., G.A.P., and G.I.A.; and supervision, G.A.P. and P.F. All authors have read and agreed to the published version of the manuscript.

Funding: This research received no external funding.

Institutional Review Board Statement: This study was approved by the General Assembly of the Veterinary Faculty of Aristotle University of Thessaloniki and was part of the PhD research project of D.G. The General Assembly of the Veterinary Faculty of Aristotle University of Thessaloniki approved a specific PhD protocol during its decision: 55/27-5-2015.

Informed Consent Statement: Not applicable.

Data Availability Statement: Data are available by the corresponding author upon request.

Conflicts of Interest: The authors declare no conflict of interest.

References

1. Dunne, C.; Siettou, C. UK Consumers' Willingness to Pay for Laying Hen Welfare. *Br. Food J.* **2020**, *122*, 2867–2880. [CrossRef]
2. Larsen, H.; Hemsworth, P.H.; Cronin, G.M.; Gebhardt-Henrich, S.G.; Smith, C.L.; Rault, J.L. Relationship between Welfare and Individual Ranging Behaviour in Commercial Free-Range Laying Hens. *Animals* **2018**, *12*, 2356–2364. [CrossRef] [PubMed]
3. Sosnówka-Czajka, E.; Herbut, E.; Skomorucha, I. Effect of Different Housing Systems on Productivity and Welfare of Laying Hens. *Ann. Anim. Sci.* **2010**, *10*, 349–360.
4. Pettersson, I.C.; Weeks, C.A.; Wilson, L.R.M.; Nicol, C.J. Consumer Perceptions of Free-Range Laying Hen Welfare. *Br. Food J.* **2016**, *118*, 1999–2013. [CrossRef]
5. Sutherland, M.A.; Webster, J.; Sutherland, I. Animal Health and Welfare Issues Facing Organic Production Systems. *Animals* **2013**, *3*, 1021–1035. [CrossRef] [PubMed]
6. Jung, L.; Brenninkmeyer, C.; Niebuhr, K.; Bestman, M.; Tuyttens, F.A.M.; Gunnarsson, S.; Sørensen, J.T.; Ferrari, P.; Knierim, U. Husbandry Conditions and Welfare Outcomes in Organic Egg Production in Eight European Countries. *Animals* **2020**, *10*, 2102. [CrossRef] [PubMed]
7. Englmaierová, M.; Tůmová, E.; Charvátová, V.; Skřivan, M. Effects of Laying Hens Housing System on Laying Performance, Egg Quality Characteristics, and Egg Microbial Contamination. *Czech J. Anim. Sci.* **2014**, *59*, 345–352. [CrossRef]
8. *EC REGULATION No1831/2003*; Regulation (ec) No 1831/2003 of the European Parliament and of the Council of 22 September 2003 on Additives for Use in Animal Nutrition (Text with Eea Relevance); Official Journal of the European Union: Brussels, Belgium, 2003.
9. Diaz-Sanchez, S.; D'Souza, D.; Biswas, D.; Hanning, I. Botanical Alternatives to Antibiotics for Use in Organic Poultry Production. *Poult. Sci.* **2015**, *94*, 1419–1430. [CrossRef] [PubMed]
10. Windisch, W.; Schedle, K.; Plitzner, C.; Kroismayr, A. Use of Phytogenic Products as Feed Additives for Swine and Poultry. *J. Anim. Sci.* **2008**, *86*, E140–E148. [CrossRef]
11. Forte, C.; Acuti, G.; Manuali, E.; Casagrande Proietti, P.; Pavone, S.; Trabalza-Marinucci, M.; Moscati, L.; Onofri, A.; Lorenzetti, C.; Franciosini, M.P. Effects of Two Different Probiotics on Microflora, Morphology, and Morphometry of Gut in Organic Laying Hens. *Poult. Sci.* **2016**, *95*, 2528–2535. [CrossRef]
12. Puvača, N.; Lika, E.; Cocoli, S.; Kika, T.S.; Bursić, V.; Vuković, G.; Simin, M.T.; Petrovic, A.; Cara, M. Use of Tea Tree Essential Oil (Melaleuca Alternifolia) in Laying Hen's Nutrition on Performance and Egg Fatty Acid Profile as a Promising Sustainable Organic Agricultural Tool. *Sustainability* **2020**, *12*, 3420. [CrossRef]
13. Brenes, A.; Roura, E. Essential Oils in Poultry Nutrition: Main Effects and Modes of Action. *Anim. Feed. Sci. Technol.* **2010**, *158*, 1–14. [CrossRef]
14. Burt, S. Essential Oils: Their Antibacterial Properties and Potential Applications in Foods—A Review. *Int. J. Food Microbiol.* **2004**, *94*, 223–253. [CrossRef]
15. Mitsch, P.; Zitterl-Eglseer, K.; Köhler, B.; Gabler, C.; Losa, R.; Zimpernik, I. The Effect of Two Different Blends of Essential Oil Components on the Proliferation of Clostridium Perfringens in the Intestines of Broiler Chickens. *Poult. Sci.* **2004**, *83*, 669–675. [CrossRef] [PubMed]
16. Johny, A.K.; Darre, M.J.; Donoghue, A.M.; Donoghue, D.J.; Venkitanarayanan, K. Antibacterial Effect of Trans-Cinnamaldehyde, Eugenol, Carvacrol, and Thymol on Salmonella Enteritidis and Campylobacter Jejuni in Chicken Cecal Contents in Vitro. *J. Appl. Poult. Res.* **2010**, *19*, 237–244. [CrossRef]
17. Grashorn, M.A. Use of Phytobiotics in Broiler Nutrition—An Alternative to Infeed Antibiotics? *J. Anim. Feed Sci.* **2010**, *19*, 338–347. [CrossRef]
18. Bozkurt, M.; Hippenstiel, F.; Abdel-Wareth, A.A.A.; Kehraus, S.; Küçükyilmaz, K.; Südekum, K.H. Effects of Selected Herbs and Essential Oils on Performance, Egg Quality and Some Metabolic Activities in Laying Hens—A Review. *Eur. Poult. Sci.* **2014**, *78*, 1–15. [CrossRef]
19. Ommati, M.M.; Zamiri, M.J.; Akhlaghi, A.; Atashi, H.; Jafarzadeh, M.R.; Rezvani, M.R.; Saemi, F. Seminal Characteristics, Sperm Fatty Acids, and Blood Biochemical Attributes in Breeder Roosters Orally Administered with Sage (*Salvia Officinalis*) Extract. *Anim. Prod. Sci.* **2013**, *53*, 548–554. [CrossRef]
20. Piesova, E.; Makova, Z.; Levkut, M.; Faixova, Z.; Pistl, J.; Marcin, A.; Levkut, M. The Effects of Sage Extract Feed Supplementation on Biochemical Parameters, Weight of Internal Organs and Salmonella Counts in Chickens. *Res. Vet. Sci.* **2012**, *93*, 1307–1308. [CrossRef] [PubMed]
21. Levkut, M.; Marcin, A.; Lenhardt, L.; Porvaz, P.; Revajová, V.; Šoltysová, B.; Blanár, J.; Ševčíková, Z.; Pist, J. Effect of Sage Extract on Alkaline Phosphatase, Enterocyte Proliferative Activity and Growth Performance in Chickens. *Acta Vet. Brno* **2010**, *79*, 177–183. [CrossRef]

22. Farhadi, M.; Hedayati, M.; Manafi, M.; Khalaji, S. Influence of Using Sage Powder (*Salvia Officinalis*) on Performance, Blood Cells, Immunity Titers, Biochemical Parameters and Small Intestine Morphology in Broiler Chickens. *Iran. J. Appl. Anim. Sci.* **2020**, *10*, 509–516.
23. Ryzner, M.; Takáčová, J.; Čobanová, K.; Plachá, I.; Venglovská, K.; Faix, Š. Effect of Dietary *Salvia Officinalis* Essential Oil and Sodium Selenite Supplementation on Antioxidative Status and Blood Phagocytic Activity in Broiler Chickens. *Acta Vet. Brno* **2013**, *82*, 43–48. [CrossRef]
24. Rasouli, B.; Movahhedkhah, S.; Seidavi, A.; Haq, Q.M.I.; Kadim, I.; Laudadio, V.; Mazzei, D.; Tufarelli, V. Effect of Sage (*Salvia Officinalis* L.) Aqueous Leaf Extract on Performance, Blood Constituents, Immunity Response and Ileal Microflora of Broiler Chickens. *Agrofor. Syst.* **2020**, *94*, 1179–1187. [CrossRef]
25. Placha, I.; Ryzner, M.; Cobanova, K.; Faixova, Z.; Faix, S. Effects of Dietary Supplementation with Sage (*Salvia Officinalis* L.) Essential Oil on Antioxidant Status and Duodenal Wall Integrity of Laying Strain Growers. *Pol. J. Vet. Sci.* **2015**, *18*, 741–749. [CrossRef] [PubMed]
26. Van den Dool, H.; Kratz, P.D. A Generalization of the Retention Index System Including Linear Temperature Programmed Gas-Liquid Partition Chromatography. *J. Chromatogr. A* **1963**, *11*, 463–471. [CrossRef]
27. Masada, Y. *Analysis of Essential Oils by Gas Chromatography and Mass Spectrometry*; Halsted Press, a Division of John Wiley & Sons, Inc.: New York, NY, USA, 1976.
28. Adams, R.P. *Identification of Essential Oil Components by Gas Chromatography/Mass Spectrometry*, 4th ed.; Allured Publishing Corporation: Carol Stream, IL, USA, 2007.
29. Galobart, J.; Barroeta, A.C.; Baucells, M.D.; Codony, R.; Ternes, W. Effect of Dietary Supplementation with Rosemary Extract and α-Tocopheryl Acetate on Lipid Oxidation in Eggs Enriched with Ω3-Fatty Acids. *Poult. Sci.* **2001**, *80*, 460–467. [CrossRef] [PubMed]
30. Botsoglou, N.A.; Fletouris, D.J.; Papageorgiou, G.E.; Vassilopoulos, V.N.; Mantis, A.J.; Trakatellis, A.G. Rapid, Sensitive, and Specific Thiobarbituric Acid Method for Measuring Lipid Peroxidation in Animal Tissue, Food, and Feedstuff Samples. *J. Agric. Food Chem.* **1994**, *42*, 1931–1937. [CrossRef]
31. International Organization of Standardization (ISO). *ISO 10272-1: Microbiology of the Food Chain-Horizontal Method for the Detection and Enumeration of Campylobacter spp.—Part 1:Detection Method*; ISO—International Organization of Standardization: Geneva, Switzerland, 2017.
32. International Organization of Standardization (ISO). *ISO 6579-1: Microbiology of the Food Chain-Horizontal Method for the Detection and Enumeration and Serotyping of Salmonella spp.—Part 1:Detection of Salmonella spp.*; ISO—International Organization of Standardization: Geneva, Switzerland, 2017.
33. International Organization of Standardization (ISO). *ISO 21528-2: Microbiology of the Food Chain-Horizontal Method for the Detection and Enumeration of Enterobacteriaceae—Part 2: Colony Count Technique*; ISO—International Organization of Standardization: Geneva, Switzerland, 2017.
34. Van der Heide, M.E.; Johansen, N.F.; Kidmose, U.; Nørgaard, J.V.; Hammershøj, M. The Effect of Deshelled and Shell-Reduced Mussel Meal on Egg Quality Parameters of Organic Laying Hens under Commercial Conditions. *J. Appl. Poult. Res.* **2021**, *30*, 100119. [CrossRef]
35. Rodenburg, T.B.; de Reu, K.; Tuyttens, F.A.M. Performance, Welfare, Health and Hygiene of Laying Hens in Non-Cage Systems. In *Alternative Systems for Poultry: Health, Welfare and Productivity*; Sandilands, V., Hocking, P., Eds.; CABI Publishing: Wallingford, UK, 2012. [CrossRef]
36. Bestman, M.; Wagenaar, J.-P. Health and Welfare in Dutch Organic Laying Hens. *Animals* **2014**, *4*, 374. [CrossRef] [PubMed]
37. Leenstra, F.; Maurer, V.; Galea, F.; Bestman, M.; Amsler-Kepalaite, Z.; Visscher, J.; Vermeij, I.; van Krimpen, M. Leistung von Legehennen in Unterschiedlichen Produktionssystemen—Wo Liegen Die Unterschiede Und Wie Kann Man Sie Reduzieren? Resultate von Gruppendiskussionen Mit Legehennenhaltern in Den Niederlanden, Der Schweiz Und Frankreich, Vergleich von Eckdaten Und Modellrechnungen. *Eur. Poult. Sci.* **2014**, *78*, 1–10. [CrossRef]
38. Mashaly, M.M.; Hendricks, G.L.; Kalama, M.A.; Gehad, A.E.; Abbas, A.O.; Patterson, P.H. Effect of Heat Stress on Production Parameters and Immune Responses of Commercial Laying Hens. *Poult. Sci.* **2004**, *83*, 889–894. [CrossRef] [PubMed]
39. Bozkurt, M.; Küçükyilmaz, K.; Çatli, A.U.; Çınar, M.; Bintaş, E.; Çöven, F. Performance, Egg Quality, and Immune Response of Laying Hens Fed Diets Supplemented with Mannan-Oligosaccharide or an Essential Oil Mixture under Moderate and Hot Environmental Conditions. *Poult. Sci.* **2012**, *91*, 1379–1386. [CrossRef] [PubMed]
40. Bölükbasi, S.C.; Erhan, M.K.; Kaynar, Ö. The effect of feeding thyme, sage and rosemary oil on laying hen performance, cholesterol and some proteins ratio of egg yolk and Escherichia coli count in feces. *Europ. Poult. Sci.* **2008**, *72*, 231–237.
41. Loetscher, Y.; Kreuzer, M.; Messikommer, R.E. Late Laying Hens Deposit Dietary Antioxidants Preferentially in the Egg and Not in the Body. *J. Appl. Poult. Res.* **2014**, *23*, 647–660. [CrossRef]
42. Çabuk, M.; Bozkurt, M.; Alçiçek, A.; Akbaþ, Y.; Küçükyýlmaz, K. Effect of a Herbal Essential Oil Mixture on Growth and Internal Organ Weight of Broilers from Young and Old Breeder Flocks. *South Afr. J. Anim. Sci.* **2006**, *36*, 135–141. [CrossRef]
43. Özek, K.; Wellmann, K.; Ertekin, B.; Tarım, B. Effects of Dietary Herbal Essential Oil Mixture and Organic Acid Preparation on Laying Traits, Gastrointestinal Tract Characteristics, Blood Parameters and Immune Response of Laying Hens in a Hot Summer Season. *J. Anim. Feed Sci.* **2011**, *20*, 575–586. [CrossRef]

44. Lolli, S.; Hidalgo, A.; Alamprese, C.; Ferrante, V.; Rossi, M. Layer Performances, Eggshell Characteristics and Bone Strength in Three Different Housing Systems. *Biotechnol. Anim. Husb.* **2013**, *29*, 591–606. [CrossRef]
45. Lara, L.; Rostagno, M. Impact of Heat Stress on Poultry Production. *Animals* **2013**, *3*, 356. [CrossRef] [PubMed]
46. Rahimi, S.; Zadeh, Z.T.; Torshizi, M.A.K.; Omidbaigi, R.; Rokni, H. Effect of the Three Herbal Extracts on Growth Performance, Immune System, Blood Factors and Intestinal Selected Bacterial Population in Broiler Chickens. *J. Agric. Sci. Tech.* **2011**, *13*, 527–539. [CrossRef]
47. Hashemipour, H.; Kermanshahi, H.; Golian, A.; Veldkamp, T. Effect of Thymol and Carvacrol Feed Supplementation on Performance, Antioxidant Enzyme Activities, Fatty Acid Composition, Digestive Enzyme Activities, and Immune Response in Broiler Chickens. *Poult. Sci.* **2013**, *92*, 2059–2069. [CrossRef]
48. Lee, K.-W.; Everts, H.; Kappert, H.J.; Frehner, M.; Losa, R.; Beynen, A.C. Effects of Dietary Essential Oil Components on Growth Performance, Digestive Enzymes and Lipid Metabolism in Female Broiler Chickens. *Br. Poult. Sci.* **2003**, *44*, 450–457. [CrossRef]
49. Jang, I.S.; Ko, Y.H.; Yang, H.Y.; Ha, J.S.; Kim, J.Y.; Kim, J.Y.; Kang, S.Y.; Yoo, D.H.; Nam, D.S.; Kim, D.H.; et al. Influence of Essential Oil Components on Growth Performance and the Functional Activity of the Pancreas and Small Intestine in Broiler Chickens. *Asian-Australas. J. Anim. Sci.* **2004**, *17*, 394–400. [CrossRef]
50. Franz, C.; Baser, K.; Windisch, W. Essential Oils and Aromatic Plants in Animal Feeding—A European Perspective. A Review. *Flavour Fragr. J.* **2010**, *25*, 327–340. [CrossRef]
51. Nadia, R.; Hassan, R.A.; Qota, E.M.; Fayek, H.M. Effect of Natural Antioxidant on Oxidative Stability of Eggs and Productive and Reproductive Performance of Laying Hens. *Int. J. Poult. Sci.* **2008**, *7*, 134–150. [CrossRef]
52. Akbarian, A.; Michiels, J.; Degroote, J.; Majdeddin, M.; Golian, A.; de Smet, S. Association between Heat Stress and Oxidative Stress in Poultry; Mitochondrial Dysfunction and Dietary Interventions with Phytochemicals. *J. Anim. Sci. Biotechnol.* **2016**, *7*, 1–14. [CrossRef] [PubMed]
53. Bozkurt, M.; Tokuşoğlu, Ö.; Küçükyilmaz, K.; Akşit, H.; Çabuk, M.; Uğur Çatli, A.; Seyrek, K.; Çinar, M. Effects of Dietary Mannan Oligosaccharide and Herbal Essential Oil Blend Supplementation on Performance and Oxidative Stability of Eggs and Liver in Laying Hens. *Ital. J. Anim. Sci.* **2012**, *11*, e41. [CrossRef]
54. Parisi, M.A.; Northcutt, J.K.; Smith, D.P.; Steinberg, E.L.; Dawson, P.L. Microbiological Contamination of Shell Eggs Produced in Conventional and Free-Range Housing Systems. *Food Control* **2015**, *47*, 161–165. [CrossRef]
55. Jones, D.R.; Anderson, K.E.; Guard, J.Y. Prevalence of Coliforms, Salmonella, Listeria, and Campylobacter Associated with Eggs and the Environment of Conventional Cage and Free-Range Egg Production. *Poult. Sci.* **2012**, *91*, 1195–1202. [CrossRef] [PubMed]
56. Hannah, J.F.; Wilson, J.L.; Cox, N.A.; Cason, J.A.; Bourassa, D.V.; Musgrove, M.T.; Richardson, L.J.; Rigsby, L.L.; Buhr, R.J. Comparison of Shell Bacteria from Unwashed and Washed Table Eggs Harvested from Caged Laying Hens and Cage-Free Floor-Housed Laying Hens. *Poult. Sci.* **2011**, *90*, 1586–1593. [CrossRef] [PubMed]

animals

MDPI

Communication

Impact of Microbial Protease Enzyme and Dietary Crude Protein Levels on Growth and Nutrients Digestibility in Broilers over 15–28 Days

Abdul Jabbar [1,2], Muhammad Tahir [1], Ibrahim A. Alhidary [3], Mutassim A. Abdelrahman [3], Hani Albadani [3], Rifat Ullah Khan [3,4], Maria Selvaggi [5], Vito Laudadio [6,*] and Vincenzo Tufarelli [6]

1 Department of Animal Nutrition, Faculty of Animal Husbandry and Veterinary Sciences, The University of Agriculture, Peshawar 25000, Pakistan; abduljabbar@upr.edu.pk (A.J.); tahir065@yahoo.co.uk (M.T.)
2 Faculty of Veterinary and Animal Sciences, University of Poonch, Rawalakot 12350, Pakistan
3 Department of Animal Production, College of Food and Agriculture Sciences, King Saud University, Riyadh 11414, Saudi Arabia; ialahidary@ksu.edu.sa (I.A.A.); amutassim@ksu.edu.sa (M.A.A.); halbadani@ksu.edu.sa (H.A.); rukhan@aup.edu.pk (R.U.K.)
4 Faculty of Animal Husbandry and Veterinary Sciences, College of Veterinary Sciences, The University of Agriculture, Peshawar 25000, Pakistan
5 Department of Agricultural and Environmental Science, University of Bari "Aldo Moro", 70125 Bari, Italy; maria.selvaggi@uniba.it
6 Department of DETO, Section of Veterinary Science and Animal Production, University of Bari "Aldo Moro", 70010 Bari, Italy; vincenzo.tufarelli@uniba.it
* Correspondence: vito.laudadio@uniba.it

Citation: Jabbar, A.; Tahir, M.; Alhidary, I.A.; Abdelrahman, M.A.; Albadani, H.; Khan, R.U.; Selvaggi, M.; Laudadio, V.; Tufarelli, V. Impact of Microbial Protease Enzyme and Dietary Crude Protein Levels on Growth and Nutrients Digestibility in Broilers over 15–28 Days. *Animals* **2021**, *11*, 2499. https://doi.org/10.3390/ani11092499

Academic Editors: Youssef A. Attia, Natalie Morgan and Velmurugu Ravindran

Received: 21 July 2021
Accepted: 24 August 2021
Published: 25 August 2021

Simple Summary: Protein is considered as the most expensive nutrient in animal diet and requires exogenous proteolytic enzymes for proper digestion in all living organisms. Thus, the present study evaluated different crude protein levels (CP) with or without exogenous protease in broilers diet over 15–28 days on their growth and nutrient digestibility. Based on the obtained findings, it is recommended a reduction in protein contents of diets from 21 to 19% when the protease enzyme is supplemented, to produce a beneficial impact on the growth and digestibility of broilers and also to reduce feeding costs.

Abstract: In this trial, a 3 × 2 factorial design with different dietary crude protein levels (CP, 17, 19 and 21%) and two levels of exogenous protease (0 and 30,000 IU/kg) was used. A total of 540 two-week old broilers (Ross-308) was randomly allocated to experimental diets over 15–28 days of age. The interaction between dietary protein levels and enzyme supplementation showed that body weight gain was significantly ($p < 0.05$) higher in birds fed CP-19 (1114.7 g) and CP-21 (1108.8 g) with enzymes supplementation. Feed intake was higher ($p < 0.05$) in broilers fed with CP-17 than CP-19 with supplementation of the protease enzyme. Results also revealed that the feed conversion ratio (FCR) was significantly ($p < 0.05$) improved in birds fed with CP-19 and CP-21 and protease supplementation. Total tract N retention was lower ($p < 0.05$) in birds fed CP-17 with no enzyme than the other dietary groups. Similarly, the gross energy (GE) was significantly ($p < 0.05$) lower in birds fed CP-17 with or without the protease enzyme. Abdominal fat was higher ($p < 0.05$) in CP-17 (0.96%) without the protease enzyme. It was concluded that a diet at 19% CP with the protease enzyme improved the performance and nutrient digestibility in broilers over 15–28 days.

Keywords: broiler; protease; dietary protein; growth; nutrient utilization

1. Introduction

The inclusion of exogenous protease may be an effective nutritional tool to improve growth and increase digestibility of crude protein and metabolizable energy in broiler diet [1]. Endogenous enzymes are secreted in animal body, but their concentration is not always sufficient for efficient protein digestion [2], especially during the growing phase, where feed intake substantially increases. Past experiments indicated that most of the crude protein (CP) went undigested without being completely utilized [3,4]. This undigested protein may be better utilized in the presence of the exogenous protease enzyme. Therefore, it is speculated that exogenous supplementation of enzymes may optimize the CP digestibility and growth performance of broiler chickens [5,6]. Previous studies have also demonstrated that nutrient density is more important for optimum growth and feed efficiency in broilers during the finisher phase than the starter phase [7–9].

To minimize the feed cost, a low-protein diet is recommended in broilers [10]; however, there is likelihood of less chance to obtain sufficient amount of essential amino acids, which may also compromise the conversion of non-essential amino acids [2]. Several exogenous enzymes are commercially available and being used in SBM-based diets to improve the growth of broiler chickens. Exogenous enzymes improve not only feed efficiency, but also reduce proteolytic fermentation, nutrients excretion in feces and bacterial toxins [11]. Moreover, the protease enzyme cleaves anti-nutritional factors, such as trypsin inhibitors, thus resulting in enhanced utilization of amino acids [2]. Earlier studies have also demonstrated that protease supplementation increased ileal digestibility, resulting in improved growth performance in broilers fed on low-protein diets [12]. In addition, little attention has been given to the digestibility of energy of nutrients in protease-supplemented birds. Many studies have documented inconclusive findings in broilers fed with different levels of CP [9,11,12]. Law et al. [12] concluded that digestibility of nutrients was higher in broilers fed with 19% CP than 16% co-supplemented with the protease enzyme during the finishing phase. Jabbar et al. [13] reported best results in term of performance and digestibility of nutrients in response to protease enzyme supplementation in broilers fed with 21% CP. Another study reported that exogenous protease supplementation in low energy-protein diets is helpful in achieving enhanced growth performance and lower abdominal fat in broilers in heat stress conditions [9]. In the published literature, discrepancies exist on the level of CP in broilers diet. We hypothesized that addition of the protease enzyme could reduce the level of CP with no compromise on the growth performance of broiler. Therefore, the objective of the present study is to evaluate different levels of CP with and without exogenous protease in broiler diet over 15–28 days on their growth and nutrient digestibility.

2. Materials and Methods

2.1. Ethical Approval

The study was approved by the Committee on Ethical and Animal Welfare, Faculty of Animal Husbandry and Veterinary Sciences, University of Agriculture, Peshawar.

2.2. Experimental Procedures

Day-old chicks were reared under standard protocol of feeding (CP 22%, 2700 Kcal/kg), environment temperature (35 °C in initial week and gradual decrease in the following week) and lighting (23:1 light and dark cycle). Wood shavings were used as bedding material with ad libitum access to drinking water. Using a 3 × 2 factorial design having three dietary protein levels (CP, 17,19 and 21%) and two levels of the protease enzyme (0 and 30,000 IU/kg), a total of 540 day-old male broiler chicks (Ross-308) were randomly allotted to 36 floor pens (n = 15 chicks/pen) from 15 to 28 days of age. There were six replicate pens per crude protein-enzyme supplementation subgroups. All the diets (pellet form) were isocaloric, having digestible amino acids level that met or exceeded the requirements [14]. The stocking density was 10 birds/m^2. The house temperature was maintained at 25 ± 2 °C and humidity was 61%. Feed supply and consumption were weighed during the entire period. Weight gain in broilers was recorded on a weekly basis for each group at the end of experimental period. The diet was mixed with the exogenous protease enzyme (Pro Plus, Medixacell, Lahore, Pakistan) at two dose levels (0 and 30,000 IU/kg). Protease was derived from *Bacillus subtilis* and capable to break down protein and anti-nutritional factors in plants with the strength activity of 600,000 PROT units/kg. The protease activity of this enzyme was measured in PROT units, which was defined as the concentration of enzyme that releases 1 μmol of ρ-nitroaniline from 1 μM of substrate/min at 37 °C and pH of 9. Three diets having 17, 19 and 21% CP with same level of metabolizable energy were prepared as reported in Table 1.

Table 1. Ingredients and chemical composition of diets over 15–28 days.

Ingredients (%)	Diet 17% CP	Diet 19% CP	Diet 21% CP
Corn	62.45	59.62	57.18
Soybean meal (44% CP)	9.27	11.08	14.51
Canola meal	7.00	7.00	8.00
Sunflower meal	5.71	6.00	5.00
Gluten meal (60% CP)	4.50	2.50	3.00
Poultry by-products	2.00	3.00	3.00
Gluten meal (40% CP)	1.21	2.60	2.81
Vegetable oil	1.00	2.19	2.04
Bone meal	1.00	1.00	1.00
Dicalcium Phosphate	1.00	1.00	0.65
Limestone	1.00	0.86	1.02
Salt	0.45	0.45	0.45
VitaVit-minerals premix [1]	0.12	0.12	0.12
L-Lys HCl	0.55	0.42	0.23
DL-Met	0.28	0.20	0.14
Valine	0.99	0.71	0.46
Threonine	0.72	0.66	0.34
Isoleucine	0.43	0.36	0.00
Arginine	0.32	0.22	0.00
Chemical composition			
ME, kcal/kg	2950	2930	2920
Crude protein, %	17.00	19.00	21.00
Ca, %	1.00	1.00	1.00
Available P, %	0.45	0.45	0.45
Met + Cys, %	0.90	0.90	0.90
Methionine, %	0.59	0.56	0.53
Lysine, %	1.15	1.15	1.15
Valine, %	0.96	0.68	0.99
Threonine, %	0.83	0.72	0.89
Isoleucine, %	0.93	0.85	0.88
Arginine, %	1.19	1.09	1.21

[1] Vitamin–mineral premix/kg contains the following per kg: vitamin A, 2,400,000 IU; vitamin D, 1,000,000 IU; vitamin E, 16,000 IU; vitamin K, 800 mg; vitamin B1, 600 mg; vitamin B_2, 1600 mg; vitamin B_6, 1000 mg; vitamin B_{12}, 6 mg; niacin, 8000 mg; folic acid, 400 mg; pantothenic acid, 3000 mg; biotin 40 mg; antioxidant, 3000 mg; cobalt, 80 mg; copper, 2000 mg; iodine, 400; iron, 1200 mg; manganese, 18,000 mg; selenium, 60 mg; zinc, 14,000 mg.

2.3. Total Tract Digestibility Assay

Total tract N retention was determined on day 28 of the experiment using acid-insoluble ash as digestibility marker. On day 25 of the experiment, the marker was mixed in the experimental rations and fed to the birds (five birds per replicate), separated in metallic cages, till the end of the trial. On day 28 of the experiment, excreta samples were collected twice a day in plastic bags and stored at −20°C. The grinded feed and excreta samples were passed through a 0.5 mm sieve and stored at −20°C. Nitrogen (N) was determined through the Kjeldahl technique and multiplied by 6.25 to find out total tract N retention. Gross energy (GE) was determined using a bomb calorimeter [13]. On day 28, five birds per replicate were randomly selected to determine abdominal fat. Birds were sacrificed (cut the neck with sharp knife) and defeathered and abdominal fat was removed from around the abdominal organs and belly. Fat pads were weighed as percentage of live body weight.

2.4. Statistical Analysis

Data were analyzed by analysis of variance (ANOVA) using the general linear model (GLM) procedure of the statistical analysis system of SAS [15] to identify the main effects of diets, enzyme and their interaction by factorial design. When the interaction was found to be significant, Tukey's test was applied to separate the means. A p-value less than 0.05 was found to be statistically significant.

3. Results

Data regarding the effects of different dietary protein and protease on broiler chick's performance are presented in Table 2. The results show that body weight gain was significantly ($p < 0.05$) lower in the group fed CP-17, with no significant ($p > 0.05$) difference in feed intake. As a consequence, FCR was significantly ($p < 0.05$) higher in birds fed with CP-17. Similarly, body weight gain was significantly ($p < 0.05$) higher in birds treated with enzymes, while FCR was significantly ($p < 0.05$) lower in the same group. The interaction between protein levels and enzyme supplementation showed that body weight gain was significantly ($p < 0.05$) higher in birds fed with CP-19 and CP-21 with enzymes supplementation. Feed intake was significantly ($p < 0.05$) higher in birds fed with CP-17 than CP-19 with supplementation of protease enzyme. The results also revealed that FCR was significantly ($p < 0.05$) lower in birds with protease supplementation.

Table 2. Body weight gain, feed intake and feed conversion ratio (FCR) of broilers fed different levels of crude protein and protease enzyme over 15–28 days.

Item	Body Weight Gain (g)	Feed Intake (g)	FCR (g/g)
	CP [1]		
CP-17	938.0 ^{b}	1895.0	2.02 ^{a}
CP-19	1098.4 ^{a}	1857.1	1.69 ^{b}
CP-21	1096.0 ^{a}	1873.2	1.71 ^{b}
Pooled SEM	76.33	102.45	0.11
	Enzyme [2]		
Without enzyme	1021.0 ^{b}	1867.2	1.84 ^{a}
CP + enzyme	1067.0 ^{a}	1883.1	1.77 ^{b}
Pooled SEM	45.66	124.00	0.12
	CP × Enzyme		
$CP\text{-}17_0$	899.0 ^{d}	1875.0 ^{ab}	2.08 ^{a}
$CP\text{-}17_E$	977.3 ^{c}	1915.4 ^{a}	1.96 ^{b}
$CP\text{-}19_0$	1082.2 ^{b}	1858.0 ^{ab}	1.71 ^{c}
$CP\text{-}19_E$	1115.0 ^{a}	1856.4 ^{b}	1.66 ^{d}
$CP\text{-}21_0$	1082.4 ^{b}	1869.0 ^{ab}	1.72 ^{c}
$CP\text{-}21_E$	1109.00 ^{ab}	1877.4 ^{ab}	1.69 ^{cd}
Pooled SEM	78.33	106.55	0.11
	p-Value		
CP	<0.001	0.1799	<0.001
Enzyme	0.0001	0.3301	0.0001
CP × Enzyme	0.0456	0.0383	0.0238

Means in the same column with different superscripts differ significantly ($p < 0.05$). [1] Diets containing different levels of CP: CP-1 = 17% CP; CP-2 = 19% CP; CP-3 = 21% CP (all these diets were with or without protease enzyme). [2] Protease enzyme at 30,000 UI/kg.

The effects of the different levels of protein and enzyme supplementation on digestibility indices and abdominal fat are given in Table 3. Results revealed that total tract N retention was significantly ($p < 0.05$) higher in birds fed with CP-19 and CP-21. The GE was significantly ($p < 0.05$) higher in the diet with CP-19 and CP-21 than CP-17. Abdominal fat was significantly ($p < 0.05$) lower in CP-19 and CP-21 than CP-17. Protease supplementation significantly ($p < 0.05$) improved total tract N retention and GE; however, abdominal fat (0.62%) was reduced ($p < 0.05$) in broiler chickens. The interaction showed that CP digestibility was lower ($p < 0.05$) in birds fed with CP-17 with no enzyme supplementation than the rest of the groups. Similarly, GE was significantly ($p < 0.05$) higher in birds fed with CP-19 and 21 plus protease supplementation than CP-17 without protease enzyme. Further, abdominal fat was significantly ($p < 0.05$) higher in CP-17 without protease enzyme than the other treatments.

Table 3. Total tract N retention, gross energy (GE) and abdominal fat of broilers fed different levels of crude protein and protease enzyme over 15–28 days.

Item	Total Tract N Retention (%)	GE (Kcal/kg)	Abdominal Fat (%)
	CP [1]		
CP-17	66.00 ^{b}	2889.0 ^{b}	0.87 ^{a}
CP-19	68.00 ^{a}	2928.0 ^{a}	0.59 ^{b}
CP-21	68.25 ^{a}	2930.0 ^{a}	0.55 ^{c}
Pooled SEM	1.74	103.55	0.01
	Enzyme [2]		
Without enzyme	66.08 ^{b}	2904.0	0.72 ^{a}
CP + enzyme	68.29 ^{a}	2927.1	0.62 ^{b}
Pooled SEM	2.42	203.45	0.01
	CP × Enzyme		
$CP\text{-}17_0$	62.71 ^{d}	2876.0 ^{b}	0.96 ^{a}
$CP\text{-}17_E$	64.00 ^{c}	2902.0 ^{ab}	0.77 ^{b}
$CP\text{-}19_0$	67.00 ^{b}	2919.0 ^{ab}	0.63 ^{c}
$CP\text{-}19_E$	68.19 ^{ab}	2936.5 ^{a}	0.55 ^{d}
$CP\text{-}21_0$	68.18 ^{ab}	2917.0 ^{ab}	0.57 ^{d}
$CP\text{-}21_E$	68.36 ^{a}	2943.2 ^{a}	0.53 ^{d}
Pooled SEM	3.33	107.00	0.01
	p-value		
CP	<0.001	<0.001	0.0348
Enzyme	0.0007	<0.001	0.0919
CP × Enzyme	0.0339	0.0009	0.0052

Means in the same column with different superscripts differ significantly ($p < 0.05$). [1] Diets contained different levels of CP: CP-1 = 17% CP; CP-2 = 19% CP; CP-3 = 21% CP (all these diets were with or without protease enzyme). [2] Protease enzyme at 30,000 UI/kg.

4. Discussion

Enzyme supplementation in broiler diet is well known for its economic, environmental and nutritional advantages. In the present study, the body weight of birds was significantly higher when fed with CP-19 plus protease enzyme, while FCR and abdominal fat were significantly lower in the same group. In contrast, the growth performance, feed intake and feed efficiency were severely affected in birds fed with CP-17 with or without enzymes. This outcome was expected, since the lower CP accumulated body fat and resulted in decreased feed efficiency and utilization. Our findings are partially similar to the previous studies reporting that low CP diets negatively affect the performance and feed efficiency of broilers [2,16]. In the present study, it was obvious that CP-19 with enzyme supplementation was superior in terms of higher growth performance in than CP-21 without enzyme supplementation. There is also an economic aspect to these findings, since the broiler producers would opt to use CP-19 with protease supplementation to reduce feed-cost. Previous studies have attempted to decrease the CP content of broiler and ducks, which resulted in some noxious effects [17,18]. In the present study, total tract N retention and GE were significantly improved in birds fed with CP-19 with supplementation of protease. Previously, it has been reported that protease enzyme supplementation improved CP digestibility in broiler chickens [3,11,19,20]. Freitas et al. [21] reported that digestibility of protein is higher in high-CP than low-CP diets, in response to the exogenous protease enzyme. Although this behavior of the protease enzyme has resulted in inconsistent growth performance in some of the experiments in broilers [22], this was not the case in our study. In this experiment, the improvement in the digestibility indices resulted in a parallel increase in the growth performance of broilers. Furthermore, in the current study, GE was the highest in birds fed a diet with CP-19 and CP-21 and protease enzyme. Freitas et al. [21] reported improved digestibility of CP in broilers in response to low-CP in the diet. Similar to our findings, Vieira et al. [23] reported improved CP digestibility in protease-supplemented broilers, when compared with a low-CP diet. Moreover, Favero et al. [24] also reported a 5% improved digestibility of energy in broilers in the highest level of CP with protease supplementation. In contrast to our results, the metabolizable energy was higher in birds given a low-CP diet [3].

In the current study, the highest fat percentage was found in broilers fed with CP-17 with no protease enzyme supplementation. It is clear from this study that low protein diet promotes abdominal fat deposition. Abdominal fat of the carcass is an unfavorable trait and its presence decreases the consumers' acceptability. It is one of the major drawbacks of the low-CP diet. In an experiment, fat content was reduced in broiler fed with low-CP diet [9]. Previous studies have reported a significantly higher percentage of abdominal fat in broiler at the end of experimental period fed with low protein diet [17,20], which has been attributed to higher calorie: protein ratio. It is inferred that exogenous protease helps to digest fats, thus increasing the availability of metabolizable energy, as recorded in the current study and reported previously [5]. Thus, it seems that excess energy beyond protein deposition is mainly accumulated as abdominal fat. In the current study, it was also clear that there was a slight difference in growth and digestibility in CP-19 and 21, elucidating that birds performed well under these two levels of CP when protease was supplemented in the diets. Previous studies have recommended a slight reduction in protein contents of the diets when the protease enzyme was supplemented to produce a beneficial impact on the growth and digestibility of broilers [9,13]. Further, it is more economical for farmers to use CP-19 than CP-21.

Therefore, based on the present findings, it was concluded that fed a diet at 19% CP with the protease enzyme improved the growth performance and nutrients digestibility during a trial period of 15–28 days.

Author Contributions: Conceptualization, A.J., M.T.; methodology, R.U.K. and I.A.A.; formal analysis, A.J. and M.T.; investigation, A.J., M.T., M.A.A. and H.A.; data curation, R.U.K. and I.A.A.; writing—original draft preparation, A.J., R.U.K. and V.T.; writing—review and editing, R.U.K., M.S., V.L. and V.T. All authors have read and agreed to the published version of the manuscript.

Funding: The authors would like to extend their sincere appreciation to the Deanship of Scientific Research at King Saud University for funding this work through the research group project No. RG.1436-021.

Institutional Review Board Statement: The study was conducted according to the guidelines of the Declaration of Helsinki and approved by the Ethics Committee of the Faculty of Animal Husbandry and Veterinary Sciences, University of Agriculture, Peshawar (No. 127/AN/FAHVS/2019).

Acknowledgments: The authors are thankful to the respective academic institutes.

Conflicts of Interest: No potential conflict of interest exists.

References

1. Hu, J.; Ingale, S.; Rathi, P.; Zhang, J.Y.; Kim, I.H. Influence of exogenous acid protease in broiler chickens fed corn–soybean meal-based diets. *J. Anim. Physiol. Anim. Nutr.* **2020**, *104*, 204–211. [CrossRef] [PubMed]
2. Ndazigaruye, G.; Kim, D.H.; Kang, C.W.; Kang, K.R.; Joo, Y.J.; Lee, S.R.; Lee, K.W. Effects of low-protein diets and exogenous protease on growth performance, carcass traits, intestinal morphology, cecal volatile fatty acids and serum parameters in broilers. *Animals* **2019**, *9*, 226. [CrossRef]
3. Angel, C.R.; Saylor, W.; Vieira, S.L.; Ward, N. Effects of a mono-component protease on performance and protein utilization in 7-to 22-day-old broiler chickens. *Poult. Sci.* **2011**, *90*, 2281–2286. [CrossRef] [PubMed]
4. Ding, X.M.; Li, D.D.; Li, Z.R.; Wang, J.P.; Zeng, Q.F.; Bai, S.P.; Su, Z.W.; Zhang, K.Y. Effects of dietary crude protein levels and exogenous protease on performance, nutrient digestibility, trypsin activity and intestinal morphology in broilers. *Livest. Sci.* **2016**, *193*, 26–31. [CrossRef]

5. Cowieson, A.J.; Roos, F.F. Toward optimal value creation through the application of exogenous mono-component protease in the diets of non-ruminants. *Anim. Feed Sci. Technol.* **2016**, *221*, 331–340. [CrossRef]
6. Saleh, A.A.; Dawood, M.M.; Badawi, N.A.; Ebeid, T.A.; Amber, K.A.; Azzam, M.M. Effect of supplemental serine-protease from *Bacillus licheniformis* on growth performance and physiological change of broiler chickens. *J. Appl. Anim. Res.* **2020**, *48*, 86–92. [CrossRef]
7. Kamran, Z.; Sarwar, M.; Nisa, M.; Nadeem, M.; Mahmood, S.; Babar, M.; Ahmed, S. Effect of low-protein diets having constant energy-to-protein ratio on performance and carcass characteristics of broiler chickens from one to thirty-five days of age. *Poult. Sci.* **2008**, *87*, 468–474. [CrossRef]
8. Cowieson, A.; Zaefarian, F.; Knap, I.; Ravindran, V.R. Interactive effects of dietary protein concentration, a mono-component exogenous protease and ascorbic acid on broiler performance, nutritional status and gut health. *Anim. Prod. Sci.* **2017**, *57*, 1058–1068. [CrossRef]
9. Law, F.L.; Zulkifli, I.; Soleimani, A.F.; Liang, J.B.; Awad, E.A. Effects of protease supplementation of low protein and/or energy diets on growth performance and blood parameters in broiler chickens under heat stress condition. *Ital. J. Anim. Sci.* **2019**, *18*, 679–689.
10. Polat, I.; Yurtseven, S.; Sarica, S. Supplementing diets for broilers that are low in crude protein and amino acids with protease. *S. Afr. J. Anim. Sci.* **2020**, *50*, 521–528.
11. Mahmood, T.; Mirza, M.A.; Nawaz, H.; Shahid, M. Effect of different exogenous proteases on growth performance, nutrient digestibility, and carcass response in broiler chickens fed poultry by-product meal-based diets. *Livest. Sci.* **2017**, *200*, 71–75. [CrossRef]
12. Law, L.; Zulkifli, I.; Soleimani, F.A.; Hossain, M.; Liang, J. Nutrient digestibility of broiler chickens fed on a low-protein diet supplemented with mono-component proteases. *Eur. Poult. Sci.* **2015**, *79*, 1–15.
13. Jabbar, A.; Tahir, M.; Khan, R.U.; Ahmad, N. Interactive effect of exogenous protease enzyme and dietary crude protein levels on growth and digestibility indices in broiler chickens during the starter phase. *Trop. Anim. Health Prod.* **2021**, *53*, 23. [CrossRef]
14. Aviagen, W. Ross 308: Broiler Nutrition Specification. Available online: https://tmea.aviagen.com/assets/Tech_Center/Ross_Broiler/RossBroilerNutritionSpecs2019-EN.pdf (accessed on 30 September 2019).
15. SAS Institute Inc. *SAS/STAT User's Guide*, 9.1 ed.; SAS Institute Inc.: Cary, NC, USA, 2002.
16. Yang, H.; Yang, Z.; Wang, Z.; Wang, W.; Huang, K.; Fan, W.; Jia, T. Effects of early dietary energy and protein dilution on growth performance, nutrient utilization and internal organs of broilers. *Ital. J. Anim. Sci.* **2015**, *14*, 3729. [CrossRef]
17. Namroud, N.F.; Shivazad, M.; Zaghari, M. Effects of fortifying low crude protein diet with crystalline amino acids on performance, blood ammonia level, and excreta characteristics of broiler chicks. *Poult. Sci.* **2008**, *87*, 2250–2258. [CrossRef] [PubMed]
18. Wang, Q.D.; Zhang, K.Y.; Zhang, Y.; Bai, S.P.; Ding, X.M.; Wang, J.P.; Peng, H.W.; Tian, G.; Xuan, Y.; Su, Z.W.; et al. Effects of dietary protein levels and protease supplementation on growth performance, carcass traits, meat quality, and standardized ileal digestibility of amino acid in Pekin ducks fed a complex diet. *Poult. Sci.* **2020**, *99*, 3557–3566. [CrossRef]
19. Ghazi, S.; Rooke, J.A.; Galbraith, H.; Bedford, M.R. The potential for the improvement of the nutritive value of soya-bean meal by different proteases in broiler chicks and broiler cockerels. *Br. Poult. Sci.* **2002**, *43*, 70–77. [CrossRef] [PubMed]
20. Law, F.L.; Zulkifli, I.; Soleimani, A.F.; Liang, J.B.; Awad, E.A. The effects of low-protein diets and protease supplementation on broiler chickens in a hot and humid tropical environment. *Asian-Australas. J. Anim. Sci.* **2018**, *8*, 1291–1300. [CrossRef]
21. Freitas, D.M.; Vieira, S.L.; Angel, C.R.; Favero, A.; Maiorka, A. Performance and nutrient utilization of broilers fed diets supplemented with a novel mono-component protease. *J. Appl. Poult. Res.* **2011**, *20*, 322–334. [CrossRef]
22. Walk, C.L.; Pirgozliev, V.; Juntunen, K.; Paloheimo, M.; Ledoux, D.R. Evaluation of novel protease enzymes on growth performance and apparent ileal digestibility of amino acids in poultry: Enzyme screening. *Poult. Sci.* **2018**, *97*, 2123–2138. [CrossRef] [PubMed]
23. Vieira, S.L.; Angel, C.R.; Miranda, D.J.A.; Favero, A.; Cruz, R.F.A.; Sorbara, J.O.B. Effects of a monocomponent protease on performance and protein utilization in 1-to 26-day-of-age turkey poults. *J. Appl. Poult. Res.* **2013**, *22*, 680–688. [CrossRef]
24. Favero, A.; Maiorka, A.; Rocha, C.; Appelt, M.D.; Sorbara, J.O.B. Effect of protease enzyme on performance and ileal digestibility of broilers grown to 42 days of age in floor pens. *Int. Poult. Sci. Forum Atlanta Ga. Abstr. Pap. M28* **2009**, *28*, 9.

animals

MDPI

Article

Quercetin Dietary Supplementation Advances Growth Performance, Gut Microbiota, and Intestinal mRNA Expression Genes in Broiler Chickens

Mervat A. Abdel-Latif [1,*], Ahmed R. Elbestawy [2], Ali H. El-Far [3], Ahmed E. Noreldin [4], Mohamed Emam [1], Roua S. Baty [5], Ghadeer M. Albadrani [6], Mohamed M. Abdel-Daim [7,8,*] and Hatem S. Abd El-Hamid [2]

[1] Department of Nutrition and Veterinary Clinical Nutrition, Faculty of Veterinary Medicine, Damanhour University, Damanhour 22511, Egypt; vetdr_2006@yahoo.com
[2] Department of Poultry and Fish Diseases, Faculty of Veterinary Medicine, Damanhour University, Damanhour 22511, Egypt; ahmed.elbestawy@vetmed.dmu.edu.eg (A.R.E.); drhatem_deltavet@yahoo.co.uk (H.S.A.E.-H.)
[3] Department of Biochemistry, Faculty of Veterinary Medicine, Damanhour University, Damanhour 22511, Egypt; ali.elfar@damanhour.edu.eg
[4] Histology and Cytology Department, Faculty of Veterinary Medicine, Damanhour University, Damanhour 22511, Egypt; ahmed.elsayed@damanhour.edu.eg
[5] Department of Biotechnology, College of Science, Taif University, P.O. Box 11099, Taif 21944, Saudi Arabia; rsbaty@tu.edu.sa
[6] Department of Biology, College of Science, Princess Nourah bint Abdulrahman University, Riyadh 11474, Saudi Arabia; gmalbadrani@pnu.edu.sa
[7] Department of Pharmaceutical Sciences, Pharmacy Program, Batterjee Medical College, P.O. Box 6231, Jeddah 21442, Saudi Arabia
[8] Pharmacology Department, Faculty of Veterinary Medicine, Suez Canal University, Ismailia 41522, Egypt
* Correspondence: mervat.abdellatif@vetmed.dmu.edu.eg (M.A.A.-L.); abdeldaim.m@vet.suez.edu.eg (M.M.A.-D.)

Citation: Abdel-Latif, M.A.; Elbestawy, A.R.; El-Far, A.H.; Noreldin, A.E.; Emam, M.; Baty, R.S.; Albadrani, G.M.; Abdel-Daim, M.M.; Abd El-Hamid, H.S. Quercetin Dietary Supplementation Advances Growth Performance, Gut Microbiota, and Intestinal mRNA Expression Genes in Broiler Chickens. *Animals* **2021**, *11*, 2302. https://doi.org/10.3390/ani11082302

Academic Editors: Natalie Morgan and Youssef A. Attia

Received: 1 July 2021
Accepted: 30 July 2021
Published: 4 August 2021

Simple Summary: The biological activity of quercetin is diverse, particularly antioxidant, antimicrobial, and antibacterial. The impacts of quercetin nutritional supplementations on growth performance, humoral immunity, gut microbiota and mRNA in broiler chickens were recorded.

Abstract: Quercetin was fed to groups of broiler chickens at concentrations of 200, 400, and 800 ppm, and a control group was supplemented with a basal diet. Results revealed that quercetin dietary supplementation numerically improved the growth performance traits and significantly increased ($p < 0.05$) the European production efficiency factor (EPEF) in the 200 ppm group. The total coliforms and *Clostridium perfringens* were decreased ($p < 0.05$) in quercetin-supplemented groups. Conversely, *Lactobacillus* counts were increased ($p < 0.05$), due to improvement of the gut microbiota environment in quercetin-supplemented groups. Moreover, the mRNA expression of intestinal Cu/Zn-superoxide dismutase (*SOD1*), glutathione peroxidase (*GSH-Px*) and nutritional transporters, including glucose transporter 2 (*GLUT2*), peptide transporter 1 (*PEPT1*), and fatty acid synthase (*FAS*) genes, were significantly upregulated in quercetin-supplemented groups. Quercetin enhanced intestinal morphometry. We can suggest quercetin supplementation in broiler chickens by levels between 200 and 400 ppm to enhance their development and gut environment.

Keywords: antioxidant; broilers; growth performance; gut microbiota; quercetin

1. Introduction

The microbiome plays a major role in the gastrointestinal tract health, immune system, and productivity of broiler chickens [1]. The link between intestinal health and overall health is raised by the quality of feed consumed by broiler chickens [2]. Many studies have been conducted to enhance the intestinal environment by adding herbs, probiotics, and exogenous enzymes to the diet of broiler chickens to increase their efficiency and productivity [3–8].

The global banning of antibiotic growth promotors has encouraged researchers to find alternative solutions. Flavonols can regulate feed intake, contribute to eubiosis, and exhibit antimicrobial, immunomodulatory, anti-inflammatory, and antioxidant properties in monogastric animals [9]. Quercetin (3,3′,4′,5,7-pentahydroxyflavone) is a constituent of flavonols, a sub-group of flavonoids that are present in some fruit (apples, berries, and grapes), herbs, and some vegetables (onions and broccoli) [10]. Quercetin is a powerful, anti-inflammatory, antimicrobial, anti-obesity, anti-hypercholesterolemic, antioxidant, anti-aging, and anticancer agent [11]. Having antioxidative action is the primary core of quercetin's biological activities. Quercetin is generally one of the most commonly used bioflavonoids for metabolic and inflammatory diseases [12].

Regarding the role of quercetin in broiler diets, the study of Goliomytis et al. [13] stated that quercetin may extend the shelf-life of meat by decreasing the lipid oxidation rate and may lead to enhanced animal health. Quercetin also increases the immune responses in broiler chickens [14]. The current research examines the impact of various quercetin supplementation concentrations on the performance, gut microbiota, humoral immunity, and some mRNAs of broiler chickens.

2. Materials and Methods

2.1. Ethics Statement

The research was endorsed by the Local Experimental Animal Care Committee of the University of Damanhour, Egypt, Faculty of Veterinary Medicine (VMD: 15/2018).

2.2. Animals, Management and the Experimental Design

One hundred and twenty-eight one-day-old Ross 308 chicks were obtained from a commercial hatchery and randomly allocated into four equal groups of mixed-sex chicks (32 birds of equal body weights per group (41 gm/chick)). Each group was subdivided into four replicates (8 birds per replicate) with an equal number of males in each replicate (2 males and 6 females) and raised on wire-floored cages of the same dimensions, and numbers of nipple drinkers and feed hoppers. They received an experimental diet for five consecutive weeks. The birds were allowed to consume feed and water ad libitum and were kept under daily observation. The environmental temperature of the 1st week was 32 °C and progressively reduced to 26 °C by the 3rd week of age, and the chicks were exposed to 23 h light. Chicks were allocated into control group 1 (control, fed on a commercial basal diet); group 2 (Q200), group 3 (Q400), and group 4 (Q800) were fed a commercial basal diet containing 200, 400, and 800 ppm quercetin (Sigma-Aldrich Chemical Co., St. Louis, MO, USA), respectively. The basal diets (corn–soybean based diet) were formulated according to the nutrient requirements for broiler chickens (Ross 308) [15]. The nutrient content of the ingredients was evaluated following the instructions of AOAC [16]. The ingredients' percentage and calculated composition analysis of the basal diet are shown in Table 1.

Broiler chickens were vaccinated under the following program in all cages: inactivated avian influenza subtype H5N1 vaccine (MeFluvac®, MEVAC, Cairo, Egypt) by subcutaneous injection, and bivalent live Newcastle disease and infectious bronchitis vaccine (Nobilis® Clone 30 + Ma5, MSD, Boxmeer, The Netherlands) at 7 days of age; live Gumboro intermediate plus (Bursine Plus® vaccine, Zoetis Inc., Florham Park, NJ, USA) at 14 days of age, and finally vaccinated with live Newcastle disease (Nobilis® ND LaSota, MSD, Boxmeer, The Netherlands) vaccine at 18 days of age. Eye drops were used to administer all live vaccines.

Table 1. Ingredients' percentage and calculated composition analysis of the experimental starter and grower diets (%, as-fed basis).

Ingredients %	Starter (0–10 d)	Grower (11–21 d)	Finisher (22–35 d)
Yellow corn	54.78	58.88	63.90
Soybean meal (44%)	33.5	29.4	24
Corn gluten (60%)	5	5	5
Corn oil	2	2.65	3.15
Dicalcium phosphate	1.73	1.6	1.5
Lime stone	1.35	1	1
Salt	0.4	0.4	0.4
DL-methionine *	0.15	0.12	0.1
HCl-lysine **	0.35	0.3	0.3
Vitamins and minerals premix ***	0.3	0.3	0.3
Antimycotoxin	0.2	0.2	0.2
Sodium bicarbonate	0.1	0.1	0.1
Choline chloride	0.05	0.05	0.05
	Calculated composition		
ME, Kcal/Kg diet	3005	3100	3195
CP%	23	21.5	19.5
Ca%	1	0.87	0.82
Avail. P%	0.47	0.44	0.41
Methionine%	0.56	0.51	0.47
Lysine%	1.44	1.29	1.14
Meth. + Cyst.%	0.93	0.86	0.78
Na%	0.20	0.20	0.20

SBM = soybean meal, ME = metabolizable energy, CP = crude protein, Av. (P) = available phosphorous. * DL—methionine 99% feed grade China. ** L—lysine 99% feed grade. *** Vitamin and mineral premix (Hero mix) produced by Hero pharm and composed (per 3 kg) of vitamin A 12,000,000 IU, vitamin D3 2,500,000 IU, vitamin E 10,000 mg, vitamin K3 2000 mg, vitamin B1 1000 mg, vitamin B2 5000 mg, vitamin B6 1500 mg, vitamin B12 10 mg, niacin 30,000 mg, biotin 50 mg, folic acid 1000 mg, pantothenic acid 10,000 mg, manganese 60,000 mg, zinc 50,000 mg, iron 30,000 mg, copper 4000 mg, iodine 300 mg, selenium 100 mg, and cobalt 100 mg.

2.3. Growth Performance

The performance parameters include the body weight (BW), feed intake (FI), body weight gain (BWG), feed conversion ratio (FCR), and protein efficiency ratio (PER). The viability percentage was assessed weekly over the whole experimental period. In addition, the European production efficiency factor (EPEF) was evaluated at days 21 and 35 of the study [17] and calculated according to the following equation:

$$\text{EPEF} = [(\text{viability}\% \times \text{body weight per kg}) \div (\text{age per day} \times \text{FCR})] \times 100 \quad (1)$$

2.4. Sample Collection

At days 21 and 35, fecal samples were collected and tested for total coliform, *C. perfringens* and *Lactobacillus* counts. At the end of the experiment, day 35, blood samples (n = 5) were collected from the wing vein without anticoagulant for serum separation. The collected samples were centrifuged at 1435× g for 15 min at 4 °C to obtain clear sera for HI test against avian influenza subtype H5N1. Five birds from each group were euthanized through anesthesia with intravenous injection of sodium pentobarbital (50 mg/kg) and immediately necropsied. Five intestinal (ileum) samples of 2 cm in length (5 cm proximal to the ileo—cecal junction) for mRNA gene expressions and histological analysis were taken from each group, processed, and analyzed as previously described [5].

2.5. Total Fecal Bacterial Count

The total coliform, *C. perfringens*, and *Lactobacillus*, counts were evaluated as previously described [5].

Briefly, ten-fold dilutions (10^{-1} to 10^{-7}) of each sample were performed with BPW and directly inoculated on MacConkey's agar for total coliform counting and incubated aerobically at 37 °C for 24 h. All red colonies within the range of 15–150 μm were selected for counting.

C. perfringens were subcultured on Perfringens agar base (Oxoid; Table 2) mixed with 400 mg of D-cycloserine per liter by the dilutions from 10^{-1} to 10^{-7} and incubated anaerobically at 37 °C, using gas generating kits (Oxoid) for 48 h. Plates with black colonies within the range of 25–250 μm were counted.

Table 2. Composition of Perfringens agar.

Ingredients	Amount (g)
Tryptose	15.0
Soya peptone	5.0
Yeast extract	5.0
Sodium metabisulphite	1.0
Ferric ammonium citrate	1.0
Agar	19.0
Distilled water added to make 1 liter	

The *Lactobacillus* count was conducted using Rogosa agar (Table 3) plates and cultured by dilutions from 10^{-1} to 10^{-7}, then incubated at 37 °C in 5% CO_2. All whitish colonies that appeared after 48 h of incubation were counted.

Table 3. Composition of Rogosa agar.

Ingredients	Amount (g)
Tryptone	10.0
Yeast extract	5.0
Glucose	20.0
Sodium acetate, anhydrous	17.0
Ammonium citrate	2.0
Potassium dihydrogen phosphate	6.0
Magnesium sulfate	0.575
Manganese sulfate	0.120
Ferrous sulfate	0.034
Bacteriological Agar	20
Tween 80	1
Distilled water added to make 1 liter	

2.6. Hemagglutination Inhibition (HI) Assay

Antibody titers for avian influenza subtype H5N1 were determined using a standard H5N1 antigen; the positive titers were the highest dilutions of serum causing complete inhibition of 4 hemagglutination units (4 HAU) of antigen [18,19].

2.7. RNA Extraction and RT-PCR

The total RNA was extracted and purified from intestinal samples (n = 5) of all groups using QIAamp RNeasy Mini kit (Qiagen, GmbH, Hilden, Germany); then, RT-PCR was done with QuantiTect SYBR Green PCR Master Mix (Qiagen, GmbH, Dusseldorf, Germany). Primers are listed in Table 4 as previously practiced [5].

2.8. Histology

The fixed samples were processed with the conventional paraffin embedding technique and stained with hematoxylin and eosin (H&E) as described by Bancroft and Layton [20]. Sectioning and slide preparation were done according to Saeed et al. [21]. Three sections were utilized from each intestinal segment (one section from serial ten sections). From every section, five well oriented complete villi were selected for the investigation. So, fifteen values were estimated for each intestinal sample. Slides were examined under a light microscope (Leica DM500) at 4× magnification, using a digital camera (Leica EC3, Leica, Germany). Measurements of the villi height (VH), villi width in the middle of the individual villus (VW), crypt depth (CD), and VH:CD ratio for each villus in the control and supplemented groups were made by using ImageJ software (NIH, Bethesda, MD, USA) [22].

2.9. Statistical Analysis

Statistical calculations were made with the SPSS programming tool (IBM SPSS. 20®) (SPSS Inc., Chicago, IL, USA) using one-way ANOVA followed by Duncan's multiple range tests. Data of the HI assay, RT-PCR, and total fecal bacterial counts were analyzed with one-way ANOVA and Tukey's multiple range tests with Graphpad prism 5. All significant deviations were based on $p < 0.05$.

Table 4. Primer sequences, target genes, amplicon sizes and cycling conditions for SYBR green RT-PCR.

Target Gene	Primers Sequences	Reverse Transcription	Primary Denaturation	Amplification (40 Cycles)			Dissociation Curve (1 Cycle)			Reference
				Secondary Denaturation	Annealing (Optics on)	Extension	Secondary Denaturation	Annealing	Final Denaturation	
β. actin	F: ATTGTCCACCGCAA ATGCTTC	50 °C 30 min	94 °C 5 min	94 °C 15 s	60 °C 30 s	72 °C 30 s	94 °C 1 min	60 °C 1 min	94 °C 1 min	[23]
	R: AAATAAAGCCATGC-CAATCTCGTC									
SOD1	F: AGGGGGTCATCCACTTCC				60 °C 30 s			60 °C 1 min		[24]
	R: CCCATTTGTGTTGTCTCCAA									
GSH-PX	F: TTGTAAACATCAGGGGCAAA									
	R: ATGGGCCAAGATCTTTCTGTAA									
GLUT2	F: CACACTATGGGCGCATGCT				60 °C 30 s			60 °C 1 min		[25]
	R: ATTGTCCCTGGAGGTGTTGGTG									
PEPT1	F: CCCCTGAGGAGGATCACTGTT									
	R: CAAAAGAGCAGCAGCAACGA									
FAS	F: CTATCGACACAGCCTGCTCCT				62 °C 30 s			62 °C 1 min		[26]
	R: CAGAATGTTGACCCCTCCTACC									

3. Results

3.1. Growth Performance and Survival Percentages

The initial live body weight between the distinct experimental groups non-significantly varied (Table 5). In Q200, Q400, and Q800, growth was enhanced during the experimental period by quercetin supplementation, compared with the control, by 3.27, 3.18, and 2.32%, respectively, and feed intake values were similar to the control group. In comparison with the control groups, the body weight gain, feed conversion ratios (FCR), and protein efficiency ratio (PER) values were not substantially improved; however, the European production efficiency factor (EPEF) was increased ($p < 0.05$) in Q200 as compared with the control group. In addition, the quercetin-supplemented groups expressed no mortality, compared to 3.13% of the control.

Table 5. Effect of dietary quercetin supplementation on growth performance and mortality rate of broilers.

	Control	Quercetin Supplementation			
		Q200	Q400	Q800	*p-Value*
Initial weight, g	41.46 ± 0.47	41.67 ± 0.49	41.67 ± 0.49	41.67 ± 0.49	0.99
[1] fBwt, g	1768.10 ± 31.56	1826.59 ± 35.88	1824.29± 31.97	1809.13 ± 25.77	0.54
[2] BWG, g	1727.14 ± 31.36	1785 ± 35.58	1782.86 ± 31.60	1767.61 ± 25.45	0.54
BWG/day, g	49.35 ± 0.90	51 ± 1.02	50.94 ± 0.90	50.50 ± 0.73	0.54
[3] FI, g	2887.24± 25.32 ^{ab}	2848.31 ± 28.45 ^{b}	2894.72 ± 6.11 ^{ab}	2926.58 ± 11.29 ^{a}	0.05
[4] FCR	1.67 ± 0.03	1.60 ± 0.033	1.63 ± 0.03	1.66 ± 0.02	0.32
[5] PER	2.80 ± 0.05	2.93 ± 0.06	2.87 ± 0.05	2.81 ± 0.04	0.27
[6] EPEF	296.54 ± 10.12 ^{b}	331.06 ± 13.03 ^{a}	323.29 ± 11.54 ^{ab}	313 ± 8.32 ^{ab}	0.01
Mortality%	3.13	0	0	0	

Note: Means within each column for each division with no common superscript letters are significantly different ($p \leq 0.05$). Abbreviations. [1] Final body weight. [2] Body weight gain. [3] Voluntary feed intake. [4] Feed conversion ratio. [5] Protein efficiency ratio. [6] EPEF = [(viability % × body weight per kg) ÷ (age per day × FCR)] × 100.

3.2. Total Fecal Bacterial Count

Total coliform counts decreased significantly ($p < 0.001$) by 21 and 35 days in all quercetin-supplemented groups (Q200, Q400, and Q800) (Figure 1A,B), compared with the control group. Total *Clostridium perfringens* counts were decreased in all quercetin supplements ($p < 0.001$) at 35 days of age (9.41 log10 CFU/g), compared with the control (Figure 1C,D). Conversely, the total number of *Lactobacillus* in groups supplemented with quercetin was enhanced ($p < 0.001$), compared with the control at both ages (21 and 35 days) (Figure 1E,F).

3.3. Hemagglutination Inhibition Test

In the quercetin-supplemented groups and controls, no significant differences were found among HI Titer values of the H5N1 avian influenza subtype; however, the titers of Q200, Q400, and Q800 were below the control level at 5.4, 5.4, 5.8 vs. 6.4 log2, respectively (Figure 2).

3.4. Antioxidant Enzymes' Gene Expressions

As shown in Figure 3A, the expressions of the intestinal Cu/Zn-superoxide dismutase (SOD1) in the Q400 and Q800 groups were significantly increased ($p < 0.001$), compared with the control group, and were higher ($p < 0.05$ and $p < 0.001$, respectively) than Q200. Moreover, the SOD1 gene expression was increased ($p < 0.001$) in Q800, compared with Q400. Data in Figure 3B show the mRNA expression of glutathione peroxidase (GSH-Px); the quercetin-supplemented groups had higher ($p < 0.001$) levels than the control. The fold change of GSH-Px expression in Q800 and Q400 was increased ($p < 0.001$), compared with Q200, while Q800 was also increased ($p < 0.001$) in comparison with Q400.

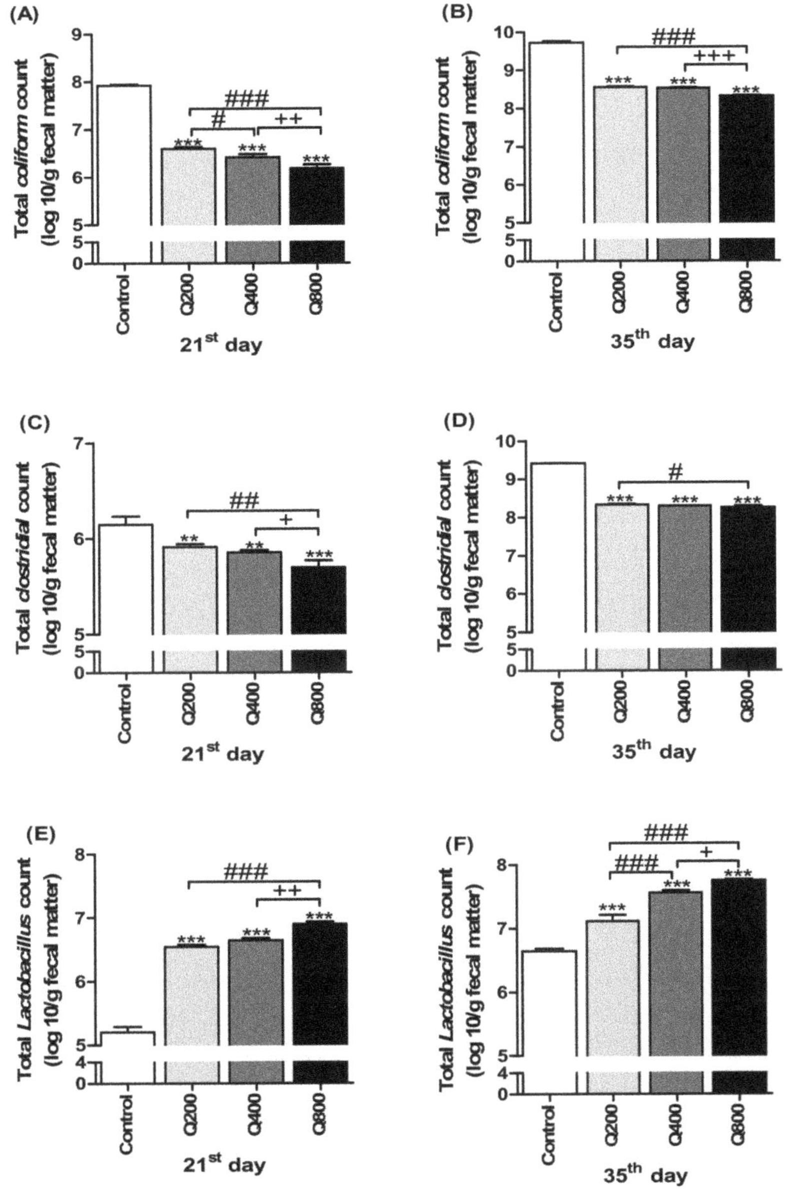

Figure 1. Total coliform (**A**,**B**), *C. perfringens* (**C**,**D**), and *Lactobacillus* (**E**,**F**) counts in fecal samples. ** $p < 0.01$ and *** $p < 0.001$ vs. control. # $p < 0.05$, ## $p < 0.01$, and ### $p < 0.001$ vs. Q200. + $p < 0.05$, ++ $p < 0.01$, and +++ $p < 0.001$ vs. Q400. Statistical analysis was performed using one-way ANOVA and Tukey's post hoc test for multiple comparisons. Q200, birds fed 200 ppm. Q400, birds fed 400 ppm. Q800, birds fed 800 ppm.

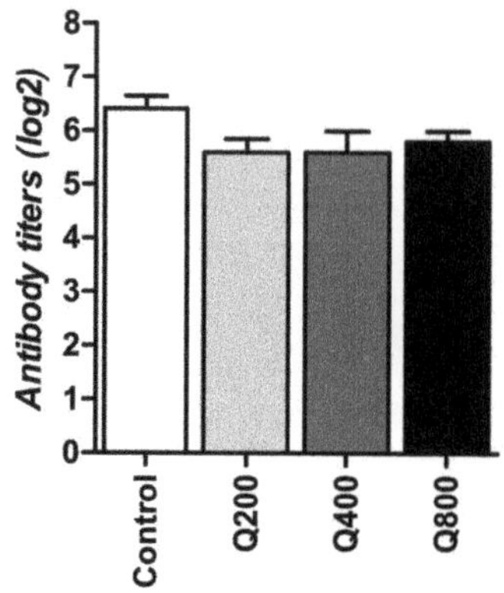

Figure 2. Antibody titer of H5N1 (log2). Statistical analysis was performed using one-way ANOVA and Tukey's post hoc test for multiple comparisons. Q200, birds fed 200 ppm. Q400, birds fed 400 ppm. Q800, birds fed 800 ppm.

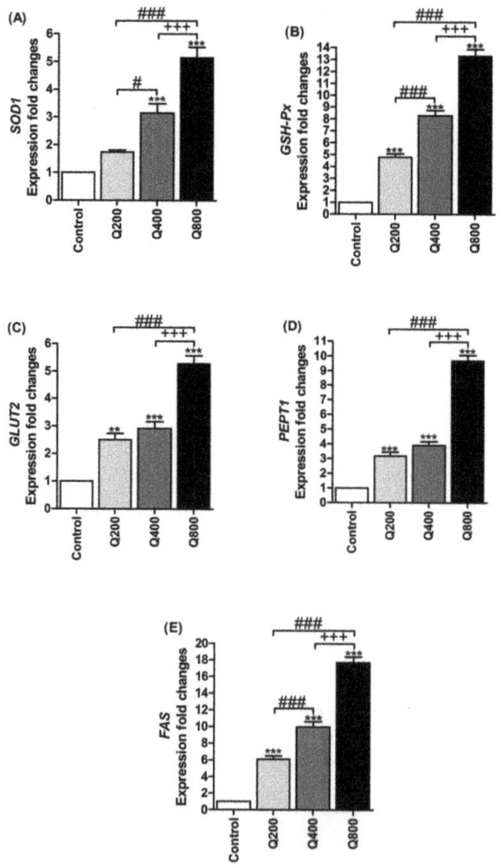

Figure 3. RT-PCR validation of the *SOD1* (**A**), *GSH-Px* (**B**), *GLUT2* (**C**), *PEPT1* (**D**), and *FAS* (**E**) genes. ** $p < 0.01$ and *** $p < 0.001$ vs. control. # $p < 0.05$ and ### $p < 0.001$ vs. Q200. +++ $p < 0.001$ vs. Q400. Statistical analysis was performed using one-way ANOVA and Tukey's post hoc test for multiple comparisons. Q200, birds fed 200 ppm. Q400, birds fed 400 ppm. Q800, birds fed 800 ppm.

3.5. Nutrients Transporter Gene Expressions

Figure 3C,D shows the expression of intestinal glucose transporter 2 (GLUT2) and peptide transporter 1 (PEPT1), respectively. GLUT2 and PEPT1 expressions were increased ($p < 0.001$) in quercetin-supplemented groups, compared with the control group. Additionally, their expressions increased ($p < 0.001$) in Q800, compared with Q400 and Q200. The same results were found for the intestinal fatty acid synthase (FAS) gene (Figure 3E) with a significant increase ($p < 0.001$) in Q400, compared with the Q200 group.

3.6. Histology

The height and width of the villi in the quercetin-supplemented groups compared with the control group significantly increased. The best supplementation for the villi area was Q200. On the other hand, Q800 had the best effect on the crypt depth. Villi height/crypt depth had the highest value in the Q200 group, due to the highest effect of this supplementation on the villi (Figure 4).

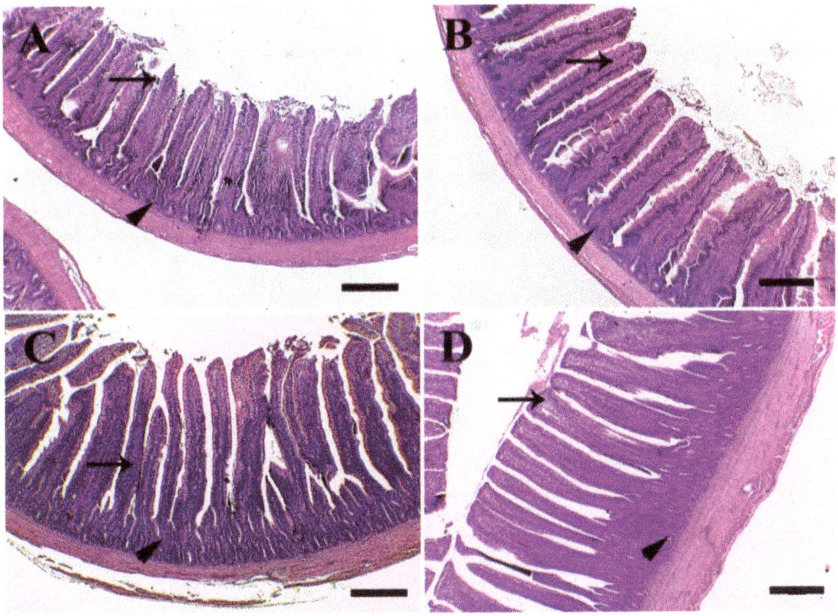

Figure 4. *Cont.*

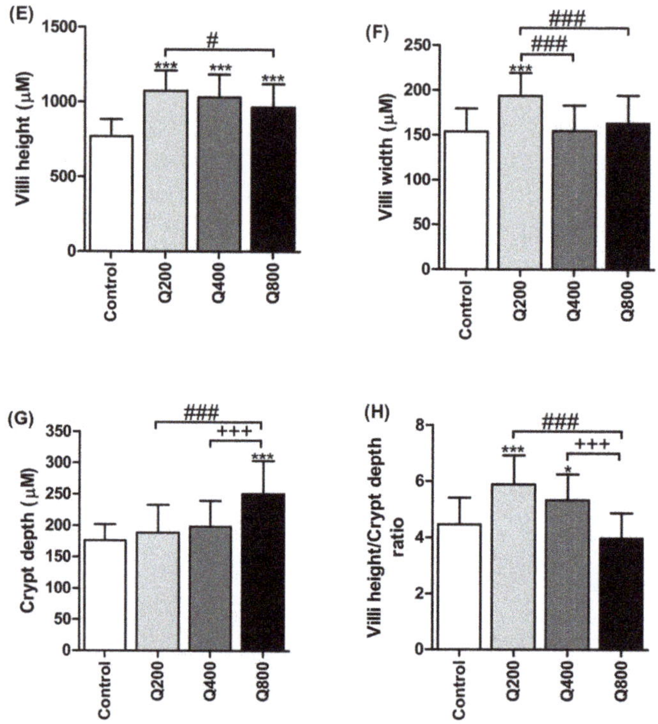

Figure 4. Light microscopic images of small intestine. (**A**) Control. (**B**) Q200 shows the highest villi height and width and the lowest crypt depth. (**C**) Q400 shows the moderate villi height, width and crypt depth. (**D**) Q800 shows the lowest villi height and width and the greatest crypt depth. (**E**) Villi height (μm). (**F**) Villi width (μm). (**G**) Crypt depth (μm). (**H**) Villi height/crypt depth. * $p < 0.05$ and *** $p < 0.001$ vs. control. $^{\#}$ $p < 0.05$ and $^{\#\#\#}$ $p < 0.001$ vs. Q200. $^{+++}$ $p < 0.001$ vs. Q400. Statistical analysis was performed using one-way ANOVA and Tukey's post hoc test for multiple comparisons. Q200, birds fed 200 ppm. Q400, birds fed 400 ppm. Q800, birds fed 800 ppm. The black arrows refer to villi. The black arrows' heads refer to crypts. Scale bar = 400 μm.

4. Discussion

Quercetin is a bioavailable glycone in mammals. Quercetin glycosides can be hydrolyzed by lactase phlorizin hydrolase in the lumen or, once entered into the enterocyte through the sodium dependent glucose transporter (SGLT1), quercetin–glucoside is hydrolyzed by β-glucosidase [27–29]. In addition, in chickens, Rupasinghe et al. [30] detected seven quercetin metabolites in the excreta of quercetin-supplemented birds, indicating that quercetin was absorbed and excreted. Alternatively, the presence of metabolites in broiler chicken excreta could be due to the action of intestinal microbiota, rather than de novo metabolism. Other studies confirmed the absorption of quercetin in pigs [31] and horses [32].

Functional feeds are used to improve poultry efficiency by regulating pathogens and improving beneficial bacteria in the intestines [33]. This results in improved weight gain, FCR, and uniformity [34]. Here, quercetin improved performance parameters, such as body weight (BW), body weight gain (BWG), FCR, PER, and EPEF, while voluntary feed intake (VFI) in the quercetin-supplemented groups did not differ significantly from the control group. In the Q200 group, EPEF increased noticeably, compared with the control group. In the same context, Liu et al. [35] supplemented hens with varying quercetin levels and found an enhanced laying rate and FCR when hens were fed 0.2 and 0.4 g quercetin per kg of body weight.

The improvement of performance parameters in our research was attributed to the metabolic prebiotic effect of quercetin [35] with its ability to modulate gut microbiota through increasing the number of beneficial bacteria as Lactobacilli and decreasing the number of *C. perfringens* and total coliform counts (selective action) that beneficially affect the broiler chickens' health and performance. Quercetin induces their antibacterial activity by acting as DNA gyrase on various cell targets [36], bacterial membrane and motility [37], type II fatty acid biosynthesis (FAS II) pathway [38], and D-alanine:d-alanine ligase (Ddl) enzyme inhibitor, acting as a bacteriostatic, preventing harmful bacterial growth [39]. Unfortunately, the fecal materials were collected without caecum voiding, so there was no differentiation between fecal and cecal droppings [40]; however, this is a very critical limitation point and it will be considered in future work.

Villus height and crypt depth from the jejunum of the quercetin fed birds increased, compared with the control. The height of villi shows the gut's absorbing ability [41]. The supplemented treatment deviated significantly from the control in the villus height and villus/crypt ratio in the jejunum at Q200, although the villus/crypt ratio of the birds fed Q800 tended to be lower than other treatments, compared with the control. These results indicate that feeding quercetin to broilers may boost intestinal morphology, which shows elevated absorption and enhanced intestinal health. Similarly, in rats, the administration of methotrexate-treated quercetin resulted in a higher villus height in the jejunum and ileum crypts [42]. Moreover, quercetin dietary supplementation upregulated the expression of nutrient transporter genes (GLUT2, PEPT1, and FAS), which play a vital role in the nutrient's metabolism. Intestinal GLUT2 is primarily a protein sensor for glucose and glucose homeostasis [43]. In the absorption of small peptides, the intestinal PEPT1 plays a key role [44], while the intestinal FAS retains the palmitoylation of the mucin 2, intestinal mucus barrier, which prevents the bowel pathogen [45]. In addition, the status of the bowel antioxidant is a key protection indicator for broiler chickens. In the present study, SOD1 and GSH-Px mRNA expressions in quercetin-supplemented groups were significantly increased dose-dependently, compared with the control. SOD enzymes catalyze the transformation of superoxide anion into less dangerous free radical hydrogen peroxide (H_2O_2) [46], and GSH-Px attacked the generated H_2O_2 [47]. These antioxidant effects of quercetin provided the birds with a strong defense with a healthy intestinal environment [5]. Iskender et al. [48] reported significant increases in antioxidant enzymes, including the activities of GSH-Px, SOD1, and glutathione levels in erythrocytic lysates of laying hens fed a diet containing 0.5 g/kg quercetin.

Immune response to vaccination is a key method to show the well-functioning of the immune system of broiler chickens [49]. Here, we investigated the impact of quercetin supplementation on the healthy intestinal immunity and, accordingly, on the overall humoral immune response, e.g., serological immune response to inactivated avian influenza vaccine (H5N1) at the end of the experiment; the quercetin-supplemented groups did not differ significantly in relation to controls. However, intestinal injury models are badly needed in the future to explore the close relationship between the gastrointestinal tract (GIT) microflora and development and/or maintenance of a functional intestinal immune system during quercetin supplementation.

5. Conclusions

Quercetin dietary supplementation in broiler chickens of 200 and 400 ppm, but not 800 ppm, enhanced their growth, intestinal and gene expression in levels of antioxidant enzymes and nutrient transportation. Quercetin is, therefore, regarded as a promising natural feed additive to broiler chickens at levels below 800 ppm.

Author Contributions: Conceptualization, M.A.A.-L., H.S.A.E.-H., and A.H.E.-F.; methodology, M.A.A.-L., A.R.E., A.E.N., M.E., and A.H.E.-F.; software, A.E.N.; validation, M.A.A.-L.; formal analysis, M.A.A.-L. and A.H.E.-F.; investigation, A.R.E.; resources, H.S.A.E.-H.; data curation, M.A.A.-

L., and A.H.E.-F.; writing—original draft preparation, M.A.A.-L., G.M.A., A.R.E., and A.H.E.-F.; writing—review and editing, M.A.A.-L., A.R.E., M.E., R.S.B., M.M.A.-D., and A.H.E.-F.; visualization, H.S.A.E.-H., and M.M.A.-D.; supervision, M.A.A.-L. and A.H.E.-F.; project administration, H.S.A.E.-H.; funding acquisition, R.S.B. All authors have read and agreed to the published version of the manuscript.

Funding: This research work was funded by the Deanship of Scientific Research at Taif University Researchers Supporting Project number TURSP-2020/269, Taif University, Taif, Saudi Arabia.

Institutional Review Board Statement: The study was conducted according to the guidelines of the Local Experimental Animal Care Committee of the University of Damanhour, Egypt, Faculty of Veterinary Medicine (VMD: 15/2018).

Data Availability Statement: The data presented in this study are available on request from the corresponding author.

Acknowledgments: The authors acknowledge the financial support of Taif University Researchers Supporting Project number TURSP-2020/269, Taif University, Taif, Saudi Arabia. The authors extend thanks to their respected institutes and universities.

Conflicts of Interest: The authors have no conflict of interest.

References

1. Clavijo, V.; Flórez, M.J.V. The gastrointestinal microbiome and its association with the control of pathogens in broiler chicken production: A review. *Poult. Sci.* **2018**, *97*, 1006–1021. [CrossRef]
2. Schokker, D.; Jansman, A.J.M.; Veninga, G.; De Bruin, N.; Vastenhouw, S.A.; De Bree, F.M.; Bossers, A.; Rebel, J.M.J.; Smits, M.A. Perturbation of microbiota in one-day old broiler chickens with antibiotic for 24 hours negatively affects intestinal immune development. *BMC Genom.* **2017**, *18*, 241. [CrossRef]
3. Ayoub, M.; El-Far, A.; Taha, N.; Korshom, M.; Mandour, A.; Abdel-Hamid, H.; El-Neweshy, M. The biochemical protective role of some herbs against aflatoxicosis in ducklings: II. Nigella sativa. *Lucr. Științ. Univ. Științe Agric. Med. Vet. Ser. Zooteh.* **2011**, *55*, 68–77.
4. Abdel-Latif, M.A.; Abd El-Hack, M.E.; Swelum, A.A.; Saadeldin, I.M.; Elbestawy, A.R.; Shewita, R.S.; Ba-Awadh, H.A.; Alowaimer, A.N.; Abd El-Hamid, H.S. Single and Combined Effects of Clostridium butyricum and Saccharomyces cerevisiae on Growth Indices, Intestinal Health, and Immunity of Broilers. *Animals* **2018**, *8*, 184. [CrossRef] [PubMed]
5. Abdel-Latif, M.A.; El-Far, A.H.; Elbestawy, A.R.; Ghanem, R.; Mousa, S.A.; Abd El-Hamid, H.S. Exogenous dietary lysozyme improves the growth performance and gut microbiota in broiler chickens targeting the antioxidant and non-specific immunity mRNA expression. *PLoS ONE* **2017**, *12*, e0185153. [CrossRef]
6. Abd El-Hack, M.E.; Samak, D.H.; Noreldin, A.E.; El-Naggar, K.; Abdo, M. Probiotics and plant-derived compounds as eco-friendly agents to inhibit microbial toxins in poultry feed: A comprehensive review. *Environ. Sci. Pollut. Res. Int.* **2018**, *25*, 31971–31986. [CrossRef]
7. Abd El-Hack, M.E.; Chaudhry, M.T.; Mahrose, K.M.; Noreldin, A.; Emam, M.; Alagawany, M. The efficacy of using exogenous enzymes cocktail on production, egg quality, egg nutrients and blood metabolites of laying hens fed distiller's dried grains with solubles. *J. Anim. Physiol. Anim. Nutr.* **2018**, *102*, e726–e735. [CrossRef] [PubMed]
8. El-Far, A.H.; Ahmed, H.A.; Shaheen, H.M. Dietary Supplementation of Phoenix dactylifera Seeds Enhances Performance, Immune Response, and Antioxidant Status in Broilers. *Oxid. Med. Cell. Longev.* **2016**, *2016*, 5454963. [CrossRef]
9. Wenk, C. Herbs and botanicals as feed additives in monogastric animals. *Asian Aust. J. Anim. Sci.* **2003**, *16*, 282–289. [CrossRef]
10. Mlcek, J.; Jurikova, T.; Skrovankova, S.; Sochor, J. Quercetin and Its Anti-Allergic Immune Response. *Molecules (Basel Switzerland)* **2016**, *21*, 623. [CrossRef] [PubMed]
11. David, A.V.A.; Arulmoli, R.; Parasuraman, S. Overviews of Biological Importance of Quercetin: A Bioactive Flavonoid. *Pharmacogn. Rev.* **2016**, *10*, 84–89. [CrossRef]
12. Lakhanpal, P.; Rai, D.K. Quercetin: A versatile flavonoid. *Internet J. Med. Update* **2007**, *2*, 22–37. [CrossRef]
13. Goliomytis, M.; Tsoureki, D.; Simitzis, P.; Charismiadou, M.; Hager-Theodorides, A.; Deligeorgis, S. The effects of quercetin dietary supplementation on broiler growth performance, meat quality, and oxidative stability. *Poult. Sci.* **2014**, *93*, 1957–1962. [CrossRef]
14. Saeed, M.; Naveed, M.; Arain, M.; Arif, M.; El-Hack, M.A.; Alagawany, M.; Siyal, F.; Soomro, R.; Sun, C. Quercetin: Nutritional and beneficial effects in poultry. *World's Poult. Sci. J.* **2017**, *73*, 355–364. [CrossRef]
15. NRC. *Nutrient Requirements of Poultry*; National Academies Press: Washington, DC, USA, 1994.
16. AOAC. *Official Methods of Analysis of the Association of Analytical Chemists International*; AOAC: Gaithersburg, MD, USA, 2005.
17. Marcu, A.; Vacaru-Opriş, I.; Dumitrescu, G.; Ciochina, L.P.; Marcu, A.; Nicula, M.; Peţ, I.; Dronca, D.; Kelciov, B. The influence of the genotype on economic efficiency of broiler chickens growth. *Sci. Pap. Anim. Sci. Biotechnol.* **2013**, *46*, 339–346.

18. Takátsy, G. The use of spiral loops in serological and virological micro-methods. *Acta Microbiol. Immunol. Hung.* **1955**, *50*, 369–382. [CrossRef] [PubMed]
19. Brugh, M., Jr. A simple method for recording and analyzing serological data. *Avian Dis.* **1978**, *22*, 362–365. [CrossRef]
20. Bancroft, J.D.; Layton, C. The hematoxylin and eosin, connective and mesenchymal tissues with their stains. In *Bancroft s Theory and Practice of Histological Techniques*, 7th ed.; Kim suvarna, S., Layton, C., Bancroft, J.D., Eds.; Churchill Livingstone: Philadelphia, PA, USA, 2013; pp. 173–186.
21. Saeed, M.; Yatao, X.; Hassan, F.U.; Arain, M.A.; Abd El-Hack, M.E.; Noreldin, A.E.; Sun, C. Influence of Graded Levels of l-Theanine Dietary Supplementation on Growth Performance, Carcass Traits, Meat Quality, Organs Histomorphometry, Blood Chemistry and Immune Response of Broiler Chickens. *Int. J. Mol. Sci.* **2018**, *19*, 462. [CrossRef] [PubMed]
22. Schneider, C.A.; Rasband, W.S.; Eliceiri, K.W. NIH Image to ImageJ: 25 years of image analysis. *Nat. Methods* **2012**, *9*, 671–675. [CrossRef] [PubMed]
23. Hassan, F.A.M.; Roushdy, E.M.; Kishawy, A.T.Y.; Zaglool, A.W.; Tukur, H.A.; Saadeldin, I.M. Growth Performance, Antioxidant Capacity, Lipid-Related Transcript Expression and the Economics of Broiler Chickens Fed Different Levels of Rutin. *Animals* **2018**, *9*, 7. [CrossRef]
24. Akbarian, A.; Michiels, J.; Golian, A.; Buyse, J.; Wang, Y.; De Smet, S. Gene expression of heat shock protein 70 and antioxidant enzymes, oxidative status, and meat oxidative stability of cyclically heat-challenged finishing broilers fed Origanum compactum and Curcuma xanthorrhiza essential oils. *Poult. Sci.* **2014**, *93*, 1930–1941. [CrossRef]
25. Ebrahimi, R.; Faseleh Jahromi, M.; Liang, J.B.; Soleimani Farjam, A.; Shokryazdan, P.; Idrus, Z. Effect of Dietary Lead on Intestinal Nutrient Transporters mRNA Expression in Broiler Chickens. *BioMed Res. Int.* **2015**, *2015*, 149745. [CrossRef] [PubMed]
26. Zhou, M.; Zeng, D.; Ni, X.; Tu, T.; Yin, Z.; Pan, K.; Jing, B. Effects of Bacillus licheniformis on the growth performance and expression of lipid metabolism-related genes in broiler chickens challenged with Clostridium perfringens-induced necrotic enteritis. *Lipids Health Dis.* **2016**, *15*, 48. [CrossRef]
27. Graefe, E.U.; Wittig, J.; Mueller, S.; Riethling, A.K.; Uehleke, B.; Drewelow, B.; Pforte, H.; Jacobasch, G.; Derendorf, H.; Veit, M. Pharmacokinetics and bioavailability of quercetin glycosides in humans. *J. Clin. Pharmacol.* **2001**, *41*, 492–499. [CrossRef]
28. Crespy, V.; Morand, C.; Besson, C.; Manach, C.; Demigne, C.; Remesy, C. Quercetin, but not its glycosides, is absorbed from the rat stomach. *J. Agric. Food Chem.* **2002**, *50*, 618–621. [CrossRef]
29. Day, A.J.; Gee, J.M.; DuPont, M.S.; Johnson, I.T.; Williamson, G. Absorption of quercetin-3-glucoside and quercetin-4′-glucoside in the rat small intestine: The role of lactase phlorizin hydrolase and the sodium-dependent glucose transporter. *Biochem. Pharmacol.* **2003**, *65*, 1199–1206. [CrossRef]
30. Rupasinghe, H.P.; Ronalds, C.M.; Rathgeber, B.; Robinson, R.A. Absorption and tissue distribution of dietary quercetin and quercetin glycosides of apple skin in broiler chickens. *J. Sci. Food Agric.* **2010**, *90*, 1172–1178. [CrossRef] [PubMed]
31. Bieger, J.; Cermak, R.; Blank, R.; de Boer, V.C.; Hollman, P.C.; Kamphues, J.; Wolffram, S. Tissue distribution of quercetin in pigs after long-term dietary supplementation. *J. Nutr.* **2008**, *138*, 1417–1420. [CrossRef] [PubMed]
32. Wein, S.; Wolffram, S. Oral bioavailability of quercetin in horses. *J. Equine Vet. Sci.* **2013**, *33*, 441–445. [CrossRef]
33. Sallam, K.I.; Ishioroshi, M.; Samejima, K. Antioxidant and antimicrobial effects of garlic in chicken sausage. *Lebensm. Wiss. Technol.* **2004**, *37*, 849–855. [CrossRef]
34. Willis, W.L.; Goktepe, I.; Isikhuemhen, O.S.; Reed, M.; King, K.; Murray, C. The effect of mushroom and pokeweed extract on salmonella, egg production, and weight loss in molting hens. *Poult. Sci.* **2008**, *87*, 2451–2457. [CrossRef] [PubMed]
35. Liu, H.N.; Liu, Y.; Hu, L.L.; Suo, Y.L.; Zhang, L.; Jin, F.; Feng, X.A.; Teng, N.; Li, Y. Effects of dietary supplementation of quercetin on performance, egg quality, cecal microflora populations, and antioxidant status in laying hens. *Poult. Sci.* **2014**, *93*, 347–353. [CrossRef]
36. Ohemeng, K.; Schwender, C.; Fu, K.; Barrett, J. DNA gyrase inhibitory and antibacterial activity of some flavones (1). *Bioorga. Med. Chem. Lett.* **1993**, *3*, 225–230. [CrossRef]
37. Mirzoeva, O.; Grishanin, R.; Calder, P. Antimicrobial action of propolis and some of its components: The effects on growth, membrane potential and motility of bacteria. *Microbiol. Res.* **1997**, *152*, 239–246. [CrossRef]
38. Tasdemir, D.; Lack, G.; Brun, R.; Rüedi, P.; Scapozza, L.; Perozzo, R. Inhibition of Plasmodium f alciparum Fatty Acid Biosynthesis: Evaluation of FabG, FabZ, and FabI as Drug Targets for Flavonoids. *J. Med. Chem.* **2006**, *49*, 3345–3353. [CrossRef]
39. Wu, D.; Kong, Y.; Han, C.; Chen, J.; Hu, L.; Jiang, H.; Shen, X. D-Alanine: D-alanine ligase as a new target for the flavonoids quercetin and apigenin. *Int. J. Antimicrob. Agents* **2008**, *32*, 421–426. [CrossRef]
40. Rychlik, I. Composition and Function of Chicken Gut Microbiota. *Animals* **2020**, *10*, 103. [CrossRef]
41. Teirlynck, E.; Bjerrum, L.; Eeckhaut, V.; Huygebaert, G.; Pasmans, F.; Haesebrouck, F.; Dewulf, J.; Ducatelle, R.; Van Immerseel, F. The cereal type in feed influences gut wall morphology and intestinal immune cell infiltration in broiler chickens. *Br. J. Nutr.* **2009**, *102*, 1453–1461. [CrossRef] [PubMed]
42. Sukhotnik, I.; Moati, D.; Shaoul, R.; Loberman, B.; Pollak, Y.; Schwartz, B. Quercetin prevents small intestinal damage and enhances intestinal recovery during methotrexate-induced intestinal mucositis of rats. *Food Nutr. Res.* **2018**, *62*. [CrossRef]
43. Thorens, B. GLUT2, glucose sensing and glucose homeostasis. *Diabetologia* **2015**, *58*, 221–232. [CrossRef]
44. Wang, C.Y.; Liu, S.; Xie, X.N.; Tan, Z.R. Regulation profile of the intestinal peptide transporter 1 (PepT1). *Drug Des. Dev. Ther.* **2017**, *11*, 3511–3517. [CrossRef] [PubMed]

45. Wei, X.; Yang, Z.; Rey, F.E.; Ridaura, V.K.; Davidson, N.O.; Gordon, J.I.; Semenkovich, C.F. Fatty acid synthase modulates intestinal barrier function through palmitoylation of mucin 2. *Cell Host Microbe* **2012**, *11*, 140–152. [CrossRef]
46. Wang, Y.; Branicky, R.; Noë, A.; Hekimi, S. Superoxide dismutases: Dual roles in controlling ROS damage and regulating ROS signaling. *J. Cell Biol.* **2018**, *217*, 1915–1928. [CrossRef]
47. Nagababu, E.; Chrest, F.J.; Rifkind, J.M. Hydrogen-peroxide-induced heme degradation in red blood cells: The protective roles of catalase and glutathione peroxidase. *Biochim. Biophys. Acta* **2003**, *1620*, 211–217. [CrossRef]
48. Iskender, H.; Yenice, G.; Dokumacioglu, E.; Kaynar, O.; Hayirli, A.; Kaya, A. The effects of dietary flavonoid supplementation on the antioxidant status of laying hens. *Braz. J. Poult. Sci.* **2016**, *18*, 663–668. [CrossRef]
49. Korver, D. Implications of changing immune function through nutrition in poultry. *Anim. Feed Sci. Technol.* **2012**, *173*, 54–64. [CrossRef]

animals

Article

Dietary Supplementation of Shredded, Steam-Exploded Pine Particles Decreases Pathogenic Microbes in the Cecum of Acute Heat-Stressed Broilers

Akshat Goel [1,2], Beom-June Kim [1], Chris-Major Ncho [1], Chae-Mi Jeong [1,3], Vaishali Gupta [1,3], Ji-Young Jung [2,4], Si-Young Ha [4], Dong-Hwan Lee [4], Jae-Kyung Yang [2,4] and Yang-Ho Choi [1,2,3,*]

1 Department of Animal Science, Gyeongsang National University, Jinju 52828, Korea; genesakshat@gnu.ac.kr (A.G.); premix@naver.com (B.-J.K.); major159@gnu.ac.kr (C.-M.N.); wjdcoa@gnu.ac.kr (C.-M.J.); vaishali2020@gnu.ac.kr (V.G.)
2 Institute of Agriculture and Life Sciences, Gyeongsang National University, Jinju 52828, Korea; charmhanjjy@nate.com (J.-Y.J.); jkyang@gnu.ac.kr (J.-K.Y.)
3 Division of Applied Life Sciences (BK21 Plus Program), Gyeongsang National University, Jinju 52828, Korea
4 Department of Environmental Materials Science, Gyeongsang National University, Jinju 52828, Korea; hellohsy2@gmail.com (S.-Y.H.); ehdxos@naver.com (D.-H.L.)
* Correspondence: yhchoi@gnu.ac.kr

Simple Summary: The importance of the gut in poultry can be explained by the microbiome action. Prebiotics has gained attention as potential substances for improving gut health. The presence of insoluble dietary fiber from a cheap source such as wood specially prepared by the steam explosion that facilitates the depolymerization of hemicellulose could be an added advantage for microbial existence. Heat stress (HS) has been known to have drastic effects on chickens. In this study, we investigated the effect of dietary steam-exploded pine particle (SPP) supplementation and subsequent acute HS on productive performance and cecum microbiome in broilers. The HS tends to decrease the percent difference in body weight and rectal temperature. Metagenome analysis revealed similar richness and diversity in microbial communities. Firmicutes and Bacteroidetes were the most abundant phylum and were inversely correlated with each other. Furthermore, Firmicutes was also inversely correlated with unfavorable bacterial phyla. Supplementation of SPP in diets helped by enhancing favorable and reducing unfavorable bacterial genus in the cecum of the HS chickens. Although a clear advantage of using SPP on production parameters in broilers was not revealed, this study provided useful information to understand the modulation of microbiota during HS in dietary SPP supplemented chickens.

Abstract: The gut microbiome stimulates nutrient metabolism and could effectively generate heat tolerance in chickens. This study investigates the effects of dietary steam-exploded pine particle (SPP) supplementation and subsequent acute heat stress on productive performance and cecum microbiome in broilers. Eight-day Ross 308 broilers were distributed in three groups with 0%, 1%, and 2% SPP in diets. On the 41st day, forty birds were allocated to four groups with ten birds each. The treatments were control diet at thermoneutral temperature (0% NT) and acute heat-stressed (HS) birds fed control (0% HS), 1% (1% HS), and 2% (2% HS) SPP. Parameters recorded were body weight (BW), feed intake (FI), rectal temperature (RT), relative organ weight, and metagenome analysis from cecum samples. Percent difference in BW, FI, and RT was decreased in HS birds. Metagenome analysis revealed similar richness and diversity in microbial communities. The relative abundance of the bacterial genus such as *Limosilactobacillus*, *Drancourtella*, and *Ihubacter* was increased while that of *Alistipes*, *Alkalibacter*, *Lachnotalea*, and *Turicibacter* was decreased in SPP supplemented HS birds. Concludingly, the production performance of broilers is negatively influenced during HS, and 2% dietary SPP supplementation may reduce the adverse effects of HS by modifying the microbiota in chickens.

Keywords: acute heat stress; broilers; cecum metagenome; insoluble fiber; performance; steam-exploded pine

Citation: Goel, A.; Kim, B.-J.; Ncho, C.-M.; Jeong, C.-M.; Gupta, V.; Jung, J.-Y.; Ha, S.-Y.; Lee, D.-H.; Yang, J.-K.; Choi, Y.-H. Dietary Supplementation of Shredded, Steam-Exploded Pine Particles Decreases Pathogenic Microbes in the Cecum of Acute Heat-Stressed Broilers. *Animals* **2021**, *11*, 2252. https://doi.org/10.3390/ani11082252

Academic Editors: Natalie Morgan and Youssef A. Attia

Received: 2 July 2021
Accepted: 26 July 2021
Published: 30 July 2021

1. Introduction

The consistent increase in the environmental temperature is a serious threat to livestock production. Broiler production contributes significantly to the livestock industry and is severely affected under high ambient temperature conditions. Heat stress (HS) leads to drastic effects on the production performances and immunocompetence that enhances the mortality rate in chickens [1,2]. Various dietary supplements have been utilized to attempt to minimize the drastic effects of HS. However, due to the extensive competitive nature of the poultry industry, it is critical to keep feed costs at the lower end. The utilization of agricultural waste in the context of the cheap availability of additives could be one of the strategies for selection. Furthermore, choosing a feed additive with heat mitigating properties may provide extra benefits.

Prebiotics are non-digestible fibers that improve health by positively modifying the favorable intestinal microflora [3]. Microorganisms present in the gut have a crucial role in maintaining intestinal homeostasis [4]. Supplements that contribute to modulating the microflora can be of great importance. In light of the above facts, using non-digestible fibers from low-cost agriculture byproducts could be a good solution. Wood powder is rich in non-digestible fibers and can act as the source for the growth of the good microorganisms in the gut, and can be used as a prebiotic.

Further processing of wood powder from different sources such as pine trees may enhance its benefits. Thermal treatments may increase the utilization of insoluble dietary fiber sources from wood. Among the different existing thermal treatments, steam explosion, for example, is one of the most commonly used for the fractionation of biomass components. Steam explosion pretreatment is a simple, low-cost, and environmentally friendly technology that causes the depolymerization of hemicellulose and lignin into soluble oligomers, taking advantage of its high-temperature profile [5]. Thus, steam explosion facilitates the breakdown of lignocellulosic biomass structure [6,7] and has been used to apply their end products to a variety of purposes such as feed [8–10].

The gut is a vital organ of the digestive system, and most of the undigested fibrous material reaches the large intestine. It has previously been demonstrated that the breakdown of fibrous material takes place in the cecum by anaerobic fermentation of undigested fibrous material such as cellulose, starch, and other resistant polysaccharides through microbial action [11]. Thus, the importance of evaluating the microbial load in the cecum is high. Dietary supplementation of additives with prebiotic properties may help the proliferation of useful microbes and simultaneously reduce the propagation of pathogens in the gut [12]. However, due to the scarce use of wood powder from a pine tree in chicken diets, little information is available regarding its modulatory role on the gut microbiota of chickens, especially under HS conditions.

The present study evaluated the effect of increasing amounts of dietary supplementation of pine tree (*Pinus densiflora*) particles on the performance and cecum microbiota in acute heat-stressed broiler chickens. We hypothesize that birds fed on diets containing pine tree particles may increase good bacteria, reduce the pathogenic load in the cecum during HS conditions, and further help mitigate the harmful effects of HS on production performances in chickens.

2. Materials and Methods

The present experiment was conducted at the animal research facility of the Gyeongsang National University, Korea. The Animal Ethics Committee approved all the experimental procedures of the Gyeongsang National University (GNU-200916-C0057).

2.1. Animal Housing and Treatments

In the study, 260-day-old straight run (mixed sex) Ross 308 broiler chicks were procured from a local hatchery (Ohsung Hatchery, Seongju, Korea) and raised in a controlled environment with continuous lighting. For the first seven days, all chicks were distributed in a total of thirteen cages, with 20 chicks in each cage. A commercial feed and water were supplied ad libitum. On the 8th day, each chick was weighed ($n = 216$) and assigned to one of three different treatment groups containing twelve replicates of each treatment and six birds in each cage. The experimental diets contained 0%, 1%, and 2% shredded, steam-exploded pine particles (SPP) passing through a 10-mesh sieve replacing corn in their feed ingredients. The selection of dosage was made based on the results of the previous studies [10]. The preparation of SPP was done by exploding the pinewood chips of approximately $2 \times 2 \times 0.5$ cm^3 with steam at 200 °C for 11.5 min and stored at 20 °C until use. On the 41st day of age, a total of 40 birds (0%: 20 birds, 1%: 10 birds, 2%: 10 birds) were randomly selected and distributed into four groups of five replicates each treatment and two birds per cage. One group (0% SPP) was kept at thermoneutral temperature (21.0 °C) served as control while the other three groups (0, 1, and 2% SPP) were heat-stressed in a separate room by gradually increasing the temperature of the room to 31 °C within the first three hours and then maintained the same temperature for another 3 h. The total HS period was 6 h (Figure 1). Finally, there were a total of four treatments: Control diet (0% SPP) at thermoneutral temperature (0% NT); Control diet with acute HS (0% HS), 1% SPP-supplemented diet at acute HS (1% HS), and 2% SPP-supplemented diet at acute HS (2% HS).

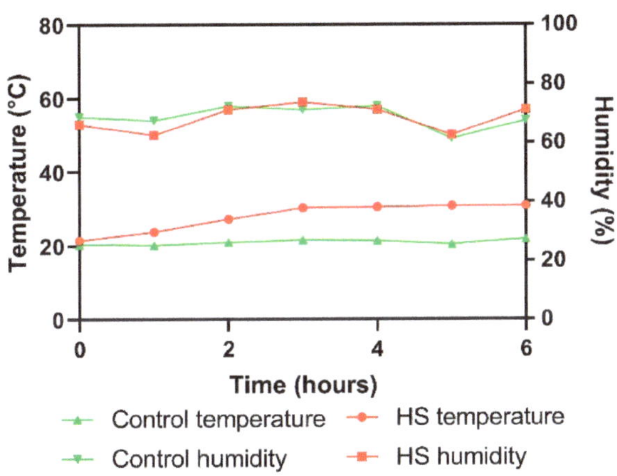

Figure 1. The temperature of thermoneutral control and heat stress rooms during the period of the experiment.

2.2. Sampling and Data Collection

At the end of the experiment, all the birds were weighed using digital balance, and rectal temperature (RT) was recorded by inserting a digital thermometer (HI 91610, Hanna instruments Inc., Padova, Italy) to approximately 3 cm inside the cloaca. Residual feed from each cage was also recorded to calculate the net feed intake. A total of 6 birds from each treatment were randomly selected and euthanized using carbon dioxide on 41 days of age. The liver, bursa of Fabricius, and spleen were dissected free, weighed, and presented as absolute and relative to body weight (BW). Cecum samples were immediately collected in sterilized 50 mL falcon tubes and stored at −80 °C.

2.3. Microbial Analysis

Total DNA containing the microbial communities from the cecum sample was extracted using a DNeasyPowerSoil Kit (Qiagen, Hilden, Germany) according to the manufacturer's instructions, followed by quantification using Quant-IT PicoGreen (Invitrogen, Waltham, MA, USA). To evaluate the metagenome of the cecum DNA samples, a 16S metagenomic sequencing library was constructed using a Herculase II Fusion DNA Polymerase Nextera XT Index Kit V2 (Illumina, San Diego, CA, USA) following manufacturer instruction. The library was then sequenced with the Illumina platform at Macrogen, Inc. (Seoul, Korea). FASTQ files were created for each sample which was then subjected to quality profiling, adapter trimming, and read filtering using fastp program [13]. The paired-end reads were assembled into one sequence using FLASH (v1.2.11) software [14], and the assembled sequences less than 400 bp and more than 500 bp were removed. The number of operational taxonomic units (OTUs) was determined by de novo clustering with a 97% sequence identity cutoff using the CD-HIT-EST program [15]. Each sequence OTU was checked for its taxonomic similarity using BLAST+ (v2.9.0) program [16] against the reference database (NCBI 16S Microbial), and the identical coverage of less than 85% was identified as not defined. The microbial communities in each sample in terms of OTU abundance and taxonomic information were analyzed using QIIME (v1.9) software. The species diversity and homogeneity among the microbial community in a sample were evaluated through Shannon, Goods Coverage, and Inverse Simpson Index. Alpha diversity was presented through the rarefaction curve of the Chao1 and observed OTUs. Beta diversity was evaluated using Weighted/Unweighted UniFrac distances.

2.4. Statistical Analysis

The data relating to animal studies such as growth performances, RT, absolute and relative organ weights were analyzed using one-way ANOVA, followed by Duncan's multiple range test. Alpha diversity (community richness and diversity) and taxonomic analysis (phylum and genus) were done using the Kruskal–Wallis test and adjusted with Bonferroni correction. Differences were considered statistically significant at $p < 0.05$ unless otherwise stated. All the data were expressed as mean ± standard error of the mean (SEM). The beta diversity analysis includes principal coordinate analysis (PCoA), calculated by weighted Unifrac and unweighted Unifrac. Spearman's correlation was used to evaluate the correlation among the bacteria of the phylum. Analysis was conducted using IBM SPSS Statistics package 25.0 (IBM software, Chicago, IL, USA). Graph pad Prism software was used to draw the graphs.

3. Results

There was no difference in the initial and final BW of chickens measured before and after six hours of HS. The percent difference in BW was significantly affected, and HS birds had a lower ($p < 0.05$) percent difference in BW irrespective of diets compared to birds kept at control temperature. Feed intake was also decreased ($p < 0.05$) in dietary SPP (1% and 2% SPP) supplemented HS birds in comparison to birds kept at thermoneutral temperature (Table 1).

Table 1. Effects of dietary steam-exploded pine particles supplementation on the growth performances of thermoneutral and heat-stressed broiler chickens.

Treatments	Initial BW	Final BW	% Difference in BW	Feed Intake
0% NT	2689.5 ± 86.5	2721.5 ± 84.2	$1.23^{b} \pm 0.37$	$55.60^{a} \pm 2.2$
0% HS	2631 ± 111.2	2608.5 ± 109.8	$-0.87^{a} \pm 0.45$	$43.20^{a,b} \pm 5.2$
1% HS	2645.5 ± 39.7	2631.5 ± 39.5	$-0.52^{a} \pm 0.73$	$40.00^{a} \pm 5.9$
2% HS	2670 ± 50.7	2648.5 ± 49.7	$-0.80^{a} \pm 0.24$	$38.60^{a} \pm 2.1$
p-value	0.952	0.746	0.022	0.044

Chickens were fed with diets containing 0% (control), 1%, and 2% shredded, steam-exploded pine particles (SPP) from 8th day to 41st day of age. On 41st day, birds were either kept at thermoneutral temperature (21.0 °C) and provided control diet (0% NT) or heat-stressed at 31.0 °C for six hours and supplemented with 0% (0% HS), 1% (1% HS) and 2% (2% HS) SPP in diets. Data show mean ± SEM ($n = 10$). ^{a,b}: different letters indicate significant differences ($p < 0.05$). Abbreviations: BW, body weight; NT, normal temperature; HS, heat stress.

Compared with those in the birds kept at thermoneutral temperature, rectal temperature (RT) was significantly increased ($p < 0.001$) in HS birds having 0, 1, or 2% SPP in diets, indicating an HS effect (Table 2).

Table 2. Effects of dietary steam-exploded pine particles supplementation on the rectal temperature of thermoneutral and heat-stressed broiler chickens.

Treatments	Rectal Temperature	
	Before	After
0% NT	41.73 ± 0.03	41.88 [a] ± 0.05
0% HS	41.87 ± 0.04	43.75 [b] ± 0.25
1% HS	41.78 ± 0.05	43.52 [b] ± 0.23
2% HS	41.80 ± 0.04	43.30 [b] ± 0.20
p-value	0.157	0.001

Chickens were fed with diets containing 0% (control), 1%, and 2% shredded, steam-exploded pine particles (SPP) from the 8th day to the 41st day of age. On 41st day, birds were either kept at thermoneutral temperature (21.0 °C) and provided control diet (0% NT) or heat-stressed at 31.0 °C for six hours and supplemented with 0% (0% HS), 1% (1% HS) and 2% (2% HS) SPP in diets. "Before" and "After" indicates rectal temperature taken before starting and after completing six hours of heat stress respectively. Data show mean ± SEM ($n = 6$). [a,b]: Different letters indicate significant differences ($p < 0.05$). Abbreviations: NT, normal temperature; HS, heat stress.

The absolute and relative organ weights of the liver, bursa of Fabricius, and spleen were similar, and no differences ($p > 0.05$) were observed among different treatments (Table 3).

Table 3. Effects of dietary steam-exploded pine particles supplementation on the absolute organ weight (g) and relative organ weight (% body weight) of thermoneutral and heat-stressed broiler chickens.

Treatments	Absolute Organ Weight (g)			Relative Organ Weight (%)		
	Liver	Bursa of Fabricius	Spleen	Liver	Bursa of Fabricius	Spleen
0% NT	65.0 ± 5.56	5.02 ± 0.44	2.92 ± 0.25	2.42 ± 0.18	0.19 ± 0.02	0.11 ± 0.01
0% HS	62.9 ± 2.92	4.18 ± 0.34	3.38 ± 0.61	2.46 ± 0.16	0.16 ± 0.01	0.13 ± 0.02
1% HS	62.3 ± 4.33	4.18 ± 0.63	3.44 ± 0.29	2.43 ± 0.16	0.16 ± 0.03	0.14 ± 0.01
2% HS	69.2 ± 3.60	4.57 ± 0.56	3.00 ± 0.46	2.59 ± 0.08	0.17 ± 0.02	0.11 ± 0.02
p-value	0.653	0.614	0.805	0.844	0.811	0.651

Chickens were fed with diets containing 0% (control), 1%, and 2% shredded, steam-exploded pine particles (SPP) from 8th day to 41st day of age. On 41st day, birds were either kept at thermoneutral temperature (21.0 °C) and provided control diet (0% NT) or heat-stressed at 31.0 °C for six hours and supplemented with 0% (0% HS), 1% (1% HS) and 2% (2% HS) SPP in diets. Data show mean ± SEM ($n = 6$). Abbreviations: NT, normal temperature; HS, heat stress.

A total of 24 cecum samples (six each treatment group) were used to generate 574,031 sequences ranging from 19,503 to 29,051 sequences for each sample after a 97% sequence similarity and removal of chimeric reads for quality control. A total of 5522 operational taxonomic units (OTU) were generated via clustering analysis with an abundance greater than 0.005% ranged from 196 to 260 OTUs for each sample.

Rarefaction curves for chao1 and observed OTUs (Figure 2) became flattered to the right and reached a plateau indicating that a reasonable number of reads were used in the analysis and reveals most of the bacterial community in the cecum samples of chickens.

Community richness was analyzed using OTUs and Chao1 and is presented in Figure 3. The OTUs and Chao1 were similar and were not affected among the treatment groups. The diversity in the community was analyzed using Shannon, Inverse Simpson, and Goods coverage and is presented in Figure 3. Shannon index has shown increasing trend ($p = 0.071$) in 2% HS and 0% HS against its counterparts. No variation was observed in Inverse Simpson and Goods coverage and was similar among the treatments group.

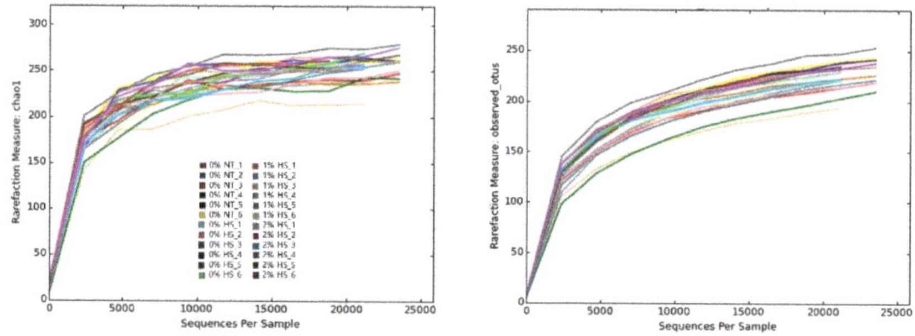

Figure 2. Rarefaction curve of Chao1 and observed OTUs for each sample. There are twenty-four samples represented by different colored lines that belong to four treatment groups of six replicates each. The treatments were control diet containing 0% steam-exploded pine particles (SPP) at thermoneutral temperature (0% NT), control diet with acute heat stress (0% HS), 1% SPP-supplemented diet at acute heat stress (1% HS), and 2% SPP-supplemented diet at acute heat stress (2% HS). The thermoneutral birds were maintained at 21.0 °C while the temperature of the heat-stressed room was raised to 31 °C within the first three hours and then maintained for another three hours that made the total HS period of six hours.

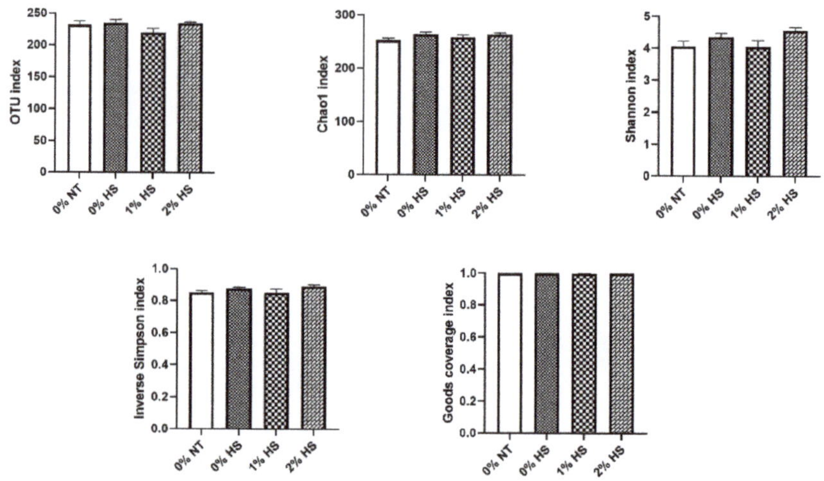

Figure 3. Effects of dietary shredded, steam-exploded pine particles (SPP) supplementation on the community richness and diversity of cecum microflora in broiler chickens exposed to either thermoneutral or heat stress conditions. The treatments were control diet containing 0% SPP at thermoneutral temperature (0% NT), control diet with acute heat stress (0% HS), 1% SPP-supplemented diet at acute heat stress (1% HS), and 2% SPP-supplemented diet at acute heat stress (2% HS). The thermoneutral birds were maintained at 21.0 °C while the temperature of the heat-stressed room was raised to 31 °C within the first three hours and then maintained for another three hours that made the total HS period of six hours.

The PCoA based on unweighted unifrac distances were similar, while weighted unifrac distances showed scattered plots (Figure 4).

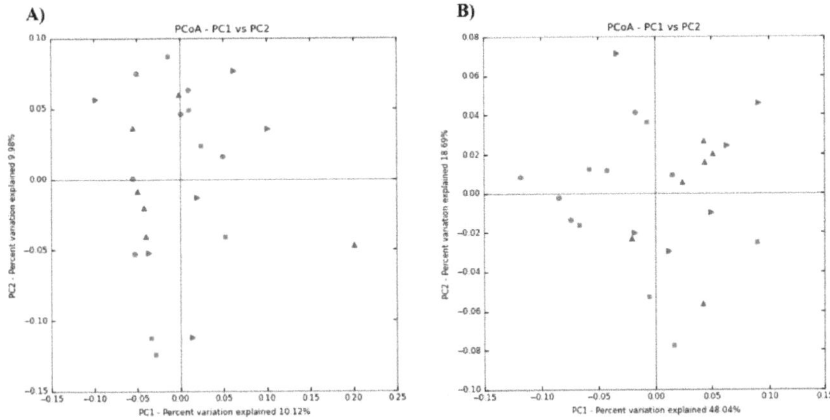

Figure 4. The composition changes of cecum microbiota based on principal coordinate analysis (PCoA) including unweighted (**A**) and weighted (**B**) unifrac distances in the dietary steam-exploded pine particles (SPP) supplemented chickens kept at either thermoneutral or at heat-stress. The treatments were control diet containing 0% SPP at thermoneutral temperature (0% NT), control diet with acute heat stress (0% HS), 1% SPP-supplemented diet at acute heat stress (1% HS), and 2% SPP-supplemented diet at acute heat stress (2% HS). The thermoneutral birds were maintained at 21.0 °C while the temperature of the heat-stressed room was raised to 31 °C within the first three hours and then maintained for another three hours that makes the total HS period of six hours. The red triangle indicates 0% NT, the blue triangle indicates 0% HS, the orange square indicates 1% HS, and the green circle indicates 2% HS treated chickens.

The phylum microbiota of cecum samples from different treatments is presented in Figure 5. At the phylum level the cecum microbiota was dominated by Firmicutes (0% NT: 60.48%, 0% HS: 65.35, 1% HS: 70.40, 2% HS: 73.17) and Bacteroidetes (0% NT: 34.58%, 0% HS: 31.99, 1% HS: 27.62, 2% HS: 24.86) followed by Verrucomicrobia, Tenericutes, Actinobacteria, Proteobacteria and Candidatus Melainabacteria.

The two predominant phyla, Firmicutes and Bacteroidetes are strongly correlated inversely (Spearman $R = -0.950$, $p = 0.001$) in terms of their abundances (Figure 6). Similar inverse correlation of Firmicutes was also observed with Verrucomicrobia (Spearman $R = -0.525$, $p = 0.008$) and Proteobacteria (Spearman $R = -0.366$, $p < 0.079$). However, Firmicutes was positively correlated (Spearman $R = 0.397$, $p = 0.054$) and Bacteroidetes was inversely correlated (Spearman $R = -0.427$, $p = 0.038$) with Actinobacteria.

The Firmicutes to Bacteroidetes ratio has shown an increasing trend with increasing concentration of SPP (Figure 7).

Figure 8 presents the cecum microbiota at the genus level in different treatment groups. The analysis of genus microbiota revealed that *Bacteroides*, *Faecalibacterium*, and *Blautia* are the most dominant genera followed by *Limosilactobacillus*, *Alistipes*, *Lactobacillus*.

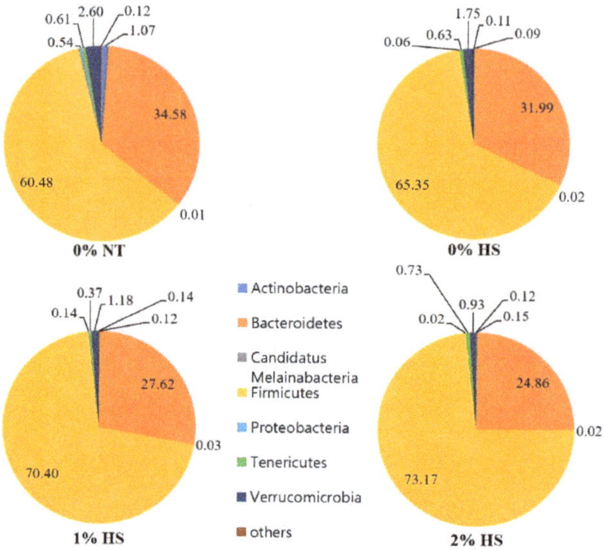

Figure 5. Effects of dietary steam-exploded pine particles supplementation to the chickens kept at either thermoneutral or heat stress on the composition of cecum microflora at Phylum level. The treatments were control diet containing 0% steam-exploded pine particles (SPP) at thermoneutral temperature (0% NT), control diet with acute heat stress (0% HS), 1% SPP-supplemented diet at acute heat stress (1% HS), and 2% SPP-supplemented diet at acute heat stress (2% HS). The thermoneutral birds were maintained at 21.0 °C while the temperature of the heat-stressed room was raised to 31 °C within the first three hours and then maintained for another three hours that makes the total HS period of six hours.

Among the top five dominant genera, *Alistipes* and *Limosilactobacillus* were significantly different among treatments. The relative abundance of *Alistipes* was decreased ($p = 0.010$), and that of *Limosilactobacillus* ($p = 0.042$) was increased by 2% HS in comparison to 0% HS (Figure 9). Five more genera were significantly modified by increasing the dietary concentration of SPP and subsequent HS in the cecum of broiler chickens (Figure 9). For instance, the abundance of *Ihubacter* was increased ($p = 0.049$) in 2% HS compared to 1% HS while the abundance of *Alkalibacter* was decreased ($p = 0.003$) in 1% HS and 2% HS in comparison to 0% NT. The abundance of *Lachnotalea* was increased ($p = 0.049$) in 0% HS compared to 0% NT but was similar and showed numerically lower values in 2% HS compared to 0% NT. The abundance of *Drancourtella* was increased ($p = 0.039$) in 2% HS in comparison to 0% NT. The abundance of *Turicibacter* was increased ($p = 0.042$) in 0% HS in comparison to 2% HS. However, not significant, but a numerically higher abundance of *Turicibacter* was also seen in 0% HS than that of 0% NT.

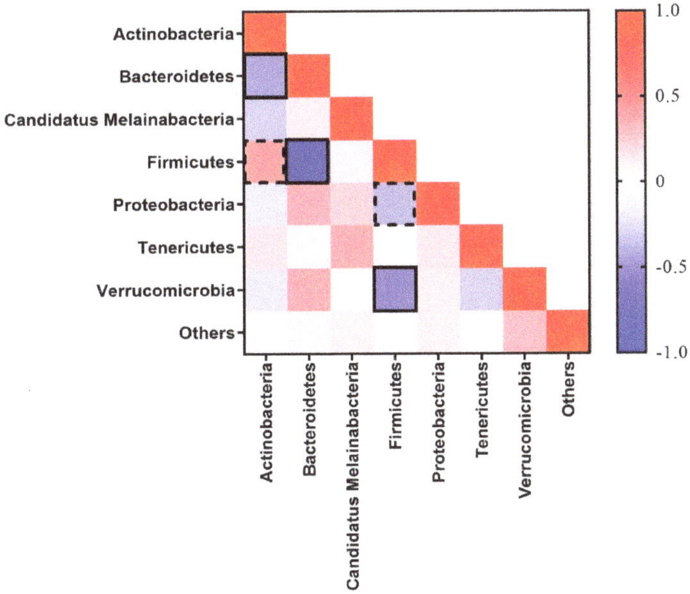

Figure 6. Correlation among the bacterial phylum of the cecum in broiler chickens supplemented with steam-exploded pine particles and kept at either thermoneutral or heat stress conditions. Values from each phylum were used to determine the correlation among each other. Zero indicates similarity without any possibility of correlation, while plus and minus values indicate positive and negative correlation, respectively. The continuous line and dash line indicate significance at $p < 0.05$ and $p < 0.10$, respectively.

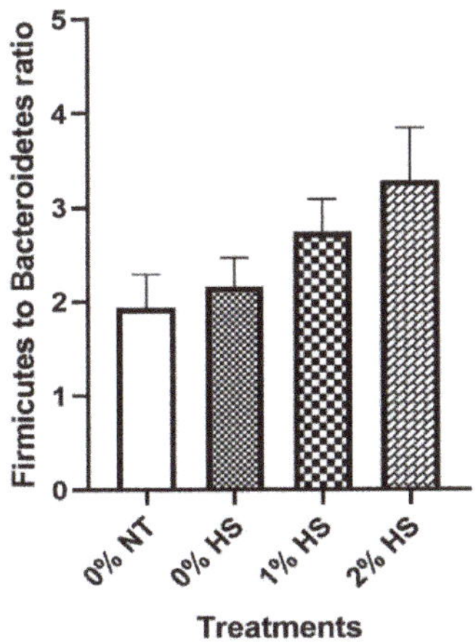

Figure 7. Effects of dietary steam-exploded pine particles supplementation on the Firmicutes to Bacteroidetes ratio in broiler chickens kept at either thermoneutral or heat stress conditions. The treatments were control diet containing 0% steam-exploded pine particles (SPP) at thermoneutral temperature (0% NT), control diet with acute heat stress (0% HS), 1% SPP-supplemented diet at acute heat stress (1% HS), and 2% SPP-supplemented diet at acute heat stress (2% HS). The thermoneutral birds were maintained at 21.0 °C while the temperature of the heat-stressed room was raised to 31 °C within the first three hours and then maintained for another three hours that makes the total HS period of six hours.

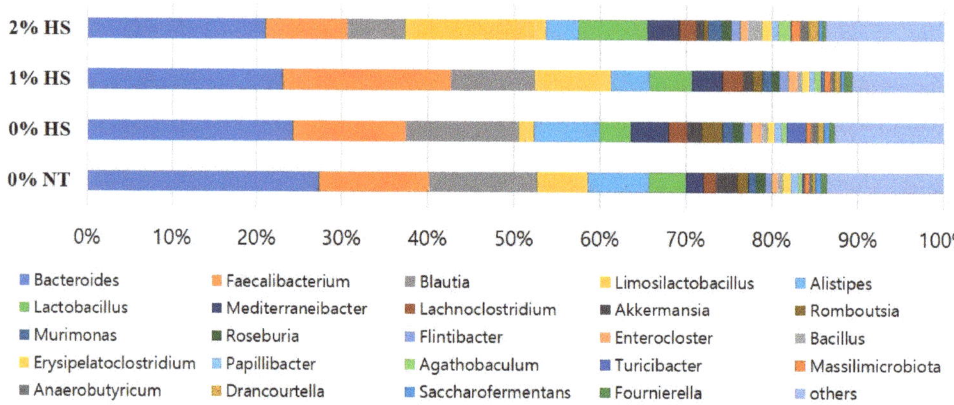

Figure 8. Effects of dietary steam-exploded pine particles supplementation and subsequent heat stress on the composition of cecum microflora of chicken at the genus level. Data represent the top 24 abundant microflorae. The treatments were control diet containing 0% steam-exploded pine particles (SPP) at thermoneutral temperature (0% NT), control diet with acute heat stress (0% HS), 1% SPP-supplemented diet at acute heat stress (1% HS), and 2% SPP-supplemented diet at acute heat stress (2% HS). The thermoneutral birds were maintained at 21.0 °C while the temperature of the heat-stressed room was raised to 31 °C within the first three hours and then maintained for another three hours that makes the total HS period of six hours.

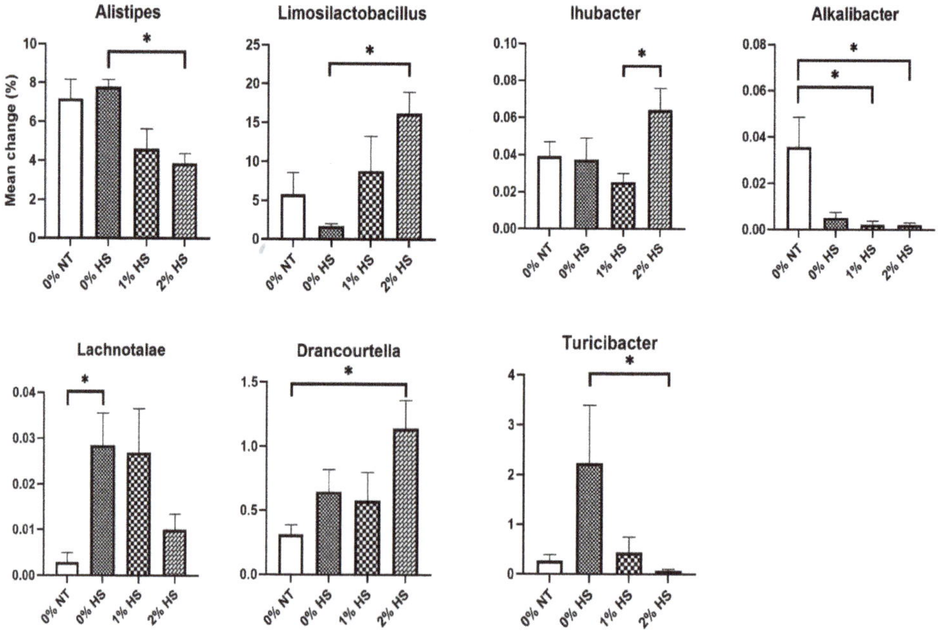

Figure 9. Effects of dietary steam-exploded pine particles supplementation and subsequent heat stress on the significantly modified genus of the cecum in broilers. The treatments were control diet containing 0% steam-exploded pine particles (SPP) at thermoneutral temperature (0% NT), control diet with acute heat stress (0% HS), 1% SPP-supplemented diet at acute heat stress (1% HS), and 2% SPP-supplemented diet at acute heat stress (2% HS). The thermoneutral birds were maintained at 21.0 °C while the temperature of the heat-stressed room was raised to 31 °C within the first three hours and then maintained for another three hours that makes the total HS period of six hours. * indicates a significant difference ($p < 0.05$).

4. Discussion

Broilers are meat-type chickens and are mostly reared for livestock production. Therefore, continuous improvement in the performance of broilers is the prerequisite of the poultry industry. Several stressors were identified that negatively influence broiler performances. HS is among the environmental stressors known to have drastic effects on production performance in chickens. The present study was conducted to evaluate the adverse effect of acute HS and to identify if supplementation of SPP during HS can lead to overcoming some adverse effects of HS. Although no effect was observed on the initial and final BW of the broilers among the treatment groups, a decrease in the percent difference in BW in the HS birds suggested the harmful effects of HS on chickens. A decrease in the percent difference in BW could be attributed to a reduction in feed intake. An increase in respiration rates [17] due to the absence of sweat glands may reduce the feed intake of birds. Furthermore, secretion of anorexic hormone under HS could also result in decreasing feed intake [18]. The present study also reported a decrease in feed intake in HS (1 and 2% SPP) birds. This corroborates the previous studies where acute HS resulted in decreasing the BW gain and feed intake when birds were exposed to 31 °C for ten h [19]. Contrary to this, in the present study, no variation was found in the feed intake of birds kept at thermoneutral or HS when fed a 0% SPP diet. The difference in the results could be attributed to the time of HS for which it was conducted.

High ambient temperature tends to increase the RT in chickens. In the present study, RT was increased in birds exposed to HS in comparison to birds kept at thermoneutral temperatures. Our results are consistent with the previous reports where the HS challenge at 35 °C even for 1.5 and 3 h enhances RT in chicken [20]. An increase in the RT could be related to respiration rates. Due to the absence of sweat glands, chickens are less able to dissipate heat in the surroundings, thus increasing respiration rates [17], which may further increase RT.

Different organs have specific roles in the body. The liver is vital for nutrition and metabolism in the body, whereas the bursa of the Fabricius and spleen are mainly responsible for imparting immunity in chickens. Stress is known to induce modulation in organ development, but inflection in weight will depend on the severity of stress to which the chickens are exposed. The weights of the liver, bursa of Fabricius, and spleen were similar and not affected among the treatment groups. These results correlate with the previous study where lymphoid organ weights (thymus, spleen, and bursa of Fabricius) were not affected when birds were exposed to 31 °C for 10 h [19]. Contrary to this, a decrease in the lymphoid organ weight was reported when birds were exposed to 31 °C or 36 °C for ten h from 35 days to 41 days of age [1]. Together, the difference in the results could be related to the time of exposure and the intensity of HS.

The progression of microbes in the gut starts as early as after hatch and may be influenced by many factors such as diets and other environments. It has been suggested that dietary prebiotic supplementation may influence growth performance by positively modifying the gut microbiota in chickens [21]. Since prebiotics cannot be absorbed or digested in the upper gastrointestinal tract (GIT), it passes to the lower GIT where it acts as a source of food for the good bacteria reducing the pathogenic bacteria attachment. However, HS has been associated with the degradation of intestinal integrity and GIT impairment, leading to the penetration of pathogenic microbes in chickens [22]. It is expected that dietary SPP supplementation as a source of prebiotic may help in reducing the adverse effect of HS due to the growth of beneficial bacteria and reduction of pathogenic bacteria.

To analyze the community richness, OTUs and Chao1 estimators were evaluated, and no difference was observed among the treatment groups. This suggests that the number of species in all the treatment groups was similar and was not affected by either HS or SPP supplementation. A similar trend was also observed in estimating the community diversity as Inverse Simpson, and Goods coverage was not affected among the treatment groups. However, the Shannon index did show an increasing trend in 2% HS and 0% HS treatments.

The SPP acts as a prebiotic and may help in increasing the abundance of beneficial bacteria in the gut [12], thus showing increasing trends of microbial diversity in 2% HS treatment. The increasing trend of Shannon diversity in 0% HS treatment could be attributed to the invasion of pathogenic microbes due to challenged intestinal gut integrity under HS [22].

In general, unweighted and weighted unifrac distances are used in microbial ecology, accounting for the presence or absence of observed organisms and their abundance, respectively. The PCoA based on unweighted unifrac distances showed scattered plots indicating similar microbiota. However, for weighted unifrac distances, dimension one contributed 48.04% of the inertia while dimension two contributed 18.69%, cumulatively contributing to almost 66% variability in caecum microbiota of SPP supplemented thermoneutral and heat-stressed chickens. Higher variability in weighted unifrac distances could be attributed to the addition of SPP-containing fiber in diets which may enhance the abundance of microflora in the cecum in a concentration-dependent manner. In the present study, Firmicutes and Bacteroidetes were found to be the most abundant microbiota in the chicken cecum, which was similar to the previous studies [23]. Furthermore, our results also showed that these two most abundant phylum microbiotas are inversely correlated with each other, and the Firmicutes percentage increases while Bacteroidetes percentage decreases with increasing concentration of dietary SPP supplementation. This could be related to the negative correlation of Bacteroidetes with BW along with higher cecum Firmicutes to Bacteroidetes ratio in obese hosts [24]. Although we did not find any significant differences in BW and Firmicutes to Bacteroidetes ratio, increasing Firmicutes to Bacteroidetes ratio trends with increasing concentration of dietary SPP supplementation indicates fattening of chicken due to the prebiotic effects. The reason behind the nonsignificant results could be related to the short term of HS. Long-term HS has been associated with growth retardation in chickens [1]. Higher differences with significantly positive variations in the SPP treatment groups could be expected if HS persists for an extended period of time and opens the way for future studies. The Bacteroidetes phylum was also found to be inversely correlated with another phylum of Actinobacteria. The exact reason behind this not clear. However, it could be due to the positive trend among the correlation between Firmicutes and Actinobacteria. A similar trend was also reported in chickens on the 9th and 18th day of age after hatching [25].

The bacteria of phylum Verrucomicrobia are known to possess mucin degrading ability and, if found in abundance, are associated with several diseases [26]. Proteobacteria is another gram-negative bacteria from the phylum that has been associated with pathogenicity and inflammation [27]. Verrucomicrobia was found to be negatively correlated with the Firmicutes in the present study, and its number was decreased with the increasing concentration of SPP supplementation in HS birds. A similar negative correlation trend was also observed among Firmicutes and Proteobacteria. However, a numerically higher number of Firmicutes was found with increasing SPP supplemented chickens. This indicates that SPP supplementation may help in maintaining gut health by reducing the colonization of bacteria associated with the degradation of intestinal barrier integrity and thus reducing the penetration of pathogenic microbes.

Due to a lack of information about the specific role of *Alistipes* in chickens, Biasato and his group suggested importance for its characterization, but they looked upon its increased abundance in their treatment group as a positive effect [28]. Parker and his group discussed the protective effects of *Alistipes* against liver fibrosis, colitis, cancer immunotherapy, and cardiovascular disease and suggested its pathogenic effects in colorectal cancer and mental depression [29]. However, in the present study, the relative abundance of *Alistipes* was decreased by 2% HS in comparison to 0% HS. However, the exact reason behind this is not clear since *Alistipes* belongs to phylum Bacteroidetes which has shown a decreasing trend with increasing dietary SPP concentration in the present study and could be the reason for the significant reduction of *Alistipes*.

Limosilactobacillus reuteri produces antimicrobials, including organic acids, ethanol, and reuterin, obstructing the pathogenic microbial colonization [30]. Numerically lower values of *Limosilactobacillus* in 0% HS in comparison to 0% NT might be due to the enhanced penetration of pathogens under HS. Furthermore, a significant increase in the *Limosilactobacillus* abundance in 2% HS in comparison to 0% HS proposes the beneficial effects of SPP supplementation under HS conditions. The presence of higher *Limosilactobacillus* may toughen the intestinal barrier and reduces the chances of pathogen penetration.

Gut microbiota has been associated with the production of trimethylamine *N*-oxide (TMAO), having a protective effect against adverse conditions such as temperature, salinity, and hydrostatic pressure [31]. Although its increased levels have been associated with the adverse effect on cardiovascular health in humans [32], it has an important physiological role in lower animals [31]. In chickens, the preferred level of TMAO is not yet evaluated and its adverse effects on cardiovascular health are yet to be determined. *Ihubacter massiliensis* is one of the critical bacteria responsible for the production of TMAO [33]. In the present study, the abundance of *Ihubacter* was numerically higher in 2% HS against its counterparts (0% NT and 0% HS) and was significantly increased in 2% HS compared to 1% HS. We assume that SPP supplementation may have a protective effect on heat-stressed chickens by modifying the abundance of TMAO producing *Ihubacter*. We did not evaluate the TMAO levels in chicken, however, its correlation with *Ihubacter* to modify heat tolerance capacity could be interesting for future studies.

The abundance of *Alkalibacter* was significantly decreased in dietary SPP supplemented chickens (1% HS and 2% HS) in comparison to 0% NT. Prebiotic selectively stimulates the growth of beneficial bacteria, enhancing the production of short-chain fatty acids and lactic acid as fermentation products, reducing the intestine's pH [34,35]. *Alkalibacter* being alkaliphilic might be reduced in the cecum of SPP supplemented chickens due to its prebiotic effect. Although the effect of HS in decreasing the abundance of *Alkalibacter* in this study could not be ignored, the relationship between the increase in temperature with pH is yet to be established.

High ambient temperature produces oxidative stress and is associated with ROS generation [22]. Malondialdehyde (MDA) is the secondary product generated after the degradation of lipids during lipid peroxidation [36]. Thus, the MDA is expected to be increased under HS. However, the increase in MDA differs in tissues and depends on many factors such as intensity, time, and temperature of HS exposure. The abundance of *Lachnotalea glycerini* is positively correlated to serum MDA in mice [37]. An increase in *Lachnotalea* in 0% HS in comparison to 0% NT indicates the effect of HS. Furthermore, decreasing trend with numerically lower abundance of *Lachnotalea* in the 2% HS in comparison to 0% HS suggested the role of SPP supplementation to recover from HS.

Drancourtella has been isolated from fresh human stool [38] and its role is not clear; however, an increase in its abundance in 2% HS in comparison to 0% NT indicates its beneficial effects in reducing the harmful effect of HS in the chicken cecum. Due to lack of information, future studies are warranted to confirm its role in the chicken gut.

Turicibacter is considered harmful for health due to its inverse correlation with tight junction in mice [39]. The presence of *Turicibacter* in monogastric animals may cause subclinical infection and can negatively modulate the microbiota of the gut [40]. The contagious activity of *Turicibacter* may perhaps be due to the intestinal damage caused by the pathogenic invasion. The increment in the abundance of *Turicibacter* in *Salmonella* infected chickens also supports the above notion [4]. The association of challenged gut health under HS conditions has already been explained previously [22]. The present study shows a numerically higher abundance of *Turicibacter* in 0% HS in comparison to 0% NT indicates the occurrence of pathogenic bacterial invasion due to the intestinal injury under HS. Significant reduction in the *Turicibacter* abundance in 2% HS treatment confirms the role of SPP supplementation as prebiotics to retain intestinal health and maintaining microbial homeostasis.

5. Conclusions

In conclusion, acute HS negatively influences broiler chickens' production performance and rectal temperature. HS tends to increase the Shannon diversity index indicating pathogenic microbial filtration. Nevertheless, increased Shannon diversity index in dietary SPP supplemented chickens along with fewer pathogenic phylum indicates improved gut health. An increase in the abundance of the favorable genus such as *Limosilactobacillus* and *Ihubacter* while decreasing unfavorable genus such as *Lachnotalea and Turicibacter* in SPP supplemented diets during HS suggested its role in modifying gut health.

Author Contributions: Conceptualization, J.-K.Y. and Y.-H.C.; methodology, A.G., B.-J.K., C.-M.N., C.-M.J., V.G., J.-Y.J., S.-Y.H., D.-H.L., J.-K.Y. and Y.-H.C.; software, A.G. and C.-M.N.; validation, A.G. and Y.-H.C.; writing—original draft preparation, A.G.; writing—review and editing, A.G., J.-Y.J., J.-K.Y. and Y.-H.C.; supervision, Y.-H.C.; funding acquisition, J.-K.Y. and Y.-H.C. All authors have read and agreed to the published version of the manuscript.

Funding: This work was supported in part by the Forest Science Technology R&D Program (2020193C10-2022-BA01) provided by Korea Forest Service (Korea Forestry Promotion Institute), and by the National Research Foundation of Korea (NRF) grant funded by the Korean government (MSIT) (2019H1D3A1A01071142).

Institutional Review Board Statement: All animal handling procedures complied with the Animal Ethics Committee at Gyeongsang National University (GNU-200916-C0057).

Data Availability Statement: The data presented in this study are available on request from the corresponding author. The data are not publicly available due to other manuscripts under review.

Acknowledgments: The authors are thankful to the Korea Forest Service (Korea Forestry Promotion Institute), and the National Research Foundation of Korea (NRF) funded by the Korean government (MSIT) for their support.

Conflicts of Interest: The authors declare no conflict of interest.

References

1. Quinteiro-Filho, W.M.; Ribeiro, A.; Ferraz-de-Paula, V.; Pinheiro, M.L.; Sakai, M.; Sa, L.R.; Ferreira, A.J.; Palermo-Neto, J. Heat stress impairs performance parameters, induces intestinal injury, and decreases macrophage activity in broiler chickens. *Poult. Sci.* **2010**, *89*, 1905–1914. [CrossRef] [PubMed]
2. Awad, E.A.; Najaa, M.; Zulaikha, Z.A.; Zulkifli, I.; Soleimani, A.F. Effects of heat stress on growth performance, selected physiological and immunological parameters, caecal microflora, and meat quality in two broiler strains. *Asian-Australas J. Anim. Sci.* **2019**. [CrossRef]
3. Kim, S.A.; Jang, M.J.; Kim, S.Y.; Yang, Y.; Pavlidis, H.O.; Ricke, S.C. Potential for Prebiotics as Feed Additives to Limit Foodborne Campylobacter Establishment in the Poultry Gastrointestinal Tract. *Front. Microbiol.* **2019**, *10*, 91. [CrossRef] [PubMed]
4. Oh, J.K.; Pajarillo, E.A.B.; Chae, J.P.; Kim, I.H.; Yang, D.S.; Kang, D.K. Effects of Bacillus subtilis CSL2 on the composition and functional diversity of the faecal microbiota of broiler chickens challenged with Salmonella Gallinarum. *J. Anim. Sci. Biotechnol.* **2017**, *8*, 1. [CrossRef] [PubMed]
5. Li, J.; Henriksson, G.; Gellerstedt, G. Lignin depolymerization/repolymerization and its critical role for delignification of aspen wood by steam explosion. *Bioresour. Technol.* **2007**, *98*, 3061–3068. [CrossRef]
6. Ibrahim, M.M.; Agblevor, F.A.; El-Zawawy, W.K. Isolation and characterization of cellulose and lignin from steam-exploded lignocellulosic biomass. *BioResources* **2010**, *5*, 397–418.
7. Obame, S.N.; Ziegler-Devin, I.; Safou-Tchima, R.; Brosse, N. Homolytic and Heterolytic Cleavage of β-Ether Linkages in Hardwood Lignin by Steam Explosion. *J. Agric. Food Chem.* **2019**, *67*, 5989–5996. [CrossRef] [PubMed]
8. Liu, Y.; Ren, X.; Wu, H.; Meng, Q.; Zhou, Z. Steam Explosion Treatment of Byproduct Feedstuffs for Potential Use as Ruminant Feed. *Animals* **2019**, *9*, 688. [CrossRef] [PubMed]
9. Chen, X.-j.; Lin, Q.-m.; Rizwan, M.; Zhao, X.-r.; Li, G.-t. Steam explosion of crop straws improves the characteristics of biochar as a soil amendment. *J. Integr. Agric.* **2019**, *18*, 1486–1495. [CrossRef]
10. Jung, J.Y.; Heo, J.M.; Yang, J.-K. Effects of steam-exploded wood as an insoluble dietary fiber source on the performance characteristics of Broilers. *BioResources* **2019**, *14*, 1512–1524. [CrossRef]
11. Svihus, B.; Choct, M.; Classen, H.L. Function and nutritional roles of the avian caeca: A review. *World's Poult. Sci. J.* **2013**, *69*, 249–264. [CrossRef]
12. Patel, S.; Goyal, A. The current trends and future perspectives of prebiotics research: A review. *3 Biotech* **2012**, *2*, 115–125. [CrossRef]

13. Chen, S.; Zhou, Y.; Chen, Y.; Gu, J. fastp: An ultra-fast all-in-one FASTQ preprocessor. *Bioinformatics* **2018**, *34*, i884–i890. [CrossRef] [PubMed]
14. Magoč, T.; Salzberg, S.L. FLASH: Fast length adjustment of short reads to improve genome assemblies. *Bioinformatics* **2011**, *27*, 2957–2963. [CrossRef]
15. Li, W.; Fu, L.; Niu, B.; Wu, S.; Wooley, J. Ultrafast clustering algorithms for metagenomic sequence analysis. *Brief. Bioinform.* **2012**, *13*, 656–668. [CrossRef]
16. Zhang, Z.; Schwartz, S.; Wagner, L.; Miller, W. A greedy algorithm for aligning DNA sequences. *J. Comput. Biol.* **2000**, *7*, 203–214. [CrossRef]
17. Mack, L.A.; Felver-Gant, J.N.; Dennis, R.L.; Cheng, H.W. Genetic variations alter production and behavioral responses following heat stress in 2 strains of laying hens. *Poult. Sci.* **2013**, *92*, 285–294. [CrossRef]
18. He, X.; Lu, Z.; Ma, B.; Zhang, L.; Li, J.; Jiang, Y.; Zhou, G.; Gao, F. Effects of chronic heat exposure on growth performance, intestinal epithelial histology, appetite-related hormones and genes expression in broilers. *J. Sci. Food Agric.* **2018**, *98*, 4471–4478. [CrossRef]
19. Quinteiro-Filho, W.M.; Rodrigues, M.V.; Ribeiro, A.; Ferraz-de-Paula, V.; Pinheiro, M.L.; Sa, L.R.; Ferreira, A.J.; Palermo-Neto, J. Acute heat stress impairs performance parameters and induces mild intestinal enteritis in broiler chickens: Role of acute hypothalamic-pituitary-adrenal axis activation. *J. Anim. Sci.* **2012**, *90*, 1986–1994. [CrossRef] [PubMed]
20. Hartanto, S.; Ko, H.S.; Jee, S.H.; Kang, J.U.; Seo, J.S.; Kang, Y.H.; Kim, H.N.; Ohh, S.J. Effect of dietary nutmeg oil on heat-stress tolerance-related parameters in Korean native chicken reared under hot temperature. *J. Anim. Physiol. Anim. Nutr.* **2019**, *103*, 1160–1167. [CrossRef] [PubMed]
21. Adhikari, P.A.; Kim, W.K. Overview of prebiotics and probiotics: Focus on performance, gut health and immunity—A review. *Ann. Anim. Sci.* **2017**, *17*, 949–966. [CrossRef]
22. Goel, A.; Ncho, C.M.; Choi, Y.-H. Regulation of gene expression in chickens by heat stress. *J. Anim. Sci. Biotechnol.* **2021**, *12*, 11. [CrossRef]
23. Shi, D.; Bai, L.; Qu, Q.; Zhou, S.; Yang, M.; Guo, S.; Li, Q.; Liu, C. Impact of gut microbiota structure in heat-stressed broilers. *Poult. Sci.* **2019**, *98*, 2405–2413. [CrossRef]
24. Ley, R.E.; Turnbaugh, P.J.; Klein, S.; Gordon, J.I. Microbial ecology: Human gut microbes associated with obesity. *Nature* **2006**, *444*, 1022–1023. [CrossRef]
25. Thomas, M.; Wongkuna, S.; Ghimire, S.; Kumar, R.; Antony, L.; Doerner, K.C.; Singery, A.; Nelson, E.; Woyengo, T.; Chankhamhaengdecha, S.; et al. Gut Microbial Dynamics during Conventionalization of Germfree Chicken. *mSphere* **2019**, *4*, e00035-00019. [CrossRef] [PubMed]
26. Geerlings, S.Y.; Kostopoulos, I.; de Vos, W.M.; Belzer, C. Akkermansia muciniphila in the Human Gastrointestinal Tract: When, Where, and How? *Microorganisms* **2018**, *6*, 75. [CrossRef]
27. Rizzatti, G.; Lopetuso, L.R.; Gibiino, G.; Binda, C.; Gasbarrini, A. Proteobacteria: A Common Factor in Human Diseases. *Biomed. Res. Int.* **2017**, *2017*, 9351507. [CrossRef] [PubMed]
28. Biasato, I.; Ferrocino, I.; Grego, E.; Dabbou, S.; Gai, F.; Gasco, L.; Cocolin, L.; Capucchio, M.T.; Schiavone, A. Gut Microbiota and Mucin Composition in Female Broiler Chickens Fed Diets including Yellow Mealworm (*Tenebrio molitor*, L.). *Animals* **2019**, *9*, 213. [CrossRef]
29. Parker, B.J.; Wearsch, P.A.; Veloo, A.C.M.; Rodriguez-Palacios, A. The Genus Alistipes: Gut Bacteria With Emerging Implications to Inflammation, Cancer, and Mental Health. *Front. Immunol.* **2020**, *11*, 906. [CrossRef]
30. Mu, Q.; Tavella, V.J.; Luo, X.M. Role of Lactobacillus reuteri in Human Health and Diseases. *Front. Microbiol.* **2018**, *9*, 757. [CrossRef] [PubMed]
31. Velasquez, M.T.; Ramezani, A.; Manal, A.; Raj, D.S. Trimethylamine N-Oxide: The Good, the Bad and the Unknown. *Toxins* **2016**, *8*, 326. [CrossRef] [PubMed]
32. Wang, Z.; Klipfell, E.; Bennett, B.J.; Koeth, R.; Levison, B.S.; Dugar, B.; Feldstein, A.E.; Britt, E.B.; Fu, X.; Chung, Y.M.; et al. Gut flora metabolism of phosphatidylcholine promotes cardiovascular disease. *Nature* **2011**, *472*, 57–63. [CrossRef] [PubMed]
33. Wu, W.K.; Panyod, S.; Liu, P.Y.; Chen, C.C.; Kao, H.L.; Chuang, H.L.; Chen, Y.H.; Zou, H.B.; Kuo, H.C.; Kuo, C.H.; et al. Characterization of TMAO productivity from carnitine challenge facilitates personalized nutrition and microbiome signatures discovery. *Microbiome* **2020**, *8*, 162. [CrossRef]
34. Poeker, S.A.; Geirnaert, A.; Berchtold, L.; Greppi, A.; Krych, L.; Steinert, R.E.; de Wouters, T.; Lacroix, C. Understanding the prebiotic potential of different dietary fibers using an in vitro continuous adult fermentation model (PolyFermS). *Sci. Rep.* **2018**, *8*, 4318. [CrossRef]
35. Ricke, S.C.; Lee, S.I.; Kim, S.A.; Park, S.H.; Shi, Z. Prebiotics and the poultry gastrointestinal tract microbiome. *Poult. Sci.* **2020**, *99*, 670–677. [CrossRef]
36. Ayala, A.; Muñoz, M.F.; Argüelles, S. Lipid peroxidation: Production, metabolism, and signaling mechanisms of malondialdehyde and 4-hydroxy-2-nonenal. *Oxid. Med. Cell. Longev.* **2014**, *2014*, 360438. [CrossRef] [PubMed]
37. Zhang, J.; Yi, C.; Han, J.; Ming, T.; Zhou, J.; Lu, C.; Li, Y.; Su, X. Novel high-docosahexaenoic-acid tuna oil supplementation modulates gut microbiota and alleviates obesity in high-fat diet mice. *Food Sci. Nutr.* **2020**, *8*, 6513–6527. [CrossRef] [PubMed]

38. Durand, G.A.; Lagier, J.C.; Khelaifia, S.; Armstrong, N.; Robert, C.; Rathored, J.; Fournier, P.E.; Raoult, D. Drancourtella massiliensis gen. nov., sp. nov. isolated from fresh healthy human faecal sample from South France. *New Microbes New Infect.* **2016**, *11*, 34–42. [CrossRef] [PubMed]
39. Liu, A.; Lv, H.; Wang, H.; Yang, H.; Li, Y.; Qian, J. Aging Increases the Severity of Colitis and the Related Changes to the Gut Barrier and Gut Microbiota in Humans and Mice. *J. Gerontol. Ser. A Biol. Sci. Med. Sci.* **2020**, *75*, 1284–1292. [CrossRef]
40. Rettedal, E.; Vilain, S.; Lindblom, S.; Lehnert, K.; Scofield, C.; George, S.; Clay, S.; Kaushik, R.S.; Rosa, A.J.; Francis, D.; et al. Alteration of the ileal microbiota of weanling piglets by the growth-promoting antibiotic chlortetracycline. *Appl. Environ. Microbiol.* **2009**, *75*, 5489–5495. [CrossRef] [PubMed]

animals

Article

Impacts of Supplementing Broiler Diets with Biological Curcumin, Zinc Nanoparticles and *Bacillus licheniformis* on Growth, Carcass Traits, Blood Indices, Meat Quality and Cecal Microbial Load

Mohamed E. Abd El-Hack [1,*], Bothaina A. Alaidaroos [2], Reem M. Farsi [2], Diaa E. Abou-Kassem [3], Mohamed T. El-Saadony [4], Ahmed M. Saad [5], Manal E. Shafi [2], Najah M. Albaqami [2], Ayman E. Taha [6] and Elwy A. Ashour [1]

1 Poultry Department, Faculty of Agriculture, Zagazig University, Zagazig 44511, Egypt; elwynurition@yahoo.com
2 Department of Biological Sciences, Faculty of Science, King Abdulaziz University, Jeddah 21577, Saudi Arabia; b.alaidaros@kau.edu.sa (B.A.A.); rfarsi@kau.edu.sa (R.M.F.); meshafi@kau.edu.sa (M.E.S.); nalbogami@kau.edu.sa (N.M.A.)
3 Animal and Poultry Production Department, Faculty of Technology and Development, Zagazig University, Zagazig 44519, Egypt; alaa1.aa837@gmail.com
4 Department of Agricultural Microbiology, Faculty of Agriculture, Zagazig University, Zagazig 44511, Egypt; m_tlatelsado@yahoo.com
5 Biochemistry Department, Faculty of Agriculture, Zagazig University, Zagazig 44511, Egypt; ahmedm4187@gmail.com
6 Department of Animal Husbandry and Animal Wealth Development, Faculty of Veterinary Medicine, Alexandria University, Edfina 22756, Egypt; a.taha@alexu.edu.eg
* Correspondence: m.ezzat@zu.edu.eg

check for updates

Citation: Abd El-Hack, M.E.; Alaidaroos, B.A.; Farsi, R.M.; Abou-Kassem, D.E.; El-Saadony, M.T.; Saad, A.M.; Shafi, M.E.; Albaqami, N.M.; Taha, A.E.; Ashour, E.A. Impacts of Supplementing Broiler Diets with Biological Curcumin, Zinc Nanoparticles and *Bacillus licheniformis* on Growth, Carcass Traits, Blood Indices, Meat Quality and Cecal Microbial Load. *Animals* **2021**, *11*, 1878. https://doi.org/10.3390/ani11071878

Academic Editors: Natalie Morgan and Youssef A. Attia

Received: 7 June 2021
Accepted: 22 June 2021
Published: 24 June 2021

Simple Summary: The present study aimed to investigate the beneficial effects of zinc nanoparticles (ZnNPs) and curcumin nanoparticles (CurNPs) as well as *Bacillus licheniformis* (Bl) supplementation on broiler growth, chemical blood indices, and cecal microbes. The results showed considerable antimicrobial activity against pathogenic bacteria and fungi with ZnNPs and CurNPs supplementations. At the same time, ZnNPs, CurNPs, and Bl improved broiler performance, carcass traits, meat quality traits, and some blood indices. Therefore, the inclusion of ZnNPs, CurNPs, and Bl is recommended for broiler feeding regimens to improve the performance and health status.

Abstract: The current study aimed to investigate the influence of dietary zinc nanoparticles (ZnNPs), curcumin nanoparticles (CurNPs), and *Bacillus licheniformis* (Bl) on the growth, carcass, blood metabolites, and the count of some cecal microorganisms of Indian River (IR) broilers. Chicks were allotted into seven experimental groups: control group, 1st, 2nd and 3rd groups were given diets enriched with ZnNPs, CurNPs and Bl (3.0, 5.0 and 2.0 cm^3/kg diet, respectively). The 4th, 5th and 6th groups were given diets supplemented with ZnNPs (3.0) + Bl (2.0) (ZP); ZnNPs (3.0) + CurNPs (5.0) (ZC) and ZnNPs (3.0) + CurNPs (5.0) + Bl (2.0) (ZCP) cm^3/kg diet, respectively. The results revealed that ZnNPs and CurNPs exhibited a considerable antimicrobial activity against pathogenic bacteria and fungi. They also inhibited the growth of microbes in a range of 50–95 μg/mL. The diet supplemented with ZnNPs, CurNPs, and Bl increased the body weight compared to the control after five weeks of age. Additionally, values of daily feed intake increased in these groups; however, the feed conversion ratio decreased. All values of carcass traits were better than that of the control. The treatments led to decreased abdominal lipids compared to the control. The activity of liver enzymes and malondialdehyde (MDA) activity decreased in the treated groups. In a converse trend, the levels of oxidative enzymes, amylase, protease, lipase and immunoglobulin were higher than that of the control. Meat quality properties were improved and cecal microbial counts were decreased. In conclusion, the ZnNPs, CurNPs, and Bl improved the broiler's weights, carcass traits, meat quality traits, as well as some blood indices and cecal microbial load. Therefore, the inclusion of ZnNPs, CurNPs, or Bl is recommended for broiler feeding regimens to improve the performance and health status.

Keywords: ZnNPs; CurNPs; *Bacillus*; broilers; growth; microbial aspects

1. Introduction

In the poultry industry, the main goal is to supply safer feed to enhance performance and health [1]. In 2006, the European Union banned the addition of antibiotics to farm animal diets to avoid microbial resistance to antibiotics [2]. The use of natural feed additives in broiler diets such as plants and their derivatives and the study of their effect on the quantity and quality of poultry are now global trends [3–7].

Nanoparticles (NPs) can be synthesized using physical, chemical, and biological methods. The biological synthesis is safe, clean, biocompatible, eco-friendly, and accomplishes fast reduction of metal ions at room temperature, unlike physical or chemical methods that consume extensive energy or use toxic solvents, respectively [8–10]. Microbes are potent eco-friendly nano-factories and can control the size and shape of biological nanoparticles [11]. Nanoparticles offer excellent properties, such as a large surface area, increased catalytic activity, and powerful adsorption capacity [12]. There are signs that nanoparticles can raise the absorption of nutrients [13].

Zinc (Zn) is an important microelement that affects many biological processes in birds, i.e., carbohydrates, protein and fat metabolism, immunity, hormone building, DNA and protein synthesis, and antioxidant properties [14]. According to the National Research Council (NRC), poultry needs 40 ppm of zinc per day [15]. On a commercial scale, feed industrialists add an extra amount of zinc ranging from 100–120 ppm to their feed to obtain rapid development in chicks' growth [16]. This increase in zinc leads to a rise in feed production costs and an increase in the Zn excretion in feces, which leads to environmental pollution. It affects the balance of other microelements and reduces vitamins. However, increasing the bioavailability of zinc may fix these problems, and zinc bioavailability increases when it is present in nano form. Recently, nano-zinc has been used as a feed additive because of the beneficial effects on the metabolism and health of birds. It improves immunity through its antibacterial activity [17–19]. Various studies stated that utilizing ZnNPs as a feed supplement enhances the following properties: body weight gain, feed conversion ratio, meat quality, and egg quantity. Additionally, it was also observed that it affected cecal microbiota and enhanced the immune system [20,21].

Curcumin has many pharmacological activities in treating various diseases [22]. It also increases the activity of digestive enzymes [22] and inhibits lipid oxidation [23]. Curcumin has antioxidant activity and controls the cecal microbiota. Curcumin alleviates oxidative liver injury by modulating the disruption of the cecum microbiota and lipid metabolism induced by ochratoxin A. Curcumin is recommended as a prophylactic measure to prevent ochratoxin A (OTA)-induced hepatic oxidative injury [24]. The nano form of curcumin increases its beneficial influences. It is easy for the nanoparticles to pass into the cell membranes and interact with the cell contents [25]. Hence, curcumin nanoparticles (CurNPs) increase the availability and the intake of curcumin [26]. A significant positive effect on chick's performance was observed in the diet supplemented with CurNPs [27]. Sayrafi, et al. [28] stated that reducing liver enzyme activity after CurNPs addition might be due to its antioxidant properties. Moreover, Partovi, et al. [29] reported that supplementing 300 mg of CurNPs/kg diet was a useful nutritional source, which can improve carcass parameters, protein content, redness and oxidative stability of broiler chicken breast meat infected with *Eimeria* species and decrease drip loss and cooking loss while did not have negative effect on texture profile of the chicken broiler meat.

The use of probiotics in poultry diets has steadily increased across the years because of the highest demand for antibiotic-free poultry [30]. The probiotic market in 2018 had a profit of 80 million United States Dollar (USD) and the addition of probiotics in the poultry diet maintains the global probiotic market, as its profits are expected to reach 125 million USD by 2025, with an annual increase of 7.7% [30]. Probiotics in broiler diets increase the growth and laying outcomes, reduce pathogenic bacteria in the gut, and raise beneficial microbiota and immunity [31]. *Bacillus* sp. microbes are among the most extensively used, direct-fed growth promoters. These bacteria serve as an alternative to antibiotics. A broiler diet supplemented with *B. licheniformis* can significantly improve BWG and FCR despite *C. perfringens* infection [32,33]. These effects are mainly attributed to *B. licheniformis* can enhance nutrient digestion and utilization in broilers by producing several enzymes, such as lipase, protease, and amylase. Recently, Abou-Kassem, et al. [34] found that all of the growth and carcass aspects were significantly influenced by dietary probiotic addition compared to the control group in quail. Some previous studies have used single chemical and physical synthesized nanoparticle additives in broiler feed to assess their effects on broiler performance; other studies that included more than one feed additive reported synergistic effects due to their active components on different parameters of broiler performance [3,7]. Here, we hypothesized that the novel combination of biological synthesized nanoparticles and probiotics and their combinations might lead to promising and synergetic effects on most broiler performance traits. Thus, this study aimed to evaluate the antimicrobial activity of ZnNPs, CurNPs, and probiotic (*B. licheniformis*) and their synergetic effects on growth performance, carcass properties, blood indices, meat quality, and cecal microbial load.

2. Materials and Methods

2.1. Nanoparticles Biosynthesis and Antimicrobial Properties

2.1.1. Bacterial Isolates, Biosynthesis and Characterization of ZnNPs and CurNPs

Bacillus subtilis LA4 and *Bacillus subtilis* AM12 were used in biosynthesize curcumin nanoparticles (CurNPs) and zinc nanoparticles (ZnNPs). These bacterial strains were isolated from soil samples collected from different regions in Zagazig city, Egypt [9,35–37]. The CurNPs were fabricated as follows, 40 mL of *Bacillus subtilis* LA4 supernatant was mixed with 60 mL of curcumin (0.27 mmol) in a 250 mL screw bottle then placed in a shaking incubator (160 rpm) at 30 °C for 72 h, pH 6, and nutrient broth media (NB) [25]. Other bottles containing 40 of NB and 60 mL of curcumin were incubated under the same conditions to prove the curcumin biotransformation occurred by *Bacillus subtilis* LA4 supernatant by the differences in color observation [25]. On the other hand, ZnNPs were produced by mixing 10 mL of *Bacillus subtilis* AM12 supernatant with 90 mL of zinc nitrate (1 mmol), pH 7, then was incubated in a shaking incubator (130 rpm) at 30 °C for 72 h. Zinc nitrate in NB was used as a control. Visible changes in the colorless mixture to white color indicate the biotransformation of zinc nitrate to ZnNPs by *Bacillus subtilis* AM12 supernatant [18]. Several methods were used to investigate the behavior of ZnNPs and CurNPs. The size, shape, and aggregation were assessed using transmission electron microscopy (TEM), the hydrodynamic size distribution was measured using dynamic light scattering (DLS), and the stability of the surface charge was estimated using zeta potential [38–41]. The obtained CurNPs were of spherical shape with a mean diameter of 65–80 nm measured by transmission electron microscopy (TEM) (JEOL 1010, JEOL Ltd., Tokyo, Japan), and had a negative charge of −25.3 mV by zeta potential analysis (Nano Z2 Malven, Malvern Hills, UK). The obtained ZnNPs were spherical with a mean diameter of 22–43 nm measured by TEM and a negative charge of −28.7 mV by zeta potential analysis (Figure 1).

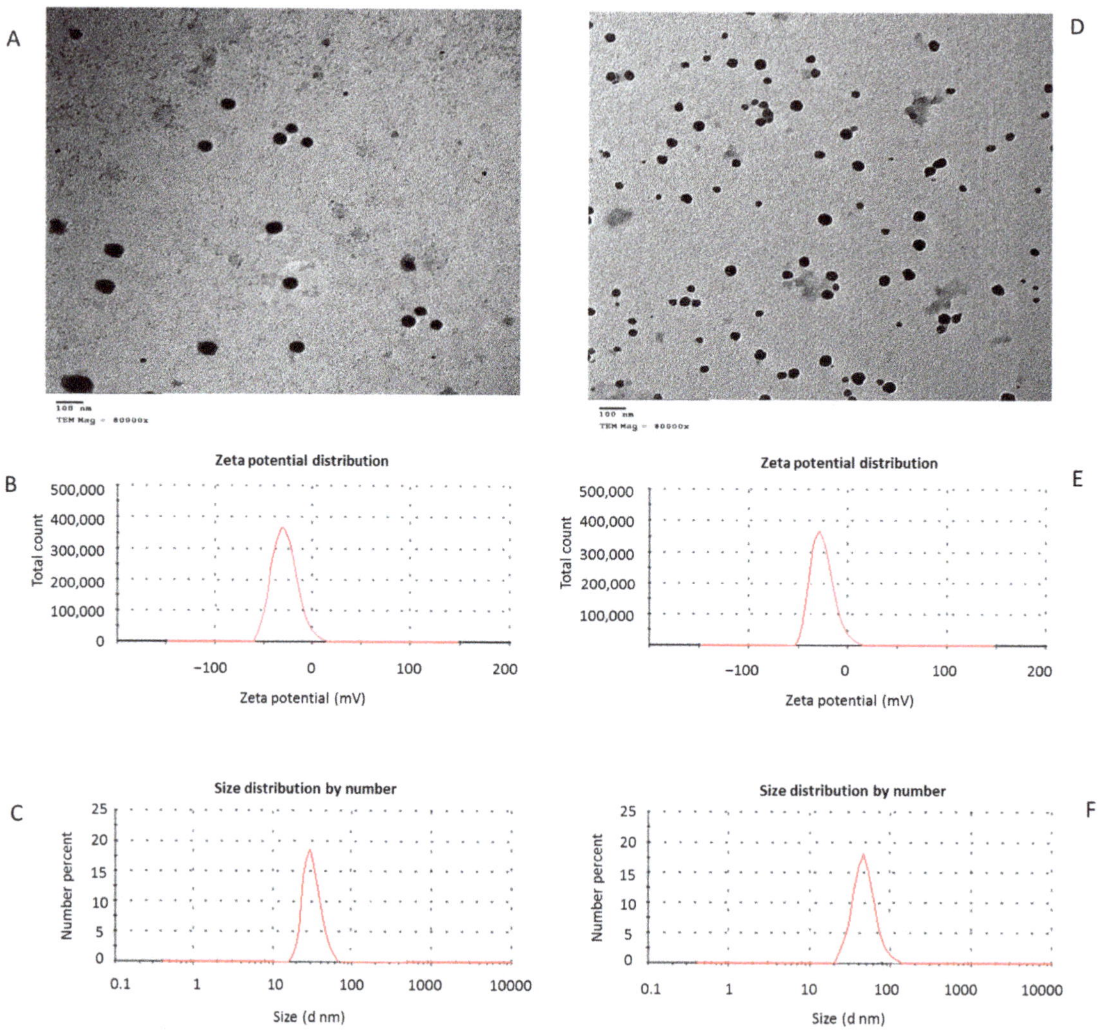

Figure 1. Characterization of zinc nanoparticles (ZnNPs) and curcumin nanoparticles (CurNPs); (**A–C**) shape, charge, and size of ZnNPs accessed by TEM, zeta potential, and zeta sizer, respectively. (**D–F**) shape, charge, and size of CurNPs accessed by TEM, zeta potential, and zeta sizer, respectively.

2.1.2. Antimicrobial Activity of CurNPs and ZnNPs

The bacterial isolates (Bacillus cereus, Listeria monocytogenes, and Staphylococcus pyogenes as gram-positive bacteria, Escherichia coli, Salmonella typhi, and Pseudomonas aeruginosa as gram-negative bacteria), and fungal isolates (Alternaria alternate, Aspergillus flavus, Fusarium oxysporum, Aspergillus niger, Penicillium solitum, and Penicillium crustosum) were used in this study to determine the antimicrobial activity of CurNPs and ZnNPs and obtained from Agricultural Microbiology department, Faculty of Agriculture, Zagazig University, Egypt. The antimicrobial activity of CurNPs and ZnNPs were estimated by the disc diffusion method [18,25,36]. Mueller Hinton agar (MHA) plates were inoculated with 0.1 mL of fresh bacterial inoculum and spread on the plate surface. Sabouraud dextrose agar (SDA) plates were inoculated with fungal mycelium disc in the plate center. The inoculated MHA and SDA plates were loaded with paper discs previously saturated with different concentrations of CurNPs and ZnNPs (100, 150, 200, 250, and 300 μg/mL). The paper discs were put on the sides of the plates. The MHA plates were incubated at 37 °C for a day and SDA plates at 28 °C for five days. The obtained inhibition zones diameters (mm) were measured [42,43].

The minimum inhibitory concentration (MIC) was estimated by the micro-dilution broth assay described in European Committee on Antimicrobial Susceptibility Testing [44]. In brief, tubes containing 9 mL of Muller Hinton broth (MHB) for bacteria or Sabourad dextrose broth (SDB) for fungi were inoculated with 0.1 mL of bacterial inoculum or standard fungal spore suspension (3×10^3 CFU/mL), then 0.05 mL of CurNPs and ZnNPs at different concentrations (100, 150, 200, 250, and 300 μg/mL) were added. Free CurNPs and ZnNPs tubes were used as controls. The MHB and SDB tubes were incubated at 37 °C for a day and 28 °C for five days, respectively. The MIC was the lowest concentration of CurNPs and ZnNPs that inhibited bacterial and fungal growth. On the other hand, minimum bactericidal concentration (MBC) and minimum fungicidal concentration (MFC) were estimated according to CLSI [45]. A loop of MIC tubes were spread over new MHA or SDA plates and incubated at previous conditions, then observed the bacterial or fungal growth. The lowest concentration that kills the bacterial or fungal growth was considered (MBC) or (MFC) [46,47].

2.2. Birds, Experimental Design and Diets

This study was performed in Poultry Research Farm, Department of Poultry, Faculty of Agriculture, Zagazig University, Zagazig, Egypt. All procedures were carried out following the guidelines of the local committee for experimental animal care and confirmed by the ethics of the Institutional Council of the Poultry Department, Faculty of Agriculture, Zagazig University, Zagazig, Egypt. A total of 420 unsexed—one week old Indian River (IR) broilers with an initial body weight of 104.40 ± 0.12 g were used in a completely randomized design experiment involving seven groups with 60 birds per group six replicates of 10 animals each. Chicks were purchased from a commercial hatchery. Chicks were housed in floor pens with clean pine shavings-based litter (50 cm rise × 100 cm wide × 100 cm height; 10 chicks each) and exposed to near-continuous photoperiod length a 23 L:1D with rotation [48]. The ZnNPs, CurNPs (500 mg/L), and BL (1.5×10^8 CFU/mL) were obtained from the Department of Agricultural Microbiology, Faculty of Agriculture, Zagazig University, Zagazig, Egypt and *B. licheniformis* was tested as probiotic bacteria [49]. The experimental groups were: the control group which was fed the basal diet. The first, second and third groups were given diets supplemented with ZnNPs, CurNPs and probiotics at 3.0, 5.0 and 2.0 mL/kg feed. The fourth, fifth and sixth groups were fed a diet supplemented with a combination of ZnNPs (3.0) + *B. licheniformis* (Bl) (2.0) (ZP); ZnNPs (3.0) + CurNPs (5.0) (ZC) and ZnNPs (3.0) + CurNPs (5.0) + *B. licheniformis* (BL) (2.0) (ZCP) mL/kg feed, respectively. Firstly, the nanoparticle of zinc and curcumin was added to the diet, then we did the pelletization process, after that, we top-dressed the probiotics on the pelletized feed to avoid the inactivation process of bacteria. The basal diets were formulated to meet the nutritional specifications of Indian River® Broiler Management guide [50]. All chicks were given the diets in the pellets form for 1–5 weeks of age, as shown in Table 1. The diets were given in two stages: starter (1–3 weeks) and finisher (3–5 weeks). All chicks were managed in the same ecological, managerial and hygienic conditions.

Table 1. Composition and chemical analysis of the starter and finisher basal diets (as fed).

Items	Starter (1–3 Weeks)	Finisher (3–5 Weeks)
	Ingredients %	
Yellow corn	55.89	57
Soybean meal 44%	31.5	29.5
Gluten meal 60%	6.5	4.83
Dicalcium phosphate	1.7	1.7
Limestone	1.24	1.15
Vit-min Premix *	0.3	0.3
NaCl	0.3	0.3
DL-Methionine	0.13	0.0
L-Lysine HCl	0.24	0.18
Choline 50%	0.2	0.2
Soybean oil	2.0	4.84
Total	100	100
	Calculated analysis:	
Dry matter %	91.72	90.43
Crude protein %	23.00	20.94
Metabolizable energy (kcal/kg diet)	2996.30	3150.70
Calcium %	1.00	0.96
Phosphorous (Available) %	0.44	0.44
Lysine %	1.3	1.17
Methionine + Cysteine %	0.90	0.70
Crude fiber %	3.52	3.38

* Vitamin-mineral premix provided per kg diet: vitamin A, 12,000 IU; vitamin D_3, 5000 IU; vitamin. E, 16.7 g; vitamin K, 0.67 g; vitamin B_1, 0.67 g; vitamin B_2, 2 g; vitamin B_6, 0.67 g; vitamin B_{12}, 0.004 g; nicotinic acid, 16.7 g; pantothenic acid, 6.67 g; biotin, 0.07 g; folic acid, 1.67 g; choline chloride, 400 g; Zn, 23.3 g; Mn, 10 g; Fe, 25 g; Cu, 1.67 g; I, 0.25 g; Se, 0.033 g and Mg, 133.4 g.

2.3. *Traits Measured*

2.3.1. Performance, Carcass, and Blood Biochemical Parameters

The parameters were measured once a week. Average day feed intake (FI), body weight gain (BWG), and feed conversion ratio (FCR) were calculated. Forty-two chicks were randomly selected from different pens within each treatment (one bird from each replicate within the group) and slaughtered at 35 days. The carcasses were weighed, and the edible parts (liver, gizzards, and hearts) and spleen, bursa, and abdominal fat were weighted as g/kg of the slaughter weight (SW). Carcass and dressed weights were expressed as (dressed weight = carcass weight + edible weight)/live body weight. The mortality rate was recorded weekly for each group and cumulatively calculated for the entire period of the experiment (1–5 weeks).

Blood sampling was performed during slaughtering from six birds randomly selected from different pens within each treatment (one bird from each replicate within the group). Samples were immediately centrifuged (Janetzki, T32c, 5000 rpm, Wall-hausen, Germany) at 2146.56× g for 15 min. The obtained serum was then frozen at −25 °C till the biochemical tests [51,52]. Hematological parameters (WBCs: white blood cells; LYM: lymphocytes; GRA: granulocytes; RBCs: red blood cells; HGB: hemoglobin; HCT: hematocrit; MCV: Mean corpuscular volume; MCH: Mean corpuscular hemoglobin; PLT: Platelet count) were measured according to Salvaggio et al. [53]. The levels of total protein, albumin, glucose, alkaline phosphatase (ALP), alanine amino-transferase (ALT), aspartate amino-transferase (AST), uric acid, creatinine, urea-N were measured as [53,54] triglycerides (TG), total cholesterol (TC), low-density lipoprotein (LDL), very-low-density lipoprotein (VLDL) and high-density lipoprotein (HDL) were measured using kits according to the protocol provided by the manufacturer (Spinreact, Ctra.Santa Coloma, Spain) [55]. Commercial kits from Biodiagnostic Company (Giza, Egypt) determined the immunity parameters IgG, A and M., antioxidant enzymes: glutathione (GSH), malondialdehyde (MDA). The activities of glutathione reductase (GSR), glutathione S-transferase (GST), superoxide dismutase (SOD), catalase (CAT) and glutathione peroxidase (GPx) were measured by a colorimetric method using kits according to the protocol provided by the manufacturer (Cell Biolabs Inc., San Diego, CA, USA).

2.3.2. Breast Meat Quality and Sensory Evaluation

The color parameters [L^* (lightness), a^* (redness), and b^* (yellowness)] of raw and cooked meat samples (cubes, 2 cm) (n = 6/treatment) were measured by Hunter Lab colorimeter (Color Flex EZ, Reston, VA, USA) following the procedure described in [56,57]. The shear force value of cooked meat cubes of 2 cm was measured by a texture analyzer (Compac-100 model, Sun Scientific Co., Tokyo, Japan) equipped with a cross head and a load cell. The speed of the cross head was set at 240 mm/min and a load cell of 10 kg was used. The cutting force was vertically applied to the meat fibers. The peak value profile of shear force was reported as the value of shear force. The lipid oxidation was measured by a 2-thiobarbituric acid test (TBA) [58]. Total volatile bases nitrogen (TVBN) was estimated according to Botta et al. [59]. The pH value of minced meat samples was assessed using a pH meter (pH 211 HANNA instruments Inc., Woonsocket, RI, USA). The chemical composition of meat was also estimated [57]. Moisture content was determined by oven method [60]; protein was determined by Kjeldahl method [60]; fat was estimated by the Soxhlet apparatus method [60]; a muffin assessed ash at 600 °C [60].

Sensory evaluation: the cooked meat samples (n = 6/treatment) were cut into cubes (2 cm) [61]. Eight experienced panelists have received meat samples in foam plate coded with random 3-digits. The sensory panel followed the descriptive sensory assessment carried out using a variation of the Sow and Grongnet [62] and Zhuang and Savage [63] process. The panelists have evaluated the following attributes (color, flavor, appearance, and juiciness) using a 7-point hedonic scale, where 1 = strongly dislike and 7 = strongly like. Tap water was provided between sessions to alter the mouth feel.

2.3.3. Microbial Count in Diet and Cecal Samples

The dietary samples (n = 6/treatment) were microbiologically examined at intervals of 0, 7, 14, and 21 days. Dietary samples were mixed with sterile saline peptone water (1 g/L peptone and 8.5 g/L NaCl) at a screw bottle and homogenized for ten minutes. Different media were used to enumerate the microbial count. Total bacterial count (TBC) was counted at plate count agar after incubation at 30 °C for two days. The total yeasts and molds count (TYMC) were estimated on Rose Bengal Chloramphenicol agar after incubation for five days at 25 °C. Total coliforms were counted on Violet Red Bile Agar (Biolife, Italy) after incubation at 37 °C for 24 h [64]. *Escherichia coli* were counted at Tryptone Bile Glucuronide Agar after incubation at 37 °C for 24 h. Additionally, the microbial count in broiler cecum was estimated as in diet. Five cecal samples were obtained from each group randomly and then were homogenized in a sterilized screw bottle with sterile saline peptone water (1 g/L peptone and 8.5 g/L NaCl). Decimal serial dilutions up to 10^7 were prepared. The different microorganisms in this study were counted on specific media [51,65–67]. The total bacterial count was enumerated as per Sheiha et al. [68] and Reda et al. [25] on Plate Count Agar (PCA) after incubation at 30 °C for two days. Violet Red Bile Agar (Biolife, Italy) was used for counting coliform after incubation at 37 °C for 24 h. *Escherichia coli* was counted at Tryptone Bile Glucuronide Agar after incubation at 37 °C for 24 h [69]. *Salmonella* spp. was counted on S.S. agar as per Edwards and Hilderbrand [70]. The yeasts and molds were enumerated as per Kurtzman et al. [71]. MRS-medium was used to count Lactic acid bacteria, according to Argyri et al. [72]. *Enterococcus* spp. was counted on Chromocult enterococci agar; red colonies indicated that it was found [73].

2.4. Statistical Analysis

The SPSS v 20 (IBM Corp., Armonk, NY, USA) and one-way ANOVA test were used to analyze the replicated data using GLM procedures. The statistical model used was:

$$Y_{ij} = \mu + T_i + e_{ij} \quad (1)$$

where Y_{ij} = observed value; μ = overall mean; T_i = treatment effect (control, and 1–6); and e_{ij} = random error. Differences among recorded means were estimated by the test of Student–Newman–Keuls. The SEM and mean values were reported. The differences between groups are considered significant at $p < 0.05$.

GLM model analyzed the data of the antimicrobial activity according to the following model:

$$Y_{aijk} = \mu + M_a + N_i + C_j + N_i{*}C_j + e_{aijk} \quad (2)$$

where Y_{ijk} = observed value; μ = overall mean; M_a = fixed effect of microorganism (bacteria or fungal), N_i = nanoparticle effect (zinc and curcumin), C_j = concentration of nanoparticles (100, 150, 200, 250 and 300 µg/mL for each nanoparticle type); $N_i{*}C_j$ = the interaction between nanoparticles and the dose of the nanoparticles; e_{aijk} = random error. The differences between means were compared by LSD at 5% level of probability.

GLM model analyzed the data of the microbiological activity of dietary samples at intervals according to the following model:

$$Y_{ijk} = \mu + T_i + S_j + N_i{*}S^j + e_{ijk} \quad (3)$$

where Y_{ijk} = observed value; μ = overall mean; N_i = treatment effect (control, and 1–6), S_j = day of sampling (0, 7, 14, and 21 days); $T_i{*}S_j$ = the interaction between treatment and the days of sampling; e_{ijk} = random error. The differences between means were compared by LSD at 5% level of probability.

3. Results

3.1. Antimicrobial Activity of CurNPs and ZnNPs

Table 2 showed the antibacterial effect of CurNPs and ZnNPs against six bacterial isolates. The inhibition zones diameters (IZDs) of CurNPs and ZnNPs were increasing in concentration-dependent manner. The results showed that CurNPs exhibited more antibacterial activity than ZnNPs; the CurNPs IZDs were in the range of 13.7–25.6 mm, the ZnNPs IZDs were in the range of 12.7–24.5. The largest IZD was observed against *S. pyogenes* with 25.6 and 24.5 for CurNPs and ZnNPs. Therefore, *S. pyogenes* was the most sensitive Gram-positive bacteria to the tested nanoparticles. On the other hand, *P. aeruginosa* was the most resistant Gram-negative bacteria to CurNPs and ZnNPs concentrations. Likely; the CurNPs had higher IZDs against tested fungi in a range of 17.9–30.2 mm compared to ZnNPs. *A. niger* was the most sensitive fungi to nanoparticles and *P. crustosum* was the most resistant. The CurNPs and ZnNPs inhibited the tested microbial growth with MIC values of 45–90 and 50–95 µg/mL, respectively. No microbial growth was detected at concentrations between 85–155, and 90–170 µg/mL for CurNPs, and ZnNPs. on the other hands, the combination between CurNPs and ZnNPs increased the antimicrobial activity achieving no bacterial or fungal growth in the range of 80–140 µg/mL (data not shown).

Table 2. Antimicrobial activity of ZnNPs and CurNPs against the tested bacteria and fungi represented by the inhibition zones diameters (mm), MIC, MFC and MBC (μg/mL). ($n = 6$).

Microorganisms	**ZnNPs (μg/mL)**					**Means**	**CurNPs (μg/mL)**					**Means**	**MIC**		**MBC**	
Bacteria	**100**	**150**	**200**	**250**	**300**		**100**	**150**	**200**	**250**	**300**		**ZnNPs**	**CurNPs**	**ZnNPs**	**CurNPs**
B. cereus	17.26 b	18.22 b	20.91 b	21.65 b	22.84 b	20.18 ^{AB}	18.32 b	19.62 b	21.41 b	22.22 b	23.43 b	21.00 ^{AB}	60.12 d	55.23 d	110.00 e	100.31 e
L. monocytogenes	16.55 c	17.53 c	19.52 c	20.54 c	23.42 c	19.5 ^{B}	17.43 c	18.93 c	20.95 c	23.43 c	24.22 c	20.99 ^{B}	75.21 cd	69.24 cd	130.01 d	120.21 d
S. pyogenes	18.36 a	19.45 a	21.83 a	22.92 a	24.53 a	21.42 ^{A}	19.55 a	20.34 a	22.53 a	23.65 a	25.64 a	22.34 ^{A}	50.00 e	45.33 e	90.23 f	85.43 f
E. coli	15.28 d	16.96 d	18.81 d	19.57 d	20.67 d	18.26 ^{BC}	16.28 d	18.38 d	19.24 d	20.56 d	21.48 d	19.19 ^{C}	80.13 c	71.11 c	140.14 c	130.24 c
S. typhi	14.38 e	15.28 e	17.24 e	18.48 e	19.88 e	17.05 ^{C}	15.29 e	16.97 e	18.56 e	19.41 e	20.69 e	18.18 ^{CD}	85.36 b	80.10 b	150.62 b	140.78 b
p. aeruginosa	12.79 f	14.82 f	16.46 f	17.55 f	18.52 f	16.03 ^{D}	13.41 f	15.59 f	17.44 f	18.22 f	19.52 f	16.84 ^{D}	95.61 a	90.34 a	170.14 a	160.14 a
Means	15.77 ^{D}	17.04 ^{C}	19.13 ^{B}	20.12 ^{AB}	21.64 ^{A}		16.71 ^{D}	18.31 ^{C}	20.02 ^{B}	21.25 ^{AB}	22.50 ^{A}		74.41 ^{A}	68.56 ^{B}	131.86 ^{A}	122.85 ^{B}
SEM	0.23	0.56	0.37	0.67	0.50		0.36	0.44	0.40	0.70	0.49		0.56	0.60	0.88	0.99
p-value																
Bacteria (B) Concentration (C) B × C			<0.001 <0.001 <0.001						<0.001 <0.001 <0.001				<0.001	<0.001	<0.001	<0.001
Fungi						Means						Means	MIC		MFC	
A. alternate	20.91 c	21.51 c	22.92 c	24.21 c	25.25 c	22.96 ^{B}	21.42 c	22.24 c	23.54 c	25.02 c	26.42 c	23.73 ^{B}	70.12 c	65.33 c	120.23 d	110.24 d
A. flavus	21.81 b	22.32 b	24.23 b	26.52 b	28.22 b	24.62 ^{AB}	22.32 b	23.05 b	24.82 b	27.33 b	29.42 b	25.39 ^{AB}	65.13 cd	60.27 cd	110.24 e	105.34 e
F. oxysporum	19.62 d	20.55 d	21.94 d	23.24 d	24.23 d	21.92 ^{C}	20.13 d	21.22 d	22.55 d	24.05 d	25.43 d	22.68 ^{BC}	75.52 bc	70.31 bc	140.29 c	130.24 c
A. niger	22.43 a	24.84 a	25.79 a	27.65 a	29.14 a	25.97 ^{A}	22.95 a	25.52 a	26.38 a	28.46 a	30.24 a	26.71 ^{A}	55.15 d	50.51 d	100.23 f	90.34 f
p. solitum	18.52 e	19.57 e	20.88 e	22.67 e	23.42 e	21.01 ^{C}	19.06 e	20.21 e	21.42 e	23.47 e	24.68 e	21.77 ^{C}	80.33 b	75.36 b	150.14 b	140.22 b
p. crustosum	16.43 f	18.22 f	19.54 f	21.48 f	22.61 f	19.66 ^{D}	17.92 f	18.93 f	20.11 f	22.22 f	23.87 f	20.61 ^{D}	95.56 a	90.25 a	170.34 a	155.31 a
Means	19.95 ^{D}	21.17 ^{C}	22.55 ^{B}	24.30 ^{AB}	25.48 ^{A}		20.63 ^{D}	21.86 ^{C}	23.14 ^{B}	25.09 ^{AB}	26.68 ^{A}		73.64 ^{A}	68.67 ^{B}	131.91 ^{A}	121.95 ^{B}
SEM	0.25	0.60	0.33	0.74	0.43		0.32	0.41	0.43	0.77	0.40		0.51	0.62	0.90	0.83
p-value																
Fungi (F) Concentration (C) F × C			<0.001 <0.001 <0.001						<0.001 <0.001 <0.001				<0.001	<0.001	<0.001	<0.001

a–f and $^{A–D}$ Mean in the same column within bacterial or fungal isolates within the same nanoparticle concentration with different lowercase letters are significantly different at $p < 0.05$; differences between means of the same bacterial or fungal isolates within the row and same concentration of different nanoparticles considered significant at $p < 0.05$; MIC: minimum inhibitory concentration; MBC: minimum bactericidal concentration; MFC: minimum fungicidal concentration. SEM: standard errors means.

3.2. Growth Performance

The effects of CurNPs, ZnNPs and Bl on live body weight (LBW) and BWG of chicks are shown in Table 3. Tests of variance offered a remarkable ($p < 0.001$) impact of treatments on LBW, BWG and FI over the various experimental groups at the starter (1–3 weeks), finisher (3–5 weeks) and only effects on LBW and BWG at the whole cycle (1–5 weeks) compared to the control, respectively. For the whole cycle, the best LBW was for T5 and T6; and the best BWG was for all the groups with ZnNPs (T1, T4, T5 and T6).

Table 3. Live body weight (g), body weight gain (g), feed intake (g), feed conversion ratio and mortality rate (%) of broilers as affected by dietary supplementation of ZnNPs, CurNPs, and *Bl*. (n = 60/treatment).

Items	Treatments							SEM	*p*-Value
	Control	1	2	3	4	5	6		
Starter period (1–3 weeks)									
LBW (g)	667.41 b	642.36 d	645.36 d	652.45 c	668.18 b	665.95 b	683.73 a	4.03	0.038
BWG (g/day)	40.53 a	36.40 b	33.15 c	35.04 bc	38.02 b	37.25 b	40.12 a	0.59	<0.001
FI (g/day)	70.74 a	55.42 b	53.53 b	54.13 b	56.87 b	57.52 b	58.08 b	1.33	<0.001
FCR	1.74 a	1.52 c	1.61 b	1.55 c	1.50 c	1.54 c	1.45 d	0.02	0.004
Finisher period (3–5 weeks)									
LBW (g)	1843.30 e	2161.91 b	2062.36 de	2084.15 d	2143.43 c	2177.60 a	2190.54 a	26.26	<0.001
BWG (g)	81.05 c	117.61 a	112.38 ab	109.27 b	112.30 ab	118.72 a	114.54 ab	2.74	<0.001
FI (g)	116.48 d	148.56 a	150.57 a	131.66 c	136.11 c	150.47 a	141.66 b	3.13	0.007
FCR	1.44 a	1.26 c	1.34 b	1.20 d	1.21 d	1.27 c	1.24 c	0.02	0.034
Whole cycle (1–5 weeks)									
LBW (g)	1843.30 e	2161.91 b	2062.36 de	2084.15 d	2143.43 c	2177.60 a	2190.54 a	26.26	<0.001
BWG (g/day)	56.08 c	65.22 a	61.69 b	63.05 b	64.87 a	65.52 a	66.28 a	0.78	<0.001
FI (g/day)	87.69	89.84	84.65	84.02	86.88	90.63	88.31	0.75	0.133
FCR	1.56 a	1.39 b	1.40 b	1.35 c	1.35 c	1.39 b	1.33 c	0.02	0.001
Mortality rate (%)	2.30	2.72	2.72	0.00	1.82	2.72	0.00	0.82	0.542

Basal diet (Control); T1: basal diet + 3.0 cm^3 ZnNPs; T2: basal diet + 5.0 cm^3 CurNPs; T3: basal diet + 2.0 cm^3 BL; T4: basal diet + 3.0 cm^3 ZnNPs + 2 cm^3 BL; T5: basal diet +3.0 cm^3 ZnNPs + 5.0 cm^3 CurNPs and T6: basal diet +3.0 ZnNPs + 5.0 CurNPs + 2.0 cm^3 BL. Live body weight (LBW); body weight gain (BWG); feed intake (FI); feed conversion ratio (FCR). SEM: standard error mean. a–e, Different letters within one row are significantly different ($p < 0.05$).

The FI decreased significantly in all treated groups compared to control broilers at the starter period (1–3 weeks), while FI increased significantly in T1, T2, and T5 compared to other treated and control birds at the finisher period (3–5 weeks). On the other hand, FI did not exhibit any significant differences between all treated and control groups throughout the cycle (1–5 weeks) (Table 3).

The influences of biological ZnNPs, CurNPs and BL on FCR are illustrated in Table 3. Generally, diets supplemented with ZnNPs, CurNPs and *B. licheniformis* (Bl) gave a better FCR than the control. The T6 group showed the best FCR at the started period, whereas T3 and T4 presented the best FCR in the finisher period. Regarding the whole cycle T3, T4 and T6 presented the best FCR values.

3.3. Carcass Traits

As shown in Table 4, all carcass traits were significantly ($p < 0.05$) impacted by the dietary treatment except for % of the bursa. The best carcass and dressing values are for T6 and T4. However, the lowest value for carcass was for control, and the lowest for dressing was control and T2. All dietary treatments, except for T2 (increased), reduced abdominal fat percentage compared to control.

Table 4. Carcass traits (%) of broilers as affected by dietary supplementation of ZnNPs, CurNPs and *Bl.* ($n = 6$).

Carcass Traits (as a % of Pre-Slaughter Weight)							Items
Bursa	Spleen	Abdominal Fat	Dressing	Giblets	Carcass	Pre-Slaughter Weight	
0.10	0.12 b	1.11 b	77.29 d	3.28 e	74.00 d	2010.00 bc	Control
0.12	0.08 c	0.81 d	78.69 b	4.20 a	74.49 c	2047.50 b	T1
0.08	0.09 c	1.32 a	77.51 d	3.88 b	73.63 e	1960.00 c	T2
0.14	0.12 b	0.38 e	78.85 b	3.76 c	75.09 b	1995.00 bc	T3
0.13	0.16 a	0.94 c	79.19 ab	3.60 d	75.58 a	2240.00 a	T4
0.15	0.12 b	0.84 d	78.08 c	3.76 c	74.32 c	2020.00 bc	T5
0.17	0.12 b	0.79 d	79.55 a	3.71 c	75.85 a	2075.00 b	T6
0.01	0.01	0.07	0.23	0.11	0.71	21.45	SEM
0.135	0.017	0.002	0.006	<0.001	0.010	0.001	*p*-value

Basal diet (Control); T1: basal diet + 3.0 cm^3 ZnNPs; T2: basal diet + 5.0 cm^3 CurNPs; T3: basal diet + 2.0 cm^3 BL; T4: basal diet + 3.0 cm^3 ZnNPs + 2 cm^3 BL; T5: basal diet + 3.0 cm^3 ZnNPs + 5.0 cm^3 CurNPs and T6: basal diet + 3.0 ZnNPs + 5.0 CurNPs + 2.0 cm^3 BL. SEM: standard error mean. a–e Different letters within one column are significantly different ($p < 0.05$).

3.4. Blood Biochemical Indices

Compared to the control group; the dietary supplementation of CurNPs, ZnNPs, and *B. licheniformis* (Bl) combination in T5 and T6 significantly increased hemoglobin content, MCH, MCV, MCHC, RBCs count, monocytes %, albumen content and IgA level as well as GSH, GSR and SOD enzymes. T6 resulted in significant increase in PCV, blood platelets count, and percentages of lymphocytes, neutrophils, basophils, and eosinophils. Moreover, a significant increase in calcium level and GST, duodenal amylase, protease and lipase enzymatic activity was reported compared to the control group. Additionally, lipid indicators and parameters were significantly lower in the T6 group than in the control and other groups. A significant increase in SOD and GSH, immunoglobulins, and hydrolysis enzymes were observed in the T6 group compared with the control group and other groups (Table 5).

3.5. Meat Quality

Data in Table 6 reported that the addition of CurNPs, ZnNPs, and Bl (5:3:2 cm^3/kg) to the broiler diet significantly increased meat moisture (T5 and T6), protein content (T6) and pH (T6) compared to control group. Moreover, the lipid content of meat was significantly lower in T6 compared to other treated and control groups. Furthermore, the T6 significantly enhanced the yellowness (b*), juiciness, tenderness and taste of the meat compared to the control group.

3.6. Microbial Count in Diet and Cecal Samples

Generally, the microbial count was significantly ($p \leq 0.05$) lower in the treated groups than that of the control. However, the interaction effect showed that T6 excelled the other treatment groups in reducing the microbial count in diet samples with a relative decrease of 35% in total bacterial count (TBC), 45% in total yeasts and molds count (TYMC), 63% in *E. coli* count and 50% of coliform count compared to control. The microbial count increased with feeding period. The T6 followed T5 in the microbial count reduction (Table 7). Additionally, total bacterial count, total yeasts and molds, coliform, *E. coli*, *Salmonella* spp., *Enterococcus* spp., and lactic acid bacteria were counted in the cecum. Data in Table 8 showed that the T6 group followed by T5 caused a significant reduction in all microbial counts in the broiler's cecum with a relative decrease of 18–30%. Conversely, the count of lactic acid bacteria was higher than the control by 25%. *Salmonella* did not exist in the cecum of the treated groups. The different treatments significantly reduced the count of *Enterococcus* spp.

Table 5. Blood hematology, biochemical parameters, serum antioxidants, duodenal enzymes and immunity of broilers as affected by dietary supplementation of ZnNPs, CurNPs, and *Bacillus licheniformis*. (n = 6).

Serum Parameters	Control	T1	T2	T3	T4	T5	T6	SEM	p Value
Hematological									
Hemoglobin (g/dL)	6.90 bc	7.00 b	7.20 ab	7.00 b	7.00 b	7.20 ab	7.30 a	0.051	<0.0001
MCH (P/g)	22.90 c	23.00 bc	23.80 b	23.00 bc	22.90 c	24.00 ab	25.10 a	0.279	<0.0001
MCV (Fl)	63.00 d	63.50 c	64.00 bc	63.80 c	64.50 b	65.00 ab	66.50 a	0.405	<0.0001
MCHC (g/d)	33.22 cd	33.90 c	34.50 b	34.00 bc	33.50 c	35.10 ab	35.50 a	0.285	<0.0001
PCV (%)	22.25 c	22.50 bc	25.20 ab	21.50 cd	23.00 b	24.50 b	25.90 a	0.579	<0.0001
RBC ($\times10^6$/U L)	3.30 b	3.40 b	3.60 ab	2.50 c	3.40 b	3.50 ab	3.80 a	0.144	<0.0001
WBC ($\times10^3$ j/L)	4.00 a	4.00 a	3.25 c	2.10 d	3.50 b	3.12 bc	4.10 a	0.251	<0.0001
Platelets (10^3 j/L)	210 cd	212 c	215 bc	199 e	219 b	205 d	225 a	3.027	<0.0001
Lymphocytes (%)	72.00 bc	67.10 d	73.10 b	72.10 bc	72.12 bc	70.31 c	75.12 a	0.877	<0.0001
Neutrophils (%)	21.00 d	21.10 d	27.12 b	25.60 c	29.12 ab	23.10 cd	30.12 a	1.286	<0.0001
Basophils (%)	0.00	0.00	0.00	1.10 b	1.20 b	0.90 bc	1.50 a	0.228	<0.0001
Monocytes (%)	0.30 d	0.40 d	1.90 bc	0.90 c	2.30 b	2.50 ab	2.80 a	0.363	<0.0001
Eosinophils (%)	4.30 b	4.50 ab	3.80 c	4.30 b	4.50 ab	4.10 c	5.13 a	0.135	<0.0001
Biochemical									
Glucose (mmol/L)	6.12 bc	6.11 bc	6.22 bc	6.30 ab	6.10 b	6.50 a	6.01 bc	0.069	<0.0001
Total protein (g/dL)	4.50 d	4.90 d	5.80 b	6.50 a	5.12 c	5.10 c	6.30 ab	0.266	<0.0001
Albumin (g/dL)	3.20 cd	3.50 c	4.10 b	4.10 b	4.00 b	4.20 ab	4.50 a	0.153	<0.0001
Globulin (g/dL)	2.10 d	2.50 c	2.60 c	2.40 cd	2.50 c	3.05 b	3.60 a	0.171	<0.0001
Chloride (mmol/L)	105 a	99 d	95 d	100 c	103 b	100 c	90.00 e	1.755	<0.0001
Calcium (mmol/L)	2.20 c	2.40 c	3.50 a	3.20 ab	3.05 b	3.05 b	3.50 a	0.177	<0.0001
AST (U/I)	23.12 c	25.42 bc	25.12 bc	35.12 a	27.05 b	34.22 ab	20.10 c	1.948	<0.0001
ALT (U/I)	20.14 c	16.10 d	19.01 c	25.22 ab	23.23 b	26.14 a	18.10 cd	1.309	<0.0001
Urea (mmol/L)	5.12 a	4.20 b	3.90 c	5.01 a	4.25 b	3.99 c	4.10 bc	0.161	<0.0001
Triglycerides (mmol/L)	1.20 a	1.10 ab	0.90 c	1.05 b	1.01 b	0.98 bc	0.70 c	0.056	<0.0001
Cholesterol (mmol/L)	4.10 ab	4.20 a	4.05 b	4.20 a	4.04 c	4.15 bc	4.12 c	0.030	<0.0001
LDL (mmol/L)	3.40 ab	3.50 a	1.70 c	3.40 ab	1.70 c	2.40 b	1.60 c	0.310	<0.0001
VLDL (mmol/L)	0.44 ab	0.45 a	0.30 b	0.28 c	0.17	0.32 ab	0.25 c	0.035	<0.0001

Table 5. *Cont.*

Serum Parameters	Control	T1	T2	T3	T4	T5	T6	SEM	p Value
Oxidative enzymes									
GSH	1.10 d	1.30 c	1.30 c	1.20 cd	1.50 b	1.70 ab	1.80 a	0.091	<0.0001
GSR	1.30 d	1.50 cd	1.60 c	1.35 cd	1.80 b	1.90 ab	2.00 a	0.095	<0.0001
GST	1.40 cd	1.60 c	1.60 c	1.31 cd	1.80 b	2.00 b	2.30 a	0.121	<0.0001
SOD	1.50 e	1.70 d	1.80 cd	1.90 c	2.40 b	2.50 ab	2.70 a	0.160	<0.0001
MDA	17.90 a	17.10 ab	16.80 b	16.20 bc	16.20 bc	14.50 c	13.10 d	0.575	<0.0001
Duodenal Enzyme activity									
Amylase	2830 g	2935 f	3040 e	3145 d	3750 c	4022 b	4155 a	192.42	<0.0001
Protease	130 fg	140 f	170 e	180 d	190 c	225 b	234 a	13.72	<0.0001
Lipase	95.00f	100 e	111 de	120 d	130 c	140 b	145 a	6.75	<0.0001
Immunoglobulin level									
IgA	7.50 cd	7.80 c	7.90 bc	8.00 bc	8.50 b	9.70 ab	10.20 a	0.362	<0.0001
IgM	2.50 cd	2.90 c	3.00 cd	3.10 cd	3.50 c	3.90 b	4.50 a	0.237	<0.0001
IgG	13.00 cd	14.30 c	15.90 bc	16.00 bc	16.00 bc	17.00 b	18.90 a	0.658	<0.0001

Basal diet (Control); T1: basal diet + 3.0 cm^3 ZnNPs; T2: basal diet + 5.0 cm^3 CurNPs; T3: basal diet + 2.0 cm^3 Bl; T4: basal diet + 3.0 cm^3 ZnNPs + 2 cm^3 BL; T5: basal diet + 3.0 cm^3 ZnNPs + 5.0 cm^3 CurNPs and T6: basal diet + 3.0 ZnNPs + 5.0 CurNPs + 2.0 cm^3 Bl. MCH: mean corpuscular hemoglobin; MCV: mean corpuscular volume; MCHC: mean corpuscular hemoglobin concentration; PCV: packed cell volume; RBCs: red blood cells; WBCs: white blood cells; AST: aspartate aminotransferase; ALT: alanine aminotransferase; LDL: low-density lipoprotein; VLDL: very low-density lipoprotein; GSH: glutathione; GSR: glutathione reductase; GST: glutathione-S-transferase; SOD: superoxide dismutase; MDA: malondealdehyde. IgA: immunoglobulin A; IgM: immunoglobulin M; IgG; immunoglobulin G. SEM: standard error mean. a–g Different letters within one raw are significantly different ($p < 0.05$).

Table 6. Chemical, color parameters, and raw and cooked (sensorial) meat quality of broiler affected by dietary supplementation of ZnNPs, CurNPs, and *Bacillus licheniformis*. ($n = 6$).

Quality Parameters	Control	T1	T2	T3	T4	T5	T6	SEM	*p*-Value
				Chemical					
Moisture	65.9 cd	64.30 d	68.00 b	68.20 b	67.10 c	70.00 ab	71.00 a	0.803	<0.0001
Protein	19.45 d	20.12 c	21.40 c	22.00 b	21.20 c	21.22 c	23.00 a	0.408	<0.0001
Lipid	14.1 ab	15.00 a	10.20 c	9.30 cd	11.50 b	9.00 d	6.00 e	1.087	<0.0001
Ash	0.89 b	1.10 a	0.90 b	1.00 ab	1.00 ab	0.80 c	0.30 d	0.092	<0.0001
pH	5.5 de	6.00 c	5.80 d	6.10 c	6.30 b	6.20 bc	6.80 a	0.143	<0.0001
TBVN	6.4 a	5.90 b	5.90 b	5.60 bc	5.50 bc	5.10 c	4.80 cd	0.187	<0.0001
TBA	0.60 a	0.60 a	0.5 ab	0.4 b	0.3 b	0.2 c	0.2 c	0.061	<0.0001
				Color					
*L**	60.10 ab	58.20 c	60.20 ab	60.00 ab	59.60 b	59.0 bc	61.20 a	0.334	<0.0001
*a**	6.00 bc	6.50 ab	6.00 bc	6.00 bc	6.70 a	6.40 b	5.80 c	0.116	<0.0001
*b**	15.00 bc	15.00 bc	15.90 b	14.10 cd	14.80 c	14.20 cd	16.10 a	0.268	<0.0001
				Sensorial					
Juiciness	4.20 bc	4.05 c	4.35 b	4.30 b	4.4 b	4.56 ab	4.80 a	0.085	<0.0001
Tenderness	4.90 b	4.75 bc	4.70 c	4.95 ab	4.70 c	4.96 ab	5.20 a	0.063	<0.0001
Taste	4.35 c	4.20 cd	4.44 b	4.34 c	4.20 cd	4.56 ab	4.80 a	0.075	<0.0001
Aroma	4.50 ab	4.35 c	4.50 ab	4.55 ab	4.56 ab	4.46 b	4.70 a	0.037	<0.0001

Basal diet (Control); T1: basal diet + 3.0 cm^3 ZnNPs; T2: basal diet + 5.0 cm^3 CurNPs; T3: basal diet + 2.0 cm^3 Bl; T4: basal diet + 3.0 cm^3 ZnNPs + 2 cm^3 BL; T5: basal diet + 3.0 cm^3 ZnNPs + 5.0 cm^3 CurNPs and T6: basal diet + 3.0 ZnNPs + 5.0 CurNPs + 2.0 cm^3 Bl; TVBN: total volatile basic nitrogen; TBA: thiobarbituric acid; *L**: lightness; *a**: redness; *b**: yellowness. SEM: standard error mean. a–e Different letters within one raw are significantly different ($p < 0.05$).

Table 7. Dietary microbiota (total bacteria, yeast and molds, *E. coli*, and coliform) presented (Log CFU/mL) in broiler during feeding period of 0–21 days as affected by dietary supplementation of ZnNPs, CurNPs, and *Bacillus licheniformis*. ($n = 6$).

Samples/Microbial Count	TBC				*p* **Value**	TYMC				*p* **Value**
Feeding Period (day)	**0**	**7**	**14**	**21**		**0**	**7**	**14**	**21**	
Control	5.80 a,D	6.01 a,C	6.33 a,B	6.70 a,A	<0.001	3.81 a,D	4.00 a,C	4.30 a,B	4.82 a,A	<0.001
T1	5.54 ab,D	5.72 b,C	6.12 ab,B	6.51 ab,A	<0.001	3.52 ab,D	3.72 ab,C	4.03 ab,B	4.55 b,A	<0.001
T2	5.20 bD	5.41 bcC	5.93 bB	6.22 b,A	<0.001	3.14 b,D	3.51 b,C	3.82 b,B	4.12 bc,A	<0.001
T3	4.86 c,D	5.14 c,C	5.50 bc,B	5.90 bc,A	<0.001	2.82 bc,D	3.14 c,C	3.44 c,B	3.84 c,A	<0.001
T4	4.60 d,D	4.95 cd,C	5.32 c,B	5.74 c,A	<0.001	2.58 c,D	2.90 cd,C	3.15 cd,B	3.69 cd,A	<0.001
T5	4.22 d,D	4.72 d,C	5.11 cd,B	5.38 cd,A	<0.001	2.30 cd,D	2.71 d,C	2.82 d,B	3.33 d,A	<0.001
T6	3.93 e,A	4.11 e,C	4.85 d,B	5.09 d,A	<0.001	2.10 d,D	2.42 de,C	2.68 de,B	3.15 de,A	<0.001
SEM	0.25	0.27	0.61	0.44		0.46	0.64	0.09	0.14	
p value	<0.001	<0.001	<0.001	<0.001		<0.001	<0.001	<0.001	<0.001	
Samples/Microbial Count	*E. coli*				*p***Value**	Coliform				*p***Value**
Feeding Period (day)	**0**	**7**	**14**	**21**		**0**	**7**	**14**	**21**	
Control	2.22 a,C	2.56 a,B	2.92 a,AB	3.12 a,A	<0.001	2.91 a,D	3.33 a,C	3.62 a,B	4.02 a,A	<0.001
T1	2.01 ab,D	2.32 ab,C	2.62 ab,B	2.81 ab,A	<0.001	2.72 ab,D	3.05 ab,C	3.43 ab,B	3.81 ab,A	<0.001
T2	1.88 b,D	2.11 b,C	2.31 b,B	2.54 b,A	<0.001	2.44 b,D	2.84 b,C	3.11 b,B	3.51 b,A	<0.001
T3	1.56 bc,D	1.92 c,C	2.15 bc,B	2.37 bc,A	<0.001	2.21 bc,D	2.58 bc,C	2.94 bc,B	3.23 c,A	<0.001
T4	1.34 c,C	1.73 c,B	1.96 c,A	2.08 c,A	<0.001	1.94 c,C	2.11 c,C	2.77 c,B	3.07 cd,A	<0.001
T5	1.12 cd,D	1.54 cd,C	1.64 cd,B	1.82 cd,A	<0.001	1.66 cd,C	1.84 d,C	2.52 cd,B	2.88 d,A	<0.001
T6	0.85 d,C	1.29 d,B	1.48 d,A	1.53 d,A	<0.001	1.48 d,D	1.62 de,C	2.29 d,B	2.57 e,A	<0.001
SEM	0.21	0.23	0.59	0.40		0.42	0.60	0.07	0.10	
p value	<0.001	<0.001	<0.001	<0.001		<0.001	<0.001	<0.001	<0.001	

Means followed by different lower case in the same column indicate significant difference ($p < 0.05$), different uppercase letters in same raw within each parameter indicate significant differences ($p < 0.05$); T1: basal diet + 3.0 cm^3 ZnNPs; T2: basal diet + 5.0 cm^3 CurNPs; T3: basal diet + 2.0 cm^3 Bl; T4: basal diet + 3.0 cm^3 ZnNPs + 2 cm^3 Bl; T5: basal diet +3.0 cm^3 ZnNPs + 5.0 cm^3 CurNPs and T6: basal diet +3.0 ZnNPs + 5.0 CurNPs + 2.0 cm^3 Bl; TBC; Total bacterial count; TYMC: total yeast-mold count; *E. coli*: *Escherichia coli*; SEM: standard error mean. a–e, A–D: Different letters within one column are significantly different ($p < 0.05$). SEM: standard error means.

Table 8. Cecal microbiota (total bacteria, yeasts and molds, *E. coli*, coliform, *Salmonella* spp., *Enterococcus* spp., and lactic acid bacteria) are represented (Log CFU/g) broiler during feeding as affected by dietary supplementation of ZnNPs, CurNPs, and *Bacillus licheniformis*. ($n = 5$).

Samples	Total Bacteria	Total Yeasts and Molds	*E. coli*	*Salmonella* spp.	*Enterococcus* spp.	Coliform	Lactic Acid Bacteria
Control	9.20 a	3.92 a	5.70 a	1.58	5.82 a	6.70 a	4.51 d
T1	8.40 b	3.59 ab	5.50 ab	ND	5.60 b	6.51 ab	4.81 cd
T2	7.83 bc	3.52 b	5.30 b	ND	5.50 b	6.23 b	5.20 c
T3	7.65 c	3.41 bc	5.11 c	ND	5.21 bc	6.25 c	5.42 c
T4	7.40 cd	3.21 c	4.82 cd	ND	5.12 c	5.84 c	5.62 b
T5	7.25 d	3.00 cd	4.62 d	ND	4.90 c	5.55 cd	5.81 ab
T6	6.50 e	2.80 d	4.44 d	ND	4.71 d	5.26 d	6.18 a
SEM	0.17	0.22	0.14	0.20	0.30	0.19	0.20
p value	0.001	0.004	0.001	0.001	0.012	0.021	0.002

Basal diet (Control); T1: basal diet + 3.0 cm^3 ZnNPs; T2: basal diet + 5.0 cm^3 CurNPs; T3: basal diet + 2.0 cm^3 Bl; T4: basal diet + 3.0 cm^3 ZnNPs + 2 cm^3 BL; T5: basal diet +3.0 cm^3 ZnNPs + 5.0 cm^3 CurNPs and T6: basal diet + 3.0 ZnNPs + 5.0 CurNPs + 2.0 cm^3 Bl; *E. coli*: *Escherichia coli*; SEM: standard error mean. a–e: Different letters within one column are significantly different ($p < 0.05$).

4. Discussion

The antimicrobial and antioxidant activities of CurNPs and ZnNPs are attributable to their small size and their phenolic compounds content on the surface of nanoparticles (NPs) [74]. These active compounds may be the reason for the beneficial effects of NPs additive on the growth, carcass properties, biochemical blood indicators, meat quality and microbial status of birds and the combination between CurNPs and ZnNPs increased the antimicrobial and antioxidant activities. For the antimicrobial activity of curcumin nanoparticles against tested pathogenic bacteria, Bhawana, et al. [75] confirmed the antibacterial activity of CurNPs on *S. aureus*, *B. subtilis*, *E. coli*, and *p. aeruginosa*. The authors added that the effect of CurNPs was more effective on Gram-positive than Gram-negative bacteria. Additionally, Narayanan, et al. [76] reported that the IDZs of ZnONPs (40 µg/mL) against *S. aureus* and *E. coli* were 19 and 14 mm, respectively. The IDZ of ZnONPs (50 µg/mL) on these bacteria were 18 and 16 mm [77]. These results are in agreement with the obtained results. Sirelkhatim, et al. [78] and Hassani Sangani, et al. [79] showed that the MIC and MBC values of ZnONPs were in the range of 158–325 µg/mL against 15 isolates of *p. aeruginosa* as well. The MIC was 50 µg/mL against *E. coli*, *Bacillus subtilis*, and *Staphylococcus aureus* and 25 µg/mL against *Vibrio cholera* and *Clostridium botulinum*.

The CurNPs and ZnNPs exhibited broad-spectrum antifungal activity in the current study. Arciniegas-Grijalba, et al. [80] found that zinc oxide nanoparticles in vitro antifungal activity against *Erythricium salmonicolor* causal of pink disease was through inhibiting the fungal mycelia. Additionally, Alnashi and Fattah [81] revealed that curcumin nanoparticles produced by turmeric methanolic extract have antifungal activity against *Candida albicans* ATCC 10,231 and *Geotricum candidum* NRRL Y-552 and mold strains, including: *Aspergillus niger* ATCC 102, *Aspergillus flavus* ATCC 247 and *Fusarium moniliform* ATCC 206 with MIC range of 15–25µg/mL. No available studies discussed the combination between CurNPs and ZnNPs. The beneficial impacts of probiotics on bird's performance may occur via changing the gut medium and raising the immunization of helpful gastric microorganisms. The competitive action reduces the harmful bacteria and the excitation of the immune system [31,82]. Probiotics settle the beneficial bacteria in the intestine and the competition with pathogenic bacteria leaves no area for hurtful bacteria to live in or set up. Additionally, probiotics stimulate the digestive enzyme secretion such as β galactosidase, α amylase, etc., which helps enhance the performance of animals [83].

The combinations of both NPs (T5) and the triple combination in the study (T6) presented the best LWG among groups, whereas all treatments supplemented with ZnNPs showed the best BWG values (T1, T4, T5 and T6) throughout the whole cycle. This positive effect may be due to the role of ZnNPs in raising the intestinal absorption ability by increasing the mucosal efficiency [84]. The higher adsorption capacity of Zn nanoparticles improves Zn bioavailability [21,85]. Previous studies indicated that utilizing ZnNPs as feed additives could improve LBW and FCR, reduce gut microbial populations, and boost the immunity system [20,21]. Our results agree with Mahmoud, et al. [86], who reported that ZnNPs (10 ppm) safely improved the body weight gain and FCR in broilers. As well, Fathi, et al. [87] observed that birds fed diets supplemented with nano-ZnO had higher ($p< 0.05$) body weight gain and lower FCR than the control. Several studies assured that diets supplemented with zinc increased growth rate and improved feed efficiency in broilers [21,88–90]. The different combinations between ZnNPs and Bl (T4), ZnNPs and CurNPs (T5), or ZnNPs, CurNPs, and Bl (T6) gave a reasonable growth rate compared to the control. This finding may depend on the role of each additive and their synergistic effect after combination. As reported in Table 3, the best growth performance (LBW, BWG, and FCR) was given by the T6 group, which had the triple combinations (ZnNPs, CurNPs, and Bl).

The nano form of curcumin increases its bioavailability and can raise its absorption [91], therefore, CurNPs can be applied as a safe and natural feed supplement [29]. The administration of CurNPs in chicks drinking water has also been reported to develop body weight and FCR. It has been reported that curcumin enhances the excretion of bile acids and stimulates protease, lipase, amylase, trypsin and chymotrypsin enzymes [92]. Therefore, the beneficial impact of curcumin on broiler growth may be due to the raised secretion of these enzymes. Recently, Reda et al. [25] showed that nano curcumin improved ($p < 0.0001$) LBW, BWG and FCR of growing quails at 5 and 1–5 weeks of age. The authors added that the FI ($p < 0.0001$) was reduced in birds fed CurNPs rations (0.1, 0.3, and 0.4 g/kg) compared to the control from 1 to 5 wks. The enhancements in the broiler performance given a diet supplemented with curcumin was probably due to improvements in the intestinal morphology of the broiler [27]. The considerable impact of curcumin might belong to its antibacterial, antioxidant and anti-inflammatory activities [93]. The method of using CurNPs in broilers' diets has been recently examined. It may have a potential pathway to activate the physiological and health status of broilers.

For carcass traits, the enhancement in some carcass traits in birds received T6 diet may be attributed to the antimicrobial activity of ZnNPs that reduces the pathogenic microbes' load and improves gut health [94]. Previous investigations showed that dietary ZnNPs (40 to 90 ppm) supplementation increased dressing percentage and carcass yield [5,93,95]. Moreover, Mahmoud et al. [86] confirmed that relative weights of spleen and bursa in birds fed diets supplemented with 10, 20, or 40 ppm of ZnNP's were higher ($p < 0.05$) than the untreated groups. However, Abdel-Moneim et al. [96] found no statistical differences in carcass traits of broilers fed *bifidobacteria* administrated groups.

The results obtained from Tables 5–8 showed an essential impact for feed supplementation concerning the biochemical indices, enzyme secretion, immunity status, increasing beneficial microorganisms, and reducing pathogenic bacteria, that achieving the best health and performance. No available studies investigate the beneficial impact of a similar combination; however, the findings of the current study were similar to the obtained results in a single addition Premavalli, et al. [97] and Abdel-Moneim, et al. [98] on growing Japanese quail and Jin, et al. [99] and Abd El-Moneim et al. [96] on broilers. The latter authors showed the valuable action of probiotics on bird's viability and healthy digestive tract. Zinc oxide nanoparticles can impact the birds' metabolic potency and health status due to their anti-bacterial and immune-modulation properties [18,19]. In addition, various studies revealed that higher dosages of ZnNPs such as 30–80 ppm [20,90] could enhance the broiler performance [85]. This is because zinc is an important microelement and a portion of more than 300 enzymes participating in controlling nucleic acid and protein metabolism [100,101]. Furthermore, ZnNPs can adjust the broiler's metabolism by raising the actions of insulin and growth hormone genes [102]. El-Katcha, et al. [103] mentioned that dietary addition of ZnNPs at 15 ppm enhanced the weight gain of broilers. In line, Zhao et al. [21] found that ZnNPs have beneficial effects on enhancing broiler performance.

The increasing in immunoglobulin levels in blood and cells in T6 may be attributed to synergistic effects of the triple combinations of ZnNPs, CurNPs and *Bacillus licheniformis* through the additivity of their antioxidant properties that resulted in the enhancement of the immune systems of birds. The antioxidant activity of CurNPs has a modulating effect on blood indices. The addition of CurNPs to the broiler's diet lowered AST levels in the blood [28]. Conversely, increased LDH levels and improved liver functions were reported in broilers fed a diet supplemented with turmeric (5 g/kg) [104]. Previous reports confirmed that feed supplemented with CurNPs (400 mg/kg diet) significantly decreased the lipid profile and reduced blood cholesterol as a diagnostic marker of lipid metabolism [105–107]. Emadi and Kermanshahi [104] observed a decrease in HDL levels and increased LDL levels in chickens that consumed a diet supplemented with turmeric. This led to improved liver function because of the inhibition of the HMGCR enzyme responsible for TC production in hepatic tissues [108]. Curcumin is a powerful antioxidant [109]. It reduces oxidative stress by modifying hepatic nuclear transcription factors and reducing lipid peroxidation in serum and muscles [110].

Diet supplemented with turmeric roots increased the levels of SOD, GSH and decreased levels of MDA [111]. Dietary curcumin supplementation reduced blood MDA and increased CAT, SOD, and GSH levels compared to controls [110,112]. It has been proven that the turmeric plant activates the immune cells (B and T) [113]. On the other hand, the addition of ZnNPs to broiler's diet led to an increase in the levels of blood lipids according to Fathi et al. [87]; Al-Daraji and Amen [114]. The latter authors also stated that a diet supplemented with ZnNPs (20 mg/kg) led to increased blood cholesterol. This increase is due to the role of zinc as the main part of many lipid enzymes. However, it stimulates SOD activity to scavenge free radicals [21]. Probiotics did not affect serum parameters except serum calcium and glucose Alkhalf, et al. [115].

The obtained results showed that ZnNPs, CurNPs, and Bl combinations gave the best improvements in meat quality. It increased the moisture and protein content and increased the tenderness and juiciness of the meat. Likewise, adding some natural extracts such as green tea extract and grape seeds to cooked beef patties did not affect the meat's sensory properties [116,117]. Additionally, the triple combination had an increased pH value that may differ due to the basic nature of the combination among ZnNPs, CuNPs, and Bl. However, TBVN and TBA values were significantly decreased, indicating the inhibitory effect of this combination against lipid oxidation and protein deterioration of meat by microbial enzymes [21].

The intestinal microbial load plays an important role in poultry health. The addition of CurNPs to the broiler diet modified the microbial balance in the intestine by increasing lactic acid bacteria count and reducing the pathogenic bacteria count, i.e., *S. aureus* and *E. coli* compared to the control [118,119]. Gupta, et al. [120] found that turmeric extracts inhibit pathogenic bacteria. Moreover, they reduce the intestinal bacterial count [121]. On the other hand, the diet supplemented with biological ZnNPs (100 mg/kg) caused increased bacterial count except for *E. coli* and *Enterococcus* spp. While the dietary addition of 200 and 400 mg ZnNPs/kg diet significantly decreased the bacterial count except for the intestinal bacteria. The concentration of 300 mg ZnNPs/kg diet significantly increased the bacterial load except for *Salmonella* spp.

The combination of ($ZnNPs^+$ $CurNPs^+$ Bl) achieved synergism, and their production was economic-effective. The synergism resulted in improving growth performance, blood indices, meat quality, and antioxidants parameters. The combination between CurNPs, ZnNPs, and BL exhibited more antimicrobial activity than the individual additions. Sequentially, the pathogenic microorganism reduced and increased lactic acid bacteria compared to single additions. Conclusively, the cost of production of biologically-synthesized CurNPs and ZnNPs is very low compared to the other methods (chemical or physical) with higher safety levels. In our study, the nanoparticles were used in very small doses compared to the original additives of the curcumin and zinc in broiler feed used in higher doses, which is very cost-effective. Additionally, BL is considered a very safe and cheap probiotic compared to other feed additives included in broiler feed, such as antibiotics.

5. Conclusions

The results of the current study assured a considerable antimicrobial activity against pathogenic bacteria and fungi with ZnNPs and CurNPs supplementation. The combination of (ZnNPs+ CurNPs+ Bl) achieves synergistic effects on enhancing the broiler's weights, performance, carcass traits, digestive enzymes, meat quality traits, blood indices, and cecal microbial load and the antioxidants capacity. Therefore, the inclusion of ZnNPs, CurNPs, Bl, and their combinations is recommended for broiler feeding regimens to improve the performance and health status with economic benefits.

Author Contributions: Conceptualization, M.E.A.E.-H., D.E.A.-K. and E.A.A.; methodology, M.T.E.-S. and A.M.S.; software, M.E.A.E.-H. and A.E.T.; formal analysis, A.E.T.; investigation, E.A.A., M.E.S. and N.M.A.; data curation, M.E.A.E.-H. and A.E.T.; writing—original draft preparation, M.E.A.E.-H., B.A.A. and R.M.F.; writing—review and editing, M.E.A.E.-H.; visualization, M.E.A.E.-H.; funding acquisition, B.A.A. and R.M.F. All authors have read and agreed to the published version of the manuscript.

Funding: This research received no external funding.

Institutional Review Board Statement: The authors confirm that the ethical policies of the journal, as noted on the journal's author guidelines page, have been adhered to and the appropriate ethical review committee approval has been received. The authors confirm that they have followed EU standards for the protection of animals used for scientific purposes. Ethical code number ZU-IACUC/2/F/95/2018.

Informed Consent Statement: Not applicable.

Data Availability Statement: Not applicable.

Conflicts of Interest: The authors declare that they have no conflict of interest.

References

1. Khan, R.U.; Naz, S.; Javdani, M.; Nikousefat, Z.; Selvaggi, M.; Tufarelli, V.; Laudadio, V. The use of turmeric (*Curcuma longa*) in poultry feed. *Worlds Poult. Sci. J.* **2012**, *68*, 97–103. [CrossRef]
2. European Commission. *Ban on Antibiotics as Growth Promoters in Animal Feed Enters into Effect*; European Commision: Brussels, Belgium, 2005.
3. Alagawany, M.; Elnesr, S.; Farag, M. Use of liquorice (*Glycyrrhiza glabra*) in poultry nutrition: Global impacts on performance, carcass and meat quality. *Worlds Poult. Sci. J.* **2019**, *75*, 293–304. [CrossRef]
4. Alagawany, M.; Elnesr, S.S.; Farag, M.R.; Abd El-Hack, M.E.; Khafaga, A.F.; Taha, A.E.; Tiwari, R.; Yatoo, M.; Bhatt, P. Use of licorice (*Glycyrrhiza glabra*) herb as a feed additive in poultry: Current knowledge and prospects. *Animals* **2019**, *9*, 536. [CrossRef]
5. Khafaga, A.F.; Abd El-Hack, M.E.; Taha, A.E.; Elnesr, S.S.; Alagawany, M. The potential modulatory role of herbal additives against Cd toxicity in human, animal, and poultry: A review. *Environ. Sci. Pollut. Res.* **2019**, *26*, 4588–4604. [CrossRef] [PubMed]
6. Abd El-Hack, M.; Ashour, E.; Elaraby, G.M.; Osman, A.; Arif, M. Influences of dietary supplementation of peanut skin powder (*Arachis Hypogaea*) on growth performance, carcass traits, blood chemistry, antioxidant activity and meat quality of broilers. *Anim. Prod. Sci.* **2018**, *58*, 965–972. [CrossRef]
7. Ashour, E.A.; Abd El-Hack, M.E.; Swelum, A.A.; Osman, A.O.; Taha, A.E.; Alhimaidi, A.R.; Ismail, I.E. Does the dietary graded levels of herbal mixture powder impact growth, carcass traits, blood indices and meat quality of the broilers? *Ital. J. Anim. Sci.* **2020**, *19*, 1226–1235. [CrossRef]
8. El-Saadony, M.T.; Alkhatib, F.M.; Alzahrani, S.O.; Shafi, M.E.; Abdel-Hamid, S.E.; Taha, T.F.; Aboelenin, S.M.; Soliman, M.M.; Ahmed, N.H. Impact of mycogenic zinc nanoparticles on performance, behavior, immune response, and microbial load in *Oreochromis niloticus*. *Saudi J. Biol. Sci.* **2021**, in press. [CrossRef]
9. Akl, B.; Nader, M.M.; El-Saadony, M.T. Biosynthesis of silver nanoparticles by *Serratia marcescens* ssp *sakuensis* and its antibacterial application against some pathogenic bacteria. *J. Agric. Chem. Biotechnol.* **2020**, *11*, 1–8. [CrossRef]
10. El-Saadony, M.T.; Saad, A.M.; Najjar, A.A.; Alzahrani, S.O.; Alkhatib, F.M.; Shafi, M.E.; Selem, E.; Desoky, E.-S.M.; Fouda, S.E.; El-Tahan, A.M.; et al. The use of biological selenium nanoparticles to suppress *Triticum aestivum* L. crown and root rot diseases induced by *Fusarium* species and improve yield under drought and heat stress. *Saudi J. Biol. Sci.* **2021**, in press. [CrossRef]
11. Li, X.; Xu, H.; Chen, Z.; Chen, G. Biosynthesis of nanoparticles by microorganisms and their applications. *J. Nanomater.* **2011**, 1–16. [CrossRef]

12. Wijnhoven, S.W.; Peijnenburg, W.J.; Herberts, C.A.; Hagens, W.I.; Oomen, A.G.; Heugens, E.H.; Roszek, B.; Bisschops, J.; Gosens, I.; Van De Meent, D. Nano-silver–a review of available data and knowledge gaps in human and environmental risk assessment. *Nanotoxicology* **2009**, *3*, 109–138. [CrossRef]
13. Liao, C.-D.; Hung, W.-L.; Jan, K.-C.; Yeh, A.-I.; Ho, C.-T.; Hwang, L.S. Nano/sub-microsized lignan glycosides from sesame meal exhibit higher transport and absorption efficiency in Caco-2 cell monolayer. *Food Chem.* **2010**, *119*, 896–902. [CrossRef]
14. Jahanian, R.; Moghaddam, H.N.; Rezaei, A. Improved broiler chick performance by dietary supplementation of organic zinc sources. Asian-australas. *J. Anim. Sci.* **2008**, *21*, 1348–1354.
15. NRC. *National Research Council. Nutrient Requirements of Poultry*, 9th ed.; National Academy Press: Washington, DC, USA, 1994.
16. Feng, J.; Ma, W.; Niu, H.; Wu, X.; Wang, Y. Effects of zinc glycine chelate on growth, hematological, and immunological characteristics in broilers. *Biol. Trace Elem. Res.* **2010**, *133*, 203–211. [CrossRef] [PubMed]
17. Sagar, P.D.; Mandal, A.; Akbar, N.; Dinani, O. Effect of different levels and sources of zinc on growth performance and immunity of broiler chicken during summer. *Int. J. Curr. Microbiol. Appl. Sci.* **2018**, *7*, 459–471. [CrossRef]
18. Reda, F.M.; El-Saadony, M.T.; El-Rayes, T.K.; Attia, A.I.; El-Sayed, S.A.; Ahmed, S.Y.; Madkour, M.; Alagawany, M. Use of biological nano zinc as a feed additive in quail nutrition: Biosynthesis, antimicrobial activity and its effect on growth, feed utilisation, blood metabolites and intestinal microbiota. *Ital. J. Anim. Sci.* **2021**, *20*, 324–335. [CrossRef]
19. Akhavan-Salamat, H.; Ghasemi, H. Effect of different sources and contents of zinc on growth performance, carcass characteristics, humoral immunity and antioxidant status of broiler chickens exposed to high environmental temperatures. *Livest. Sci.* **2019**, *223*, 76–83. [CrossRef]
20. Ahmadi, F.; Ebrahimnezhad, Y.; Sis, N.M.; Ghiasi, J. The effects of zinc oxide nanoparticles on performance, digestive organs and serum lipid concentrations in broiler chickens during starter period. *Int. J. Biosci.* **2013**, *3*, 23–29. [CrossRef]
21. Zhao, C.-Y.; Tan, S.-X.; Xiao, X.-Y.; Qiu, X.-S.; Pan, J.-Q.; Tang, Z.-X. Effects of dietary zinc oxide nanoparticles on growth performance and antioxidative status in broilers. *Biol. Trace Elem. Res.* **2014**, *160*, 361–367. [CrossRef]
22. Toghyani, M.; Toghyani, M.; Gheisari, A.; Ghalamkari, G.; Eghbalsaied, S. Evaluation of cinnamon and garlic as antibiotic growth promoter substitutions on performance, immune responses, serum biochemical and haematological parameters in broiler chicks. *Livest. Sci.* **2011**, *138*, 167–173. [CrossRef]
23. Rukkumani, R.; Aruna, K.; Varma, P.S.; Rajasekaran, K.N.; Menon, V.P. Comparative effects of curcumin and an analog of curcumin on alcohol and PUFA induced oxidative stress. *J. Pharm. Pharm. Sci.* **2004**, *7*, 274–283.
24. Zhai, S.; Ruan, D.; Zhu, Y.; Li, M.; Ye, H.; Wang, W.; Yang, L. Protective effect of curcumin on ochratoxin A–induced liver oxidative injury in duck is mediated by modulating lipid metabolism and the intestinal microbiota. *Poult. Sci.* **2020**, *99*, 1124–1134. [CrossRef]
25. Reda, F.M.; El-Saadony, M.T.; Elnesr, S.S.; Alagawany, M.; Tufarelli, V. Effect of dietary supplementation of biological curcumin nanoparticles on growth and carcass traits, antioxidant status, immunity and caecal microbiota of Japanese quails. *Animals* **2020**, *10*, 754. [CrossRef]
26. Shaikh, J.; Ankola, D.; Beniwal, V.; Singh, D.; Kumar, M.R. Nanoparticle encapsulation improves oral bioavailability of curcumin by at least 9-fold when compared to curcumin administered with piperine as absorption enhancer. *Eur. J. Pharm. Sci.* **2009**, *37*, 223–230. [CrossRef] [PubMed]
27. Rajput, N.; Muhammah, N.; Yan, R.; Zhong, X.; Wang, T. Effect of dietary supplementation of curcumin on growth performance, intestinal morphology and nutrients utilization of broiler chicks. *J. Poult. Sci.* **2012**, 0120065. [CrossRef]
28. Sayrafi, R.; Hosseini, S.; Ahmadi, M. The protective effects of nanocurcumin on liver toxicity induced by salinomycin in broiler chickens. *Rev. Méd. Vét.* **2017**, *168*, 136–142.
29. Partovi, R.; Seifi, S.; Pabast, M.; Babaei, A. Effects of dietary supplementation with nanocurcumin on quality and safety of meat from broiler chicken infected with Eimeria species. *J. Food Saf.* **2019**, *39*, e12703. [CrossRef]
30. Ahuja, K.; Mamtani, K. Poultry Probiotic Ingredients Market Size by Product. In *(Lactobacilli, Bifidobacterium, Streptococcus, Bacillus), by Application (Broilers, Layers, Turkeys, Breeders, Chicks & Poults), Regional Outlook, Application Potential, Price Trends, Competitive Market Share & Forecast, 2019–2025*; Global Market Insights: Selbyville, DE, USA, 2019.
31. Jha, R.; Das, R.; Oak, S.; Mishra, P. Probiotics (direct-fed microbials) in poultry nutrition and their effects on nutrient utilization, growth and laying performance, and gut health: A systematic review. *Animals* **2020**, *10*, 1863. [CrossRef]
32. Attia, Y.; Zeweil, H.; Alsaffar, A.; El-Shafy, A. Effect of non-antibiotic feed additives as an alternative to flavomycin on productivity, meat quality and blood parameters in broilers. *Arch. Geflügelk* **2011**, *75*, 40–48.
33. Santoso, U.; Tanaka, K.; Ohtani, S.; Sakaida, M. Effect of fermented product from Bacillus subtilis on feed conversion efficiency, lipid accumulation and ammonia production in broiler chicks. *Asian Australas. J. Anim. Sci.* **2001**, *14*, 333–337. [CrossRef]
34. Abou-Kassem, D.; Elsadek, M.; Abdel-Moneim, A.; Mahgoub, S.; Elaraby, G.; Taha, A.; Elshafie, M.; Alkhawtani, D.; Abd El-Hack, M.; Ashour, E. Growth, carcass characteristics, meat quality, and microbial aspects of growing quail fed diets enriched with two different types of probiotics (*Bacillus toyonensis* and *Bifidobacterium bifidum*). *Poult. Sci.* **2021**, *100*, 84–93. [CrossRef]
35. El-Saadony, M.T.; Abd El-Hack, M.E.; Taha, A.E.; Fouda, M.M.; Ajarem, J.S.; Maodaa, S.N.; Allam, A.A.; Elshaer, N. Ecofriendly synthesis and insecticidal application of copper nanoparticles against the storage pest *Tribolium castaneum*. *Nanomaterials* **2020**, *10*, 587. [CrossRef] [PubMed]

36. El-Saadony, M.T.; El-Wafai, N.A.; El-Fattah, H.I.A.; Mahgoub, S.A. Biosynthesis, optimization and characterization of silver nanoparticles using a soil isolate of *Bacillus pseudomycoides* MT32 and their antifungal activity against some pathogenic fungi. *Adv. Anim. Vet. Sci.* **2019**, *7*, 238–249. [CrossRef]
37. Kannan, M.N.; Sethi, S.; Badoni, A.; Chamoli, V.; Bahuguna, N.C. Isolation and characterization of bacterial isolates from agriculture field soil of Roorkee region. *J. Pharm. Phytochem.* **2018**, *5*, 108–110.
38. El-Saadony, M.T.; Desoky, E.-S.M.; Saad, A.M.; Eid, R.S.; Selem, E.; Elrys, A.S. Biological silicon nanoparticles improve *Phaseolus vulgaris* L. yield and minimize its contaminant contents on a heavy metals-contaminated saline soil. *J. Environ. Sci.* **2021**, *106*, 1–14. [CrossRef]
39. Elbeshehy, E.K.; Elazzazy, A.M.; Aggelis, G. Silver nanoparticles synthesis mediated by new isolates of *Bacillus* spp., nanoparticle characterization and their activity against Bean Yellow Mosaic Virus and human pathogens. *Front. Microbiol.* **2015**, *6*, 453. [CrossRef]
40. Saad, A.M.; El-Saadony, M.T.; El-Tahan, A.M.; Sayed, S.; Moustafa, M.A.; Taha, A.E.; Taha, T.F.; Ramadan, M.M. Polyphenolic extracts from pomegranate and watermelon wastes as substrate to fabricate Sustainable Silver nanoparticles with larvicidal effect against *Spodoptera littoralis*. *Saudi J. Biol. Sci.* **2021**, in press. [CrossRef]
41. Mishra, M.; Paliwal, J.S.; Singh, S.K.; Selvarajan, E.; Subathradevi, C.; Mohanasrinivasan, V. Studies on the inhibitory activity of biologically synthesized and characterized zinc oxide nanoparticles using *lactobacillus sporogens* against *Staphylococcus aureus*. *J. Pure. Appl. Microbiol.* **2013**, *7*, 1–6.
42. Gulluce, M.; Sahin, F.; Sokmen, M.; Ozer, H.; Daferera, D.; Sokmen, A.; Polissiou, M.; Adiguzel, A.; Ozkan, H. Antimicrobial and antioxidant properties of the essential oils and methanol extract from *Mentha longifolia* L. ssp. *longifolia*. *Food Chem.* **2007**, *103*, 1449–1456. [CrossRef]
43. Rahmoun, N.M.; Ziane, H.; Boucherit-Otmani, Z. Antibacterial and antifungal screening of four medicinal plants. *J. Coast. Life Med.* **2014**, *2*, 975–979.
44. Eucast. Determination of minimum inhibitory concentrations (MICs) of antimicrobial agents by broth dilution. *Clin. Microbiol. Infect.* **2003**, *9*, 1–7.
45. CLSI. *Performance standards for Antimicrobial Susceptibility Testing, 18th Informational Supplement. M100-S18*; Clinical and Laboratory Standards Institute: Wayne, PA, USA, 2016.
46. Ali, S.S.; Morsy, R.; El-Zawawy, N.A.; Fareed, M.F.; Bedaiwy, M.Y. Synthesized zinc peroxide nanoparticles (ZnO_2-NPs): A novel antimicrobial, anti-elastase, anti-keratinase, and anti-inflammatory approach toward polymicrobial burn wounds. *Int. J. Nanomed.* **2017**, *12*, 6059. [CrossRef]
47. Usman, J.G.; Sodipo, O.; Sandabe, U. In vitro antimicrobial activity of *Cucumis metuliferus* E. Mey. Ex. Naudin fruit extracts against Salmonella gallinarum. *Int. J. Phytomed.* **2014**, *6*, 268.
48. Fidan, E.D.; Nazlıgül, A.; Türkyılmaz, M.K.; Aypak, S.Ü.; Kilimci, F.S.; Karaarslan, S.; Kaya, M. Effect of photoperiod length and light intensity on some welfare criteria, carcass, and meat quality characteristics in broilers. *Rev. Bras. Zootec.* **2017**, *46*, 202–210. [CrossRef]
49. Trela, J.; Kierończyk, B.; Hautekiet, V.; Józefiak, D. Combination of bacillus licheniformis and salinomycin: Effect on the growth performance and git microbial populations of broiler chickens. *Animals* **2020**, *10*, 889. [CrossRef] [PubMed]
50. IR-BroilerHandbook2018-EN.pdf. 2018. A.I.R.B.M.H. Available online: http://www.aviagen.com (accessed on 27 May 2018).
51. Abdelnour, S.A.; Swelum, A.A.; Salama, A.; Al-Ghadi, M.Q.; Qattan, S.Y.; Abd El-Hack, M.E.; Khafaga, A.F.; Alhimaidi, A.R.; Almutairi, B.O.; Ammari, A.A.; et al. The beneficial impacts of dietary phycocyanin supplementation on growing rabbits under high ambient temperature. *Ital. J. Anim. Sci.* **2020**, *19*, 1046–1056. [CrossRef]
52. Sitohy, M.; Osman, A.; Gharib, A.; Chobert, J.-M.; Haertlé, T. Preliminary assessment of potential toxicity of methylated soybean protein and methylated β-lactoglobulin in male Wistar rats. *Food Chem. Toxicol.* **2013**, *59*, 618–625. [CrossRef]
53. Salvaggio, A.; Periti, M.; Miano, L.; Tavanelli, M.; Marzorati, D. Body mass index and liver enzyme activity in serum. *Clin. Chem.* **1991**, *37*, 720–723. [CrossRef]
54. Owen, J.; Iggo, B.; Scandrett, F.; Stewart, C. The determination of creatinine in plasma or serum, and in urine; A critical examination. *Biochem. J.* **1954**, *58*, 426–437. [CrossRef]
55. Abdel-Hamid, M.; Osman, A.; El-Hadary, A.; Romeih, E.; Sitohy, M.; Li, L. Hepatoprotective action of papain-hydrolyzed buffalo milk protein on carbon tetrachloride oxidative stressed albino rats. *J. Dairy Sci.* **2020**, *103*, 1884–1893. [CrossRef]
56. Wattanachant, S.; Benjakul, S.; Ledward, D. Composition, color, and texture of Thai indigenous and broiler chicken muscles. *Poult. Sci.* **2004**, *83*, 123–128. [CrossRef] [PubMed]
57. Wattanachant, S.; Benjakul, S.; Ledward, D. Microstructure and thermal characteristics of Thai indigenous and broiler chicken muscles. *Poult. Sci.* **2005**, *84*, 328–336. [CrossRef] [PubMed]
58. Fernandez-Lopez, J.; Zhi, N.; Aleson-Carbonell, L.; Pérez-Alvarez, J.A.; Kuri, V. Antioxidant and antibacterial activities of natural extracts: Application in beef meatballs. *Meat Sci.* **2005**, *69*, 371–380. [CrossRef] [PubMed]
59. Botta, J.; Lauder, J.; Jewer, M. Effect of methodology on total volatile basic nitrogen (TVB-N) determination as an index of quality of fresh Atlantic cod (*Gadus morhua*). *J. Food Sci.* **1984**, *49*, 734–736. [CrossRef]
60. AOAC. *Official Methods of Analysis*, 18th ed.; Association of Official Analytical Chemists: Arlington, VA, USA, 2005.
61. Zhao, C.; Liu, Y.; Lai, S.; Cao, H.; Guan, Y.; San Cheang, W.; Liu, B.; Zhao, K.; Miao, S.; Riviere, C. Effects of domestic cooking process on the chemical and biological properties of dietary phytochemicals. *Trends Food Sci. Technol.* **2019**, *85*, 55–66. [CrossRef]

62. Sow, T.; Grongnet, J.-F. Sensory characteristics and consumer preference for chicken meat in Guinea. *Poult. Sci.* **2010**, *89*, 2281–2292. [CrossRef]
63. Zhuang, H.; Savage, E.M. Effect of postmortem deboning time on sensory descriptive flavor and texture profiles of cooked boneless skinless broiler thighs. *LWT Food Sci. Technol.* **2011**, *44*, 2087–2090. [CrossRef]
64. Harrigan, W.F.; McCance, M.E. *Laboratory Methods in Food and Dairy Microbiology*; Academic Press Inc.: London, UK, 1976.
65. Alagawany, M.; El-Saadony, M.; Elnesr, S.; Farahat, M.; Attia, G.; Madkour, M.; Reda, F. Use of lemongrass essential oil as a feed additive in quail's nutrition: Its effect on growth, carcass, blood biochemistry, antioxidant and immunological indices, digestive enzymes and intestinal microbiota. *Poult. Sci.* **2021**, *100*, 101172. [CrossRef]
66. Alagawany, M.; Madkour, M.; El-Saadony, M.T.; Reda, F.M. *Paenibacillus polymyxa* (LM31) as a new feed additive: Antioxidant and antimicrobial activity and its effects on growth, blood biochemistry, and intestinal bacterial populations of growing Japanese quail. *Anim. Feed Sci. Technol.* **2021**, *276*, 114920. [CrossRef]
67. Ashour, E.A.; Abd El-Hack, M.E.; Shafi, M.E.; Alghamdi, W.Y.; Taha, A.E.; Swelum, A.A.; Tufarelli, V.; Mulla, Z.S.; El-Ghareeb, W.R.; El-Saadony, M.T. Impacts of green coffee powder supplementation on growth performance, carcass characteristics, blood indices, meat quality and gut microbial load in broilers. *Agriculture* **2020**, *10*, 457. [CrossRef]
68. Sheiha, A.M.; Abdelnour, S.A.; Abd El-Hack, M.E.; Mohamed, E.; Khafaga, A.F.; Metwally, K.A.; Ajarem, J.S.; Maodaa, S.N.; Allam, A.A.; El-Saadony, M.T. Effects of dietary biological or chemical-synthesized nano-selenium supplementation on growing rabbits exposed to thermal stress. *Animals* **2020**, *10*, 430. [CrossRef]
69. Szabo, R.A.; Todd, E.C.; Jean, A. Method to isolate *Escherichia coli* O157: H7 from food. *J. Food Prot.* **1986**, *49*, 768–772. [CrossRef] [PubMed]
70. Edwards, E.A.; Hilderbrand, R.L. Method for identifying Salmonella and Shigella directly from the primary isolation plate by coagglutination of protein A-containing *staphylococci* sensitized with specific antibody. *J. Clin. Microbiol.* **1976**, *3*, 339–343. [CrossRef]
71. Kurtzman, C.; Fell, J.W.; Boekhout, T. *The Yeasts: A Taxonomic Study*; Elsevier: Amsterdam, The Netherlands, 2011.
72. Argyri, A.A.; Zoumpopoulou, G.; Karatzas, K.-A.G.; Tsakalidou, E.; Nychas, G.-J.E.; Panagou, E.Z.; Tassou, C.C. Selection of potential probiotic lactic acid bacteria from fermented olives by in vitro tests. *Food Microbiol.* **2013**, *33*, 282–291. [CrossRef] [PubMed]
73. Miranda, J.; Franco, C.; Vázquez, B.; Fente, C.; Barros-Velázquez, J.; Cepeda, A. Evaluation of Chromocult® enterococci agar for the isolation and selective enumeration of *Enterococcus* spp. in broilers. *Lett. Appl. Microbiol.* **2005**, *41*, 153–156. [CrossRef]
74. El-Saadony, M.T.; Sitohy, M.Z.; Ramadan, M.F.; Saad, A.M. Green nanotechnology for preserving and enriching yogurt with biologically available iron (II). *Innov. Food Sci. Emerg. Technol.* **2021**, 102645. [CrossRef]
75. Bhawana, B.R.; Buttar, H.S.; Jain, V.; Jain, N. Curcumin nanoparticles: Preparation, characterization, and antimicrobial study. *J. Agric. Food Chem.* **2011**, *59*, 2056–2061. [CrossRef]
76. Narayanan, P.; Wilson, W.S.; Abraham, A.T.; Sevanan, M. Synthesis, characterization, and antimicrobial activity of zinc oxide nanoparticles against human pathogens. *BioNanoScience* **2012**, *2*, 329–335. [CrossRef]
77. Azam, A.; Ahmed, A.S.; Oves, M.; Khan, M.S.; Habib, S.S.; Memic, A. Antimicrobial activity of metal oxide nanoparticles against Gram-positive and Gram-negative bacteria: A comparative study. *Int. J. Nanomed.* **2012**, *7*, 6003. [CrossRef] [PubMed]
78. Sirelkhatim, A.; Mahmud, S.; Seeni, A.; Kaus, N.H.M.; Ann, L.C.; Bakhori, S.K.M.; Hasan, H.; Mohamad, D. Review on zinc oxide nanoparticles: Antibacterial activity and toxicity mechanism. *Nano Micro Lett.* **2015**, *7*, 219–242. [CrossRef]
79. Hassani Sangani, M.; Nakhaei Moghaddam, M.; Forghanifard, M.M. Inhibitory effect of zinc oxide nanoparticles on pseudomonas aeruginosa biofilm formation. *Nanomed. J.* **2015**, *2*, 121–128.
80. Arciniegas-Grijalba, P.; Patiño-Portela, M.; Mosquera-Sánchez, L.; Guerrero-Vargas, J.; Rodríguez-Páez, J. ZnO nanoparticles (ZnO-NPs) and their antifungal activity against coffee fungus Erythricium salmonicolor. *Appl. Nanosci.* **2017**, *7*, 225–241. [CrossRef]
81. Alnashi, B.; Fattah, A.A. Antimicrobial activity of raw and nano turmeric powder extracts. *Middle East. J. Appl. Sci.* **2016**, *6*, 787–796.
82. Rhayat, L.; Jacquier, V.; Brinch, K.; Nielsen, P.; Nelson, A.; Geraert, P.-A.; Devillard, E. Bacillus subtilis s train specificity affects performance improvement in broilers. *Poult. Sci.* **2017**, *96*, 2274–2280. [CrossRef] [PubMed]
83. Jadhav, K.; Sharma, K.; Katoch, S.; Sharma, V.; Mane, B. Probiotics in broiler poultry feeds: A review. *J. Anim. Nutr. Physiol.* **2015**, *1*, 4–16.
84. Hafez, A.; Hegazi, S.; Bakr, A.; Shishtawy, H. Effect of zinc oxide nanoparticles on growth performance and absorptive capacity of the intestinal villi in broiler chickens. *Life Sci. J.* **2017**, *14*, 67–72.
85. Abedini, M.; Shariatmadari, F.; Torshizi, M.K.; Ahmadi, H. Effects of a dietary supplementation with zinc oxide nanoparticles, compared to zinc oxide and zinc methionine, on performance, egg quality, and zinc status of laying hens. *Live Sci.* **2017**, *203*, 30–36. [CrossRef]
86. Mahmoud, U.T.; Abdel-Mohsein, H.S.; Mahmoud, M.A.; Amen, O.A.; Hassan, R.I.; Abd-El-Malek, A.M.; Rageb, S.M.; Waly, H.S.; Othman, A.A.; Osman, M.A. Effect of zinc oxide nanoparticles on broilers' performance and health status. *Trop. Anim. Health Prod.* **2020**, *52*, 2043–2054. [CrossRef]
87. Fathi, M.; Haydari, M.; Tanha, T. Effects of zinc oxide nanoparticles on antioxidant status, serum enzymes activities, biochemical parameters and performance in broiler chickens. *J. Livest. Sci. Technol.* **2016**, *4*, 7–13.

88. Sandoval, M.; Henry, P.; Luo, X.; Littell, R.; Miles, R.; Ammerman, C. Performance and tissue zinc and metallothionein accumulation in chicks fed a high dietary level of zinc. *Poult. Sci.* **1998**, *77*, 1354–1363. [CrossRef]
89. Roberson, K.D.; Edwards, H.M., Jr. Effects of 1,25-dihydroxycholecalciferol and phytase on zinc utilization in broiler chicks. *Poult. Sci.* **1994**, *73*, 1312–1326. [CrossRef] [PubMed]
90. Fazilati, M. Investigation toxicity properties of zinc oxide nanoparticles on liver enzymes in male rat. *Eur. J. Exp. Biol.* **2013**, *3*, 97–103.
91. Hani, U.; Shivakumar, H. Solubility enhancement and delivery systems of curcumin a herbal medicine: A review. *Curr. Drug Deliv.* **2014**, *11*, 792–804. [CrossRef]
92. Platel, K.; Srinivasan, K. Influence of dietary spices and their active principles on pancreatic digestive enzymes in albino rats. *Food Nahr.* **2000**, *44*, 42–46. [CrossRef]
93. Chattopadhyay, I.; Biswas, K.; Bandyopadhyay, U.; Banerjee, R.K. Turmeric and curcumin: Biological actions and medicinal applications. *Curr. Sci.* **2004**, 44–53.
94. Sahoo, A.; Swain, R.; Mishra, S.K. Effect of inorganic, organic and nano zinc supplemented diets on bioavailability and immunity status of broilers. *Int. J. Adv. Res.* **2014**, *2*, 828–837.
95. Lina, T.; Jianyang, J.; Fenghua, Z.; Huiying, R.; Wenli, L. Effect of nano-zinc oxide on the production and dressing performance of broiler. *Chin. Agric. Sci. Bull.* **2009**, *2*, 318.
96. Abd El-Moneim, A.E.-M.E.; El-Wardany, I.; Abu-Taleb, A.M.; Wakwak, M.M.; Ebeid, T.A.; Saleh, A.A. Assessment of in ovo administration of *Bifidobacterium bifidum* and *Bifidobacterium longum* on performance, ileal histomorphometry, blood hematological, and biochemical parameters of broilers. *Probiotics Antimicro Proteins* **2019**, 1–12. [CrossRef]
97. Premavalli, K.; Sangilimadan, K.; Omprakash, A. Effect of supplementation of multi-species probiotic on production performance of Japanese quail. *Inter. J. Chem. Stud.* **2018**, *6*, 2164–2166.
98. Abdel-Moneim, A.-M.E.; Selim, D.A.; Basuony, H.A.; Sabic, E.M.; Saleh, A.A.; Ebeid, T.A. Effect of dietary supplementation of *Bacillus subtilis* spores on growth performance, oxidative status, and digestive enzyme activities in Japanese quail birds. *Rop. Anim. Health Prod.* **2020**, *52*, 671–680. [CrossRef]
99. Jin, L.; Ho, Y.; Abdullah, N.; Ali, M.; Jalaludin, S. Effects of adherent Lactobacillus cultures on growth, weight of organs and intestinal microflora and volatile fatty acids in broilers. *Anim. Feed Sci. Technol.* **1998**, *70*, 197–209. [CrossRef]
100. Badawi, M.; Ali, M.; Behairy, A. Effects of zinc sources supplementation on performance of broiler chickens. *J. Am. Sci.* **2017**, *13*, 35–40.
101. Jarosz, M.; Olbert, M.; Wyszogrodzka, G.; Młyniec, K.; Librowski, T. Antioxidant and anti-inflammatory effects of zinc. Zinc-dependent NF-κB signaling. *Inflammopharmacology* **2017**, *25*, 11–24. [CrossRef] [PubMed]
102. Ibrahim, D.; Ali, H.A.; El-Mandrawy, S.A. Effects of different zinc sources on performance, bio distribution of minerals and expression of genes related to metabolism of broiler chickens. *Zagazig Vet. J.* **2017**, *45*, 292–304. [CrossRef]
103. El-Katcha, M.; Soltan, M.A.; El-Badry, M. Effect of dietary replacement of inorganic zinc by organic or nanoparticles sources on growth performance, immune response and intestinal histopathology of broiler chicken. *Alex. J. Vet. Sci.* **2017**, *55*. [CrossRef]
104. Emadi, M.; Kermanshahi, H. Effect of turmeric rhizome powder on the activity of some blood enzymes in broiler chickens. *Int. J. Poult. Sci* **2007**, *6*, 48–51. [CrossRef]
105. Gholiyan, A.; Jafari, M.R. Effect of different levels of nanoparticle curcumin on egg quality, blood parameters, immune response and digestibility in laying hens. *Res. Anim. Prod. Sci. Res.* **2018**, *9*, 26–34.
106. Unnikrishnan, M.; Rao, M. Inhibition of nitrite induced oxidation of hemoglobin by curcuminoids. *Die Pharm.* **1995**, *50*, 490–492.
107. De Beer, M.; McMurtry, J.; Brocht, D.; Coon, C. An examination of the role of feeding regimens in regulating metabolism during the broiler breeder grower period. 2. Plasma hormones and metabolites. *Poult. Sci.* **2008**, *87*, 264–275. [CrossRef] [PubMed]
108. AL-Kassie, G.A.; Mohseen, A.M.; Abd-AL-Jaleel, R.A. Modification of productive performance and physiological aspects of broilers on the addition of a mixture of cumin and turmeric to the diet. *Res. Opin. Anim. Vet. Sci.* **2011**, *27*, 847–854.
109. Das, K.C.; Das, C.K. Curcumin (diferuloylmethane), a singlet oxygen (1O2) quencher. *Biochem. Biophys. Res. Commun.* **2002**, *295*, 62–66. [CrossRef]
110. Sahin, K.; Orhan, C.; Tuzcu, Z.; Tuzcu, M.; Sahin, N. Curcumin ameloriates heat stress via inhibition of oxidative stress and modulation of Nrf2/HO-1 pathway in quail. *Food Chem. Toxicol.* **2012**, *50*, 4035–4041. [CrossRef] [PubMed]
111. Wang, D.; Huang, H.; Zhou, L.; Li, W.; Zhou, H.; Hou, G.; Liu, J.; Hu, L. Effects of dietary supplementation with turmeric rhizome extract on growth performance, carcass characteristics, antioxidant capability, and meat quality of Wenchang broiler chickens. *Ital. J. Anim. Sci.* **2015**, *14*, 3870. [CrossRef]
112. Nawab, A.; Li, G.; Liu, W.; Lan, R.; Wu, J.; Zhao, Y.; Kang, K.; Kieser, B.; Sun, C.; Tang, S. Effect of dietary curcumin on the antioxidant status of laying hens under high-temperature condition. *J. Therm. Biol.* **2019**, *86*, 102449. [CrossRef]
113. Jagetia, G.C.; Aggarwal, B.B. "Spicing up" of the immune system by curcumin. *J. Clin. Immunol.* **2007**, *27*, 19–35. [CrossRef] [PubMed]
114. Al-Daraji, H.J.; Amen, M.H. Effect of dietary zinc on certain blood traits of broiler breeder chickens. *Int. J. Poult. Sci.* **2011**, *10*, 807–813. [CrossRef]
115. Alkhalf, A.; Alhaj, M.; Al-Homidan, I. Influence of probiotic supplementation on blood parameters and growth performance in broiler chickens. *Saudi J. Biol. Sci.* **2010**, *17*, 219–225. [CrossRef]

116. Bech-Larsen, T.; Scholderer, J. Functional foods in Europe: Consumer research, market experiences and regulatory aspects. *Trends Food Sci. Technol.* **2007**, *18*, 231–234. [CrossRef]
117. Banon, S.; Díaz, P.; Rodríguez, M.; Garrido, M.D.; Price, A. Ascorbate, green tea and grape seed extracts increase the shelf life of low sulphite beef patties. *Meat Sci.* **2007**, *77*, 626–633. [CrossRef]
118. Ahmed, I.; El-Rayes, T.; Ahmed, A.I. Assessment of dietary supplementation of turmeric (curcuma longa) as a phytobiotic on broiler performance and bacterial count. *Egypt. J. Nutr. Feeds* **2018**, *21*, 519–528. [CrossRef]
119. Feng, W.; Wang, H.; Zhang, P.; Gao, C.; Tao, J.; Ge, Z.; Zhu, D.; Bi, Y. Modulation of gut microbiota contributes to curcumin-mediated attenuation of hepatic steatosis in rats. *Biochim. Biophys. Acta (BBA) Gen. Subj.* **2017**, *1861*, 1801–1812. [CrossRef] [PubMed]
120. Gupta, A.; Mahajan, S.; Sharma, R. Evaluation of antimicrobial activity of Curcuma longa rhizome extract against Staphylococcus aureus. *Biotechnol. Rep.* **2015**, *6*, 51–55. [CrossRef] [PubMed]
121. Niamsa, N.; Sittiwet, C. Antimicrobial activity of Curcuma longa aqueous extract. *J. Pharmacol. Toxicol.* **2009**, *4*, 173–177. [CrossRef]

animals

MDPI

Article

Effect of Different Dietary Betaine Fortifications on Performance, Carcass Traits, Meat Quality, Blood Biochemistry, and Hematology of Broilers Exposed to Various Temperature Patterns

Ahmed A. Al-Sagan [1,*], Abdullah H. Al-Yemni [2], Alaeldein M. Abudabos [3], Abdulaziz A. Al-Abdullatif [3] and Elsayed O. Hussein [3]

1 King Abdulaziz City for Science & Technology, P.O. Box 6086, Riyadh 11442, Saudi Arabia
2 Arabian Agriculture Services Company (ARASCO), Riyadh 11593, Saudi Arabia; abdullah@arasco.com
3 Department of Animal Production, College of Food and Agriculture Sciences, King Saud University, Riyadh 11451, Saudi Arabia; aabudabos@ksu.edu.sa (A.M.A.); aalabdullatif@ksu.edu.sa (A.A.A.-A.); shessin@ksu.edu.sa (E.O.H.)
* Correspondence: abdeen@kacst.edu.sa

Simple Summary: Enhancing production of broilers in hot climates is essential to overcoming high-heat problems. High temperature during the summer season is one of the most influential factors affecting poultry production. Heat stress has long been identified as dampening growth performance and increasing vulnerability to diseases. Several methods have been used to alleviate the adverse effects of heat stress. One of these is the dietary fortification of betaine. Betaine (Trimethylglycine) naturally occurs in various plants, and it can be extracted from beetroot. The current experiment examined the effects of dietary betaine fortification on broiler chickens at 1–40 days of age. At 18 days of age, half of the birds were kept under thermos-neutral temperature (22–24 °C), while the other half were kept under high temperature (35 °C). The production efficiency factor was best ($p < 0.05$) for birds that received 0.10% betaine. Betaine fortification improved the growth, feed utilization, and production index. Betaine fortification at 0.1% betaine supplementation with an 84 g increase in growth, 4.6 points improvement in feed utilization, and 24 points enhancement in production index compared with no betaine fortification. Betaine fortification (0.1%) during heat stress reduced the negative impact on performance and improved production efficiency, suggesting that betaine may be as a practical tool for improving poultry production in hot regions.

Abstract: Improving broilers' production in the hot region is essential to overcome heat-stress challenges. The current experiment examined the effects of betaine's fortification (0.0, 0.075, 0.10, and 0.15%) to broiler chickens during days 1–40 of age. The growth period was divided into the starter (1–18 d) and growing-finishing (19–40 d). During the starter period, there was no heat challenge, and all birds were kept under the same conditions. At 18 days of age, half of the birds were kept under thermos-neutral temperature (TN, 22–24 °C), while the other half were kept under high temperature (HT, 35 °C). However, the production efficiency factor (PEF) was the best ($p < 0.05$) for birds that received 0.10% betaine. Betaine fortification improved ($p < 0.05$ and 0.01) body weight gain (BWG), feed conversion ratio (FCR), and production efficiency factor (PEF) in the cumulative finisher heat-stress challenge period (19–40 d). The best performance was achieved at 0.1% betaine fortification with 84 g gain, 4.6 points improvement in FCR, and 24 points improvements in PEF as compared to no betaine fortification. The heat-stressed group consumed less feed (239 g), gained less weight (179 g), converted feed less efficiently (2.6 points), and, as a result, had lower FEF (29 points) as compared to the TN group. Conclusively, heat challenge had a powerful effect on growth performance, meat characteristics, and blood parameters, especially during the grower-finisher period. Betaine fortification (0.1%) during heat stress reduced the negative impact on performance and improved production efficiency, suggesting that betaine is a useful nutritional tool under stress conditions that deserves further investigation.

Keywords: broilers; heat stress; betaine; performance; meat quality; blood constituents

Citation: Al-Sagan, A.A.; Al-Yemni, A.H.; Abudabos, A.M.; Al-Abdullatif, A.A.; Hussein, E.O. Effect of Different Dietary Betaine Fortifications on Performance, Carcass Traits, Meat Quality, Blood Biochemistry, and Hematology of Broilers Exposed to Various Temperature Patterns. *Animals* **2021**, *11*, 1555. https://doi.org/10.3390/ani11061555

Academic Editors: Natalie Morgan and Youssef A. Attia

Received: 28 April 2021
Accepted: 12 May 2021
Published: 26 May 2021

1. Introduction

Recently, poultry production's husbandry and feeding protocols have become crucial issues for farmers, particularly in hot areas, as they can lessen the comparative heat burden [1–5]. High temperature during the summer season is one of the most significant factors affecting poultry production [6–8]. Heat stress has long been identified as dampening growth performance and egg production and increasing vulnerability to diseases [4,9–11].

Given the substantial importance of poultry products and high-quality and affordable animal protein to consumers, there is an urgency to study the factors that hinder poultry production. One of the major factors that influence poultry production is heat stress, particularly in desert regions with scorching climates, such as Saudi Arabia, which has an average annual temperature of about 43.4 °C.

Several approaches have been used to alleviate the adverse effects of heat stress [2–8,10]. One of these is the dietary fortification of betaine [12–16]. Betaine (Trimethyl glycine) naturally occurs in various plants, and it can be extracted from beetroot [12]. It is a key source of the methyl group (CH_3) in poultry additives [1,17], and it is created by choline and glycine in cell mitochondria [12,16]. Betaine plays two valuable roles in the metabolism of the animals [18,19]. Its initial role is to provide methyl activists, and its other role is linked to its zwitterionic feature of Osmo protectant or osmolyte, which aids in preserving cell water homeostasis [20,21] without influencing cell metabolism. Osmolytes, which act as chaperones, stop or postpone stress-induced denaturation and aid previously denatured proteins [22,23]. Additionally, the methyl group is fundamental in the metabolism of poultry species, as they cannot create the methyl group and therefore require it in their diets [18,24]. Betaine is known to amplify patient sensitivity to hypertonic heat stress and reduce the demand for inducible heat-shock protein expression [17,23]. The recommended dose of betaine for broiler chickens is disputed in the literature and requires further verification, particularly under normal and heat-stress conditions. Therefore, this study aimed to determine the effect of different dietary betaine fortifications on performance, carcass traits, meat quality, blood biochemistry, and hematology of broilers exposed to various temperature patterns after 18 days of age.

2. Materials and Methods

2.1. Study Ethics and Area

The research was conducted according to the ethics and guidelines of the King Abdulaziz City for Science and Technology under project no. LGP—35-269 at the Research Station of the King Abdulaziz City for Science and Technology, Al-Muzahmiyya, which is about 50 km away from Riyadh City, Suadi Arabia.

2.2. Birds and Experimental Design

During the starter period, a total of 480 1-day-old broiler chicks (Ross 308) were allotted to 96 cages (60 W × 50 D × 40 H cm), with five animals per cage. Isocaloric and isonitrogenous starter (1–18 d) and grower-finisher (19–40 d) diets, in mashed form, were formulated according to the Ross 308 recommendation guide (Table 1). During the starter period (1–18 d), the animals were divided into four experimental groups, with each consisting of 24 replicates of five chicks each: the control group feed was a supplemented basal diet, while the other three groups received increasing levels of betaine (0.075%, 0.10%, and 0.15%). Betaine is an anhydrous betaine (96%), and it is a product of Selko feed additives, which is a part of Trouw Nutrition, belonging to Nutreco Company, Rotterdam, the Netherlands. Selko's recommended dose of betaine is 1–2 g/kg for complete feed.

At 18 days of age and until the slaughter age (40 d), birds were placed in two different rooms with 48 cages per room and 12 cages per group in each room. One room was set as thermoneutral (TN), and the other was that of high-temperature (HT) treatment. For the first week, both rooms were on a 24 h light schedule. Then, they were maintained on a light–dark cycle of 20 h: 4 h. An HT cycle was applied in the HT room. The daily temperature of the HT room fluctuated between 24 and 35 °C. However, the TN room temperature was kept at 24 °C during the remaining part of the tested period. The HT chicks were exposed to 35 °C between 8:00 to 15:00. Then, the temperature was gradually decreased to 24 °C. The average outdoor temperature and relative humidity during the experimental period were 35 ± 4 and 31 ± 5, respectively.

Table 1. Dietary ingredients and chemical composition of starter (1–18 days of age) and finisher (19–40 days of age) diets.

Ingredients	1–18 d	19–40 d
	Basal Starter	Basel Finisher
	%	
Corn	59.105	62.00
Soybean meal	27.15	25.35
Corn gluten meal	6.70	4.275
Corn oil	2.53	3.72
Dicalcium phosphate	2.20	2.03
Limestone	0.70	0.57
Salt	0.40	0.40
VM Mix [1]	0.50	0.50
DL-Methionine	0.20	0.76
Lysine-HCL	0.33	0.24
Threonine	0.135	0.105
Choline chloride	0.05	0.05
Total	100	100
Calculated [1] and determined [2] composition		
ME, kcal/kg [1]	3050	3150
Crude protein, % [2]	22.0	21.3
Crude fat,% [2]	35.3	34.6
Crude fiber,% [2]	35.5	46.2
Crude ash,% [2]	60.3	54.2
Lysine, % [1]	1.22	1.10
Sulfur amino acids, % [2]	0.89	0.80
Threonine, % [2]	0.80	0.72
Calcium, % [2]	0.94	0.85
Phosphorus, % [2]	0.45	0.42

[1] Vitamin–mineral premix contains the following per kg: vitamin A, 2,400,000 IU; vitamin D, 1,000,000 IU; vitamin E, 16,000 IU; vitamin K, 800 mg; vitamin B1, 600 mg; vitamin B_2, 1600 mg; vitamin B_6, 1000 mg; vitamin B_{12}, 6 mg; niacin, 8000 mg; folic acid, 400 mg; pantothenic acid, 3000 mg; biotin 40 mg; antioxidant, 3000 mg; cobalt, 80 mg; copper, 2000 mg; iodine, 400; iron, 1200 mg; manganese, 18,000 mg; selenium, 60 mg, and zinc, 14,000 mg; 25% lysine; 32% methionine; [2] Determined values.

Thus, during the growing period, the experimental groups of 12 replicates of five birds each received one of the following eight treatments as 4 betaine levels x 2 temperatures factorial arrangement in a completely randomized design. The treatments were as follows: control diet in the TN room, control diet in the HT room, control + 0.075% betaine in the TN room, control + 0.075% betaine in the HT room, control + 0.10% betaine in the TN room, control + 0.10% betaine in the HT room, control + 0.15% betaine in the TN room, and control + 0.15% betaine in the HT room.

2.3. *Measurements*

2.3.1. Performance Measurements

Body weight gain (BWG) and feed intake (FI) per cage/replicate were recorded every 6 days, while the feed conversion ratio was adjusted in accordance with mortality and computed using the feed intake and BWG of each cage. The production efficiency factor (PEF) was calculated weekly by using the following formula: PEF = ((Livability × Live body weight (kg))/((Age in days × FCR)) × 100.

2.3.2. Carcass and Meat Characteristic Measurements

At day 40, 12 birds were randomly selected per treatment to represent all treatment replicates (a total 96 birds) for measuring carcass characteristics and meat quality traits after being fasted overnight. After euthanasia, the jugular vein was cut, and the heads, feathers, nicks, and shanks were cut. Then, the left parts of the carcasses were cut into two quarters of leg and breast, and they were eventually weighed. The yield percentage of each section was obtained on a dressed weight basis. For the small-intestine measurements, the entire empty gastrointestinal tract was removed aseptically, the ceca and small intestine were firstly weighed, and their total length was determined. After that, the small intestine was separated into the jejunum, ileum, and duodenum, and their lengths and weights were determined. The intestine percentage was determined as the ratio between the total intestinal weight (g) and dressed weight (g).

The right and left breasts from each bird were used to determine meat quality. The breast muscle's pH was measured at 10 min and 24 h postmortem by a microprocessor pH-meter (Model pH 211; Hanna Instruments, Woonsocket, RI, USA). The breast muscle temperature was measured at 15 min postmortem, using a portable digital thermocouple (EcoScan Temp JKT; Thermo Scientific, Waltham, MA, USA). Using a chroma meter (CR-400; Konica Minolta, Tokyo, Japan), the L* (lightness), a* (redness), and b* (yellowness) classes for the breast muscles were determined at 15 min and 24 h after slaughtering fin accordance with the Commission International de l'Eclairage (1976) [25]. The breasts were later stored at −80 °C, till analysis, to determine broilers' meat quality traits. The frozen muscles were thawed overnight, at 4 °C, before the examination. Then the myofibril fragmentation index (MFI) of the breast muscle was determined [26], as well as the water-holding capacity (WHC) [27] and its modification [28]. Cooking loss (CL) was measured [29]. Cooked meat samples were then used to evaluate shear force (SF) by following Wheeler et al. [30].

Blood samples were taken via brachial venipuncture from 12 randomly chosen chicks from each treatment and placed into plain tubes for analysis at 35 d of age. The samples were then centrifuged at 5 °C and 3000 rpm for 10 min, and the serum was collected and stored at −80 °C until further analysis. Albumin, protein, globulin, triglycerides (TGs), cholesterol, and glucose analysis was conducted by using enzymatic colorimetric kits that were bought from Bio-diagnostic, Cairo, Egypt. The globulin concentration was recorded as the difference between albumin and protein concentrations.

Hematological parameters were assessed by following Attia and Hassan [2] and Attia et al. [3]. One hundred leukocytes were counted per slide for each bird. The ratio of heterophile to lymphocyte was determined by using differential counts between heterophils and lymphocytes and the respective percentages of each cell.

2.4. Statistical Analysis

Data for the starter period were analyzed by using analysis of variance (ANOVA) in a completely randomized design, using the general linear model (GLM) procedure in Statistical Analysis System (SAS) software [31]. When the means were significantly different t $p < 0.05$, Tukey's test was applied to separate means. For the finisher period, all data were analyzed by using the SAS [31] for a 2 × 4 factorial experiment. The replicate was the experimental unit. Percentage data were log10 transformed before conducting ANOVA. Before running the statistical analyses, the normality of both the error and data distributions were tested by using the Shapiro–Wilk test of normality in SAS® [31]. The ANOVA assumptions were validated according to the random selection of the samples and normal error distribution, and the data distribution and variance homogeneity.

3. Results

3.1. Experimental Diets, and Performance during the Starter Period

The basal experimental diets fed during the starter and finisher periods are displayed in Table 1. The diets were isocaloric and iso nitrogenous and supplemented with different betaine levels 0, 0.075, 0.10 and 0.15%. The diets meet the nutritional requirements of broiler chickens, depending on the starter and finisher periods.

The cumulative FI, BWG, FCR, and PEF of Ross 308 for the starter period are shown in Table 2. No significant differences in FI, BWG, and FCR were found for the cumulative period. However, the PEF was highest ($p < 0.05$) for birds that had received 0.10% betaine compared with their counterparts. The result indicated a little benefit of supplementing the chicken diets with betaine during the starter period when there was no heat stress.

Table 2. Cumulative feed intake (FI), body weight gain (BWG), feed conversion ratio (FCR), and production efficiency factor (PEF) of broiler chickens given experimental diets at 0–18 days of age.

Treatment	Betaine Added	Performance FI (g)	BWG (g)	FCR (g:g)	PEF
1	0.0%	761	577	1.320	267 ^{b}
2	0.075%	761	581	1.311	270 ^{b}
3	0.10%	772	601	1.286	285 ^{a}
4	0.15%	748	576	1.299	271 ^{b}
SEM		8.96	8.21	0.009	4.92
p value		NS	NS	NS	*

^{a,b} Means within a column within the same factor with different superscript letters differ significantly (* $p < 0.05$, NS: not significant). NS, not significant; SEM, standard error of the mean.

3.2. Performance during the Grower-Finisher Period

Performance data for the grower-finisher period (19–40 d) are presented in Table 3. Betaine fortification caused significant differences ($p < 0.05$ and 0.01) in BWG, FCR, and PEF. Birds that received 0.10% betaine gained 84 g of body weight, and their FCR and PEF improved by 4.6 and 24 points, respectively, compared with birds without betaine fortification. The temperature had an evident effect on all performance parameters over the cumulative grower-finisher period. Broilers in the TN treatment consumed more feed (239 g), gained more weight (179 g), converted feed more efficiently (2.6 points), and showed higher FEF (29 points) than those in the HT treatment (Table 3). However, the interaction effect between betaine and temperature was insignificant on the growth performance indices.

Table 3. The cumulative effect of betaine fortifications and a heat challenge of 35 °C on feed intake (FI), live weight gain (BWG), feed conversion ratio (FCR), and production efficiency factor (PEF) of broiler chickens during 19–40 days of age.

Treatment	Betaine	Temperature	Performance FI (g)	BWG (g)	FCR (g:g)	PEF
	Interaction between betaine and temperature					
1	0.0%	TN	2981	1851	1.611	373
2	0.0%	HT	2704	1658	1.635	346
3	0.075%	TN	2887	1853	1.559	384
4	0.075%	HT	2706	1705	1.589	363
5	0.10%	TN	3004	1917	1.570	397
6	0.10%	HT	2781	1758	1.583	370
7	0.15%	TN	2972	1906	1.561	393
8	0.15%	HT	2697	1690	1.597	352
SEM±			35.84	30.05	0.014	7.76
Betaine average						
0.0%			2843	1754 ^{b}	1.623 ^{a}	359 ^{b}
0.075%			2797	1779 ^{ab}	1.574 ^{b}	374 ^{ab}
0.10%			2893	1838 ^{a}	1.577 ^{b}	383 ^{a}
0.15%			2834	1798 ^{ab}	1.579 ^{b}	372 ^{ab}
Temperature average						
Temperature average						
TN			2961 ^{a}	1882 ^{a}	1.575 ^{a}	387 ^{a}
HT			2722 ^{b}	1703 ^{b}	1.601 ^{b}	358 ^{b}
Statistical probabilities						
Betaine			NS	*	**	*
Temperature			***	***	*	***
Betaine × Temperature			NS	NS	NS	NS

^{a,b} Means within a column within the same factor with different superscripts differ significantly (* $p < 0.05$, ** $p < 0.01$, *** $p < 0.001$; NS, not significant). SEM: standard error of the mean.

3.3. Body Parts and Meat Characteristics

The mean percentage yield of carcass parts in different treatments at 40 d of age is documented in Table 4. The betaine level did not cause any differences in the percentage yield of carcass parts. However, temperature affected the breast yield ($p < 0.01$) and fat percentage ($p < 0.05$). The broilers that were exposed to TN had higher breast yields and lower fat percentage than the HT group. A two-way interaction (Betaine x Temperature) was significant ($p < 0.05$) for the relative intestine weight (IRW). For the TN treatment, the 0.10% betaine level significantly decreased the IRW compared with the unsupplemented TN group. Under HT conditions, betaine fortifications did not affect IRW. In the 0.10% betaine and HT treatment, the IRW was significantly higher than TN treatment.

Table 4. Effect of betaine fortifications and temperature on carcass part yields as percentages of broiler dressed weight of 40-day-old broiler chickens.

Treatment	Betaine	Temperature	Dressing (%)	Breast (%)	Leg (%)	Fat (%)	Liver (%)	IRW (%)
		Interaction between betaine and temperature						
1	0.0%	TN	72.5	39.5	29.9	1.7	2.3	3.8 ^{a}
2	0.0%	HT	72.9	38.4	30.5	2.3	2.2	3.5 ^{ab}
3	0.075%	TN	72.9	39.5	30.2	1.7	2.2	3.6 ^{ab}
4	0.075%	HT	72.9	38.5	30.1	1.9	2.3	3.7 ^{a}
5	0.10%	TN	73.5	40.0	28.9	1.8	2.3	3.4 ^{b}
6	0.10%	HT	72.9	38.0	30.8	1.7	2.1	3.8 ^{a}
7	0.15%	TN	73.1	40.4	29.8	1.7	2.2	3.6 ^{ab}
8	0.15%	HT	73.3	40.1	29.6	2.0	2.0	3.5 ^{ab}
SEM±			0.38	39.5	0.52	0.16	0.10	0.11
	Betaine average							
0.0%			72.7	38.9	30.2	2.0	2.3	3.6
0.075%			73.0	39.0	30.1	1.8	2.2	3.7
0.10%			73.2	39.0	29.8	1.7	2.2	3.6
0.15%			73.2	40.3	29.7	1.8	2.1	3.5
	Temperature average							
TN			73.0	39.9 ^{a}	29.7	1.7 ^{b}	2.2	3.6
HT			73.0	38.7 ^{b}	30.3	2.0 ^{a}	2.1	3.6
	Statistical probabilities							
	Betaine		NS	NS	NS	NS	NS	NS
	Temperature		NS	**	NS	*	NS	NS
Betaine × Temperature			NS	NS	NS	NS	NS	*

^{a,b} Means within a column within the same factor with different superscripts differ significantly (* $p < 0.05$, ** $p < 0.01$; NS, not significant). SEM: standard error of the mean.

3.4. Breast Quality Characteristics

Betaine did not affect the pH value when measured 10 min postmortem (Table 5). However, the temperature did: chicken breasts from the TN group had a higher pH (6.57; $p < 0.01$) than those from the HT group (6.49). Breast temperature was altered ($p < 0.001$) by the betaine level, temperature, and their interaction. Higher temperatures were obtained from the breast meat of HT-exposed broilers fed 0.10% and 0.15% betaine, while the lowest temperature was found for the unsupplemented HT group.

Yellowness (b*) and lightness (L*) values of the breast changed ($p < 0.05$) with temperature 10 min postmortem (Table 5). Higher (b*) and (L*) values were obtained for the TN group than the HT group. The values of a* (redness) were impacted ($p < 0.05$) by the betaine x temperature interaction. The results showed that the redness of meat was higher for the breast of HT-exposed broilers fed 0.15% betaine, while the lowest redness score was found for the unsupplemented, 0.15% betaine-supplemented TN group, and 0.075 betaine-supplemented HT group. The (L*) score was higher ($p < 0.05$) for the TN group than the HT group 24 h postmortem (Table 5).

Table 5. Effect of betaine fortifications and temperature on meat parameters of 40-day-old broiler chickens.

Treatment	Betaine	Temperature	pH-10 min.	Temp.-10 min. (°C)	Color-10 min.			pH-24 h	Color-24 h			WHC	MFI	CL	SF (Kgf/g)
					L*	a*	b*		L*	a*	b*				
Interaction between betaine and temperature															
1	0.0%	TN	6.60	31.9 ^{b}	47.5	3.64 ^{b}	8.67	6.44 ^{a}	50.9	3.83	9.13	0.24	62.1	39.7	1.57
2	0.0%	HT	6.48	29.2	44.7	4.44 ^{ab}	7.90	6.34 ^{c}	47.1	4.60	10.31	0.23	57.2	35.7	1.81
3	0.075%	TN	6.56	31.8 ^{b}	45.4	4.02 ^{ab}	8.87	6.37 ^{bc}	48.8	4.22	11.10	0.22	63.9	35.1	1.28
4	0.075%	HT	6.52	30.5 ^{c}	45.9	3.24 ^{b}	8.54	6.38 ^{bc}	49.1	3.78	10.32	0.19	65.0	30.6	1.50
5	0.10%	TN	6.61	32.2 ^{b}	48.8	4.09 ^{ab}	8.30	6.40 ^{ab}	49.4	4.06	10.55	0.23	58.8	30.7	1.99
6	0.10%	HT	6.45	33.1 ^{a}	46.2	3.69 ^{ab}	7.16	6.35 ^{c}	47.3	5.00	9.48	0.22	52.9	31.3	1.72
7	0.15%	TN	6.50	31.7 ^{b}	47.8	3.14 ^{b}	9.71	6.35 ^{c}	48.3	3.92	11.59	0.23	62.3	33.9	1.63
8	0.15%	HT	6.49	32.9 ^{a}	45.4	4.97 ^{a}	8.25	6.38 ^{bc}	46.2	5.13	8.77	0.21	63.6	28.7	1.54
SEM±			0.03	0.18	1.11	0.46	0.55	0.018	1.23	0.47	0.075	0.014	2.31	1.64	0.13
Betaine average															
	0.0%		6.54	30.6 ^{c}	46.1	4.03	8.29	6.39	49.0	4.21	9.72	0.24	59.7 ^{bc}	36.2 ^{a}	1.69 ^{ab}
	0.075%		6.54	31.1 ^{b}	45.7	3.63	8.71	6.37	48.9	3.99	10.71	0.21	64.5 ^{a}	32.8 ^{b}	1.39 ^{c}
	0.10%		6.53	32.6 ^{a}	47.5	3.89	7.74	6.38	48.4	4.53	10.01	0.22	55.8 ^{c}	31.0 ^{b}	1.86 ^{a}
	0.15%		6.49	32.3 ^{a}	46.6	4.06	8.98	6.36	47.2	4.53	10.18	0.22	62.9 ^{ab}	31.3 ^{b}	1.59 ^{bc}
Temperature average															
		TN	6.57 ^{a}	31.9 ^{a}	47.4 ^{a}	3.72	8.89 ^{a}	6.39 ^{a}	49.3 ^{a}	4.00	10.59	0.23	61.8	34.1 ^{a}	1.62
		HT	6.49 ^{b}	31.4 ^{b}	45.5 ^{b}	4.08	7.97 ^{b}	6.36 ^{b}	47.4 ^{b}	4.62	9.72	0.22	59.7	31.6 ^{b}	1.64
Statistical probabilities															
Betaine			NS	***	NS	NS	NS	NS	NS	NS	NS	NS	**	**	**
Temperature			**	***	*	NS	*	*	*	NS	NS	NS	NS	*	NS
Betaine × Temperature			NS	***	NS	*	NS	**	NS	NS	NS	NS	NS	NS	NS

pH-10, Temp.-10, and color-10 min = pH, temperature, and color measurements of the breast 10 min postmortem, respectively; pH-24 h, Temp.-24 h, and color-24 h = pH, temperature, and color measurements of the breast 24 h post mortem, respectively; CL, cooking loss; WHC, water-holding capacity; MFI, myofibril fragmentation index; SF, shearing force. ^{abc} The means within a coulumn within the same factor with different superscripts differ significantly (* $p < 0.05$, ** $p < 0.01$, *** $p < 0.001$; NS, not significant). SEM: standard error of the mean.

Betaine did not affect the pH value when measured at 24 h postmortem (Table 5). Breasts from the TN group had a higher pH (6.39) than the HT group (6.36). The pH value at 24 h was influenced ($p < 0.01$) by the betaine x temperature interaction. After 24 h, the breast pH was higher for NS-exposed broilers fed an unsupplemented diet than most of the experimental groups, except for the TN group fed a 0.010% betaine-supplemented diet. Betaine supplementation at 0.10% decreased the breast pH of the HT group compared with the TN group.

The WHC was not influenced by the examined factors. However, the MFI increased in the 0.075% and 0.15% betaine treatments compared with the other groups. The percentage CL was highest ($p < 0.01$) for the control group breasts (Table 5). Additionally, the CL was higher ($p < 0.05$) in the TN group than the HT group. The lowest SF has observed in the breasts of birds fed 0.075% betaine, while the highest SF was obtained for the chickens that consumed 0.0% and 0.10% betaine (Table 5).

3.5. Blood Biochemical and Hematological Parameters

Table 6 shows the effect of the applied treatments on some serum biochemical parameters of broilers. Betaine supplementation changed ($p < 0.05$) only TGs and albumin concentrations. All serum biochemicals, except albumin, varied by temperature ($p < 0.05$), and all previous parameters were lower in the TN group than the HT group, except for glucose. The two-way interaction (Betaine × Temperature) was only significant ($p < 0.05$) for glucose concentration (Table 6). The results showed that betaine supplementation at 0.15% to the HT group increased the serum glucose level compared with the other groups.

Table 6. Effect of betaine fortifications and temperature on pooled blood parameters of 35-day-old broiler chickens.

Treatment	Betaine	Temperature	Glucose (mg/dL)	TGs [1] (mg/dL)	Cholesterol (mg/dL)	TP [2] (g/dL)	Albumin (g/dL)	Globulin (g/dL)	A:G [3]
Interaction between betaine and temperature									
1	0.0%	TN	191.0 [c]	78.9	95.0	2.9	1.7	1.2	1.6
2	0.0%	HT	218.4 [b]	71.2	102.0	3.1	1.8	1.3	1.6
3	0.075%	TN	210.2 [bc]	72.6	89.5	2.7	1.7	1.0	1.9
4	0.075%	HT	221.6 [b]	63.2	92.4	3.1	1.5	1.6	1.1
5	0.10%	TN	202.6 [bc]	68.2	90.6	3.0	1.7	1.3	1.5
6	0.10%	HT	198.7 [bc]	64.2	95.8	3.1	1.8	1.3	1.4
7	0.15%	TN	198.7 [bc]	73.3	92.0	3.0	1.7	1.3	1.5
8	0.15%	HT	242.9 [a]	63.3	95.5	2.9	1.5	1.5	1.2
SEM±			8.3	3.33	3.02	0.09	0.08	0.11	0.19
Betaine average									
0.0%			204.7	75.0 [a]	98.5	3.0	1.8 [a]	1.2	1.6
0.075%			215.9	67.9 [b]	91.0	2.9	1.6 [ab]	1.3	1.5
0.10%			200.6	66.2 [b]	93.2	3.1	1.7 [ab]	1.3	1.5
0.15%			220.8	68.3 [b]	93.8	2.9	1.6 [b]	1.4	1.3
Temperature average									
TN			200.6 [b]	73.3 [a]	91.8 [b]	2.9 [b]	1.7	1.2 [b]	1.6 [a]
HT			220.4 [a]	65.5 [b]	96.0 [a]	3.1 [a]	1.6	1.4 [a]	1.3 [b]
Statistical probabilities									
Betaine			NS	*	NS	NS	*	NS	NS
Temperature			**	**	*	*	NS	**	*
Betaine × Temperature			*	NS	NS	NS	NS	NS	NS

[a,b,c] Means within a column within the same factor with different superscripts differ significantly (* $p < 0.05$, ** $p < 0.01$; NS, not significant). SEM, standard error of the mean; [1] TGs, triglycerides, [2] TP, total protein, [3] A:G ratio, albumin to globulin ratio.

However, the lowest glucose level was for the TN group's unsupplemented control. Within betaine levels, HT increased the glucose level in the unsupplemented control and the 0.15% betaine-supplemented group. There were no changes in serum glucose within the other betaine levels due to temperature.

Table 7 shows the effect of the different treatments on the studied hematological parameters of the broilers. Heterophils, lymphocytes, and their ratio (H:L ratio) were altered ($p < 0.001$) by the HT treatment. The number of heterophils and the H:L ratio were lower in the TN group than the HT group, while the number of lymphocytes was higher in the HT group. There were no changes in the hematological blood characteristics due to the interaction between temperature and dietary betaine levels.

Table 7. Effect of betaine fortifications and temperature on pooled blood parameters of 35-day-old broiler chickens.

Treatment	Betaine	Temperature	Hetophile (%)	Lymphocyte (%)	H:L [1] Ratio
Interaction between Betaine and temperature					
1	0.0%	TN	12.0	88.0	0.14
2	0.0%	HT	27.0	71.0	0.42
3	0.075%	TN	14.0	86.0	0.16
4	0.075%	HT	29.0	70.0	0.44
5	0.10%	TN	15.0	85.0	0.18
6	0.10%	HT	28.0	71.0	0.42
7	0.15%	TN	12.0	87.0	0.14
8	0.15%	HT	25.0	73.0	0.36
SEM±			2.10	2.33	0.05
Betaine average					
0.0%			20.0	79.0	0.28
0.075%			21.0	78.0	0.30
0.10%			21.0	78.0	0.30
0.15%			19.0	80.0	0.25
Temperature average					
TN			13.0 ^{b}	86.0 ^{a}	0.16 ^{b}
HT			27.0 ^{a}	71.0 ^{b}	0.41 ^{a}
Statistical probabilities					
Betaine			NS	NS	NS
Temperature			***	***	***
Betaine × Temperature			NS	NS	NS

^{a,b} The means within a column within the same factor with different superscripts differ significantly (*** $p < 0.001$; NS, not significant). SEM, standard error of the mean; [1] H:L ratio, heterophil to lymphocyte ratio.

4. Discussion

Alleviating heat stress improves broiler performance, and determining the effectiveness of using dietary betaine supplementation to alleviate heat stress was the primary goal of the present trial. Through the first 18 days of age, in the absence of heat stress, dietary supplementation of betaine at 0.10% improved the PEF without affecting the FI, BWG, and FCR, which agrees with Uzunoğlu and Yalçin [16]. We found that 0.10% betaine supplementation was adequate for broiler chickens of 1–18 days of age. It should be mentioned that the experimental diets fed containing adequate nutrients, particularly methionine and choline, to avoid nutritional deficiency and declare the impact of betaine without, confound results with methyl donor agents (1).

It should be mentioned that, although several studies have been reported previously in the same topic, confirmation studies are needed in different parts of the world due to different enviromental conditions among various regions of the world. In our study, heat stress impaired growth performance and caused several changes in the metabolic and antioxidants status of the chickens, as also indicated for other poultry species [2–6]. It is known that the response to heat stress is influenced by exposure time, the temperature-humidity index, strain, and breed [4,7,8,10,32]. The current work exhibited reduced FI and impaired FCR in the HT group during the grower-finisher period, demonstrating that the birds did not adapt to temperatures that were above optimal. Heat stress is augmented by radiating temperature loss through the relocation of blood from the center to the periphery of the body, where it can be released to the environment [2–7]. This relocation of the blood during heat stress possibly causes hypoxia and tissue harm inside the gastrointestinal tract [22], influencing the digestive and absorptive system of the body and possibly compromising feed efficiency. Additionally, damage to the intestinal barrier can cause a gateway for pathogens, increasing the risk of disease [33,34].

One chief outcome of the current work was that betaine supplementation at 0.10% enhanced BWG, FCR, and PEF in broilers exposed to heat stress, suggesting that betaine may be a useful antistress nutritional tool. It should be noticed that betaine performs a fundamental function by controlling the cellular osmotic atmosphere during heat stress, reducing dryness by intensifying the WHC of the cell, and improving antioxidant balance [21,22,35]. Furthermore, betaine acts as a methyl group contributor, which is beneficial for the creation of protein in the body [1,17,23]. As an osmolyte, betaine lets proteins preserve their conformational steadiness in the existence of high uric acid levels and variations in cell salinity. Numerous tissues rely on betaine as an osmolyte, and it is found in the kidneys, brain, liver, intestines, and leukocytes [12].

Dietary betaine supplementation at three different concentrations (0.10%, 0.15%, and 0.20%) significantly improved FI, BWG, and FCR in heat-stressed broilers, especially at higher betaine levels [36]. Dietary betaine supplementation augmented BWG and enhanced FCR during 21–42 days of age, and they had no significant impact on the FI of broilers, as indicated by Shakeri et al. [22]. In growing rabbits, betaine enhanced BW and significantly boosted the FCR [37] in heat-stress environments. The incorporation of betaine in broiler feeds was efficient at advancing their performance [12,19]. However, Nutautaitė [17] reported that dietary betaine supplementation could not change BW and BWG, but it altered FCR in broilers. In contrast to our results, Uzunoğlu and Yalçin [16] showed that betaine dietary supplementation did not affect the growth performance indices of broilers, which could be due to different experimental conditions and the supplemented dose of betaine between the two studies. Different experimental results suggested the need for further experiments to escape the gap in contradictory results due to different levels of betaine used, animal species and environmental conditions, and composition of the experimental diets.

Our results showed that betaine supplementation did not affect broiler carcass traits, but heat stress decreased breast-meat percentage and increased fat percentages. It was imagined that dietary betaine would affect carcass weight and the relative weights of different carcass parts depending on its methyl group. This would boost methionine and cystine availability for protein synthesis [1,38] and glycine for protein creation. Similarly, it would contribute to decreasing fat precipitation in the body across numerous metabolic means [12]. However, this mode of action was not recorded herein, which agrees with Nutautaitė [17] and is contradictory to Nofal et al. [39]. In this regard, Attia et al. [32] stated that heat stress decreased the percentage of dressed carcasses, liver, and giblets. The same authors suggested that supplementing diets with 1 g betaine/kg caused complete retrieval of the adverse consequences. In contrast, El-Shinnawy [19] revealed that broilers that consumed betaine-supplemented diets had increased carcass and carcass part yield percentages.

In our study, the HT treatment decreased breast muscle colors L* and b*, CL, and pH after 24 h. Betaine supplementation changed the MFI, CL, and shearing force, but it did not affect the WHC and breast pH after 24 h. These findings contrast with those by Zhang et al. [40], who reported that breast meat from broilers raised under high temperature exhibited greater L* and CL than those reared under normal temperatures. The latter authors demonstrated that high temperatures could encourage rigor mortis, causing a quicker pH decline and higher L* value, leading to whitish, exudative meat traits [40]. Comparable results were also reported by Attia et al. [32], who indicated that heat stress and dietary betaine supplementation did not affect meat pH and color. In contrast, betaine addition at 0.5 g/kg diet lessened the WHC when compared with the control group, which could be due to increased water retention and osmolality of the cells. Nutautaitė et al. [17]

showed that betaine dietary supplementation did not influence breast-muscle colors L*, b*, and the WHC in broilers. Additionally, broilers fed diets supplemented with 0.10% and 0.20% betaine had the highest CL, while birds consuming the control diet had the highest SF [17]. Additionally, Liu [41] found comparable findings to ours. In their trial, which aimed to reduce the impacts of transport stress on broilers, Chen et al. [24] reported that carcasses of broilers fed diets supplemented with betaine (0.05% or 0.10%) showed similar results for pH and CL compared with the control. Moreover, Al-Tamimi et al. [23] found that betaine addition did not alter the pH, CL, and shearing force of broiler carcass meat.

The escalation in blood albumin attributable to betaine incorporation was linked with its capability as a methyl group contributor, which is necessary for protein metabolism [1,36] and its role in improving immunity [16]. Similar outcomes were observed by Uzunoğlu and Yalçin [16], who stated that dietary betaine supplementation significantly reduced TGs without affecting the other blood parameters studied. Attia et al. [32] found that there was a decline in plasma glucose and total protein due to heat stress, although their results for plasma TGs were contradictory. Additionally, it was found that supplementing 1 g of betaine/kg diet resulted in complete relief in plasma glucose and incomplete relief in TGs and total protein [32]. Attia et al. [42] concluded that plasma glucose decreased in laying hens that were raised under heat-stress conditions and consumed a diet with no supplementation compared with their counterparts. Furthermore, El-Shinnawy [19] indicated that broilers fed diets supplemented with betaine had higher levels of TGs than those in the control group. In the present results, heat-stress impacted levels of glucose, TGs, cholesterol, and total protein. Similar findings were also reported by Ghasemi and Nari [42], who found significant differences in these biochemicals due to heat stress. They stated that dietary betaine supplementation did not impact the investigated blood biochemicals, except TGs [43].

Hematological parameters were negatively affected by the HT treatment, showing an increased number of heterophils, decreased number of lymphocytes, and an increased stress index (heterophil/lymphocyte ratio). However, there was no interaction between heat stress and betaine, which agrees with Abdelsattar et al. [44], who showed that betaine-supplemented diets in growing lambs receiving saline water did not cause a change in lymphocyte numbers.

The results of the heat-stress period (19–40 days of age) revealed that the heat challenge of 35 °C had a powerful effect on performance, meat characteristics, and blood parameters. Betaine fortification during days 19–40 of age during heat challenge reduced the negative impact on performance and improved production efficiency. This indicates that the supplementation of 0.10% betaine during the heat challenge was adequate for broilers. Further studies are required to test betaine's ability, especially with different levels of methionine and choline, to reduce methionine supplementation and make betaine supplementation cost-effective.

Among the four tested levels of betaine, the practical level was found to be 0.1%, suggesting that betaine may be a useful nutritional tool under different stress conditions that deserve further investigation.

5. Conclusions

The best performance was achieved at 0.1% betaine fortifications with 84 g gain, 4.6 points improvement in FCR, and 24 points improvement in PEF compared to no betaine supplementation. The heat-stressed group consumed less feed (239 g), gained less weight (179 g), converted feed less efficiently (2.6 points), and, as a result, had lower FEF (29 points) as compared to the TN group. Conclusively, heat challenge had a powerful effect on growth performance, meat characteristics, and blood parameters, especially during the grower-finisher period. Betaine fortification (0.1%) during heat challenge reduced the negative impact on performance and improved production efficiency. As an antistress agent, the role of betaine should be addressed in further research and explore its application benefits.

Author Contributions: A.A.A.-S., experimental design, field work, and 1st draft of the manuscript; A.H.A.-Y., A.A.A.-A., and A.M.A., data collection, statistical analyses, and contribution to experimental setup; E.O.H., lab work and proofreading of the manuscript. All authors have read and agreed to the published version of the manuscript.

Funding: This article was funded by King Abdulaziz City for Science and Technology (KACST) under project number LGP-35-269. Therefore, the authors appreciate the technical and financial support of the DSR.

Institutional Review Board Statement: The research was approved on 15 September 2015 under protocol LG-35-269 according to the ethics and guidelines of King Abdulaziz City for Science and Technology.

Data Availability Statement: All data are presented in the text and tables of this manuscript.

Acknowledgments: The authors express their sincere gratitude and appreciation to King Abdulaziz City for Science and Technology (KACST) for allowing the use of its facilities to conduct this study.

Conflicts of Interest: The authors declare no conflict of interest.

References

1. Attia, Y.A.; Hsaan, R.A.; Shehatta, M.H.; Abd El-Hady, S. Growth, carcass quality and serum constituents of slow growing chicks as affected by betaine addition to diets containing 2 different levels of methionine. *Int. J. Poult. Sci.* **2005**, *4*, 856–865.
2. Attia, Y.A.; Hassan, S.S. Broiler tolerance to heat stress at various dietary protein/energy levels. *Eur. Poult. Sci.* **2017**, *81*. [CrossRef]
3. Attia, Y.A.; Al-Harthi, M.A.; Elnaggar, A.S. Productive, physiological and immunological responses of two broiler strains fed different dietary regimens and exposed to heat stress. *Ital. J. Anim. Sci.* **2018**, *17*, 686–697. [CrossRef]
4. Farghly, M.F.A.; Mahrose, K.M.; Galal, A.E.; Ali, R.M.; Ahmad, E.A.M.; Rehman, Z.U.; Ullah, Z.; Ding, C. Implementation of different feed withdrawal times and water temperatures in managing turkeys during heat stress. *Poult. Sci.* **2018**, *97*, 3076–3084. [CrossRef]
5. Farghly, M.F.A.; Mahrose, K.H.M.; Mahmoud, G.B.; Ali, R.M.; Daghash, W.; Metwally, K.A.; Abougaba, M.S. Lighting programs as an appliance to improve growing New Zealand white rabbit's performance. *Int. J. Biometeorol.* **2020**, *64*, 1295–1303. [CrossRef]
6. Farghly, M.F.A.; Mahrose, K.M.; Ullah, Z.; Rehman, Z.U.; Ding, C. Influence of swimming time in alleviating the deleterious effects of hot summer on growing Muscovy duck performance. *Poult. Sci.* **2017**, *96*, 3912–3919. [CrossRef]
7. Rizk, Y.S.; Fahim, H.N.; Beshara, M.M.; Mahrose, K.M.; Awad, A.L. Response of duck breeders to dietary L-Carnitine supplementation during summer season. *An. Acad. Bras. Ciências* **2019**, *91*, e20180907. [CrossRef]
8. Awad, A.; Fahim, H.; El-Shhat, A.E.; Mahrose, K.H.; Shazly, S. Dietary *Echinacea purpurea* administration enhanced egg laying performance, serum lipid profile, antioxidant status and semen quality in duck breeders during summer season. *J. Anim. Physiol. Anim. Nutr.* **2021**. [CrossRef] [PubMed]
9. De Silva, P.; Kalubowila, A. Influence of feed withdrawal for three hour time period on growth performance and carcass parameters of later stage of male broiler chickens. *Iran. J. Appl. Anim. Sci.* **2012**, *2*, 191–197.
10. Hassan, F.A.; Mahrose, K.M.; Basyony, M.M. Effects of grape seed extract as a natural antioxidant on growth performance, carcass characteristics and antioxidant status of rabbits during heat stress. *Arch. Anim. Nutr.* **2016**, *70*, 141–154. [CrossRef]
11. Abd El-Hack, M.E.; Mahrose, K.H.M.; Arif, M.; Chaudhry, M.T.; Saadeldin, I.M.; Saeed, M.; Soomro, R.N.; Abbasi, I.H.R.; Rehman, Z.U. Alleviating the environmental heat burden on laying hens by feeding on diets enriched with certain antioxidants (vitamin E and selenium) individually or combined. *Environ. Sci. Pollut. Res.* **2017**, *24*, 10708–10717. [CrossRef]
12. Sakomura, N.K.; Barbosa, N.A.A.; Longo, F.A.; da Silva, E.P.; Bonato, M.A.; Fernandes, J.B.K. Effect of dietary betaine supplementation on the performance, carcass yield, and intestinal morphometrics of broilers submitted to heat stress. *Braz. J. Poult. Sci.* **2013**, *15*, 105–112. [CrossRef]
13. Ratriyanto, A.; Indreswari, R.; Nuhriawangsa, A.M.P. Effects of dietary protein level and betaine supplementation on nutrient digestibility and performance of Japanese quails. *Braz. J. Poult. Sci.* **2017**, *19*, 445–454. [CrossRef]
14. Attia, Y.A.; El-Hamid, A.E.A.; Abdallah, A.A.; Berika, M.A.; El-Gandy, M.F.; Sahin, K.; Abou-Shehema, B.M. Effect of betaine, vitamin C, and vitamin E on egg quality, hatchability, and markers of liver and renal functions in dual-purpose breeding hens exposed to chronic heat stress. *Eur. Poult. Sci.* **2018**, 82. [CrossRef]
15. Attia, Y.A.; El-Naggar, A.S.; Abou-Shehema, B.M.; Abdella, A.A. Effect of Supplementation with Trimethylglycine (Betaine) and/or Vitamins on Semen Quality, Fertility, Antioxidant Status, DNA Repair and Welfare of Roosters Exposed to Chronic Heat Stress. *Animals* **2019**, *9*, 547. [CrossRef] [PubMed]
16. Uzunoğlu, K.; Yalçin, S. Effects of dietary supplementation of betaine and sepiolite on performance and intestinal health in broilers. *Ank. Üniv. Vet. Fak. Derg.* **2019**, *66*, 221–229.
17. Nutautaitė, M.; Alijošius, S.; Bliznikas, S.; Šašytė, V.; Vilienė, V.; Pockevičius, A.; Racevičiūtė-Stupelienė, A. Effect of betaine, a methyl group donor, on broiler chicken growth performance, breast muscle quality characteristics, oxidative status and amino acid content. *Ital. J. Anim. Sci.* **2020**, *19*, 621–629. [CrossRef]

18. Summers, J.D. Effect of Choline or Betaine Supplementation on Broilers Exposed to Different Temperature Treatments. Master's Thesis, University of Tennessee, Knoxville, TN, USA, 2013.
19. El-Shinnawy, A.M. Effect of betaine supplementation to methionine adequate diet on growth performance, carcass characteristics, some blood parameters and economic efficiency of broilers. *J. Anim. Poult. Prod.* **2015**, *6*, 27–41. [CrossRef]
20. Willingham, B.D.; Ragland, T.J.; Ormsbee, M.J. Betaine supplementation may improve heat tolerance: Potential mechanisms in humans. *Nutrients* **2020**, *12*, 2939. [CrossRef]
21. Wasti, S.; Sah, N.; Mishra, B. Impact of heat stress on poultry health and performances, and potential mitigation strategies. *Animals* **2020**, *10*, 1266. [CrossRef]
22. Shakeri, M.; Cottrell, J.J.; Wilkinson, S.; Ringuet, M.; Furness, J.B.; Dunshea, F.R. Betaine and antioxidants improve growth performance, breast muscle development and ameliorate thermoregulatory responses to cyclic heat exposure in broiler chickens. *Animals* **2018**, *8*, 162. [CrossRef] [PubMed]
23. Al-Tamimi, H.; Mahmoud, K.; Al-Dawood, A.; Nusairat, B.; Bani Khalaf, H. Thermotolerance of broiler chicks ingesting dietary betaine and/or creatine. *Animals* **2019**, *9*, 742. [CrossRef] [PubMed]
24. Chen, R.; When, C.; Gu, Y.; Wang, C.; Chen, Y.; Zhuang, S.; Zhou, Y. Dietary betaine supplementation improves meat quality of transported broilers through altering muscle anaerobic glycolysis and antioxidant capacity. *J. Sci. Food Agric.* **2020**, *100*, 2656–2663. [CrossRef]
25. Commission International de l'Eclairage. Official Recommendations on Uniform Colour Space, Colour Difference Equations and Metric Colour Terms. *Color Res. Appl.* **1977**, *2*. [CrossRef]
26. Culler, R.D.; Parrish, F.C., Jr.; Smith, G.C.; Cross, H.R. Relationship of myofibril fragmentation index to certain chemical, physical and sensory characteristics of bovine longissimus muscle. *J. Food Sci.* **1978**, *43*, 1177–1180. [CrossRef]
27. Hamm, R. Biochemistry of meat hydration. *Adv. Food Res.* **1960**, *10*, 355–463. [PubMed]
28. Wilhelm, A.E.; Maganhini, M.B.; Hernández-Blazquez, F.J.; Ida, E.I.; Shimokomaki, M. Protease activity and the ultrastructure of broiler chicken PSE (pale, soft, exudative) meat. *Food Chem.* **2010**, *119*, 1201–1204. [CrossRef]
29. Attia, Y.A.; Al-Harthi, M.A.; Korish, M.M.; Shiboob, M.M. Evaluation of the broiler meat quality in the retail market: Effects of type and source of carcasses. *Rev. Mex. Cienc. Pecu.* **2016**, *7*, 321–339. [CrossRef]
30. Wheeler, T.L.; Cundiff, L.V.; Shackelford, S.D.; Koohmaraie, M. Characterization of biological types of cattle (Cycle VII): Carcass, yield, and longissimus palatability traits. *J. Anim. Sci.* **2005**, *83*, 196–207. [CrossRef] [PubMed]
31. SAS Institute. *SAS/SRATI User's Guide, Statistics*; SAS Institute Inc.: Cary, NC, USA, 2004.
32. Attia, Y.A.; Hassan, R.A.; Qota, E.M. Recovery from adverse effects of heat stress on slow-growing chicks in the tropics 1: Effect of ascorbic acid and different levels of betaine. *Trop. Anim. Health Prod.* **2009**, *41*, 807–818. [CrossRef] [PubMed]
33. Keita, Å.V.; Söderholm, J.D. The intestinal barrier and its regulation by neuroimmune factors. *Neurogastroenterol. Motil.* **2010**, *22*, 718–733. [CrossRef]
34. Dunshea, F.R.; Leury, B.J.; Fahri, F.T.; Digiacomo, K.; Hung, A.; Chauhan, S.S.; Clarke, I.J.; Collier, R.J.; Little, S.; Baumgard, L.H.; et al. Amelioration of thermal stress impacts in dairy cows. *Anim. Prod. Sci.* **2013**, *53*, 965–975. [CrossRef]
35. Ratriyanto, A.; Mosenthin, R. Osmoregulatory function of betaine in alleviating heat stress in poultry. *J. Anim. Physiol. Anim. Nutr.* **2018**, *102*, 1634–1650. [CrossRef]
36. Chand, N.; Naz, S.; Maris, H.; Khan, R.U.; Khan, S.; Qureshi, M.S. Effect of betaine supplementation on the performance and immune response of heat stressed broilers. *Pak. J. Zool.* **2017**, *49*, 1857–1862. [CrossRef]
37. Hassan, R.A.; Ebeid, T.A.; Abd El-Lateif, A.I.; Ismail, N.B. Effect of dietary betaine supplementation on growth, carcass and immunity of New Zealand White rabbits under high ambient temperature. *Livest. Sci.* **2011**, *135*, 103–109. [CrossRef]
38. Mcdevitt, R.M.; Mack, S.; Wallis, I.R. Can betaine partially replace or enhance the effect of methionine by improving broiler growth and carcass characteristics. *Br. Poult. Sci.* **2000**, *41*, 473–480. [CrossRef] [PubMed]
39. Nofal, M.E.; Magda, A.; Galal, S.M.M.; Mousa, M.; Doaa, M.M.; Yassein, M.; Bealsh, A.M.A. Effect of dietary betaine supplementation on productive, physiological and immunological performance and carcass characteristic of growing developed chicks under the condition of heat stress. *Egypt. Poult. Sci. J.* **2015**, *35*, 237–259.
40. Zhang, Z.Y.; Jia, G.Q.; Zuo, J.J.; Zhang, Y.; Lei, J.; Ren, L.; Feng, D.Y. Effects of constant and cyclic heat stress on muscle metabolism and meat quality of broiler breast fillet and thigh meat. *Poult. Sci.* **2012**, *91*, 2931–2937. [CrossRef] [PubMed]
41. Liu, W.; Yuan, Y.; Sun, C.; Balasubramanian, B.; Zhao, Z.; An, L. Effects of dietary betaine on growth performance, digestive function, carcass traits, and meat quality in indigenous yellow-feathered broiler under long-term heat stress. *Animals* **2019**, *9*, 506. [CrossRef]
42. Attia, Y.A.; El-Hamid, A.E.-H.E.A.; Abedalla, A.A.; Berika, M.A.; Al-Harthi, M.A.; Kucuk, O.; Sahin, K.; Abou-Shehema, B.M. Laying performance, digestibility and plasma hormones in laying hens exposed to chronic heat stress as affected by betaine, vitamin C, and/or vitamin E supplementation. *Springerplus* **2016**, *5*, 1–12. [CrossRef] [PubMed]
43. Ghasemi, H.A.; Nari, N. Effect of supplementary betaine on growth performance, blood biochemical profile, and immune response in heat-stressed broilers fed different dietary protein levels. *J. Appl. Poult. Res.* **2020**, *29*, 301–313. [CrossRef]
44. Abdelsattar, M.M.; Hussein, A.M.A.; Haridy, M.; Abd El-Ati, M.N.; Saleem, A.M.; Zhang, N. Betaine counteracts the harmful effects of saline water induced to growing lambs. *Egypt. J. Sheep Goat Sci.* **2019**, *14*, 19–31.

animals

MDPI

Article

Myostatin and Related Factors Are Involved in Skeletal Muscle Protein Breakdown in Growing Broilers Exposed to Constant Heat Stress

Xiumei Li, Minhong Zhang *, Jinghai Feng and Ying Zhou

State Key Laboratory of Animal Nutrition, Institute of Animal Sciences, Chinese Academy of Agricultural Sciences, Beijing 100193, China; llxiumei93@163.com (X.L.); fjh6289@126.com (J.F.); 15624955881@163.com (Y.Z.)

* Correspondence: zhangminhong@caas.cn

Simple Summary: Our results showed that constant heat stress could affect the expression of myostatin and related factors involved in skeletal muscle protein breakdown in growing broilers, resulting in a decrease in muscle protein deposition. These findings suggest a new strategy for regulating muscle protein breakdown in growing broilers, which could benefit the modern broiler industry in combating constant heat stress.

Abstract: Heat stress has an adverse effect on the development of poultry farming, which has always aroused great concern. This study was carried out to investigate the protein breakdown mechanism responsible for the suppressive effect of constant heat stress on muscle growth in growing broilers. A total of 96, 29-day-old, Arbor Acres male broilers were randomly divided into two groups, a thermoneutral control (21 ± 1 °C, TC) and a heat stress (31 ± 1 °C, HS) group, with six replicates in each group and eight birds in each replicate. The trial period lasted for 14 d, and the trial was performed at 60 ± 7% relative humidity, a wind speed of <0.5 m/s and an ammonia level of <5 ppm. The results showed that the average daily feed intake and average daily gain in the HS group were distinctly lower than those in the TC group ($p < 0.05$), whereas the HS group showed a significantly increased feed conversion ratio, nitrogen excretion per weight gain and nitrogen excretion per feed intake compared to the TC group ($p < 0.05$). In addition, the HS group showed a significantly reduced breast muscle yield and nitrogen utilization in the broilers ($p < 0.05$). The HS group showed an increase in the serum corticosterone level ($p < 0.05$) and a decrease in the thyroxine levels in the broiler chickens ($p < 0.05$) compared to the TC group, whereas the HS group showed no significant changes in the serum 3,5,3′-triiodothyronine levels compared to the TC group ($p > 0.05$). Moreover, the HS group showed increased mRNA expression levels of myostatin, Smad3, forkhead box O 4, muscle atrophy F-box and muscle ring-finger 1, but reduced mRNA expression levels of the mammalian target of rapamycin, the protein kinase B and the myogenic determination factor 1 ($p < 0.05$). In conclusion, the poor growth performance of birds under constant heat stress may be due to an increased protein breakdown via an mRNA expression of myostatin and related factors.

Keywords: chicken; high temperature; regulatory factors; protein degradation; growth

Citation: Li, X.; Zhang, M.; Feng, J.; Zhou, Y. Myostatin and Related Factors Are Involved in Skeletal Muscle Protein Breakdown in Growing Broilers Exposed to Constant Heat Stress. *Animals* **2021**, *11*, 1467. https://doi.org/10.3390/ani11051467

Academic Editors: Natalie Morgan, Youssef A. Attia and Velmurugu Ravindran

Received: 26 March 2021
Accepted: 12 May 2021
Published: 20 May 2021

1. Introduction

Heat stress (HS) is a recognized long-term problem in the poultry farming industry, and its effect on poultry production performance is a common challenge in tropical and subtropical climatic conditions and even during the summer in temperate regions [1]. It is well known that HS can decrease the production of birds [2–5]. According to the United States Department of Agriculture's statistics, global chicken consumption reached 101.486 million tons in 2018. Poultry meat has seen a rise in its global demand because of its relatively high nutritional value, its low price and the lack of cultural or religious obstacles for its consumption [6]. Almost 50% of a broiler's total body mass is skeletal muscle, which exhibits a strong metabolic activity as the largest protein source in the body [7]. Heat stress has a significant effect on the growth of breast muscle, which can greatly reduce the breast muscle yield [8–12] and result in a substantial economic loss.

The essence of muscle growth is the accumulation of protein, and the balance between the rates of protein synthesis and protein breakdown influences the muscle mass [13]. At present, studies on protein synthesis are mainly focused on the insulin-like growth factor-1 (*IGF-1*)/protein kinase B (*Akt*) pathway in mammals and birds [14–16]. One of the principal protein degradation systems in skeletal muscle is the ubiquitin–proteasome pathway [17]. Forkhead box O (*FoxO*) transcription factors play an important role in muscle wastage through the regulation of ubiquitin E3 ligases, muscle atrophy F-box (*MAFbx*) and muscle ring-finger 1 (*MuRF1*) [14].

Myostatin, a transforming growth factor-beta family member, is a potent negative regulator of skeletal muscle growth [18]. Many previous in vitro studies have revealed that myostatin affects mammalian muscle growth. For example, blocking the myostatin activity in mice has applications in the promotion of muscle growth [19], and myostatin can inhibit the *Akt* activation in human skeletal muscle cells [20]. In addition, McFarlane et al. (2006) reported that in cultured C2C12 muscle cells, myostatin treatment blocked the *IGF-1*/phosphatidylinositol 3-kinase/*Akt* pathway and activated *FoxO1*, leading to an increased expression of *MAFbx* and *MuRF-1*. Smad3 transcription factors are found downstream of myostatin type II receptors and can be activated by the interaction of myostatin with its receptors [21].

In birds, a recent study revealed that myostatin significantly increased the phosphorylation rate of Smad2 and the mRNA levels of *MAFbx* in a chick's embryonic myotubes cultured at 37 °C for 2 h in vitro [22]. Several previous studies of broilers have shown that HS affected muscle growth via the expression of several genes. Acute (24 h) HS decreased the expression of the *IGF-1* and phosphatidylinositol 3-kinase R1 genes in the liver and increased the cathepsin L2 and *MAFbx* gene expression in male broilers (Cobb 500) [23]. Chronic HS decreased the muscle protein synthesis in Arbor Acres male broilers by downregulating the *IGF*-mammalian target of rapamycin (*mTOR*) signaling pathway [16]. Furukawa et al. (2016) [24] reported that short-term (0, 0.5 and 3 d) HS treatment induced a superoxide production in the muscle mitochondrial of Ross male broilers as well as a *MAFbx* gene expression and affected the signaling pathway governing the *FoxO3* activity and expression. Zuo et al. (2015) [1] reported that constant HS reduced the skeletal muscle protein deposition in broilers by decreasing the *IGF-1*, phosphatidylinositol 3-kinase and p70S6 kinase expression and increasing the *MuRF1* and *MAFbx* expression. However, whether the myostatin and related factors involved in skeletal muscle protein breakdown in growing broilers are affected by constant HS remains unclear. Therefore, the present study aimed to investigate the effects of constant HS on the growth performance, breast muscle yield, nitrogen utilization and mRNA expression of myostatin and related factors in growing broilers.

2. Materials and Methods

2.1. Birds and Treatments

One-day-old male Arbor Acres broilers were kept in one-tier cages and were maintained with administrative procedures and a standard corn–soybean-meal diet consistent with the Nutrient Requirements of Poultry (1994) for Arbor Acres broilers. At 29 d old, a total of 96 healthy Arbor Acres male chicks with similar BWs (1000 ± 70 g) were selected and randomly divided into two groups, the thermoneutral control (TC) and the HS group, which were raised in two environmentally controlled chambers. There were 6 cages (one-tier, 0.80 m × 0.80 m × 0.40 m) with 8 birds per cage in each environmentally controlled chamber (each cage served as a replicate). From the age of 29 to 42 d, the birds in the TC group were reared at a constant temperature of 21 ± 1 °C, whereas those in the HS group were reared at a constant temperature of 31 ± 1 °C. The two chambers were maintained at

60 ± 7% RH with a wind velocity of <0.5 m/s, an ammonia level of <5 ppm and a 24-h light. All broilers had ad libitum access to feed and water.

2.2. Sampling Collection and Chemical Analysis

Growth performance and breast muscle yield. At 42 d of age, 12 broilers from each group (two sampled birds per replicate) were randomly selected and euthanized by cervical dislocation. Their feed intake was recorded daily to calculate the average daily feed intake (ADFI). Their average daily gain (ADG) was calculated as the difference between the values of the body weight of all the birds at the beginning (29 d) and the end (42 d) of the experiment. The feed conversion ratio (FCR) was calculated as the ratio of the ADG to the ADFI. The breast muscle yield was expressed as the ratio of the breast muscle mass to the eviscerated carcass weight.

2.2.1. Nitrogen Utilization and Nitrogen Excretion

The randomly selected feed from the two groups was mixed, reduced to 200 g using the quartering method and crushed. Then, the feed was placed in a sealed bag prior to being tested to detect its nitrogen intake. The excreta of all the birds in each cage was collected (each cage was a replicate) on the 14th day of the experiment (every 4 h for a total of 6 times) to detect nitrogen in the excreta. Sundries in the excreta were removed, and the excreta was weighed and then nitrogen-fixed with 10 mL of 10% H_2SO_4 per 100 g excreta sample. Finally, the excreta collected from one cage was mixed and 200 g of the excreta was selected using the quartering method, dried at 65–70 °C until it reached a constant weight, collected and crushed. The content of nitrogen in the feed and the excreta was determined using the Kjeldahl method [25]. The following equations were used for the calculation:

$$\text{Nitrogen utilization (\%)} = [(\text{nitrogen intake-nitrogen excretion})/\text{nitrogen intake}] \times 100 \quad (1)$$

$$\text{Nitrogen excretion per weight gain (\%)} = (\text{nitrogen excretion}/\text{daily gain}) \times 100 \quad (2)$$

$$\text{Nitrogen excretion per feed intake (\%)} = (\text{nitrogen excretion}/\text{daily feed intake}) \times 100 \quad (3)$$

2.2.2. Blood Measurements

Serum samples were obtained by blood centrifugation at 3000× *g* for 20 min at 4 °C, and then stored at −20 °C until the analysis took place. The serum corticosterone, 3,5,3′-triiodothyronine (T3) and thyroxine (T4) levels were measured by radioimmunoassay using a gamma radioimmunoassay counter (GC-2010, Anhui Ustczonkia Scientific Instruments Co., Ltd., Anhui, China). All procedures were conducted by following the manufacturer's instructions.

2.2.3. Regulatory Factors Gene Expression

Breast muscle samples were collected and stored at −80 °C for further analysis. Total RNA was isolated from each breast muscle sample using a TRIzol reagent (Invitrogen, Carlsbad, CA, USA) as described in the manufacturer's instructions. A real-time quantitative PCR was carried out using a LightCycler 96 system (LightCycler 96 system, Roche, Basel, Switzerland) according to a common real-time quantitative PCR method. The mRNA levels of myostatin, Smad3, *MAFbx*, *MuRF1*, *mTOR*, *FoxO4*, myogenic determination factor 1 (*MyoD*) and *Akt* in the breast muscle were examined. The primers for the target genes were designed and confirmed based on the sequences described in GenBank, which are listed in Table 1. The *β-actin* gene was used as an internal control for normalization. The mRNA expression data were analyzed using the $2^{-\Delta\Delta Ct}$ method [26].

Table 1. Primers used for quantitative RT-PCR.

Primer Name [1]	Primer Sequence [2] 5′-3′	Product Size (bp)	GenBank Accession Number
β-actin	F: TGCTGTGTTCCCATCTATCG R: TTGGTGACAATACCGTGTTCA	150	NM_205518
mTOR	F: AAGGATGCTGACAAACGCTATGGA R: ACTGACTGACTGGCTGAGTAGGAG	225	XM_417614
myostatin	F: TACCCGCTGACAGTGGATTTC R: GCCTCTGGGATTTGCTTGG	153	NM_001001461
MyoD	F: GGAGAGGATTTCCACAGACAACTC R: CTCCACTGTCACTCAGGTTTCCT	113	NM_204214
Akt	F: GCTGGCATTGTTTGGCAAGATGT R: GCGGTTCCACTGGCTGAATAGG	215	NM_205055
FoxO4	F: GCTCTTCTCACACCTGGCTCTC R: TGGTTCTGCCTGCTGCTCTG	186	XM_015278657
Smad3	F: GCGTTCTGGTGCTCCATATCCTAC R: TCCTCTTCCGATGTGCCGTCTC	192	NM_204475
MAFbx	F: CAGTGAGCCAGCCTCTTGTGATG R: TTCAGCCAGTGTGACAGTCTCAGT	114	NM_001030956
MuRF1	F: GCGAGCAGGAGGACAAGACAAG R: CAAGACTGACTGTGAAGGCATCCA	240	XM_424369

[1] *β-actin*, beta-actin; *mTOR*, mammalian target of rapamycin; *MyoD*, myogenic determination factor 1; *Akt*, protein kinase B; *FoxO4*, forkhead box O 4; *MAFbx*, muscle atrophy F-box; *MuRF1*, muscle ring-finger 1. [2] F, forward; R, reverse.

2.3. Statistical Analysis

The data from the present study were analyzed using a one-way ANOVA by SAS 9.2 (SAS Institute Inc., Cary, NC, USA). The ADG, ADFI, FCR, nitrogen utilization, nitrogen excretion per weight gain and nitrogen excretion per feed intake were analyzed using the cage as the experimental unit, and other indexes were analyzed by determining the mean of two sampled birds per replicate as the experimental unit ($n = 6$). The data are expressed as the means ± SD. Statistical significance was indicated at $p < 0.05$.

3. Results

3.1. Growth Performance, Breast Muscle Yield, Nitrogen Utilization and Nitrogen Excretion

During the trial period, none of the birds suffered from clinical diseases, and there was no mortality. As shown in Table 2, the ADFI and ADG in the HS group were significantly lower than those in the TC group ($p < 0.05$), and the HS group had a significantly increased FCR compared to the TC group ($p < 0.05$). As shown in Figure 1A, the nitrogen utilization in the HS group was significantly lower than that in the TC group ($p < 0.05$), whereas the nitrogen excretion per weight gain (Figure 1B) and the nitrogen excretion per feed intake (Figure 1C) in the HS group were significantly higher than those in the TC group ($p < 0.05$). As shown in Figure 2, compared to those in the TC group, the broilers in the HS group showed a significantly reduced breast muscle yield ($p < 0.05$).

Table 2. Effects of heat stress on the growth performance of broilers [1].

Items	TC [2]	HS [2]	*p*-Value
ADFI (g)	$153.30^{a} \pm 1.17$	$127.30^{b} \pm 1.76$	0.02
ADG (g)	$82.08^{a} \pm 0.53$	$62.07^{b} \pm 1.01$	<0.01
FCR	$1.87^{b} \pm 0.01$	$2.05^{a} \pm 0.01$	<0.01

ADFI = average daily feed intake; ADG = average daily gain; FCR = ADFI/ADG. [1] All the means are reported as the means ± SD ($n = 6$). [2] TC, thermoneutral control group; HS, heat stress group. [a,b] Values with different superscripts differ significantly at $p < 0.05$.

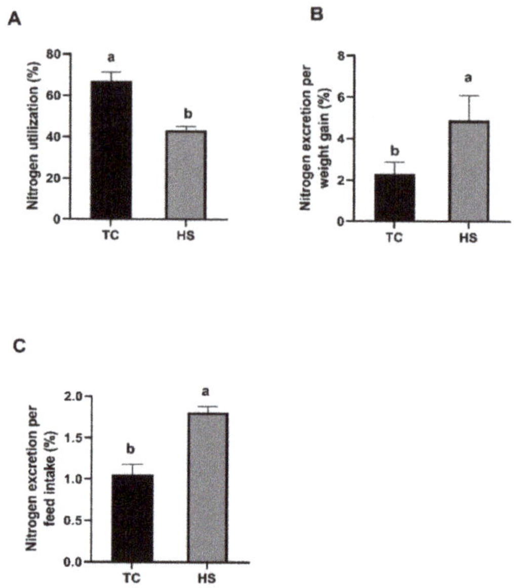

Figure 1. Effects of heat stress on the nitrogen utilization (**A**), nitrogen excretion per weight gain (**B**) and nitrogen excretion per feed intake (**C**) of broilers. TC, thermoneutral control group; HS, heat stress group. Each bar presents the means ± SD (n = 6). ^{a,b} Values with different superscripts differ significantly at $p < 0.05$.

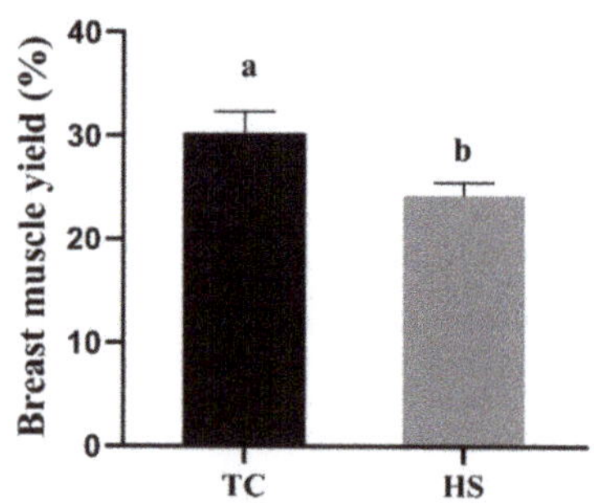

Figure 2. Effects of heat stress on the breast muscle yield of broilers. TC, thermoneutral control group; HS, heat stress group. Each bar presents the means ± SD (n = 6). ^{a,b} Values with different superscripts differ significantly at $p < 0.05$.

3.2. Blood Biochemical Indexes

We tested the effects of HS on the blood biochemical indexes of broilers, and the results of the serum corticosterone, T4 and T3 levels in broilers are presented in Figure 3. As shown in Figure 3A, the serum corticosterone levels in the broilers in the HS group were significantly higher ($p < 0.05$) than the serum corticosterone levels in the broilers in the TC group, whereas the levels of T4 in the serum from the broiler chickens in the HS group were significantly less than the serum T4 levels in the TC group ($p < 0.05$, Figure 3B). No significant differences in the levels of T3 were observed between the TC and HS groups ($p > 0.05$, Figure 3C).

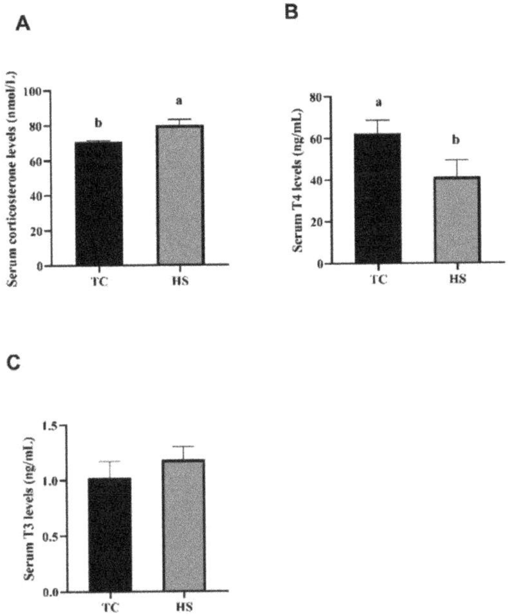

Figure 3. Effects of heat stress on the serum corticosterone (**A**), thyroxine (T4) (**B**) and 3,5,3′-triiodothyronine (T3) (**C**) levels in broilers. TC, thermoneutral control group; HS, heat stress group. Each bar presents the means ± SD (n = 6). [a,b] Values with different superscripts differ significantly at $p < 0.05$.

3.3. Regulatory Factors Gene Expression

The gene expression of breast muscle growth-related regulatory factors in broilers exposed to constant HS was determined, and the results of the myostatin, Smad3, *FoxO4*, *MAFbx*, *MuRF1*, *Akt*, *MyoD* and *mTOR* mRNA expression levels are presented in Figure 4. As shown in Figure 4, the HS group showed significantly increased mRNA expression levels of myostatin compared to the TC group ($p < 0.05$), and the HS group showed significantly increased mRNA expression levels of Smad3, *FoxO4*, *MAFbx* and *MuRF1* as well, whereas the HS group showed significantly reduced mRNA expression levels of *Akt*, *MyoD* and *mTOR* ($p < 0.05$).

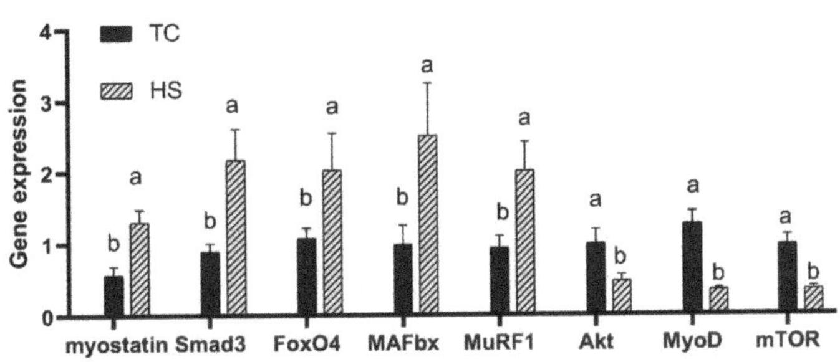

Figure 4. Effects of heat stress on the mRNA expression of breast muscle growth-related regulatory factors in broilers. TC, thermoneutral control group; HS, heat stress group. Each bar presents the means ± SD (n = 6). [a,b] Values with different superscripts differ significantly at $p < 0.05$. *FoxO4* = forkhead box O 4; *MAFbx* = muscle atrophy F-box; *MuRF1* = muscle ring-finger 1; *Akt* = protein kinase B; *MyoD* = myogenic determination factor 1; *mTOR* = mammalian target of rapamycin.

4. Discussion

Previous studies indicated that HS could affect productivity, reduce the body weight and feed intake, and increase the FCR expression of broilers [2,9,27,28]. The data in the present study revealed that the HS group showed significantly reduced ADFI and ADG values and increased FCR values compared to the TC group. These results were consistent with the findings of previous studies. HS significantly decreased nitrogen utilization and increased the nitrogen excretion per weight gain and nitrogen excretion per feed intake in this study, which revealed that HS reduced protein utilization. Kumar et al. (2017) [29] reported that a reduced nitrogen excretion is a result of an increased digestibility of protein and an increased deposition of protein in broilers. A significant increase was observed in the nitrogen excretion per weight gain and the nitrogen excretion per feed intake, showing that HS directly led to muscle protein degradation. The breast muscle is a main part of the total body, and HS directly impairs broiler production. It has been reported that HS decreased the proportion of breast muscle [10,30,31]; as expected, our results were the same as those of previous studies and indicated that HS significantly reduced the breast muscle yield of broilers. Muscle growth is the result of a rate of protein synthesis greater than the rate of protein breakdown; in heat-stressed broilers, both protein synthesis and breakdown are affected by heat exposure [32]. Based on the decreases in the ADFI, ADG, breast muscle yield and nitrogen utilization and the increases in the FCR, nitrogen excretion per weight gain and nitrogen excretion per feed intake, HS reduced protein deposition and promoted protein breakdown in the broilers. Lin et al. (2004) [33] pointed out that the reduced growth rate was related to the proteolysis that was induced by corticosterone, which could cause a reduction in animal body weight gain. T3 and T4 are the major hormones that are required to support normal growth and are known to impact almost every physiological process in chickens [34]. Studies have shown that HS reduced the plasma concentrations of T4 and T3 [35]; based on the present results, HS significantly increased serum corticosterone levels but reduced the levels of T4, as observed in previous studies [36–38]. HS has no effect on serum T3 levels, probably because the effect of HS on thyroid hormones is determined by many factors [39]. These endocrinological changes, which are consistent with the changes in growth performance caused by HS, could accelerate protein hydrolysis in vivo.

Many studies have reported that a high ambient temperature decreases muscle protein content [10,16,40], and Yunianto et al. (1997) [8] demonstrated that HS decreases muscle protein synthesis and accelerates protein breakdown. In the past few years, the *IGF-1*-phosphatidylinositol 3-kinase-*mTOR* signaling pathway, which is responsible for regulating the protein synthesis pathways, has been defined and studied [15,41]. Ma et al. (2018) [16] reported that chronic heat stress decreased muscle protein synthesis by downregulating the *IGF–mTOR* signaling pathway. *Akt* can phosphorylate a series of protein substrates to activate its downstream *mTOR* channels once it is activated [1]. The mammalian target of rapamycin can mediate protein synthesis through its downstream targets, and the *mTOR* pathway is known as a key signaling pathway that regulates the muscular hypertrophy process in vivo [42]. It has been reported that a member of the transforming growth factor-beta superfamily, myostatin, has a dramatically negative effect on muscle growth by binding to *Akt* in order to elicit its biological effects [18]. Forbes et al. (2006) [43] indicated that the R-Smads, Smad2 and Smad3, could be activated by myostatin to transduce signaling, which could cause the formation of complexes in the nucleus to regulate the expression of target genes through interactions with transcription coactivators or repressors. In normal chicken myotubes cultured in vitro, Smad controlled the myostatin expression [22]. The present study showed an increase in Smad3 mRNA levels in growing broilers under constant heat stress, indicating that there is a relationship between Smad3 and myostatin expression changes under constant HS, but the specific molecular changes remain to be further studied. Intracellular protein degradation occurs mainly through the ubiquitin–protein enzyme (proteasome) process, which is associated with *FoxO* transcription factors and two muscle-specific ubiquitin ligases (E3s), *MAFbx* and *MuRF1* [1,44]. The *FoxO* transcription factors play an important role in muscle atrophy. Sandri et al. (2004) [14] showed that the decreased activity of the *Akt* signaling pathway seemed to lead to an increase in the hypophosphorylated active forms of the *FoxO* transcription factors, and the *FoxO* transcription factors resulted in skeletal muscle atrophy by regulating atrophy-related genes, including *MAFbx*. In the current study, the HS group showed increased mRNA expression levels of myostatin, *FoxO4*, *MAFbx* and *MuRF1*, and reduced mRNA expression levels of *mTOR* and *Akt*. These data implied that HS may decrease broiler breast muscle growth by increasing the activity of myostatin and related factors, thus promoting muscle protein degradation. The myogenic determination factor 1 is involved in myoblast differentiation and is required for fast fiber formation [45,46]. Moreover, it has been reported that myostatin negatively regulates *MyoD* expression in muscle [47]. In the present study, we found that the mRNA expression levels of *MyoD* in breast muscle were reduced in the HS group, which indicated that HS affected breast muscle differentiation and hypertrophy, and thereby inhibited muscle growth.

5. Conclusions

The change in myostatin mRNA expression in growing broilers exposed to constant HS was preliminarily studied for the first time in this experiment. Therefore, we speculated that in growing broilers under high temperature conditions, protein deposition may be affected not only by the *IGF-1-Akt* signaling pathway, but also by myostatin and related factors involved in skeletal muscle protein breakdown. Moreover, further studies are required to clarify the molecular mechanism of myostatin and related factors in growing broilers under constant HS.

Author Contributions: Conceptualization, M.Z., X.L. and J.F.; methodology, X.L. and Y.Z.; software, X.L.; validation, M.Z.; formal analysis, X.L.; investigation, X.L., J.F. and Y.Z.; resources, M.Z.; data curation, X.L.; writing—original draft preparation, X.L.; writing—review and editing, X.L., M.Z., J.F. and Y.Z.; funding acquisition, M.Z. All authors have read and agreed to the published version of the manuscript.

Funding: This research was supported by the National Key Research and Development Program of China (2016YFD0500509). This work was also supported by the Independent Research Program of State Key Laboratory of Animal Nutrition (2004DA125184G1609) and the Science and Technology Innovation Project of the Chinese Academy of Agricultural Sciences (ASTIP-IAS09).

Institutional Review Board Statement: The study was conducted according to the guidelines of the Declaration of Helsinki and approved by the Animal Experimental Welfare and Ethical Inspection Committee of Institute of Animal Science, Chinese Academy of Agricultural Sciences (permit number: IAS2019-43).

Data Availability Statement: Not Applicable.

Conflicts of Interest: The authors declare no conflict of interest.

References

1. Zuo, J.; Xu, M.; Abdullahi, Y.A.; Ma, L.; Zhang, Z.; Feng, D. Constant heat stress reduces skeletal muscle protein deposition in broilers. *J. Sci. Food Agric.* **2015**, *95*, 429–436. [CrossRef] [PubMed]
2. Sahin, K.; Orhan, C.; Tuzcu, M.; Sahin, N.; Hayirli, A.; Bilgili, S.; Kucuk, O. Lycopene activates antioxidant enzymes and nuclear transcription factor systems in heat-stressed broilers. *Poult. Sci.* **2016**, *95*, 1088–1095. [CrossRef] [PubMed]
3. Zhai, W.; Corzo, A.; Mejia, L.; Zumwalt, C.D.; Peebles, E.D. Effects of dietary amino acid density and metabolizable energy level on the growth and meat yield of summer-reared broilers. *J. Appl. Poult. Res.* **2014**, *23*, 501–515. [CrossRef]
4. Awad, A.E.; Idrus, Z.; Farjam, A.S.; Bello, A. Growth performance, duodenal morphology and the caecal microbial population in female broiler chickens fed glycine-fortified low protein diets under heat stress conditions. *Br. Poult. Sci.* **2018**, *59*, 340–348. [CrossRef]
5. Sifa, D.; Bai, X.; Zhang, D.; Hu, H.; Wu, X.; Wen, A.; He, S.; Zhao, L. Dietary glutamine improves meat quality, skeletal muscle antioxidant capacity and glutamine metabolism in broilers under acute heat stress. *J. Appl. Anim. Res.* **2018**, *46*, 1412–1417. [CrossRef]
6. Petracci, M.; Cavani, C. Muscle Growth and Poultry Meat Quality Issues. *Nutrients* **2012**, *4*, 1–12. [CrossRef]
7. Matsakas, A.; Patel, K. Skeletal muscle fibre plasticity in response to selected environmental and physiological stimuli. *Histol. Histopathol.* **2009**, *24*, 611–629.

8. Yunianto, V.D.; Hayashit, K.; Kaiwda, S.; Ohtsuka, A.; Tomita, Y. Effect of environmental temperature on muscle protein turnover and heat production in tube-fed broiler chickens. *Br. J. Nutr.* **1997**, *77*, 897–909. [CrossRef]
9. Dai, S.F.; Gao, F.; Zhang, W.H.; Song, S.X.; Xu, X.L.; Zhou, G.H. Effects of dietary glutamine and gamma-aminobutyric acid on performance, carcass characteristics and serum parameters in broilers under circular heat stress. *Anim. Feed Sci. Technol.* **2011**, *168*, 51–60. [CrossRef]
10. Zhang, Z.; Jia, G.; Zuo, J.; Zhang, Y.; Lei, J.; Ren, L.; Feng, D. Effects of constant and cyclic heat stress on muscle metabolism and meat quality of broiler breast fillet and thigh meat. *Poult. Sci.* **2012**, *91*, 2931–2937. [CrossRef]
11. Akşit, M.; Yalçin, S.; Ozkan, S.; Metin, K.; Ozdemir, D. Effects of Temperature During Rearing and Crating on Stress Parameters and Meat Quality of Broilers. *Poult. Sci.* **2006**, *85*, 1867. [CrossRef]
12. Rosa, P.; Filho, D.F.; Dahlke, F.; Vieira, B.; Macari, M.; Furlan, R. Performance and carcass characteristics of broiler chickens with different growth potential and submitted to heat stress. *Rev. Bras. De Ciência Avícola* **2007**, *9*, 181–186. [CrossRef]
13. Li, Y.; Li, F.; Duan, Y.; Guo, Q.; Wang, W.; Wen, C.; Huang, X.; Yin, Y. The Protein and Energy Metabolic Response of Skeletal Muscle to the Low-Protein Diets in Growing Pigs. *J. Agric. Food Chem.* **2017**, *65*, 8544–8551. [CrossRef]
14. Sandri, M.; Sandri, C.; Gilbert, A.; Skurk, C.; Calabria, E.; Picard, A.; Walsh, K.; Schiaffino, S.; Lecker, S.H.; Goldberg, A.L. Foxo transcription factors induce the atrophy-related ubiquitin ligase atrogin-1 and cause skeletal muscle atrophy. *Cell* **2004**, *117*, 399–412. [CrossRef]
15. Lozier, N.R.; Kopchick, J.J.; Lacalle, S.D. Relative Contributions of Myostatin and the GH/IGF-1 Axis in Body Composition and Muscle Strength. *Front. Physiol.* **2018**, *9*, 1418. [CrossRef]
16. Ma, B.; He, X.; Lu, Z.; Lin, Z.; Li, J.; Yun, J.; Zhou, G.; Gao, F. Chronic heat stress affects muscle hypertrophy, muscle protein synthesis and uptake of amino acid in broilers via insulin like growth factor-mammalian target of rapamycin signal pathway. *Poult. Sci.* **2018**, *97*, 4150–4158. [CrossRef]
17. Gomes, M.D.; Lecker, S.H.; Jagoe, R.T.; Navon, A.; Goldberg, A.L. Atrogin-1, a muscle-specific F-box protein highly expressed during muscle atrophy. *Proc. Natl. Acad. Sci. USA* **2001**, *98*, 14440–14445. [CrossRef]
18. Rebbapragada, A.; Benchabane, H.; Wrana, J.; Celeste, A.; Attisano, L. Myostatin signals through a transforming growth factor β-like signaling pathway to block adipogenesis. *Mol. Cell. Biol.* **2003**, *23*, 7230–7242. [CrossRef]
19. Lee, S.J.; Mcpherron, A.C. Regulation of myostatin activity and muscle growth. *Proc. Natl. Acad. Sci. USA* **2001**, *98*, 9306–9311. [CrossRef]
20. Trendelenburg, A.U.; Meyer, A.; Rohner, D.; Boyle, J.; Hatakeyama, S.; Glass, D.J. Myostatin reduces Akt/TORC1/p70S6K signaling, inhibiting myoblast differentiation and myotube size. *Am. J. Physiol. Cell Physiol.* **2009**, *296*, C1258–C1270. [CrossRef]
21. Gao, L.; Yang, M.; Wei, Z.; Gu, M.; Li, G. MSTN Mutant Promotes Myogenic Differentiation by Increasing Demethylase TET1 Expression via the SMAD2/SMAD3 Pathway. *Int. J. Biol. Sci.* **2020**, *16*, 1324–1334. [CrossRef]
22. Saneyasu, T.; Honda, K.; Kamisoyama, H. Myostatin Increases Smad2 Phosphorylation and Atrogin-1 Expression in Chick Embryonic Myotubes. *J. Poult. Sci.* **2019**, *56*, 224–230. [CrossRef]
23. Del Vesco, A.P.; Eliane, G.; de Oliveira Grieser, D.; Zancanela, V.; Soares, M.A.M.; de Oliveira Neto, A.R. Effects of methionine supplementation on the expression of oxidative stress-related genes in acute heat stress-exposed broilers. *Br. J. Nutr.* **2015**, *113*, 549–559. [CrossRef]
24. Furukawa, K.; Kikusato, M.; Kamizono, T.; Toyomizu, M. Time-course changes in muscle protein degradation in heat-stressed chickens: Possible involvement of corticosterone and mitochondrial reactive oxygen species generation in induction of the ubiquitin-proteasome system. *Gen. Comp. Endocrinol.* **2016**, *228*, 105–110. [CrossRef]
25. Helrich, K. *Official Methods of Analysis of the Association of Official Analytical Chemists*; Association of Official Analytical Chemists: Washington, DC, USA, 1990.
26. Livak, K.J.; Schmittgen, T.D. Analysis of Relative Gene Expression Data using Real-Time Quantitative PCR. *Methods* **2002**, *25*, 402–408. [CrossRef] [PubMed]
27. Zulkifili, I.; Akmal, A.F.; Soleimani, A.F.; Hossain, M.A.; Awad, E.A. Effects of low-protein diets on acute phase proteins and heat shock protein 70 responses, and growth performance in broiler chickens under heat stress condition. *Poult. Sci.* **2018**, *97*, 1306–1314. [CrossRef]
28. Luo, J.; Song, J.; Liu, L.; Xue, B.; Tian, G.; Yang, Y. Effect of epigallocatechin gallate on growth performance and serum biochemical metabolites in heat-stressed broilers. *Poult. Sci.* **2018**, *97*, 599–606. [CrossRef]
29. Kumar, P.; Patra, A.K.; Mandal, G.P.; Samanta, I.; Pradhan, S. Effect of black cumin seeds on growth performance, nutrient utilization, immunity, gut health and nitrogen excretion in broiler chickens. *J. Sci. Food Agric.* **2017**, *97*, 3742–3751. [CrossRef]
30. Hadad, Y.; Cahaner, A.; Halevy, O. Featherless and feathered broilers under control versus hot conditions. 2. Breast muscle development and growth in pre-and posthatch periods. *Poult. Sci.* **2014**, *93*, 1076–1088. [CrossRef]
31. Piestun, Y.; Patael, T.; Yahav, S.; Velleman, S.G.; Halevy, O. Early posthatch thermal stress affects breast muscle development and satellite cell growth and characteristics in broilers. *Poult. Sci.* **2017**, *96*, 2877–2888. [CrossRef] [PubMed]
32. Lin, H.; Sui, S.J.; Jiao, H.C.; Buyse, J.; Decuypere, E. Impaired development of broiler chickens by stress mimicked by corticosterone exposure. *Comp. Biochem. Physiol. Part. A Mol. Integr. Physiol.* **2006**, *143*, 400–405. [CrossRef] [PubMed]
33. Lin, H.; Decuypere, E.; Buyse, J. Oxidative stress induced by corticosterone administration in broiler chickens (Gallus gallus domesticus) 1. Chronic exposure. *Comp. Biochem. Physiol. Part. B Biochem. Mol. Biol.* **2004**, *139*, 737–744. [CrossRef]

34. Xiao, Y.; Wu, C.; Li, K.; Gui, G.; Zhang, G.; Yang, H. Association of growth rate with hormone levels and myogenic gene expression profile in broilers. *J. Anim. Sci. Biotechnol.* **2017**, *8*, 43. [CrossRef] [PubMed]
35. Kataria, N.; Kataria, A.K.; Gahlot, A.K. Ambient temperature associated variations in serum hormones and interrelated analytes of broiler chickens in arid tract. *Slov. Vet. Res.* **2008**, *45*, 127–34616612.
36. Sohail, M.U.; Hume, M.E.; Byrd, J.A.; Nisbet, D.J.; Ijaz, A.; Sohail, A.; Shabbir, M.Z.; Rehman, H. Effect of supplementation of prebiotic mannan-oligosaccharides and probiotic mixture on growth performance of broilers subjected to chronic heat stress. *Poult. Sci.* **2012**, *91*, 2235. [CrossRef]
37. Lara, L.; Rostagno, M. Impact of Heat Stress on Poultry Production. *Animals* **2013**, *3*, 356–369. [CrossRef]
38. Loyau, T.; Bedrani, L.; Berri, C.; Métayer-Coustard, S.; Praud, C.; Coustham, V.; Mignon-Grasteau, S.; Duclos, M.J.; Tesseraud, S.; Rideau, N. Cyclic variations in incubation conditions induce adaptive responses to later heat exposure in chickens: A review. *Animal* **2015**, *9*, 76–85. [CrossRef]
39. Jiang, S.; Mohammed, A.A.; Jacobs, J.A.; Cramer, T.A.; Cheng, H.W. Effect of synbiotics on thyroid hormones, intestinal histomorphology, and heat shock protein 70 expression in broiler chickens reared under cyclic heat stress-ScienceDirect. *Poult. Sci.* **2020**, *99*, 142–150. [CrossRef]
40. Geraert, P.A.; Padilha, J.; Guillaumin, S. Metabolic and endocrine changes induced by chronic heatexposure in broiler chickens: Growth performance, body composition and energy retention. *Br. J. Nutr.* **1996**, *75*, 195–204. [CrossRef]
41. Mcfarlane, C.; Plummer, E.; Thomas, M.; Hennebry, A.; Ashby, M.; Ling, N.; Smith, H.; Sharma, M.; Kambadur, R. Myostatin induces cachexia by activating the ubiquitin proteolytic system through an NF-κB-independent, FoxO1-dependent mechanism. *J. Cell. Physiol.* **2006**, *209*, 501–514. [CrossRef]
42. Bodine, S.C.; Stitt, T.N.; Gonzalez, M.; Kline, W.O.; Yancopoulos, G.D. Akt/mTOR pathway is a crucial regulator of skeletal muscle hypertrophy and can prevent muscle atrophy in vivo. *Nat. Cell Biol.* **2001**, *3*, 1014–1019. [CrossRef] [PubMed]
43. Forbes, D.; Jackman, M.; Bishop, A.; Thomas, M.; Kambadur, R.; Sharma, M. Myostatin auto-regulates its expression by feedback loop through Smad7 dependent mechanism. *J. Cell. Physiol.* **2006**, *206*, 264–272. [CrossRef] [PubMed]
44. Wen, C.; Yueping, C.; Ping, W.; Wang, T.; Zhou, Y. MSTN, mTOR and FoxO4 are involved in the enhancement of breast muscle growth by methionine in broilers with lower hatching weight. *PLoS ONE* **2014**, *9*, e114236. [CrossRef] [PubMed]
45. Kitamura, T.; Kitamura, Y.I.; Funahashi, Y.; Shawber, C.J.; Castrillon, D.H.; Kollipara, R.; De Pinho, R.A.; Kitajewski, J.; Accili, D. A Foxo/Notch pathway controls myogenic differentiation and fiber type specification. *J. Clin. Investig.* **2007**, *117*, 2477–2485. [CrossRef]
46. Yi, C.; Yao, Z.; Sarkar, D.; Lawrence, M.; Sanchez, G.J.; Parker, M.H.; Macquarrie, K.L.; Davison, J.; Morgan, M.T.; Ruzzo, W.L. Genome-wide MyoD Binding in Skeletal Muscle Cells: A Potential for Broad Cellular Reprogramming. *Dev. Cell* **2010**, *18*, 662–674.
47. Hennebry, A.; Berry, C.; Siriett, V.; O'Callaghan, P.; Chau, L.; Watson, T.; Sharma, M.; Kambadur, R. Myostatin regulates fiber-type composition of skeletal muscle by regulating MEF2 and MyoD gene expression. *Am. J. Physiol. Cell Physiol.* **2009**, *296*, C525–C534. [CrossRef]

animals MDPI

Article

Effects of Heat Stress on Gut-Microbial Metabolites, Gastrointestinal Peptides, Glycolipid Metabolism, and Performance of Broilers

Guangju Wang, Xiumei Li, Ying Zhou, Jinghai Feng and Minhong Zhang *

State Key Laboratory of Animal Nutrition, Institute of Animal Sciences, Chinese Academy of Agricultural Sciences, Beijing 100193, China; 82101185163@caas.cn (G.W.); llxiumei93@163.com (X.L.); 15624955881@163.com (Y.Z.); fjh6289@126.com (J.F.)

* Correspondence: zmh66@126.com; Tel.: +86-10-6289-5517

Simple Summary: In the summer, heat stress is a main factor that causes poor performance in broilers. Broilers are more susceptible to high temperature environments than mammals due to their lack of sweat glands and being covered in feathers. Heat stress can alter the regulation of glycolipid metabolism, which is manifested by unstable levels of blood glucose, insulin, total cholesterol, and triglyceride. Heat stress also affects the structure of gut microbes and gastrointestinal peptides. However, the relationship among microbiota, gastrointestinal peptides, glycolipid metabolism, and production performance under heat stress is still unclear. Moreover, exploring these mechanisms can help in the development of strategies that alleviate the negative effects of performance by heat stress. Our results suggest that the poor production performance of broilers under heat stress may be related to short chain fatty acids fermented by intestinal microbiota involved in regulating metabolic disorders.

Citation: Wang, G.; Li, X.; Zhou, Y.; Feng, J.; Zhang, M. Effects of Heat Stress on Gut-Microbial Metabolites, Gastrointestinal Peptides, Glycolipid Metabolism, and Performance of Broilers. *Animals* **2021**, *11*, 1286. https://doi.org/10.3390/ani11051286

Academic Editor: Natalie Morgan

Received: 31 March 2021
Accepted: 28 April 2021
Published: 30 April 2021

Abstract: This paper investigated the effects of heat stress on gut-microbial metabolites, gastrointestinal peptides, glycolipid metabolism, and performance of broilers. Thus, 132 male Arbor Acres broilers, 28-days-old, were randomly distributed to undergo two treatments: thermoneutral control (TC, 21 °C) and high temperature (HT, 31 °C). The results showed that the average daily gain (ADG), average daily feed intake (ADFI), and gastric inhibitory polypeptide (GIP) concentration in the jejunum significantly decreased the core temperature, feed conversion ratio (FCR), and ghrelin of the hypothalamus, and cholecystokinin (CCK) in jejunum, and serum significantly increased in the HT group ($p < 0.05$). Exploration of the structure of cecal microbes was accomplished by sequencing 16S rRNA genes. The sequencing results showed that the proportion of *Christensenellaceae* and *Lachnospiraceae* decreased significantly whereas the proportion of *Peptococcaceae* increased at the family level ($p < 0.05$). *Ruminococcus* and *Clostridium* abundances significantly increased at the genus level. Furthermore, the content of acetate in the HT group significantly increased. Biochemical parameters showed that the blood glucose concentration of the HT group significantly decreased, and the TG (serum triglycerides), TC (total cholesterol), insulin concentration, and the insulin resistance index significantly increased. Nonesterified fatty acid (NEFA) in the HT group decreased significantly. In conclusion, the results of this paper suggest that the poor production performance of broilers under heat stress may be related to short-chain fatty acids (SCFAs) fermented by intestinal microbiota involved in regulating metabolic disorders.

Keywords: heat stress; microbial metabolites; gastrointestinal peptides; glycolipid metabolism

1. Introduction

Chicken meat is considered an important source of dietary protein worldwide. As per the Food and Agriculture Organization (FAO), total global chicken production was 118.0 million tons in 2019, accounting for a large proportion of the whole meat production.

Broilers are severely affected by heat stress; it can cause slow growth rate, low feed intake, and decreased immunity, leading to economic losses [1,2]. Studies have shown that heat stress leads to glucolipid metabolism disorder [3–5]. Heat stress accelerates decomposition of glycogen and inhibits glycogen production [6]. When ambient temperature is higher than the tolerable limit of broilers, the energy produced cannot meet the needs, resulting in insulin resistance and glucolipid metabolism disorders [7]. Previous studies investigated the impacts of heat stress on broilers, which was found to cause blood glucose instability [3,8,9], decrease insulin sensitivity [10], serum cholesterol, and triglycerides [11], increase body fat in the abdomen, and decrease plasma non-esterified fatty acid (NEFA) concentrations [12]. Recently, several studies have shown that heat stress affects the metabolites and structures of the intestinal microflora [13,14] and gastrointestinal peptide [15] in broilers. However, to date, there are no specific studies on the effects of heat stress on broiler microflora and its metabolites, gastrointestinal peptide, glycolipid metabolism, and the association among the three.

Multiple studies were conducted on the associations among bacterial metabolites, gastrointestinal peptides, and glucose and lipid metabolism on metabolic disorders in mice and humans. Intestinal microbiota are involved in metabolic processes and energy homeostasis [16]. Previous studies have shown that heat stress can alter the gut microbiota of mice [17]. The Firmicutes to Bacteroidetes ratio decreased significantly under 30 °C [18]. Short-chain fatty acids (SCFAs) are metabolites formed by gut microbes from complex dietary carbohydrates. Researchers found that SCFAs stimulated mouse L cells to produce glucagon-like peptide 1 (GLP-1), peptide YY (PYY), and other intestinal anorexia hormones, reducing appetite [19]. GLP-1 and GIP are related to insulin secretion, and approximately 70% of β-cell insulin secretion is controlled by GIP and GLP-1 [20]. Insulin secretion disorders lead to metabolic disorders, such as diabetes and obesity [21]. As mentioned above, heat stress causes metabolic disorders in broilers. However, the association between metabolic disorder and microbiota under heat stress is still unclear in mice and humans.

Thus, the purpose of this study was to investigate the effects of heat stress on the microbiota and its metabolites, gastrointestinal peptides, and glycolipid metabolism in broilers, and explore the relationship among them to provide a scientific basis for reconstructing the intestinal flora, to alleviate the decline in production performance caused by heat stress.

2. Materials and Methods

2.1. Animals and Experimental Design

A total 132 male, one day old, Arbor Acres broilers were purchased from commercial hatcheries and housed in three-layer (8400 cm^2 per layer) metal cages at ambient temperatures that decreased with age. All birds had free access to feed and water (room temperature) ad libitum. The broilers were fed on crumble diets (Table 1). Then, birds were divided into a thermoneutral control group (TC, 21 °C) or a high temperature group (HT, 31 °C), with six biological replicates per group, 11 birds per replicate, at 21 days old. Birds were transferred to the environmental temperature control chamber, while maintaining a temperature of 21 °C, a humidity of 60%, for 7 days. The experiment started at 28 days old. Temperature control was adjusted by an artificial environmental control chamber developed by the Institute of Animal Science of the Chinese Academy of Agricultural Sciences (CAAS). There was no significant difference in the initial body weight of broilers in the two groups. The temperature of the two groups was constant and humidity remained at 60% until the end of the experiment, which lasted for 21 days. The lighting program was continuous throughout the experimental (fluorescent light, 40 W). To reduce stress, irrelevant personnel were prohibited from entering the artificial environment control chamber. This program was approved by the Experiment Animal Welfare and Ethical at the Institute of Animal Science of CAAS.

Table 1. Composition and nutrient levels of the basal diet.

Items	Content (%)
Ingredients	
Corn	56.51
Soybean meal	35.52
Soybean oil	4.50
Na Cl	0.30
Limestone	1.00
Ca HPO_4	1.78
d L-Met	0.11
Premix (1)	0.28
Total	100.00
Nutrient levels (2)	
ME/(MJ/Kg)	12.73
CP	20.07
Ca	0.90
AP	0.40
Lys	1.00
Met	0.42
Met + Cys	0.78

(1) Premix provided the following per kg of the diet: VA 10,000 IU, VD3 3400 IU, VE 16 IU, VK3 2.0 mg, VB1 2.0 mg, VB2 6.4 mg, VB6 2.0 mg, VB12 0.012 mg, pantothenic acid calcium 10 mg, nicotinic acid 26 mg, folic acid 1 mg, biotin 0.1 mg, choline 500 mg, Zn($ZnSO_4 \cdot 7H_2O$) 40 mg, Fe($FeSO_4 \cdot 7H_2O$) 80 mg, Cu($CuSO_4 \cdot 5H_2O$) 8 mg, Mn($MnSO_4 \cdot H_2O$) 80 mg, I(KI) 0.35 mg, Se(Na_2SeO_3) 0.15 mg. (2) Calculated values.

2.2. Sample and Data Collection

During the experiment, we used an 0.01 g sensitive electronic body weight scale (manufacturer: Mettler Toledo, PL2002) to record the initial feed weight, final feed weight, and body weight of the bird replicates, then calculated the average feed intake, average body weight, average daily weight gain, and feed efficiency for each replicate. This experiment used a rectal probe thermometer to measure the body core temperature of one bird replicate, which was randomly selected from each replicate, and measured four times a day during the experiment. At the end of the experiment, one bird, after 12 h fasting, was randomly selected from each replicate for insulin and blood glucose determination. Feed then continued for two hours, and one bird was randomly selected from each replicate to collect samples. The blood was immediately collected through the wing vein into heparinized tubes and centrifuged at 10,000× *g* for 4 min at 4 °C. Then the plasma was collected and stored at −20 °C until the analysis of the concentrations of PYY, ghrelin, CCK, GIP, and GLP-1. Immediately after the blood samples were obtained, the birds were humanely sacrificed, and the tissues of the cecum contents, intestinal mucosa, pancreas, and hypothalamus were collected. The samples were placed in a cryopreservation tube and stored in a −80 °C refrigerator.

2.3. Determination of Gastrointestinal Peptides

The concentrations of CCK, Ghrelin, GLP-1, GIP, and PYY in the intestinal mucosa and serum were determined by the enzyme-linked immunosorbent assay (ELISA). The intestinal mucosa needed to be grinded, weighed, diluted with PBS buffer at 1:9, and centrifuged for 20 min (2000–3000 rpm). The supernatant was then carefully taken for testing. The standard wells were set up, the samples added (diluent and enzyme label reagent in sequence), and were incubated for 60 min. The ELISA plate was washed with a washing solution for more than five times, and the color reagent and stop solution were added. Finally, an enzyme-labeled instrument was used to measure the absorbance of each well (OD value). The contents of each peptide in the sample were calculated through the standard curve.

2.4. DNA Extraction and PCR Amplification

Microbial community genomic DNA was extracted from cecal content samples using the E.Z.N.A.® soil DNA Kit (Omega BioTek, Norcross, GA, USA) according to the manufacturer's instructions. The DNA extract was checked on 1% agarose gel, and DNA concentration and purity were determined with NanoDrop 2000 UV-vis spectrophotometer (Thermo Scientific, Wilmington, DE, USA). The hypervariable region V3-V4 of the bacterial 16S rRNA gene was amplified with primer pairs 338F (5′-ACTCCTACGGGAGG-CAGCAG-3′) and 806R (5′-GGACTACHVGGGTWTCTAAT -3′) by an ABI GeneAmp® 9700 PCR thermocycler (Applied Biosystems, Carlsbad, CA, USA). The PCR amplification of 16S rRNA gene was performed as follows: initial denaturation at 95 °C for 3 min, followed by 27 cycles of denaturing at 95 °C for 30 s, annealing at 55 °C for 30 s and extension at 72 °C for 45 s, and single extension at 72 °C for 10 min, and end at 10 °C. The PCR mixtures contain 5 × TransStart FastPfu buffer 4 μL, 2.5 mM dNTPs 2 μL, forward primer (5 μM) 0.8 μL, reverse primer (5 μM) 0.8 μL, TransStart FastPfu DNA Polymerase 0.4 μL, template DNA 10 ng, and finally ddH2O up to 20 μL. PCR reactions were performed in triplicate. The PCR product was extracted from 2% agarose gel and purified using the AxyPrep DNA Gel Extraction Kit (Axygen Biosciences, Union City, CA, USA), according to manufacturer's instructions, and quantified using Quantus™ Fluorometer (Promega, Madison, WI, USA).

2.5. Illumina MiSeq Sequencing

Purified amplicons were pooled in equimolar and paired-end sequences on an Illumina MiSeq PE300 platform/NovaSeq PE250 platform (Illumina, San Diego, CA, USA), according to the standard protocols by Majorbio Bio-Pharm Technology Co. Ltd. (Shanghai, China).

2.6. Processing of Sequencing Data

The raw 16S rRNA gene sequencing reads were demultiplexed, quality-filtered by fastp version 0.20.0 [22] and merged by FLASH version 1.2.7 [23] with the following criteria: (i) the 300 bp reads were truncated at any site receiving an average quality score of <20 over a 50 bp sliding window, and the truncated reads shorter than 50 bp were discarded. Reads containing ambiguous characters were also discarded. (ii) Only overlapping sequences longer than 10 bp were assembled according to their overlapped sequences. The maximum mismatch ratio of overlap region was 0.2. Reads that could not be assembled were discarded. (iii) Samples were distinguished according to the barcode and primers, and the sequence direction was adjusted, exact barcode matched, and two nucleotides mismatched in the primer matching. Operational taxonomic units (OTUs) with 97% similarity cutoff [24,25] were clustered using UPARSE version 7.1 [24], and chimeric sequences were identified and removed. The taxonomy of each OTU representative sequence was analyzed by RDP Classifier version 2.2 [26] against the 16S rRNA database using confidence threshold of 0.7.

2.7. Determination of SCFAs

We accurately weighed 1 g of cecal content, added 5 mL of ultra-pure water, shook, and mixed for 30 min, overnight at 4 °C, then centrifuged at 10,000 rpm for 10 min, and transferred the supernatant to a 10 mL cuvette. We added 4 mL of ultra-pure water to the precipitation, shook and mixed for 30 min, then centrifuged at 10,000 rpm to transfer the supernatant into a 10 mL cuvette for constant volume. We transferred the liquid in the colorimetric tube to a 10 mL centrifuge tube, centrifuged at 12,000 rpm for 15 min, and then transferred the supernatant to a 2 mL centrifuge tube, according to V:V = 9:1 (900 μL supernatant + 100 μL 25% metaphosphoric acid), mixed well, and let it stand at room temperature for 3–4 h centrifugation, 45 um microporous membrane (nylon series) filtration. We added the machine bottle (more than 600 μL) to be tested. Chromatographic conditions: db-ffap column, 30 m * 250 μm * 0.25 μm; carrier gas: high purity nitrogen (99.999%), flow rate: 0.8 mL/min; auxiliary gas: high purity hydrogen (99.999%), detector FID temperature: 280 °C, injection port temperature: 250 °C, split ratio: 50:1, injection volume: 1 μL; temperature programming: initial temperature: 60 °C, the temperature rose to 220 °C at the rate of 20 °C/min, and held for 1 min.

2.8. Determination of Serum Biochemical Parameters

The concentrations of blood glucose, serum insulin, triglycerides, total cholesterol, and NEFA were measured using the kits provided by Nanjing Jiancheng Bioengineering Institute (Jiangsu, China). The insulin resistance index was calculated using a formula as previously described: insulin resistance index = insulin/(22.5$e^{-lnglucose}$) [27].

2.9. Statistical Analysis

All statistical analyses were conducted using GraphPad Prism 8 software (GraphPad Software, Inc. La Jolla, CA, USA). An independent sample t-test (unpaired Student's *t* test and Mann–Whitney test) was used for the comparison of the two treatments. Replicate served as the experimental unit. The confidence interval was 95%, and $p < 0.05$ indicated a significant difference in the treatment effect; values were expressed as the mean ± SEM.

3. Results

3.1. Effect of High Temperature on Performance

The effect of heat stress on performance is shown in Table 2. Compared with the TC group, ADG and ADFI in the HT group decreased significantly ($p < 0.05$). Compared with the TC group, the core temperature of birds and FCR in the HT group significantly increased ($p < 0.05$).

Table 2. Effects of thermal environment on performance indices of broilers during experiment period.

	Treatments			
Item	**TC**	**HT**	**SEM**	***p* Value**
IABW (g)	1427.31	1432.10	19.35	>0.05
FABW (g)	2840.22 ^{a}	2567.78 ^{b}	24.32	<0.05
ADG (g/d)	80.58 ^{a}	70.70 ^{b}	5.03	<0.05
ADFI (g/d)	157.51 ^{a}	148.95 ^{b}	7.16	<0.05
FCR (g/g)	1.95 ^{b}	2.11b ^{a}	0.11	<0.05
CT (°C)	41.49 ^{b}	42.66 ^{a}	1.28	<0.05

Values are means ± SEM. TC, thermoneutral control group; HT, high temperature group; IABW, initial average body weight (29 d); FABW, final average body weight (43 d); ADG, average daily gain; ADFI, average daily feed intake; FCR, feed conversion rate; CT, core temperature. ^{a,b} Means within the same line with different superscript differ significantly.($p < 0.05$).

3.2. Effects of High Temperature on Gastrointestinal Peptide

The effects of heat stress on the neuropeptides of gut and hypothalamus concentrations in six-week-old chickens were examined and the results are presented in Figures 1–3. Compared with the TC group, GIP concentration in the jejunum of the HT group significantly decreased ($p < 0.05$). The concentration of CCK in the jejunum and serum and ghrelin of the hypothalamus in the HT group was significantly higher than that in the TC group ($p < 0.05$). The GLP-1 in the ileum, PYY in the pancreas, and ghrelin in the jejunum in the HT group were not significantly different from that of the TC group.

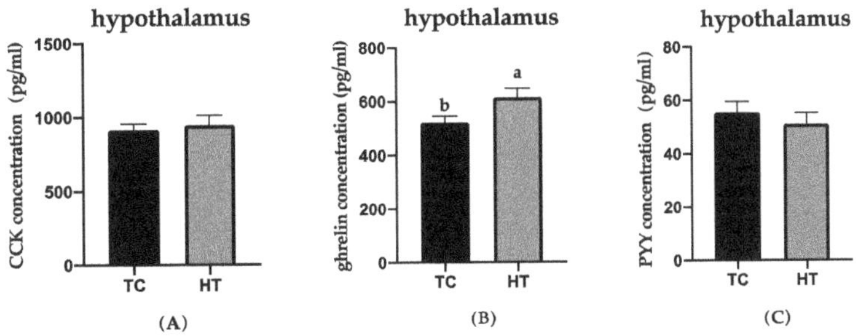

Figure 1. Effects of high temperatures on CCK, ghrelin, and PYY concentrations in the hypothalamus. (**A**), cholecystokinin; (**B**), ghrelin; (**C**), peptide YY. a,b, Means with different letters within columns indicates significant differences ($p < 0.05$).

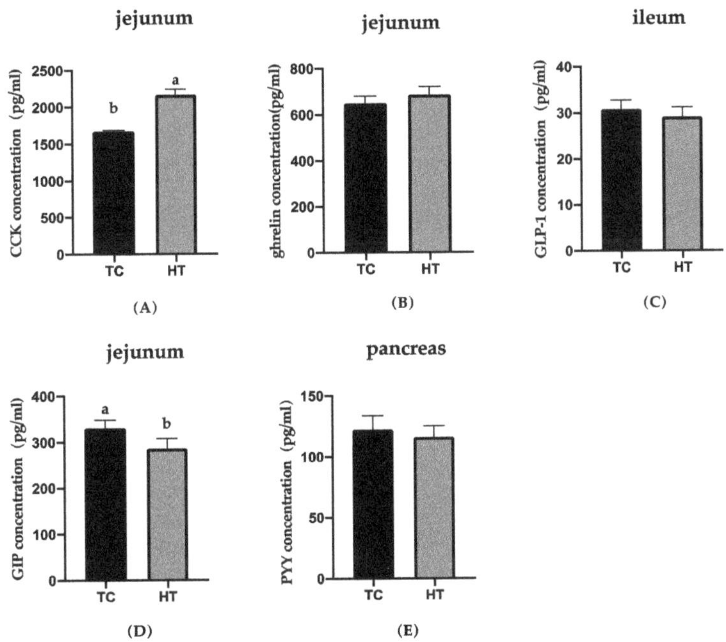

Figure 2. Effects of high temperature on concentrations of CCK, ghrelin, GIP in jejunum, GLP-1 in ileum and PYY in pancreas. (**A**), cholecystokinin; (**B**), ghrelin; (**C**), glucagon-like peptide-1; (**D**), glucose-dependent insulinotropic polypeptide; (**E**), peptide YY. a,b, Means with different letters within columns indicates significant differences ($p < 0.05$).

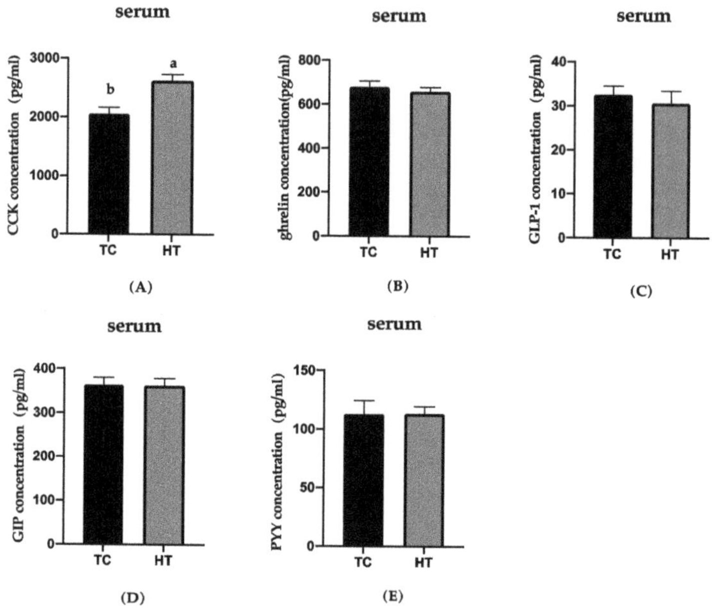

Figure 3. Effects of high temperature on concentration of CCK, ghrelin, GLP-1, GIP and PYY in serum. (**A**), cholecystokinin; (**B**), ghrelin; (**C**), glucagon-like peptide-1; (**D**), glucose-dependent insulinotropic polypeptide; (**E**), peptide YY. a,b Means with different letters within columns indicates significant differences ($p < 0.05$).

3.3. *Effects of High Temperature on Cecal Microbial Composition*

The intestinal microbial community of broilers is mainly concentrated in the cecum; it is the main place where microorganisms participate in the regulation of body metabolism. Therefore, we analyzed the composition of the caecal flora of two groups of broilers at the levels of phylum, family, and genus. The compositions of cecal microbiota at phylum, family, and genus levels are provided in Table 3. At the phylum level, the two groups of flora were mainly composed of *Firmicutes*, *Bacteroides*, and *Actinomycetes*; however, there was no significant difference in abundance. At the family level, the dominant bacteria were mainly *Ruminococcaceae*; the proportion of *Christensenellaceae* and *Lachnospiraceae* decreased significantly, and increased the proportion of *Peptostreptococcaceae* ($p < 0.05$). It may show that heat stress significantly affects the dominant intestinal flora at the family level. At the genus level, the dominant bacteria were mainly *Faecalibacterium*, *Romboutsia*. *Ruminococcus* and *Clostridium* abundances significantly increased, indicating that the thermal environment significantly increased the proportion of them.

Table 3. Effects of the thermal environment on cecum digesta microbiota composition.

Level	Species Name	Treatments		SEM	*p* Value
		TC	HT		
phylum	*Firmicutes* (%)	89.33	87.49	2.18	>0.05
	Bacteroides (%)	5.69	5.18	0.74	>0.05
	Actinomycetes (%)	3.78	6.50	1.12	>0.05
family	*Ruminococcaceae* (%)	51.32	51.14	1.25	>0.05
	Lachnospiraceae (%)	18.68 [a]	10.15 [b]	0.80	<0.05
	Christensenellaceae (%)	1.20 [a]	0.51 [b]	0.07	<0.05
	Peptococcaceae (%)	0.21 [b]	0.54 [a]	0.05	<0.05
genus	*Faecalibacterium* (%)	28.32	31.63	1.18	>0.05
	Romboutsia (%)	6.91	11.29	0.45	>0.05
	Ruminococcus (%)	0.08 [b]	0.15 [a]	0.02	<0.05
	Clostridium (%)	2.55 [b]	5.16 [a]	0.40	<0.05
	Faecalibacterium (%)	28.32	31.63	1.18	>0.05

Values are means ± SEM. TC, thermoneutral control group; HT, high temperature group. [a,b] Means within the same line with different superscript differ significantly ($p < 0.05$).

3.4. Effects of High Temperature on SCFAs Concentration

The concentration of short chain fatty acids in cecal contents of the two treatment groups is shown in Table 4. Compared with the TC group, the content of acetate in the HT group significantly increased, but there was no significant difference in propionic acids and butyric acids.

Table 4. Effects of thermal environment on the content of SCFAs in the cecum.

Item	Treatments		SEM	*p* Value
	TC	HT		
acetate (µg/mL)	572.9 [b]	741.1 [a]	16.79	<0.05
propionic acids (µg/mL)	56.45	63.96	5.89	>0.05
butyric acids (µg/mL)	262.9	252.7	29.71	>0.05

Values are means ± SEM. SCFAs, short chain fatty acids; TC, thermoneutral control group; HT, high temperature group. [a,b] Means within the same line with different superscript differ significantly. ($p < 0.05$).

3.5. Effects of High Temperatures on Serum Parameters

The blood glucose, insulin, insulin resistance index, TG, TC, and NEFA concentrations are shown in Figure 4. Compared with the TC group, the blood glucose and NEFA concentrations in the HT group significantly reduced, the insulin, TC, and TG concentrations significantly increased. Compared with the TC group, the insulin resistance index of broilers under heat stress increased significantly, resulting in severe insulin resistance.

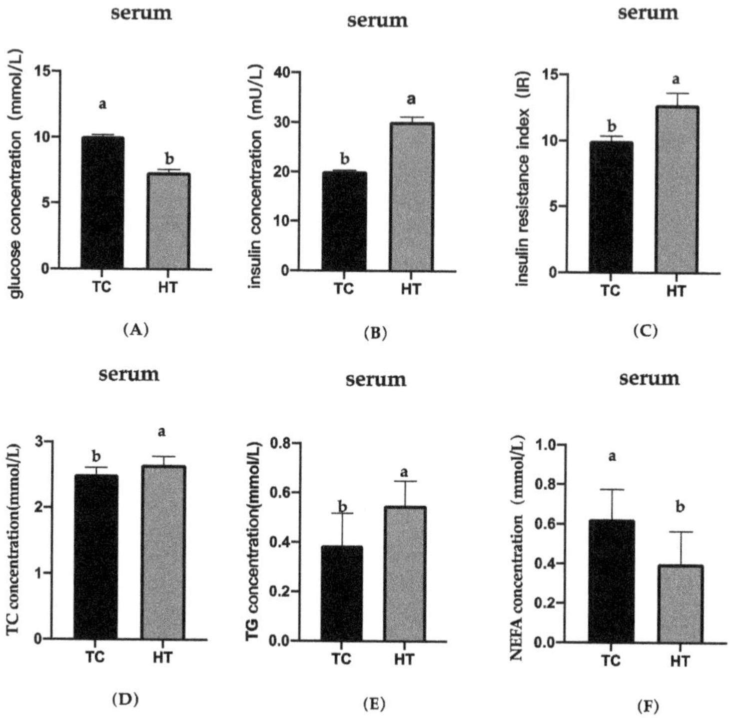

Figure 4. Effect of high temperature on blood glucose, insulin, insulin resistance index, TC, TG, and NEFA in serum. (**A**), blood glucose; (**B**), insulin; (**C**), insulin resistance index; (**D**), total cholesterol; (**E**), triglyceride; (**F**), non-esterified fatty acid. ^{a,b} means with different letters within columns indicates significant differences ($p < 0.05$).

4. Discussion

Thermal environment severely impairs the performance [28,29], changes the composition of gut microbiota [30], and affects glycolipid metabolism [31] of broilers. However, previous studies have not been able to account for the associations among them. The present study clearly established a significant outcome of a thermal environment on performance, gut microbiota, gastrointestinal peptides, and glycolipid metabolism in broilers, and developed a preliminary understanding of the relationship among them.

Significant elevations were detected in the core temperatures of the birds, meaning that heat stress was produced. Previous literature demonstrated that heat stress could negatively impact ADFI, ADG, and FCR in broilers [2,29,32–34]. The present study showed that ADFI, ADG, and FCR were seriously affected by high temperatures of 31 °C compared with the 21 °C group, which agreed with the results of previous studies [2,35]. These results are widely accepted, they imply that heat exposure will directly impair productive performance of birds. In general, the reason for the decline in productive performance is largely due to the lower feed intake; birds reduce feed intake to minimize excess metabolic energy production to keep themselves cool [36]. Another important reason for the weight gain reduction observed in birds experiencing heat stress would be metabolism disorder. Previous studies have shown that thermal environment can disrupt metabolic homeostasis, accelerating protein catabolism [37], increasing abdominal fat deposition [38], changing blood glucose levels, decreasing insulin sensitivity, and causing a negative energy balance [39]. Birds suffering from heat stress may choose to use more energy to resist the damage caused by high temperatures rather than to grow or reproduce. Therefore, the underlying cause of poultry production performance degradation may be metabolic disorder. The current study observed that heat stress led to elevated levels of insulin, serum triglycerides, and total cholesterol, and a reduction in serum blood glucose and NEFA. Changes in blood glucose in broilers exposed to high temperatures showed mixed results, as there was evidence for it to increase [3], decrease, or remain unchanged [9]. Numerous factors influence blood glucose and the reasons for the differences include different physiological statuses and experimental designs. The results of elevated levels of insulin [40], serum triglycerides, total cholesterol, and reduced NEFA [41] in this study are consistent with previously reported studies. It is possible that insulin levels were elevated due to compensatory secretions to maintain glucose tolerance. The increased basal insulin levels may explain the lack of an increase in basal NEFA levels because insulin is a potent antilipolytic hormone. We also calculated that the insulin resistance index increased significantly, meaning insulin resistance occurs under heat stress. In summary, these observations imply that heat stress alters carbohydrate and lipid metabolism, leading to metabolic disorder.

Gut microbes have previously been susceptible to heat stress in both mice and livestock. The present study found that heat stress changed the colony composition at the family level, reduced the proportions of *Christensenellaceae* and *Lachnospiraceae*, and increased the proportion of *Peptococcaceae*. It was reported that *Christensenellaceae* is related to healthy glucose metabolism and reduces the risk of obesity [42]. A previous study reported gut dysbiosis in mice that underwent chronic water avoidance stress; in particular, the abundance of *Lachnospiraceae* declined [43]. Similar results were obtained in this paper. The increase of *Peptococcaceae* was mainly related to histopathological infection, indicating that heat stress increased the threat factors of gut health. At the genus level, heat stress increased the proportion of *Ruminococcus* and *Clostridium*. In addition, our results showed that heat stress increased the content of acetate in cecal contents. Since *Ruminococcus* and *Clostridium* are the main acetate-producing microorganisms, the increase of acetate content in cecal contents may be associated to the increase of *Ruminococcus* and *Clostridium*.

Gastrointestinal peptides are mainly secreted by intestinal endocrine cells, including ghrelin, CCK, GIP, PYY, GLP-1, which are directly involved in the regulation of gastrointestinal peristalsis, sensation, and secretion. They have the dual functions of promoting the secretion of neurotransmitters and hormones, and regulating feeding behavior, nutrient absorption, energy metabolism, intestinal peristalsis, and emptying [44]. The results we obtained show that heat stress significantly increased CCK concentration in the jejunum and circulation—similar to the results obtained by a previous study [45]. CCK proved to be an anorexic hormone in various poultry studies, which can gather short-term post-eating satiety signals and inhibit broiler feeding [46]. It was reported that exogenous injection of CCK could significantly reduce the feed intake of poultry [47]. Therefore, the significant decrease in feed intake of broilers in the HT group may be due to the large amount of CCK secreted by the intestines. Several studies have shown that a certain dose of ghrelin is injected into the cerebral ventricle of chicks, and the results are found to strongly inhibit the feed intake of chicks [48,49]. In the HT group, a significant increase in hypothalamic ghrelin levels was observed, which is consistent with the results of a previous study [36]. There were no significant changes in other peptides except for ghrelin in the hypothalamus; no significant changes in PYY, ghrelin, and GLP-1 were observed in the intestinal tract and circulation. This is consistent with previous research [50]. It was found that heat stress significantly reduced the plasma GIP concentration of poultry and pigs [51]. The present study found that heat stress significantly reduced the GIP concentration in the jejunum of broilers, and the serum GIP concentration had a decreasing trend. This is supported by previous results. The decreased GIP concentration may be caused by the decreased number of intestinal K cells or the decreased GIP expression in K cells. Another possible explanation for this is that blood glucose was significantly reduced because GIP secretion is glucose-dependent [52].

The gut microbiota participating in host metabolism is frequently reported on [53–55]. According to various studies, such a process is associated with bacterial metabolites, such as SCFAs [56,57]. In the current study, the proportion of acetate-producing bacteria, such as *Ruminococcus* and *Clostridium*, increased, leading to an increase in acetate production. Studies have found that oral prebiotics may affect the secretion of related gastrointestinal hormones, such as GLP-1 and ghrelin, through the short-chain fatty acids produced by microbiota fermentation [55]. SCFAs have also been shown to cross the blood–brain barrier, directly into the brain, to control feed intake [58]. Furthermore, it was reported that SCFAs can stimulate mouse intestinal L cells to secrete PYY, CCK, and ghrelin, by activating G protein coupled receptors, such as Gpr41 [59]. A recent study suggested that SCFAs inhibit the secretion of GIP by activating the FFAR3 receptor of K cells [60]. This evidence implies that the increased CCK and decreased GIP relate to the SCFAs in the current study. Therefore, it is not surprising that the increase in CCK secretion by L cells and the decrease in GIP secretion by K cells may be due to the increased acetate content in the cecal contents.

5. Conclusions

This paper argues that the poor production performance of broilers under heat stress may be related to SCFAs fermented by gut microbiota involvement in regulating metabolic disorders. Our data will provide valuable insight for future studies on humans in tropical environments who suffer from metabolic disorders, such as diabetes and obesity.

Author Contributions: Conceptualization, G.W. and M.Z.; methodology, G.W., X.L., and M.Z.; software, G.W.; validation, G.W., X.L., and J.F.; formal analysis, G.W., Y.Z., and X.L.; investigation, G.W., Y.Z., and X.L.; resources, G.W.; data curation, G.W., X.L.; writing—original draft preparation, G.W.; writing—review and editing, G.W., X.L., J.F., and M.Z.; visualization, G.W.; supervision, M.Z.; project administration, J.F.; funding acquisition, M.Z. All authors have read and agreed to the published version of the manuscript.

Funding: This study was supported by the National Key Research and Development Program of China (2016YFD0500509). This research was also supported by the Science and Technology Innovation Project of the Chinese Academy of Agricultural Sciences (ASTIP-IAS09).

Institutional Review Board Statement: The study was conducted according to the guidelines of the Declaration of Helsinki, and approved by the Institutional Ethics Committee of Experiment Animal Welfare and Ethical at the Institute of Animal Science of CAAS (NO:IAS2021-75; 8 April 2021).

Data Availability Statement: Not Applicable.

Conflicts of Interest: The authors declare no conflict of interest.

References

1. St-Pierre, N.R.; Cobanov, B.; Schnitkey, G. Economic Losses from Heat Stress by US Livestock Industries. *J. Dairy Sci.* **2003**, *86*, E52–E77. [CrossRef]
2. Hosseini-Vashan, S.; Golian, A.; Yaghobfar, A.; Zarban, A.; Afzali, N.; Esmaeilinasab, P. Antioxidant Status, Immune System, Blood Metabolites and Carcass Characteristic of Broiler Chickens Fed Turmeric Rhizome Powder under Heat Stress. *Afr. J. Biotechnol.* **2012**, *11*, 16118–16125.
3. Garriga, C.; Hunter, R.R.; Amat, C.; Planas, J.M.; Mitchell, M.A.; Moretó, M. Heat Stress Increases Apical Glucose Transport in the Chicken Jejunum. *Am. J. Physiol.-Regul. Integr. Comp. Physiol.* **2006**, *290*, R195–R201. [CrossRef] [PubMed]
4. Fan, Y.; Croom, J.; Christensen, V.; Black, B.; Bird, A.; Daniel, L.; Mcbride, B.; Eisen, E. Jejunal Glucose Uptake and Oxygen Consumption in Turkey Poults Selected for Rapid Growth. *Poult. Sci.* **1997**, *76*, 1738–1745. [CrossRef] [PubMed]
5. Attia, Y.A.; Al-Harthi, M.A.; Sh Elnaggar, A. Productive, Physiological and Immunological Responses of Two Broiler Strains Fed Different Dietary Regimens and Exposed to Heat Stress. *Ital. J. Anim. Sci.* **2018**, *17*, 686–697. [CrossRef]
6. Afsal, A.; Sejian, V.; Madiajagan, B.; Krishnan, G. Heat Stress and Livestock Adaptation: Neuro-Endocrine Regulation. *Int. J. Vet. Anim. Med.* **2018**, *1*, 1–8.
7. Xiong, Y.; Yi, H.; Wu, Q.; Jiang, Z.; Wang, L. Effects of Acute Heat Stress on Intestinal Microbiota in Grow-finishing Pigs, and Associations with Feed Intake and Serum Profile. *J. Appl. Microbiol.* **2020**, *128*, 840–852. [CrossRef]
8. Akşit, M.; Yalçin, S.; Ozkan, S.; Metin, K.; Ozdemir, D. Effects of Temperature During Rearing and Crating on Stress Parameters and Meat Quality of Broilers. *Poult. Sci.* **2006**, *85*, 1867. [CrossRef]
9. Lin, H.; Decuypere, E.; Buyse, J. Acute Heat Stress Induces Oxidative Stress in Broiler Chickens. *Comp. Biochem. Physiol. A. Mol. Integr. Physiol.* **2006**, *144*, 11–17. [CrossRef]

10. Lu, Z.; He, X.F.; Ma, B.B.; Zhang, L.; Li, J.L.; Jiang, Y.; Zhou, G.H.; Gao, F. Increased Fat Synthesis and Limited Apolipoprotein B Cause Lipid Accumulation in the Liver of Broiler Chickens Exposed to Chronic Heat Stress. *Poult. Sci.* **2019**, *98*, 3695–3704. [CrossRef]
11. Sands, J.; Smith, M. Effects of Dietary Manganese Proteinate or Chromium Picolinate Supplementation on Plasma Insulin, Glucagon, Glucose and Serum Lipids in Broiler Chickens Reared under Thermoneutral or Heat Stress Conditions. *Int. J. Poult. Sci.* **2002**, *1*, 145–149.
12. Geraert, P.A.; Padilha, J.C.F.; Guillaumin, S. Metabolic and Endocrine Changes Induced by Chronic Heatexposure in Broiler Chickens: Growth Performance, Body Composition and Energy Retention. *Br. J. Nutr.* **1996**, *75*, 195–204. [CrossRef] [PubMed]
13. Song, J.; Xiao, K.; Ke, Y.L.; Jiao, L.F.; Hu, C.H.; Diao, Q.Y.; Shi, B.; Zou, X.T. Effect of a Probiotic Mixture on Intestinal Microflora, Morphology, and Barrier Integrity of Broilers Subjected to Heat Stress. *Poult. Sci.* **2014**, *93*, 581–588. [CrossRef]
14. Park, S.-O.; Hwangbo, J.; Ryu, C.-M.; Park, B.-S.; Chae, H.-S.; Choi, H.-C.; Seo, O.-S.; Choi, Y.-H. Effects of Extreme Heat Stress on Growth Performance, Lymphoid Organ, IgG and Cecum Microflora of Broiler Chickens. *Int. J. Agric. Biol.* **2013**, *15*, 5.
15. Ramiah, S.K.; Atta Awad, E.; Hemly, N.I.M.; Ebrahimi, M.; Joshua, O.; Jamshed, M.; Saminathan, M.; Soleimani, A.F.; Idrus, Z. Effects of Zinc Oxide Nanoparticles on Regulatory Appetite and Heat Stress Protein Genes in Broiler Chickens Subjected to Heat Stress. *J. Anim. Sci.* **2020**, *98*, skaa300. [CrossRef]
16. Tremaroli, V.; Bäckhed, F. Functional Interactions between the Gut Microbiota and Host Metabolism. *Nature* **2012**, *489*, 242–249. [CrossRef]
17. Ducray, H.A.G.; Globa, L.; Pustovyy, O.; Morrison, E.; Vodyanoy, V.; Sorokulova, I. Yeast Fermentate Prebiotic Improves Intestinal Barrier Integrity during Heat Stress by Modulation of the Gut Microbiota in Rats. *J. Appl. Microbiol.* **2019**, *127*, 1192–1206. [CrossRef]
18. Giles, K.; Pluvinage, B.; Boraston, A.B. Structure of a Glycoside Hydrolase Family 50 Enzyme from a Subfamily That Is Enriched in Human Gut Microbiome Bacteroidetes: Structure of a Bacteroides GH50. *Proteins Struct. Funct. Bioinform.* **2017**, *85*, 182–187. [CrossRef]
19. Cani, P.D.; Dewever, C.; Delzenne, N.M. Inulin-Type Fructans Modulate Gastrointestinal Peptides Involved in Appetite Regulation (Glucagon-like Peptide-1 and Ghrelin) in Rats. *Br. J. Nutr.* **2004**, *92*, 521–526. [CrossRef]
20. Baggio, L.L.; Drucker, D.J. Biology of Incretins: GLP-1 and GIP. *Gastroenterology* **2007**, *132*, 2131–2157. [CrossRef]
21. Højlund, K. Metabolism and Insulin Signaling in Common Metabolic Disorders and Inherited Insulin Resistance. *Dan. Med. J.* **2014**, *61*, B4890. [PubMed]
22. Chen, S.; Zhou, Y.; Chen, Y.; Gu, J. Fastp: An Ultra-Fast All-in-One FASTQ Preprocessor. *Bioinformatics* **2018**, *34*, i884–i890. [CrossRef]
23. Magoc, T.; Salzberg, S.L. FLASH: Fast Length Adjustment of Short Reads to Improve Genome Assemblies. *Bioinformatics* **2011**, *27*, 2957–2963. [CrossRef] [PubMed]
24. Edgar, R.C. UPARSE: Highly Accurate OTU Sequences from Microbial Amplicon Reads. *Nat. Methods* **2013**, *10*, 996–998. [CrossRef] [PubMed]
25. Stackebrandt, E.; Goebel, B.M. Taxonomic Note: A Place for DNA-DNA Reassociation and 16S RRNA Sequence Analysis in the Present Species Definition in Bacteriology. *Int. J. Syst. Evol. Microbiol.* **1994**, *44*, 846–849. [CrossRef]
26. Wang, Q.; Garrity, G.M.; Tiedje, J.M.; Cole, J.R. Naïve Bayesian Classifier for Rapid Assignment of RRNA Sequences into the New Bacterial Taxonomy. *Appl. Environ. Microbiol.* **2007**, *73*, 5261–5267. [CrossRef] [PubMed]
27. Matthews, D.R.; Hosker, J.P.; Rudenski, A.S.; Naylor, B.A.; Treacher, D.F.; Turner, R.C. Homeostasis Model Assessment: Insulin Resistance and Beta-Cell Function from Fasting Plasma Glucose and Insulin Concentrations in Man. *Diabetologia* **1985**, *28*, 412. [CrossRef]
28. Bartlett, J.R.; Smith, M.O. Effects of Different Levels of Zinc on the Performance and Immunocompetence of Broilers Under Heat Stress. *Poult. Sci.* **2003**, *82*, 1580–1588. [CrossRef]
29. Niu, Z.Y.; Liu, F.Z.; Yan, Q.L.; Li, W.C. Effects of Different Levels of Vitamin E on Growth Performance and Immune Responses of Broilers under Heat Stress. *Poult. Sci.* **2009**, *88*, 2101–2107. [CrossRef]
30. Shi, D.; Bai, L.; Qu, Q.; Zhou, S.; Yang, M.; Guo, S.; Li, Q.; Liu, C. Impact of Gut Microbiota Structure in Heat-Stressed Broilers. *Poult. Sci.* **2019**, *98*, 2405–2413. [CrossRef]
31. Lu, Z.; He, X.; Ma, B.; Zhang, L.; Li, J.; Jiang, Y.; Zhou, G.; Gao, F. Chronic Heat Stress Impairs the Quality of Breast-Muscle Meat in Broilers by Affecting Redox Status and Energy-Substance Metabolism. *J. Agric. Food Chem.* **2017**, *65*, 11251–11258. [CrossRef] [PubMed]
32. Mujahid, A.; Yoshiki, Y.; Akiba, Y.; Toyomizu, M. Superoxide Radical Production in Chicken Skeletal Muscle Induced by Acute Heat Stress. *Poult. Sci.* **2005**, *84*, 307–314. [CrossRef] [PubMed]
33. Quinteiro-Filho, W.M.; Ribeiro, A.; Ferraz-De-Paula, V.; Pinheiro, M.L.; Sakai, M.; Sa, L.; Ferreira, A.; Palermo-Neto, J. Heat Stress Impairs Performance Parameters, Induces Intestinal Injury, and Decreases Macrophage Activity in Broiler Chickens. *Poult. Sci.* **2010**, *89*, 1905–1914. [CrossRef] [PubMed]
34. Zeferino, C.P.; Komiyama, C.M.; Pelícia, V.C.; Fascina, V.B.; Aoyagi, M.M.; Coutinho, L.L.; Sartori, J.R.; Moura, A.S.A.M.T. Carcass and Meat Quality Traits of Chickens Fed Diets Concurrently Supplemented with Vitamins C and E under Constant Heat Stress. *Animal* **2016**, *10*, 163–171. [CrossRef] [PubMed]

35. Han, G.; Yang, H.; Wang, Y.; Haraguchi, S.; Miyazaki, T.; Bungo, T.; Tashiro, K.; Furuse, M.; Chowdhury, V.S. L-Leucine Increases the Daily Body Temperature and Affords Thermotolerance in Broiler Chicks. *Asian-Australas. J. Anim. Sci.* **2019**, *32*, 842–848. [CrossRef] [PubMed]
36. Song, Z.; Liu, L.; Sheikhahmadi, A.; Jiao, H.; Lin, H. Effect of Heat Exposure on Gene Expression of Feed Intake Regulatory Peptides in Laying Hens. *J. Biomed. Biotechnol.* **2012**, *2012*, 1–8. [CrossRef]
37. Yunianto, V.D.; Hayashit, K.; Kaiwda, S.; Ohtsuka, A.; Tomita, Y. Effect of Environmental Temperature on Muscle Protein Turnover and Heat Production in Tube-Fed Broiler Chickens. *Br. J. Nutr.* **1997**, *77*, 897–909. [CrossRef]
38. Lu, Q.; Wen, J.; Zhang, H. Effect of Chronic Heat Exposure on Fat Deposition and Meat Quality in Two Genetic Types of Chicken. *Poult. Sci.* **2007**, *86*, 1059–1064. [CrossRef]
39. Toghyani, M.; Toghyani, M.; Shivazad, M.; Gheisari, A.; Bahadoran, R. Chromium Supplementation Can Alleviate the Negative Effects of Heat Stress on Growth Performance, Carcass Traits, and Meat Lipid Oxidation of Broiler Chicks without Any Adverse Impacts on Blood Constituents. *Biol. Trace Elem. Res.* **2012**, *146*, 171–180. [CrossRef]
40. Yuan, L.; Lin, H.; Jiang, K.J.; Jiao, H.C.; Song, Z.G. Corticosterone Administration and High-Energy Feed Results in Enhanced Fat Accumulation and Insulin Resistance in Broiler Chickens. *Br. Poult. Sci.* **2008**, *49*, 487–495. [CrossRef]
41. Lu, Z.; He, X.; Ma, B.; Zhang, L.; Li, J.; Jiang, Y.; Zhou, G.; Gao, F. Serum Metabolomics Study of Nutrient Metabolic Variations in Chronic Heat-Stressed Broilers. *Br. J. Nutr.* **2018**, *119*, 771–781. [CrossRef]
42. Goodrich, J.K.; Waters, J.L.; Poole, A.C.; Sutter, J.L.; Koren, O.; Blekhman, R.; Beaumont, M.; Van Treuren, W.; Knight, R.; Bell, J.T.; et al. Human Genetics Shape the Gut Microbiome. *Cell* **2014**, *159*, 789–799. [CrossRef]
43. Zhang, J.; Song, L.; Wang, Y.; Liu, C.; Zhang, L.; Zhu, S.; Liu, S.; Duan, L. Beneficial Effect of Butyrate-producing Lachnospiraceae on Stress-induced Visceral Hypersensitivity in Rats. *J. Gastroenterol. Hepatol.* **2019**, *34*, 1368–1376. [CrossRef]
44. Demidova, T.Y.; Lobanova, K.G.; Oynotkinova, O.S. Gut Microbiota Is an Endocrine Organ. *Obes. Metab.* **2020**, *17*, 299–306. [CrossRef]
45. He, X.; Lu, Z.; Ma, B.; Zhang, L.; Li, J.; Jiang, Y.; Zhou, G.; Gao, F. Chronic Heat Stress Alters Hypothalamus Integrity, the Serum Indexes and Attenuates Expressions of Hypothalamic Appetite Genes in Broilers. *J. Therm. Biol.* **2019**, *81*, 110–117. [CrossRef]
46. Melo-Duran, D.; Gonzalez-Ortiz, G.; Sola-Oriol, D.; Martinez-Mora, M.; Perez, J.F.; Bedford, M.R. Relationship between Peptide YY, Cholecystokinin and Fermentation Products in Fasted, Re-Fed and Ad Libitum Fed Broiler Chickens. *Anim. Feed Sci. Technol.* **2019**, *247*, 141–148. [CrossRef]
47. Louie, D.S.; May, D.; Miller, P.; Owyang, C. Cholecystokinin Mediates Feedback Regulation of Pancreatic Enzyme Secretion in Rats. *Am. J. Physiol. Gastrointest. Liver Physiol.* **1986**, *250*, G252–G259. [CrossRef] [PubMed]
48. Furuse, M.; Tachibana, T.; Ohgushi, A.; Ando, R.; Yoshimatsu, T.; Denbow, D.M. Intracerebroventricular Injection of Ghrelin and Growth Hormone Releasing Factor Inhibits Food Intake in Neonatal Chicks. *Neurosci. Lett.* **2001**, *301*, 123–126. [CrossRef]
49. Xu, P.; Siegel, P.B.; Denbow, D.M. Genetic Selection for Body Weight in Chickens Has Altered Responses of the Brain's AMPK System to Food Intake Regulation Effect of Ghrelin, but Not Obestatin. *Behav. Brain Res.* **2011**, *221*, 216–226. [CrossRef]
50. Liu, L.; Song, Z.; Jiao, H.; Lin, H. Glucocorticoids Increase NPY Gene Expression via Hypothalamic AMPK Signaling in Broiler Chicks. *Endocrinology* **2014**, *155*, 2190–2198. [CrossRef] [PubMed]
51. Pearce, S. Evaluation of the Chronological Impact Heat Stress Has on Swine Intestinal Function and Integrity. Ph.D. Thesis, Iowa State University, Ames, IA, USA, 2014.
52. Ding, K.-H.; Shi, X.-M.; Zhong, Q.; Kang, B.; Xie, D.; Bollag, W.B.; Bollag, R.J.; Hill, W.; Washington, W.; Mi, Q.-S.; et al. Impact of Glucose-Dependent Insulinotropic Peptide on Age-Induced Bone Loss. *J. Bone Miner. Res.* **2007**, *23*, 536–543. [CrossRef] [PubMed]
53. Flint, H.J.; Duncan, S.H.; Scott, K.P.; Louis, P. Links between Diet, Gut Microbiota Composition and Gut Metabolism. *Proc. Nutr. Soc.* **2015**, *74*, 13–22. [CrossRef] [PubMed]
54. Holmes, E.; Li, J.V.; Marchesi, J.R.; Nicholson, J.K. Gut Microbiota Composition and Activity in Relation to Host Metabolic Phenotype and Disease Risk. *Cell Metab.* **2012**, *16*, 559–564. [CrossRef]
55. Cani, P.D.; Lecourt, E.; Dewulf, E.M.; Sohet, F.M.; Pachikian, B.D.; Naslain, D.; De Backer, F.; Neyrinck, A.M.; Delzenne, N.M. Gut Microbiota Fermentation of Prebiotics Increases Satietogenic and Incretin Gut Peptide Production with Consequences for Appetite Sensation and Glucose Response after a Meal. *Am. J. Clin. Nutr.* **2009**, *90*, 1236–1243. [CrossRef]
56. den Besten, G.; Bleeker, A.; Gerding, A.; van Eunen, K.; Havinga, R.; van Dijk, T.H.; Oosterveer, M.H.; Jonker, J.W.; Groen, A.K.; Reijngoud, D.-J.; et al. Short-Chain Fatty Acids Protect Against High-Fat Diet–Induced Obesity via a PPARγ-Dependent Switch From Lipogenesis to Fat Oxidation. *Diabetes* **2015**, *64*, 2398–2408. [CrossRef]
57. Chambers, E.S.; Viardot, A.; Psichas, A.; Morrison, D.J.; Murphy, K.G.; Zac-Varghese, S.E.K.; MacDougall, K.; Preston, T.; Tedford, C.; Finlayson, G.S.; et al. Effects of Targeted Delivery of Propionate to the Human Colon on Appetite Regulation, Body Weight Maintenance and Adiposity in Overweight Adults. *Gut* **2015**, *64*, 1744–1754. [CrossRef]
58. Klosterbuer, A.; Sanders, L.; Potter, S.; Thomas, W.; Slavin, J. A Blend of Soluble Fiber and Resistant Starch Promotes Feelings of Fullness in Humans. *FASEB J.* **2010**, *24*, 220–224.
59. Samuel, B.S.; Shaito, A.; Motoike, T.; Rey, F.E.; Backhed, F.; Manchester, J.K.; Hammer, R.E.; Williams, S.C.; Crowley, J.; Yanagisawa, M.; et al. Effects of the Gut Microbiota on Host Adiposity Are Modulated by the Short-Chain Fatty-Acid Binding G Protein-Coupled Receptor, Gpr41. *Proc. Natl. Acad. Sci. USA* **2008**, *105*, 16767–16772. [CrossRef]
60. Lee, E.-Y.; Zhang, X.; Miyamoto, J.; Kimura, I.; Taknaka, T.; Furusawa, K.; Jomori, T.; Fujimoto, K.; Uematsu, S.; Miki, T. Gut Carbohydrate Inhibits GIP Secretion via a Microbiota/SCFA/FFAR3 Pathway. *J. Endocrinol.* **2018**, *239*, 267–276. [CrossRef]

animals

MDPI

Review

Protein and Amino Acid Metabolism in Poultry during and after Heat Stress: A Review

Mohammed M. Qaid [1,2,*] and Maged A. Al-Garadi [1,2,*]

1 Animal Production Department, College of Food and Agriculture Sciences, King Saud University, Riyadh 11451, Saudi Arabia
2 Veterinary Medicine Faculty, Thamar University, Dhamar 13020, Yemen
* Correspondence: mqaid@ksu.edu.sa (M.M.Q.); malgaradi@ksu.edu.sa (M.A.A.-G.); Tel.: +966-509-844-024 (M.M.Q.); +966-533429411 (M.A.A.-G.)

Simple Summary: Broilers must be reared under thermoneutral conditions and comfort zones; therefore, any deviation from the neutral thermal zone causes stress and a consequent disturbance in the turnover or the metabolism of nutrients. This review addressed the biosynthesis of amino acids and/or protein metabolism under normal conditions and heat stress conditions. In addition, hormonal responses to stress and the role of endocrine hormones in protein metabolism have been reviewed. In addition, the aim of this review is to summarize the studies related to the assessment of heat stress, the physiological stress regulation mechanism, and the nutritional strategies for the prevention of heat stress in poultry.

Abstract: This review examined the influence of environmental heat stress, a concern facing modern broiler producers, on protein metabolism and broiler performance, as well as the physiological mechanisms that activate and control or minimize the detrimental impacts of stress. In addition, available scientific papers that focused on amino acids (AA) digestibility under stress conditions were analyzed. Furthermore, AA supplementation, a good strategy to enhance broiler thermotolerance, amelioration, or stress control, by keeping stress at optimal levels rather than its elimination, plays an important role in the success of poultry breeding. Poultry maintain homeothermy, and their response to heat stress is mainly due to elevated ambient temperature and the failure of effective heat loss, which causes a considerable negative economic impact on the poultry industry worldwide. Reduced feed intake, typically observed during heat stress, was the primary driver for meat production loss. However, accumulating evidence indicates that heat stress influences poultry metabolism and endocrine profiles independently of reduced feed intake. In conclusion, high ambient temperatures significantly reduced dietary AA intake, which in turn reduced protein deposition and growth in broilers. Further studies are required to determine the quantity of the AA needed in warm and hot climates and to introduce genetic tools for animal breeding associated with the heat stress in chickens.

Keywords: amino acids; broiler; heat stress; heat tolerance; protein metabolism

check for updates

Citation: Qaid, M.M.; Al-Garadi, M.A. Protein and Amino Acid Metabolism in Poultry during and after Heat Stress: A Review. *Animals* **2021**, *11*, 1167. https://doi.org/10.3390/ani11041167

Academic Editor: Natalie Morgan

Received: 11 March 2021
Accepted: 12 April 2021
Published: 19 April 2021

1. Introduction

Poultry meat is an essential source of dietary protein, and the industry has developed high grade poultry because of improved farming techniques, automation equipment, and comprehensive and balanced feeding, and other new technologies [1]. In the past decade, broiler production has increased rapidly in tropical and subtropical areas and is expected to sustain robust growth in the future. However, according to a review [2], high environmental temperatures are one of the greatest challenges of poultry and live stock performance, leading to the decline in production efficiency in these countries. Furthermore, modern commercial broilers are more sensitive to heat stress than previous generations due to their higher performance, growth rate, and feed conversion efficiency [3]. Indeed, commercial poultry strains can reach a high production yield, but their body metabolism, being comparatively accelerated, has poor thermoregulation and is poorly adapted to the living environments compared to native backyard chickens [1]. The higher growth rate of broilers has several consequences, such as higher feed consumption and metabolism and elevated production of internal heat. To reduce the heat load and avoid heat-induced mortality in birds, heat loss and/or lowering heat production could be achieved through reduced feed consumption, resulting in a depressed growth rate and lower final body weight, reduced breast meat weight, and lower egg quality, size, and rate in hens. In addition, the decreased feed consumption could result in nutrient deficiency, such as proteins, AA, and energy [4–6].

Among the environmental factors, heat stress negatively affects feed consumption, body weight gain [7], and carcass characteristics [2]. Additionally, HS may cause oxidative stress in the body and develop many free radicals, stimulating membrane lipid peroxidation, and hence the degradation of DNA and protein membranes [8,9]. In poultry, mitochondrial superoxide production as oxidative stress was observed on exposure to HS [9]. Besides the role of AA as protein and peptide components, some AA (e.g., glutamine, cysteine, leucine, arginine, tryptophan, and proline) are involved in the regulation of metabolic pathways, thereby affecting growth, protein accumulation, maintenance, immunity, and health [10].

The ambient temperature impacts the protein turnover rate in broiler skeletal muscle [11]. Not only protein anabolism, but also protein catabolism are energetically expensive. The growth depression of heat-exposed chickens showed lower protein gain and retention. High protein sources have beneficial effects through the improved growth of heat-stressed broilers [12]. They indicated that protein synthesis in broilers was more affected than protein breakdown with HS, resulting in reduced protein deposition in the skeletal muscles. A review [11] showed that the high ambient temperatures and dietary protein consumption affected muscle protein turnover in broilers. In broiler chickens, HS alters muscle protein and AA metabolism and accelerates liver gluconeogenesis for energy supply [13]. Dietary approaches, such as modifications of energy and protein content of the diet, are the most practical and preferred ways to alleviate heat distress in poultry and enhance broiler performance under these conditions [14,15]. Improving the overall equilibrium of the dietary AA was more effective than increasing total protein consumption [16].

Limited studies are available that address the effects of HS on protein metabolism in broilers. Therefore, it is necessary to detect mechanisms or methods that allow producers to effectively reduce the detrimental influences of environmental HS on broilers, in particular on protein metabolism via protein accretion or degradation of muscle. The present review will focus on the effects of environmental heat stress on protein metabolism and broiler performance, as well as the physiological mechanisms and nutritional strategies that mitigate the negative effects of heat stress, particularly the role of AA in reducing HS in stressed broilers.

2. Amino Acid and/or Protein Metabolism

AA are required for most biological activities. The AA transport into the apical membrane and out of the lateral basal membrane of enterocytes. Their transport relies on sodium-dependent symporters, proton-motive forces, antiporters, and the gradient of other AA. The metabolic fate of absorbed AA mainly depends on nutrient availability [17]. AA moving through catabolic pathways ultimately serve as precursors of gluconeogenesis [13] and contribute to 40% of the total AA loss in fasted animals. Proteins are synthesized from free AA, which become available either from dietary (the end product of digestion) or from metabolic origins as the result of AA biosynthesis within the body. These AA, either circulating via the blood or accumulating within tissues, form pools. The AA concentrations within these pools are based on the equilibrium between gains and losses [18]. Dietary AA are used to build protein for muscle growth, membrane glycoproteins, and enzymes involved in numerous biochemical processes, and act as precursors for the synthesis of DNA/RNA [10]. The AA catabolize in the liver to integrate into protein, which supplies peripheral tissues [13]. Protein turnover refers to the equilibrium between the anabolism and catabolism of protein. The metabolic utilization of AA is equally diverse. Anabolism or protein synthesis facilitates dietary AA to fuse into proteins, or biosynthesize in the body tissues. Catabolism occurs through the breakdown of proteins to build amino groups that produce urea or further protein. In addition, to produces carbon skeleton molecules for glucose production (glucogenesis) or fatty acids (lipogenesis), carbon dioxide, and the release of energy. The role of endocrine hormones in protein metabolism is shown in Table 1.

Table 1. Role of endocrine hormones in protein metabolism and summary of potential methods for assessing stress.

Hormones	Protein Synthesis	Proteolysis
Insulin	Stimulated	Inhibited
Glucagon	Inhibited	Stimulated
Epinephrine	Inhibited	Stimulated
Glucocorticoids: ACTH *, CS, and Cortisol	Inhibited	Stimulated (gluconeogenesis)
Thyroid hormones T_4 and T_3	Accelerated skeletal muscle protein turnover and heat production under the hot conditions	
Growth hormone	Stimulated	Inhibited
	Potential Methods for Assessing Stress	
Behavioral/Physiological	**Endocrine**	**Metabolic Systems**
Activity/sleep patterns	Catecholamines	Immune function
Posture/stereotypes	ACTH/CRH, glucocorticoids	Disease state
Feed and water intake	Gonadotrophin/sex steroids	Growth performance
Heart rate and blood pressure	Endorphin (β), renin and prolactin	Reproductive performance

* abbreviations: ACTH: adrenocorticotropin, CS: corticosterone, CRH: Corticotropin-releasing hormone, T_3: triiodothyronine, and T_4: thyroxine, adapted from [19,20].

Nitrogen excretion can be used to determine protein balance through the measurement of nitrogen losses during protein catabolism or recycling [18]. Endogenous or dietary proteins hydrolyze the previous absorption. The tissue proteins of birds are renewed frequently with the liberation of endogenous AA. Furthermore, there are many metabolic reactions converting metabolites into nonessential AA [18].

Recently, AA are applied not only as signaling molecules of the cell and the protein phosphorylation cascade, but also as regulators of gene expression. Moreover, AA are fundamental precursors for hormone synthesis and other nitrogenous elements that have considerable biological significance. Normal levels of AA and their metabolites, such as glutathione, polyamines, taurine, nitric oxide, serotonin, and thyroid hormones are needed for their functions. Nevertheless, elevated levels of AA and their metabolites, such as ammonia, asymmetric dimethylarginine, and homocysteine are considered pathogenic for the body, and lead to oxidative stress, and cause diseases and disorders of the cardiovascular and neurological systems. Therefore, an ideal balance of AA in the feed and bloodstream is crucial for the homeostasis of the body. AA not only have a role as the building blocks of polypeptides and proteins, but also regulate the fundamental metabolic routes that are essential for growth, maintenance, immunity, and reproduction. These functional AA include glutamine, leucine, proline, arginine, cysteine, and tryptophan [10].

3. Biosynthesis of Amino Acids

Birds and all vertebrates, dissimilar to plants and many bacteria, are unable to synthesize some AA, so these are termed essential AA and are required for tissue renewal through protein synthesis. Thus, essential AA must be supplemented in the diet. For protein synthesis, all AA are similarly essential owing to the absence of any AA interfere with the anabolic processes. However, nutritionally AA are classified into three groups [18]. Essential AA must be provided by feed and may be classified into two groups. One group is strictly essential because they cannot be synthesized, even from AA metabolic intermediates, such as glucogenic that yield intermediates of glycolysis pathway or ketogenic that yield intermediates of acetyl-CoA or acetoacetate. The transaminases of that group are absent, for instance methionine, lysine, tryptophan, threonine, and phenylalanine. The other group may be insufficiently synthesized from their precursors, for example, glycine, leucine, isoleucine, valine, arginine, histidine, and proline. Semiessential AA may be synthesized from essential AA. Tyrosine and cysteine originate from phenylalanine and methionine, respectively. Cysteine is synthesized from serine (nonessential AA) and methionine (essential AA). Nonessential AA are easily synthesized from intermediary metabolites or similarly nonessential AA: alanine, serine, aspartic and glutamic acids in the former group; and asparagine and glutamine in the latter group [18].

4. Effect of HS on Protein Metabolism or Turnover

Heat as a stress factor affects protein metabolism during the postabsorptive stage as muscle breakdown and changes in the quantity of lean tissue may occur in different species [21]. In the muscle protein, the RNA/DNA synthesis capacity is reduced by HS [22]. During environmental hyperthermia, muscle tissue catabolism is increased due to increased plasma markers during muscle breakdown. In lactating cows, HS increases plasma urea nitrogen concentration [23]; however, whether this elevation stems from reduced plasma volume, increased protein degradation, or other reasons remains unknown. Thus, blood urea nitrogen (BUN) is used as an indicator of muscle catabolism or breakdown, because tissue degradation results in an increase in BUN [21]. A review [24] reviewed that uric acid excretion is increased in stressed poultry owing to corticosterone -driven gluconeogenesis. Other indicators or measures of protein breakdown (muscle catabolism) include increased plasma creatinine, Nt-methyl histidine, creatine, and creatine kinase (CK) concentrations. An increase in these markers has been detected during heat load in chickens, turkeys [25], cows [26], pigs [27], and humans [28]. The increased level of these parameters indicates enhanced muscle protein catabolism.

Insulin stimulates protein synthesis or accretion. However, during heat-load, increased muscle protein degradation causes the liver to utilize available AA as gluconeogenic substrates from the carbon skeleton through the gluconeogenesis pathway [19]. Under stress conditions, the corticoid hormones (CS, ACTH) suppress the synthesis of tissue proteins and boost proteolysis, as catabolic action is elevated in the blood stream. Glycerol produced from lipid degradation is one of the gluconeogenic substrates and accounts for 20% of the glucose production. Therefore, the other products of protein catabolism are used as substrates for glucose production. First, not only the heart and lung tissue proteins are enhanced by the catabolic transformations of protein composition, but also all tissues except for the nervous system. The muscle tissues (muscle protein) that have the highest body nitrogen content are more sensitive to corticosterone administration, resulting in decreased muscle mass and growth retardation in stressed chickens [29]. A study [30] found that feed deprivation reduced protein synthesis in the liver of starved chickens, as well as plasma albumin and total protein levels. A study [31] indicated that the depletion of plasma free AA, elevated blood uric acid concentration, reduced protein synthesis possibly reflected reduced N retention and more active protein catabolism in broilers challenged by very short-term high temperatures. However, chronic exposure to HS decreased protein digestion, decreased feed digestibility, reduced protein breakdown, reduced protein synthesis in the muscles, and decreased most plasma free AA (especially branched-chain AA and sulfur) [32], whereas the serum levels of glutamic acid, aspartic acid, and phenylalanine increased [33]. It was found that protein synthesis and N deposition were depressed and proteolysis increased during HS [34]. AA catabolism was enhanced under chronic HS [13]; thus, all plasma free AA concentrations decreased, except for glutamic acid, aspartic acid, and phenylalanine. Based on these studies, protein breakdown may increase rapidly in very short-term HS, resulting in a decrease in protein synthesis and an increase in plasma uric acid levels, but then decrease protein breakdown and maintain uric acid levels around normal concentrations as the thermal stress continues.

Two previous studies, one on chickens and the other on turkeys, found that heat stress reduced uric acid levels in the blood, which could be attributed to a lower level of total protein as a result of hypotonic overhydration [25]. Although no sodium concentration was determined in these studies, water intoxication due to excessive water intake causes overhydration when the amount of water intake exceeds that of water excretion in the kidney. As a result, the sodium level in the blood is diluted, resulting in hyponatremia. As a result, hyponatremia is the most common electrolyte disorder that must be carefully managed [35].

There is little knowledge about the renal function of broilers in hot climates, especially in terms of compensating for water and electrolyte loss. During acute heat exposure, there were variable changes in urinary electrolyte excretion in chickens. Reduced glomerular filtration rates (GFR), tubular sodium reabsorption rates, and filtered water amounts may help heat-acclimated birds reduce the metabolic heat load associated with active solute recovery from the glomerular ultrafiltrate. When heat-acclimated birds consume excessive water intake to support evaporative cooling, these changes in kidney function are thought to reduce urinary fluid and solute loss [36]. More research is needed, however, to better explain how various factors may contribute to this evidence.

In addition, a study [37] demonstrated increased uric acid levels in heat stressed chickens. Hence, the application of high-protein diets in HS broilers leads to increased blood plasma uric acid and relieved oxidative stress. Furthermore, the activity of enzymes during AA or protein metabolism under stress conditions has been analyzed. Aspartate aminotransferase (AST) and alanine aminotransferase (ALT) are intracellular enzymes produced in the liver, skeletal muscles, and heart of poultry, and used as indicators of the liver, muscle, and heart damage [38].

The rate of protein accretion is always a constant balance between breakdown (protein-lysis) and synthesis (protein-genesis) [18]. The reduction in protein accretion under conditions of chronic HS is because the rate of protein-genesis is more greatly affected than the rate of proteolysis [33]. Protein synthesis was reduced more in the breast muscles than that in leg muscles; this may be related to higher oxidative metabolism of the leg muscles and increased glycolytic metabolism of the breast muscle [39]. Increasing dietary protein content from 20% to 25% at 32 °C did not affect the rate of protein synthesis but did increase muscle protein deposition, possibly by reducing protein breakdown [33]. The authors of [40] suggested that energy for protein synthesis at the molecular level may be limited at high temperatures; glucose supplementation improves the growth rate at high ambient temperatures. The effects of HS on protein turnover are controversial but may be related to the magnitude and duration of the heat load producing either a detrimental or therapeutic effect. Both HS and pair feeding reduced the muscle mass of rats; however, pair-fed animals had higher protein degradation, leading to a more severe loss of skeletal muscle that might be attributed to protein preservation triggered by heat exposure [41].

5. Heat Shock Proteins

Heat stress produces the over-expression of heat shock factors and heat shock proteins (HSP) in bird tissues. HSP regulate multiple molecular pathways in cells in response to stress conditions and change the homeostasis of cells and tissues [1]. HSP affect mediators of inflammation and infection. HSP are molecular chaperones during increased heat, and offer defense. HSP possess mediated responses to endotoxin stimulated synthesis of cytokine, and [42] reviewed that HSP 70 overlap with NFκB transcription, leading to the deactivation of the inflammatory response. Intestinal permeability offers new targets for HS remedy. A study [43] reported that when any living organisms are exposed to HS, the synthesis of most proteins is delayed; however, a group of highly conserved proteins, HSP, is rapidly synthesized. HS causes an increase in HSP synthesis, and are also known as stress proteins [44].

A study [45] indicated that HS and subsequent elevated HSP might inhibit muscle mass increase, even with unchanged feed intake. Glutamine seems to have a protective effect on heat-shocked skeletal myotubes by inhibiting protein degradation [46] and this effect might be mediated by HSPs (primarily HSP70 and HSP25/27), independently of glutamine metabolism based on a nonsufficient-metabolizable glutamine analog to mimic the HSP enhancing effect [47]. Additionally, a review [48] reported that increased HSPs defend cells from damage and protect them from apoptosis. HSP 70 is the most common family of HSPs and considered the most conservative, and is plentiful in most living organisms and increases synthesis after cell stress [49]. Glutamine supplementation has been found to increase HSP expression and improve the stress response [47].

6. Physiological Mechanism of Stress Regulation in Poultry

According to a review [50], physiological stress regulation mechanisms are classified into three stages: alarm reaction (neurogenic system), resistance or adaptation (endocrine system), and exhaustion. Under HS in fowls, heat generation and metabolizable energy (ME) intake are decreased, which might be owing to reduced thyroid hormones and corticosterone concentration since those endocrine hormones are related to protein turnover acceleration in muscle and thermogenesis [51]. During HS in birds, abnormal pathways occur, including gluconeogenesis, and as protein catabolism increases the efficiency of energy absorption is decreased because of the increased energy retention. Therefore, during periods of stress, it is possible for decreases in growth to be accompanied by increases in body fat deposition [52]. Poultry's normal body temperature is around 41–42 °C, and the thermoneutral temperature for maximum growth is between 18–21 °C [53]. The environmental temperature, the thermal neutral zone, and the influence of the ambient temperature on heat production and body temperature are shown in Figure 1.

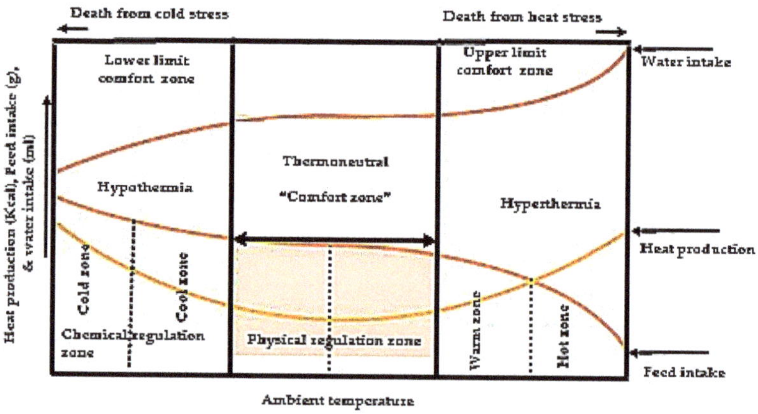

Figure 1. Feed and water intake, and body temperature production related to ambient temperature. Modified after [54].

Poultry produce heat through muscular activity and metabolic processes. The optimum or ideal temperature for performance is 19–22 °C in laying hens, and 18–22 °C in broilers [55]. Heat produced in the body is lost through conduction, convection, radiation, evaporation, and fecal excretion. Heat loss falls into two main categories. First, sensible heat loss occurs through convection, conduction, and radiation when hens are in a comfortable environment of 21–25 °C, and show optimum growth rate, egg quality and size, quality of egg shell, egg production, and hatchability. Second, insensible heat loss occurs through panting (evaporative heat loss), and begins when the temperature reaches 26.67 °C [56]. In addition, birds can increase respiration rates up to 10× normal. Additionally, chickens diminish heat by raising and spreading their wings and separating themselves from others. HS has a negative impact on both physiological and behavioral activities. Monitoring these criteria during rearing is critical for identifying HS properties and taking appropriate actions to mitigate the effects of HS while developing high-quality poultry through physiological and management strategies such as heat stress acclimation and poultry housing facilities.

7. Hormonal Responses to Stress and the Hormonal Control of Protein Metabolism

Hormone signaling plays a vital role in regulating homeostasis, which includes growth, metabolism, reproduction, and immunity. The overall responses to stress shown in Figure 2. Rapid endocrine responses are mediated by the sympathetic nervous system activation of the adrenal medulla (SA system). However, the long-term effects are due to the activation of the hypothalamic–pituitary–adrenal cortex axis (HPA axis) and the production of glucocorticoids for long time.

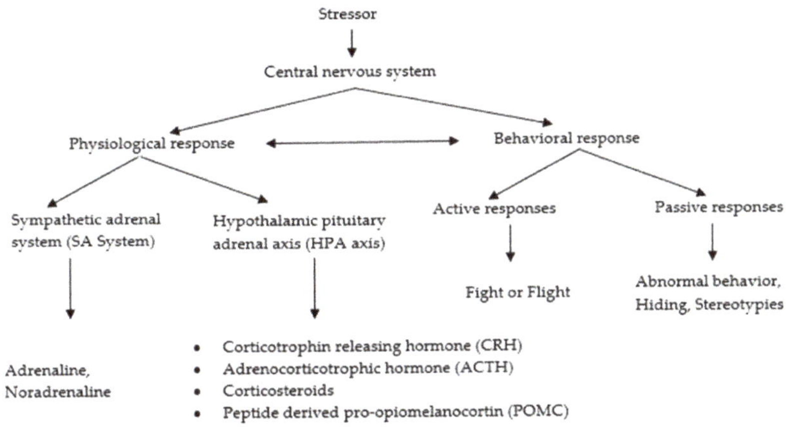

Figure 2. Scheme of overall responses to stress.

The effects of stress on growth performance and reproduction through stress hormonal axis, the reproductive axis, and their interaction are shown in (Figures 3 and 4). The effects of HS on appetite and reproductive hormones are negative. Monitoring appetite and reproductive hormone regulation during rearing are critical for mitigating the negative effects of HS and developing high-quality poultry through hormonal strategies.

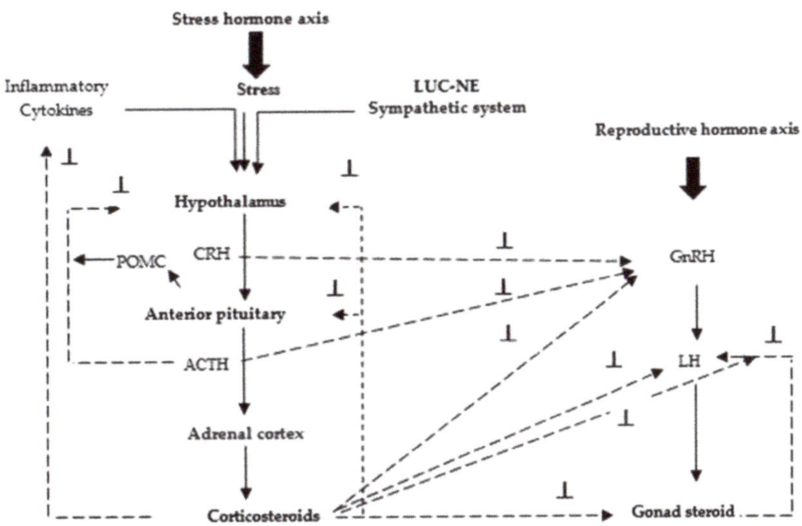

Figure 3. Hypothalamic–pituitary–adrenal axis and its impact on animal reproduction (stress hormone

axis include CRH: Corticotropin-releasing hormone; ACTH: Adrenocorticotropic hormone; corticosteroids contain glucocorticoids and mineralocorticoids; POMC: pro-opiomelanocortin; LUC-NE: locus ceruleus neurons which secrete noradrenaline; reproductive hormone axis includes GnRH Gonadotropin-releasing hormone; LH: luteinizing hormone; gonad steroid: testosterone, estradiol, progesterone; ⊥: inhibition. Modified after [20].

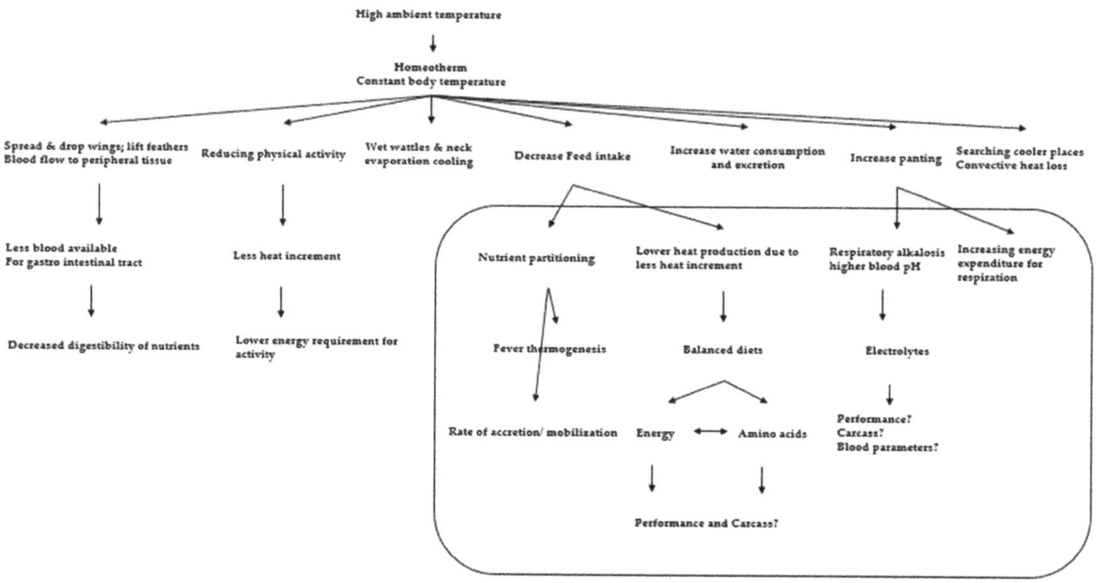

Figure 4. Behavioral and physiological adjustments of chickens at high ambient temperatures and its effects on nutrient intake and utilization. Modified after [57].

8. Assessment of Stress

HS has a negative impact on production performance, intestinal health, body temperature, immune responses, appetite hormone regulation, and oxidative properties. It is critical to monitor these criteria during rearing in order to identify HS possessions and take timely action to mitigate the negative effects of high ambient temperature. Stress can be an assessment by three potential methods through behavioral/physiological, endocrine, and metabolic systems measurements. These have been suggested as possible indicators of animal well-being (Table 1) [20]. In addition, neuropeptide Y (NPY) expression is increased in heat-exposed chick brains. NPY has a hypothermic action through the body temperature and heat stress regulation in chicks [58].

9. Nutritional Strategies for Preventing HS in Poultry

Nutritional strategies targeted to alleviate and overcome the adverse effects of HS in domestic fowl [59], include preserving feed consumption, electrolytes, water balance, or even by adding vitamins (as ascorbic acid) and minerals [4,5]. Primary strategies in changing the diet formulation of broilers under constant or cycling high-temperature conditions include the suitable use of protein-rich ingredients (AA and crude protein) [12]. It is necessary to ensure the balance of certain AA, especially, the arginine: lysine ratio, and the supplementation synthetic methionine to correct any nutritional shortages [60]. A review [61] found that when protein is the source of energy, the heat increment or specific dynamic action is much greater than when fat or carbohydrate are the sources of energy. Consequently, there are concerns regarding diet-induced heat production related to protein in hyperthermic broilers. Some authors have mentioned the harmful effects of feeding high protein diets [62], leading to the recommendation of a reduced protein diet to control further higher thermogenesis [63]. However, higher dietary crude protein (CP) can compensate reduced AA consumption in stressed broilers, thus it seems to be beneficial in hot conditions, resulting in an improved growth rate [64]. In addition, a review [65] reported that limited protein supplementation decreased water drinking under HS and limits broilers' performance. Therefore, the AA balance plays a chief role in the scientific conflict regarding the proteins needed for hyperthermic poultry, and it is necessary to determine the AA required for thermoneutrality. Additionally, the protein needed can change gradually after HS exposure, depending on the time exposed. Moreover, a study [66] found lower protein degradation with HS could be normalized with thyroxine supplementation.

Dietary supplementation with one or a mixture of functional AA (glutamine, leucine, proline, arginine, cysteine, and tryptophan) is possibly beneficial. First, for ameliorating or reducing health threats during different periods of the life cycle, such as the metabolic syndrome, fetal growth limitation, weaning-associated wasting syndrome and intestinal dysfunction, neonatal morbidity and mortality, diabetes, obesity, infertility and cardiovascular disease. Second, for improving or optimizing the efficiency of metabolic transformations to boost muscle development, meat and egg quality, and milk production, and reducing adiposity by inhibiting excess fat deposition. Thus, AA has important functions in both health and nutrition [10].

Dietary glutamine supplementation alleviates heat stress, resulting in improved performance and humoral immune response in poultry [67]. In addition, glutamine minimizes the HS effects in heat-stressed chickens in the first weeks of life [68]. Besides it plays several roles in the metabolism and homeostasis of tissues. A study [69] reported that glutamic acid and glutamine supplementation, as a conditionally essential AA in broilers under stress conditions, could be beneficial in improving the growth performance and health. For optimal broiler performance, the use of a high-fat diet (fat is less thermogenic than carbohydrates) with adequate levels of essential AA [70] has been suggested. However, high lysine or Arg:Lys ratio during HS did not reduce the adverse effects of heat stress or even improve the growth rate. Consequently, there is a further challenge to determine the best nutrient during feeding in many fowls during HS [6]. The addition of appropriate feed additives may be beneficial in improving intestinal absorption and minimizing the negative effects of HS. The addition of active substances during incubation is the most recent advancement. By instilling thermotolerance in newly hatched birds, these methods are expected to have an impact on the poultry industry. The physiology, production, and immunological response of broilers under heat stress are all affected by the feeding regimen, which should be tailored to the Ross-308 and Cobb-500 strains [71]. It is necessary to monitor nutritional strategies during nutrition applications in order to prevent HS and produce healthy and comfortable poultry with maintaining feed consumption, dietary adjustments, and appropriate diet formulation. For example, dietary protein-rich ingredients, AA balance, or dietary supplementation with one or a combination of functional AA are all important. Electrolytes, vitamins (such as ascorbic acid), and mineral drinking water supplementation, as well as acid–base balance, are also suggested.

10. Effect of HS on Amino Acids

The breakdown of dietary protein results in highly elevated heat generation than that of the catabolism of carbohydrates and fats in poultry under a thermoneutral zone (Table 2).

Table 2. The biochemical efficiency of absorbed nutrients for ATP and lipid synthesis; reviewed in [72].

Nutrients	Calorific Value (kJ/g)	ATP Production (%)	Lipid Synthesis (%)
Starch	17.7	68	74
Protein	23.8	58	53
Fatty acids	39.8	66	90

A study [73] found that feeding broilers more protein than their nutrient requirements did not improve performance at 33 °C. Low protein diets, on the other hand, had a negative impact on broiler performance at high ambient temperatures [64]. A study [74] attributed these effects to lower feed consumption, decreased consumption of AA, and therefore poor body weight gain and feed efficiency. According to a review [75], HS reduced AA levels in the birds including citrulline in chicks' plasma and leucine in the embryonic brain and liver. As a result, oral L-citrulline increased thermotolerance and decreased body temperature in layer chicks. A review [72] reported that under HS conditions, broilers aged 21 d–49 d should be fed diets containing 90 to 100 percent of the National Research Council (NRC) [76] recommended levels of AA and protein in diets containing 13.4 MJ ME/kg. According to previous studies, nutritionists did not compensate for diminished consumption in the hot ambient temperature by elevating protein and AA levels. Therefore, the final impact on growth relies on optimal or ideal protein quantity. The ideal AA composition for the maintenance or production varies with ambient temperature and among species, which can be attributed to metabolic stress alterations (Table 3).

Table 3. Estimated ideal protein ratio for a starting hen, broiler, and pig, expressed as a lysine needed percentage [70].

Amino Acid	Hen Turkeys	Broiler Chicken	Pigs
Lysine	100	100	100
Methionine + Cystine	59	72	60
Threonine	55	67	65
Valine	76	77	68
Arginine	105	105	NA[1]
Histidine	36	31	32
Isoleucine	69	67	60
Leucine	124	100	111
Phenylalanine + Tyrosine	105	105	95
Tryptophan	16	16	18

NA[1] = not available.

The optimal or ideal AA for maintenance varies from the ideal AA for production. Birds require higher methionine and cystine, threonine, and fewer leucines than turkeys and pigs, relative to lysine. Some authors reported that greater lysine or Arg:Lys ratios in broiler diets have a beneficial impact, whereas others showed an adverse effect at HT on gain and breast yield [72]. Therefore, dietary AA influenced heat generation [77] and improved broiler performance under high temperatures while decreasing nitrogen excretion by 21% between 28 and 49 days of age [78].

11. Future Perspectives and Conclusions

In conclusion, this review discusses the impact and consequences of HS in poultry. In addition, previous work was summarized, and some recommendations for developing high-quality and comfortable poultry through physiological (including HSP regulation), hormonal, and nutritional strategies were provided. Although the influence of HS on protein metabolic conversions in poultry can be concluded from this review, the scientific and medical evidence is inconclusive. Thus, further molecular studies are necessary to determine efficient HS regulation strategies, to better clarify the mechanisms involved in HS tolerance, to understand the HSP family as a useful biomarker for detecting HS. Then, for improved production efficacy in poultry, it is necessary to manage heat stress optimally. Recently, researchers interested in exploring a new generation of genetic tools that are capable of clarifying the molecular pathways associated with the heat stress in chickens, are offering new perspectives for the use of these tools in animal breeding.

Author Contributions: Funding acquisition, M.A.A.-G.; project administration, M.A.A.-G.; writing—original draft, M.M.Q.; writing—review and editing, M.M.Q. and M.A.A.-G. All authors have read and agreed to the published version of the manuscript.

Funding: This research was funded by Deanship of Scientific Research at King Saud University, grant number IFKSURP-322.

Institutional Review Board Statement: Not applicable.

Data Availability Statement: Not applicable.

Acknowledgments: The authors extend their appreciation to the Deputyship for Research and Innovation, "Ministry of Education" in Saudi Arabia for funding this research work through the project number IFKSURP-322. The authors extend their thanks to the RSSU at King Saud University for their technical support.

Conflicts of Interest: The authors declare no conflict of interest.

References

1. Perini, F.; Cendron, F.; Rovelli, G.; Castellini, C.; Cassandro, M.; Lasagna, E. Emerging genetic tools to investigate molecular pathways related to heat stress in chickens: A review. *Animals* **2021**, *11*, 46. [CrossRef] [PubMed]
2. Liu, L.; Ren, M.; Ren, K.; Jin, Y.; Yan, M. Heat stress impacts on broiler performance: A systematic review and meta-analysis. *Poult. Sci.* **2020**, *99*, 6205–6211. [CrossRef] [PubMed]
3. Zaboli, G.; Huang, X.; Feng, X.; Ahn, D.U. How can heat stress affect chicken meat quality? A review. *Poult. Sci.* **2019**, *98*, 1551–1556. [CrossRef] [PubMed]
4. Toplu, H.D.O.; Oral, D.; Nazligul, A.; Karaarslan, S.; Kaya, M.; Yagin, O. Effects of heat conditioning and dietary ascorbic acid supplementation on growth performance, carcass and meat quality characteristics in heat-stressed broilers. *Ankara Üniversitesi Veteriner Fakültesi Dergisi* **2014**, *61*, 295–302. [CrossRef]
5. Majekodunmi, B.; Ogunwole, O.; Sokunbi, O. Effect of supplemental electrolytes and ascorbic acid on the performance and carcass characteristics of broiler raised during high temperature period in Nigeria. *Int. J. Poult. Sci.* **2012**, *11*, 125. [CrossRef]
6. Syafwan, W. Effects of Dietary Changes on Heat Stress in Broiler and Kampung Chickens. Ph.D. Thesis, Wageningen University, Wageningen, The Netherlands, 2012.
7. Wasti, S.; Sah, N.; Mishra, B. Impact of heat stress on poultry health and performances, and potential mitigation strategies. *Animals* **2020**, *10*, 1266. [CrossRef]
8. Aitken, R.J. Impact of oxidative stress on male and female germ cells: Implications for fertility. *Reproduction* **2020**, *159*, R189–R201. [CrossRef] [PubMed]
9. Akbarian, A.; Michiels, J.; Degroote, J.; Majdeddin, M.; Golian, A.; de Smet, S. Association between heat stress and oxidative stress in poultry; Mitochondrial dysfunction and dietary interventions with phytochemicals. *J. Anim. Sci. Biotechnol.* **2016**, *7*, 1–14. [CrossRef] [PubMed]
10. Wu, G. Functional amino acids in nutrition and health. *Amino Acids* **2013**, *45*, 407–411. [CrossRef] [PubMed]
11. Maharjan, P.; Mullenix, G.; Hilton, K.; Beitia, A.; Weil, J.; Suesuttajit, N.; Martinez, D.; Umberson, C.; England, J.; Caldas, J. Effects of dietary amino acid levels and ambient temperature on mixed muscle protein turnover in Pectoralis major during finisher feeding period in two broiler lines. *J. Anim. Physiol. Anim. Nutr.* **2020**, *104*, 1351–1364. [CrossRef] [PubMed]
12. Attia, Y.A.; Hassan, S.S. Broiler tolerance to heat stress at various dietary protein/energy levels. *Europ. Poult. Sci.* **2017**, *81*, 1–15.
13. Ma, B.; Zhang, L.; Li, J.; Xing, T.; Jiang, Y.; Gao, F. Heat stress alters muscle protein and amino acid metabolism and accelerates liver gluconeogenesis for energy supply in broilers. *Poult. Sci.* **2021**, *100*, 215–223. [CrossRef] [PubMed]
14. Attia, Y.; Hassan, R.; Tag El-Din, A.; Abou-Shehema, B. Effect of ascorbic acid or increasing metabolizable energy level with or without supplementation of some essential amino acids on productive and physiological traits of slow-growing chicks exposed to chronic heat stress. *J. Anim. Physiol. Anim. Nutr.* **2011**, *95*, 744–755. [CrossRef] [PubMed]
15. Awad, E.; Zulkifli, I.; Soleimani, A.; Law, F.; Ramiah, S.; Mohamed-Yousif, I.; Hussein, E.; Khalil, E. Response of broilers to reduced-protein diets under heat stress conditions. *World Poult. Sci. J.* **2019**, *75*, 583–598. [CrossRef]
16. Sahin, N.; Onderci, M.; Sahin, K.; Gursu, M.; Smith, M. Ascorbic acid and melatonin reduce heat-induced performance inhibition and oxidative stress in Japanese quails. *Br. Poult. Sci.* **2004**, *45*, 116–122. [CrossRef] [PubMed]
17. Selle, P.; Liu, S.Y. The relevance of starch and protein digestive dynamics in poultry. *J. Appl. Poult. Res.* **2019**, *28*, 531–545. [CrossRef]
18. Dublecz, K. *Poultry Nutrition*; Pannon Egyetem: Egyetem, Hungary, 2011; pp. 18–22.
19. Qaid, M.M.; Abdelrahman, M.M. Role of insulin and other related hormones in energy metabolism—A review. *Cogent Food Agric.* **2016**, *2*, 1267691. [CrossRef]
20. Squires, E.J. *Applied Animal Endocrinology*; CABI: Wallingford, UK, 2010.
21. Baumgard, L.H.; Rhoads, R.P., Jr. Effects of heat stress on postabsorptive metabolism and energetics. *Annu. Rev. Anim. Biosci.* **2013**, *1*, 311–337. [CrossRef]

22. Sanz Fernandez, M. Effects of heat stress on post-absorptive metabolism. In Proceedings of the ADSA-ASAS Midwest Meeting, Des Moines, IA, USA, 17–19 March 2014.
23. Wheelock, J.; Rhoads, R.; Vanbaale, M.; Sanders, S.; Baumgard, L. Effects of heat stress on energetic metabolism in lactating Holstein cows. *J. Dairy Sci.* **2010**, *93*, 644–655. [CrossRef]
24. Virden, W.; Kidd, M. Physiological stress in broilers: Ramifications on nutrient digestibility and responses. *J. Appl. Poult. Res.* **2009**, *18*, 338–347. [CrossRef]
25. Ognik, K.; Sembratowicz, I. Stress as a factor modifying the metabolism in poultry. A review. *Annales UMCS Zootech.* **2012**, *30*, 34–43. [CrossRef]
26. Kamiya, M.; Kamiya, Y.; Tanaka, M.; Oki, T.; Nishiba, Y.; Shioya, S. Effects of high ambient temperature and restricted feed intake on urinary and plasma 3-methylhistidine in lactating Holstein cows. *Anim. Sci. J.* **2006**, *77*, 201–207. [CrossRef]
27. Pearce, S. The Effects of Heat Stress and Nutritional Status on Metabolism and Intestinal Integrity in Growing Pigs. Master's Thesis, Iowa State University, Ames, IA, USA, 2011.
28. Santiago, L.Â.M.; Lima Neto, L.G.; Pereira, G.B.; Leite, R.D.; Mostarda, C.T.; Navarro, F. Influence of creatine kinase on C-reactive protein in muscle adaptation. *Rev. Bras. Med. Esporte* **2019**, *25*, 413–417. [CrossRef]
29. Hara, T.; Ohtsuka, A.; Hayashi, K. Role of Ca^{2+} in corticosterone-induced muscle growth retardation. *Anim. Sci. J.* **2002**, *73*, 383–387.
30. Yaman, M.A.; Kita, K.; Okumura, J.-I. Various macronutrient intakes additively stimulate protein synthesis in liver and muscle of food-deprived chicks. *J. Nutr.* **2000**, *130*, 70–76. [CrossRef]
31. Ostrowski-Meissner, H.T. The physiological and biochemical responses of broilers exposed to short-term thermal stress. *Comp. Biochem. Physiol. Part A Physiol.* **1981**, *70*, 1–8. [CrossRef]
32. Siddiqui, S.H.; Kang, D.; Park, J.; Khan, M.; Shim, K. Chronic heat stress regulates the relation between heat shock protein and immunity in broiler small intestine. *Sci. Rep.* **2020**, *10*, 1–11. [CrossRef] [PubMed]
33. Temim, S.; Chagneau, A.-M.; Peresson, R.; Tesseraud, S. Chronic heat exposure alters protein turnover of three different skeletal muscles in finishing broiler chickens fed 20 or 25% protein diets. *J. Nutr.* **2000**, *130*, 813–819. [CrossRef] [PubMed]
34. Gonzalez-Esquerra, R.; Leeson, S. Effects of acute versus chronic heat stress on broiler response to dietary protein. *Poult. Sci.* **2005**, *84*, 1562–1569. [CrossRef] [PubMed]
35. Joo, M.A.; Kim, E.Y. Hyponatremia caused by excessive intake of water as a form of child abuse. *Ann. Pediatr. Endocrinol. Metab.* **2013**, *18*, 95. [CrossRef] [PubMed]
36. Chaiyabutr, N. Physiological reactions of poultry to heat stress and methods to reduce its effects on poultry production. *Thai J. Vet. Med.* **2004**, *34*, 17–30.
37. Tsahar, E.; Arad, Z.; Izhaki, I.; Guglielmo, C.G. The relationship between uric acid and its oxidative product allantoin: A potential indicator for the evaluation of oxidative stress in birds. *J. Comp. Physiol. B* **2006**, *176*, 653–661. [CrossRef] [PubMed]
38. Pijarska, I.; Czech, A.; Malec, H.; Tymczyna, L. Effect of road transportation of chicks on blood biochemical indices and productive results of broilers. *Vet. Med.* **2006**, *62*, 408–410.
39. Temim, S.; Chagneau, A.-M.; Guillaumin, S.; Michel, J.; Peresson, R.; Geraert, P.-A.; Tesseraud, S. Effects of chronic heat exposure and protein intake on growth performance, nitrogen retention and muscle development in broiler chickens. *Reprod. Nutr. Dev.* **1999**, *39*, 145–156. [CrossRef] [PubMed]
40. Hayashi, K.; Taniguchi, N.; Kinjou, T.; Ohtsuka, A. Heat stress causes death via glucose metabolism in broiler. In Proceedings of the 13th European Symposium on Poultry Nutrition and Ascites Workshop, Blankenberge, Belgium, 30 September–4 October 2001.
41. Samuels, S.; McAllister, T.; Thompson, J. Skeletal and heart muscle protein turnover during long-term exposure to high environmental temperatures in young rats. *Can. J. Physiol. Pharmacol.* **2000**, *78*, 557–564. [CrossRef]
42. Kim, J.Y.; Yenari, M.A. The immune modulating properties of the heat shock proteins after brain injury. *Anat. Cell Biol.* **2013**, *46*, 1–7. [CrossRef] [PubMed]
43. Al-Aqil, A.; Zulkifli, I. Changes in heat shock protein 70 expression and blood characteristics in transported broiler chickens as affected by housing and early age feed restriction. *Poult. Sci.* **2009**, *88*, 1358–1364. [CrossRef]
44. Youssef, S.; Selim, N.; Galal, M. Increasing the activity of antioxidant system in broiler chicks during summer season: 1-effect of some anti-heat stress agents on broiler performance, immune response and heat shock protein 70. *Egypt. Poult. Sci. J.* **2016**, *36*. [CrossRef]
45. Frier, B.C.; Locke, M. Heat stress inhibits skeletal muscle hypertrophy. *Cell Stress Chaperones* **2007**, *12*, 132–141. [CrossRef]
46. Zouhal, H.; Jacob, C.; Delamarche, P.; Gratas-Delamarche, A. Catecholamines and the effects of exercise, training and gender. *Sports Med.* **2008**, *38*, 401–423. [CrossRef]
47. Wischmeyer, P.E. Glutamine and heat shock protein expression. *Nutrition* **2002**, *18*, 225–228. [CrossRef]
48. Khan, R.; Naz, S.; Nikousefat, Z.; Selvaggi, M.; Laudadio, V.; Tufarelli, V. Effect of ascorbic acid in heat-stressed poultry. *World Poult. Sci. J.* **2012**, *68*, 477. [CrossRef]
49. Ming, J.; Xie, J.; Xu, P.; Liu, W.; Ge, X.; Liu, B.; He, Y.; Cheng, Y.; Zhou, Q.; Pan, L. Molecular cloning and expression of two HSP70 genes in the Wuchang bream (*Megalobrama amblycephala Yih*). *Fish Shellfish Immunol.* **2010**, *28*, 407–418. [CrossRef]
50. Mohan, J. *Physiology of Stress in Poultry*; Central Avian Research Intitute: Bareilly, India, 2005; Volume 11.

51. Yunianto, V.D.; Hayashit, K.; Kaiwda, S.; Ohtsuka, A.; Tomita, Y. Effect of environmental temperature on muscle protein turnover and heat production in tube-fed broiler chickens. *Br. J. Nutr.* **1997**, *77*, 897–909. [CrossRef] [PubMed]
52. Siegel, H.V.; van Kampen, M. Energy relationships in growing chickens given daily injections of corticosterone. *Br. Poult. Sci.* **1984**, *25*, 477–485. [CrossRef] [PubMed]
53. Naga Raja Kumari, K.; Narendra Nath, D. Ameliorative measures to counter heat stress in poultry. *World Poult. Sci. J.* **2018**, *74*, 117–130. [CrossRef]
54. Donkoh, A.; Atuahene, C. Management of environmental temperature and rations for poultry production in the hot and humid tropics. *Int. J. Biometeorol.* **1988**, *32*, 247–253. [CrossRef]
55. Shakeri, M.; Oskoueian, E.; Le, H.H.; Shakeri, M. Strategies to combat heat stress in broiler chickens: Unveiling the roles of selenium, vitamin E and vitamin C. *Vet. Sci.* **2020**, *7*, 71. [CrossRef]
56. Anderson, P. Further explorations into the avian-human bond. In Proceedings of the 14th Annual International Society for Anthrozoology Conference, Niagara Falls, NY, USA, 11–12 July 2005.
57. Veldkamp, T. Heat Stress and Diet Utilization in Male Turkeys: The Role of Dietary Energy and Amino Acids. Ph.D. Thesis, Wageningen University, Wageningen, The Netherlands, 2002.
58. Chowdhury, V.S. Heat stress biomarker amino acids and neuropeptide afford thermotolerance in chicks. *J. Poult. Sci.* **2019**, *56*, 1–11. [CrossRef]
59. Yahav, S. Domestic fowl-strategies to confront environmental conditions. *Poult. Avian Biol. Rev.* **2000**, *11*, 81–95.
60. Gonzalez-Esquerra, R.; Leeson, S. Physiological and metabolic responses of broilers to heat stress-implications for protein and amino acid nutrition. *World Poult. Sci. J.* **2006**, *62*, 282–295. [CrossRef]
61. Singh, P.; Kesharwani, R.K.; Keservani, R.K. Protein, carbohydrates, and fats: Energy metabolism. In *Sustained Energy for Enhanced Human Functions and Activity*; Elsevier: Amsterdam, The Netherlands, 2017; pp. 103–115.
62. Cheng, T.K.; Hamre, M.L.; Coon, C.N. Responses of broilers to dietary protein levels and amino acid supplementation to low protein diets at various environmental temperatures. *J. Appl. Poult. Res.* **1997**, *6*, 18–33. [CrossRef]
63. Teeter, R.G.; Belay, T. Broiler management during acute heat stress. *Anim. Feed Sci. Technol.* **1996**, *58*, 127–142. [CrossRef]
64. Furlan, R.L.; de Faria Filho, D.; Rosa, P.; Macari, M. Does low-protein diet improve broiler performance under heat stress conditions? *Rev. Bras. Ciência Avícola* **2004**, *6*, 71–79. [CrossRef]
65. Van der Klis, J.; de Lange, L. Water intake in poultry. In Proceedings of the 19th European Symposium on Poultry Nutrition, Postdam, Germany, 26–29 August 2013; pp. 102–107.
66. Harris, C.; Milne, G.; McDiarmid, R. The retention and metabolism of N^{τ}-methylhistidine by cockerels: Implications for the measurement of muscle protein breakdown determined from the excretion of N^{τ} methylhistidine in excreta. *Br. J. Nutr.* **1987**, *57*, 467–478. [CrossRef]
67. Dai, S.; Wang, L.; Wen, A.; Wang, L. Protection of glutamine supplementation to performance, intestinal enzyme activity and morphosis in broiler under heat stress. *J. Chin. Cereals Oils Assoc.* **2009**, *24*, 103–107.
68. Porto, M.; Givisiez, P.; Saraiva, E.; Costa, F.; Moreira Filho, A.; Andrade, M.; Brandão, P.; Guerra, R. Glutamic acid improves body weight gain and intestinal morphology of broiler chickens submitted to heat stress. *Revista Brasileira de Ciência Avícola* **2015**, *17*, 355–362. [CrossRef]
69. Olubodun, J.O.; Zulkifli, I.; Farjam, A.S.; Hair-Bejo, M.; Kasim, A. Glutamine and glutamic acid supplementation enhances performance of broiler chickens under the hot and humid tropical condition. *Ital. J. Anim. Sci.* **2015**, *14*, 3263. [CrossRef]
70. Firman, J.D.; Boling, S. Lysine: Ideal protein in turkeys. *Poult. Sci.* **1998**, *77*, 105–110. [CrossRef]
71. Attia, Y.A.; Al-Harthi, M.A.; Elnaggar, A.S. Productive, physiological and immunological responses of two broiler strains fed different dietary regimens and exposed to heat stress. *Ital. J. Anim. Sci.* **2018**, *17*, 686–697. [CrossRef]
72. Syafwan, S.; Kwakkel, R.; Verstegen, M. Heat stress and feeding strategies in meat-type chickens. *World Poult. Sci. J.* **2011**, *67*, 653–674. [CrossRef]
73. Cheng, T.K.; Hamre, M.L.; Coon, C.N. Effect of environmental temperature, dietary protein, and energy levels on broiler performance. *J. Appl. Poult. Res.* **1997**, *6*, 1–17. [CrossRef]
74. Alleman, F.; Leclercq, B. Effect of dietary protein and environmental temperature on growth performance and water consumption of male broiler chickens. *Br. Poult. Sci.* **1997**, *38*, 607–610. [CrossRef] [PubMed]
75. Chowdhury, V.S.; Han, G.; Eltahan, H.M.; Haraguchi, S.; Gilbert, E.R.; Cline, M.A.; Cockrem, J.F.; Bungo, T.; Furuse, M. Potential role of amino acids in the adaptation of chicks and market-age broilers to heat stress. *Front. Vet. Sci.* **2020**, *7*, 610541. [CrossRef] [PubMed]
76. NRC. *Nutrient Requirements of Poultry*; National Academy Press: Washington, DC, USA, 1994.
77. Zarate, A.; Moran, E.; Burnham, D. Exceeding essential amino acid requirements and improving their balance as a means to minimize heat stress in broilers. *J. Appl. Poult. Res.* **2003**, *12*, 37–44. [CrossRef]
78. Attia, Y.A.; Bovera, F.; Wang, J.; Al-Harthi, M.A.; Kim, W.K. Multiple amino acid supplementations to low-protein diets: Effect on performance, carcass yield, meat quality and nitrogen excretion of finishing broilers under hot climate conditions. *Animals* **2020**, *10*, 973. [CrossRef] [PubMed]

animals

MDPI

Review

Date (*Phoenix dactylifera* L.) by-Products: Chemical Composition, Nutritive Value and Applications in Poultry Nutrition, an Updating Review

Adel I. Attia [1], Fayiz M. Reda [1], Amlan K. Patra [2], Shaaban S. Elnesr [3], Youssef A. Attia [4,5,6,*] and Mahmoud Alagawany [1,*]

1 Poultry Department, Faculty of Agriculture, Zagazig University, Zagazig 44511, Egypt; adelatti159195@gmail.com (A.I.A.); fayizreda@yahoo.com (F.M.R.)
2 Department of Animal Nutrition, West Bengal University of Animal and Fishery Sciences, Belgachia, Kolkata 700037, India; patra_amlan@yahoo.com
3 Department of Poultry Production, Faculty of Agriculture, Fayoum University, Fayoum 63514, Egypt; ssn00@fayoum.edu.eg
4 Agriculture Department, Faculty of Environmental Sciences, King Abdulaziz University, Jeddah 21589, Saudi Arabia
5 The Strategic Center to Kingdom Vision Realization, King Abdulaziz University, Jeddah 21589, Saudi Arabia
6 Department of Animal and Poultry Production, Faculty of Agriculture, Damanhour University, Damanhour 22516, Egypt
* Correspondence: yaattia@kau.edu.sa (Y.A.A.); dr.mahmoud.alagwany@gmail.com (M.A.)

check for updates

Citation: Attia, A.I.; Reda, F.M.; Patra, A.K.; Elnesr, S.S.; Attia, Y.A.; Alagawany, M. Date (*Phoenix dactylifera* L.) by-Products: Chemical Composition, Nutritive Value and Applications in Poultry Nutrition, an Updating Review. *Animals* **2021**, *11*, 1133. https://doi.org/10.3390/ani11041133

Academic Editors: Ilias Giannenas and Jan Jankowski

Received: 3 March 2021
Accepted: 12 April 2021
Published: 15 April 2021

Simple Summary: The demand and price of feed/feed stuffs was greatly altered after the COVID-19 crisis, showing great interest on depends on local feed stuffs and a decrease in the imported ones. Date palm (*Phoenix dactylifera* L.) is a common crop in arid and semiarid regions for production of date fruits. This updating review provides the recent knowledge on the use of date palm by-products and improving their utilization in poultry rations. This article focuses on the chemical composition, amino and fatty acids, mineral content, growth performance, economic aspects and nutritional potential in comparison with widely used corn and barley grains in poultry.

Abstract: Several by-products, such as whole cull dates, date stones (also called pits, seeds, kernels), sugar-extracted date pulp, are produced from date fruit processing industries. These by-products, particularly date stone meal represent 10 to 20% of the whole dates are wasted, causing environmental issues. However, the date stone is rich in various nutrients, such as nitrogen-free extract, fiber, fat, and minerals, which could be used as an alternative feed source in poultry nutrition. However, the high fiber content in date stone meal restricts its use in poultry diets. Whole date wastes and date pulp have lower protein, fiber and fat content than those in date stone meal. Several studies have investigated the use of date stone meal and other by-products as a replacement of dietary corn and barley on feed intake, growth, feed efficiency, and nutrient utilization in poultry. The compilation of results obtained from different studies indicates that date fruit by-products, particularly date stone meal, could be incorporated up to 10% levels, replacing corn or barley grains without hampering production performance, and this could reduce the production cost. Moreover, the use of date stone meal at lower levels (5–10%) sometimes shows better growth performance, probably due to the presence of bioactive principles such as antioxidants and phenolic compounds. The use of date stone meal in poultry diets may be practically important under certain circumstances not only under the COVID-19 crisis due to the lockdown of airports, dry ports, ports and traffic restrictions, but also in countries with a limited supply of classical feed resources and which depend on imported ones. Finally, it can recommend that using date stone meal (DSM) up to 5–10% in poultry diets positively affect the productive performance. Due to the variability in the nutritive value of the DSM, developing a formula for metabolizable energy (ME) of DSM = a (intercept) − b (slope) × crude fiber (CF) content should be considered to improve the precision of feed formula for poultry. However, the nutritional value of DSM might not only depend on crude fiber but also on other compounds.

Keywords: date by-product; date stone; poultry; nutritive value; production performance

1. Introduction

Several agricultural and agro-industrial by-products are produced during the production or processing of fruits and vegetable crops, which may cause environmental pollution. These by-products were of great interest during the COVID-19 crisis due to lockdown, close of dry borders and ports and traffic restrictions [1,2]. The use of alternative feedstuffs and locally available feed ingredients would improve a possible solution. Date by-products are the best-known by-products that are produced from the arid and desert regions, which are sometimes used in livestock and poultry diets.

In arid and semiarid regions of the world, date palm (*Phoenix dactylifera* L.) constitutes one of the main crops. In countries of the Middle East, date cultivation is common, as it is where about 70% of the world date production occurs [3]. The world date production has been reported to have increased from 1.8 million tons in 1961 to 2.8 million tons in 1985 and to 5.4 million tons in 2001 [4]. The fruit of date palm is comprised of a fleshy pericarp and an inedible seed. A number of date by-products, such as whole cull dates, date stones (also called date pits, seeds, kernels or pips), sugar-extracted fruit date pulp and pressed cakes from date processing are available for date industries [5]. The date stone is a hard oblong body, ventrally grooved and present in the midst of the date fruit, and its length ranges from 12 to 36 mm, while the width ranges from 6 to 14 mm. The weight of the seeds represents 10–20% of the date fruit [6]. The size and weight of the seeds depend upon the variety, maturity and growing conditions. At the back of the seeds, there is a micropyle, where the embryo is located. The embryo is a small oblong body surrounded by the endosperm. The latter is a group of cells inside their walls. The hemicellulose is accumulated, giving it thickness and hardness. Date stone meal (DSM) is rich in carbohydrates and fat along with crude protein content compared to barley and corn grains depending upon the varieties of date palm, which is used as a feed ingredient for livestock feeding. However, DSM contains high amounts of crude fiber, which is the constraint for its use in the diets of monogastric animals, especially for poultry [7]. Tabook et al. [8] pointed out that date fiber can be included at a level of 5% in a broiler diet without affecting the performance of broiler chickens. On the other hand, Slominski [9] reported that enzyme supplementation improved nutrient digestibility and nutrient utilization and allowed use of the agro-industry by-product in poultry nutrition. Moreover, DSM also has functional properties due to the presence of mannans, phenolic compounds and antioxidant activity [5]. A number of studies have been conducted to utilize date fruit by-products as a feed ingredient of poultry diets. The DSM can give a potential alternative for energy source in diet ingredients of poultry. Dietary inclusion of DSM significantly improved the performance and feed utilization of poultry [3,10]. These low-cost feed ingredients have emerged as potential alternatives in poultry that can reduce the cost of rations, but the effects of these by-products on production performance varied among the studies. Therefore, this review was undertaken to compile the findings of the experiments on date fruit by-products for the chemical composition, nutritive values, growth performance, carcass traits and economic aspects in comparison with widely used corn and barley grains in poultry.

2. Chemical Composition and Feeding Values of Date Waste

2.1. Proximate Analysis

The proximate analysis of DSM was studied by several investigators and is shown in Table 1. The chemical composition (%) ranges widely between 88.2 and 92.0, 5.27 and 10.5, 2.89 and 9.43, 8.22 and 28.9, 58.0 and 72.4 and 0.98 and 9.08 for dry matter (DM), crude protein (CP), ether extract (EE), crude fiber (CF), nitrogen free extract (NFE) and ash, respectively. Additionally, the DSM contains 3.3 MJ/kg DM true metabolizable energy (TME) for poultry. Based on the chemical composition of DM, the nutritional value is very low for TME, due to the very high fiber content and the low protein content, resulting in metabolizable energy levels as low as 2.9 MJ/kg. It can be used in poultry diets provided that this low energy level is taken into account in diet formulation and compensated by supplementation of oil or fat [11].

The chemical composition of dried date press cake (excluding the pits) varies, particularly for CF, CP and NFE depending upon the sources and varieties of dates but are within the range of 87.7 to 95.2 for DM, 4.4 to 8.1 for CP, 9.1 to 21.8 for CF, 1.8 to 2.7 for EE, 2 to 3.5 for ash and 55.3 to 72.7 for NFE on a dry matter basis [5]. The chemical composition of date wastes is usually close to the chemical composition of corn and barley, except for CF values which are higher than those of corn and barley (8.22–16.6 vs. 2.3 and 5.0%, respectively). Higher fiber content is present in DSM or whole date fruit due to the presence of greater fiber content in the outer layer of the date seeds. The pulp of the dates has less fiber, low CP and higher NFE. The sugar extracted date fruit pulp was reported to contain 5.5% CP, 11.8% CF, 0.4% EE, 79.6% NFE and 2.7% ash [12]. Thus, pulp can be added to broiler diets at levels up to 20%. As date pulp contains less CP, whole date wastes have a low CP (2.9%), which is much lower than for DSM, maize and barley [13]. The fruit flesh is rich in NFE (76.2%), which is closer to values of barley (79.0%) but greater than that of DSM. The amino acid in flesh, mainly methionine (0.06%) and lysine (0.10%) are lower than those in the DSM, maize and barley due to lower CP concentration in the flesh [14].

Date is found to be low in protein and fat but highly rich in sugars, especially glucose and fructose [15]. Date is a good source of energy, where 100 g of date can provide about 1.31 MJ. In addition, minerals were reported, such as selenium, potassium, copper, and magnesium. The daily consumption of 100 g can cover about 15% of the recommended requirement of these minerals. Date contains about 8.0 g/100 g date, but insoluble dietary fiber is the major part of dietary fiber in this fruit. Regarding the vitamin content, vitamins C and B-complex are the major vitamins in dates. It is also a good source of antioxidants, mainly phenolic compounds and carotenoids. The seeds of date contain considerable concentrations of fat (9.0 g/100 g) and protein (5.1 g/100 g) when compared to the flesh. Additionally, seeds are high in phenolics (3942 mg/100 g), dietary fiber (73.1 g/100 g), and antioxidants (80,400 mmol/100 g) [16].

Shaba et al. [17] reported that date palm fruits contain low amounts of anti-nutritional factors, such as oxalate, tannin, saponin, alkaloid and cyanide. This indicates that the date seeds can be used effectively as the anti-nutritional composition is low and there would be no interference with the nutrient-like minerals and protein in the body. Depending on different varieties of date by-products, date pits contain significant amounts of micronutrients and macronutrients, but all varieties are excellent sources of fiber and maybe used as important sources of functional foods [18–20].

Table 1. Chemical composition (%) of different varieties of date stone meal compared with yellow corn and barley.

References	Variety	DM	CP	EE	CF	NFE	Ash	TME (MJ/kg DM)
[19]	Sewi	91.4	6.44	8.16	14.2	70.2	0.98	ND
[21]	ND	90.9	10.5	9.43	8.60	69.5	2.01	ND
[22]	ND	90.5	6.4	6.70	28.9	58.0	1.9	3.30
Yellow corn *	-	88.0	7.7	3.8	2.3	84.8	1.4	ND
Barley *	-	88.0	11.9	2.00	5.00	78.8	2.3	ND

* Central laboratory of food and feed, CLFF; ND: not defined or not determined; DM = dry matter; CP = crude protein; EE = ether extract; CF = crude fiber; NFE = nitrogen free extract, TME = true metabolizable energy.

2.2. Digestion Coefficients and Feeding Values of Date Waste

The digestibility values of CP, CF and NFE, in general, were lower than those of barley and corn, while the digestion coefficient of EE was better than those of barley and corn (Table 2). Metabolic energy values of DSM were lower than that of corn (14.14 MJ/kg). More studies in this regard revealed the nutritional value of DSM. For example, Sawaya et al. [23]

reported that the protein digestibility and calculated protein efficiency ratios were 64.1, 1.67 for Ruzeiz, 62.9, 1.65 for Sifri date, respectively, with an average of 63.5, and 1.65 for Ruzeiz and Sifri compared to 90.0 and 2.50% for casein. Attia and Al-Harthi [12] evaluated date waste in broiler chickens fed at 0 to 200 g/kg diet replacing wheat bran containing similar energy and protein concentration and observed that digestibility of nutrients was not affected except decreased CP digestibility at 200 g/kg. In addition, El-Deek et al. [24] reported that inclusion (0 to 150 g/kg diet at 6 levels) of inedible date waste (85% fruit and 15% pits) in broiler chickens did not affect nutrient digestibility. Al-Homidan [3] reported ME values of 7.23 and 4.71 MJ/kg, respectively, for broiler chickens. In future, developing a formula for ME of DSM = a − b × CF seems essential due to variability in the nutritive value of DSM to improve the accuracy of feed formula for poultry. Nonetheless, the nutritional value might not only depend on crude fiber but also on other compounds.

Table 2. Digestion coefficients (%) and metabolizable energy (ME, MJ/kg) of different varieties of date stone meal compared with yellow corn and barley.

References	Species	Variety	DM	OM	CP	CF	EE	NFE	ME, MJ/kg
[14]	Broiler	Sewi	45.0	44.2	42.9	3.45	84.2	49.7	9.57

DM = dry matter; OM = organic matter; CP = crude protein; EE = ether extract; CF = crude fiber; NFE = nitrogen free extract; ME = metabolizable energy.

2.3. Fiber Fractions

Fiber fractions of DSM as reported by several workers, compared with yellow corn and barley, are shown in Table 3. The values ranged from 39.6 to 57.5, 51.6 to 75.0, 12.0 to 17.5, 26.1 to 42.5 and 7.21 to 11.0 for acid detergent fiber (ADF), neutral detergent fiber (NDF), hemicellulose, cellulose, and lignin, respectively, which were higher than those of both corn and barley. Fiber fractions of DSM were reported to be 57.6, 46.4 and 13.7 for NDF, ADF and ADL, respectively [25]. These results indicated that the low-energy values of DSM compared with corn and barley could be due to higher fiber fractions [11].

Table 3. Fiber fractions (% of dry matter) of date stone meal compared with yellow corn and barley.

Fiber Fraction	Date Stone Meal [5]	Yellow Corn ***	Barley ***
ADF *	57.5	2.20	6.00
NDF **	75.0	9.00	17.0
Hemicellulose	17.5	ND	ND
Cellulose	42.5	2.00	4.00
Lignin	11.0	1.00	2.00

* Acid detergent fiber (cellulose + lignin); ** Neutral detergent fiber (hemicellulose + cellulose + lignin); *** Central Laboratory of food, CLFF [22]; ND: Not defined or not determined.

2.4. Fatty Acid Content

Oil extracted from date seeds is a pale yellowish–green in color and has a pleasant odor [5,26,27]. Abdel-Nabey [19] reported that the crude oil extracted from date pits has a pale-yellow color and is semi-solid at refrigerated temperature (−40 °C). The main median characteristics are as follows: an acidity value of 1.11, 0.96% unsaponifiable matter, a refracted index of 1.46, an iodine value of 49.7 and a saponification value of 219.6. The date seed oil is a natural source of medium-chain fatty acids (MCFAs) such as lauric acid (10.36–35.31%) and capric acid (0.2–0.8%). These MCFAs are effective against a wide range of microorganisms in chicken intestines and can improve the growth performance and immunity of poultry [28].

The values (% of fat) ranged between 12.20 and 23.06, 9.70 and 11.30, 10.11 and 12.70, 1.56 and 3.56, 35.1 and 45.80, 8.10 and 11.00, and 0.37 and 0.80 for lauric, myristic, palmitic, stearic, oleic, linoleic and linolenic acids, respectively. The values were close to the corn, except for myristic and oleic fatty acids, which were higher than those of corn were. Corn contained higher linoleic and linolenic than DSM (51.5 and 1.7% vs. 8.10 to 11.0% and

0.08 to 0.37%, respectively). Abdel-Nabey [19] reported that date pit oil has a very low content of linolenic acid (0.37) in comparison with many vegetable oils, i.e., cottonseed, sunflower and maize oils. The ratio of unsaturated to saturated fatty acids varies from 1.10 to 1.20 with an average of 1.17 because the total amount of unsaturated fatty acids is slightly higher than those of saturated fatty acids. This ratio, however, is lower than those of the common edible vegetable oils in which their degree of unsaturation is quite high.

2.5. Amino Acid Content

Amino acid content of DSM as indicated by several studies, compared to corn and barley, are shown in Table 4. The highest amino acid contents of DSM are almost close to those of corn grain, while they are lower than those of barley, except for arginine which is higher than that of corn and close to that of barley (0.36–0.60 vs. 0.17 and 0.52%).

Table 4. Amino acid content (% of dry matter) of date stone meal compared with yellow corn and barley.

Amino Acid	[29]	[30]
	Date Stone Meal	Barley
Aspartic	0.52	ND
Threonine	0.17	0.37
Serine	0.20	0.46
Glutamic	0.98	ND
Proline	0.19	ND
Glycine	0.27	0.44
Alanine	0.25	ND
Valine	0.28	0.52
Methionine	0.09	0.18
Isoleucine	0.20	0.37
Leucine	0.33	0.76
Tyrosine	0.06	0.35
Phenylalanine	0.20	0.56
Histidine	0.12	0.27
Lysine	0.27	0.40
Arginine	0.60	0.52
Cystine	0.10	0.24
Protein	5.70	11.0

ND: not determined.

2.6. Mineral Content

Studies showing both major and trace elements of DSM in comparison to corn and barley are presented in Tables 5 and 6. Date stone meal contents of Mg, K and Na are close to those of both corn and barley. The level of P is close to that of corn, but lower than that of barley, while the content of Ca is higher than those of both corn and barley. The trace elements (Zn, Cu, Mn and Fe) of DSM are in general higher than those of corn but close to those of barley. Calcium and phosphorus levels (0.76 and 0.52%, respectively) are present in higher amounts than in maize and barley. The protein content (8.1%) in DSM is higher than that of whole date but is nearly close to that of maize grains (8.6%) [14].

Table 5. The major mineral element content (% of dry matter) in different varieties date stone meal compared with yellow corn and barley.

References	Variety	Major Elements				
		Ca	P	Mg	K	Na
[30]	Amhate	0.046	0.085	0.145	0.471	0.025
Yellow corn [31]	-	0.02	0.28	0.12	0.30	0.02
Barley [31]	-	0.03	0.36	0.14	0.48	0.04

Table 6. The trace element content (mg/kg of dry matter) of different varieties of date stone meal compared with yellow corn and barley.

References	Variety	Major elements			
		Zn	Cu	Mn	Fe
[30]	Amhate	24.5	5.50	5.7	191
Yellow corn [31]	-	18.0	3.00	7.00	45.0
Barley [31]	-	30.0	10.0	18.0	78.0

3. Effect of Date Waste on Poultry

3.1. *Effect of Date Waste on Growth Performance*

Several studies recommended that DSM up to 10% in broiler diets can be incorporated without negative effects on growth performance [1,4]. El-Deek et al. [24] tested six levels (0 to 150 g/kg diet) of inedible date waste (85% fruit and 15% pits) in broiler chickens and noted that body weight, feed intake and feed efficiency were similar at all levels of date waste. Date palm pollen is used to improve fertility and ovulation in men and women due to it contains fatty acids, amino acids, saponins, flavonoids and sterols. Egg mass, egg weight and egg production of laying hens treated with different levels of date palm pollen were increased when compared to the control [32]. Jaffer and Jassim [33] used 0, 5, 10 and 15 of % date by-product with or without enzyme instead of maize and they found body weight, feed intake, and feed conversion as well as economic feasibility were not affected by this by-product.

As DSM contains high fiber content, which limits its use in poultry diets, some processing of DSM has been performed to improve feed utilization efficiency and performance of chickens. For example, Al-Saffar et al. [34] conducted two experiments to investigate the effect of phytase and/or multienzymes (xylanase, β-glucanase, cellulase, hemicellulose, protease and amyloglucosidase) on improving the utilization of DSM in laying hens. In experiment 1, DSM completely replaced corn in isocaloric and isonitrogenous diets, but with added phytase (500, units of phytase (FTU)/kg) and/or multienzymes (0.1%). In experiment 2, DSM was added at 0, 15, and 30% in isocaloric and isonitrogenous diets, and diets containing DSM were fed without or with phytase and/or 0.1% multienzymes. In experiment 1, laying performance and eggshell quality of hens reduced significantly due to complete substitution of corn, but hatchability and fertility were not influenced. Supplementation of phytase and multienzymes did not restore the laying performance compared to the control level. In experiment 2, the results showed that DSM could be included in diets up to 30% when supplemented with multienzymes without affecting the laying performance of hens.

Hussein and Alhadrami [35] used date pits at 0, 10, 15, and 20% with or without enzyme supplementation (Avizyme 1500) to the starter and finisher diets, while grower diets contained 10% uncooked or cooked date pits. The results showed that adding date pits with or without enzyme to the starter and finisher broiler diet had no effect on broiler performance. Body weight gain of birds who received 10% uncooked date pits were significantly higher compared to the control group birds. Groups fed a diet containing 10% date pits had no effects on feed intake or feed conversion ratio. Barreveld [5] reported that date stone could successfully replace 10% barley in chick diets. The increased live gain over the control is related to more feed intake and may be due to the existence of a growth promoting substance in date stone. Similarly, date waste (0 to 200 g/kg diet replacing wheat bran containing similar energy and protein concentration) fed to broiler chickens did not affect body weight gain up to 200 g/kg diet [36].

Al-Homidan [3] determined the effect of adding date waste (whole date waste meal and date pits meal) in starter and finisher broiler diets. Seven levels of date waste representing 0, 8, 16, 24% date waste meal and 5, 10, and 15% date pits meal were used in the starting period, while 0, 8, 16, 24% date waste meal and 6, 12 and 18% date pits meal were used during the finishing period. Results indicated that body weight and daily gain were improved when chicks were fed diets containing 8% date waste meal or 10% date pits meal during the starting period (0–4 weeks). During the finishing period (5–7 weeks), chicks fed a diet containing 8% and 16 date waste meal or 18% date pits meal had better body weight and daily gain than the control. Feed conversion improved significantly when the level of date pits meal was 10% during the growing period. During the finishing period, feed conversion was better when diets included 18% date pits meal compared with the control.

In broiler chickens, date waste (0 to 200 g/kg diet replacing wheat bran) reduced feed intake and improved feed efficiency and European feed production efficiency at 50 g/kg, whereas these variables in other date waste levels were similar to the control [12]. Kamel et al. [14] fed diets containing date pits at 5, 10 and 15%, and whole zahdi dates at 5, 10, 30 and 47.7% to broiler chicks. They found that the incorporation of 47.7% whole date as a total replacement of corn resulted in a slight decrease in feed conversion of chicks.

In some studies, the use of DSM improved production performance when used at low levels compared with the corn or barley grains, which is attributed to the presence of bioactive compounds present in it [34–37]. In addition, DSM may be effective against the aflatoxicosis. Abdel-Sattar et al. [36] reported that the inclusion of DSM (2 and 4%) in diets of broiler chickens ameliorated the negative effects of aflatoxins and provided a partial protective effect against aflatoxicosis, which was dose-related as 2% supplementation provided better protection than 4% supplementation. Moreover, supplementation of DPS in the diets reduced malondialdehyde content in muscle.

El-Far et al. [37] conducted a study on broiler chickens fed diets containing 2, 4 and 6% DSM. Results indicated that DSM significantly increased serum reduced glutathione content for all DSM, and interferon-gamma and interleukin-2 levels were significantly higher in chicks fed 2% DSM. Therefore, the authors concluded that increase growth performance, immunity, and antioxidant status of chickens due to DSM supplementation might be related to the antioxidant and immune-stimulant constituents in DSM.

The productivity and qualitative traits of eggs of laying hens were not adversely affected by the inclusion of date pits in the diets [38]. Hermes and Al-Homidan [39] clarified that the best performance was seen when 16% date waste meal or 10% date pits meal could be replaced by yellow corn without any adverse effect on productive performance, egg quality. Up to 20% date seeds was tested without an adverse effect on egg production and feed efficiency, but it decreased egg quality [40]. Al-Harthi et al. [41] recommended that using date pits as alternative feedstuff enhanced maturity of Lohmann pullets and reproduction performance of layers. Ghasemi et al. [42] stated that date pits at the level of 10% could be partly used as alternative feedstuffs in laying hens' diets, without negatively affecting the productive and health parameters but more than 10% feed intake was reduced due to the presence of non-starch polysaccharides which increase the viscosity of gut contents.

Generally, the age of the bird affects the benefit from adding dates to the diets, as old birds benefit from the fibers more than young birds. Due to the high percentage of fiber in date wastes, it should be added to finisher diets with a higher percentage than the starter diets. Additionally, ducks and laying hens benefit from the diets containing date waste more than broiler chickens. Finally, from the previous studies that were reviewed, it can be concluded that using DSM up to 5–10% in diets of poultry (broiler, laying hen, ducks, quail or turkey) positively affects the productive performance. We summarized effects of date by-products on performance of poultry in Table 7.

Table 7. Effects of Date by-products on performance of poultry.

Sources and Levels	Species	Results	Reference
Date pits (0, 10, 20, 30 and 40%),	Broiler chicks	Adding date pits at 10% had no effect on broiler performance. carcass characteristics and blood parameters.	[43]
DW (0 to 200 g/kg diet)	Broiler chicks	No effect on BWG	[36]
In the starting period: DWM (0, 8, 16, 24%) and DPM (5, 10, and 15%) In the finishing period: DWM (0, 8, 16, 24%) and DPM (6, 12 and 18%)	Broiler chicks	In the starting period, BW and BWG were improved when chicks were fed diets containing 8% DWM or 10% DPM, and 10% DPM improved FCR. In the finishing period, chicks fed a diet containing 8% and 16 DWM or 18% DPM had better BW and BWG than the control. 18% DPM improved FCR	[3]
DW (50 and 200 g/kg diet)	Broiler chicks	DW (50 g/kg diet) reduced FI and improved feed efficiency and European feed production efficiency.	[12]
Date pits (5, 10 and 15%) and whole zahdi dates (5, 10 and 47.7%)	Broiler chicks	47.7% whole date as a total replacement of corn resulted in a slight decrease in FCR	[14]

DW = date waste; DWM = date waste meal; DPM = date pits meal; BW = body weight; BWG = body weight gain; FCR = feed conversion ratio; FI = feed intake.

3.2. *Effect of Date Waste on Carcass Traits*

Kamel et al. [14] fed broiler chicks with diets containing date pits at 0, 5, 10 and 15% or whole Zahdi dates at 0, 5, 10, 30 and 47.7%. The gross examination of various internal organs (liver, heart, spleen, pancreas and intestines) in both date by-products revealed abnormalities. The weight of inedible parts (including feather) of carcass as well as their proportional weight showed an insignificant response to the effect of date seeds levels [4]. The absolute and percentage weights of total edible meat differed significantly between 5% date seeds (1188 g) and the control group (1347 g). Differences for carcass weight between the 5% group (1077 g) and the control group (1229 g) were also significant, while differences between 5% and other groups were not significant. Percentage of carcass weights of the control group (66.4%) was higher than the other groups except 10% (67.2%). In addition, giblet weights were not significant for all groups.

Osman et al. [44] studied the extent of using DSM in starter and grower Pekin duckling diets. The results indicated that dressing percentage was not significantly affected by the level of DSM in the diets, but gizzard percentage increased when increasing the level of DSM in the diets. Moreover, there were pronounced trends for increased absolute and relative digestive tract weight and length of digestive tract and caecum when increasing level of DSM in the diets. This could be attributed to the gritty nature and to the high fiber content of DSM. Such increase in gizzard percentages may be an adaptation of the birds to digest the gritty and fibrous feed at the high levels of DSM. In addition, both small intestine and caecum length increased gradually with the increasing of DSM, but fasting body weight, carcass, liver and heart percentages were not affected by the level of DSM in the broiler diets.

Al-Homidan [3] determined the effect of addition of date wastes (whole date waste meal, and date pits meal) to starter and finisher broiler diets containing 0, 8, 16 and 24% whole date waste meal, or 5, 10 and 15% date pit meal in the starting period and 0, 8, 16 and 24% whole date waste meal and 6, 12 and 18% date pit meal during finishing period. Results indicated that the corresponding values for dressing percentage ranged between 68.7 and 70.5%, while ready to cook (hot carcass weight and giblets weight) percentages ranged between 73.3 and 75.6%, without significant differences among the levels.

Attia and Al-Harthi [12] reported that date waste in broiler chickens fed at 0 to 200 g/kg diet did not affect dressing percentage, weight (%) of liver, gizzard, but altered many other carcass traits such as weight (%) of heart, spleen, pancreas and intestine. Similarly, inedible date waste up to 150 g/kg diet of broiler chickens did not affect dressing percentage, chemical composition of meat along with water holding capacity, pH, tenderness, and color, but weight (%) of heart, pancreas and spleen differed among the inclusion levels [37].

3.3. Effect of Date Waste on Mortality

Regarding mortality, Kamel et al. [14] found no effect due to DSM levels on broiler mortality. They reported that there was no link between chick mortality and the use of date pits in the diets. The effect of using DSM as a substitute for yellow corn in chickens quail diets up to 24% levels on the mortality rate of Japanese quails was studied [36]. These studies reported that up to 10% DSM can be fed to chickens without affecting growth performance. Al-Homidan [3] found no significant effect on broiler mortality rate (the mortality rate was 2.2% during total experimental period of 0 to 7 weeks) due to using date waste (0, 8, 16, 24% whole date waste meal) or (5, 10, and 15% DSM) during the starter period, or (0, 8, 16, and 24% whole date waste meal) and (6, 12 and 18% DSM) during the finishing period. Date waste up to 200 g/kg diet replacing wheat bran did not affect mortality of chickens [37].

3.4. Effect of Date Waste on Feed Cost and Economic Efficiency

The net profits in broilers fed 0%, 1%, 2%, 3% and 4% of date palm kernel were 27.01, 32.77, 36.78, 43.47 and 44.51 (Rs.) per broiler, respectively. It was reported that the maximum net profit (44.51/bird) was recorded from the birds fed 4.0% date palm kernel, closely followed by those birds fed 3.0% date palm kernel (43.47/bird) [45]. Furthermore, in broilers, the feed cost per kg live weight gain was statistically less in birds fed date palm kernel with or without enzymes compared to the basal diet [46].

Al-Homidan [3] found that using date wastes (whole date waste meal and DSM) at 5, 10 and 15% in broiler diets reduced the cost of the diet compared with the control group. Application and beneficial uses of date by-product and wastes in poultry nutrition are illustrated in Figure 1.

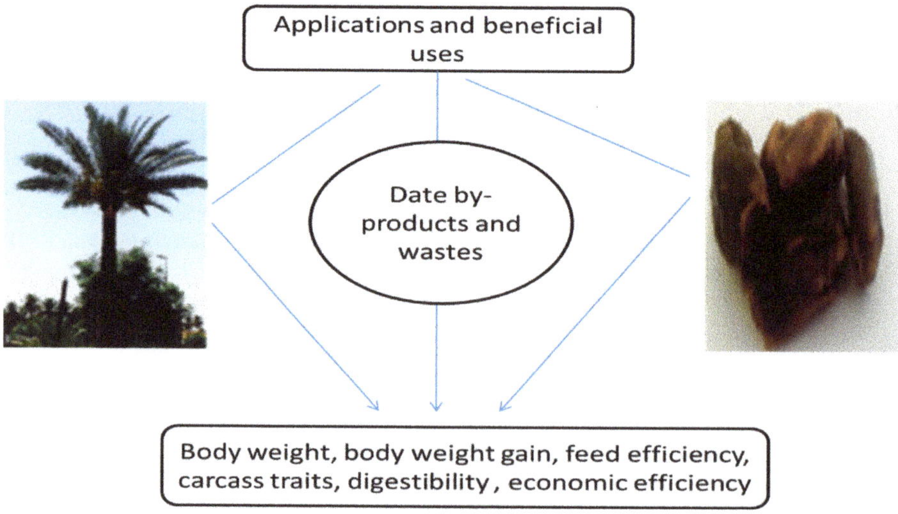

Figure 1. Application and beneficial uses of date by-products and wastes in poultry nutrition.

4. Recent Knowledge and Future Direction

Recently, there have been trends to utilize the phytochemical properties of industrial fruit or herbs wastes in animal feeding, which will not only reduce environmental burdens, but also improve animal production and health [47,48]. Similarly, phytochemical or nutraceutical prosperities of date pit or waste have been explored or attempts have been made to improve them using solid state degradation by *Trichoderma reesei* L. and their utilization in animal feeds [49–53]. Alyileili et al. [49] reported that solid state degradation of date pit with *T. reesei* enhanced mannan oligosaccharide content, phenolic (142 g gallic equivalent/kg) and flavonoid (117 g rutin equivalent/kg) content of degraded date pits and antioxidant activities. Due to the presence of probiotic, phenolics and flavonoids along with high antioxidant activities, the degraded date pits might provide beneficial impacts on gut health and barrier function [54,55]. Therefore, extra-nutritional effects of *T. reesei*-mediated solid state degraded date pit were explored in poultry [56,57]. The inclusion of 10% degraded date pits (phenolics, flavonoids and mannan–oligosaccharide) in the diets of broiler chickens increased catalase, superoxide dismutase and glutathione peroxidase activities in the serum, liver, and intestine, and decreased malondialdehyde content compared to the control group [50]. It also increased pancreatic enzyme activities, the villus length, and the villus/crypt ratio, and decreased the crypt depth of the intestine [51]. The number of pathogenic bacteria such as *Escherichia coli, Enterobacteriacea, Salmonella*, and *Shigella* were reduced, but beneficial *bacteria such as Bifidobacterium* and *Lactobacillus* were increased in the gut of poultry by 10% degraded date pits [52]. Additionally, immunoglobulin A and G content in serum and intestinal tissue, gene expressions of mucin-2, cathelicidin-1, beta defensins, and liver expressed antimicrobial peptide-2 (which prevents the attachment of invading pathogens in intestinal epithelia) in jejunum of chickens fed degraded date pits were increased [52]. However, body weight, feed intake, and feed conversion ratio were not influenced by 10% degraded date pits compared with the control [52]. Alyileili et al. [53] compared 5 and 10% non-degraded date pits vs. degraded date pits in chickens. The populations of *Escherichia coli, Salmonella, Campylobacter* and *Shigella* spp. were reduced in both degraded and non-degraded date pit-diets compared with the control, but degraded date pit diets had a stronger effect than the non-degraded date pit diets. Body weight gains were not affected, but feed conversion ratio was higher for non-degraded date pits diets than control and degraded date pit diets. European production efficiency index was greater with 5% and 10% degraded date pits than non-degraded date pits and the control [53]. Overall, these results suggested that date pits can improve gut health by reducing pathogenic bacteria, enhancing beneficial bacteria, gene expressions of antimicrobial peptides in the jejunum, and immunoglobulin concentrations in the intestine and serum while degraded date pits likely have greater effects than the non-degraded date pits. Future studies may also investigate the gut barrier and electrophysiological functions such as mRNA expressions of tight junction proteins, permeability, especially during a pathogenic microbial challenge models due to the presence of plant bioactive principles (such as phenolics and flavonoids) in non-degraded or degraded date pits because these plants bioactive have been shown to improve the gut barrier function in animals including in poultry [56,57].

Dates by-products are currently used in the feeding of animals, poultry and fish industries. Due to the presence of a large amount of total fiber, they are considered to have potential health benefits for human as prebiotics [58]. The dietary fiber concentrate of date seeds/kernels presented the potential to be used as a novel source of prebiotic feed additives, by increasing the numbers of *Lactobacillus paracasei* as probiotics [58]. Fiber concentrate of date fruit has been converted into another product rich in antioxidant soluble fiber by enzymatic hydrolysis [59].

Based on the nutritional values and chemical composition of the protein, carbohydrate, fat and minerals content in the date kernels, it has been reported that they could be used as a novel source of functional products with interesting technological functionalities that could also be applied as an excellent source of prebiotic due to the high content of dietary fiber [16,60]. All the above information will enhance the value of date by-products, fruits, seeds, etc. as excellent and cheap sources of a natural diet that can act as bioactive and nutritive ingredients in the pharmaceutical industries, food sector, and for other aspects [61,62].

5. Conclusions

Date fruit by-products, particularly DSM, have good nutritional values in poultry diets. The date has high concentrations of oleic acid and lysine and methionine content which are comparable to the corn and barley grains. Considering variability in the nutritive value of the DSM, developing a formula for ME of DSM = a (intercept) − b (slope) × crude fiber content should be considered to improve the feed formula accuracy. However, the nutritional value might not only depend on crude fiber but also on other compounds.

The results obtained from different studies indicate that date fruit by-products, particularly DSM, could be incorporated up to 5–10% levels, replacing corn or barley grains without hampering feed intake, growth, feed efficiency and nutrient utilization in poultry (broiler, laying hen, ducks, quail and turkey), and this could reduce the production cost. Maximum levels of DMS should be 5% in young birds and 10% in grower–finishers. Higher levels could be acceptable in slow-growing birds due to low-nutritional requirements. It also seems that young birds are less tolerant of DSM than older birds. In layers, up to 20% DMS was evaluated without any adverse effect on feed efficiency and egg production, but it decreased egg quality. The balance of protein and energy in the diet should always be established carefully. These agriculture-by-products become important if circularity becomes more important, not only in the Covid-19 crisis, but also in countries with a limited supply of classical feed resources and dependence on imported ones.

Author Contributions: All authors contributed equally to this review, and have read and agreed to the published version of the manuscript.

Funding: This research received no external funding.

Institutional Review Board Statement: Not applicable.

Data Availability Statement: All data are published in the cited literature and reported in the text of this manuscript.

Conflicts of Interest: The authors declare no conflict of interest.

References

1. Hafez, H.M.; Attia, Y.A. Challenges to the Poultry Industry: Current Perspectives and Strategic Future after the COVID-19 Outbreak. *Front. Vet. Sci.* **2020**, *7*. [CrossRef] [PubMed]
2. Hashem, N.M.; González-Bulnes, A.; Rodriguez-Morales, A.J. Animal Welfare and Livestock Supply Chain Sustainability under the COVID-19 Outbreak: An Overview. *Front. Vet. Sci.* **2020**, *7*, 582528. [CrossRef]
3. Al–Homidan, A.H. Date waste (whole dates and date pits) as ingredients in Broiler diets. *Egypt. Poult. Sci.* **2003**, *23*, 15–35.
4. Zaid, A.; Arias-Jiménez, E.J. "Date Palm Cultivation", Food and Agriculture Organization, Plant Production and Protection, Paper 156, Rome. 2002, p. 292. Available online: http://www.fao.org/docrep/006/Y4360E/y4360e00.htm (accessed on 20 March 2020).
5. Barreveld, W.H. "Date palm products" FAO, Agricultural Services Bulletin. 1993, p. 101. Available online: http://www.fao.org/3/t0681E/t0681E00.htm (accessed on 20 March 2020).
6. Diab, K.A.; Aboul-Ela, E.I. In vivo comparative studies on antigenotoxicity of date palm (phoenix dactylifera l.) pits extract against dna damage induced by n-nitroso-n-methylurea in mice. *Toxicol. Inter.* **2012**, *19*, 279–286. [CrossRef] [PubMed]
7. Zanu, H.K.; Abangiba, J.; Badoo, W.A.; Akparibo, A.D.; Sam, R. Laying chickens response to various levels of palm kernel cake in diets. *Int. J. Livest. Prod.* **2012**, *3*, 12–16. [CrossRef]
8. Tabook, N.; Kadim, I.; Mahgoub, O.; Al-Marzooqi, W. The effect of date fibre supplemented with an exogenous enzyme on the performance and meat quality of broiler chickens. *Br. Poult. Sci.* **2006**, *47*, 73–82. [CrossRef]
9. Daghir, N.J. *Feedstuffs Used in Hot Regions" (Daghir, NJ) in "Poultry Production in Hot Climates*; CABI: Wallingford, UK, 2008; pp. 160–196.
10. Masoudi, A.; Bojarpour, M.; Chaji, M.; Eslami, M.; Mirzadeh, K. Economic Value of Date Pits Replaced with Maize in Broiler Chicken Diet. *J. Anim. Vet. Adv.* **2010**, *9*, 1578–1581. [CrossRef]
11. Zaghari, M.; Taherkhani, R.; Ghasemi, M.; Shivazad, M. Estimation of metabolisable energy content of date pit and its effect on lipid and protein oxidation in broiler chicks. *J. Sci. Food Agric.* **2009**, *89*, 2336–2341. [CrossRef]
12. Attia, Y.A.; Al-Harthi, M. Effect of supplementation of date waste to broiler diets on performance, nutrient digestibility, carcass characteristics and physiological parameters. *Effect Suppl. Date Waste Broiler Diets Perform. Nutr. Dig. Carcass Charact. Physiol. Parameters* **2015**, *79*, 1–10. [CrossRef]
13. Gohl, B. *Tropical Feeds*; Food and Agriculture Organization of the United Nations: Rome, Italy, 1975; p. 614.

14. Kamel, B.S.; Diab, M.F.; Ilian, M.A.; Salman, A.J. Nutritional Value of Whole Dates and Date Pits in Broiler Rations. *Poult. Sci.* **1981**, *60*, 1005–1011. [CrossRef]
15. Chaira, N.; Ferchichi, A.; Mrabet, A.; Maheur, S. Chemical composition of the flesh and the pit of date palm fruit and radical scavenging activity of their extracts. *Pak. J. Biol. Sci.* **2007**, *10*, 2202–2207. [CrossRef]
16. Al-Farsi, M.A.; Lee, C.Y. Nutritional and functional properties of dates: A review. *Crit. Rev. Food Sci. Nutr.* **2008**, *48*, 877–887. [CrossRef]
17. Shaba, E.Y.; Ndamitso, M.M.; Mathew, J.T.; Etsunyakpa, M.B.; Tsado, A.N.; Muhammad, S.S. Nutritional and anti-nutritional composition of date palm (Phoenix dactylifera L.) fruits sold in major markets of Minna Niger State, Nigeria. *Afr. J. Pure Appl. Chem.* **2015**, *9*, 167–174. [CrossRef]
18. Habib, H.M.; Ibrahim, W.H. Nutritional quality evaluation of eighteen date pit varieties. *Int. J. Food Sci. Nutr.* **2009**, *60*, 99–111. [CrossRef] [PubMed]
19. Abdel-Nabey, A.A. Chemical composition and oil characteristics of date pits of six Egyptian cultivars. *Alex. J. Agric. Res.* **1999**, *44*, 127–141.
20. Gheisari, H.R.; Heydari, S.; Basiri, S. The effect of date versus sugar on sensory, physicochemical, and antioxidant properties of ice cream. *Iran J. Vet. Res.* **2020**, *21*, 9–14. [PubMed]
21. Abd EL-Rahman, G.A.; Abou El-Nassr, H.M.; Fayed, A.M.; Nassr, M.S. Utilization of some agro-Industrial by-products in fattening lambs on the natural ranges in the south of valley. *Egypt. J. Nutr. Feed.* **2003**, *6*, 815–865.
22. Heuzé, V.; Tran, G.; Delagarde, R.; Bastianelli, D.; Lebas, F.; Date Palm Seeds. Feedipedia, a Programme by INRAE, CIRAD, AFZ and FAO. 2015. Available online: https://www.feedipedia.org/node/687 (accessed on 11 June 2020).
23. Sawaya, W.N.; Khalil, J.K.; Safi, W.J. Chemical Composition and Nutritional Quality of Date Seeds. *J. Food Sci.* **1984**, *49*, 617–619. [CrossRef]
24. El-Deek, A.A.; Attia, Y.A.; Al-Harthi, M.A. Including whole inedible date in grower-finisher broiler diets and the impact on productive performance, nutrient digestibility and meat quality. *Animal* **2010**, *4*, 1647–1652. [CrossRef]
25. Mansouri, A.; Embarek, G.; Kokkalou, E.; Kefalas, P. Phenolic profile and antioxidant activity of the Algerian ripe date palm fruit (*Phoenix dactylifera*). *Food Chem.* **2005**, *89*, 411–420. [CrossRef]
26. Shanoon, A.Q.; Jassim, M.S.; Mohamed, A.H.; Latef, M.S.; Abad AL-Raheem, A.M. The effect of using different levels from date palm pollen in diet on productive performance and some eggs quality measurements for layer hens lohman. *Anim. Vet. Sci.* **2015**, *3*, 1–4. [CrossRef]
27. Mrabet, A.; Jiménez-Araujo, A.; Guillén-Bejarano, R.; Rodríguez-Arcos, R.; Sindic, M. Date Seeds: A Promising Source of Oil with Functional Properties. *Foods* **2020**, *9*, 787. [CrossRef] [PubMed]
28. Çenesiz, A.A.; Çiftci, I. Modulatory effects of medium chain fatty acids in poultry nutrition and health. *World's Poult. Sci. J.* **2020**, *76*, 234–248. [CrossRef]
29. Belal, I.E. Evaluating fungi-degraded date pits as a feed ingredient for Nile tilapia *Oreochromis niloticus* L. *Aquac. Nutr.* **2008**, *14*, 445–452. [CrossRef]
30. Kholif, A.M.; Abo El-nor, S.A. Date Seeds as a New Component in Diets for Dairy Animals. Fifth Arab Conference of Food Science & Technology. December 1993. Available online: https://www.clocate.com/conference/international-conference-on-nutrition-food-science-and-technology/69386/ (accessed on 20 June 2020).
31. National Research Council, NRC. *Nutrient Requirements of Poultry*; National Academy of Sciences: Washington, DC, USA, 1994.
32. Saleh, M.; Kokoszyński, D.; Mousa, M.; Abuoghaba, A. Effect of Date Palm Pollen Supplementation on the Egg Production, Ovarian Follicles Development, Hematological Variables and Hormonal Profile of Laying Hens. *Animals* **2021**, *11*, 69. [CrossRef]
33. Jassim, J.M. Effect of Using Date By-product with Enzyme on Performance and Blood Parameter of Broiler. *Int. J. Poult. Sci.* **2010**, *9*, 895–897. [CrossRef]
34. Al-Saffar, A.E.; Attia, Y.A.; Mahmoud, M.B.; Zewell, H.S.; Bovera, F. Productive and reproductive performance and egg quality of laying hens fed diets containing different levels of date pits with enzyme supplementations. *Trop. Anim. Health Prod.* **2012**, *45*, 327–334. [CrossRef]
35. Hussein, A.S.; Alhadrami, G.A. Effect of enzyme supplementation and diets containing date pits on growth and feed utilization of broiler chicks. *Agric. Mar. Sci.* **2003**, *8*, 67–71.
36. Abdel-Sattar, W.M.; Sadek, K.M.; Elbestawy, A.R.; Mourad, D.M. Biochemical protective role of Phoenix dactylifera seeds against aflatoxicosis in broilers. *Damanhour J. Vet. Sci.* **2019**, *1*, 5–12.
37. El-Far, A.H.; Ahmed, H.A.; Shaheen, H.M. Dietary Supplementation of Phoenix dactylifera Seeds Enhances Performance, Immune Response, and Antioxidant Status in Broilers. *Oxidative Med. Cell. Longev.* **2016**, *2016*, 1–9. [CrossRef]
38. Perez, J.F.; Gernat, A.G.; Murillo, J.G. Research notes: The effect of different levels of palm kernel meal in layer diets. *Poult. Sci.* **2000**, *79*, 77–79. [CrossRef] [PubMed]
39. Hermes, I.H.; Al-Homidan, A.H. Effects of using date waste (whole dates and date pits) on performance, egg components and quality characteristics of Baladi Saudi and Leghorn laying hens. *Egypt. J. Nutr. Feed.* **2004**, *7*, 223–241.
40. Mahmoud, H.A. Using date seed meal in Dandarawi laying hens and growing cockerels diets. *Assiut. Vet. Med. J.* **2008**, *54*, 11–23.
41. Al-Harthi, M.A.; El-Deek, A.A.; Yakout, H.M.; AL-Refaee, M. The nutritive value of date waste meal as a feedstuff for Lohmann brown pullets and layers. *J. Poult. Sci.* **2009**, *46*, 303–312. [CrossRef]

42. Ghasemi, R.; Torki, M.; Ghasemi, H.A.; Zarei, M. Single or combined effects of date pits and olive pulps on productive traits, egg quality, serum lipids and leucocytes profiles of laying hens. *J. Appl. Anim. Res.* **2014**, *42*, 103–109. [CrossRef]
43. Masoudi, A.; Chaji, M.; Bojarpour, M.; Mirzadeh, K. Effects of different levels of date pits on performance, carcass characteristics and blood parameters of broiler chickens. *J. Appl. Anim. Res.* **2011**, *39*, 399–405. [CrossRef]
44. Osman, A.M.A.; Toson, M.A.; El-Bogdady, A.H. Effect of replacing part of starter and grower diets by date stone meal on performance of pekin duckling. *J. Agric. Sci. Mansoura Univ.* **1995**, *20*, 2151–2159.
45. Tareen, M.H.; Wagan, R.; Siyal, F.A.; Babazadeh, D.; Bhutto, Z.A.; Arain, M.A.; Saeed, M. Effect of various levels of date palm kernel on growth performance of broilers. *Vet. World* **2017**, *10*, 227–232. [CrossRef]
46. Shakila, S.; Reddy, P.S.; Reddy, P.V.V.; Ramana, J.V.; Ravi, A. Effect of palm kernel meal on the performance of broilers. *Tamilnadu J. Vet. Anim. Sci.* **2012**, *8*, 227–234.
47. Singh, P.; Hundal, J.S.; Patra, A.K.; Wadhwa, M.; Sharma, A. Sustainable utilization of Aloe vera waste in the diet of lactating cows for improvement of milk production performance and reduction of carbon footprint. *J. Clean. Prod.* **2021**, *288*, 125118. [CrossRef]
48. Singla, A.; Hundal, J.S.; Patra, A.K.; Wadhwa, M.; Nagarajappa, V.; Malhotra, P. Effect of dietary supplementation of Emblica officinalis fruit pomace on methane emission, ruminal fermentation, nutrient utilization, and milk production performance in buffaloes. *Environ. Sci. Pollut. Res.* **2021**, *28*, 18120–18133. [CrossRef]
49. Alyileili, S.R.; Hussein, A.S.; Ibrahim, W.; El-Tarabily, K.A. Phytochemical Composition and Antioxidant Activity of Trichoderma reesei Degraded Date (*Phoenix dactylifera* L.) Pits. *Curr. Bioact. Compd.* **2020**, *16*, 528–536. [CrossRef]
50. Alyileili, S.R.; El-Tarabily, K.A.; Belal, I.E.H.; Ibrahim, W.H.; Sulaiman, M.; Hussein, A.S. Effect of Trichoderma reesei Degraded Date Pits on Antioxidant Enzyme Activities and Biochemical Responses of Broiler Chickens. *Front. Vet. Sci.* **2020**, *7*. [CrossRef]
51. Alyileili, S.R.; El-Tarabily, K.A.; Belal, I.E.H.; Ibrahim, W.H.; Sulaiman, M.; Hussein, A.S. Intestinal Development and Histomorphometry of Broiler Chickens Fed Trichoderma reesei Degraded Date Seed Diets. *Front. Vet. Sci.* **2020**, *7*. [CrossRef]
52. Alyileili, S.R.; El-Tarabily, K.; Ibrahim, W.H.; Sulaiman, M.; Hussein, A.S. Effect of Trichoderma reesei Degraded Date Pits Supplementation on Growth Performance, Immunoglobulin Levels, and Intestinal Barrier Functions of Broiler Chickens. *Recent Pat. Food Nutr. Agric.* **2020**, *11*, 168–181. [CrossRef]
53. Alyileili, S.; Belal, I.; Hussein, A.; El-Tarabily, K. Effect of Inclusion of Degraded and Non-Degraded Date Pits in Broilers' Diet on their Intestinal Microbiota and Growth Performance. *Animals* **2020**, *10*, 2041. [CrossRef]
54. Abdollahi, M.R.; Hosking, B.J.; Ning, D.; Ravindran, V. Influence of Palm Kernel Meal Inclusion and Exogenous Enzyme Supplementation on Growth Performance, Energy Utilization, and Nutrient Digestibility in Young Broilers. *Asian Australas. J. Anim. Sci.* **2016**, *29*, 539–548. [CrossRef]
55. Rodríguez, L.G.R.; Gasga, V.M.Z.; Pescuma, M.; Van Nieuwenhove, C.; Mozzi, F.; Burgos, J.A.S. Fruits and fruit by-products as sources of bioactive compounds. Benefits and trends of lactic acid fermentation in the development of novel fruit-based functional beverages. *Food Res. Int.* **2021**, *140*, 109854. [CrossRef] [PubMed]
56. Patra, A.K.; Amasheh, S.; Aschenbach, J.R. Modulation of gastrointestinal barrier and nutrient transport function in farm animals by natural plant bioactive compounds—A comprehensive review. *Crit. Rev. Food Sci. Nutr.* **2018**, *59*, 3237–3266. [CrossRef]
57. Patra, A.K. Influence of Plant Bioactive Compounds on Intestinal Epithelial Barrier in Poultry. *Mini Rev. Med. Chem.* **2020**, *20*, 566–577. [CrossRef]
58. Al-Thubiani, A.S.; Khan, M.S.A. The prebiotic properties of date palm (*Phoenix dactylifera* L.) seeds in stimulating probiotic lactobacillus. *J. Pure Appl. Microbiol.* **2017**, *11*, 1675–1686. [CrossRef]
59. Mrabet, A.; Rodríguez-Gutiérrez, G.; Rubio-Senent, F.; Hamza, H.; Rodríguez-Arcos, R.; Guillén-Bejarano, R.; Sindic, M.; Jiménez-Araujo, A. Enzymatic conversion of date fruit fiber concentrates into a new product enriched in antioxidant soluble fiber. *LWT* **2017**, *75*, 727–734. [CrossRef]
60. Niazi, S.; Khan, I.M.; Pasha, I.; Rasheed, S.; Ahmad, S.; Shoaib, M. Date palm: Composition, health claim and food applications. *Int. J. Pub. Health Sys.* **2017**, *2*, 9–17.
61. Idowu, A.T.; Igiehon, O.O.; Adekoya, A.E.; Idowu, S. Dates palm fruits: A review of their nutritional components, bioactivities and functional food applications. *AIMS Agric. Food* **2020**, *5*, 734–755. [CrossRef]
62. Maqsood, S.; Adiamo, O.; Ahmad, M.; Mudgil, P. Bioactive compounds from date fruit and seed as potential nutraceutical and functional food ingredients. *Food Chem.* **2020**, *308*, 125522. [CrossRef]

animals MDPI

Article

Research on the Effects of Gender and Feeding Geese Oats and Hybrid Rye on Their Slaughter Traits and Meat Quality

Dariusz Lisiak [1], Piotr Janiszewski [1,*], Eugenia Grześkowiak [1], Karol Borzuta [1], Beata Lisiak [1], Łukasz Samardakiewicz [1], Tomasz Schwarz [2], Krzysztof Powałowski [1] and Krzysztof Andres [2]

1 Department of Meat and Fat Technology, Prof. Wacław Dąbrowski Institute of Agricultural and Food Biotechnology, 60-111 Poznań, Poland; dariusz.lisiak@ibprs.pl (D.L.); eugenia.grzeskowiak@ibprs.pl (E.G.); karol.borzuta@ibprs.pl (K.B.); beata.lisiak@ibprs.pl (B.L.); lukasz.samardakiewicz@ibprs.pl (Ł.S.); krzysztof.powalowski@ibprs.pl (K.P.)

2 Faculty of Animal Science, University of Agriculture in Kraków, 31-120 Kraków, Poland; tomasz.schwarz@urk.edu.pl (T.S.); krzysztof.andres@urk.edu.pl (K.A.)

* Correspondence: piotr.janiszewski@ibprs.pl

Simple Summary: The research showed the influence of the gender of geese on some slaughter value traits and meat quality of geese slaughtered at the age of 17 weeks. It was found that ganders had a greater body and carcass weight than females but no significant differences in slaughter yield nor in carcass element share were observed. Gender did not have an effect on the majority of the studied breast meat quality traits. The studied feeding model did not have a significant influence on goslings' body weight; however, the birds fed hybrid rye had a lower slaughter yield as compared with those fed oats, but their meat had better physical and chemical characteristics (lower fat content, lower drip loss, higher protein content and, in the female goslings, also better sensory quality). Hybrid rye may be used in geese feed because it does not have a negative effect on pre-slaughter body weight and has a positive effect on some meat quality traits as compared with feeding geese oats.

Abstract: The aim of the study was to determine the effect of feeding Zatorska variety geese hybrid rye, oats, or a mixture of both grains (1:1) on slaughter value and meat quality. At 14 weeks old, the birds were separated into three feeding groups (n = 12) and were fed between 15 and 17 weeks of age with hybrid rye, oats, or a mixture of these two grains. The research proved the effect of gender and feeding on some slaughter value traits and meat quality of the goslings' breast meat. It was found that the ganders had a 10% to 15% higher body and carcass weight than the females. No significant differences were observed between the genders within the majority of the physical and chemical characteristics as well as the sensory traits. The feeding type did not have a significant effect on the goslings' body weight and carcass element share. The birds fed hybrid rye had a 2 percentage points lower slaughter yield than those birds fed oats which was combined with a lower share of subcutaneous fat (measured as the weight of the tissue coming from dissection) in birds fed hybrid rye. The meat of the birds fed hybrid rye had some better physical, chemical characteristics and, in the female goslings, also better sensory quality. The results indicated that hybrid rye may be used in feeding goslings at the end of the growing period, because it did not cause any negative effects on the pre-slaughter body weight and had a positive effect on some meat quality traits, such as better sensory estimation results, higher protein content, and lower drip losses.

Keywords: geese; feeding; rye; oats; slaughter value; quality traits

Citation: Lisiak, D.; Janiszewski, P.; Grześkowiak, E.; Borzuta, K.; Lisiak, B.; Samardakiewicz, Ł.; Schwarz, T.; Powałowski, K.; Andres, K. Research on the Effects of Gender and Feeding Geese Oats and Hybrid Rye on Their Slaughter Traits and Meat Quality. *Animals* **2021**, *11*, 672. https://doi.org/10.3390/ani11030672

Academic Editor: Ilias Giannenas

Received: 9 December 2020
Accepted: 25 February 2021
Published: 3 March 2021

1. Introduction

The share of geese in the poultry production sector in Poland is at a low level, amounting to just 1.5%. Approximately 95% of the geese population in mass production are the White Kołudzka geese bred in the Experimental Center of the National Research Institute of Animal Production in Kołuda Wielka [1–4], originating from the white Italian geese.

Purebred animals may still be found on some breeding farms especially in the flocks under the protection program.

The results of many studies have proven that genotype has a significant influence on tissue content of carcass and chemical content of meat, as well as the muscle structure [3,5]. For instance, geese from rare breed flocks have a lower body weight, better musculature, and a lower fat content as compared with the popular white Kołudzka geese [2,3,5,6].

The Zatorska variety geese, kept in the Experimental Center of the Agricultural Academy in Cracow, belong to the Southern varieties which are used in the protection program.

In a study by Kapkowska et al. [7], the results of fattening the Zatorska and white Kołudzka geese by feeding oats between 14 and 17 weeks were compared, proving that the Zatorska geese body weight (both sexes) in the 17th week of age (5648 g) was significantly lower than the white Kołudzka geese body weight (6814 g). Moreover, the Zatorska geese had a lower slaughter yield and higher thigh and shank muscle share. Gomułka et al. [8] estimated the microstructure of the pectoralis superficalis and biceps femoris muscles and determined the technological parameters of the birds' meat at the age of 17 weeks after fattening by feeding oats. The muscle microstructure seemed to be similar in the Zatorska and white Kołudzka geese apart from the higher content of type I muscle fibres in the M. pectoralis superficialis in the Zatorska geese. Similar values for the majority of the physical traits for the muscles of both groups of geese were observed [8].

Geese fed oats have been the subject of many studies and publications [9–12]. Rye is traditionally regarded as a grain of limited suitability in poultry feeding especially in young slaughter birds. The reason for these limitations is the negative impact on the digestive system because of higher food content viscosity, the lower speed rate of the movement of intestinal digestion, and the lower digestibility of nutrients. Such observations may lead to negative effects in terms of poultry production and welfare factors [11]. In recent years hybrid varieties of rye have appeared on the market such as Brasetto which has better agrotechnical features. Furthermore, hybrid rye has a lower anti-nutritional substance content especially with low starch polysaccharides (NPS). NPS present in grains are among others pentosans (xylans and arabinians) as well as beta glucan. Rye grain is especially rich in arabinoxylans (AX) that are not digested by monogastric animals due to the lack of suitable enzymes [13,14]. The AX structure varies because of many factors, i.e., the grain type and variety. The various AX structures have an impact on the functional features and anti-nutritional effect level [13,14].

The research of Świątkiewicz and Arczewska-Włosek [14] showed that the Brasetto hybrid rye grain can be used in the feeding of young chickens, since it is a useful energy and protein source. Introducing 15–20% rye into the feeding mixture for older birds (from 22 days of age) contributed to obtaining the desired production results, while introducing Brasetto rye added to the feed mixture for smallest chickens (1–21 days) had a negative effect on daily gain [13].

The aim of this study was to determine the effect of gender and feed mixture on slaughter value and physical, chemical, and sensory traits of the meat of birds fed hybrid rye, oats, or a mixture of these two grains.

2. Material and Methods

2.1. Experimental Material

The experiment was performed according to the guidelines issued by the Ethics Commission (Regulation 22/2016 of 20 January 2016r ILKE in Cracow). The experiment was performed on the Zatorska variety geese. This is a Polish rare breed meat type that is under the genetic resource protection program.

The geese were hatched from hatching eggs (Brinsea Ova-Easy 380) that belonged to the Experimental and Education Centre of WHBiZ UR (Faculty of the Animal Science University of Agriculture, Cracow, Poland). The experimental flock was made up of 300 young geese that were reared according to the slaughter geese rearing regulations in

Poland. Geese reared in Poland in the last 3 weeks of fattening are fed only oats and water. Oats, due the specific chemical composition, characterized by a high fat content and the profile of fatty acids profitable for human health, increase the value of goose meat and fat [2].

After hatching on 18 July 2017 and after gendering, the young geese were moved to a nursery where they were kept on straw bedding with natural light and with additional artificial lighting for 24 h in the first week of life.

For the first 3 weeks, the geese did not have access to pasture and the density was 7 geese per m^2. The temperature in the nursery was lowered from 28 °C to 22 °C within this period and the relative humidity ranged between 65–70%.

2.2. Feeding

The plan of the experiment is presented in Table 1 below.

Table 1. The experimental plan of the Zatorska geese breed fed different kinds of cereals.

Weeks of Being Reared	Sample Size (n)	Feeding Method
1 to 3	100 nestlings	Feed and water ad libitum, composition of the concentrate and its nutritive value is shown in Table 2
4 to 14	100 birds	Feed and water ad libitum, access in pasture for at least 8 h per day, composition of the concentrate and its nutritive value is shown in Table 2, at the end of 14th week the birds were weighed and divided into 3 feed groups of males and 3 groups of females.
15 to 17	Group A, males n = 12 and females n = 12	Fed only hybrid rye Brasetto cultivare, birds fed ad libitum with access to straw aviary
	Group B, males n = 12 and females n = 12	Fed oats, birds fed ad libitum with access to straw aviary
	Group C, males n = 12 and females n = 12	Concentrate of oats and hybrid rye Brasetto cultivare (1:1, by weight), birds fed ad libitum with access to straw aviary

Table 2. Composition of concentrates and chemical composition of feed used in geese feeding (%).

Feeding Component %	Feeding Period	
	0–3 Weeks	4–14 Weeks
Feed phosphate	1.0	1.0
Limestone	1.8	1.2
Maize	40.0	35.0
Premix	1.2	1.1
Wheat	27.5	24.2
Wheat bran	-	10.0
Soy bean meal	28.5	23.5
Sunflower extracted meal	-	4.0
Crude protein	19.5	19.2
Crude fibre	2.8	3.8
Vegetable oils and crude fat	2.5	2.6
Crude ash	5.4	5.1
Lysine	0.97	0.91
Methionine	0.48	0.40
Calcium	0.94	0.73
Sodium	0.17	0.17
Available phosphorus	0.36	0.44
Metabolic energy, MJ/kg feed	11.50	10.20

For the first 3 weeks, the geese had access to feed and water through the adjusted to their age troughs and semi-automatic drinking bowls. The geese were fed dry complete feed of the proper content and nutrients which are presented in Table 2.

Between the 4th and 14th week of age, the geese were kept on straw bedding with 3.6 goslings/m^2 stocking density in a building with windows. The geese had access to pasture, with 12.5 m^2 space for each bird, and they stayed on the pasture for at least 8 h a day. The temperature in the nursery building was 20–22 °C and the relative humidity ranged between 65–70%. Between the 4th and 14th week of age, the geese were fed a mixture similar to the feed provided to geese between 0–3 weeks of life but additionally enriched with sunflower meal and wheat bran (Table 2).

After reaching 14 weeks, the birds were weighed and divided into groups according to gender and randomly divided into 3 feeding groups for each sex (n = 12). The birds of each group were fed ad libitum with grains according to the following schemes:

Group A: Brasetto hybrid rye with the chemical composition according to Świątkiewicz and Arczewska Włosek [13];

Group B: Oats, according to the composition and nutritional recommendations for geese [14,15];

Group C: A mixture of the Brasetto hybrid rye and oats (1:1 by weight).

The fattening was performed in pens of 1 bird/m^2 density on the floor system, in a building with windows, on straw bedding, with 8 h access to straw pasture (stocking density 1.2 m^2/bird). The temperature in the building was 12–18 °C. and the relative humidity ranged between 65–70%.

In the 17th week of life after being fattened by feeding grain, the birds were marked and not fed for 10 h because emptying the intestines has an impact on the quality and shelf life of carcasses during storage. After no food for 10 h, each bird was weighed individually, and then slaughtered in the commercial slaughter plant. The carcasses, after gutting, were chilled using the blowing and spraying method. Then, each carcass was weighed individually. The slaughter yield was calculated according to the following formula: MT/MC × 100%, where MT is a weight of chilled carcass and MC body weight before being slaughtered.

2.3. *Dissection*

The cut-up of the carcasses was performed according to the Ziołecki and Doruchowski [16] method for determining yield of the following parts: breast muscles, leg muscles, skin with fat, abdominal fat (from the bottom part of the abdomen), neck with fat (cut between the last cervical vertebrae and the first thoracic vertebrae), and wings (cut at the shoulder joint). Each carcass part was weighed and its share in the total chilled carcass was calculated.

2.4. *Physical and Chemical Characteristics of Breast Muscles*

Twenty-four hours after slaughter, the pH was measured using a Mettler Toledo 1140 type pH meter with a Mettler Toledo electrode (Mettler Toledo, OH 43240, D.C., US). The meat colour was determined on the muscle cross-section using a Konica Minolta Chroma CR 400 tool (Konica Minolta, Tokyo, Japan). Colour was classified according to the CiE lab determining L* (lightness), a* (redness), and b* (yellowness), with the following measurement parameters: light source D65; observer 2°; measuring head slot 8 mm; and calibration on the white tile L* = 97.83, a* = 0.45, and b* = 1.88.

The samples of the final products were taken for lab analysis. The water content was measured according to the ISO 1442 (2000) [17]. Approximately 3 g of the minced meat was put on a weighing dish, weighed, and dried at a temperature of 105 °C up to the moment when the stable mass was reached. The water content expressed in % was calculated as a difference between the sample weight before and after drying.

The intramuscular fat content was established according to the ISO 1444 (2000) procedure [18]. The dried and weighed sample was placed in an extraction tube and the fat substances were extracted with paraffin oil in a Soxtherm device produced by the Gerhardt Laboratory System (Gerhard GmbH & Co, Konigswinter, Germany). The fat content was calculated as the difference between the sample weight before and after extraction. The protein content was established according to the Polish norm PN/A-04018 [19] with a Kjeltec System 1002 Distilling Unit (Foss, Hilleroed, Denmark), according to the manufacturer's instructions. The sodium chloride in the final products was established according to the ISO 1841-2 (2002) procedure.

Apart from measuring the colour with a device it was also determined visually using a pattern with a 1 to 5 scale (1, light pink colour to 5, dark red colour) [20].

Similarly, meat marbling was estimated on the muscle cross-section according to a pattern with a 1 to 5 scale (1, slight fat content to 5, high fat content [20].

Drip loss was determined as follows: A muscle sample of approximately 100 g was weighed, placed in a plastic bag, and left in a refrigerator at a temperature of 4 °C for 48 h. The drip loss was calculated and based on the weight difference before and after storage.

Cooking loss was determined as follows: A muscle sample of approximately 150 g was weighed. Next, it was heated in water up to a temperature of 75 °C in the geometric sample centre. After cooling, the sample was weighed, and the weight loss was calculated based on the difference between the weights before and after cooking [21].

2.5. Sensory Test

A sensory test of cooked meat was performed by a team of 5 people trained in terms of sensory sensitivity with the Baryłko-Pikielna and Matuszewska method [21]. The estimation was performed in daylight at room temperature. Meat flavour, juiciness, tenderness, and palatability were determined using a 5-point scale [22]. In the cooked meat samples, meat tenderness was measured by using the Warner–Bratzler shear force method in a Zwick Roell Z 0.5 device (Zwick Roell, Ulm, Germany) with 500 (Kilonewton—kN) force and 100 mm/min of head movement.

2.6. Statistical Analyse

The results were statistically developed by calculating mean values (x) and standard error of the arithmetic mean (SEM). The statistical significance of the differences between the mean values of groups was verified by using a two factors variance analysis according to Anova procedure using the Statistica program Version 13 (StatSoft Hamburg, Germany).

3. Results

The geese body weight before slaughter in all of the feeding groups was at a similar level (Table 3). Clear differences were found between the genders ($p \leq 0.01$). The ganders reached a higher body weight than the young female geese (by approximately 10–15%). A similar effect was observed for carcass weight, but the effects of feeding was confirmed here in the group of young female geese that were heavier in the group fed oats than in the group fed hybrid rye ($p \leq 0.05$). A significantly lower slaughter yield was observed in both genders fed hybrid rye than those fed oats and the difference between the groups was over 2 percentage points (pp). The slaughter yield of those geese fed a mixture of both grains did not differ significantly as compared with the birds that were fed hybrid rye or oats only.

The weight and the percentage share of the basic carcass parts are presented below in Table 4. The weight of these parts was significantly different between the sexes ($p \leq 0.01$), except for body frame weight in the oats group. The higher weight of the other parts was observed in the ganders due to their higher carcass weight. The feeding method had a significant impact on the leg weights of the young female geese which were heavier in the group fed oats than those fed hybrid rye and a mixture of both grains ($p \leq 0.01$). The carcass parts yield did not depend on the gender and the feeding method, except for the higher share of wings in the ganders fed hybrid rye.

Table 3. Mean of the slaughter traits of geese fattened by feeding hybrid rye or oats.

Specification	Fattening Groups			SEM	p-Value for	
	Rye A	Oats B	Rye/Oats C		Fattening	Interaction
Pre-slaughter weight (g)						
Males	5341.25	5444.58	5460.42	70.25	0.764	0.597
Females	4617.92	4905.00	4715.00	57.14	0.105	
SEM	109.38	102.56	100.92			
p-Value for sex	0.000	0.006	0.000			
Carcass weight (g)						
Males	3368.80	3580.00	3500.20	54.88	0.292	0.800
Females	2931.50 b	3226.30 a	3036.80	47.35	0.031	
SEM	74.06	82.43	63.41			
p-Value for sex	0.001	0.028	0.000			
Dressing percentage (%)						
Males	63.02 B	65.64 A	64.12	0.35	0.006	0.949
Females	63.43 b	65.68 a	64.60	0.39	0.039	
SEM	0.34	0.48	0.34			
p-Value for sex	0.561	0.963	0.505			

N = 12 in every experimental group males and females; the letters A, B, C mean the fattening groups, i.e., (A) birds fed hybrid rye, (B) birds fed oats, (C) birds fed a mixture of hybrid rye/oats (1:1); SEM, standard error of the means; means marked with a letter (A, B, C or a, b, c) represents the fattening group with which it differs statistically, means marked by lowercase (a, b, c) are significant at $p \leq 0.05$ and means marked by uppercase (A, B, C) are significant at $p \leq 0.01$, the lack of a letter means that the mean does not statistically differ with any other fattening group.

Table 4. Weight and the proportion of parts in geese carcasses.

Specification	Fattening Groups			SEM	p-Value for	
	Rye A	Oats B	Rye/Oats C		Fattening	Interaction
Neck with skin (g)						
Males	209.90	212.9	211.10	4.52	0.963	0.992
Females	172.80	177.1	175.70	3.22	0.867	
SEM	6.07	6.05	6.26			
p-Value for sex	0.001	0.001	0.002			
Neck with skin (%)						
Males	6.23	5.95	6.02	0.09	0.417	0.758
Females	5.91	5.49	5.78	0.08	0.108	
SEM	0.12	0.09	0.11			
p-Value for sex	0.180	0.001	0.284			
Wing (g)						
Males	537.80	538.80	536.10	8.68	0.992	0.677
Females	451.60	475.90	457.20	6.53	0.289	
SEM	12.97	12.18	11.72			
p-Value for sex	0.000	0.007	0.000			
Wing (%)						
Males	15.98 b	15.09 a	15.30	0.15	0.040	0.718
Females	15.40	14.79	15.06	0.11	0.089	
SEM	0.13	0.19	0.15			
p-Value for sex	0.022	0.435	0.436			
Legs (g)						
Males	687.40	730.80	682.40	12.97	0.253	0.808
Females	578.70 b	640.50 ac	567.3 b	10.89	0.009	
SEM	18.28	16.67	17.49			
p-Value for sex	0.001	0.004	0.000			

Table 4. *Cont.*

Specification	Fattening Groups			SEM	*p*-Value for	
	Rye A	Oats B	Rye/Oats C		Fattening	Interaction
Legs (%)						
Males	20.39	20.46	19.52	0.25	0.236	0.933
Females	19.79	19.94	18.67	0.28	0.135	
SEM	0.31	0.32	0.33			
p-Value for sex	0.350	0.431	0.202			
Body frame (g)						
Males	839.30	885.20	861.80	20.22	0.665	0.921
Females	724.90	783.30	734.30	15.90	0.281	
SEM	21.48	28.31	24.79			
p-Value for sex	0.005	0.071	0.007			
Body frame (%)						
Males	24.92	24.79	24.59	0.44	0.957	0.970
Females	24.74	24.24	24.24	0.40	0.847	
SEM	0.37	056	0.60			
p-Value for sex	0.812	0.635	0.777			

N = 12 in every experimental group males and females; the letters A, B, C mean the fattening groups, i.e., (A) birds fed hybrid rye, (B) birds fed oats, (C) birds fed a mixture of hybrid rye/oats (1:1); SEM, standard error of the means; means marked with a letter (A, B, C or a, b, c) represents the fattening group with which it differs statistically, means marked by lowercase (a, b, c) are significant at $p \leq 0.05$ and means marked by uppercase (A,B,C) are significant at $p \leq 0.01$, the lack of a letter means that the mean does not statistically differ with any other fattening group. Additionally, if the mean is signed by two letters "ac", that means significant difference as compared with means of both signed fattening groups a and c.

The average weight and share of the chosen muscles and fat is presented below in Table 5. No significant differences between the dissection elements were found in terms of sex and feeding method, except for leg muscle weight and skin with fat weight. The muscle weight from the ganders' legs was approximately 17% higher than in the young female geese ($p \leq 0.01$). The weight of the meat from the legs was significantly higher in the female geese fed oats ($p \leq 0.05$) than those fed hybrid rye or a grain mixture. The weight of skin with the fat of females was lower in those birds fed hybrid rye only, but the share of this element was the same in all groups. A no gender not feeding method effect was confirmed on the weight and percentage share of breast muscles and abdominal fat. The only exception was the group of birds fed a mixture of both grains in which the breast muscle weight of the ganders occurred to be higher than in the female geese ($p \leq 0.05$) and the abdominal fat share was higher than in the female geese (2.04% and 2.94%, respectively).

The results of the breast muscle physical parameters measurements are presented below in Table 6. Lower average pH_{24} values were reported only in the muscles of geese fed oats (5.88). No significant effect of sex and feeding method on colour lightness L* and colour parameters (a* and b*) was observed. These parameter values were similar with the following ranges: L* from 35.60 to 36.52, a* from 18.39 to 20.15, and b* from −1.76 to −0.95. Generally, these measurements were confirmed by a visual colour estimation of the muscles of the birds, both genders, fed hybrid rye or oats (3.12 to 3.38 points). However, the muscles of the female geese fed a mixture of both grains were lighter (2.97 points) than those fed hybrid rye (3.28 points).

The meat marbling of both sexes was significantly influenced by the feeding method ($p \leq 0.01$). Greater marbling was observed in the muscles of birds in both sexes fed oats (1.6 points) than those fed hybrid rye (1.32 points). In addition, birds fed a mixture of both grains had greater marbling than those fed hybrid rye ($p \leq 0.05$) but only in the case of female geese (1.53 points). The drip losses in both sexes were influenced by the feeding method ($p \leq 0.01$). No drip losses were observed in the muscles of both sexes fed hybrid rye, which may have been associated with the higher pH_{24} value, which was above 6.0.

Table 5. Mean of weight and percentage of chosen muscles and fat obtained from the dissection of geese carcasses.

Specification	Fattening Groups			SEM	*p*-Value for	
	Rye A	Oats B	Rye/Oats C		Fattening	Interaction
Breast muscle (g)						
Males	520.20	579.20	571.4	14.28	0.184	0.631
Females	489.30	537.90	495.3	14.35	0.330	
SEM	15.21	20.67	17.11			
p-Value for sex	0.322	0.321	0.023			
Breast muscle %						
Males	15.40	16.12	16.36	0.26	0.295	0.497
Females	16.66	16.65	16.29	0.36	0.903	
SEM	0.33	0.37	0.46			
p-Value for sex	0.058	0.488	0.949			
Legs muscles (g)						
Males	431.60	461.60	442.10	9.66	0.450	0.700
Females	357.50 b	394.60 c	350.70 b	7.75	0.040	
SEM	13.43	12.66	13.21			
p-Value for sex	0.003	0.005	0.000			
Legs muscles %						
Males	12.79	12.91	12.65	0.21	0.885	0.721
Females	12.24	12.30	11.54	0.23	0.332	
SEM	0.27	0.29	0.29			
p-Value for sex	0.313	0.300	0.042			
Skin with subcutaneous fat (g)						
Males	473.80	537.4	557.1	16.46	0.094	0.877
Females	435.80 b,c	515.7 a	509.4 a	14.56	0.040	
SEM	16.47	20.97	16.39			
p-Value for sex	0.258	0.615	0.150			
Skin with subcutaneous fat %						
Males	14.08	14.98	15.92	0.26	0.295	0.497
Females	14.84	15.93	16.73	0.36	0.903	
SEM	0.43	0.47	0.44			
p-Value for sex	0.393	0.321	0.374			
Abdominal fat (g)						
Males	90.80	85.60	71.80	5.62	0.371	0.119
Females	70.10	83.40	89.30	4.93	0.274	
SEM	7.51	5.32	6.51			
p-Value for sex	0.174	0.844	0.185			
Abdominal fat (%)						
Males	2.70	2.34	2.04	0.16	0.249	0.082
Females	2.37	2.57	2.94	0.15	0.319	
SEM	0.22	0.13	0.21			
p-Value for sex	0.472	0.408	0.034			

N = 12 in every experimental group males and females; the letters A, B, C mean the fattening groups, i.e., (A) birds fed hybrid rye, (B) birds fed oats, (C) birds fed a mixture of hybrid rye/oats (1:1); SEM, standard error of the means; means marked with a letter (A, B, C or a, b, c) represents the fattening group with which it differs statistically, means marked by lowercase (a, b, c) are significant at $p \leq 0.05$ and means marked by uppercase (A,B,C) are significant at $p \leq 0.01$, the lack of a letter means that the mean does not statistically differ with any other fattening group. Additionally, if the mean is signed by two letters e.g., "ac", that means significant difference as compared with means of both signed fattening groups a and c.

Table 6. Selected physical parameters of breast muscles.

Specification	Fattening Groups			SEM	*p*-Value	
	Rye A	Oats B	Rye/Oats C		Fattening	Interaction
pH$_{24}$						
Males	6.04	5.95	6.02	0.02	0.071	0.632
Females	6.03 B	5.88 A	5.96	0.021	0.010	
SEM	0.03	0.01	0.02			
p-Value for sex	0.858	0.019	0.121			
Colour L*						
Males	36.52	36.49	35.60	0.32	0.424	0.440
Females	35.67	36.28	36.48	0.44	0.748	
SEM	0.60	0.64	0.39			
p-Value for sex	0.369	0.784	0.453			
a* redness						
Males	19.19	19.29	20.15	0.27	0.299	0.671
Females	18.96	18.39	19.10	0.29	0.570	
SEM	0.32	0.34	0.38			
p-Value for sex	0.725	0.188	0.170			
b* yellowness						
Males	−1.41	−0.95	−1.37	0.20	0.603	0.854
Females	−1.76	−1.69	−1.21	0.33	0.768	
SEM	0.33	0.29	0.39			
p-Value for sex	0.605	0.203	0.835			
Colour, points						
Males	3.38	3.26	3.30	0.06	0.669	0.413
Females	3.28 c	3.12	2.97 a	0.05	0.045	
SEM	0.06	0.07	0.07			
p-Value for sex	0.439	0.332	0.012			
Marbling, points						
Males	1.32 B	1.59 Ac	1.43 b	0.03	0.001	0.543
Females	1.32 Bc	1.60 A	1.53 a	0.04	0.005	
SEM	0.03	0.04	0.04			
p-Value for sex	0.990	0.896	0.185			
Drip loss %						
Males	0.00 BC	1.34 A	1.55 A	0.19	0.001	0.495
Females	0.00 BC	1.92 A	1.86 A	0.19	0.000	
SEM	0.00	0.26	0.16			
p-Value for sex		0.264	0.341			
Cooking loss %						
Males	37.32 B	32.07 Ac	35.89 b	0.70		0.000
Females	33.73 B	38.30 AC	33.04 B	0.66		
SEM	0.89	1.01	0.53			
p-Value for sex	0.040	0.001	0.004			

N = 12 in every experimental group males and females; the letters A, B, C mean the fattening groups, i.e., (A) birds fed hybrid rye, (B) birds fed oats, (C) birds fed a mixture of hybrid rye/oats (1:1); SEM, standard error of the means; means marked with a letter (A, B, C or a, b, c) represents the fattening group with which it differs statistically, means marked by lowercase (a, b, c) are significant at $p \leq 0.05$ and means marked by uppercase (A,B,C) are significant at $p \leq 0.01$, the lack of a letter means that the mean does not statistically differ with any other fattening group. Additionally, if the mean is signed by two letters e.g., "ac", that means significant difference as compared with means of both signed fattening groups a and c.

Muscle loss in cooking was significantly influenced by the sex–feed type interaction ($p \leq 0.01$). Clearly higher loss was observed in the muscles of female geese fed with oats (38.3%) than those fed hybrid rye or a mixture of both grains. Whereas in the ganders' group this dependence was the opposite, ganders fed oats had a significantly lower loss rate in cooking than the ganders of the other feeding groups.

The results of the basic chemical composition (Table 7) showed a significant effect of feeding on the fat and protein level ($p \leq 0.01$). Clearly a higher fat content was observed in the muscles of ganders and female geese fed a mixture of both grains (5.10% and 4.83%, respectively) as compared with birds fed hybrid rye only or oats only (3.6% on average).

The higher protein content was confirmed in the muscles of both sexes of birds fed hybrid rye than those fed oats (by approximately 0.9 pp) or the grain mixture (by approximately 1.8 pp).

Table 7. Basic chemical composition of the breast muscles.

Traits	Fettening Groups			SEM	*p*-Value for	
	Rye A	Oats B	Rye/Oats C		Fattening	Interaction
Water content (%)						
Males	69.75	70.72	70.34	0.22	0.193	0.976
Females	69.94	70.96	70.69	0.22	0.136	
SEM	0.304	0.242	0.213			
p-Value for sex	0.764	0.632	0.429			
Fat content(%)						
Males	3.68 C	3.60 C	5.10 AB	0.160	0.000	0.710
Females	3.76 C	3.57 C	4.83 AB	0.157	0.001	
SEM	0.13	0.13	0.17			
p-Value for sex	0.764	0.919	0.466			
Total protein content (%)						
Males	25.01 C	24.14	23.07 A	0.244	0.003	0.981
Females	24.77 C	23.92	22.96 A	0.250	0.009	
SEM	0.340	0.250	0.160			
p-Value for sex	0.727	0.665	0.755			

N = 12 in every experimental group males and females; the letters A, B, C mean the fattening groups, i.e., (A) birds fed hybrid rye, (B) birds fed oats, (C) birds fed a mixture of hybrid rye/oats (1:1); SEM, standard error of the means; means marked with a letter (A, B, C or a, b, c) represents the fattening group with which it differs statistically, means marked by lowercase (a, b, c) are significant at $p \leq 0.05$ and means marked by uppercase (A,B,C) are significant at $p \leq 0.01$, the lack of a letter means that the mean does not statistically differ with any other fattening group. Additionally, if the mean is signed by two letters e.g., "ac", that means significant difference as compared with means of both signed fattening groups a and c.

The sensory test results of the cooked breast muscles are presented in Table 8. A statistically significant sex–feed type interaction was observed in the flavour test which indicated that the muscle flavour of geese fed the hybrid rye and oat mixture was less intense (4.05 points) than that of both sexes fed either hybrid rye only or oats only (on average approximately 4.35 points). Feeding had an impact on the remaining organoleptic features, i.e., juiciness, tenderness, and palatability but only in the young female geese. The lowest values for these traits were in the group fed the grain mixture. The sheer force measurements using device did not confirm the differences in the meat tenderness sensory tests results because they ranged between 15.32 and 17.49 N in all of the tested groups. An interesting fact was observed in terms of gender effect on the sensory test results. Feeding the geese hybrid rye or oats showed no differences between the genders in terms of sensory traits but, in the group fed a mixture of these two grains, female geese muscles obtained lower scores than ganders' muscles in the meat flavour, taste, juiciness, and tenderness tests.

Table 8. Sensory evaluation results (points) and shear force measurements of cooked breast.

Traits	Fattening Groups			SEM	*p*-Value for	
	Rye A	Oats B	Rye/Oats C		Fattening	Interaction
Flavour (points)						
Males	4.35	4.31	4.35	0.03		0.000
Females	4.46 C	4.38 C	4.05 AB	0.04		
SEM	0.03	0.04	0.04			
p-Value for sex	0.130	0.421	0.000			
Juiciness (points)						
Males	3.75	3.88	3.76	0.06	0.648	0.064
Females	3.88 bC	3.61 a	3.54 A	0.04	0.004	
SEM	0.06	0.08	0.05			
p-Value for sex	0.325	0.092	0.037			
Tenderness (points)						
Males	3.89	3.95	3.95	0.04	0.813	0.069
Females	4.03 c	3.93	3.70 a	0.05	0.036	
SEM	0.07	0.06	0.06			
p-Value for sex	0.287	0.860	0.025			
Palatability (points)						
Males	4.10	4.05	4.02	0.04	0.687	0.092
Females	4.09 C	4.05 C	3.76 AB	0.04	0.001	
SEM	0.05	0.05	0.04			
p-Value for sex	0.925	1.00	0.002			
Shear force (N)						
Males	17.09	17.49	16.62	0.44	0.738	0.754
Females	15.32	16.01	15.84	0.40	0.755	
SEM	0.55	0.64	0.36			
p-Value for sex	0.111	0.251	0.358			

N = 12 in every experimental group males and females; the letters A, B, C mean the fattening groups, i.e., (A) birds fed hybrid rye, (B) birds fed oats, (C) birds fed a mixture of hybrid rye/oats (1:1); SEM, standard error of the means; means marked with a letter (A, B, C or a, b, c) represents the fattening group with which it differs statistically, means marked by lowercase (a, b, c) are significant at $p \leq 0.05$ and means marked by uppercase (A,B,C) are significant at $p \leq 0.01$, the lack of a letter means that the mean does not statistically differ with any other fattening group. Additionally, if the mean is signed by two letters e.g., "ac", that means significant difference as compared with means of both signed fattening groups a and c.

4. Discussion

4.1. The Effect of Gender and Feeding on the Slaughter Value and Physical Traits of Meat

The results showed that ganders' weight before slaughter as well as carcass weight was 10–15% higher than in female geese but gender did not have an effect on the slaughter yield, which ranged among the groups from 63% to 65%. No significant differences between the carcass tissues obtained from dissection according to gender and feeding method were found, except for leg muscle weight and skin with fat weight. Similar results were obtained by Kapkowska et al. [7] who reported a higher body weight of the male Zatorska and white Kołudzka varieties fed oats up to 17 weeks of life. The authors did not find any significant differences in slaughter yield between the genders, reaching approximately 64%. Nevertheless, in this study, the birds of both genders fed oats obtained a significantly higher slaughter yield by approximately 2 pp as compared with the group fed hybrid rye.

A higher ganders' carcass weight was followed by significantly higher weight of carcass elements but there were no differences in their percentage share in the carcass between the two sexes. The meat weight from legs was higher in ganders than in female geese for all groups ($p \leq 0.01$) and in females it was only higher if they were fed oats rather than fed hybrid rye or the mixture of both grains ($p \leq 0.01$). In the group of female geese fed the grain mixture, the abdominal fat content was significantly higher ($p \leq 0.05$) than in ganders' group. The share of meat from the leg was lower only in those female geese fed the grain mixture ($p \leq 0.05$). No significant differences were found in the percentage share of breast muscles between the sexes or in skin with a fat share. The results obtained by Kapkowska et al. [7] confirmed a higher yield of breast muscles in male carcasses as compared with female carcasses in both of the Zatorska and white Kołudzka geese.

No significant differences between the sexes were reported in terms of the tested physical breast muscle features such as colour lightness and its' a* and b* parameters, meat marbling and drip loss in all feeding groups. A significantly higher pH_{24} ($p \leq 0.05$) value was observed only in ganders from the group fed oats than in female geese, reaching 5.95 and 5.88, respectively. There were no one-way differences between the sexes in meat cooking losses (significant sex–feed type interaction, $p \leq 0.01$). In the groups fed hybrid rye or the grain mixture, ganders showed higher thermal losses than geese in the group fed oats, which is difficult to explain and justify. The fairly high pH_{24} of goose meat was confirmed by the research of Kapkowska et al. [7] in which the value in both sexes of the Zatorska variety reached 6.09 on average. Additionally, meat colour was pretty dark (L = 38.5), similar to the results obtained in this research (L = approximately 36–37). In addition, the muscles of the geese of the southern varieties presented similar L* values but with a better, lower pH_{24}, reaching 5.78 on average [3].

No significant gender effect on the basic chemical composition of breast muscle was reported in any of the feeding groups. The average water content was approximately from 70 to 70.7%, fat content was approximately 3.6–3.8% (except for the group fed the grain mixture, which was between 4.8% and 5.1%) and protein content was approximately 23–25%. In a study by Biesiada-Drzazga [23] on the meat of White Kołudzka geese, the following parameters were observed in birds which were 10 weeks old: protein content of 21% and fat content between 3.1% and 5.1%, depending on the feeding group. The authors of other publications [6,21,24] depending on the breed, origin, and the diet reported on average 19% to 24% protein and 2.3% to 6.3% fat content, in the breast muscle.

Gender did not have any impact on all of the studied sensory traits (flavour, taste, juiciness, and tenderness) of breast muscle in groups fed hybrid rye or oats, but it did have an effect in the case of hybrid rye and oat mixture feeding. The ganders in this feeding group obtained better scores in flavour ($p \leq 0.01$), juiciness ($p \leq 0.05$), taste ($p \leq 0.01$), and tenderness ($p \leq 0.05$) than in females and the difference in scores was approximately 0.3 points. The average results of the sensory tests for the studied features ranged from 3.80 to 4.35 points. Better results in the sensory tests of the southern variety geese meat were reported by Lewko et al. [11] who obtained 4.8 points in cooked breast muscles.

A high score, approximately 4 points, in meat tenderness estimated by the sensory test was confirmed by measurements performed with a device measuring the shear force with low values (from 15.30 to 17.49 N), which did not depend on the gender. In contrast, Kapkowska et al. [7] observed significantly higher shear force values (43.2 to 50.2 N) but, as confirmed in the literature, they were still within the range of good meat tenderness, i.e., approximately 50 N [21].

4.2. The Effect of Feeding on Slaughter Traits and Geese Meat Quality

The different feeding groups did not show differences in terms of body weight, but the geese fed hybrid rye obtained approximately a 2 pp lower slaughter yield than those geese fed oats ($p \leq 0.01$ in males and $p \leq 0.05$ in females). The higher slaughter yield in geese fed oats could have been influenced by the higher weight of fat with skin in female geese by approximately 80 g and in ganders by 63 g as compared with the group fed hybrid rye.

Basically, the feeding method did not have any impact on body weight or on the share of carcass elements. The only exception was the weight of geese legs ($p \leq 0.01$) and the percentage share of wings in males ($p \leq 0.05$) from the group fed oats. The highest leg weight of female geese was observed in the oat group, whereas the wing share of males was the highest in the hybrid rye group. The feeding method did not have an effect on the weight and percentage share of breast muscles and share of leg muscles, as well as fat with skin percentage and share and weight of abdominal fat. Other authors have obtained different results. According to a study by Kapkowska et al. [7], the carcasses of Zatorska and White Kołudzka geese had a higher fat content with skin share as compared with the results obtained in this study. Additionally, in a study by Biesiada-Drzazga [2], the abdominal fat share in White Kołudzka geese fed a concentrated feed ranged from 4.2% to 5.8% which indicated a higher fat share than that observed in the geese within this research (2.04% to 2.94% on average). This was also confirmed by Karwowska et al. [25] where the fat share ranged between 6% and 8.5% in the carcasses of White Kołudzka geese fed corn silage and beet pulp, as well as another study by Kokoszyński et al. [26] that applied feeding a corn mixture with 20% addition of oats.

Among the estimated physical features, feeding had an influence on the lower pH_{24} value of breast muscles of female geese fed oats, meat marbling (the highest observed in feeding oats, the lowest in feeding hybrid rye), drip loss (no drip with hybrid rye feeding and approximately 1.3% to 2.0% with oats and mixed grain feeding), as well as the thermal losses of breast muscle samples (the lowest in the oat fattening of ganders and the highest in female geese of this oats group). The changes were not directed one-way (i.e., significant sex–feed type interaction). No effect of the feeding method on colour lightness L* and redness and yellowness parameters was observed but this meat colour should be described as fairly dark (L* = approximately 36), which was also confirmed by Kapkowska et al. [7] and Okruszek et al. [27] who reported average L* = 38.5 for geese breast muscles. However, Lewko et al. [11] observed a lighter colour (L = 44.2) in the breast muscles of southern geese varieties fed oats. Low drip loss was also reported by Kapkowska et al. [7] in 17-week-old geese after feeding oats (0.5%), as well as by Biesek et al. [1] in geese fed a mixture with lupin (0.33% to 0.63%), which are also confirmed in this study. Slightly lower thermal losses in cooking (approximately 30% to 32%), as in this investigation, were observed by Gumułka et al. [8] and Kapkowska et al. [7] in the breast muscles of Zatorska geese after being fed oats.

A basic chemical composition analysis proved the significant effect of feeding on fat and protein content. The highest fat content was observed in the muscles of geese and ganders, which were both fed a grain mixture (4.83% and 5.13%, respectively) as compared with those fed hybrid rye or oats (3.60% on average). These observations were confirmed by the results of visual marbling estimation.

Biesiada-Drzazga [24] reported that, in the breast muscle of White Kołudzka geese fed until 10 weeks of age a concentrated feed with soy and sunflower meal, fat content levels reached 5.1%, 4.3%, and 3.1%, in different experimental groups, and the protein content was 21% on average; the significant effect of feeding on protein content was observed in the studied population. In the muscles of birds fed hybrid rye, a significantly higher protein content, approximately 2 pp, was observed in both genders as compared with birds fed a mixture of two both grains. The muscle protein content in female geese and ganders fed oats was similar (24% on average).

The authors of other publications [6,21,22,25] depending on the geese breed, origin, and diet have reported protein contents on average from 19% to 24% and fat contents from 2.3% to 6.3% in breast muscle. In this study, we confirmed that feeding oats or hybrid rye did not affect the chemical composition of the meat but caused the relatively high protein content and lower fat content.

The results of sensory tests in terms of flavour, juiciness, tenderness, and palatability of cooked meat ranged from 3.8 to 4.35 points. Among these features, there was one statistically significant sex–feed type interaction in the flavour estimation ($p \leq 0.01$). The muscles of female geese had the lowest score (4.05), whereas the goslings in three feeding groups presented, on average, the same values within this parameter (approximately 4.3 points). A statistically significant ($p \leq 0.01$) feeding impact on meat juiciness and the palatability of female geese was observed. Better scores were observed in the meat of female geese fed hybrid rye (over 4 points). Lewko et al. [11] estimated the sensory traits of southern varieties of goose meat depending on the origin, gender, and diet. Cooked breast muscles obtained 4.8 points for the estimated sensory features.

Meat tenderness measurements using devices did not show the effect of feeding on the shear force that fit within 15.3 and 17.49 N. These data proved the good tenderness of the tested meat. Kapkowska et al. [7] reported a higher shear force for the breast muscle of the Zatorska and White Kołudzka varieties, ranging from 43.2 to 50.2 N and Karwoska et al. [9] obtained, for the same musles of the White Kołudzka variety, a mean of 49.8 N.

In summary, it should be stated based on available studies [28,29] the direction for the development of rye genetics should be to create the new cultivars with reduced levels of antinutritional substances. The high level of monosaccharides, as the best source of metabolic energy, could also contribute to an increase in feed intake. The favourable proportions of carbohydrates and lipids in hybrid rye could also determine its nutritional value and possibility have a positive effect on carcass quality. Additionally, the using of hybrid rye in animal feeding could bring economic benefits, because the price of the rye is lower than the other grains [27].

5. Conclusions

Our results show the impact of gender on some slaughter value traits and meat quality of young geese slaughtered at 17 weeks old. We found that ganders' carcass weight and their live weight before slaughter was 10–15% higher than that of female geese. The slaughter yield and the share of cuts and same dissection elements such as breast muscles and skin with fat did not differ between sex. No significant differences were found between genders in colour lightness and a* and b* colour parameters, as well as marbling, drip losses, and basic chemical composition of breast muscle. The sensory estimation results of ganders were better than young female geese only in the case of those ganders fed hybrid rye and a mixture of oats and hybrid rye.

During the last three weeks before slaughter, the type of feeding (hybrid rye, oats, or hybrid rye and oats mixture) did not have a significant effect on geese body weight, but the birds fed hybrid rye had approximately a 2 pp lower slaughter yield than the geese fed oats. The feeding type did not have an influence on the male carcass weight and the share of majority cuts in both sexes, but it had an effect on pH_{24} (it was lower in female geese fed oats), drip losses (none in hybrid rye feeding), marbling (higher in feeding with oats), cooking losses (significant sex–feed type interaction), fat content (higher in mixture feeding), and protein content (higher in hybrid rye feeding). Better sensory estimation results of breast muscles were observed in the young female geese fed hybrid rye or oats.

The obtained results indicate that hybrid rye may be used in the feeding of young geese without causing any negative effects on the final body weight and as a result improving some meat quality traits.

Author Contributions: Conceptualization, T.S. and D.L.; methodology, K.B., D.L., K.A. and P.J.; software, D.L.; validation, E.G., B.L., Ł.S. and K.P.; formal analysis, T.S. and D.L.; investigation resources, D.L.; data curation, E.G., P.J., B.L., Ł.S. and K.P. writing—original draft preparation, E.G. and K.B.; writing—review and editing, P.J. and K.B. visualization D.L., T.S. and K.A.; supervision, T.S., D.L. and K.B. project administration, T.S.; funding acquisition, T.S. and D.L. All authors have read and agreed to the published version of the manuscript.

Funding: This research was funded by the National Centre for Research and Development (https://www.ncbr.gov.pl/en/ (accessed on 24 February 2021)) in Poland and conducted within the Biostrateg program (grant "ENERGYFEED" number BIOSTRATEG2/297910/12/NCBR/2016).

Institutional Review Board Statement: All experimental procedures performed on live animals followed the EU Directive 2010/63/EU for animal experiments and the Polish law for the care of animals used in research and education. According to Polish law (Regulation of January 15, 2015 on the protection of animals used for scientific or educational purposes) the ethical approval of research is not formally required if experiments involve only the standard operating procedures typically carried out on a commercial farm. Slaughter of animal aimed to obtain tissues for laboratory analyses is not formally considered as research procedure and the ethical approval of such action is not required.

Data Availability Statement: Not applicable.

Conflicts of Interest: We certify that there is no conflict of interest with any financial organization regarding the material discussed in the manuscript.

References

1. Biesek, J.; Kuźniacka, J.; Banaszak, M.; Adamski, M. The Quality of Carcass and Meat from Geese Fed Diets with or without Soybean Meal. *Animals* **2020**, *10*, 200. [CrossRef] [PubMed]
2. Biesiada-Drzazga, B.; Janocha, A.; Koncerewicz, A. Effect of genotype and rearing system on fatness and fat quality in geese White Kołuda breead. *Postępy Nauk Technol. Przem. Rolno Spoż.* **2011**, *66*, 19–25. (In Polish)
3. Gornowicz, E.; Weglarzy, K.; Pietrzak, M.; Bereza, M. Slaughter and meat traits of domestic geese of the southern breeds. *Wiadomości Zootech.* **2012**, *4*, 5–16. (In Polish)
4. Księ́żkiewicz, J. Genetic Resources Station of Water Poultry in Animal Husbandry Institute Dworzyska—Past and contemporary period. *Wiadomości Zootech.* **2009**, *47*, 69–76. (In Polish)
5. Mazanowski, A.; Adamski, M.; Kisiel, A.; Urbanowski, M. Comparision of meat and reproductive traits in native varieties of southern and northern geese. *Roczn. Nauk. Zoot.* **2006**, *33*, 105–123. (In Polish)
6. Haraf, G. Influence of feeding and gese genotype on carcasss dissection and meat quality—The review of research. *Eng. Sci. Technol.* **2014**, *1*, 1–42. (In Polish)
7. Kapkowska, E.; Gumułka, M.; Rabsztyn, A.; Połtowicz, K.; Andres, K. Comparative study on fattening results of Zatorska and White Kołuda geese. *Ann. Anim. Sci.* **2011**, *11*, 207–217.
8. Gumułka, M.; Wojtysiak, D.; Kapkowska, E.; Połtowicz, K.; Rabsztyn, A. Microstructure and technological meat quality of geese from conservation flock and commercial hybrids. *Ann. Anim. Sci.* **2009**, *9*, 205–213.
9. Karwowska, M.; Mikołajczak, J.; Borowski, S.; Dolatowski, Z.J.; Marć-Pieńkowska, J.; Budziński, W. Effect of noise generated by the wind turbine on the quality of goose muscles and abdominal fat. *Ann. Anim. Sci.* **2014**, *14*, 441–451. [CrossRef]
10. Damaziak, K.; Stelmasiak, A.; Michalczuk, M.; Wyrwisz, J.; Moczkowska, M.; Marcinkowska-Lesiak, M.M.; Niemiec, J.; Wierzbicka, A. Analysis of storage possibility of raw and smoked breast meat of oat-fattened White Kołuda® goose based on their quality characteristics. *Poult. Sci.* **2016**, *95*, 2186–2197. [CrossRef]
11. Lewko, L.; Gornowicz, E.; Pietrzak, M.; Korol, W. The effect of origin, sex and feeding on sensory evaulation and some quality characteristics of goose meat from polish native flocks. *Ann. Anim. Sci.* **2017**, *17*, 1185–1196. [CrossRef]
12. Uhlířová, L.; Tůmová, E.; Chodová, D.; Volek, Z.; Machander, V. Fatty acid composition of goose meat depending on genotype and sex. *Asian-Australas. J. Anim. Sci.* **2019**, *32*, 137–143. [CrossRef] [PubMed]
13. Swiątkiewicz, S.; Arczewska-Włosek, A. Utilization of rye grain as a source of nutritien in feeding of poultry I. Study on chicken brojlers. *Polskie Drobiar.* **2020**, *4*, 42–45. (In Polish)
14. Swiątkiewicz, S.; Arczewska-Włosek, A. Utilization of rye grain as a source of nutrient in feeding of poultry. II Study on laying breeds of fowl. *Polskie Drobiar.* **2020**, *5*, 31–33. (In Polish)
15. Stein, H.; Lagos, L.; Casas, G. Nutritional value of feed ingredients of plant origin fed to pigs. *Anim. Feed. Sci. Technol.* **2016**, *218*, 33–69. [CrossRef]
16. Ziołecki, J.; Doruchowski, W. *Methods for Assessing Slaughter Value*; COBRO Publisher: Poznań, Poland, 1989; pp. 1–22. (In Polish)
17. ISO 1442. *Meat and Meat Products. Determination of Water Content*; ISO: Geneva, Switzerland, 2000.
18. ISO 1444. *Meat and Meat Products. Determination of Free Fat Content*; ISO: Geneva, Switzerland, 2000.
19. PN-75/A-04018. *Meat and Meat Products. Determination of Protein Content by Khejdahl Method*; FOSS: Hilleroed, Denmark, 2002.
20. Kaufman, R.G.; Cassens, R.G.; Scherer, A.; Meeker, D.L. Variations in pork quality: History, definition, extent, resolution. *Swine Health Prod. Off. J. Am. Assoc. Swine Pract.* **1993**, *1*, 1–3.
21. Baryłko Pikielna, M.; Matuszewska, I. *Sensory Research of Food*; Sci Pub. Ed. II; Scientific Publishing PSFT: Krakow, Polish, 2014; p. 375. (In Polish)
22. PN-ISO 4121: Sensory Analysis. Methodology—Evaluation of Food Products by Methods Using Scales. Available online: https://www.iso.org/standard/9871.html (accessed on 24 February 2021).
23. Biesiada-Drzazga, B. Effect of feeding with mixtures containing sunflower extracted meal and yellow lupine meal on the quality of muscular and fatty tissues in geese. *Rocz. Inst. Przem. Mięsn Tłuszcz.* **2008**, *46*, 25–34. (In Polish)
24. Biesiada-Drzazga, B. Description of selected characteristics of muscle and fat tissue of 10-week White Kołuda W31 geese. *Acta Sci. Pol. Technol. Aliment. ACTA.* **2006**, *5*, 47–54.
25. Karwowska, M.; Grabowicz, M.; Stadnik, J.; Szterk, P.; Bernacki, Z.; Dolatowski, Z.J. The effect of corn or beet pulp silage supplemented diet on production parameters, oxidative stability of muscles and fatty acid composition of abdominal fat in geese. *Ann. Anim. Sci.* **2017**, *17*, 887–902. [CrossRef]
26. Kokoszyński, D.; Bernacki, Z.; Grabowicz, M.; Stańczak, K. Effect of corn silage and quantitative feed restriction on growth performance, body measurements, and carcass tissue composition in White Kołuda W31 geese. *Poult. Sci.* **2014**, *93*, 1993–1999. [CrossRef]
27. Okruszek, J.A.; Książkiewicz, J.; Wołoszyn, J.; Haraf, G.; Orkusz, A.; Szukalski, G. Changes in selected physicochemical parameers of breast muscles of geese from Polisch conservation flocks depending on duration of the post slaughter period. *Arch. Tierz. Dummerstorf* **2008**, *3*, 255–265.

28. Makarska, E.; Schon, A.; Gardzielewska, A. The content of alkylresorcinols and trypsin inhibitors activity in translocational rye strains and parental components *Secale cerale* L. and *Daypyrum villosim* L. P. Candardy. *Ann. Univ. Mariae Curie Skłodowska* **2007**, *LXII*, 117–121.
29. Schwarz, T.; Kuleta, W.; Turek, A.; Tuz, R.; Nowicki, J.; Rudzki, B.; Bartolewski, P. Assesing the efficiency of using a modern hybryd rye cultivar for pig fattening, with emphasis on production costs and carcass quality. *Anim. Prod. Sci.* **2014**, *55*, 467–473. [CrossRef]

MDPI
St. Alban-Anlage 66
4052 Basel
Switzerland
Tel. +41 61 683 77 34
Fax +41 61 302 89 18
www.mdpi.com

Animals Editorial Office
E-mail: animals@mdpi.com
www.mdpi.com/journal/animals

www.ingramcontent.com/pod-product-compliance
Lightning Source LLC
Chambersburg PA
CBHW061546220926
51850CB00005B/261